ELECTRICAL ENERGY EFFICIENCY

ELECTRICAL ENERGY EFFICIENCY

TECHNOLOGIES AND APPLICATIONS

Andreas Sumper
BarcelonaTech (UPC), Institute for Energy Research (IREC), Spain

Angelo Baggini
University of Bergamo, Italy

A John Wiley & Sons, Ltd., Publication

This edition first published 2012
© 2012 John Wiley & Sons, Ltd

Registered office
John Wiley & Sons Ltd, The Atrium, Southern Gate, Chichester, West Sussex, PO19 8SQ, United Kingdom

For details of our global editorial offices, for customer services and for information about how to apply for permission to reuse the copyright material in this book please see our website at www.wiley.com.

Library of Congress Cataloging-in-Publication Data

Electrical energy efficiency : technologies and applications / Andreas Sumper and Angelo Baggini.
    p. cm.
  Includes bibliographical references and index.
  ISBN 978-0-470-97551-0 (hardback)
  1. Electric power–Conservation–Standards.   2. Energy conservation–Standards.   3. Energy dissipation.
4. Electric power transmission–Reliability.   I. Baggini, Angelo B.   II. Sumper, Andreas.
  TJ163.3.E39 2012
  621.31–dc23

                                                                                2012000609

A catalogue record for this book is available from the British Library.

Print ISBN: 9780470975510

Typeset in 10/12pt Times by Aptara Inc., New Delhi, India
Printed and bound in Singapore by Markono Print Media Pte Ltd

Contents

List of Contributors

Angelo Baggini
Industrial Engineering Department
University of Bergamo
Via Marconi 5 24044 Dalmine BG, Italy

Joan Bergas-Jané
Centre d'Innovació Tecnològica en
Convertidors Estàtics i Accionaments
(CITCEA)
Universitat Politècnica de Catalunya (UPC)
Escuela Técnica Superior de Ingeniería
Industrial de Barcelona
Av. Diagonal, 647. Planta 2 08028
Barcelona, Spain

Franco Bua
ECD Engineering Consulting and Design
Vai Maffi 21
27100 Pavia, Italy

Mircea Chindris
Electrical Power Systems Dept.
Technical University of Cluj-Napoca
15, C.Daicoviciu st.
400020 Cluj-Napoca, Romania

Andrei Czicker
Electrical Power Systems Dept.
Technical University of Cluj-Napoca
15, C.Daicoviciu st.
400020 Cluj-Napoca, Romania

Wim Deprez
Dept. Electrical Engineering ESAT

K.U. Leuven, Research group ELECTA
Kasteelpark Arenberg 10
3001 Heverlee, Belgium

Stefan Fassbinder
Berantung elektrotechnische Anwendungen
Deutsches Kupferinstitut
Am Bonneshof 5
D-40474 Dusseldorf, Germany

Zbigniew Hanzelka
University of Science and Technology –
AGH
30-059 Cracow, Al. Mickiewicza 30
Poland

Joris Lemmens
Dept. Electrical Engineering ESAT
K.U. Leuven, Research group ELECTA
Kasteelpark Arenberg 10
3001 Heverlee, Belgium

Annalisa Marra
ECD Engineering Consulting and Design
Vai Maffi 21
27100 Pavia, Italy

Daniel Montesinos-Miracle
Centre d'Innovació Tecnològica en
Convertidors Estàtics i Accionaments
(CITCEA)
Universitat Politècnica de Catalunya (UPC)
Escuela Técnica Superior de Ingeniería
Industrial de Barcelona

Av. Diagonal, 647. Planta 2 08028
Barcelona, Spain

Paola Pezzini
Centre d'Innovació Tecnològica en
Convertidors Estàtics i Accionaments
(CITCEA)
Universitat Politècnica de Catalunya (UPC)
Escuela Técnica Superior de Ingeniería
Industrial de Barcelona
Av. Diagonal, 647. Planta 2 08028
Barcelona, Spain

Krzysztof Piątek
University of Science and Technology –
AGH
30-059 Cracow, Al. Mickiewicza 30,
Poland

Edris Pouresmaeil
Centre d'Innovació Tecnològica en
Convertidors Estàtics i Accionaments
(CITCEA)
Universitat Politècnica de Catalunya
(UPC)
Escuela Técnica Superior de Ingeniería
Industrial de Barcelona
Av. Diagonal, 647. Planta 2 08028
Barcelona, Spain

Jaume Salom
Institut de Recerca en Energia de Catalunya
(IREC)
Jardins de les Dones de Negre 1, 2ª pl.
08930 Sant Adrià de Besòs, Spain

Antoni Sudrià-Andreu
Centre d'Innovació Tecnològica en
Convertidors Estàtics i Accionaments
(CITCEA)
Universitat Politècnica de Catalunya (UPC)
Escuela Técnica Superior de Ingeniería
Industrial de Barcelona
Av. Diagonal, 647. Planta 2 08028
Barcelona, Spain

Andreas Sumper
Centre d'Innovació Tecnològica en
Convertidors Estàtics i Accionaments
(CITCEA)
Universitat Politècnica de Catalunya (UPC)
Escola Universitària d'Enginyeria Tècnica
Industrial de Barcelona
Carrer Comte d'Urgell, 187 - 08036
Barcelona, Spain

and

Institut de Recerca en Energia de Catalunya
(IREC)
Jardins de les Dones de Negre 1, 2ª pl.
08930 Sant Adrià de Besòs, Spain

Waldemar Szpyra
University of Science and Technology –
AGH
30-059 Cracow, Al. Mickiewicza 30,
Poland

Roman Targosz
Polish Copper Promotional Centre
Plac Jana Pawla II 1-2
50-136 Wrocalw, Poland

Roberto Villafáfila-Robles
Centre d'Innovació Tecnològica en
Convertidors Estàtics i Accionaments
(CITCEA)
Universitat Politècnica de Catalunya (UPC)
Escola Universitària d'Enginyeria Tècnica
Industrial de Barcelona
Carrer Comte d'Urgell, 187 - 08036
Barcelona, Spain

Irena Wasiak
Politechnika Łódzka
Wydział Elektrotechniki, Elektroniki,
Automatyki i Informatyki
Instytut Elektroenergetyki
ul. Stefanowskiego 18/22
90-924 Łódź, Poland

Preface

Energy efficiency technologies are common technologies from different engineering fields used to reduce the energy required to provide products and services. As electricity is the most flexible energy form known to humans and one of the most important energy forms used in industry and commercial applications, a specific focus on electrical energy efficiency is required. So, electrical energy efficiency is a set of engineering technologies that are dedicated to increasing the electrical energy efficiency of applications. These engineering technologies are very widespread and can vary from power quality engineering to the thermal engineering of electrical applications, including economic aspects.

Together with electrical safety, in the coming years electrical energy efficiency should become one of the mandatory design criteria in every process, installation or building.

The difficulty of electrical energy efficiency engineering is to obtain a holistic view of an application; in most cases a specific knowledge of the technology is needed, but a deep understanding of the industrial process and the problem to be solved is necessary in order to achieve the overall efficiency goal. Often, optimal solutions for partial problems provide a moderate contribution to the overall energy efficiency of the process. Engineers should have multidisciplinary knowledge, for instance knowledge about electrical applications, power quality, control techniques and heat transfer. Also, an important aspect to consider is the ability to analyse the industrial process and to determine what efficiency actions need to be taken.

The increase in electrical energy efficiency is closely related to the evaluation of the efficiency measures to be taken, mainly by investment analysis. Efficient solutions often need higher investments and these usually need management approval. The manager also has to understand how energy efficient solutions can improve the process efficiency and therefore a higher productivity can be achieved.

In 2000 a group of academics and industrialists launched a life-long learning programme co-funded by the European Commission dedicated to Power Quality problems called Leonardo Power Quality Initiative (LPQI). This project created a network of experts in energy that created several follow-on projects such as LPQIves and Leonardo Energy. Most of the information on these programmes is available at the Leonardo Energy webpage (http://www.leonardo-energy.org). Inspired by this project, part of this working group contributed to the *Handbook of Power Quality*, edited by Angelo Baggini in 2008.

In one of the project meetings in Brussels in 2008 the idea of a comprehensive book on electrical energy efficiency was born and the content of the book was worked out during the following years.

The novel approach in this book is to give the reader a straightforward introduction to the technologies and their applications used to increase electrical energy efficiency. The reader will find efficiency aspects emphasized in this comprehensive book and an expert view given on the most important industrial and commercial fields of electrical engineering. Each chapter covers a different technology in order to achieve an efficiency goal in a wide range of application fields.

Before you begin to study this book, we would like to mention the important contributions of all the authors of the chapters from all around the world. Without their expert views, this work would not be possible. We hope that you find this book interesting reading.

Andreas Sumper, Barcelona, Spain
Angelo Baggini, Pavia, Italy

Foreword

There are no doubts that energy security and climate change are two of the most frequent topics discussed by policy makers. The oil price is now at around US$100 per barrel and, because of the increasing demand and the continuing depletion of the reserves, this price level will stay or may even increase. The human impact on climate change is not disputed anymore in the scientific community, as well as the worrying news that the irreversible impact has already started and only a drastic change in the level of CO_2 emissions will mitigate the large and very costly impact on the society.

Energy efficiency and energy conservation are gaining importance as key components in many national and international strategies to mitigate the impact of climate change, to improve security of energy supply and increase competitiveness, to preserve natural resources (energy, material and water, amongst others) and also to reduce other energy-related environmental pollution. However, investment in energy efficiency technologies from R&D to implementation, in buildings, equipment and industrial systems, is still far too less than the economics and the energy and climate change situation would suggest.

Energy efficiency policies, programmes and support schemes are still very much needed to overcome market, institutional, financial and legal barriers, and to create a favourable market for energy efficiency investments at the level that a rational economic behaviour would justify. In particular, support schemes for energy efficient technologies are very much debated as many consider that the future energy cost savings should be enough to motivate end users.

The other major issue is the awareness that what matters in climate change is to reduce the absolute energy demand if we want to mitigate the inevitable full climate change impact. Reduction in energy demand can be achieved by improving the energy efficiency of the service provided (technological aspect) and/or by realising energy savings without necessarily making technological improvements (behavioural aspect, for instance less overheating or overcooling, less driving). Energy efficiency is an important component to achieve energy savings, as it allows having the same services (e.g. lighting, cooling, heating) with less use of energy. However, improved energy efficiency – i.e. replacing a technology with a more energy efficient one – does not per se assure energy savings, and there are numerous examples where as a result of introducing a more efficient technology the actual consumption indeed increases, because of the rebound effect or because of installing larger and more numerous appliances and equipment (larger volume of appliances, more frequent usage).

There is an increased interest in energy efficiency and energy savings amongst policy makers, economists and academics (from the technology, economy, policy and human behaviour side). There is the need to further explore energy efficiency technologies (such as control systems, solid state lighting, variable speed drives and vacuum insulation) and gather new evidence on policies and socio-economic issues related to energy use, consumption and behaviour. At the same time, with increased policy activities in the energy efficiency and energy saving field, there is a new need to evaluate the past and present policies in different countries, to show the clear contribution of energy efficiency to energy security and climate change mitigation.

Paolo Bertoldi
European Commission
Joint Research Centre
Ispra Italy

1

Overview of Standardization of Energy Efficiency

Franco Bua and Angelo Baggini

Since the oil shocks of the 1970s, many countries worldwide have promoted energy efficiency improvements across all sectors of their economies. As a result of these policies and structural changes in their economies, these countries have been able to decouple primary energy use from economic growth.

The rate of decline in energy intensity has not remained constant over time; in most countries the rate of decline tended to be higher from 1970 to 1990[1].

The International Energy Agency (IEA) reports that the oil price shocks of the 1970s and the resulting energy policies have apparently been more effective in controlling the growth in energy demand and CO_2 emissions than the energy efficiency and climate policies implemented in the 1990s[2].

However, since the early 2000s, the rate of improvement in energy intensity has tended to increase, possibly in association with the increase in energy prices and greater attention to climate change issues.

It goes without saying that, these days, improving energy efficiency has become a priority in the political agenda of all countries, being key to addressing energy security and both environmental and economic challenges.

In order to support governments with their implementation of energy efficiency, many organizations have worked out a broad range of recommendations and proposed actions for

---

[1] IEA (2007), Energy Use in the New Millennium – Trends in IEA countries, OECD/IEA, Paris.
[2] IEA (2007), Energy Use in the New Millennium – Trends in IEA countries, OECD/IEA, Paris.

Electrical Energy Efficiency: Technologies and Applications, First Edition. Andreas Sumper and Angelo Baggini.
© 2012 John Wiley & Sons, Ltd. Published 2012 by John Wiley & Sons, Ltd.

well identified priority areas[3]. Each country would select the policies that best suit its efficiency commitment as well as its unique economic, social and political situation.

A classification of these policy options and measures[4] is given by the World Energy Council[5] as follows:

- Institutions and programmes
  - Institutions: agencies (national, regional and local), Ministry department
  - National programmes of energy efficiency with quantitative targets and laws
- Regulatory measures
  - Minimum efficiency standards and labels for electrical appliances (e.g. refrigerators, washing machines, AC, lamps, water heaters, motors), cars and buildings (new and existing)
  - Other regulations for designated consumers: mandatory energy managers, mandatory energy consumption reporting, mandatory energy saving and mandatory maintenance
  - Obligation of energy savings for energy companies at consumers' premises
- Financial and fiscal measures
  - Subsidies for audits by sector (industry, commercial, public, households, low income households transport)
  - Subsidies or soft loans (i.e. loans with subsidised interest rates) for energy efficiency investment and equipment by sector
- Fiscal measures
  - Tax credit
  - Accelerate depreciation
  - Tax reduction for efficiency investment, by type of tax (import, VAT, purchase, annual car registration) and by type of equipment (appliances, cars, lamps)
- Cross-cutting measures
  - Innovative communication tools
  - Voluntary agreements.

Exercises have been carried out extensively to measure how effective these energy efficiency policies are. As an example, IEA reviews the state of the art of the energy efficiency policies, highlighting strengths and areas for improvement (Table 1.1 and Table 1.2).

Despite having a huge potential, energy efficiency policies[8] are difficult to implement. Why? Energy efficiency faces pervasive barriers, including lack of access to capital for energy efficiency investments, insufficient information, and externality costs that are not reflected in energy prices. Moreover political commitment to maximizing the implementation of energy efficiency policies may also have been challenged by the current economic crisis. Energy

[3] For example IEA recommended the adoption of a set of specific energy efficiency policy measures to the last four G8 summits; for further information on the full set of recommendations, refer to http://www.iea.org/textbase/papers/2008/cd_energy_efficiency_policy/index_EnergyEfficiencyPolicy_2008.pdf

[4] A comprehensive database of energy efficiency policies and measures is provided by IEA (http//www.iea.org/textbase/pm/index_effi.asp)

[5] WEC, *Energy Efficiency: A Recipe for Success*, 2010, p. 40.

[6] IEA, *Implementing Energy Efficiency Policies*, 2009, p. 23.

[7] IEA, *Implementing Energy Efficiency Policies*, 2009, p. 33.

[8] It is worth mentioning that there is literature on most common criticisms of energy efficiency policies and programmes. These critics argue that energy efficiency policies and programmes are unwarranted or are a failure. IEA promoted the publication of a paper that compiles, categorises, and then evaluates those criticisms of energy efficiency policies (see 7).

Table 1.1 Summary of strengths and innovations in IEA member countries' energy efficiency policies in the building, industrial and transport sectors[6]

Buildings	
	• Full implementation of building certification in several EU countries
	• Policies promoting passive energy houses
	• Energy efficiency requirements in building codes
Industry	
	• High coverage of industry energy statistics in all countries
	• Policies for promoting energy management
	• Ad hoc policies for SMEs
	• Policies for cogeneration, energy efficient electric motors
Transport	
	• Policies aimed at rolling resistance of tyres
	• Fuel efficiency standards for light and heavy duty vehicles (JP only)
	• Eco drive policies
	• Scrappage schemes encouraging purchase of more efficient and less polluting new vehicles

Table 1.2 Summary of challenges and areas for improvement in IEA member countries' energy efficiency policies in the building, industrial and transport sectors[7]

Buildings	• Establish stronger energy efficiency requirements for buildings
	• Strengthen support for Passive Houses and zero-energy buildings
	• Increase promotion of energy efficiency windows and glazing
Industry	• Establish measures to optimize energy efficiency in motor driven systems
	• Set up policies and measures to assist SMEs
Transport	• Ensure the implementation fuel efficiency standards of planned policies
	• Create fuel efficiency standards for heavy duty vehicles

efficiency programmes must compete for funding with other priorities such as employment, health and social security.

1.1   Standardization

As stated above, energy efficiency faces barriers to success. Examples of such barriers include: the lack of awareness of the savings potential, inadequate performance efficiency information and metrics, the tendency to focus on the performance of individual components rather than the energy yield or consumption of complete systems, split incentives and the tendency to focus on lowest initial cost rather than life cycle cost. Standards can help in overcoming some of these barriers. Standards, for instance, can provide common measurement and test methods to assess the use of energy and the reductions attained through new technologies and processes, as well as providing a means of codifying best practices and management processes for efficient energy use and conservation.

Furthermore, standards can provide design checklists and guides that can be applied to both the design of new systems and the retrofit of existing systems; they can provide standard calculation methods so that sound comparisons of alternatives can be made in specific situations and they can help with the adaptation of infrastructure to integrate new technologies and aid interoperability.

An overview of the current standardization activities on energy efficiency is given in the following sections.[9]

1.1.1 ISO

The work of the ISO (International Organization for Standardization) on energy efficiency began in June 2007 when the ISO Council Task Force on Energy Efficiency and Renewable Energy Sources identified five areas of high priority that were deemed to have the highest potential to contribute substantially to energy savings and greenhouse gas emission reductions, namely:

• Calculation methods
• Energy management standards
• Biofuels
• Retrofitting and refurbishing
• Buildings.

In line with the Council's request[10], the Technical Management Board (TMB) established a Strategic Advisory Group (SAG) on Energy efficiency and renewable energy sources[11] for an initial period of 2 years (until February 2010). SAG E was asked to provide advice and guidance to TMB on priority standards and actions, including involving stakeholders' collaboration with other international organizations and co-ordination between ISO and TCs, etc. The goal was to speed up the process of devising a standardization programme in this field that will serve public policy objectives and market needs.

SAG-E produced an extensive report, providing 66 recommendations, which were endorsed by the TMB. SAG-E activity has been extended for another 3 years.

1.1.1.1 ISO 50001

In February 2008, the ISO Technical Management Board approved the establishment of a new project committee, ISO/PC 242, Energy management[12], building on practices and existing national or regional standards.

ISO 50001 will establish an international framework for industrial and commercial facilities, or entire companies, to manage all aspects of energy, including procurement and use. After four committee meetings, spanning a period of two years, the document was published in June 2011 and was adopted by CEN and CENELEC as ISO EN 50001 in October 2011. The

---

[9] This overview took into account standardization, directly concentrating on energy efficiency from a system approach point of view.

[10] Resolution 28/2007.

[11] ISO Technical Management Board Resolutions 22/2008.

[12] ISO Technical Management Board Resolutions 15/2008.

standard is intended to provide organizations and companies with a recognized framework for integrating energy efficiency into their management practices.

ISO 50001 will provide organizations and companies with technical and management strategies to increase energy efficiency, reduce costs, and improve environmental performance.

1.1.1.2   ISO/IEC JPC 2

In 2009, ISO and the International Electrotechnical Commission (IEC) created the joint project committee ISO/IEC JPC 2, Energy efficiency and renewable energy sources – Common terminology, whose primary objective is to develop a standard that will identify cross-cutting concepts with terms and definitions associated with energy efficiency and renewable energy sources, while taking into account terminology that has already been elaborated in sector-specific ISO and IEC technical committees.

Three working groups (WG) were established at the first meeting of ISO/IEC JPC in January 2010:

- WG 1, Energy efficiency : Concepts and diagrams, coordinated by ANSI (USA)
- WG 2, Inputs from existing reference documents, coordinated by SIS/SEK (Sweden)
- WG 3, Renewable energy sources – Terms and definitions, coordinated by AFNOR (France).

The Committee Draft (CD) step was launched in October 2011.

1.1.2   IEC

IEC's vision on Energy efficiency is outlined in its White Paper, 'Coping with the Energy Challenge'[13]. Developed by the IEC Market Strategy Board (MSB), this document maps out global energy needs and potential solutions over the next 30 years and the IEC's role in meeting the challenges.

IEC thinks that a system approach that takes into account all aspects of generating, transporting and consuming energy must be considered to cope with the energy efficiency challenges and that measurement procedures and methods of evaluating energy efficiency must be specified in order to assess potential improvements properly and to optimize technological issues (Best Available Technology, BAT).

1.1.2.1   SG1 'Energy Efficiency and Renewable Resources'

In 2007 IEC began to establish subsidiary bodies to advise its Management Board on strategic issues that would determine future technical work. Among these was the SG1, which was established on the specific topic of energy efficiency.

---

[13] http://www.iec.ch/smartenergy/

SG 1 was established at the beginning of 2007 and was tasked to:

- analyse the status quo in the field of energy efficiency and renewable resources (existing IEC standards, on-going projects)
- identify gaps and opportunities for new work in IEC's field of competence
- set objectives for electrical energy efficiency in products and systems
- formulate recommendations for further actions.

Since then experts from other groups inside the IEC and other organizations such as IEA, CIE, etc. have met to present their activities and achievements in the areas of energy efficiency and renewable resources and to provide their input to the discussions.

The main outcomes of SG1's work are 34 recommendations that were sent to SMB and TC members for comments.

1.1.2.2 SG3 'Smart Grid'

In this context it is worth mentioning another Study Group (SG) that is linked to energy efficiency: SG3 'Smart Grid'. SG3, set up in 2008, provides advice on fast-moving ideas and technologies that are likely to form the basis for new International Standards or IEC Technical Committees in the area of Smart Grid technologies.

SG3 has developed the framework and provides strategic guidance to all Technical Committees involved in Smart Grid work and has developed the Smart Grid Roadmap[14], which covers standards for interoperability, transmission, distribution, metering, connecting consumers and cyber security.

1.1.2.3 SG4 'LVDC Distribution Systems up to 1500 V DC'

SG 4 was set up in 2009 with the objective of having a global systematic approach and to align and coordinate activities in many areas where LVDC is used, such as green data centres, commercial buildings, electricity storage for all mobile products (with batteries), EVs, etc, including all mobile products with batteries, lighting, multimedia, ICT, etc. with electronic supply units.

SG4 is another example of an area of activity that is not directly dedicated to energy efficiency but whose role could be strategic in harvesting energy efficiency potential.

1.1.3   CEN and CENELEC

CEN and CENELEC were the most proactive standardization organizations as they started in 2002, to analyse the challenges of standardization in the field of energy efficiency and to elaborate general strategy.

Another interesting and valuable aspect is that CEN and CENELEC decided to start this activity jointly, thus implementing de facto an integrated system approach that is of utmost importance.

---

[14] http://www.iec.ch/smartgrid/downloads/sg3_roadmap.pdf

Table 1.3   CEN–CENELEC Joint Working Groups active in the field of Energy Management and Energy Efficiency standardization

Technical body	Scope of work
JWG1 'Energy Audits'	
JWG2 'Guarantees of origin and energy certificates'	Standardization on guarantees of origin for trading and/or disclosure/labelling of electricity and CHP and on energy certificates
JWG3 'Energy Management and related services – General requirements and qualification procedures'	To elaborate EN standards in the energy management and related services field: • Energy Management Systems: definition and requirements • Energy Service Companies (ESCO): definition, requirements and qualification procedures • Energy Managers and Experts: roles, professional requirements and qualification Procedures
JWG4 'Energy efficiency and saving calculation'	Standards for common methods of calculation of energy consumption, energy efficiencies and energy savings and for a common measurement and verification of protocol and methodology for energy use indicators

The CEN/CENELEC BT JWG 'Energy management' was set up at the beginning of 2002 to initiate a European collective view of the general strategy for improvement of energy efficiency standardization and to set an agreement between all CEN/CENELEC members on the objectives to be achieved.

The working group acted as an advisory group to CEN and CENELEC BTs on all political and strategic matters relating to standardization in the field of energy efficiency from 2002 to 2005. The main results of the work are synthesized in a report[15] that gives an overview of proposals in standardization in the field of energy management, classified into three level of priorities[16]. This document is still the basis for CEN and CENELEC standardization activity in the field of energy efficiency.

The key technical bodies involved in energy efficiency standardization are summarised in Table 1.3 together with the most important standardisation activities (Table 1.4).

1.1.3.1   SFEM

In response to the CEN/CENELEC BT JWG 'Energy management' recommendation, CEN and CENELEC have created a horizontal structure, a Sector Forum Energy Management (SFEM), dedicated to the definition of a common strategy for standardization in the field of energy management and energy efficiency. SFEM is a platform for stakeholders to share information and experiences, and to identify priorities regarding standardization in the energy sector.

---

[15] http://www.cen.eu/cen/Sectors/Sectors/UtilitiesAndEnergy/Forum/Documents/BTN7359FinalReportJWG.pdf

[16] Level A – for immediate action; Level B – that need further investigation or research before standardisation could be done; Level C – that need to be discussed in the context of a strategic and holistic view, i.e. policy questions.

Table 1.4 CEN–CENELEC standards and projects in the field of Energy Management and Energy Efficiency

Publication/Project	Title
EN 16001:2009 (pr=22320)	Energy management systems – Requirements with guidance for use
EN 15900:2010 (pr=22416)	Energy efficiency services – Definitions and requirements
prEN 16247-1:2011 (pr=23294)	Energy audits – Part 1: General requirements
prEN 50XXX (pr=23227)	Guarantees of origin related to energy – Guarantees of origin for electricity
prEN PT EEB Doc:2010 (pr=23079)	Energy efficiency benchmarking methodology
EN ISO 50001:2011 (pr=23639)	Energy management systems – Requirements with guidance for use
prEN 16212:2010 (pr=23138)	Standard on top down and bottom up methods of calculation of energy consumption, energy efficiencies and energy savings

SFEM is designed:

- to maintain and enlarge the network of partners created during the lifetime of the CEN/CENELEC BT JWG "Energy Management", especially with regards to new members;
- to initiate further investigation and to evaluate in which field or for which subject, further standardization work is needed and including subjects identified as Priority B or C by the former CEN/CENELEC BT JWG "Energy Management";
- to coordinate on-going European Standardization activities concerning Energy Management;
- to organize the CEN and CENELEC response to European legislation and Europe general strategy in the Energy Management sector;
- to maintain the exchange of information, experience and prospecting especially on the initiatives in course in the different countries or at European level.

SFEM meets twice a year, does not carry out any standardization activity and formulates recommendations to CEN and CENELEC for further actions. CEN and CENELEC usually react by setting up dedicated technical bodies (usually joint working groups) with specific scopes of work.

Further Readings

H. Geller and S. Attali, The experience with energy efficiency policies and programmes in IEA countries. Learning from the Critics, IEA Information Paper, 2005.
IEA, *Implementing Energy Efficiency Policies*, 2009, OECD/IEA, Paris.
WEC, WEC: *Energy Efficiency: A Recipe for Success*, 2010.

2

Cables and Lines

Paola Pezzini and Andreas Sumper

In distribution systems the power transmission capacity is directly given by the product of the operating voltage and the maximum current that can be transmitted. The operating voltage being a fixed value, the delivery capability of the system at a given voltage depends on the conductor's capacity if carrying current.

The delivery capacity is called the ampacity of the cable system [1] and its calculation is carried out taking into consideration both steady state and transient calculations [2]. The calculations for cables in air and for buried cables are slightly different, due to the surrounding medium with which the cable has to interact. The ampacity calculations for air cables should take into account solar radiation and the amount of wind in the area in which the cable system is installed. Ampacity calculations for buried cables should consider the soil in which the cable system is installed.

Ampacity calculations require the solution of the heat transfer equations because insulation and cable size are independent parameters, inter-related by thermal considerations. Cable ampacity calculations require the determination of the temperature of the conductor for a given current loading. The ampacity rating is directly proportional to the conductor size: the larger the conductor size (lower Joule losses) the higher the ampacity. On the other hand, the insulation requirements are determined by the operating voltage and they also directly influence the ampacity value: high insulation requirements (lower heat dissipation) mean a lower ampacity. The parameters that influence the value of the ampacity are the number and types of cable, the thermal resistance of the medium surrounding the cable (soil or air), the depth of burial in the case of buried cables and the horizontal spacing between the cables of the system. The clear relationship between the conductor current and the temperature leads to a study of how the heat generated while a current is transmitted is dissipated. The resolution of the basic heat transfer equations is the first step to achieving the cable rating calculations and cable ampacity; they depend mainly on the efficiency of the dissipation process, along with the limits imposed on the insulation temperature.

Electrical Energy Efficiency: Technologies and Applications, First Edition. Andreas Sumper and Angelo Baggini.
© 2012 John Wiley & Sons, Ltd. Published 2012 by John Wiley & Sons, Ltd.

Technical criteria alone are nowadays not enough to obtain the best sizing for a cable. In fact, the minimum admissible section obtained from the solution of the heat transfer equations does not take into account the cost of the losses that will be present during the cable's lifetime. The selection of the cable size should therefore take into consideration the sum of the initial cost and the cost of the losses: the cost of energy losses can be calculated by estimating load growth and the cost of energy. If the sum of the future costs of energy losses and the initial cost of purchase and installation are minimized, then the most economical size of conductor has been achieved. Using this minimization, the saving in overall cost is due to the reduction in the cost of the Joule losses compared with the increase in the cost of purchase.

2.1   Theory of Heat Transfer

An overview of the heat transfer theory is presented here because load current and conductor temperature are strictly related. The heat generated in the cable system and the rate of its dissipation must be calculated, for a given conductor material and for a given load, in order to determine the conductor temperature for a given current loading or to determine the tolerable load current for a given conductor temperature [2].

There are different mechanisms that explain how heat is transferred in different media: these three mechanisms are conduction, convection and radiation.

2.1.1   Conduction

For heat conduction, the rate equation used to express the transfer of heat between two media in contact is the equation known as Fourier's law, as in:

$$Q = -\frac{1}{\rho} A \frac{d\theta}{dx}. \tag{2.1}$$

The thermal power in the x direction is represented by Q [W] and it is directly proportional to the temperature gradient $\frac{d\theta}{dx}$ $\left[\frac{K}{m}\right]$. This gradient represents, for a given temperature distribution, $\theta(x)$ [K], the direction and the rate at which the temperature changes. A [m^2] is the area in which the thermal exchange occurs and ρ $\left[\frac{K\,m}{W}\right]$ is the thermal resistivity, a transport quality characteristic of the material. The minus in the equation represents the fact that heat is transferred in the direction of the decreasing temperature.

Conduction is the mechanism that acquires more importance when considering buried cables, where the conductor is in contact with other metallic parts and insulation.

2.1.2   Convection

Convection is the result of two mechanisms that work simultaneously: heat transfer conduction due to the presence of molecular motion and heat transfer due to fluid motion. The equation employed to describe convection is Newton's law:

$$Q = hA \left(\theta_s - \theta_{amb}\right). \tag{2.2}$$

Table 2.1 Range of values for the heat convection coefficient, h

Mechanism	Heat coefficient, h $\left[\text{W}/\text{K}\,\text{m}^2\right]$
Natural convection	5–25
Forced convection	
Gas	25–250
Fluids	50–20000
Boiling and condensation	2500–100 000

The convection thermal power Q [W] is proportional to the temperature difference between the surface θ_s and the ambient temperature, θ_{amb} A [m^2] is again the area of the thermal exchange surface and h $\left[\frac{W}{K\,m^2}\right]$ is the convection heat transfer coefficient.

Convection can be classified according to the fluid motion: forced convection and free convection. The first occurs when the flow is caused by external means: wind, pumps or fans. The second arises from density differences caused by temperature variations. Convection must be strongly taken into account for cables installed in air and the heat convection coefficient, h is the most important parameter to be calculated. Table 2.1 shows the typical range of values for the h coefficient.

2.1.3 Radiation

Energy transmission by radiation is a characteristic of all matter; it does not need a medium: radiation travels by means of electromagnetic waves, which can transmit energy in a vacuum. The thermal power emitted follows Stefan–Boltzmann's law:

$$Q = \varepsilon A \sigma_B \theta_s^{*4} \qquad (2.3)$$

The thermal power Q is directly proportional to the absolute temperature θ_s^* [K] of the surface, A [m^2], σ_B is Stefan–Boltzmann's constant $\left(\sigma_B = 5.67 \cdot 10^{-8} \left[\frac{W}{K^4\,m^2}\right]\right)$ and ε is the emissivity, a radiative property of the surface. Emissivity is the efficiency of a surface to emit, compared with an ideal radiator and its range of values is $0 \leq \varepsilon \leq 1$.

If radiation is incident on a surface, a portion will be absorbed according to the surface radiative property known as absorptivity, α [W/K^4 m^2], as presented in the following equation:

$$Q_{abs} = \alpha Q_{inc}, \qquad (2.4)$$

where $0 \leq \alpha \leq 1$. Cables both emit and absorb radiation. Therefore to determine the net rate the following equation is employed:

$$Q = \varepsilon A_r \sigma_B \left(\theta_S^{*4} - \theta_{amb}^{*4}\right). \qquad (2.5)$$

In cable systems installed in air, convection must also be taken into account, so finally the equation that should be applied is:

$$Q = h A_c \left(\theta_s - \theta_{amb}\right) + \varepsilon A_r \sigma_B \left(\theta_s^{*4} - \theta_{amb}^{*4}\right), \qquad (2.6)$$

where A_c[m] is the convection surface and A_r[m] is the radiation surface.

2.2   Current Rating of Cables Installed in Free Air

The permissible current rating of cables is calculated basically using four main values: permissible temperature rise, conductor resistance, losses and thermal resistivity. However, some quantities vary with cable design and material, so one needs to rely on an international standard. Moreover, as will be explained later, the quantities relating to the operating conditions may vary from one country to another.

Considering AC cables in air, the permissible current rating is [3]:

$$I = \left[\frac{\Delta\theta - W_d \left[0.5\, T_1 + n \left(T_2 + T_3 + T_4 \right) \right]}{RT_1 + nR \left(1 + \lambda_1 \right) T_2 + nR \left(1 + \lambda_1 + \lambda_2 \right) \left(T_3 + T_4 \right)} \right]^{0.5}, \tag{2.7}$$

where $\Delta\theta$ [K] is the permissible temperature rise of a conductor above ambient temperature, W_d [W/m] represents the dielectric losses per unit length per phase, n is the number of conductors in the cable, R [Ω/m] is the alternating current resistance of the conductor at its maximum operating temperature and T_i [K m/W] represents the thermal resistance, more specifically: T_1 is the thermal resistance per core between conductor and sheath, T_2 is the thermal resistance between sheath and armour, T_3 is the thermal resistance of the external serving and T_4 is the thermal resistance of the surrounding medium.

To evaluate the losses, several quantities are considered: AC resistance, dielectric losses, sheath and screen losses, armour, reinforcement and steel pipes losses. Here only AC resistance and dielectric losses are discussed, a further discussion on sheath and screen losses, armour, reinforcement and steel pipes losses can be found in the IEC 60287-1-1 [3]. Considering its maximum operating temperature, the AC resistance per unit length of the conductor is given by:

$$R = R'(1 + y_s + y_p), \tag{2.8}$$

where R [Ω/m] is the AC current resistance of conductor at maximum operating temperature, R' is the DC resistance of conductor at maximum operating temperature, y_s is the skin effect factor and y_p the proximity effect factor. The evaluation of these quantities can be done by IEC 60287-1-1 [3]. When a cable carries alternating current, the resistance is higher than when it carries direct current, mainly because of the skin effect, proximity effect, hysteresis and eddy current losses in ferromagnetic materials and the induced losses in short-circuited non-ferromagnetic materials [2]. Usually, only the skin and proximity effects are considered, except in very high voltage cables.

The dielectric losses per unit length in each phase are given by:

$$W_d = \omega C U_0^2 \tan \delta, \tag{2.9}$$

where $\omega = 2\pi f$, C [F/m] is the capacitance per unit length and U_0 [V] is the voltage to earth. Applying alternating voltage to paper and solid insulation causes charging currents to flow because the insulation acts as a large capacitor. Each time the voltage direction changes, the electrons must be realigned, expending a certain amount of work, which will produce heat and therefore a loss in real power, the dielectric loss [2].

As can be seen by its equation, the dielectric loss is voltage dependent and Table 3 of the IEC 60287-1-1 [3] gives, for the common used insulation materials, the value of U_0. The rest of the quantities in the equation for the dielectric losses can be also be found in the same table.

Finally, the internal and the external thermal resistance of cables are discussed. The thermal resistances per unit length of a cable, T_1, T_2 and T_3 are formulated separately. For single-core cables, the thermal resistance, T_1, between one conductor and its sheath is:

$$T_1 = \frac{\rho_T}{2\pi} \ln\left[1 + \frac{2\,t_1}{d_c}\right], \tag{2.10}$$

where ρ_T [K m/W] is the thermal resistivity of insulation, d_c [mm] is the diameter of the conductor and t_1 [mm] is the thickness of the insulation between conductor and sheath.

The thermal resistivity of the materials used for insulation can be found in Table 1 of IEC 60287-2-1 [4]. In the same part of the international standard, formulations of T_1 are given for belted cables, three-core cables, oil-filled cables, and SL and SA type cables. SL and SA cables are radial field single-core metallic sheath cables with electrostatic tape acting as the insulation screen. SL and SA refer to sheathing with lead and aluminium respectively [5].

The thermal resistance between sheath and armour for single-core, two-core and three-core cables, with a common metallic sheath, is represented by T_2 and it is formulated as follows:

$$T_2 = \frac{\rho_T}{2\pi} \ln\left[1 + \frac{2\,t_2}{D_s}\right], \tag{2.11}$$

where D_s [mm] is the external diameter of the sheath and t_2 [mm] is the thickness of the bedding. For SL and SA type cables, the formulation is given in IEC 60287-2-1 [3].

The thermal resistance of the outer covering, T_3, for external servings in the form of concentric layers is given by:

$$T_3 = \frac{\rho_T}{2\pi} \ln\left[1 + \frac{2\,t_3}{D'_a}\right], \tag{2.12}$$

where D'_a [mm] is the external diameter of the armour and t_3 [mm] is the thickness of the serving. Corrugated sheaths and pipe-type cables thermal resistances are further analysed in IEC 60287-2-1 [4].

The evaluation of external thermal resistance T_4 changes when considering cables protected from direct solar radiation or cables directly exposed to solar radiation. For the first case, the formulation is:

$$T_4 = \frac{1}{h\pi\,(\Delta\theta_s)^{\frac{1}{4}}\,D_e^*}, \tag{2.13}$$

where D_e^* [m] is the external diameter of the cable for corrugated sheaths and h is the heat dissipation coefficient (see [4]).

The international standard also presents methods of calculating $\Delta\theta_s$, the excess of cable surface temperature above ambient temperature. For groups of cables in free air, protected from solar radiation, a method for calculating reduction factors is given in IEC 60287-2-2 [6]. The method is valid when cables are mounted adjacent to each other and it is limited to:

- a maximum of nine cables in a square formation,
- a maximum of six circuits, each comprising three cables mounted in a trefoil, with up to three circuits placed side by side or two circuits placed one above the other.

Table 2.2   Ambient temperature at sea level

	Ambient air temperature		Ambient ground temperature at a depth of 1 m	
Climate	Min °C	Max °C	Min °C	Max °C
Tropical	25	55	25	40
Subtropical	10	40	15	30
Temperate	0	25	10	20

For the second case, cables directly exposed to solar radiation, the change is in the calculation of $\Delta\theta_s$ as detailed in IEC 60287-2-1 [4].

The specific operating conditions of cables vary depending on the country: reference ambient temperature and thermal resistivities of soil may therefore present different values for each country.

IEC 60287-3-1 [7] presents standard operating conditions when values are not provided by national tables.

National values are available for Australia, Austria, Canada, Finland, France, Germany, Italy, Japan, the Netherlands, Norway, Poland, Sweden, Switzerland, the United Kingdom and the United States of America.

Where national values are not given, the tables have to be used (Tables 2.2 and 2.3).

To complete this part, the purchaser is given a detailed list of information required by the cable manufacturer to select the appropriate type of cable. Information should be given on operating conditions and installation data.

Considering operation conditions, the information necessary to enable the selection of the appropriate type of cable is as follows:

a) the nominal voltage of the system, U;
b) the highest voltage of the three-phase system, U_m;
c) lightning overvoltage;
d) the system frequency;
e) the type of earthing;
f) environmental conditions should be provided, for example:

- the altitude above the sea level, if above 1000 m,
- indoor or outdoor installations,

Table 2.3   Thermal resistivity of soil

Thermal resistivity [K m/W]	Soil conditions	Weather conditions
0.7	Very moist	Continuously moist
1.0	Moist	Regular rainfall
2.0	Dry	Seldom rains
3.0	Very dry	Little or no rain

- excessive atmospheric pollution,
- termination in SF$_6$ switchgear;

g) the maximum rated current: for continuous operation, for cyclic operation and for emergency or overload operation, if any exists;
h) the expected symmetrical and asymmetrical short-circuit currents that may flow in the case of short circuits, both between phases and to earth;
j) the maximum time for which short-circuit currents may flow.

The installation data that are required can be divided into general data, underground cables and cables in air. Details on underground data can be found in [7]. For general data, the necessary information is:

a) the length and profile of route;
b) laying arrangements and how the metallic coverings are connected to each other and to earth;
c) special laying conditions, for example cables in water. Individual installations require special consideration.

For cables in air the requirements are as follows:

a) minimum, maximum and average ambient air temperature;
b) type of installation;
c) details of ventilation;
d) whether exposed to direct sunlight;
e) special conditions, for example fire risk.

2.3 Economic Aspects

Usually, the cost of losses during the lifetime of a cable is not taken into account when selecting a cable size. In fact, the selection leads to the minimum admissible cross-sectional area, minimizing the initial investment cost of the cable, but not taking into consideration its whole life-cycle. Therefore, the initial cost and the cost of the losses should be minimized (see IEC 60287-3-2 [8]) by estimating load growth and the cost of energy in order to provide the correct minimization of the sum of the future cost of energy losses and the initial cost of purchase and installation.

Purchase and installation costs are combined with the costs of energy losses and all these costs are expressed in comparable economic values, henceforth expressed as 'cu'. The date of purchase of the installation is therefore considered as the 'present' and future costs associated with costs of energy losses must be converted to their equivalent 'present values'. To do this, the 'discounting' process is applied so that the discounting rate is linked to the cost of borrowing money. The conditions and financial constraints of an individual installation influence the calculation of the present value of the cost of the losses. To obtain this value, it is indeed necessary to choose appropriate values for the future development of the load, annual increases in kWh price and annual discounting rates over the economic lifetime of the cable.

There are two different ways to calculate the economic sizing of a cable. In the first, for a particular installation, for each of the conductor sizes, the range of economic currents has to be calculated. Afterwards, it is necessary to select the size whose economic range contains the required value of the load. The second method is more suitable for a single installation: the optimum cross-sectional area for the required load is calculated and then the closest standard conductor size is selected. Economic aspects are not the only aspects that should be considered when finding the optimum size for a conductor. There are that should be considered in this order:

1. Calculate the economic cross-sectional area.
2. Check that the calculated size can carry the maximum load expected to occur and the temperature limits are respected.
3. Check short-circuit and earth fault currents.
4. Check voltage drop limits.
5. Consider other criteria that may affect the installation.

Finally, the supply quality of the installation should also be considered.

2.4   Calculation of the Current Rating: Total Costs

The first step in finding the optimum cable size is to express the total cost of installing and operating a cable during its economic life. All costs should be expressed in present values and the equation that represents the total cost (*CT*):

$$CT = CI + CJ, \tag{2.14}$$

where *CI* is the cost of the installed length of cable and *CJ* is the equivalent cost at the present value of the Joule losses.

2.4.1   Evaluation of CJ

This is the cost due to Joule losses and it is composed of two parts: the energy charge and the charge of the additional supply capacity to provide for the losses.

The cost due to energy charge (*CE*) is obtained considering first the energy losses (*EL*) [Wh] for the first year, expressed as:

$$EL = \left(I_{max}^2 \, R \, l \, N_p \, N_c\right) T, \tag{2.15}$$

where I_{max} is the maximum load on the cable during the first year, R is the apparent AC resistance of the conductor, considering skin, proximity effects and losses in metal screens and armour, l is the length of the cable. N_p is the number of phase conductor per circuit and N_c is the number of circuits carrying the same value and type of load. T represents the number of hours per year that the maximum current I_{max} would need to flow in order to produce the same total yearly energy losses at the actual, variable, load current. The operating time at maximum Joule loss can be found by applying:

$$T = \int_0^{8760} \frac{I(t)^2}{I_{max}^2} dt, \tag{2.16}$$

t being the time in hours and $I(t)$ the load current as a function of time. Finally, the cost of the first year's losses is represented by:

$$CE = \left(I_{max}^2 \, R \, l \, N_p \, N_c\right) TP, \tag{2.17}$$

where P is the cost of one watt-hour of energy at the relevant voltage level.

The cost of additional supply capacity (CA) [u/year] is:

$$CA = \left(I_{max}^2 \, R \, l \, N_p \, N_c\right) D, \tag{2.18}$$

where D is the demand charge per year and u is the arbitrary currency unit.

Considering CE and CA, the overall cost (OC) [cu] of the first year's losses is the sum of these two and, if the costs are paid at the end of year, the overall cost should be considered at its present value, therefore:

$$OC = \frac{\left(I_{max}^2 \, R \, l \, N_p \, N_c\right) (TP + D)}{\left(1 + {}^i/_{100}\right)}. \tag{2.19}$$

The present value of energy costs [cu] during N years of operation, discounted to the date of purchase is:

$$CJ = \left(I_{max}^2 \, R \, l \, N_p \, N_c\right) (TP + D) \cdot \frac{Q}{\left(1 + {}^i/_{100}\right)}, \tag{2.20}$$

where Q is a coefficient taking into account the increase in load, the increase in cost of energy over N years and the discount rate, i.

$$Q = \sum_{n=1}^{N} \left(r^{n-1}\right)$$

$$r = \frac{\left(1 + {}^a/_{100}\right)^2 \cdot \left(1 + {}^b/_{100}\right)}{\left(1 + {}^i/_{100}\right)}, \tag{2.21}$$

and a and b are the increase in load per year and the increase in the cost of energy per year.

The total cost is [cu] given by the sum of CI and CJ.

$$CT = CI + I_{max}^2 \, RlF_{[cu]}, \tag{2.22}$$

where F [cu/W] is expressed by

$$F = N_p N_c \, (TP + D) \frac{Q}{\left(1 + \frac{i}{100}\right)}. \tag{2.23}$$

2.5   Determination of Economic Conductor Sizes

To evaluate the economic size of a conductor, two approaches can be employed: the first one analyses the economic current range for each conductor in a series of sizes and the second one considers the economic conductor size for a given load.

2.5.1   Economic Current Range for Each Conductor in a Series of Sizes

For a given installation condition, all conductor sizes have an economic current range [A] that presents an upper and a lower limit. For a given conductor size, the upper and lower limits are:

$$I_{\text{low}}^{\text{max}} = \sqrt{\frac{CI - CI_1}{FI\,(R_1 - R)}}$$

$$I_{\text{up}}^{\text{max}} = \sqrt{\frac{CI_2 - CI}{FI\,(R - R_2)}}, \tag{2.24}$$

where CI is the installed cost of the length of cable whose conductor size is being considered, expressed in cu, an arbitrary currency unit. R [Ω/m] is the AC resistance per unit length of the conductor size being considered, CI_1 [cu] is the installed cost of the next smallest standard conductor, R_1 [Ω/m] is the AC resistance per unit length of the next smallest standard conductor, CI_2 [cu] is the installed cost of the next largest standard conductor and R_2 [Ω/m] is the AC resistance per unit length of the next largest standard conductor.

2.5.2   Economic Conductor Size for a Given Load

The economic conductor size is the cross-section that minimizes the total cost function [cu]:

$$CT(S) = CI(S) + I_{\text{max}}R(S)\,lF^2. \tag{2.25}$$

The equation for the relationship between $CI(S)$ and conductor size can be derived from known costs of standard cable sizes. The apparent conductor resistance [Ω/m] can be expressed as a function of the cross-section (see IEC 60287-1-1):

$$R(S) = \frac{\rho_{20}\left(1 + y_{\text{p}} + y_{\text{s}}\right)\left(1 + \lambda_1 + \lambda_2\right)\left[1 + \alpha_{20}\left(\theta_{\text{m}} - 20\right)\right]}{S}10^6, \tag{2.26}$$

where ρ_{20} [Ω m] is the DC resistivity of the conductor, y_{p} is the proximity effect, see IEC 60287-1-1 [3], y_{s} is the skin effect, see IEC 60287-1-1, λ_1 represents the sheath loss factor, see IEC 60287-1-1, λ_2 is the armour loss factor, see IEC 60287-1-1 [3], α_{20} [K^{-1}] is the temperature coefficient of resistivity for the particular conductor material at 20°C, θ_{m} [°C] is the conductor temperature and S [mm^2] is the cross-sectional area of the cable conductor.

If a linear model can be fitted to the values of initial cost [cu] for the type of cable and installation under consideration, then:

$$CI(S) = l(AS + C), \tag{2.27}$$

where A [cu/m, cu/mm^2] is the variable component of cost, related to the conductor size, C [cu/m] is the constant component of cost, unaffected by size of cable and l [m] represents the length of cable.

The optimum size [mm] can be obtained deriving the equation of $CT(S)$:

$$S_{ec} = 1000 \left[\frac{I_{max}^2 F \rho_{20} \left(1 + y_p + y_s\right) (1 + \lambda_1 + \lambda_2) \left[1 + \alpha_{20} (\theta_m - 20)\right]}{A} \right]^{0.5} . \qquad (2.28)$$

This section will not be exactly equal to a standard size: larger and smaller standard sizes must be calculated to choose the most economical one.

2.6   Summary

Ampacity calculations are required to establish the maximum current-capacity that a cable can tolerate without risking deterioration or damage. Technical criteria are necessary to carry out the ampacity calculations and the procedures generally used for the selection of a cable size lead to the minimum cross-sectional area. The initial cost is therefore minimized but the cost of the losses that will occur during the life of the cable is not minimized. In the last decade the cost of energy in Western Countries has been increasing fast and newer insulating materials allow operation at higher temperatures than before. Instead of just minimizing the initial cost, the sum of the initial cost of losses over the economic life of the cable should be also minimized.

References

[1] Ranasamy Natarajan, *Computer-Aided Power System Analysis*, CRC, 2002.
[2] George J. Anders, *Rating of Electric Power Cables in Unfavorable Thermal Environment*, Wiley-IEEE Press, May 2005.
[3] IEC 60287-1-1 Electric cables, Calculation of the current rating. Part 1-1: Current rating equations (100% load factor) and calculation of losses – General.
[4] IEC 60287-2-1 Electric cables, Calculation of the current rating. Part 2-2: Thermal resistance – Calculation of thermal resistance.
[5] George F. Moore, *Electric Cables Handbook*, 3rd edn, BICC Cables, 2004.
[6] IEC 60287-2-2 Electric cables, Calculation of the current rating. Part 2-2: Thermal resistance – A method for calculating reduction factors for groups of cables in free air, protected from solar radiation.
[7] IEC 60287-3-1 Electric cables, Calculation of the current rating. Part 3-1: Sections on operating conditions – Reference operating conditions and selection of cable type.
[8] IEC 60287-3-2 Electric cables, Calculation of the current rating. Part 3-2: Sections of operating conditions, Economic optimization of power cable size.

3

Power Transformers

Roman Targosz, Stefan Fassbinder and Angelo Baggini

Power transformers are an essential part of the electricity network as they convert electrical energy from one voltage level to another. After having been generated in a power station, electrical energy needs to be transported to the areas where it is consumed. This transport is more efficient at higher voltage, which is why power generated at 10 to 30 kV is converted by transformers into typical voltages of 220 kV up to 400 kV, or even higher. There are power transformers in large transmission substations, usually at major transmission nodes close to large power plants, which add flexibility to transmission channels and connect to subtransmission level. Transformers or autotransformers installed there are here referred to as grid coupling transformers.

Since the majority of electrical installations operate at lower voltages, the high voltage needs to be converted back close to the point of use. The first step down is transformation to 33–150 kV. This is often the level at which power is supplied to major industrial customers. Distribution companies then transform power further, down to the consumer mains voltage.

In this way, electrical energy passes through an average of four transformation stages before being consumed. A large number of transformers of different classes and sizes are needed in the transmission and distribution network, with a wide range of operating voltages. Large transformers for high voltages are called system transformers. The last transformation step into the consumer mains voltage (in Europe 400/230 V) is done by the distribution transformer. Distribution transformers operated and owned by electricity distribution companies are responsible for supplying about 70% of low voltage electricity to final users and represent about 80% of distribution transformers' stock. Voltage levels are classified as:

- Extra high voltage: transmission grid (>150 kV) typically 220–400 kV (ultra high > 400 kV)
- High voltage >70 kV up to 150 kV: subtransmission (the interface between TSO and DSO)
- Medium voltage >1 kV up to 70 kV (typically up to 36 kV)
- Low voltage < 1 kV (e.g. 110 V, 240 V, 690 V).

Electrical Energy Efficiency: Technologies and Applications, First Edition. Andreas Sumper and Angelo Baggini.
© 2012 John Wiley & Sons, Ltd. Published 2012 by John Wiley & Sons, Ltd.

Power transformers provide very good opportunities for energy saving. There have been numerous attempts to assess the energy saving potential of transformers. As an example, in 2005 Leonardo ENERGY estimated that there was at least 200 TWh global energy saving potential from distribution transformers only.

According to the International Energy Agency (IEA, http://data.iea.org) energy statistics, the global generation of electricity in 2008 was 20 270 TWh, of which around 10%, i.e. 2000 TWh, was generated in generators, mostly renewables, that are interconnected at non-high voltage level (below 70 kV). Consumption was reported at 16 816 TWh, while losses were 1656 TWh. Making the balance, the remaining 1798 TWh was kept by the energy industry for its own use. At the same time, the world generating capacity was 4625 GW.

Looking into the breakdown of transmission and distribution (T&D) losses. Analysis performed in SEEDT project indicated the following T&D losses from transformers in Europe by the year 2000:

- distribution transformers 25%;
- HV/MV transformers 10%;
- transmission transformers 10%.

This share might have slightly reduced in the last ten years, but so far none of the following losses, not considered before, have been included in the estimate of the saving potential:

- generator transformers;
- transformers used by the energy industry for its own needs (supplying power station installation and auxiliaries);
- distributed energy (DER) transformers, integrating mainly renewables into electricity networks (including converter transformers for HVDC to AC connections).

Transformers installed in electricity networks worldwide are responsible for about 40% of the total T&D losses, which results in about 650 TWh. This estimate is based on energy statistics with the following provisions:

- Europe's 45% transformer losses contribution from the year 2000 has decreased because of some realization in saving potential already and is about 10% higher than the world's average due to the power factor.
- At the same time, European estimates do not include losses from transformers as stated above.

How much of these losses can be saved? Transformers are already quite efficient devices and great progress in the reduction of losses has been already achieved. The potential savings are, however, still high as there is still a reserve in design and technology, and because so many voltage transformation steps are in use in the electricity system. For example US and European studies during the preparation of transformer Minimum Energy Performance Standards (MEPS) have indicated that distribution transformers can be about 1% more efficient using today's best available technologies compared with the average units in operation and 0.5% compared with MEPS. As for power transformers, the gap is lower and

Table 3.1 Typical energy efficiencies of different transformer types

Transformer	S [kVA]	Current density [A/mm^2]	Energy efficiency
Miniature transformer	0.001	7.000	45.00%
Small transformer	0.100	3.000	80.00%
Industrial transformer	40.000	3.397	96.00%
Distribution transformer	200.000		98.50%
Bulk supply point transformer	40 000.000	3.000	99.50%
Generator transformer	600 000.000		99.75%

traditional technologies make it practical to increase efficiencies by 0.1% to 0.2% compared with current levels.

On average this can be a nearly 50% improvement for all the electricity network transformers, representing a savings potential of more than 300 TWh. In this chapter the focus will be mainly on distribution transformers.

The largest energy saving potential is, in effect, represented by distribution transformers. They still have the largest reserve for saving energy, have a very long lifetime and the proportion of the lifetime cost of losses to production costs is relatively high. Larger power transformers are already more efficient and the production (including design and tools) is different from the case of distribution transformers. These decrease, beyond a scaling factor of exponents of 3/4 for weight, cost and losses, the proportion of production cost to lifetime cost of losses and decrease the attractiveness of reaching much improvement in efficiency. However they are still relevant. Smaller transformers, as seen in Table 3.1, are much less efficient than distribution or power transformers, however their capacity and availability factors are so low as the life cycle costing for these transformers is very often beneficial for low first cost (production) units, as experienced in the preparatory study to the Energy related Products Directive, when analysing earthing transformers.

3.1 Losses in Transformers

Transformer losses can be classified into two main components: no-load losses and load losses. These types of losses are common to all types of transformers, regardless of transformer application or power rating. There are, however, two other types of losses: extra losses created by the non-ideal quality of power and losses, which may apply particularly to larger transformers – cooling or auxiliary losses, caused by the use of cooling equipment such as fans and pumps.

3.1.1 No-Load Losses

These losses occur in the transformer core whenever the transformer is energized (even when the secondary circuit is open). They are also called iron losses or core losses and are constant. They are composed of the following:

- *Hysteresis losses*, caused by the frictional movement of magnetic domains in the core laminations being magnetized and demagnetized by alternation of the magnetic field. These losses depend on the type of material used to build a core. Silicon steel has much lower hysteresis than normal steel but amorphous metal has much better performance than silicon steel. Hysteresis losses can be reduced by material processing such as cold rolling, laser treatment or grain orientation. Hysteresis losses are usually responsible for more than a half of total no-load losses ($\sim$50% to $\sim$80%). This ratio was smaller in the past (due to the higher contribution of eddy current losses).

- *Eddy current losses*, caused by varying magnetic fields inducing eddy currents in the laminations and thus generating heat. These losses can be reduced by building the core from thin laminated sheets insulated from each other by a thin varnish layer to reduce eddy currents. Eddy current losses usually account for 20–50% of the total no-load losses.

There are also less significant stray and dielectric losses that occur in the transformer core, accounting usually for no more than 1% of total no-load losses.

3.1.2   Load Losses

These losses are commonly called copper losses or short-circuit losses. Load losses vary according to the transformer loading. They are composed of:

- *Ohmic heat loss*, sometimes referred to as copper loss, since this resistive component of load loss dominates. This loss occurs in transformer windings and is caused by the resistance of the conductor. The magnitude of these losses increases with the square of the load current and is proportional to the resistance of the winding. It can be reduced by increasing the cross-sectional area of the conductor or by reducing the winding length. Using copper as the conductor maintains the balance between weight, size, cost and resistance; adding an additional amount to increase conductor diameter, consistent with other design constraints, reduces losses.
- *Conductor eddy current losses*. Eddy currents, due to magnetic fields caused by alternating current, also occur in the windings. Reducing the cross-section of the conductor reduces eddy currents, so stranded conductors with the individual strands insulated against each other are used to achieve the required low resistance while controlling eddy current loss. Effectively, this means that the 'winding' is made up of a number of parallel windings. Since each of these windings would experience a slightly different flux, the voltage developed by each would be slightly different and connecting the ends would result in circulating currents, which would contribute to loss. This is avoided by the use of continuously transposed conductor (CTC), in which the strands are frequently transposed to average the flux differences and equalize the voltage.

A good example of how current density influences load losses is given in Table 3.2. The average commonly used value of 3 A/mm^2 will result in almost double the load resistive loss for only one-quarter (25%) higher current density, i.e. one-quarter less conductor material.

3.1.3   Auxiliary Losses

These losses are caused by using energy to run cooling fans or pumps, which help to cool larger transformers.

Table 3.2 Improvement in load losses by increasing the conductor cross-section

			'Energetic payback' from using more copper (operating hours at full load)			
	Specific loss in copper magnet wire				Time of operation	
at	1.00	A/mm^2	1.96	W/kg	1854	h
at	1.50	A/mm^2	4.41	W/kg	824	h
at	2.00	A/mm^2	7.84	W/kg	463	h
at	2.25	A/mm^2	9.92	W/kg	366	h
at	2.50	A/mm^2	12.25	W/kg	297	h
at	2.75	A/mm^2	14.82	W/kg	245	h
at	3.00	A/mm^2	17.64	W/kg	206	h
at	3.50	A/mm^2	24.01	W/kg	151	h
at	4.00	A/mm^2	31.35	W/kg	116	h
at	4.50	A/mm^2	39.68	W/kg	92	h
at	5.00	A/mm^2	48.99	W/kg	74	h
at	6.00	A/mm^2	70.55	W/kg	51	h
at	7.00	A/mm^2	96.02	W/kg	38	h
at	8.00	A/mm^2	125.42	W/kg	29	h

3.1.4   Extra Losses due to Harmonics, Unbalance and Reactive Power

This category of losses includes those extra losses that are caused by unbalanced harmonics and reactive power.

Power losses due to eddy currents depend on the square of frequency, so the presence of harmonic frequencies that are higher than the normal 50 Hz frequency causes extra losses in the core and windings. These additional losses deserve separate attention and are discussed below.

The reactive component of the load current generates a real loss even though it makes no contribution to useful load power. Losses are proportional to $1/(\cos \varphi)^2$ of the active power, but the power rating of the transformers always gives the apparent power. This reactive power is responsible for active power losses in supplying network coming from reactive power for transformer core magnetization and from reactive power stray losses. Low power factor loads should be avoided to reduce losses related to reactive power, also in transformers.

3.1.4.1   Harmonics

Before we start discussing harmonics considerations in transformers it should be stated that there is a non-univocal position if only current distortion and the harmonic effects on windings are worth practical consideration. Voltage harmonics as well as harmonics effects on no-load losses, according to some studies, may in some cases contribute significantly to extra losses in transformers.

Non-linear loads, such as power electronic devices, such as variable speed drives in motor systems, computers, UPS systems, TV sets and compact fluorescent lamps, cause harmonic currents on the network. Harmonic voltages are generated in the impedance of the network by the harmonic load currents. Harmonics increase both load and no-load losses due to increased skin effect, eddy current, stray and hysteresis losses.

3.1.4.2 Current Distortion

The most important of these losses is that due to eddy current losses in the winding, it can be very large and consequently most calculation models ignore the other harmonic-induced losses!

The precise impact of a harmonic current on load loss depends on the harmonic frequency and the way that the transformer is designed. In general, the eddy current loss increases by the square of the frequency and the square of the load current. So, if the load current contains 20% fifth harmonic, the eddy current loss due to the harmonic current component would be $5^2{}^*0.2^2$ multiplied by the eddy current loss at the fundamental frequency, meaning that the eddy current loss would have doubled.

To avoid excessive heating in transformers supplying harmonic currents, two approaches are used:

1. Reducing the maximum apparent power transferred by the transformer, often called derating. To estimate the required derating of the transformer, the load's derating factor may be calculated. This method, used commonly in Europe, is to estimate by how much a standard transformer should be de-rated so that the total loss on harmonic load does not exceed the fundamental design loss. This derating parameter is known as 'factor K'.

   The transformer derating factor is calculated according to the formula:

$$K = \left[1 + \frac{e}{1+e}\left(\frac{I_h}{I}\right)^2 \sum_{n=2}^{n=N}\left(n^q\left(\frac{I_n}{I_1}\right)^2\right)\right]^{0.5}, \tag{3.1}$$

   where
   e = the eddy current loss at the fundamental frequency divided by the loss due to a DC current equal to the rms value of the sinusoidal current, both at reference temperature;
   n = the harmonic order;
   I = the rms value of the sinusoidal current including all harmonics given by:

$$I = \left(\sum_{n=1}^{n=N}(I_n)^2\right)^{0.5} = I_1\left[\sum_{n=1}^{n=N}\left(\frac{I_n}{I_1}\right)^2\right]^{0.5}, \tag{3.2}$$

   where
   I_n = the magnitude of the nth harmonic;
   I_1 = the magnitude of the fundamental current;
   q = exponential constant that is dependent on the type of winding and frequency. Typical values are 1.7 for transformers with round rectangular cross-section conductors in both windings and 1.5 for those with foil low voltage windings.

2. Developing special transformer designs rated for non-sinusoidal load currents. This process requires analysis and minimizing of the eddy loss in the windings, calculation of the hot spot temperature rise, individual insulation of laminations, and/or increasing the size of the core or windings. Each manufacturer will use any or all of these techniques according to labour rates, production volume and the capability of his plant and equipment. These products are sold as 'K rated' transformers. During the transformer selection process, the

designer should estimate the K factor of the load and select a transformer with the same or higher K factor, defined as:

$$K = \sum_{n=1}^{n=n_{max}} I_n^2 n^2. \tag{3.3}$$

For example, when the current harmonic distortion factor THDI = 20%, additional losses in the transformer load circuit, caused by the presence of high harmonics in the load current, will increase by about 4% as related to the losses caused by the current fundamental harmonic.

As an example of specific transformer harmonics considerations the IEC 61378-1 deals with the specification, design and testing of power transformers and reactors that are intended for integration within semiconductor converter plants; it is not designed for industrial or public distribution of AC power in general. The scope of this standard is limited to applications of power converters, of any power rating, for local distribution, at moderate rated converter voltage, generally for industrial applications and typically with a highest voltage for equipment not exceeding 36 kV. The converter transformers covered by this standard may be of the oil-immersed or dry-type design. The oil-immersed transformers are required to comply with IEC 60076, and with IEC 60726 for dry-type transformers. Note also that EN 50464 Part 3 is dedicated on the Determination of the power rating of a transformer loaded with non-sinusoidal currents (K Factor).

3.1.4.3 Voltage Distortion

The common approach presented above assumes that although the magnetizing current does include harmonics, these are extremely small compared with the load current and their effect on the losses is minimal. As a result, in standards such as ANSI/IEEE C57.110 it is assumed that the presence of harmonics does not increase the core loss.

When not ignoring extra harmonic losses from voltage harmonics and also those generated in the transformer core, the full formula to calculate losses in transformers due to harmonics, is as follows:

$$P_T = 3 \sum_n I_n^2 \cdot R_n + P_{Fe} \sum \left(\frac{V_n}{V_1}\right)^m \cdot \frac{1}{n^{2.6}}, \tag{3.4}$$

where
P_T = losses of transformer due harmonic distortion;
P_{Fe} = fundamental frequency iron losses;
R_n = equivalent copper loss resistance of transformer at the nth order;
V_1 = fundamental component voltage;
V_n = harmonic voltage of order n;
I_n = harmonic current of order n;
n = order of harmonic;
m = exponent empiric value (assumed to be the value 2).

The second component in the above equation represents losses in the transformer core caused by voltage distortion. This is a partly empiric formula that may still underestimate core harmonic losses caused by current distortion.

Metglas, the introducers of amorphous metal into transformer cores, formulated the following theory.

Current distortion in power networks leads to increased transformer core losses, since hysteresis and eddy current losses vary as f and f^m respectively, where f is the frequency and m varies from ~1.5 to ~2, depending on the core material. The situation is worse in transformers using conventional steels with relatively higher hysteresis and eddy current losses than amorphous metals. Therefore the difference in overall transformer loss in amorphous core and conventional silicon steel core transformers widens as higher harmonic contents increase in the power distribution line.

A breakdown of the losses in transformers with conventional and amorphous cores is presented in Table 3.3.

Losses also occur in the magnetic circuit and they increase in the presence of the voltage harmonics. They are eddy current losses, associated with the frequency of high harmonics and magnetic loss in the core, caused by the voltage high harmonics.

First of all, the reason that they are ignored is the level of voltage distortion. It is usually at least one order of magnitude lower than the current distortion. However, voltage harmonics strengthen the effect of unbalanced zero sequence currents circulating in transformer delta windings and associated heat losses. Such losses also contribute to the 'additional' supplementary load losses. An experiment carried out with a small transformer can illustrate just how quickly this additional loss channel can attain a significant size. The series resistance of the delta-connected secondary winding is 0.1 Ω. A THD of only 3.2% in the primary voltage results in a circulating current of 2.3 A. The resulting $I^2 \times R$ loss is therefore about 0.5 W. Half a watt is about 1% of the total copper losses and doesn't actually sound that bad. If the voltage THD rises to 6.4%, which can occur in practice, the Joule heating loss would increase to 4% of the total copper losses or 4.6 A, which in this case would correspond to 28% of the rated current. The transformer load would therefore have to be reduced by 28% solely in order to prevent overheating of the secondary winding by the 150 Hz circulating current.

Now, let us take one more look at the voltage. There are a few particular situations in which the voltage can be so distorted that it has a detrimental effect on the performance of the transformer that it is driving. For example, it is an 'inherent characteristic' of small UPS

Table 3.3 Comparison of harmonics losses in amorphous and silicon steel transformers (THD = 25% at approx 56% transformer loading)

	Hysteresis loss [W]	Eddy current loss [W]	Load loss [W]	Total losses [W]
Non-distorted amorphous	99	33	966	1098
Distorted amorphous	99	74	1553	1726
Non-distorted CRGO	155	311	1084	1550
Distorted CRGO	155	698	1671	2524

systems (in other words, the alternatives are too expensive) that when power loss occurs, they generate a square wave, rather than a sinusoidal voltage. However, a non-stepped square wave will have a form factor that is 11% less than that of a sinusoidal waveform. The 11% is the factor linking the mean value and the RMS value. The quoted value is always the RMS value, or at least it should be. But the degree of magnetization depends on the mean value. The correct RMS value at the output side of a small UPS can cause significant overexcitation of the transformer to which it is connected. In addition, the harmonic distortion of a square wave is so high that very substantial no-load losses must be expected.

3.1.4.4   Mitigation of Extra Harmonic Losses

Despite transformers derating, which has already been described above, the effects of harmonics can largely be mitigated if transformers with different vector groups are installed across the system. Thanks to different phase shifts, such transformers would encourage harmonics with different phase angles to cancel each other. Also, zigzag transformers in which the delta-connected and the star-connected parts of the relevant winding have the same voltage, will minimize the effect of harmonics. Unfortunately such measures are rarely used as distribution network operators use the same vector grouping usually Dyn 5 or 11 to retain an option of transformers operating in parallel. Also the standardization and exchangeability of transformer stock are arguments against using different vector groups.

3.1.4.5   Unbalance

Transformers subject to negative sequence voltages transform them in the same way as positive-sequence voltages. The behaviour with respect to homopolar voltages depends on the primary and secondary connections and, more particularly, the presence of a neutral conductor. If, for instance, one side has a three-phase four-wire connection, neutral currents can flow. If at the other side the winding is delta-connected, the homopolar current is transformed into a circulating (and heat-causing) current in the delta. The associated homopolar magnetic flux passes through constructional parts of the transformer causing parasitic losses in parts such as the tank, sometimes requiring an additional derating. The indicative extra harmonic losses due to unbalance are presented in Table 3.4.

Table 3.4   Indicative unbalance losses in transformers

Ratio of neutral current to average phase currents	Transformers extra losses in %
0.5	6–8
1.0	15–20
1.5	35–50
2.0	70–90
3.0	150–200

3.2   Efficiency and Load Factor

The presented efficiencies are supposed to reflect the operating conditions (loading) of transformers. What matters a great deal is the load of a transformer. Transformers operate at their highest efficiency when the load and no-load losses are equal. This comes out from the following equations.

The ratio of the increments of losses to power should be a minimum:

$$\frac{\Delta P}{S} = \frac{\Delta P_O}{S} + \frac{\Delta P_k}{S} = \frac{\Delta P_O}{S} + \frac{\Delta P_k \left(\frac{S}{S_n}\right)^2}{S} = \frac{\Delta P_O}{S} + \frac{\Delta P_k}{S_n^2}S, \qquad (3.5)$$

where

ΔP = increments of losses;

P_k = load;

Po = no load;

S = instantaneous power;

S_n = rated.

This minimum can be calculated from derivatives. The first derivative of the above relation is zero, while the second derivative is positive so the optimum power is given by the relation:

$$S_{opt} = S_n \sqrt{\frac{P_o}{P_k}} \qquad (3.6)$$

In practice, the optimum efficiency of transformers where both losses are equal is between 25% and 50% of transformer load. Equation (3.6) shows how to find easily the loading at which a transformer reaches its highest efficiency. If, for example, the ratio of no-load losses by load losses is one by four, the square root of 0.25 is 0.5, which means optimum efficiency between 25% and 50% loading.

Examples of two transformers with different proportions of load (Cu) and no-load (Fe) losses are shown in Figure 3.1.

Table 3.5 compares losses and efficiency of different types of transformers 100% and 50% loaded.

Table 3.5   Losses and efficiency at 50% and 100% load

Transformer type	Rated power	Efficiency at		Loss at	
		100% rated load	50% rated load	100% rated load	50% rated load
	[MVA]	%	%	[kW]	[kW]
Generator transformer	1100	99.60	99.75	4400	1375
Interbus transformer	400	99.60	99.75	1600	500
Substation transformer	40	99.40	99.60	240	80
Distribution transformer	1	98.60	99.00	14	5

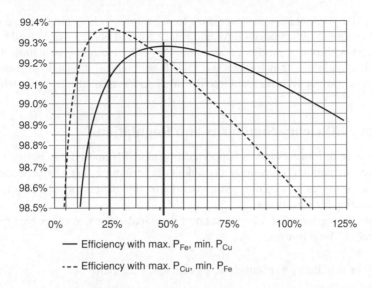

— Efficiency with max. P_{Fe}, min. P_{Cu}

--- Efficiency with max. P_{Cu}, min. P_{Fe}

Figure 3.1   Transformer efficiency versus loading

3.3   Losses and Cooling System

Transformer losses in the form of heat have to be evacuated from the inside to the outside of the transformer. Air, gas or liquid can play the role of insulator and also of the cooling agent. However cooling can be natural or forced by fans or pumps. The cooling agent and the forced flow of the coolant are important aspects of transformer efficiency. Transformers follow some physical rules that relate certain parameters to their size. These relations can be expressed as:

$$\frac{Parameter_2}{Parameter_1} = \left(\frac{S_2}{S_1}\right)^x, \tag{3.7}$$

where the left-hand side of the equation represents selected parameters of transformers 2 and 1, while S_2 and S_1 are the apparent powers of transformers 2 and 1 and x is the exponent, which in theory is $x = \frac{1}{4}$ (one-quarter) for each of the transformer's dimensions width, length and height, hence $\frac{3}{4}$ for the volume. So the volume, approximately proportional to the mass of a transformer grows less than the power throughput or, expressed inversely, the power density increases with greater power ratings.

The surface of a transformer, which is responsible for heat dissipation, thereby increases even less. The exponent will be $\frac{2}{3}$ against any of the dimensions (width, length or height) and thus $\frac{2}{3} \times \frac{3}{4} = \frac{1}{2}$ against the power rating, while the relation for weight (including active materials proportion) and losses has an exponent value of $\frac{3}{4}$ (three-quarters).

It is clear now that larger transformers needed additional cooling. The first step is to introduce liquid cooling of the transformer windings. Further cooling can then be achieved by increasing the area of the transformer cooling surfaces. This type of cooling system is known as ONAN cooling (oil natural, air natural circulation). Forced cooling is used in transformers with ratings above about 40 MVA. In this type of cooling, known as ONAF cooling (oil natural, air forced circulation), liquid cooling is augmented by cool air blown in between the oil radiators by fans.

Above about 400 MVA, it becomes necessary to use pumps to help circulate the oil coolant. This form of cooling is abbreviated OFAF and stands for 'Oil Forced, Air Forced Circulation'. In transformers with power ratings greater than 800 MVA, simply circulating the oil is no longer sufficient and these transformers use ODAF cooling (oil directed, air forced cooling) in which a jet of cooling oil is directed into the oil channels of the transformer windings.

3.4   Energy Efficiency Standards and Regulations

Energy efficiency in transformers is supported by standards and energy policy instruments. Standards are international or country documents describing either test procedures including loss tests, tolerances and guiding on transformers application including lifetime costing, loading or derating for harmonics.

Policy instruments are used more to support principal targets, such as energy efficiency. They may include the following:

- a voluntary or mandatory minimum energy efficiency standard;
- labelling;
- incentives from obligations or certificate schemes;
- other financial or fiscal incentives;
- information and motivation;
- tool-kits for buyers;
- energy advice / audits;
- co-operative procurement;
- support for R&D and pilot or demonstration projects.

Although mandatory regulations guarantee the strongest enforcement it is important to mention that energy policy should always act as a mix of instruments. Regulations usually referred to MEPS (Minimum Energy Performance Standards) for transformers have evolved in many countries over the last decade. Except for China and European proposals of MEPS for 'non-distribution' power transformers, such regulations cover distribution transformers, both liquid-immersed and dry types of transformers.

The main international normative reference is IEC 60076, Power transformers – series. The IEEE equivalent standard for IEC 60076-1 (2000) is the IEEE C57.12.00 (2006). IEC 60076 gives detailed requirements for transformers for use under defined conditions of altitude, ambient temperature for both:

- oil-immersed transformers in IEC 60076-2, and
- dry-type transformers in IEC 60076-11.

The IEC 60076 series consists of the following parts relevant to energy efficiency:

- Part 1: 1993, General – definition of terms
- Part 2: 1993, Temperature rise
- Part 3: 1980, Insulation levels and dielectric tests
- Part 5: 1976, Ability to withstand short circuit

- Part 7: 2005, Loading guide for oil-immersed power transformers. This part provides recommendations for the specification and loading of power transformers complying with IEC 60076 from the point of view of operating temperatures and thermal ageing. It also provides recommendations for loading above the nameplate rating and guidance for the planner to choose rated quantities for new installations. The use of life time is based on the hot spot temperature in the winding. An increase of the hot spot temperature with 6K is a reduction of the life time by 50%.
- Part 8: 1997, Application guide.

The most important aspects are that the maximum allowable tolerance on the total losses (sum of the load and no-load losses) is +10% of the total losses (IEC 60076-1). This standard in clause IEC 60076-1/ 7.1 stipulates that the values of the losses or the efficiency class of the transformer is not mandatory information on the rating plate of the transformer.

It is worth mentioning the initiative of Technical Committee no. 14 of IEC, which has initiated a project of new IEC 60076-XX standard: Power transformers – Part XX: Energy efficiency for distribution transformers.

This standard is intended to guide purchasers of power transformers in choosing the most appropriate level of energy efficiency, and the most appropriate method of specifying that efficiency. It will also provide a guide on the loss measurement where not provided for in other standards, and tables of standard losses for certain types of transformers.

As justification it says 'Energy efficiency is becoming more and more important as a worldwide issue for electricity transmission and distribution. A standard is needed to provide a method to calculate energy efficiency according to the way in which the transformer is to be used and the type of transformer, as the best balance between energy use and use of resources in the construction of the transformer will depend on these factors.'

The target of this standard will be:

- Calculation of energy efficiency according to the following parameters
- Type of load (inductive, reactive, resistive)
- Level of rated power
- To provide standard levels of load losses and no-load losses to suit particular efficiency requirements
- The ways in which loss measurement can be done
- The ways in which the uncertainties of measurement can be considered
- Tolerances on guarantees.

Now let us have a look into detailed MEPS in different countries.

There has been a substantial level of international activity concerning efficiency supporting instruments including MEPS for (distribution) transformers. Comparison of these international efficiency classes is not always obvious because of:

- differences in electricity distribution systems: grid voltages, grid frequencies (50 Hz versus 60Hz), etc.;
- differences in definitions for apparent power rating of the transformer (input power versus output power);

- differences in load levels at which the efficiency of the transformer is measured (50% load, 100% load, etc.);
- differences in normalized operating temperatures;
- different rated sizes of transformers.

In the process of preparing the MEPS the social, economic and technical feasibility aspects are taken into account. The common approach is to set the standards as close as possible to minimize the product life cycle. The environmental perspective is equally significant but although life cycle costing and life cycle assessment are different things, they lead to similar results for most transformers. The reason is that the use phase of a transformer, especially a distribution transformer, is usually responsible for even more than 95% of the life cycle environmental impact. The energy from transformer losses with associated emissions is the dominant component there. The potential for global warming and acidification, which are mostly energy related emissions have relative environmental impacts. All together there are seven environmental impacts, including GWP and acidification, as well as eutrophication, ozone depletion, etc. The impact of GWP and acidification are around 100 times larger than all the remaining impacts. The comparison of energy cost (including energy used in production and end of life phases) and life cycle for distribution transformers analysed in the European MEPS preparatory study is given in Figure 3.2.

The asterisks in Figure 3.2 designate amorphous core options that have not been considered as technology relevant for MEPS at the early stage. At this point it is worth delving deeper into the structure of life cycle costs.

These have three main components: transformer price, cost of load and cost of no-load losses. Here we focus more on transformer price and its relation to losses. Figure 3.3 compares product price with the capitalized cost of losses for distribution transformers. Ideally the increased price of the transformer is fully compensated for by the decreased cost of losses.

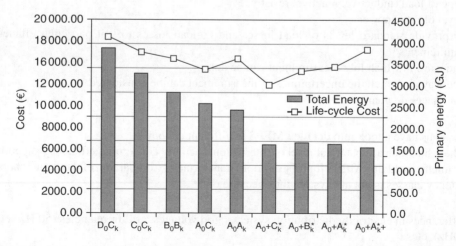

Figure 3.2 Environmental performance, life cycle costs versus primary energy for 400 kVA oil transformer

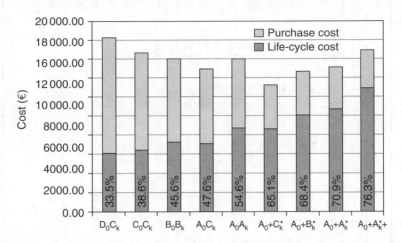

Figure 3.3 Environmental performance, price versus capitalized cost of losses for 400 kVA oil transformer

In this case the lowest life cycle cost (LLC) is at A_0C_k loss level, and this is the proposed MEPS for transformers up to 630 kVA. However, as the difference in LLC between A_0A_k and A_0C_k decreases with increase of transformer size due to higher loading and lower relative production cost, the A_0A_k loss mix is proposed for larger transformers (>630 kVA) as MEPS. At the same time, even on the given example, A_0A_k compared with A_0C_k losses provides incentive in the form of more than 100 GJ of primary energy and this value is expected to increase more than proportionally with transformer rating (the conversion factor 1 kWhe = 10.5 MJ given in the European methodology for preparatory studies was used to convert the electricity consumption).

The precise analysis of all variants that shows differences in transformer cost is presented in Table 3.6.

It is obvious that reduction of losses and associated costs comes at some expense. The US Department of Energy estimated that an increase in the energy efficiency of one percentage point increases the transformer price by 73% (DOE, 2001). The recent EuP study shows that a 400 kVA oil transformer – the size that has been selected as the most representative for the average sold distribution transformers – experiences a rate of loss reduction similar to the price increase but loss reductions at a higher efficiency level, as expected, come at a higher price increment. See the last row in Table 3.6.

MEPS are organized in the form of either maximum loss tables or efficiency tables calculated at certain loading levels; 100%, or more often 50%, which represents loading closer to real operating conditions and closer to optimum efficiency as well. The Japanese top runner scheme uses formulae to calculate efficiency for different transformers from their kVA value at 40% load. India applied an interesting idea of specifying maximum losses for two transformer loading levels 50% and 100%. This is to secure that transformers have the required proportion of no- load to load losses.

In addition to losses, European standards specify noise levels together with no-load losses as the noise in transformers is mostly related to core magnetization.

Table 3.6 Environmental performance, comparison of different options, production, losses

		DOCk	COCk		BOBk		AOCk		AOAk		A0 + Ck*	
			Relative	Absolute	Relative	Absolute	Relative	Absolute	Relative	Absolute	Relative	Absolute
Core	Core steel (kg)	468.7	1.12	524.94	1.22	571.81	1.35	632.75	1.45	679.62	1.59	747
Windings	Aluminium wire (kg)	21.44	1.06	22.73	1.44	30.87	1.37	29.37	2.07	44.38	1.59	82
	Copper wire (kg)	144.72	1.06	153.4	1.44	208.4	1.37	198.27	2.07	299.57	1.59	206
	Copper sheet (kg)	48.24	1.06	51.13	1.44	69.47	1.37	66.09	2.07	99.86	1.59	76.88
Other	Tank (kg)	266.7	1.1	293.68	1.29	343.78	1.36	361.71	1.64	438.61	1.63	434.79
	Paper (kg)	16	1.1	17.61	1.29	20.62	1.36	21.7	1.64	26.31	1.63	26.08
	Ceramic (kg)	6.02	1.1	6.63	1.29	7.76	1.36	8.16	1.64	9.9	1.63	9.81
	Oil (kg)	265.5	1.1	292.36	1.29	342.24	1.36	360.09	1.64	436.64	1.63	432.84
	Cardboard (kg)	3.65	1.1	4.02	1.29	4.71	1.36	4.96	1.64	6.01	1.63	5.96
	Plastics (kg)	2.05	1.1	2.25	1.29	2.64	1.36	2.78	1.64	3.37	1.63	3.34
	Wood (kg)	4.38	1.1	4.83	1.29	5.65	1.36	5.95	1.64	7.21	1.63	7.15
	Powder coating/Paint (kg)	5.79	1.1	6.37	1.29	7.46	1.36	7.85	1.64	9.52	1.63	9.43
	Total (kg)	1253		1380	1.29	1615	1.36	1699	1.64	2060	1.63	2041
	Volume of final product (m^3)	2.11	1.1	2.32	1.29	2.72	1.36	2.86	1.64	3.47	1.63	3.44
	P_0 (W)	750		610		520		430		430		196
	P_k (W)	4600		4600		3850		4600		3250		4554
	Efficiency at 50% load %	99.05	1.073	99.12	1.221	99.26	1.168	99.21	1.347	99.38	1.294	99.33
	Electricity losses (kWh/year)	7858		6632		5633		5055		4677		2992
	Product price (€/unit)	6122	1.05	6428	1.19	7285	1.16	7101	1.42	8693	1.41	8632
	Ratio: price increase versus total loss reduction at 50% load			98%		97%		97%		105%		109%

An overview of existing transformer efficiency schemes is given below.

3.4.1   MEPS

Main international MEPS are listed in Table 3.7.

Table 3.7   International MEPS

Country	Title
Australia	AS 2374.1.2-2003 : Power transformers – Minimum Energy Performance Standard (MEPS) requirements for distribution transformers (10-2004)
Canada	Mandatory MEPS for Transformers (01-01-2005)
EU member countries	Energy Performance Standard for Distribution and Power Transformers under preparatory work in frame of Energy related Products Directive
India	MEPS for Distribution Transformers of ratings 16, 25, 63, 100, 125, 200 kVA capacity (2010)
Israel	MEPS for Distribution Transformers – Israel
New Zealand	AS 2374.1.2 – Power Transformers Part 1.2: Minimum Energy Performance Standard (MEPS) requirements for distribution transformers (01-10-2004)
People's Republic of China	GB 20052-2006 – Minimum Allowable Values of Energy Efficiency and the Evaluating Values of Energy Conservation for Three-Phase Distribution Transformers (2006)
The United States	MEPS for Distribution Transformers (2010)

3.4.2   Mandatory Labelling

Main international efficiency labelling programmes are listed in Table 3.8.

Table 3.8   International efficiency labelling

India	Star Rating Plan – Distribution Transformer (2010)
Israel	Energy Label for Distribution Transformers – Israel
Japan	Label Display Program for Retailers – 'Top runner program' – Transformers
People's Republic of China	China Energy Label – Power Transformer (2010)

3.4.3   Voluntary Programmes

Main international voluntary schemes are listed in Table 3.9.

Table 3.9   International voluntary schemes

Chinese Taipei	Greenmark – Transformers (1992)
People's Republic of China	CQC Mark Certification – Power Transformer (2010)
Republic of Korea	Certification of high energy efficiency appliance program for Transformers (–)
The United States	ENERGY STAR – Transformers (1995)

The comparison of selected international MEPS, taking into account the lack of full equivalency as explained earlier in this section, is presented in Table 3.10.

Table 3.10 Comparison of international MEPS

1	2	3	4	5	6	7	8	9	10	11	12	13	14
Rating kVA	USA oil	USA dry	USA BAT oil	USA BAT dry MV	Australia oil	Australia dry	Australia oil High eff.	China S11	EU oil proposed	EU dry proposed	India	EU25 fleet	EU25 market
15 or 16	98.36%	97.00%	99.31%	98.75%									
25					98.28%	97.17%	98.50%				98.36%	97.12%	97.80%
30	98.62%	97.50%	99.42%	98.95%								97.58%	98.24%
45	98.76%	97.70%	99.47%	99.05%								97.56%	98.36%
50									98.54%			97.84%	98.44%
63 or 75	98.91%	98.00%	99.54%	99.17%	98.62%	97.78%	98.82%	98.67%			98.82%	98.15%	98.44%
100					98.76%	98.07%	99.00%	98.86%	98.84%	98.58%	98.98%	98.27%	98.66%
112.5	99.01%	98.20%	99.58%	99.25%									
150	99.08%	98.30%	99.61%	99.30%									
160												98.52%	98.76%
200					98.94%	98.46%	99.11%	98.96%	99.00%	98.75%	99.05%	98.67%	98.86%
225	99.17%	98.50%	99.65%	99.37%							99.12%		
250								99.08%	99.11%	98.92%	99.15%	98.88%	99.01%
300 or 315	99.23%	98.60%	99.67%	99.41%	99.04%	98.67%	99.19%		99.15%				
400								99.17%	99.21%	99.09%	99.18%	99.00%	99.15%
500	99.25%	98.70%	99.71%	99.48%	99.13%	98.84%	99.26%		99.25%				
630								99.24%	99.29%	99.12%	99.26%	99.13%	99.28%
750	99.32%	98.80%	99.66%	99.42%	99.21%	98.96%	99.32%		99.46%			99.22%	99.31%
1000	99.36%	98.90%	99.68%	99.46%	99.27%	99.03%	99.37%	99.25%	99.47%	99.29%		99.16%	99.29%
1500	99.42%		99.71%	99.51%	99.35%	99.12%	99.44%						
1600								99.34%	99.48%	99.37%		99.33%	99.39%
2000	99.46%		99.73%	99.54%	99.39%	99.16%	99.49%		99.49%	99.39%			
2500			99.74%	99.57%	99.40%	99.19%	99.50%			99.42%		99.35%	99.37%

Comments on Table 3.10

All presented efficiencies apply to three-phase transformers.

Losses in the table cannot be compared directly. They are given as in the original documents and are compared with schemes where they appear as the maximum allowable losses further recalculated to efficiency levels. One particular difference exists when comparing efficiencies in 50 Hz and 60 Hz systems. Transformers for 60 Hz tend to have higher no-load and lower load losses, if all the other parameters are kept the same. The resulting differences are about 10% at 50% load if the optimum efficiency of a transformer is close to 40%. In fact in its first attempt Australia 'recalculated' the American 60 Hz NEMA TP1 efficiency standard to Australia's 50 Hz frequency and also linearly interpolated the efficiencies at the size ratings that in Australia are different from USA. In practice a sort of check would be to deduct 0.2% from the USA standard for efficiency levels below 99% and deduct 0.1% for levels above 99% to get rough equivalency of standards. If we do so, we will observe that the USA rule is by far the most demanding scheme worldwide. In Europe levels are also quite demanding, particularly for sizes above 630 kVA, having in mind that they are dedicated to transformers with magnetic steel cores. Currently amorphous transformers are not at all a mainstream product in Europe and there is the uncertainty on the availability in the short term of amorphous material and transformer production in the EU, therefore the proposed levels are ambitious but possible to achieve with non-amorphous technology.

As comment to this, more than 95% of amorphous transformers are sold in Asia and the USA, primarily in India, China and Japan with emerging demand from the USA. Not long ago, in 2006, Spanish Endesa, a company that made some purchases of amorphous transformers reported that about 22 000 tons of amorphous steel is used. Hitachi Metals/Metglas in 2009 indicated that the production capacity of amorphous iron was 50 000 tons in 2008 and this was expected to rise up to 100 000 ton by 2010 (however not necessarily all this production will have to be dedicated to the transformer market). If we take 400 kVA as the average rating of trans-formers installed, at 600 kg core material, about 37 000 amorphous transformers are produced each year (based on 2006 figures). This is about 1.2% of the total annual sales worldwide.

Columns 4 and 5 specify DOE efficiency levels of the best available technology (BAT), which are possible to achieve if the cost is no issue. They were added to the final rule to demonstrate the gap between BAT and MEPS. Column 8 is Australian MEPS, not for ordinary but for 'high efficiency power transformers'. Finally, columns 13 and 14 are given to compare standards with current practice in Europe (source – SEEDT). Column 13 gives equivalent efficiency for the average transformers operating in Europe, while column 14 specifies average efficiencies of transformers sold in Europe in the year 2005.

A better illustration of how demanding MEPS are with the recalculation of 60 Hz frequency to 50 Hz is presented in Figure 3.4.

One extreme (the most ambitious) is USA efficiency level here referred to as BAT, in the original expressed as 'if costs were no issue'. The other extreme would be far out of scale as the South African standard SANS 780, 2004 specifies losses that at 50% loading are equivalent to an efficiency of 96.45%! It is a good example of how energy costs, which have been known to be low in South Africa, can discourage more ambitious efficiency levels.

The details of standards are presented in Section 9 of this chapter.

3.5 Life Cycle Costing

The annual energy losses of a transformer can be estimated from the following formula:

$$W_{\text{loss}} = (P_0 + P_{\text{K}} * L^2) * 8760h, \tag{3.8}$$

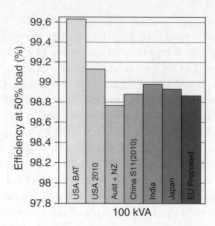

Figure 3.4   Graphical comparison of selected MEPS normalized at 50 Hz 100 kVA

where

W_{loss} is the annual energy loss in kWh;

P_0 is the no-load loss in kW – this factor is available from the transformer specifications or can be measured;

P_K is the short-circuit loss (or load loss) in kW – this factor is available from the transformer specifications or can be measured;

L is the average per-unit load on the transformer (This is absolutely precise for constant load but becomes less precise, the more the load varies);

8760 is the number of hours in a year.

To calculate the cost of these losses, they need to be converted to the moment of purchase by assigning capital values, in order to put them into the same perspective as the purchase price. This is called the Total Capitalized Cost of the losses, TCC_{loss}. This can be calculated using the following equation:

$$TCC_{loss} = W_{loss} \times \frac{(1+i)^n - 1}{i \cdot (1+i)^n} \times C \times 8760, \tag{3.9}$$

where

C = estimated average cost per kWh in each year;

i = estimated interest rate;

n = expected life time of the transformer.

3.5.1   Life Cycle Cost of Transformers

To perform the economical analysis of the transformer, it is necessary to calculate its life cycle cost, sometimes called the Total Cost of Ownership, over the life span of the transformer or, in other words, the capitalized cost of the transformer. All these terms mean the same – in one equation, costs of purchasing, operating and maintaining the transformer need to be compared, taking into account the time value of money.

The concept of the 'time value of money' is that a sum of money received today has a higher value – because it is available to be exploited – than a similar sum of money received at some future date.

In practice, some simplification can be made. While each transformer will have its own purchase price and loss factors, other costs, such as installation, maintenance and decommissioning will be similar for similar technologies and can be eliminated from the calculation. Only when different technologies are compared, e.g. air-cooled dry-type transformers with oil cooled transformers will these elements need to be taken into account.

Taking only purchase price and the cost of losses into account the Total Cost of Ownership can be calculated by the base formula:

$$TCO = PP + A * P_0 + B * P_k, \tag{3.10}$$

where
PP = purchase price of transformer;
A = assigned cost of no-load losses per watt;
P_0 = rated no-load loss;
B = assigned cost of load losses per watt;
P_k = rated load loss.

P_0 and P_k are transformer rated losses. The A and B values depend on the expected loading of the transformer, and energy prices.

The choice of the factors A and B is difficult, since they depend on the expected loading of the transformer, which is often unknown, and energy prices, which are volatile, as well as the interest rate and the anticipated economic lifetime. If the load grows over time, the growth rate must be known or estimated and the applicable energy price over the lifetime must be forecast. Typically, the value of A ranges from less than 1 to 8€/W and B is between 0.2 and 5€/W.

Below we propose a relatively simple method for determining the A and B factor for distribution transformers.

The A and B factors are calculated as follows.
No-load loss capitalization

$$A = \frac{(1+i)^n - 1}{i(1+i)^n} \times C_{kWh} \times 8760 \tag{3.11}$$

Load loss capitalization

$$B = \frac{(1+i)^n - 1}{i(1+i)^n} \times C_{kWh} \times 8760 \times \left(\frac{I_l}{I_r}\right)^2 \tag{3.12}$$

where
i = interest rate [%/year];
n = lifetime [years];
C_{kWh} = kWh price [€/kWh];
8760 = number of hours in a year [h/year];
I_l = loading current [A];
I_r = rated current [A].

These formulae assume that energy prices and the loading are constant over the transformer life. We will comment on this later.

Usually, the loss evaluation figures A and B form part of the request for a quotation and are submitted to the transformer manufacturers, who can then start the process of designing a transformer to give the required performance. The result of this open process should be to use the cheapest transformer, i.e. the one with the lowest total cost of ownership, optimized for a given application. The drawback of this process is, as mentioned, the difficulty in predicting the future load profile and electricity costs and tariffs with any confidence. On the other hand, these optimization efforts depend on the prices of materials, particularly active materials, i.e. conductor and core material. Dynamic optimization makes sense when there is the different price volatility for different materials such as aluminium and copper or high and low loss magnetic steel.

For large transformers, above a few MVA, the cost of losses are so high that transformers are custom-built, tailored to the loss evaluation figures specified in the request for a quotation for a specific project.

For distribution transformers, often bought in large batches, the process is undertaken once every few years. This yields an optimum transformer design, which is then retained for several years – less so nowadays because of the volatility of metal prices – until energy prices and load profiles have changed dramatically.

To make the capitalization more attractive, so that the use of TCO is easier, we propose the use of a graph, shown in Figure 3.5, which allows factor A to be determined.

Factor A expresses the relation between the cost of no-load losses and the following:

- electricity price;
- discount rate or company interest rate or average cost of capital;
- capitalization period or expected lifetime of the transformer.

This example illustrates that for an electricity price of 100€/MWh, an interest rate of 5% and a 10-year capitalization period, the cost of no-load loss will be 6.75 €/watt.

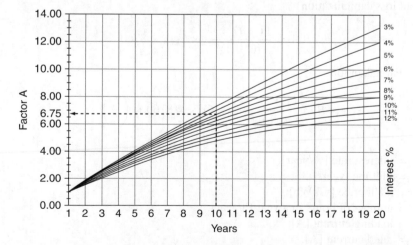

Figure 3.5   Simplified chart for calculation of factor A at electricity cost 100 €/MWh

Factor A is directly proportional to the electricity price, so it can simply be scaled to account for electricity price changes as long as the interest rate and capitalization period remain unchanged.

It is important to note that, for small interest rates, a doubling of the capitalization period will result in almost doubling the cost of losses. On the other hand, applying too high a capital rate by making, for example, too high a provision for risk, will produce a low value of loss.

Factor B, as explained previously, is simply the product of factor A and the square of the loading factor ($B = A \, (Loading)^2$). The loading factor used here is the expected average load over the lifespan of the transformer, possibly taking harmonics into account.

For larger power transformers the formula for total cost of ownership is more complex. First of all additional component should be added to the base formula, reflecting capitalized losses of auxiliary losses for fans or pumps calculated analogically as load losses. The next element treated in more detail here is peak responsibility, which is intended to compensate for the transformer peak load losses that do not occur at the system peak losses. This means that only a fraction of the peak transformer losses will contribute to the system peak demand. This relation is the ratio of two transformer loads:

- β_{syst}, transformer load at time of system peak; and
- β_s, transformer peak load.

Finally, the base equation for cost of energy is now composed of an additional element C_{inv}, which reflects system investment to the cost of generation and transmission facilities related to 1 kWh of energy transmitted through a transformer, which is necessary to supply the additional demand resulting from the transformer losses at the system peak. Since a transformer located directly at a generating station does not require an investment in transmission facilities, this value used to evaluate the losses in the generating station transformer should be less than in a transformer located at a certain location downwards transmission network.

A method for determining this value involves adding the construction cost of a recently completed or soon to be completed generating station to the cost of the transmission facilities required to connect the transformer to the plant. If power is purchased rather than self-generated, this value can be determined by dividing the demand charge by the fixed charge rate. There can be different methods to estimate such a cost, sometimes referred to as avoided cost of generation and transmission capacity. The selected method should yield the most realistic results.

In conclusion, the simplified equation that includes peak load responsibility and T&D capacity will have the following notation:

$$C = C_{kWh} + C_{inv} \times \left(\frac{\beta_{syst}}{\beta s} \right)^2. \tag{3.13}$$

The cost of the losses also depends on voltage level and less strongly on rated power, location or transformer type. Some examples or indicative values of the cost of losses in transformers are as follows.

For distribution transformers they are between 3 and 10 USD/W, sometimes even higher if very high electricity prices together with low interest rates and long lifetimes apply.

For power transformers, due to the scale, no-load losses are related to kW instead of watts and are usually between 2000 to 7000 USD/kW and average around 5000 USD/kW. There is also differentiation in accordance with generation: mostly variable fuel cost and operating costs of the transmission system.

As for load losses, the specified values are a certain fraction of no-load losses that are strictly dependent on assumed loading. In the case of distribution transformers this fraction is usually at a level of 15-20% of the cost of no-load loss (equivalent to roughly 40% loading) while for transmission transformers they are higher, usually between 30% to 50% of the cost of no-load losses (between 55% and 70% loading).

3.5.2   Detailed Considerations

Transformer loading plays a very special role in life cycle costing. Load losses vary with the square of the load.

Ideally, to calculate the load losses it would be necessary to integrate the squares of all momentary ratios of actual load to the rated load. This is practically impossible, so a methodology to analyse load losses based on the summation of energy consumed in transformers has been developed.

The formula to calculate load losses is presented below:

$$\sum P_k = \beta_S^2 \times \tau \times P_k, \tag{3.14}$$

where

$\sum P_k$ is sum of load losses in given period of time, usually one year;
β_s is the peak load of a transformer in given period of time;
τ is time duration of peak loss;
P_k is rated load loss of a transformer.

It should be noted that $(I_1/I_r)^2$ value from the B factor calculation formula can be expressed as $\beta_s^2 \times \tau$. The relationship between the time duration of the peak losses τ and the time duration of peak load T_s is shown in Figure 3.6.

T_s represents the fraction of yearly time in which energy is transformed at peak load conditions equivalent to the actual energy transformed. τ_s represents the fraction of yearly time of peak loss (which occurs at peak load) equivalent to actual load losses.

Different empirical models have been developed to define the relationship:

$$\tau_s = f(T_s)$$

and some examples are:

$$\frac{\tau_s}{8760} = \left(\frac{T_s}{8760}\right)^x \tag{3.15}$$

with 'x' varying around a value of 1.7 to 1.8 or

$$\tau_s = A \cdot \left(\frac{T_s}{8760}\right) + B \cdot \left(\frac{T_s}{8760}\right)^2 \tag{3.16}$$

with A being between 0.15 to 0.5 and the B value between 0.5 and 0.85, with additional feature of $A + B = 1$ (but there are exclusions from the last condition). The physical interpretation of all these formulae is hard and is not always proven.

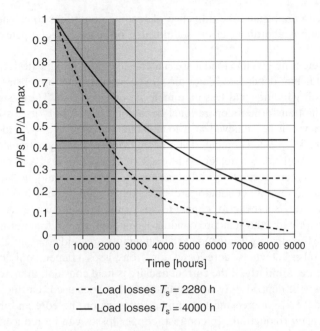

Figure 3.6 Explanation of the relationship between time of peak losses τ and time of peak load in situations when for half a year the load is 200% of the average load and for the remaining half a year is zero

In SEEDT they have compared and analysed these formulae from the accuracy point of view. The 'x' value that gives the best results is 1.73 or an A value of 0.3 and a B value of 0.7.

The average loading of distribution transformers in electricity distribution companies in the EU-27 is 18.9% and peak load is 0.53 (53%). The time of peak load is 0.36 and the time of peak loss is 0.2. Transformers in electricity distribution companies have such low loadings for many reasons, such as the anticipated high variability of load and the need to reserve capacity to provide resilience against failure of other units and sections of the network. Another reason could also be the limiting of loading on transformers that are in a poor technical condition (e.g. moist insulation and risk of its further degradation leading to failure). A further reason might be that distribution transformers are protected against short circuit, but not against overload or excessive temperature and large margins are used. These are the average figures and the situation may be quite different in different countries.

Industrial transformers are loaded higher than transformers owned by distribution companies. The average load is 37.7%, peak load above 0.7, while the times of peak load and peak loss are about 0.3 and 0.15, which means that the load (with peaks and lows) is fairly intermittent.

It is quite apparent that the τ value is around 50 to 60% of the T value. Theoretically τ is 50% of T in situations where the T curve in Figure 3.6 is a straight line between the peak load and zero (load is continuously and uniformly distributed between peak load and zero). On the other hand, when the T curve is a straight horizontal line (equal load all the year), the values T and τ will be equal.

It is necessary to understand the influence of loading conditions to calculate the B factor. In practical situations additional effort should also be made to anticipate loading changes over time.

The relative weighting given to load losses and no-load losses in the design of a transformer that is so much loading dependant, determines whether the transformer has more conductor material in the coil windings and less core material or vice versa. The design choices made will also affect the transformer's operational behaviour, particularly its losses. For instance, optimum efficiency can be achieved at a load factor of 24% or at 47% depending on the design (see Figure 3.1). When compared at constant current density, a transformer with more conductor material will exhibit greater load losses, or 'copper losses', as they are also known. Strictly speaking, a more accurate trivial name for these losses would be 'aluminium losses', as the losses in an aluminium conductor are 35% greater than those in a copper conductor of identical cross-section. But the term 'copper losses' is unlikely to change, as it reflects the fact that copper is, historically, the standard conductor material used in transformers.

If the magnetic flux density, frequency and iron quality are held constant, the no-load losses in a transformer (also known as 'core losses' or 'iron losses') depend only on the amount of iron used in the core. Similarly, if the current density is held constant, then, roughly speaking, the copper losses will depend only on the amount of copper used. On the other hand, iron losses can be reduced by increasing the number of turns on the core and thus reducing the magnetic flux density (induction). In contrast, copper losses can be reduced by operating at a higher flux density and using fewer windings on the core – but this can only be realized within strict limits, as high-quality magnetic materials have quite sharp magnetic saturation points and most conventionally designed transformers operate close to this limit. The primary means of reducing copper losses is to lower the current density, while maintaining the number of turns and the core cross-section and modifying the core in such a way that the winding window is larger and thicker wire can then be used for the windings.

A transformer that spends most of life operating under no load or minimal load conditions should therefore be designed to minimize the no-load losses, i.e. less iron and more copper. It would however be wrong to conclude from this that any transformer designed for permanent full-load operation (something that only really occurs in generator transformers in power stations and in certain industrial applications) should contain as little copper as possible. In this case, the preferred approach is to maximize the cross-section of the iron core in order to minimize the number of turns. The cross-section of the conducting wires should also be as large as possible in transformers running under continuous full load.

So far we have not included the effects coming from extra losses due to harmonics, unbalance and reactive power in life cycle costing. For harmonics and unbalance there is no easy approach and some guidance is provided in Section 3.1. In general, as an average, extra losses from harmonics are at the level of a few percent of nominal losses, however in some industrial applications and office buildings when electrical power is supplying distorting loads these losses can even double nominal losses. As for reactive power influence, this is even harder to make accurate calculation. There are some indications that the equivalent active power losses in supplying network from reactive power flowing in a transformer may be at the level of 0.15 kW/kvar.

So far we have also not considered the increase in power and energy flow in a transformer due to the connection of new customers and new loads. This may sometimes be very important as such yearly increments may lead to very high additional losses and the transformer running further and further away from the optimum (loading) point. In cases when load increase is expected to be significant, the sum of all the annual losses calculated on the basis of expected annual transformer loading should be analysed.

The last but not least of the elements to consider are energy costs. They may also increase with time as new capacities are expected to be more costly due to more expensive technologies and decreased capacity factors, as in the case of renewables or some other peak generators. There are even ideas of differentiating between the cost of energy for no-load and load losses. In the case of load losses, the higher cost of generation and increased cost of energy dispatch combined with the quadratic relation of Joule losses to load may result in the cost of energy load loss being doubled or even higher than the nominal cost of energy.

3.6 Design, Material and Manufacturing

Thanks to improved data processing with mathematical tools and models it is now possible to design a transformer using the finite-element method with the provision of electrical and mechanical strengths, heat transfer and dynamic properties, including short-circuit conditions. As a practical outcome two- or three-dimensional field plots are drawn, helping to design different transformer elements and also keep losses at desired levels with the best balance between costs and efficiency. In practice it is to minimize the influence of an out of proportion increase of one of the cost components, either the conductor, insulant, or core material for the purpose of loss optimization at the design stage.

Material and material processing technology developments have the largest influence on losses. Without evolution in material technologies the progress in transformer efficiency improvement would be impossible on so large a scale. In this section only these improvements that have effects on loss reduction are described.

Fabrication technics may incorporate some improvements; better stacking, precision in manufacturing, insulating and shielding against stray magnetic flux add smaller but still significant loss reductions.

3.6.1 Core

The materials used in both the core and the coils contribute significantly to the cost of a power transformer, whose manufacture is in any case a highly labour-intensive process.

The main milestones in core material developments have been:

- The development of cold-rolled grain-oriented (CGO) electrical steels
- The introduction of thin coatings with good mechanical properties
- The improved chemistry of the steels, e.g., Hi-B steels
- Further improvement in the orientation of the grains
- The introduction of laser-scribed and plasma-irradiated steels
- The continued reduction in the thickness of the laminations to reduce the eddy loss component of the core loss
- The introduction of amorphous ribbon (with no crystalline structure) –manufactured using rapid cooling technology – for use with distribution and small power transformers.

Chronologically these improvements are illustrated in Table 3.11.

Table 3.11 Historical development of core sheet steels

Year	Material	Thickness	Loss (50 Hz)	at flux density
1895	Iron wire		6.00 W/kg	1.0 T
1910	Warm rolled FeSi sheet	0.35 mm	2.00 W/kg	1.5 T
1950	Cold-rolled, grain-oriented	0.35 mm	1.00 W/kg	1.5 T
1960	Cold-rolled, grain-oriented	0.30 mm	0.90 W/kg	1.5 T
1965	Cold-rolled, grain-oriented	0.27 mm	0.84 W/kg	1.5 T
1970	Cold-rolled HiB sheet	0.30 mm	0.80 W/kg	1.5 T
1975–	Amorphous iron	0.03 mm	0.2 W/kg	1.3 T
–2005	Amorphous iron improved HB1	0.02 mm	0.15 W/kg	1.3 T + 10%
1980	Cold-rolled, grain-oriented	0.23 mm	0.75 W/kg	1.5 T
1980	Cold-rolled HiB sheet	0.23 mm	0.70 W/kg	1.5 T
1983	Laser treated HiB sheet	0.23 mm	0.60 W/kg	1.5 T
1985	Cold-rolled, grain oriented	0.18 mm	0.67 W/kg	1.5 T
1987	Plasma treated HiB sheet	0.23 mm	0.60 W/kg	1.5 T
1991	Chemically etched HiB sheet	0.23 mm	0.60 W/kg	1.5 T

3.6.1.1 Cold-Rolled Grain Oriented and HiB Magnetic Steel

Selecting the right sheet steel for the laminations, accurate stacking with frequent staggering (every two sheets), and minimization of the residual air gap are all key parameters in reducing open-circuit currents and no-load losses. Today, practically all core laminations are made from cold-rolled, grain-oriented steel sheet (the thinner the laminations, the lower the eddy currents) despite the significantly higher cost of this type of steel.

Applying gradually improved better grades of non-grain-oriented steel, technology of cut, decreasing laminations thickness led to reduction of these losses by approximately a factor of two over the last 30 years. When comparing these losses with the levels of the middle of the last century the factor would be more than three.

A good illustration of this progress is the SEEDT model presented in Figure 3.7. The chart presents transformers' relative age populations together with their relative shares of no-load and load losses in Europe. To help to read this picture properly one conclusion will be that, for example, replacing 10% of the oldest (and presumably based on the model, least efficient) part of the population will turn into phasing out no-load losses contributing in 21.5% to whole no-load losses. The relevant figure for load losses will be 15.2%. Similarly, taking the oldest 20% of the population out of service will save almost 35% of the total no-load losses and about 30% of total load losses.

It would not be fair if we did not mention the progress in efficiency improvement in large and very large power transformers. Although amorphous metal has not entered this part of the transformer fleet yet (and will hardly do so without major developments in amorphous cores), other improvements have resulted in about 60% no-load reductions and close to 50% load loss reductions over the last 50 years. At the same time, acoustic noise has been reduced from more than 80 dB(A) to less than 50 dB(A).

One of the key aspects in core construction is ensuring the absence of eddy current loops. It is for this reason that even in small transformers with ratings of above about 1 kVA (depending on the manufacturer), the clamping bolts are electrically insulated on one side, although these

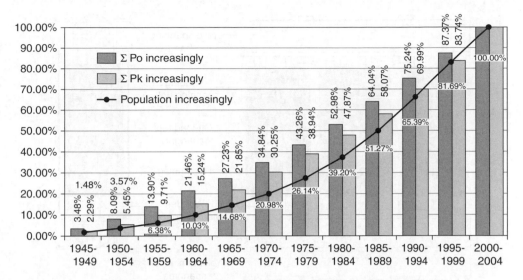

Figure 3.7   Age distribution of transformer population versus losses

benefits would also be apparent in transformers with power ratings below 100 VA. Given the advantages that insulated fastening bolts can yield in relatively small transformers, the benefits gained in much larger distribution and high-power transformers should be obvious.

In larger transformers in yoke frames made of steel pofiles rather than wood the holes have to be large enough so that an insulating bushing can be pushed over the shaft of the bolt to ensure that the bolt does not come into contact with the burred edges of the yoke plates and touches only one side of the yoke frame. If multiple contact points occur, it essentially short circuits the relevant section of the yoke. In addition, cutting bolt holes effectively reduces the cross-sectional area of the core, and eddy currents are also induced in the bolt, which, for obvious reasons, cannot be manufactured from laminated sheet. Sometimes clamping bolts made of stainless steel are chosen, because, perhaps surprisingly, stainless steel is not in fact ferromagnetic although it consists predominantly of iron and nickel – both ferromagnetic elements. The magnitude of the magnetic field in these stainless steel bolts is therefore lower, thus reducing eddy current losses. In addition, stainless steel is much better at suppressing eddy currents because its electrical conductivity is only about one-seventh that of conventional steels.

A better means of clamping the yoke laminations, though more costly than employing stainless steel bolts, is to use a clamping frame that wraps around the yoke (Figure 3.8). However, it is essential to ensure that the clamping ring does not form a closed electrical circuit that would short-circuit the yoke.

Figure 3.8 also shows transformer tap changers in a high-voltage winding, which allow for any variation in the input voltage (typically two steps of +2.5% above the nominal voltage, and two steps of −2.5% below). These are located in the central section of the winding and not at its upper or lower ends. This ensures that the effective axial height of that portion of the high-voltage coil that carries current is essentially constant as is the relative height of the HV and LV coils. Without the tight clamping, a number of windings at the upper or

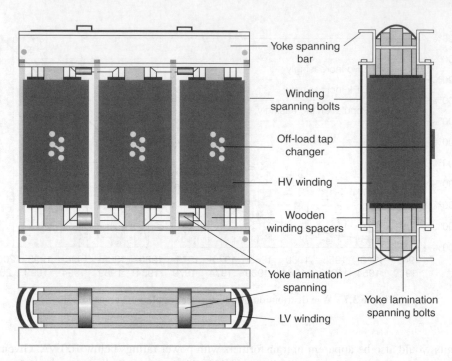

Figure 3.8   The structure of a transformer's core-and-coil assembly ('active part'). The design shown here is the more 'elegant' solution with unperforated yokes

lower end of the coil would be lost if a short circuit caused a significant force in the axial direction between the coils. In transformers that have been in service for a long time, the coils may no longer be as rigidly clamped as they were at the time of manufacture and the insulating materials may be showing signs of age. A short circuit in such a transformer or a breakdown of the insulation material as a result of a lightning strike often causes the device to fail completely. At installations where short circuits or lightning strikes occur only every few decades, a transformer can remain operational for as long as 60 years before finally having to be replaced for economic reasons.

3.6.1.2   Amorphous Steel

America's Allied-Signal admits spending more than 25 years and a great deal of R&D effort to achieve the commercial production of Metglas® amorphous alloys. The joint Hitachi/Metglas group is the world's largest promoter of amorphous technology in distribution transformers.

No-load losses can be reduced by lowering the magnetic flux density and by using special core steels. The thinner the sheet steel is, the smaller the extent of eddy current formation. Eddy currents are completely absent in core materials that do not conduct electricity (so-called ferrites), but these are reserved for radio-frequency applications as their magnetizability is too low for transformers operating at grid frequencies. Amorphous steel is a new type of core material that offers a compromise between sufficiently high magnetizability and significantly reduced core losses. While the resulting core material has a saturation magnetization of approximately 1.5 T compared with the 1.75 T exhibited by modern cold-rolled grain-oriented steels, the no-load losses in a transformer with an amorphous steel core are around 60% lower. As the saturation flux density of the core material is lower, these transformers tend to be larger and heavier and correspondingly more expensive. The transformer with an amorphous steel core is also about 6 dB louder.

Amorphous steel is made by atomizing the liquid metal and spraying it on to a rotating roller where it is quenched extremely quickly, so rapidly in fact that it cannot crystallize and remains in a disordered amorphous state, hence the name. The structure is similar to glass and therefore amorphous metals are often called metallic glass (Metglass, the licenced manufacturer). The structure is also similar to undercooled liquid. Domain walls can move freely through the random atomic structure.

The risk exists that this 'free' structure will crystallize with time and temperature. Thus crystallization limits the life time of amorphous alloys:

- 550 years when temperature is 175°C;
- 25 years when temperature is 200°C;
- 2 hours when temperature is 350°C.

So, significant overheating of amorphous core shortens its life as amorphous alloy.

The application of amorphous steel tapes is limited to distribution transformers. In the past it was commonly believed that amorphous wound cores are most suitable for smaller, single-phase units. Brittleness, mechanical sensitivity, cutting and stacking problems due to thickness in the range of only 20 to 30 μm and the necessity of annealing in the magnetic field after manufacturing seem to be prohibitive for the application in larger power transformers. Most of the advantage of the excellent loss values at low induction was compensated for by the higher stacking factor and the lower saturation point where losses increase rapidly. The recent years show that disadvantages are gradually removed, amorphous tapes may be wider and larger than 1 MVA units can be normally produced.

Amorphous metal performance offers a huge opportunity for no-load losses reduction compared with conventional transformer steel. The empiric formula to calculate magnetic losses is:

$$Magnetic\ loss = Af + B\,d^l\,f^m\,B^n/\rho,$$

where A and B are constants.

The explanation and comparison of amorphous metal and silicon steel are given in Table 3.12.

Table 3.12 Comparison of magnetic properties of amorphous metal and silicon steel

Property/Exponent	Amorphous metal	Silicon steel
ρ (resistivity)	~1.30 [μΩ m]	~50 [μΩ cm]
d (thickness)	~20 [μm]	200 [μm]
l	1–2	2
m	~1.5	~2
n	~2	~2

Metglas, as stated above, has improved amorphous material over the recent years, particularly the amorphous metals' deficiency of reduced saturation induction has been partly compensated. The new material reaches saturation at induction closer to these (even slightly above 1.4 T), characterizing traditional magnetic steel. This makes cores more compact and the transformers smaller and lighter than older amorphous designs. The makers also tried to remove other drawbacks such as brittleness and the associated difficulties during core making.

3.6.2   Windings

While the transformer core serves the magnetic circuit, the windings form the electrical circuit of a transformer. The resistive losses are directly dependent on the resistance of both primary and secondary windings. The losses will be lower if the length and the cross-section of the windings decreases. The length of the windings and the number of turns depends on the core geometry and properties of the core material. The cross-section and losses are very closely related (see Table 3.2).

According to an old rule of thumb within the transformer industry, the production cost optimum lies somewhere around a ratio of steel to copper usage of 2:1. However, it is a fairly flat optimum and, of course, varies with the ratio of steel to copper price. Independently of this, it should be taken into consideration that the operating properties of the transformer also vary when the share of metals are varied, especially with respect to losses: holding the current densities in the windings and the magnetic flux density in the core constant, the loss per kilogramme of copper or steel, respectively, will be more or less constant. So a transformer designed according to this philosophy, but with more iron and less copper, tends to have higher iron losses, and one with more copper and less iron will have higher copper losses. But this does not mean that skimping on copper and steel pays off! Rather, enhancing the core cross-section while keeping the number of turns constant will reduce core losses, and enhancing the copper cross-section, while keeping the core cross-section constant will reduce copper losses. In short: the bulkier transformer will always be more efficient, and metal prices will always be an obstacle against its implementation.

Continuously transposed conductors (Figure 3.9) are a particular development in the design of power transformers windings. Conductor design has been improved by the introduction

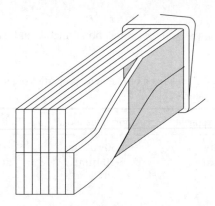

Figure 3.9   Continuously transposed conductors

of continuously transposed conductors (a single conductor subdivided into several flat sub-conductors that are regularly transposed and insulated against each other), reducing the skin effect and eddy current losses and allowing better packing density of the winding.

3.6.2.1 Superconducting (High Temperature, HTS)

In a superconducting transformer the windings, made of a high temperature superconducting material (HTS), are cooled with liquid nitrogen at about 77K, so that the resistance is almost negligible. Load losses, even after adding losses from nitrogen processing, can be still reduced by 50%. The cooling power however has to be supplied continuously at its maximum required level, thus increasing the no-load losses, despite the fact, that in practice, the transformer hardly ever runs at full load, and when it does, then only for a short time. When all these factors are taken into account, overall loss reductions turn out to be minimal.

The use of HTS transformers on a larger scale may be economically justified as cooling systems improve and the cost of liquid nitrogen, which is greatly related to cost of electrical energy production, When electricity cost falls, the benefit of loss reduction also falls. Another important factor is progress in the processing of long lengths of HTS conductors.

These transformers excluding cooling plant have lower weight and volume and are more resistant to overload but cost about 150% to 200% of the price of conventional transformers.

HTS transformers are suitable in applications where the load losses make up a high proportion of the total losses, so are not ideal in distribution transformers. The one place where superconducting transformers could be used effectively would be in railway vehicles. Once these transformers go into industrial production they will save not only weight (and therefore extra energy), but also space. The weight and space limitations in railway vehicles mean that the transformers currently in use in railway vehicles are working at their design limits and are thus significantly less efficient than comparable grid transformers. A locomotive transformer has a power rating of $\approx$5 MVA and weighs $\approx$10 t. Its efficiency is $\approx$95%. A stationary transformer of this power range would weigh some 50 t and reach $\approx$99% efficiency. During the service life of the locomotive this would yield $\approx$1 GWh of electricity savings due to the 4% loss reduction while an additional demand of $\approx$1 GWh would be needed for transporting the additional 40 t of active material (copper and magnetic steel). Obviously this would not really pay off. In such applications, transformers are much more 'squeezed' (by forced cooling) to cut the weight.

The most important advantages and disadvantages of superconducting transformers are listed in Table 3.13.

Table 3.13 Advantages and disadvantages of HTS

Pros	Cons
• Oil free, (liquid nitrogen 77K)	• 150–200% of the price of traditional transformer
• Lower weight (10–30%) excluding cooling plant	
• Slightly smaller volume	• Additional maintenance cost (cryogenic system)
• 25% overloading without accelerated ageing	
• Immediate short-circuit current limitation ability	• Installation site (extra requirements)

3.6.3   Other Developments

3.6.3.1   Gas-Insulated Transformers

Efficiency can also be improved by increasing the rate of heat dissipation from a transformer as load losses are highly temperature dependant. The example here can be gas-insulated transformer. A few power transformers use sulphur hexafluoride (SF_6) gas and they are sometimes referred to as gas-insulated transformers. Their application has strong implications on environmental issue because SF_6 has a strong impact on global warming (1 unit SF_6 = 23 600 units CO_2).

Because of the limited performance of SF_6 as an insulation and heat dissipation agent such transformers with natural or forced gas cooling have to be very efficient because the current densities need to be low by design.

This special type of transformer was (re-)developed in 1987. Gas-cooled transformers had in fact already been the subject of quite long research. When gas cooling is involved, physicists tend to think immediately of hydrogen as it has a very high heat capacity. However, heat capacity is generally expressed relative to mass, and the density (i.e. the mass per-unit volume) of hydrogen is almost one tenth that of air. If, on the other hand, the key parameter is the speed of circulation in a cooling circuit, then heat capacity per volume is more relevant, as the resistance to flow is proportional to the square of the volume flow in any given system. The gaseous material therefore selected was sulphur hexafluoride (SF6), a well-known substance that was already being used as an insulating material in switchgear and that has a density five times that of air and has a considerably better dielectric strength. Engineers carrying out electrical breakdown tests on a specially designed open-top test vessel were able to enjoy the rather unusual observation of air-filled balloons apparently floating mid-air within the test vessel. The balloons were of course resting on a bed of higher density SF6 that had been slowly and carefully filled into the container. By the way, breathing in this completely non-toxic and chemically inert gas (its inertness is directly linked to its high dielectric strength) results in the opposite of the well-known helium 'squeak'. The engineers who carried out these experiments on themselves all lived to tell the tale, proving just how harmless the gas sulphur hexafluoride is. Similar tests using hydrogen should, however, be avoided at all costs – especially by smokers! Although the heat capacity of a kilogramme of SF6 is only half that of a kilogramme of air, its heat capacity per litre is 2.5 times greater. That means that if SF6 is used as the coolant, it only needs to circulate at 40% of the speed used in air-cooled devices in order to produce the same cooling effect. As a result, the fan power can be reduced to about 32% of that needed in an equivalent forced-air cooling system. Two prototype transformers, each with a power rating of 2 MVA, a corrugated tank and internal forced cooling (i.e. cooling class GFAN – gas-forced, air natural) were built and operated in an explosion hazard area within a chemical manufacturing plant owned by the gas manufacturing company, which, it goes without saying, had an understandable interest in these trials.

3.7   Case Study – Evaluation TOC of an Industrial Transformer

Large electrical utilities are usually able to dedicate enough resources to work on economical analysis of their power distribution transformers and also have enough statistical historical data to fill databases for evaluating factors A and B. Users, on the contrary, usually do not have enough reliable data to perform the same analysis.

Table 3.14 The standard industrial consumers chosen for the calculations

Code	Annual consumption [kWh]	Average demand [kW]	Annual utilization [hours]
Ic	160.000	100	1.600
Id	1.250.000	500	2.500
Ie	2.000.000	500	4.000
If	10.000.000	2.500	4.000
Ig	24.000.000	4.000	6.000
Ih	50.000.000	10.000	5.000
Ii	70.000.000	10.000	7.000

So the scope of this case study is the evaluation of not just a single case, but a series of industrial users with different characteristics and consumptions to provide statistical results in which a user can eventually identify itself.

The study focuses on the evaluation of the present worth of the losses and on the TOC of MV/LV oil-immersed distribution transformers for an industrial customer.

Seven standard industrial consumers have been selected from the nine classified in the EUROSTAT [1] database. Only those supplied in MV have been considered; these consumers are characterized in terms of annual consumption of electrical energy, average demand and annual utilization (see Table 3.14).

The rated power and the number of transformers owned by each standard consumer has been assumed accordingly to Table 3.15.

3.7.1 Method

The *reference TOC* of the transformer(s) has been calculated for each standard consumer assuming that the transformer(s) belongs to the EN 50464-1 EoDk list. The *comparable TOC* of the transformer(s) has been calculated by considering the following features:

- oversizing the transformer in order to achieve maximum efficiency by coping with market availability;
- B_k-D_0 transformer(s);

Table 3.15 Rated power and number of transformers owned by each standard consumer

Standard consumer	Average demand [kW]	Power factor @ average demand	Number of transformers	Rated power [kVA]
Ic	100	0.90	1	160
Id	500	0.90	1	630
Ie	500	0.90	1	630
If	2500	0.90	3	1000
Ig	4000	0.90	2	2500
Ih	10 000	0.90	5	2500
Ii	10 000	0.90	5	2500

- the combination of the above solutions;
- B_k-C_0 transformer(s);
- A_k-B_0 transformer(s).

The case of the EN 50464-1 **EoDk** transformer has been assumed as the reference case for the comparison of the different solutions. The methodology consists of analysing the following outcomes:

- the TOCs of the different solutions;
- the balance between purchase cost increase and present worth of losses savings;
- the difference in the present worth of losses and TOC of the different solutions with respect to the baseline.

The TOC has been calculated on the basis of the formula given in EN 50464-1. The costs per rated watt of no-load losses (A factor) and the costs per rated watt of load losses (B factor) have been determined using the following formulae:

$$A = (12C_d + 8760C_e)F_c \quad (\text{€/kW year}), \tag{3.17}$$

$$B = (C_e h) \left(\frac{S_L}{S_r}\right)^2 F_c \quad (\text{€/kW year}), \tag{3.18}$$

$$F_c = \frac{(1+i)^n - 1}{i(1+i)^n}, \tag{3.19}$$

where
  C_d is the demand rate (€/kW);
  C_e is the cost of energy (€/kWh);
  h is the working hours (hours);
  S_L is the average apparent power of the load (kVA);
  S_r is the transformer rated power (kVA);
  F_c is the capitalization factor.

Energy prices are taken from EUROSTAT, the VAT-excluded energy price has been considered in the calculations (Figure 3.10). The demand rate has been determined on the basis of the Italian tariff structure for MV unbounded customers.

In the basic scenario, a 10-year lifetime has been assumed for a typical industrial transformer; calculations have been performed also for 15 and 20 years. Energy prices and load profile have been considered flat over the transformer lifetime.

3.7.2 Results

The major outcomes with the new ranks can be summarized as follows.

- the B_k-D_0 transformer(s) shows the lowest TOC;
- oversizing the transformer in order to reach the maximum efficiency level is not a feasible approach with respect to TOC;

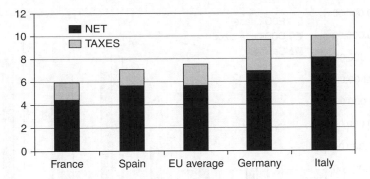

Figure 3.10   EE prices (24 GWh/y – c€/kWh – 0.1 Y/kWh)

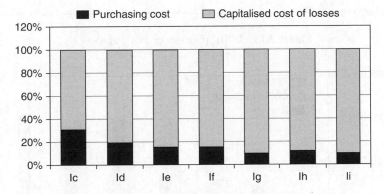

Figure 3.11   EoDk transformer composition of TOC

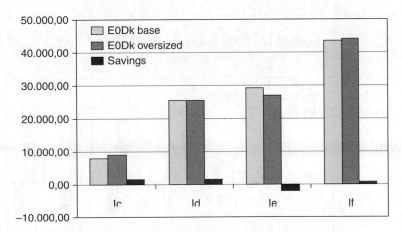

Figure 3.12   Oversized EoDk transformer TOC values (€)

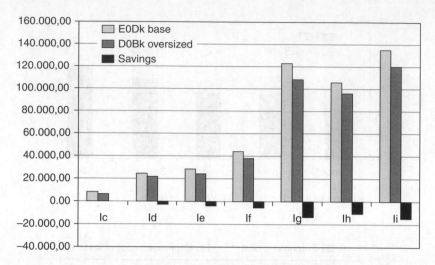

Figure 3.13   D0BK transformer TOC values (€)

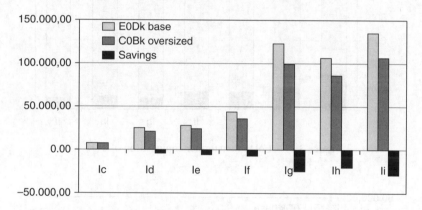

Figure 3.14   C0BK transformer TOC values (€)

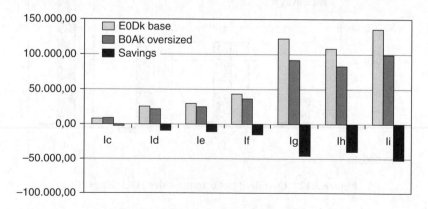

Figure 3.15   B0AK transformer TOC values (€)

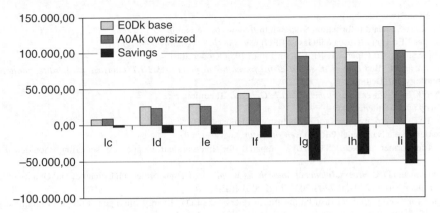

Figure 3.16   A0AK transformer TOC values (€)

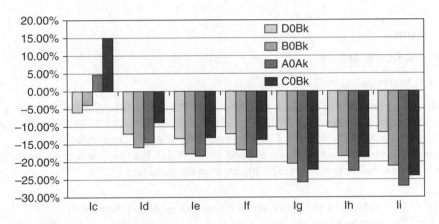

Figure 3.17   TOC saving comparison

- the combination of the two approaches seems to give no synergetic effect, on the contrary the solution B_k-D_0 seems to be more attractive;
- the solution B_k-D_0 shows the lowest payback time;
- the new A_k-B_0 list shows a comparable value of TOC savings but longer payback time, due to higher investment costs.

Figure 3.11 shows TOC composition, Figures 3.12 to 3.17 compare TOC for different industrial users and different transformer classes.

Reference

[1] *Ecodesign Preparatory Study ENTR Lot 2: Distribution and Power Transformers.* http://www.ecotransformer .org

Further Readings

M. Agostini, A. Clò, G. Goldoni: *Mercato dei trasformatori e risparmio energetico, Quaderni di Enegia n° 1.*

ANIE web site (www.anie.it);

APEC Energy Standard Information System http://www.apec-esis.org/

Autorità per l'Energia Elettrica e il Gas (AEEG) *web site (http:\\autorita.energia.it)*

R. Baehr, State of Art Paper, CIGRE Electra 13 no. 195 October 2001

A. Baggini and F. Bua *Criteri di scelta di trasformatori di potenza MT/BT industriali a perdite ridotte*, Power Technology, Ed. Delfino,

CEI EN 60076-1: *Power transformers. Part 1: General*, III edition.

ENEL *web site (www.enel.it);*

EUROSTAT – Electricity prices for EU industry on January 2002.

Gestore della Rete Nazionale (GRTN) *web site (http\\www.grtn.it);*

Global transformer markets 2002–2012 prepared for International Copper Asssociation, Goulden Reports, 2008

R. Hasegawa and D. C. Pruess, *Impact of Amorphous Metal Based* Transformers on Efficiency and Quality of Electric Power Distribution, 2004, 0-7803-7031-7, ©2001 IEEE.

Strategies for energy efficient distribution transformers – SEEDT, Energy Intelligent Europe Programme project 1985–1988, http://seedt.ntua.gr

Stefan Fassbinder, Roman Targosz, Energy Efficient Transformers, Leonardo ENERGY, 2012.

3.A   Annex

3.A.1   Selected MEPS

3.A.1.1   Australia

The Minimum Energy Performance Standards (MEPS) for distribution transformers are set out as power efficiency levels at 50% of rated load according to AS 2374.1.2 when tested in accordance with AS 2374.1 or AS 2735, as applicable (Tables 3.16 and 3.17).

Australia also has a standard that is applied for the 'High Energy Efficiency Transformers', which have an efficiency of about 20% higher than the 'standard MEPS'.

Table 3.16   Minimum power efficiency levels for oil-immersed transformers

Type	kVA	Power efficiency @ 50% load
Single phase (and SWER)	10	98.30
	16	98.52
	25	98.70
	50	98.90
Three phase	25	98.28
	63	98.62
	100	98.76
	200	98.94
	315	99.04
	500	99.13
	750	99.21
	1000	99.27
	1500	99.35
	2000	99.39
	2500	99.40

Table 3.17 Minimum power efficiency levels for dry-type transformers

Type	kVA	Power efficiency @ 50% load	
		$U_m = 12$ kV	$U_m = 24$ kV
Single phase (and SWER)	10	97.29	97.01
	16	97.60	97.27
	25	97.89	97.53
	50	98.31	97.91
Three phase	25	97.17	97.17
	63	97.78	97.78
	100	98.07	98.07
	200	98.46	98.42
	315	98.67	98.59
	500	98.84	98.74
	750	98.96	98.85
	1000	99.03	98.92
	1500	99.12	99.01
	2000	99.16	99.06
	2500	99.19	99.09

3.A.1.2 USA

The Department of Energy Minimum Efficiency Levels for Regulation of Liquid-immersed Distribution Transformers.

The USA, like Australia, have also more demanding max-tech levels for liquid-insulated (Table 3.18) transformers and dry-type transformers (Tables 3.19 and 3.20). The max-tech level represents the transformer designs that would exist if cost were no object and all design efforts were focused solely on having the highest possible efficiency level. In other words,

Table 3.18 Liguid immersed distribution transformers

Single phase		Three phase	
kVA	Efficiency (%)	kVA	Efficiency (%)
10	98.62	15	98.36
15	98.76	30	98.62
25	98.91	45	98.76
37.5	99.01	75	98.91
50	99.08	112.5	99.01
75	99.17	150	99.08
100	99.23	225	99.17
167	99.25	300	99.23
250	99.32	500	99.25
333	99.36	750	99.32
500	99.42	1000	99.36
667	99.46	1500	99.42
833	99.49	2000	99.46

Table 3.19 Medium-voltage dry-type distribution transformers at 60 Hz

	Single-phase efficiency				Three-phase efficiency		
kVA	20–45 kV BIL	46–95 kV BIL	≥96 kV BIL	kVA	20–45 kV BIL	46–95 kV BIL	≥96 kV BIL
15	98.10	97.86		15	97.50	97.18	
25	98.33	98.12		30	97.90	97.63	
37.5	98.49	98.30		45	98.0	97.86	
50	98.60	98.42		75	98.33	98.12	
75	98.73	98.57	98.53	112.5	98.49	98.30	
100	98.8	98.67	98.63	150	98.60	8.42	
167	98.96	98.83	98.80	225	9873	98.57	98.53
250	99.07	98.95	98.91	300	98.82	98.67	98.63
333	99.14	99.03	98.99	500	98.96	98.83	98.80
500	99.22	99.12	99.09	750	99.07	98.95	98.91
667	99.27	99.18	99.15	1000	99.14	99.03	98.99
833	99.31	99.23	99.20	1500	99.22	99.12	99.09
–				2000	99.27	99.18	99.15

Note: BIL means basic impulse insulation level. All efficiency values are at 50% of nameplate rated load, determined according to the DOE Test-Procedure. 10 CFR Part 431, Subpart K, Appendix A.

Table 3.20 Low-voltage dry-type distribution transformers at 60 Hz

	Single phase		Three phase
kVA	Efficiency (%)	kVA	Efficiency (%)
10	97.7	15	97.0
15	98.0	30	97.5
25	98.2	45	97.7
37.5	98.3	75	98.0
50	98.5	112.5	98.2
75	98.6	150	98.3
100	98.7	225	98.5
167	98.8	300	98.6
250	98.9	500	98.7
333	97.7	750	98.8
		1000	98.9

the Max Tech. levels represent the upper limit of efficiency values considered by the US Department of Energy in the final rule it published in October 2007. These values represent roughly between 20 to 30% higher efficiency that MEPS.

3.A.1.3   Europe

3.A.1.3.1   *European EN 50464-1 Standard – Liquid Filled*
Standardized European level of losses for liquid immersed transformers with different rated voltages and short circuit impedances are listed in the tables from Tables 3.21 to 3.23.

Table 3.21 Europe, liquid filled, voltage below 24 kV, load losses

Power rating Sn kVA		Rel. short-circuit volt u_k			Oil-immersed transformer, in watts			
		List D_K	List C_K		List B_K		List A_K	
		$\leq$24 kV	$\leq$24 kV	$\leq$36 kV	$\leq$24 kV	$\leq$36 kV	$\leq$24 kV	$\leq$36 kV
		P_K	P_K	P_K	P_K	P_K	P_K	P_K
50	4%	1350	1100	1450	875	1250	750	1050
100	4%	2150	1750	2350	1475	1950	1250	1650
160	4%	3100	2350	3350	2000	2550	1700	2150
250	4%	4200	3250	4250	2750	3500	2350	3000
315	4%	5000	3900		3250		2800	
400	4%	6000	4600	6200	3850	4900	3250	4150
500	4%	7200	5500		4600		3900	
630	4%	8400	6500	8800	5400	6500	4600	5500
630	6%	8700	6750		5600		4800	
800	6%	10 500	8400	10 500	7000	8400	6000	7000
1000	6%	13 000	10 500	13 000	9000	10 500	7600	8900
1250	6%	16 000	13 500	16 000	11 000	13 500	9500	11 500
1600	6%	20 000	17 000	19 200	14 000	17 000	12 000	14 500
2000	6%	26 000	21 000	24 000	18 000	21 000	15 000	18 000
2500	6%	32 000	26 500	29 400	22 000	26 500	18 500	22 500

Table 3.22 Europe, liquid filled, voltage below 24 kV, no-load losses

Power rating Sn kVA	No-load losses oil-immersed transformer									
	List E_0		List D_0		List C_0		List B_0		List A_0	
	$\leq$24 kV		$\leq$24 kV		$\leq$24 kV		$\leq$24 kV		$\leq$24 kV	
	P_0 W	Noise dB (A)	P_0 W	Noise dB (A)	P_0 W	Noise dB (A)	P_0 W	Noise dB (A)	P_0 W	Noise dB (A)
50	190	55	145	50	125	47	110	42	90	39
100	320	59	260	54	210	49	180	44	145	41
160	460	62	375	57	300	52	260	47	210	44
250	650	65	530	60	425	55	360	50	300	47
315	770	67	630	61	520	57	440	52	360	49
400	930	68	750	63	610	58	520	53	430	50
500	1100	69	880	64	720	59	610	54	510	51
630	1300	70	1030	65	860	60	730	55	600	52
630	1200	70	940	65	800	60	680	55	560	52
800	1400	71	1150	66	930	61	800	56	650	53
1000	1700	73	1400	68	1100	63	940	58	770	55
1250	2100	74	1750	69	1350	64	1150	59	950	56
1600	2600	76	2200	71	1700	66	1450	61	1200	58
2000	3100	78	2700	73	2100	68	1800	63	1450	60
2500	3500	81	3200	76	2500	71	2150	66	1750	63

Table 3.23   Europe, liquid filled, voltage below 36 kV, no-load losses

	No-load losses oil-immersed transformer					
	List C_{036}		List B_{036}		List A_{036}	
	$\leq$36 kV		$\leq$36 kV		$\leq$36 kV	
Power rating Sn kVA	P_0 W	Noise dB(A)	P_0 W	Noise dB(A)	P_0 W	Noise dB(A)
50	230	52	190	52	160	50
100	380	56	320	56	270	54
160	520	59	460	59	390	57
250	780	62	650	62	550	60
400	1120	65	930	65	790	63
630	1450	67	1300	67	1100	65
800	1700	68	1450	68	1300	66
1000	2000	68	1700	68	1450	67
1250	2400	70	2100	70	1750	68
1600	2800	71	2600	71	2200	69
2000	3400	73	3150	73	2700	71
2500	4100	76	3800	76	3200	73

3.A.1.3.2   *European EN 50541-1, Dry Type*

Standardized European level of losses for dry-type transformers with different rated voltages and short circuit impedances are listed in the tables from Tables 3.24 to 3.28.

Table 3.24   Europe, dry-type rated voltage $\leq$12 kV short circuit voltage 4%

	Pk	Pk	Po	Lwa	Po	Lwa	Po	Lwa
Sr	Ak	Bk	Ao		Bo		Co	
kVA	W	W	W	dB (A)	W	dB (A)	W	dB (A)
100	1800	2000	260	51	330	51	440	59
160	2500	2700	350	54	450	54	610	62
250	3200	3500	500	57	610	57	820	65
400	4500	4900	7000	60	880	60	1150	68
630	6700	7300	1000	62	1150	62	1500	70

3.A.1.3.3   *Proposed European MEPS*

The proposed European MEPS are the following ones:

- Oil-immersed transformers: units $\leq$630 kVA: A_oC_k, units >630 kVA: A_0A_k.
- Additionally in tier 1, optionally low loss core material ($\leq$0.95 W/kg at 1.7 T at 50 Hz) is proposed as minimum requirement if it is not possible to meet generic MEPS.
- Dry-type transformers: A_0A_k.

Table 3.25 Europe, dry-type rated voltage ≤12 kV short circuit voltage 6%

	Pk	Pk	Po	Lwa	Po	Lwa	Po	Lwa
Sr	Ak	Bk	Ao		Bo		Co	
kVA	W	W	W	dB (A)	W	dB (A)	W	dB (A)
100	1800	2000	260	51	330	51	440	59
160	2600	2700	350	54	450	54	610	62
250	3400	3500	500	57	610	57	820	65
400	4500	4900	700	60	880	60	1150	68
630	7100	7300	1000	62	1150	62	1500	70
800	8000	9000	1100	64	1300	65	1800	71
1000	9000	10 000	1300	65	1500	67	2100	73
1250	11 000	12 000	1500	67	1800	69	2500	75
1600	13 000	14 500	1800	68	2200	71	2800	76
2000	15 500	18 000	2200	70	2600	73	3600	78
2500	18 500	21 000	2600	71	3200	75	4300	81
3150	22 000	26 000	3150	74	3800	77	5300	83

Table 3.26 Europe, dry-type rated voltage 17.5 and 24 kV short circuit voltage 4%

	Pk	Pk	Po	Lwa	Po	Lwa	Po	Lwa	Po	Lwa
Sr	Ak	Bk	Ao		Bo		Co		Do	
kVA	W	W	W	dB (A)	W	dB (A)	W	dB (A)	W	dB (A)
100	1350	1750	330	51	360	51	400	59	600	59
160	1800	2500	450	54	490	54	580	62	870	62
250	2700	3450	640	57	660	57	800	65	1100	65
400	3800	4900	850	60	970	60	1100	68	1450	68
630	5300	6900	1250	62	1270	62	1600	70	2000	70

Table 3.27 Europe, dry-type, rated voltage 17.5 kV and 24 kV short circuit voltage 6%

	Pk	Pk	Po	Lwa	Po	Lwa	Po	Lwa
Sr	Ak	Bk	Ao		Bo		Co	
kVA	W	W	W	dB (A)	W	dB (A)	W	dB (A)
100	1800	2050	280	51	340	51	460	59
160	2600	2900	400	54	480	54	650	62
250	3400	3800	520	57	650	57	880	65
400	4500	5500	750	60	940	60	1200	68
630	7100	7600	1100	62	1250	62	1650	70
800	8000	9400	1300	64	1500	64	2000	72
1000	9000	11 000	1550	65	1800	65	2300	73
1250	11 000	13 000	1800	67	2100	67	2800	75
1600	13 000	16 000	2200	68	2400	68	3100	76
2000	16 000	18 000	2600	70	3000	70	4000	78
2500	19 000	23 000	3100	71	3600	71	5000	81
3150	22 000	28 000	3800	74	4300	74	6000	83

Table 3.28 Europe, dry-type, rated voltage 36 kV short circuit voltage 6%

	Pk	Pk	Pk	Po	Lwa	Po	Lwa	Po	Lwa
Sr	Ak	Bk	Ck	Ao		Bo		Co	
kVA	W	W	W	W	dB (A)	W	dB (A)	W	dB (A)
160	2500	2700	2900	850	57	900	62	960	66
250	3500	3800	4000	1000	59	1100	64	1280	67
400	5000	5400	5700	1200	61	1300	65	1650	69
630	7000	7500	8000	1400	63	1600	68	2200	71
800	8400	9000	9600	1650	64	1900	69	2700	72
1000	10 000	11 000	11 500	1900	65	2250	70	3100	73
1250	12 000	13 000	14 000	2200	67	2600	72	3600	75
1600	14 000	16 000	17 000	2550	68	3000	73	4200	76
2000	17 000	18 500	21 000	3000	72	3500	74	5000	78
2500	20 000	22 500	25 000	3500	73	4200	78	5800	81
3150	25 000	27 500	30 000	4100	76	5000	81	6700	83

3.A.1.4 Formulae for Losses Evaluation – American and European

Selected formulae used in analysis of transformer losses are listed in Table 3.29.

Table 3.29 Selected formulae used in analysis of transformer losses

	American		European		
Item	Definition	Name	Symbol	Definition	Name
1 Annual peak (kW)	Peak power (kW)	Peak	P	The maximum yearly power	Peak power
2 Power Factor	Load (kVA) / Power (kW)	PF	$\cos \varphi$		
3 Annual PU peak	Ratio peak load to nominal power of transformer $$Annual\ PU\ peak = peak \left/ S_B \cdot PF \right.$$	Annual PU peak	B_{av}	$$\beta_s = \frac{P}{\cos \varphi \cdot S_n}$$	Relative peak load
4 Annual PU avg. load	$$Annual\ PU\ avg.\ load = \frac{Annual\ energy}{8760 \cdot S_B \cdot PF}$$	Annual PU avg. load	B_{av}	$$\beta av = \frac{E}{8760 \cdot S_n \cdot \cos \varphi}$$	Relative average load

(continued overleaf)

Table 3.29 (*Continued*)

Item	American		European		
	Definition	Name	Symbol	Definition	Name
5 Annual Loss Factor	Ratio of losses at average load to losses at peak load $$L_sF = \left(\frac{S_{RMS}}{Peak}\right)^2$$	L_sF	No equivalent		
6 Load Factor	Ratio of average load to peak load	LF	T_w	$T_w = T_s \big/ 8760$	Relative time of peak load $T_s = E/P_s$ E – energy kWh
7	$L_sF = LF^{1.732}$ or $L_sF = 0.85 \cdot LF^2 + 015 \cdot LF$		No equivalent but: $\tau_w = T_w^{1.7}$		
8 Annual average per unit (PU) load of all transformer	$PU\,load = \dfrac{sales(MWh)}{8760 \cdot installed\,MVA \cdot PF}$	$PU\,load = S$	$\beta * av$	$\beta * av = \dfrac{\sum E[kWh]}{8760 \cdot N \cdot S_n \cdot \cos\varphi}$ S_n – rated power of transformers N – number of transformers	Average relative load

(*continued overleaf*)

Table 3.29 (Continued)

Item	American Definition	Name	European Symbol	Definition	Name
9 PU load	$PU\ load = S = \dfrac{E}{8760 \cdot S_B \cdot PF}$	S	β_{av}	$\beta av = \dfrac{E}{8760 \times S_n \times \cos\varphi}$	Average relative transformer load
10 Annual FU load factor	$PU\ load\ factor = \dfrac{E}{peak \cdot 8760 \cdot PF}$	$PU\ load/PF$	$\dfrac{T_w}{\cos\varphi}$	$\dfrac{T_w}{\cos\varphi} = \dfrac{E}{8760 P_s \times \cos\varphi} = \dfrac{T_s}{8760 \times \cos\varphi}$	Relative duration of peak load
11 Annual FU rms load	$(S_{RMS})^2 = L_sF \cdot (peak)^2]$ $\cdot 8760 \cdot \Delta P_z \cdot (S_{RMS})^2 = annual\ loss\ energy$ $S_{RMS} \approx 1,1 \cdot S$	S_{RMS} PU	Yearly energy losses: $8760 \cdot \Delta P_z \cdot \beta_s \cdot \tau_w$ and thus: $(S_{RMS})^2 = (\beta_s)^2 \cdot \tau_w$		Equivalent average yearly load that generates average yearly losses

4

Building Automation, Control and Management Systems

Angelo Baggini and Annalisa Marra

In the context of energy in major Western countries, the civil sector, which has been traditionally overshadowed by industry and transport, has become more and more important.

In Europe, for example, approximately 40% of energy consumption is related to the civil sector, with 70% and 30% related to residential and tertiary, respectively. Tertiary and residential buildings, therefore, constitute an important area for the reduction of consumption.

In addition to the most popular tools that are mainly used to address the performance of loads and components, and the thermal insulation of buildings, automation can play a major role in reducing energy consumption without impacting on the development and improvement of the service that is delivered.

The management of both the building and a process should be both efficient and effective:

1. efficient, for example, in creating such an enclosure optimized in terms of thermal insulation;
2. effective, for example, in the management of energy processes, mainly heat and electricity.

The management can be carried out by humans or by automation systems. Automation can help correct those human behaviours that lead to unnecessary consumption, assisting and guiding the user to the optimal management of facilities, saving energy and helping environmental sustainability.

To date, in contrast to what has happened with other technologies, however, building automation systems[1] have not yet seen widespread development for energy saving in buildings.

---

[1] Automation, control and management systems of the building are identified by different acronyms depending on the source and on the level: BACS Building Automation and Control System, within CEN; HBES Home and Building Electronic System, within the CENELEC; still TBM Technical Building Management at CEN. Sometimes some uncertainty in the use of such classifications are even found in the original standardization documents. In this text, even to streamline the form in the absence of specification needs and unless otherwise specified, the terms automation or automation system will generally indicate the system and, without distinction, the installation of automation, control and management of the building whether residential or tertiary.

Electrical Energy Efficiency: Technologies and Applications, First Edition. Andreas Sumper and Angelo Baggini.
© 2012 John Wiley & Sons, Ltd. Published 2012 by John Wiley & Sons, Ltd.

The causes are many. First, despite the obvious benefits, the technology of automation systems is still poorly understood: on the one hand because of the misinformation of the end users, on the other because of the resistance of some professionals to update. Also, in the residential sector, the implementation of an automation system includes a new approach requiring real design and coordination of activities with the end user.

The main reasons, however, are the higher cost of these systems compared with traditional ones, and because this energy efficiency (EE) technology is rarely counted among those that now have government economical support.

The technologies that use renewable sources, as known, are still expensive and the their huge success around the world is due mainly to the availability of public incentives.

In addition the estimation of energy and cost savings related to the adoption of an automation system is quite delicate and complex.

In practice, an automated system works better than a human for energy savings, for the following main reasons:

1. Humans don't pay enough attention to energy savings because they are focused on they other main activities.
2. The human senses are usually not sensitive enough to recognize in real time the need or the possibility of switching off a service. For example, lighting is switched on in the morning before sunrise. During the day the contribution of daylight is more than sufficient, but our senses do not immediately perceive that the light level is more than the minimum needed but are only aware of this when the level becomes really excessive.

4.1 Automation Functions for Energy Savings

Some automation functions that are often of interest with reference to comfort levels may play an important role in reducing consumption.

The basic idea is very simple: to automate and then optimize the use of existing (electrical and non-electrical) energy uses in a building, switching it off completely or partially when it is not needed.

In general, automation tasks related to energy savings are typically quite simple and less expensive than those related to other application classes (Table 4.1).

Figures 4.1 and 4.2 show the percentages of energy consumption in the tertiary and residential sectors. One can easily see that, while the energy consumption for heating is high in both cases, the use of electricity for lighting greatly predominates in the tertiary sector (offices and shopping centres require high energy consumption during working hours).

4.1.1 Temperature Control

Limiting heating or cooling periods and climatic conditions to real needs allows one to achieve significant reductions in energy requirements in the residential as well as the tertiary sector.

The most important automation functions that are useful in reducing consumption by acting on the temperature controls are:

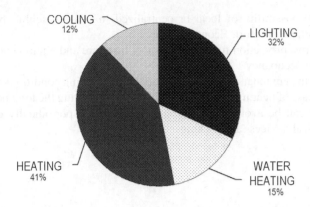

Figure 4.1 Energy consumption in the tertiary sector (ref. Italy)

1. Micro-zones with independent thermostats and valves to optimize, at the same time, comfort and consumption, instead of central or no control.
2. Indoor Air Quality (IAQ) control: a sensor analyses the air quality, activating the air exchange system only once the air becomes stale and not just periodically.
3. Air exchange scheduled at times when it is not too cold in winter and not too hot in summer, to avoid excessive temperature gradients and, therefore, losses.
4. Integration of a weather station (rain, wind, brightness) with the temperature control to avoid losses and to implement awnings, blinds, outdoor lighting and irrigation.
5. Adjusting the indoor climate conditions according to occupancy or time, switching off or controlling the opening of doors or windows.

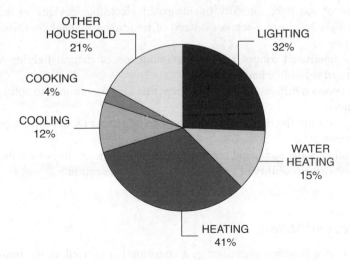

Figure 4.2 Energy consumption in the residential sector (ref. Italy)

6. Introducing the possibility of local or centralized manual switching between different modes (Comfort, Precomfort, Economy, Off).
7. Introducing a frost protection function to protect furniture and equipment in environments with occasional occupancy.
8. With reference to maintenance and the most efficient operating conditions for the mechnical installation in case of heating only, the sticking of valves during the long period of standstill in the summer can be avoided by setting a function that periodically opens and closes electromechanical devices (antisticking).

4.1.2   Lighting

Limiting the intensity of artificial lighting to when it is really needed during different periods may have an important impact on electricity consumption, especially in the service sector where the corresponding consumption is more important.

The main possibilities offered by an automation system that allows a reduction in energy consumption by acting on the lighting are as follows:

1. Automatically switching lighting on and off with all kinds of lamps (by means of timers or occupancy sensors or the level of light or twilight) instead of manually (with traditional commands or remote control).
2. Command of the power of the marker lamps or LEDs or lighting appliances in general through light sensors.
3. Automatically switching lighting on and off according to motion (high-traffic areas) or with occupancy.
4. Automatic control of external equipment and lighting showcases using a twilight criterion.
5. A temporary switch off of groups of non-critical equipment.
6. Adjustment of the light intensity by means of electronic devices to increase or decrease the light level using sensors instead of traditional manual commands of various kinds.
7. Brightness automated control for the maximum use of natural lighting with artificial lighting integrated only when needed.
8. Switching between different groups of devices that are programmed to optimize the useful life of sources.
9. The ability to count the hours of operation of an apparatus by optimizing preventive and routine maintenance.
10. The replication of the command points (individual and group) anywhere the bus cable is, will increase the probability of an effective human intervention.

4.1.3   Drives and Motors

The automation of a function for reducing a consumption as well as the introduction of a drive involves an additional consumption. But for various reasons, including laziness and lack of sensitivity of human senses, the balance can become easily positive, since, practically and statistically, human actions are not so effective.

The main functions of automation, based on drives that can lead to a reduction in consumption, are as follows:

1. Shutters for windows and door locks can be automated. To reduce energy losses the opening and closing doors can be controlled by micro switches, scenarios, timers or can be operated by remote control, or controlled by light sensors or weather stations that measure wind speed and the presence of rain. Similarly, awnings can be automated.
2. Venetian blinds: move up/down and angle adjustment of blades.
3. Integration with electric door locks.
4. Irrigation systems: scheduling start and end times to open and close water valves or depending on sunlight, avoiding unnecessary wastage of water if it rains, and preventing damage to plants.
5. Integration with systems management pools.
6. Control of shading devices (blinds, shutters, etc.) manually or automatically. Operations such as up, down, opening, closing or tilting of the slats can be made automatically depending on the position of the sun to avoid direct sunlight with less need for air conditioning in the summer (energy saving) and glare protection (greater visual comfort for workstation screens).
7. Integration of the command and control scenarios for meeting rooms, multipurpose rooms, showrooms, etc.
8. Weather alarms (rain, wind, frost) can maximize the use of control shading devices, putting them to a secure position or blocking them just when necessary.

4.1.4 Technical Alarms and Management

Technical alarms can also play an active role in limiting the consumption of a building: the detection of an abnormal condition at the outset avoids the system operating in abnormal conditions where it could be less efficient with higher losses, etc.

The main functions of automation that can lead to a reduction in consumption in a building are as follows:

1. The detection of hazardous gases such as methane, LPG, carbon monoxide (in the kitchen, garage, near fireplaces and stoves), smoke, fire, flood water.
2. The closing of solenoid valves for gas and water in a given scenario.
3. Combining an actuator, plug sockets can be switched off for selected users, eliminating the consumption of appliances in standby mode without switching off the users who require uninterrupted power.
4. Protective automatic reset in case of nuisance tripping.
5. Installation management.
6. Scheduling events, bidirectional control via phone or internet.
7. Environmental and/or electrical parameter monitoring and logging for trend analysis.

4.1.5   Remote Control

Remote control of the system can help in reducing the consumption of a building, allowing direct human intervention in the unexpected situations that require it. The ability to intervene directly when there is an unexpected situation avoids, at the outset, the operation of the system in conditions that are potentially characterized by a lower efficiency, higher losses, etc.

The main automation functions leading to a reduction in consumption based on remote control are:

1. Control of bus devices.
2. Enquiry of the status of bus devices.
3. Delivery of alarms or events.
4. Command and/or control of multiple coordinated functions.

4.2   Automation Systems

The line between traditional electrical systems and automation systems can be found in the native ability to communicate, of each device developed for the automation system:

1. Low voltage electrical systems in the traditional control device act directly on the power circuit of the controlled user.
2. In the automation systems, the control circuit is separated from the power circuit. The former consists of a network through which signal control devices and actuators, with their own electronic processing and communication, are able to exchange information in the form of messages encoded in digital form. The loads and the devices that need the power supply to operate are connected to the power network (power bus) and, for control, in addition to a signal bus. The devices that do not need a power supply to operate, like the command button, for example, are connected just to the signal bus.

The applications that are achievable with a traditional system differ, in practice, from those achievable with an automated system. This is because as soon as a bit more articulated logic is needed, the wiring complexity of the traditional system increases exponentially to the point of making the implementation impractical.

Applications built with an automation system can be classified as shown in Table 4.1.

Those of interest for energy saving are virtually all in Class 1. There is sometimes confusion over the differences between SCADA (Supervisory Control and Data Acquisition) systems and Distributed Control Systems (DCS). Generally speaking, a SCADA system always refers to a system that coordinates, but does not control processes in real time. The discussion on real-time control is muddied somewhat by newer telecommunications technology, enabling reliable, low latency, high speed communications over wide areas. Most differences between SCADA and DCS are culturally determined and can usually be ignored. As communication infrastructures with higher capacity become available, the difference between SCADA and DCS will fade. SCADA and DCS are compared below.

1. DCS is process oriented, while SCADA is data acquisition oriented.
2. DCS is process driven, while SCADA is event driven.

Table 4.1   Application classes of an automation system

	Class 1
Command	Lighting, heating, ventilation, air conditioning, actuation in general
Alarms	Rescue, intrusion, gas leak, fire, flood, technical alarms
Sound diffusion	Control

	Class 2
Sound diffusion	Speakers
Communication	Phone, etc.

	Class 3
Communication	Broadband video

3. DCS is commonly used to handle operations on a single zone while SCADA is preferred for applications that are spread over a wide geographic location.
4. DCS operator stations are always connected to its I/O, while SCADA is expected to operate despite failure of field communications.

Elements of a distributed control system may directly connect to physical equipment such as switches, pumps and valves or may work through an intermediate system such as a SCADA system.

4.2.1   KNX Systems

KNX[2] is a specific DCS system that consists of a series of input/output connected to a shared transmission medium, a bus, plus some system devices necessary for operation. Each device on the network exchanges information containing a set of datapoints, i.e. control and process variables that each device interprets. The data points can be inputs, outputs, parameters and diagnostic data.

These simple rules of communication together with the characteristics of the transmission medium constitute the architecture of the KNX system.

A KNX system is a distributed computing network (i.e. there is no device that centralizes the logic of the system) whose nodes have an individual address of 16 bits. In total therefore, in theory, up to 65 536 devices can be addressed.

4.2.1.1   Architecture

By analysing the routing rules and some electrical rules, for a KNX network you obtain the overall architecture shown in Figure 4.3,

---

[2] Of the DCS systems available on the market, KNX systems are the only ones that are standardized by an International IEC standard. In the author's opinion, standardization is a crucial issue for the possibility of the future development of automation in buildings.

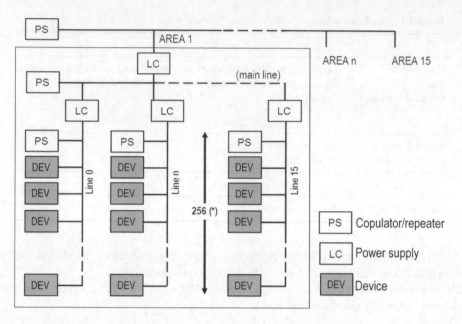

Figure 4.3 General architecture of a KNX network (*). Each line, separated by an LC, is made by a maximum of four segments each one made by a maximum of 64 devices

Area, Line Device
The KNX technology, line level, allows any topology: star, tree and bus, but not ring. A series of up to 256 devices is a line. A maximum of 15 lines can be connected by a main line, constituting an area. A KNX installation can contain up to 15 areas, connected by a backbone line.

Physical Address
Each part of the KNX network, with the exception of power, is uniquely identified by a physical address. The address consists of three numerical fields separated by a dot:

1. The 1st field defines the area of belonging (range from 0 to 15).
2. The 2nd field indicates the line (range from 0 to 15).
3. The 3rd field identifies the device (range from 0 to 255).

Group Address
The equipment in a system communicate with each other via KNX group addresses. Normally, the group addresses are structured in a hierarchy with three levels:

1. The main group, normally it is the system level (e.g. lighting, temperature control, etc.).
2. The group, a unique feature of the system considered (e.g. switches, dimmers, etc.).
3. The subgroup, devices belonging to the same function (e.g. light kitchen, bedroom window, etc.).

Group addresses are represented by separating the numeric fields that define the main group, the group and the subgroup using the forward slash (/). There is also a version of routing groups at two levels, but it is less used.

Power Supply
The power supply is the device that supplies power to devices connected to the bus line. The power supply is normally fed directly to the mains voltage and provides a DC voltage of 29 V directly to the bus terminals.

Coupler (Line, Field, Repeater)
A bus line with a power supply is generally determined by a maximum of 64 connected devices, taking into account the power consumption of the devices and the total maximum current.

KNX on each line, however, can be theoretically up to 256 devices connected to the bus if the line is structured into four segments (Figure 4.4), each with its own power supply; they are connected together by repeaters.

A repeater is just a special way of employing a coupler that galvanically separates bus lines and regenerates the signal, which prevents an electrical fault propagating to other lines and also allows one to expand the 'overall architecture' of a KNX system up to a maximum of 65 536 devices.

The device coupler can be used in different ways:

1. coupler area/region: linking the areas along the main backbone (backbone line)
2. line coupler/area: linking the lines in an area along the main line (main line)
3. repeater (booster) links two line segments, regenerating the degraded signal.

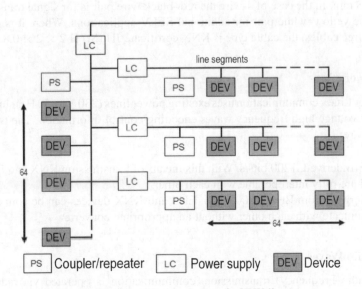

| PS | Coupler/repeater | | LC | Power supply | | DEV | Device |

Figure 4.4   Maximum size of a KNX line

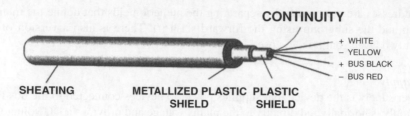

Figure 4.5   Twisted pair

4. filter frames: you can set the device coupler to block the trasmission of some telegrams, thereby avoiding when unnecessary to send them across the network. The scope is to increase the ability of communication and decrease the likelihood of errors and collisions between packages.

The couplers are addressed as any other KNX device, and can only be configured with the ETS software.

4.2.1.2   Transmission Media

Twisted Pair – TP

This is a communication cable through a bipolar twisted (helical winding), shielded and double insulated cable.

In a KNX system the bus TP-1 derived from EIB, speed 9600 bits/s is to be used. Through this transmission medium KNX and EIB can communicate with each other and are fully interoperable.

The cable to be used (Figure 4.5) shall be certified KNX as YCYM type 1 or $2 \times 2 \times 0.8 \times 2 \times 0.8$ mm. In the case of 4-wire the red–black wire pair is for signal transmission and power and the yellow–white pair for additional SELV applications. Where it is necessary to use halogen-free cables, the cable type is KNX certificate JH (St) H $2 \times 2 \times 0.8$.

4.2.1.3   Power Line

For PL (Power Line), communication uses existing power lines (230 Vac, 50 Hz) superimposing on the power voltage high frequency waves encoding control information. The two modes of transmission are:

1. PL-110: EIB derived, 1200 bits/s. With this means of transmission KNX and EIB communicate and are fully interoperable with each other.
2. PL-132: derived from EHS, 2400 bits/s. EHS and KNX devices can be connected but do not communicate with each other without an appropriate converter.

4.2.1.4   Radio Waves

With RF (Radio Frequency) transmission, communication is operated via radio waves to 868.30 MHz carrier frequency transmission inside the ISM band (Industrial Scientific

Medical), with a speed of 16.38 kbit/s. The logic states 0 and 1, needed for the digital encoding of the message are generated by a deviation from the carrier frequency.

The RF devices, which can be divided into two groups depending on the communication, may be unidirectional or bidirectional.

4.2.1.5   Ethernet

Independently of the particular transmission medium, in this case the communication is done by transferring KNX telegrams over an Ethernet network encapsulated in packages in standard protocol IP (Internet Protocol).

According to a procedure known as 'KNX/IP routing', in the largest installations, the IP network can be used as a high speed backbone (Fast-Backbone) to send KNX telegrams.

4.2.1.6   Configuration

Once the devices have been installed and the bus has been connected, two more steps are needed to put the installation into service:

1. The level of network topology, when the physical addresses are defined: in practice in this step the global architecture of the system is created.
2. The level of node, when the configuration of the application of each node are defined and the group addresses are defined (binding). The group addresses are the logical links between two or more devices connected to the KNX network. Without this link, all the devices on the network receive all the messages, but are unable to understand if the message was addressed to themselves or to another device.

Guaranteeing the consistency of the protocol and the interoperability between the devices, the KNX standard allows one to adopt devices with different configuration methods.

A KNX device can be configured in three different ways:

1. System (S-mode): it is necessary to use a PC with the ETS software.
2. Easy (E-mode): no need for a PC but other techniques for programming functions.
3. Automatic (A-mode) will not be further investigated as it is specified for the consumer market.

Table 4.2 compares the characteristics of the E-mode and S-mode configuration methods.

S-mode
The S-mode configuration is ideal when functions are required for automation and control and the various and complex system requirements are high with reference to the number of devices, the personalization features and the spatial distribution (building automation systems in the strict sense): it is based on the ETS software and requires the use of a PC.

The main advantage of this method lies in the flexibility of the design and the configuration.

Table 4.2 Comparison between E-mode and S-mode

Characteristics	E-mode	S-mode
Max. number of devices	64	>64.000
Network architecture	One line	Maximum 15 lines per area and maximum15 areas
Configuration	Configurator or ETS software	ETS software only
Protocol	KNX	KNX
Functions settable on the devices	The main one only	all

E-mode
The E-mode configuration installation is possible in two ways:

1. using a simplified method of steps leading to the reduction and simplification of all the steps necessary to put into service;
2. using the ETS software.

The E-mode devices, unlike the S-mode, already contain within them the possible configurations that, although they limit the number of functions available, on the other hand they greatly simplifiy the process of commissioning the system.

The available functions are usually sufficient to meet all the domestic application needs.

4.2.2 Scada Systems

SCADA generally refers to computer systems that monitor and control industrial, infrastructure, or facility-based processes. The term SCADA usually refers to centralized systems that monitor and control entire complex sites or buildings.

It was initially developed for:

- Infrastructure processes including electrical power transmission and distribution, wind farms, civil defence siren systems, water treatment and distribution, wastewater collection and treatment, oil and gas pipelines, and large communication systems.
- Industrial processes including those of manufacturing, production, power generation, fabrication and refining, and it may run in continuous, batch, repetitive, or discrete modes.

They have also been quickly adopted in facility processes, including buildings, airports, ships, and space stations. They monitor and control HVAC, access, and energy consumption.

A SCADA System typically consists of the following subsystems:

- A Human–Machine Interface (HMI) is the apparatus that presents process data to a human operator and, through this, the human operator monitors and controls the building or the process.
- A supervisory system, acquiring data on the building and sending commands to it.

- Remote Terminal Units (RTUs) connecting to sensors in the building, converting sensor signals to digital data and sending digital data to the supervisory system.
- Programmable Logic Controller (PLCs) used as field devices because they can be more economical, versatile, flexible, and configurable than special-purpose RTUs.
- Communication infrastructure connecting the supervisory system to the Remote Terminal Units.

Most control actions are performed automatically by Remote Terminal Units (RTUs) or by Programmable Logic Controllers (PLCs). Host control functions are usually restricted to basic overriding or supervisory level intervention. For example, a PLC may control the flow of cooling water through part of an HVAC large unit, but the SCADA system may allow operators to change the set points for the flow, and enable alarm conditions, such as loss of flow and high temperature, to be displayed and recorded. The feedback control loop passes through the RTU or PLC, while the SCADA system monitors the overall performance of the loop.

Data acquisition begins at the RTU or DSC level and includes meter readings and equipment status reports that are communicated to SCADA as required. Data is then compiled and formatted in such a way that a control room operator using the HMI can make supervisory decisions to adjust or override normal RTU (DSC) controls. Data may also be fed to a Historian, often built on a commodity Database Management System, to allow trending and other analytical auditing.

SCADA systems typically implement a distributed database, commonly referred to as a tag database, which contains data elements called tags or points. A point represents a single input or output value monitored or controlled by the system. Points can be either hard or soft. A hard point represents an actual input or output within the system, while a soft point results from logic and math operations applied to other points. Most implementations conceptually remove the distinction by making every property a 'soft' point expression, which may, in the simplest case, equal a single hard point. Points are normally stored as value-timestamp pairs: a value, and the timestamp when it was recorded or calculated. A series of value–timestamp pairs gives the history of that point. It is also common to store additional metadata with tags, such as the path to a field device or PLC register, design time comments, and alarm information.

4.2.2.1   Human–Machine Interface

A Human–Machine Interface is the apparatus that presents process data to a human operator, and through which the human operator controls the process.

An HMI is usually linked to the SCADA system's databases and software programs, to provide trending, diagnostic data, and management information, such as scheduled maintenance procedures, logistic information, detailed schematics for a particular sensor or machine, and expert-system troubleshooting guides.

The HMI system usually presents the information to the operating personnel graphically, in the form of a mimic diagram (Figure 4.6). This means that the operator can see a schematic representation of the plant being controlled. For example, a picture of a pump connected to a pipe can show the operator that the pump is running and how much fluid it is currently pumping through the pipe. The operator can then switch the pump off. The HMI software will show the flow rate of the fluid in the pipe decrease in real time. Mimic diagrams may consist of

2000-01-01 13:15

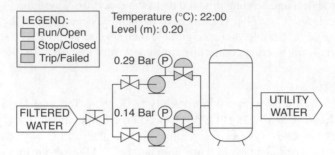

Figure 4.6 Typical Basic SCADA Animations

line graphics and schematic symbols to represent process elements, or may consist of digital photographs of the process equipment overlain with animated symbols.

The HMI package for the SCADA system typically includes a drawing program that the operators or system maintenance personnel use to change the way these points are represented in the interface. These representations can be as simple as an on-screen corridor light, which represents the state of an actual corridor light, or as complex as a multi-projector display representing the position and parameters of all the water pumps in a big building.

An important part of most SCADA implementations is alarm handling. The system monitors whether certain alarm conditions are satisfied, to determine when an alarm event has occurred. Once an alarm event has been detected, one or more actions are taken (such as the activation of one or more alarm indicators, and perhaps the generation of email or text messages so that management or remote SCADA operators are informed). In many cases, a SCADA operator may have to acknowledge the alarm event; this may deactivate some alarm indicators, whereas other indicators remain active until the alarm conditions are cleared. Alarm conditions can be explicit – for example, an alarm point is a digital status point that has either the value NORMAL or ALARM, which is calculated by a formula based on the values in other analogue and digital points – or implicit: the SCADA system might automatically monitor whether the value in an analogue point lies outside high and low limit values associated with that point. Examples of alarm indicators include a siren, a pop-up box on a screen, or a coloured or flashing area on a screen (which might act in a similar way to the 'fuel tank empty' light in a car); in each case, the role of the alarm indicator is to draw the operator's attention to the part of the system 'in alarm' so that appropriate action can be taken. In designing SCADA systems, care is needed in coping with a cascade of alarm events occurring in a short time, otherwise the underlying cause (which might not be the earliest event detected) may get lost in the noise. Unfortunately, when used as a noun, the word 'alarm' is used rather loosely in the industry; thus, depending on context, it might mean an alarm point, an alarm indicator, or an alarm event.

4.2.2.2 Remote Terminal Unit (RTU)

The RTU connects to physical equipment. Typically, an RTU converts the electrical signals from the equipment to digital values such as the open/closed status from a switch or a valve, or measurements such as pressure, flow, voltage or current. By converting and sending these electrical signals out to equipment the RTU can control the equipment, such as opening or closing a switch or a valve, or setting the speed of a pump. It can also control the flow of a liquid.

SCADA solutions often have Distributed Control System (DCS) components. Use of 'smart' RTUs or PLCs, which are capable of autonomously executing simple logic processes without involving the master computer, is increasing. A standardized control programming language, IEC 61131-3 (a suite of five programming languages including Function Block, Ladder, Structured Text, Sequence Function Charts and Instruction List) is frequently used to create programs that run on these RTUs and PLCs. Unlike a procedural language such as the C programming language or Visual Basic, IEC 61131-3 has minimal training requirements by virtue of resembling historic physical control arrays. This allows SCADA system engineers to perform both the design and implementation of a program to be executed on an RTU or PLC. A Programmable automation controller (PAC) is a compact controller that combines the features and capabilities of a PC-based control system with that of a typical PLC. PACs are deployed in SCADA systems to provide RTU and PLC functions. In many electrical substation SCADA applications, 'distributed RTUs' use information processors or station computers to communicate with digital protective relays, PACs, and other devices for I/O, and communicate with the SCADA master in lieu of a traditional RTU.

Virtually all major PLC manufacturers offer integrated HMI/SCADA systems, many of them using open and non-proprietary communications protocols. Numerous specialized third-party HMI/SCADA packages, offering built-in compatibility with most major PLCs, have also entered the market, allowing mechanical engineers, electrical engineers and technicians to configure HMIs themselves, without the need for a custom-made program written by a software developer.

4.2.2.3 Supervisory Station

The term 'Supervisory Station' refers to the servers and software responsible for communicating with the field equipment (RTUs, PLCs, etc.), and then to the HMI software running on workstations in the control room, or elsewhere. In smaller SCADA systems, the master station may be composed of a single PC. In larger SCADA systems, the master station may include multiple servers, distributed software applications, and disaster recovery sites. To increase the integrity of the system the multiple servers will often be configured in a dual-redundant or hot-standby formation providing continuous control and monitoring in the event of a server failure.

4.2.2.4 Communication Infrastructure and Methods

SCADA systems have traditionally used combinations of radio and direct serial or modem connections to meet communication requirements, although Ethernet and IP over SONET/SDH is also frequently used at large sites such as railways and power stations. The remote management or monitoring function of a SCADA system is often referred to as telemetry.

This has also come under threat with some customers wanting SCADA data to travel over their pre-established corporate networks or to share the network with other applications. The legacy of the early low-bandwidth protocols remains, though. SCADA protocols are designed to be very compact and many are designed to send information to the master station only when the master station polls the RTU. Typical legacy SCADA protocols include Modbus RTU, RP-570, Profibus and Conitel. These communication protocols are all SCADA-vendor specific but are widely adopted and used. Standard protocols are IEC 60870-5-101 or 104, IEC 61850 and DNP3. These communication protocols are standardized and recognized by all major SCADA vendors. Many of these protocols now contain extensions to operate over TCP/IP. Although some believe it is good security engineering practice to avoid connecting SCADA systems to the internet so the attack surface is reduced, many industries, such as wastewater collection and water distribution, have used existing cellular networks to monitor their infrastructure along with internet portals for end-user data delivery and modification. This practice has been ongoing for many years with no known data breach incidents to date. Cellular network data is fully encrypted, using sophisticated encryption standards, before transmission and internet data transmission, over an 'https' site, is highly secure.

RTUs and other automatic controller devices were being developed before the advent of industry-wide standards for interoperability. The result is that developers and their management created a multitude of control protocols. Among the larger vendors, there was also the incentive to create their own protocol to 'lock in' their customer base. A list of automation protocols is being compiled here.

Recently, OLE for Process Control (OPC) has become a widely accepted solution for intercommunicating between different hardware and software, even allowing communication between devices originally not intended to be part of an industrial network.

4.3 Automation Device Own Consumption

For a correct assessment in principle in an automation system the device's own consumption should be considered to estimate the total consumption (and eventually corresponding cost). However, as is easily understood, the consumption of the devices is very low compared with the energy savings introduced: it is therefore possible to ignore them without introducing a significant error.

Table 4.3 lists the most frequent consumptions by the devices used in the proposed design solutions for automation systems.

The voltage supplied from the KNX system bus is 30 VDC, the power consumption of devices can vary from 30 mW to 540 mW. So usually low power consumption does not significantly affect the total consumption and can be neglected in the calculation.

4.4 Basic Schemes

This section describes the automation functions that are useful for a rational use of energy. For easy reference, the functions have been divided on the basis of the system under control.

4.4.1 Heating and Cooling

4.4.1.1 Automatic Control of Every Room by Thermostatically Controlled Valves or Electronic Regulator (Figure 4.7)

In every room there is a controller (R) that is not coordinated with those of other environments. The device is not connected to the bus, but it has a current output to control the discharge valve

Table 4.3 Typical automation device's own consumption (ref. KNX)

Device	Own consumption (mA)
Actuator 1 channel	5
Actuator 4 channels	10
Actuator 8 channels	15
Actuator 12 channels	15
Actuator 1 channel for motor	8
Actuator 2 channels for motor	18
Actuator 8 channels for motor	18
Actuator dimmer	7
Contact interface 4 channels	9
Infrared motion sensor	5
Thermostat	5
Chronothermostat	5
Control panel*	2
Touch screen panel*	1
USB interface	5

Note: Own consumption for the control panel and touch screen panel indicate the power drawn by the only bus system, as these devices are powered directly from the mains.

of the heat transfer fluid. Alternatively, a thermostat could be used or a thermostatic valve for the environment.

Category
Emission control

Logic
Temperature can be adjusted in each individual zone, but there is no communication between the controllers.

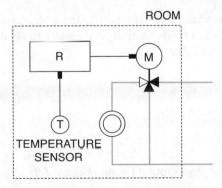

Figure 4.7 Automatic control of every room by thermostatically controlled valves or electronic regulator scheme

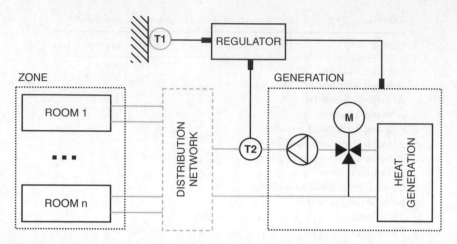

Figure 4.8 Control of water temperature with compensated supply temperature depending on the outside temperature

Main Components
Stand alone thermostats or electronic controllers
Thermostatic valves

4.4.1.2 Control of Water Temperature with Compensated Supply Temperature Depending on the Outside Temperature (Figure 4.8)

The regulator (R) sends a signal to the heat generator on the basis of which the temperature of the fluid is varied depending on the outside temperature. Temperature sensors detect the temperature of the fluid flow (T2) and the external flow temperature (T1). There are no room temperature sensor nor communication with the bus outside the thermal sub-system.

Category
Water temperature control in the distribution system (supply and return)

Logic
The temperature of the heat transfer network is adjusted with outside temperature compensation. For thermal calculations, it needs the flow temperature to obtain an average temperature inside the zone to the set point value.

Main Components
Electronic controller with:

- External temperature sensor
- Supply temperature sensor

4.4.1.3 On/Off Control of Distribution Pump (Figure 4.9)

The On/Off control on the distribution pump is based on the temperature set on the thermostat of the area, which activates or stops the flow of heat transfer fluid. In the compact boiler, pump

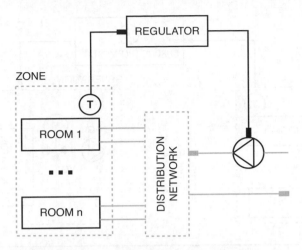

Figure 4.9   On/Off control of distribution pump

distribution is usually built inside the compact boiler. There is no communication with the bus outside the thermal sub-system.

Category
Distribution pump control

Logic
Controller (R) stops the function of the distribution pump when the temperature exceeds the set point. Operation starts again when the temperature drops below the set point.

Main Components
Regulator or thermostat

4.4.1.4   Automatic Control Function with a Fixed Time Program (Figure 4.10)

The thermostat, which is the same for all environments, measures the temperature of the area using a sensor (T) and handles the generation operation based on time and on the temperature set point. The circulation of the heat transfer fluid is started by commanding, for example, the pump (PG) and the distribution valve (VG), when the zone temperature falls below the prescribed value during the time set on the thermostat. There is no expected communication with the bus outside the thermal sub system.

Category
Generation and/or distribution and intermittent clock

Logic
The control works according to the time and the temperature profile set on an environment thermostat in anticipation of the presence of people but without considering the variations in thermal load and the actual occupancy.

Main Components
Thermostat or electronic controller

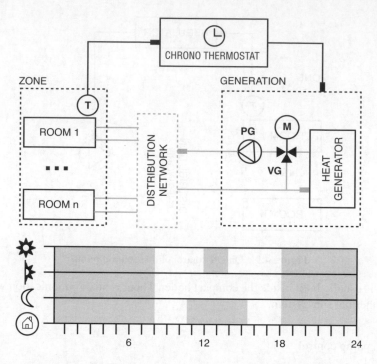

Figure 4.10   Automatic Control Function with a fixed time program

4.4.1.5   Partial Interlock (Depending on the HVAC System)

The interlock between heating and cooling systems consists of a controller (RC) that activates the heating when the outside temperature is below the heating set point (e.g. 20°C), and a controller (RF) that activates the cooling when the outside temperature is above the cooling set point (e.g. 24°C). The sensor signal TE can vary the width of the neutral zone depending on the outside temperature.

The scheme depends on the type of system used.

Category
Interlock between heating and cooling at an output and/or distribution level

Logic
The flow temperature of the two heat transfer fluids can be influenced by the outside temperature to decrease the possibility of heating in summer and cooling in winter. To minimize the possibility of simultaneous heating and cooling, the control function is achieved by providing a suitably large interval between the flow temperature of the heat transfer medium and that of the cold heat transfer (neutral zone).

Main Components
• Heating flow regulator
• Cooling flow regulator
• Outside temperature sensor

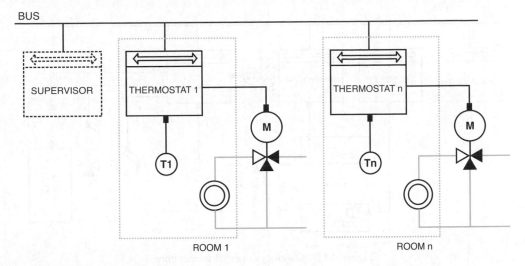

Figure 4.11 Function of automatic control of any room with communication between the regulators and toward the system BUS

4.4.1.6 Function of Automatic Control of Any Room with Communication Between the Regulators and Toward the System BUS (Figure 4.11)

The temperature of each room can be adjusted by a thermostat coordinated with regulators in other rooms.

Category
Emission control

Logic
A thermostat with a temperature sensor integrated or remote via a mixing valve regulates the flow of heat transfer fluid. The function allows the coordination of temperature adjustment between different environments and their management from any central location.

Main Components
• Thermostats
• Mixing solenoid (or intercept)
• Possible central supervisor

4.4.1.7 Control of Internal Temperature (Figure 4.12)

In any environment, the temperature is regulated by an electronic regulator that acts on the respective valve and pump. The temperature of the heat transfer fluid in any environment is continuously modified and optimized, depending on the outside temperature and the set-point. The temperature adjustment of the heat transfer in the network can be made with a mix of supply with the return, or by acting directly on the burner performance.

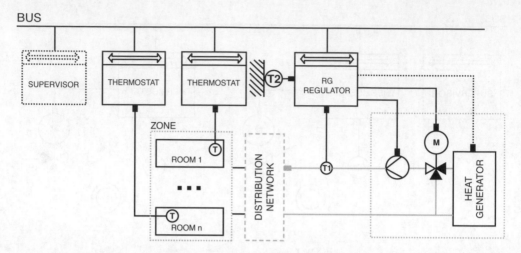

Figure 4.12   Control of internal temperature

Category

Water temperature control in the distribution system (supply and return)

Logic

The generation regulator (RG) regulates the flow water temperature by adjusting the mixing valve, or directly on the burner. The adjustment is due to the thermostat, which senses the temperature in every room, the T2 sensor that detects the external temperature and the sensor T1, which detects the flow temperature of the fluid. For each outdoor temperature, including a preset number, you get the minimum radiator hot water temperature needed to reach the set point.

Main Components

- Generation regulator
- Flow temperature sensor in the network (remote)
- Outside temperature sensor
- Room thermostats

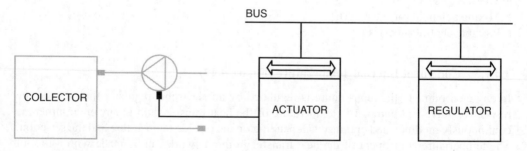

Figure 4.13   Control of the distribution pumps at a variable speed with ΔP constant

4.4.1.8   Control of the Distribution Pump at a Variable Speed with ΔP Constant (Figure 4.13)

A variable speed drive in the pump can be controlled in an On/Off way or can regulate the heat flow to load. In case of partial load, this reduces electricity consumption.

Category
Control of distribution pump

Logic
The pump head is kept constant and its scope (speed) is adjusted in proportion to the load (opening or closing of one or more hydraulic circuits).

Main Components
• Circulation pump with variable and self-regulated in ΔP constant speed
• Regulator
• Actuator (variable frequency inverter that could be incorporated in the pump or in the regulator)

4.4.1.9   Automatic Control with Optimized Start/Stop (Figure 4.14)

The activation is with reference to the start time of the period of comfort, so that the required temperature is reached by the set start time. The setting depends on the type of system controlled, i.e. the type of terminals (fan coils, radiators, floor panels, radiant ceiling, etc.), on the type of control (boiler, flow temperature) and on the type of building (ground, isolation, etc.).

The optimized stop time anticipates the off time of the plant so that the specified temperature for the end time period of comfort is not below a certain value.

Category
Intermittent control of generation and/or distribution

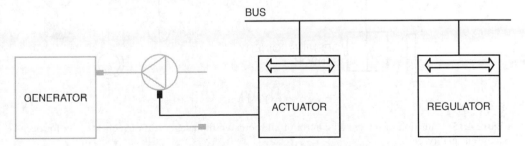

Figure 4.14   Automatic control with optimized start/stop

Logic

Starting and stopping are achieved without the use of external devices. The controller software uses the data entered in the starting phase (e.g. thermal inertia and orientation of the building, constant of time of terminals) to optimize production and distribution of energy without compromising comfort.

Main components
• System controller
• Possible central supervisor for monitoring and/or management of the system

4.4.1.10   Function of Integrated Control of All Rooms with Management of Requests (e.g. for Occupancy, Air Quality, etc.) (Figure 4.15)

The function provides for monitoring the temperature of each room with the possibility of turning off the heating or to put it on low power when there is nobody in the room or the exterior doors are open. The controller must be able to communicate with other controllers and towards the bus system.

Category
Emission control

Logic
The regulator controls the flow of hot water commanding an On/Off mode or modulating the mixing solenoid valve. The heater can be discontinued or placed in a state of Precomfort when the occupancy sensor reveals a lack of people in the room or when the sensor reveals

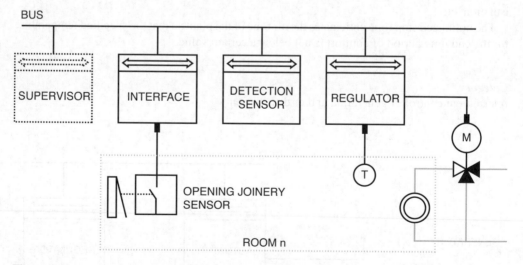

Figure 4.15   Function of integrated control of all rooms with management of requests (e.g. for occupancy, air quality, etc.)

a door opening to the outside. The controller and devices such as the occupancy sensor and the binary interface send, on the bus line, information realted to the local control (e.g. thermal load, occupation, windows/doors status, etc.).

Main Components
- Thermostat
- Motion sensor
- Environmental temperature sensor
- Doors-open sensor
- Contact interface
- On-Off solenoid/regulation

4.4.1.11   Function of Total Interlock

The central system prevents, by electrical interlock, the simultaneous starting of both hot and cold generators: it is possible to condition the system with the outside temperature to improve comfort and to properly implement the heat/cool change.

The scheme depends on the type of system used.

Category
Intermittent control of generation and/or distribution

Logic
The overal interlock between the cooling and heating ensures that, in the environment, the hot and cold generators never operate simultaneously in the same area. This depends on the type of generator used, on the distribution of heat transfer fluid and its control (central or in individual zones).

Main Components
- Flow regulator for heating
- Flow regulator for cooling
- External temperature sensor

4.4.2   Ventilation and Air Conditioning

4.4.2.1   Time Control (Figure 4.16)

A program controls the switching on and off at set times of the injection and extraction fans (V1 and V2). The room air flow is controlled by V1 and/or V2 fans and is set for the maximum load. For energy saving, when fans are on, thermal generators for room air conditioning of the AHU (air handling unit) should be turned off.

Category
Air flow control in environment

Logic
The system operates under centralized planning, controlling air flow, according to a predetermined time program.

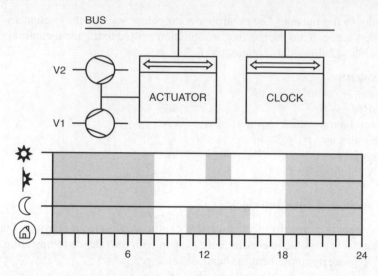

Figure 4.16   Time control

Main Components
- System clock
- Actuator

4.4.2.2   Time On/Off Control (Figure 4.17)

A program controls the switching on and off of the entry and extraction fans (V1 and V2) at regular times by an actuator. The room air flow is controlled by fans V1 and/or V2 and is set for the maximum load. For energy saving generators should be turned off for AHU

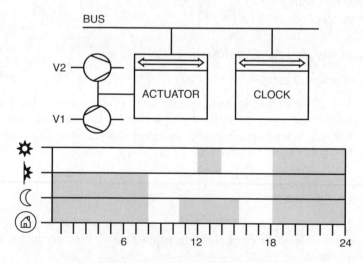

Figure 4.17   Time On/Off control

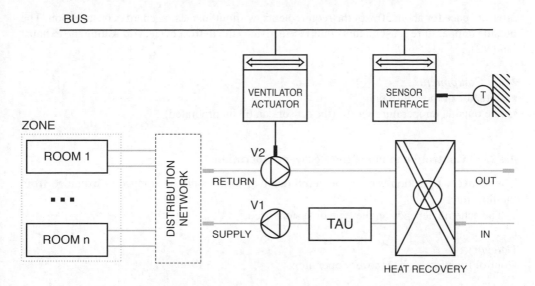

Figure 4.18 Defrost control with heat recovery

room-air conditioning when the fans are off, sending a special message to the actuator of AHU controllers.

Category
Air flow control into the air handling unit

Logic
This function affects the flow of air conditioning in every room of the local area taking as a reference only one room. The air handling system provides the flow for the maximum load to all areas during periods of occupation, prepared by a time schedule.

Main Components
- System clock
- Actuator

4.4.2.3   Defrost Control with Heat Recovery (Figure 4.18)

During the winter, ice forms in the heat recovery due to humidity and the outside air temperature. It is therefore necessary to increase the power of the exhaust fan to ensure the proper flow of fresh air in spaces, resulting in a greater consumption of electricity. By taking control of cycle defrost an increased consumption of energy is avoided.

Category
Defrost control with heat recovery

Logic
The formation of ice on the external coil determines the automatic activation of the defrost cycle, which usually lasts about 3–5 minutes. The AHU continues to provide heat to the

interior space for about 50% of the requirements, without increasing energy consumption. The outside temperature sensor and its interface activates the defrost cycle, controlling the exhaust fan (V2).

Main Components
- Actuator of the return fan
- The outside temperature sensor (the sensor can be incorporated)

4.4.2.4   Control Function of the Recovery Operation

The recovery operation can be counterproductive, so it should be stopped or excluded from the flow duct.

The scheme depends on the type of system used.

Category
Control function of the recovery operation.

Logic
The heat recovery will start automatically when the machine and temperature conditions permit it.

Main Components
The heat recovery manufacturer usually provides the temperature sensors (appropriately placed) and the related actuators to stop the machine in case of overheating. This condition may degrade or prevent the operation of the recovery in hot weather. The control components are highly dependent on their type (eg. cross-flow heat recovery, heat exchangers, spinning mass, etc.) and on the construction used, so a temperature control should be requested from the supplier of the machine.

4.4.2.5   Night Cooling (Figure 4.19)

Through the system clock, the time is established at which night cooling is started; the interface sensor detects the T1 room temperature and the T2 outside temperature.

The controller performs the following comparisons:

1. The T1 room temperature is higher than a fixed value SP1: $T1 > SP1$.
2. The T2 outside temperature is lower than a fixed value SP2: $T2 < SP2$.
3. The difference in temperature between the room and outside exceeds a fixed value SP3: $(T1 - T2) > SP3$.

And, in sequence:

4. It deactivates the air handling unit (AHU).
5. It opens the entry/release shutters and outdoor air and it closes the recirculation damper.
6. It activates the forced ventilation (V1 and V2).

Category
Natural cooling

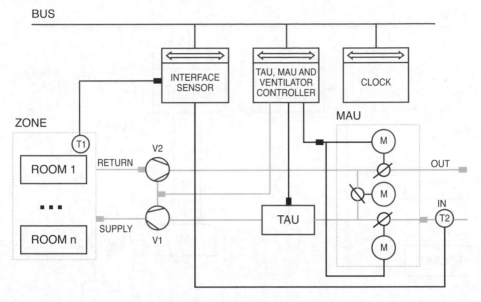

Figure 4.19   Night cooling

Logic
Hot and cold coils are disabled; the single night ventilation is active, which allows you to use, especially in spring and autumn, outside air to pre-cool the premises at night. In this way, the cooling, which would slow down due to the thermal inertia of the building, becomes faster and cheaper and air conditioning energy can be saved the following day.

Main Components
- Damper controller (UMA)
- Fan controller
- AHU controller
- System clock by programming time (day–night)
- Interface temperature sensors

4.4.2.6   Constant Set Point Control (Figure 4.20)

The flow temperature is set manually on the AHU controller while the interface temperature sensor sends the value of the measured temperature to the AHU. The deviation from the set value is cleared by the AHU, regulating hot or cold coils gradually.

Category
Flow temperature control

Logic
The flow temperature is set up manually at AHU level (set point temperature) for the maximum load provided for all controlled premises. The set point is constant but, if necessary, it can be adjusted manually. This is true for both heating and cooling.

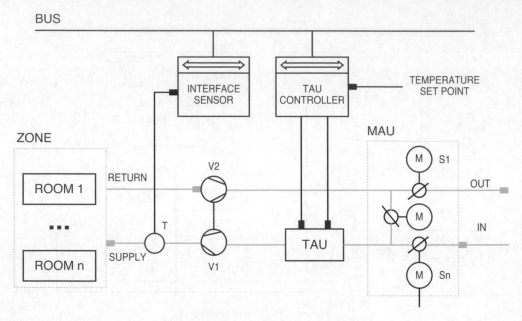

Figure 4.20   Constant set point control

Main Components
- Interface temperature sensors
- AHU controller

4.4.2.7   Humidity Limitation on the Flow of Air (Figure 4.21)

The humidistat controls the maximum relative humidity limit set for the flow by air entering, either continuously or intermittently, the AHU humidifier, so that the relative flow
humidity is maintained below the set threshold. As the air humidification causes air cooling, when TAU is on, it is necessary to heat the air, controlling the temperature with
the T sensor.

Logic
In small systems, humidity control is achieved with a humidistat that controls the humidifier
every time the humidity drops below the set value, switching it off.
    This humidification system is a typical On/Off control.

Main Components
- Humidistat
- Temperature sensor
- Regulator for AHU

4.4.2.8   Automatic Control of Pressure or Flow (Figure 4.22)

The shutters of the duct S1, Sn, open or closed, carry the load variable for air flow in the
general flow duct. The pressure sensor P detects the differential pressure variable value (e.g.
pressure difference between the duct and surrounding space). The fan actuator adjusts the

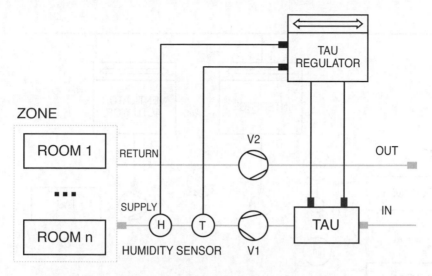

Figure 4.21 Humidity limitation on the flow of air

air pressure in the channel to a set value to compensate for the variation of pressure every time a damper changes state. The differential prescribed pressure value can be referred to the number of dampers installed: a proper compensation is achieved by varying the value prescribed according to the number of open shutters over the total number.

Logic
The flow of treated air in every room matches the demand of the load (total or partial users/ connected spaces). In this way, at a partial load, consumption is reduced to that of the unit treatment air fan.

Category
AHU flow control

Main Components
- Fan actuator
- Pressure sensor interface

4.4.2.9   Free Cooling (Figure 4.23)

When the outside weather conditions allow it, the gradual adjustment of the outdoor air dampers (S1), expulsion (S2) and recirculation (S3) enables the mixing of indoor air (circulation) with the outside air to cool the environment. The TAU controller controls the heating and cooling coils to maintain the required temperature in the room. Note that, without the intervention of the AHU, the space-air follows the conditions of the outside air temperature by adding the heat produced by electrical appliances, solar radiation, the presence of people and so on, then it always remains a bit hotter.

When the room temperature detected by the CT sensor increases, the UMA works on the dampers to mix outdoor air (cooler than the interior) with recirculating air, without reaching

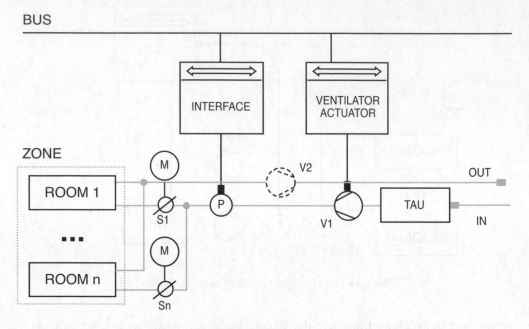

Figure 4.22   Automatic control of pressure or flow

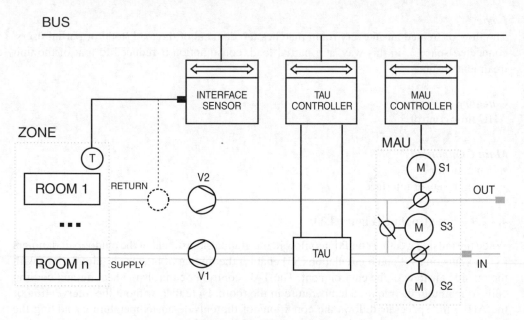

Figure 4.23   Free cooling

the activation temperature of the cooling coil. In this way free cooling is obtained by measuring only the room temperature. Instead of T sensor, a temperature sensor measuring the return air can be used (dashed in the diagram).

Category
Free mechanical cooling

Logic
In spring and autumn, the room temperature controller can be calibrated to compensate for the excess heat with a larger amount of outside air, giving free cooling. When the outdoor temperature is in the range 10–20°C for several months a year, this measure can guarantee important savings. You can refresh the space without using the AHU and heat recovery, but using the outside air when it has a temperature below that of the indoor air.

Main Components
• UMA actuator
• AHU/UMA controller
• Interface temperature sensor

4.4.2.10 Set-Point External Temperature-Dependent (Figure 4.24)

The control of the flow temperature is influenced by the outside temperature (climatic compensation). The compensation function is stored in the TAU controller that regulates the activation of hot and cold coils. This function allows the modification of winter and summer supply temperature within the predetermined ranges of outdoor temperature.

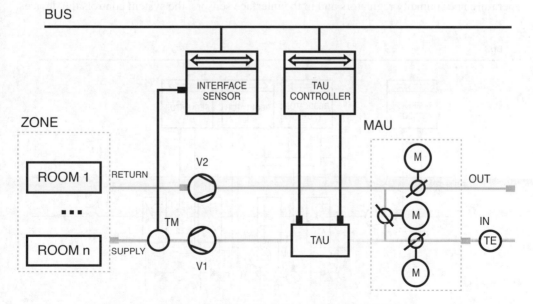

Figure 4.24 Set-point external temperature-dependent

Category
Flow temperature control

Logic
The set-point of the supply temperature is set for the total continuous load in the different zones. The flow temperature is adjusted on the basis of a simple function (e.g. linear) of the outside temperature, corresponding to the likely demand for individual rooms served. The system does not allow one to take into account requests for changes in individual rooms.

Main Components
- Interface temperature sensors
- AHU controller

4.4.2.11 Control Function of Flow Air Humidity (Figure 4.25)

A control adjusts the relative humidity of the flow air to a specified value that is constant in all rooms. The control is carried out by the use of regulators that act in sequence on all parts of the AHU in order to bring the humidity up to the specified value: the sequences of cooling, condensation, humidification and hotting causes changes in the air humidity content in absolute and/or relative terms. In order to obtain efficient adjustments (e.g. simultaneous heating and cooling) it is important to set the set-point controllers with the largest possible dead zones.

Category
Humidity control

Logic
Under the conditions of temperature T1 and humidity Ur, the set points set up in the temperature and humidity regulators and in the interface sensors, the system controls the devices

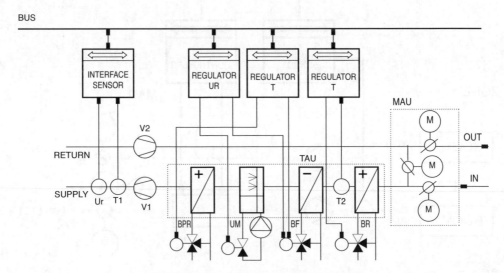

Figure 4.25 Control function of flow air humidity

BF, UM and BPR to cool or heat, humidify or dehumidify in order to maintain the relative humidity and flow air temperature to the values set. The relative humidity is detected by the sensor Ur in the flow duct. Along with humidity detection, it is still necessary to also measure the flow temperature through the T1 sensor. Dehumidification is achieved with the cold coil (Bf) that receives commands both from the temperature humidity controller and the controller (a software switch of priorities, present in the regulators, it gives precedence to the command that requires the greatest cooling capacity). The temperature controller controls the cold (BF) and heat (BPR) coils based on the signal received from the temperature sensor T1. The preheating coil BC is not involved directly in the control of the humidity but it allows to protect the coil against freezing in cold weather and to ensure a minimum temperature (12–16°C), detected by the sensor T2 and the controller connected to it, for optimal functioning of the system.

Main Components
- Air-relative humidity regulator
- Temperature regulator (T1 and T2)
- Interface sensors T/Ur
- Flow air humidity control (humidity control).

4.4.2.12 Occupancy Control Function (Figure 4.26)

Where people are present, the occupancy sensor enables the flow and activates through the flow air actuator, the circulation fans V1 and/or V2 in the spaces. For energy saving the AHU generators should be turned off when the fans are inactive.

Category
Space flow air control

Logic
The system controls the flow of supply air, set to the maximum load, according to the presence of people in the environment/area. In this way the energy loss is minimal, only occurring during periods of actual employment and at a partial load.

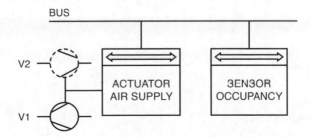

Figure 4.26 Occupancy control function

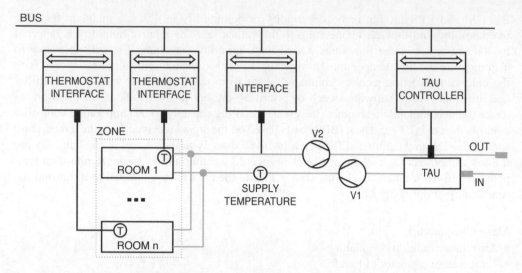

Figure 4.27   Function to set point, load-dependent

Main Components
- Supply air actuator
- Occupancy sensor
- AHU controller

4.4.2.13   Function to Set Point, Load-Dependent (Figure 4.27)

The system is able to reduce the production of heat and cold to that which is strictly necessary in order to avoid loss in production and distribution. The AHU controller collects the load data (exemplified by the value of the space temperature) that is coming from all areas/zones and optimizes the value of the flow temperature Tm with sequences of hot or cold.

Logic
The flow temperature set point is a function of the active loads in different rooms: the adjustment can be achieved with a system that centralizes the collection of data based on temperatures and/or the state of air conditioners in the rooms and sets the flow temperature accordingly.

Category
Flow temperature control

Main Components
- Interface temperature sensors
- AHU controller

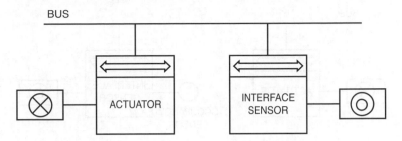

Figure 4.28 Function of switch on manual and automatic shut-off

4.4.3 Lighting

4.4.3.1 Function of Switch on Manual and Automatic Shut-Off (Figure 4.28)

A generated signal causes automatic shutdown at least once a day, typically during the evening, to inhibit unnecessary operations at night and avoid unnecessary energy consumption. The light may be turned off gradually as the switch-off warning to any user.

Category
Occupancy control

Logic
Switching on the lighting occurs by acting on the power command connected to the input device. Shutdown can be manual, but can also take place automatically after a preset time. The lighting is switched on and off manually by one or more control points with instant-closing and delayed opening.

Main Components
- Bus interface for input of manual control of lighting
- Lighting control actuator

4.4.3.2 Manual Power Control Function and Occupancy Detection Auto-On/Reduction/Off (Figure 4.29)

The occupancy sensor or the manual control command on the light through the actuator. Then, when the sensor no longer detects a presence, the lighting is automatically reduced to a certain percentage and/or switched off.

Category Occupancy Control Logic
Lighting can be turned on manually by commands installed in the illuminated area. If the users do not switch the light off manually, the light is kept on by the system automatically in a reduced state (not more than 20%) for a certain time interval following the last detection of a presence in the controlled environment (e.g. 3 minutes). If for another given period of time no presence is detected in the environment, the light is switched off completely.

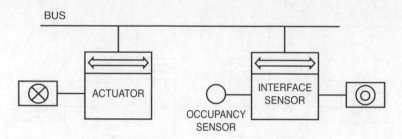

Figure 4.29   Manual power control function and occupancy detection auto-on/reduction/off

Main Components
- Occupancy sensor
- Actuator

4.4.3.3   Motorized Control with Automatic Drive Control of Sunscreens (Figure 4.30)

The blinds controller adjusts the position through the actuator, depending on the ambient light detected by the light sensor.

Category Occupancy Control Logic
A daylight control allows energy saving in winter, and protects against glare and against overheating in summer. The flow of heat lost at night can be reduced by controlling the blinds.

Main Components
- Ambient light sensor
- Curtains/Blinds controller (summer/winter)
- Actuator

4.4.3.4   Automatic Daylight Control Function (Figure 4.31)

The occupancy sensor controls the power of the lighting system through the actuator, depending on the presence of people and the ambient light.

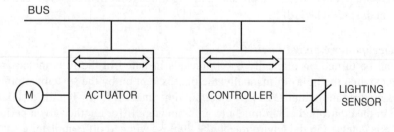

Figure 4.30   Motorized control with automatic drive control of sunscreens

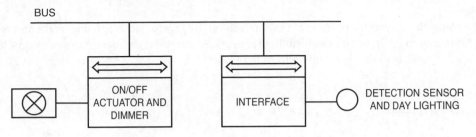

Figure 4.31   Automatic daylight control function

Category
Daylight control

Logic
The system adjusts the brightness of the lamps in the given area according to the light coming from the outside. The light shuts down after a delay after the last detection of the presence of people.

Main Components
- Occupancy and daylight sensor
- On/Off actuator and dimmer

4.4.4   Sunscreens

4.4.4.1   Control Combined Light/Blinds/HVAC Function (Figure 4.32)

The free solar energy control saves energy in winter, and provides protection against glare and overheating in summer. The flow of heat lost at night can be reduced by controlling shutters.

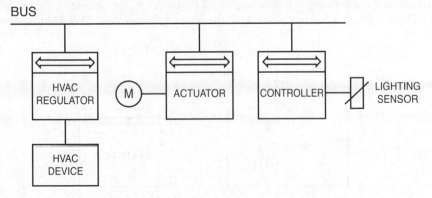

Figure 4.32   Control combined light/blinds/HVAC function

Logic

The controller adjusts the position of the blinds through the actuator, depending on the ambient light measured by the sensor (ambient light sensor). A controller regulates the operation of the HVAC according to the data coming from the sensor.

Category

Sunscreens control

Main Components
- Ambient light sensor
- Shutter actuator
- Curtain/blinds controller (summer/winter)
- HVAC controller

4.4.5 Technical Building Management

4.4.5.1 Function Centralized Control (Figure 4.33)

The system allows easy adaptation to users' needs and provides the following functions:

- centralized setting time programs and set point;
- centralized optimization of the system;
- control at regular intervals of the operation settings of the heating, cooling, ventilation and lighting and their set point in such a way that they are well prepared for the current use and meet the real needs;
- setting the operating parameters of the auditors;
- assessment, at regular intervals, of controllers of premise/area. Their set points are often modified by the user. The centralized system can detect and correct scroll settings due to user error or misunderstanding;
- interlock control (emission and distribution) between heating and cooling: it is only partial since its set point must to be regularly changed in order to minimize the simultaneous use of heating and cooling;
- the functions of alarms/faults detection and supervision are appropriate to the needs of the user and allow the settings optimization of the different controllers. This must be done with

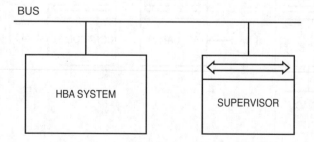

Figure 4.33   Function centralized control

easy instructions to detect abnormal operation (alarm) and simple means of recording and displaying results;
- implementation of energy management programs based on the presence or absence of personnel in the rooms;
- automatic start of automation scenarios and control based on the evolution of the occupancy parameters and of factors that affect energy consumption.

Category
Home system or building automation control scheme

Logic
It allows the system control and management to be centralized.

Main Components
- Touch screen
- PC
- Smart phone, etc.

4.4.6   Technical Installations in the Building

4.4.6.1   Function of Fault Detection, Diagnosis and Provision of Technical Support (Figure 4.34)

The TBM (Technical Building Management) system adds to the functions of the centralized automatic control:

- device diagnostics,
- device status,
- operation time,
- type of fault,
- fault detection in devices/actuators/sensors/control elements,
- capacity of the system to activate the request for technical support both for periodic maintenance and system malfunctions.

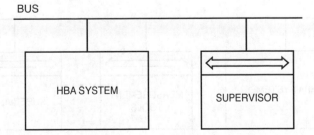

Figure 4.34   Function of fault detection, diagnosis and provision of technical support

Category
Technical building systems management (TBM) scheme

Logic
To improve the efficiency of a plant, it is particularly important that the system is able to provide a regularly diagnostic on the efficiency of the plant, its operating parameters and to identify failures. It is important to be able to link these management parameters with a technical support, whose frequency will depend on the complexity of the plant.

Main Components
• Touch-screen panel
• PC
• Smart phone, etc.

4.4.6.2 Function of the Report on Energy Use, Internal Conditions and Possibilities for Improvement (Figure 4.35)

For real optical management of a system and to monitor its effectiveness it is essential that a system includes a display of the energy consumption parameters and of the various operating conditions. This ensures continuous energy efficiency and the ability to intervene at a later time, adapting the system to changes in housing needs and alterations in the environment.

Category
Technical systems Building Management (TBM) scheme

Logic
A report on the status of the energy consumption and indoor conditions (heating, cooling, lighting, etc.) must be prepared. This informative report can cover for example:

• a building energy certificate,
• evaluating the improvement of energy and building system,
• a detection function,
• energy detection,
• monitoring the temperature of the room and internal air quality,
• monitoring the power consumption of the system.

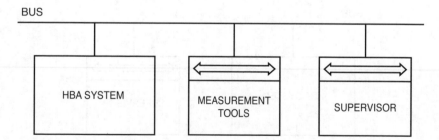

Figure 4.35   Function of the report on energy use, internal conditions and possibilities for improvement

Main Components
- Measurement of energy consumption (software and hardware)
- Software on personal computers

4.5 The Estimate of Building Energy Performance

Understandable difficulties in dealing with the impact of the integrated management, in terms of energy saving, of various functions of the building has meant that until now the automation has not been considered as well as the efficient components to achieve the goals set by official authority in saving energy consumption. Even if not yet fully mature, in the opinion of the authors, the European standard EN 15232 'Energy performance of buildings. Automation Effect, control and technical management of buildings' might make a crucial contribution in facing this situation by allowing the exploitation of the potential of the automation applications for the rational use of energy resources.

The tools and statistical data made available by the standard document should allow the designer, the public administrator and also the end user, to introduce quickly enough the basic strategic choices for achieving a good technical and economic compromise, as well as an environmental balance between reduced consumption during the life of the building and building's first cost. Available data show that the automatic adjustment of different levels of lighting or heating and cooling, depending on the absence or presence of people in a living or working context has the following advantages:

- it always brings, whatever the level of automation, real reductions compared with traditional systems in energy consumption without affecting comfort and security;
- by delegating even simple tasks to automation, such as turning off unnecessary lighting, which could potentially be done by a human user, in fact could lead to tangible and measurable results in the direction of limiting consumption. The availability of conventional and officially recognized instruments, like the standard EN 15232, could help in the process of including building automation in the legislative instruments addressed to energy saving.

4.5.1 European Standard EN 15232

In a building, good energy management is a basic and fundamental factor in reducing energy consumption and raise the quality of life (comfort) while working in an indoor environments. A practical limit in diffusion of automation, among other common tools for reducing consumption, is linked to the difficulties in evaluating ex ante the benefits in terms of energy savings. To allow a formal evaluation of the effectiveness of the investment or in comparing investments, the uncertainty of the results is crucial.

While the ex-ante quantification of benefits related to the adoption of a more efficient apparatus, such as water heaters, are easily estimated, the impact of a system, like the building automation one, involves a set of horizontal skills and modelling problems.

In this direction, in the opinion of the authors, a focal point is given by the publication of the European Standard EN 15232 'Energy performance of buildings. Automation Effect, control and technical building management'. Mostly ex ante, the standard sets just the problem of estimating the impact on energy consumption of building automation, control and management systems. It allows to concretely add to the classical approach based on the adoption of a more energy performing component (thermal insulation, heat generator, low energy light bulbs, etc.), the one of improving and optimizing the usage of a standard component. This last approach also allows the adoption on existing components and, of course, the combination with the adoption of a more performing component.

The main innovative features of the Standard EN 15232 can be summarized as follows:

- a conventional definition of four classes for classification of the technology contents of automation in terms of building energy performance;
- a list of all the automation, control and management functions related to building energy performance and a correlation of these with the performance ratings above;
- the definition of conventional methods for estimating building energy performance.

4.5.1.1   Automation Classes for Energy Efficiency

From an energy efficiency perspective, the standard EN 15232 defines four conventional classes, corresponding to the installed level of automation, control and management system (Figure 4.36):

- Class D 'NON ENERGY EFFICIENT BACS': this includes the traditional and technical systems with no automation for energy efficiency.
- Class C 'STANDARD': this corresponds to buildings with automation systems and normal controls and is considered the reference class.
- Class B 'ADVANCED': this includes buildings with automation and control systems as well as TBM (Technical Building Management: technical installations management in the building) for centralized control.

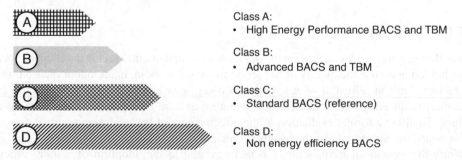

Class A:
• High Energy Performance BACS and TBM

Class B:
• Advanced BACS and TBM

Class C:
• Standard BACS (reference)

Class D:
• Non energy efficiency BACS

Figure 4.36   Graphical representation (not in the standard) of the classes of installed automation, control and management system

- Class A 'HIGH ENERGY PERFORMANCE': as Class B but with more performing automatic control, such as to ensure high performance energy system. Such an approach gives a series of communication advantages:

  o It is already qualitatively known by the user.
  o It summarizes in a single significant and simple parameter (the class) the performance result of a complex system hardly understandable by the general public.
  The definition of each class is based on the existence or not of a set of automation control and management functions without details of their implementation.

4.5.1.2   Definition of Automation Classes

A tool provided by the Standard EN 15232 is the list of all the automation, control and management functions relating to the building energy performance and from the correlation of these with the automation classes for energy efficiency in the previous section.

The functions required for the definition of belonging to a given class differ for residential buildings and tertiary sectors. Class C requires the construction of minimum functions for automation and control. Class B requires the characteristic functions of the previous class plus some additional specific features such as room controllers to communicate with the building central control. Class A requires some specific features of TBM as additional features to Class B. It may be noted that although even if tied to the rigidity introduced by the adopted approach[3], the tables can be a useful operational tool for all figures involved in the building sector, in particular:

- Owners, architects and building designers/installers for the definition of automation functions
- Public authorities for a definition of the minimum requirements relating to the efficiency of automation
- Public authorities and inspectors for the definition of assessment and verification procedures
- Public authorities for the definition and verification of procedures for calculating energy efficiency
- Electrical designers and automation programmers for the definition of the hardware components, the software and procedures for verifying the energy performance of the system.

Tables 4.4–4.6 give examples of tables for heating control, a HVAC system and lighting control.

4.5.2   Comparison of Methods: Detailed Calculations and BAC Factors

The main contents of the EN15232 standard for the quantitative estimation of efficiency and energy savings introduced by the application of automation control and management systems

---

[3] With reference to Table 4.6, we consider, for example, the case of an underground building aiming at class A, but in which obviously the adoption of an automatic control for daylight makes no sense.

Table 4.4 Examples of tables defining the functions of automation, control and management with an impact on building energy performance. Heating control

AUTOMATIC CONTROL	DEFINITION OF CLASSES							
HEATING CONTROL	Residential				Non residential			
Emission control	D	C	B	A	D	C	B	A
The control system is installed at the emitter or room level, for case 1 one system can control several rooms								
0 No automatic control	▨				▨			
1 Central automatic control	▨				▨			
2 Individual room automatic control by thermostatic valves or electronic controller	▨	▨			▨	▨		
3 Individual room control with communication between controllers and to BACS	▨	▨	▨		▨	▨	▨	
4 Integrated individual room control including demand control (by occupancy, air quality, etc.)	▨	▨		▨	▨	▨		▨
Control of distribution network hot water temperature (supply or return)								
Similar function can be applied to the control of direct electric heating networks								
0 No automatic control	▨	▨			▨	▨		
1 Outside temperature compensated control	▨	▨			▨	▨		
2 Indoor temperature control	▨	▨		▨	▨	▨		▨
Control of distribution pumps								
The controlled pumps can be installed at different levels in the network								
0 No control	▨				▨			
1 On/off control	▨	▨			▨	▨		
2 Variable speed pump control with constant Dp	▨	▨		▨	▨	▨		▨
3 Variable speed pump control with proportional Dp	▨	▨		▨	▨	▨		▨

to the various types of new or existing buildings consist of an impressive statistical work and simulations carried out by experts in the field of European legislation in the CEN/CENELEC[4].

In particular, the EN 15232 standard offers two different procedures for calculating the efficiency of automation, control and management systems (Figure 4.37):

- Detailed calculation
- Calculation based on efficiency factors (BAC factors in the English text of the standard).

The detailed calculation is best used only when the system is completely known, that is, when all functions have already been established. The method of efficiency factors, certainly characterized by a higher level of formality and accuracy, while being based on measurements and calculations performed on a large number of different types of buildings, in rooms with

---

[4] The work has also given rise to other standards of EN15000, EN12000 and EN13000, dedicated to the automation of heating, cooling, ventilation, hot water, lighting, shutter control/spaces brightness, the centralization and integrated control of different applications, diagnostics and consumption/improving detection of automation parameters.

Table 4.5 Examples of tables defining the functions of automation, control and management with an impact on energy performance. Control of the HVAC system

AUTOMATIC CONTROL	DEFINITION OF CLASSES							
VENTILATION AIR CONDITIONING CONTROL	Residential				Non residential			
Air flow control at the room level	D	C	B	A	D	C	B	A
0 No automatic control								
1 Maual control								
2 Time control								
3 Occupancy control								
4 Demand control								
Air flow control at the air humidifier level								
0 No automatic control								
1 On/off time control								
2 Automatic flow or pressure control with or without pressure control								
Heat exchange ... control								
0 Without ... control								
1 With ... control								
Heat exchange overheating control								
0 Without distrubiting control								
1 Without overhating control								
Free mechanical cooling								
0 No control								
1 Night control								
2 Free cooling								
3 ... directed control								
Supply temperature control								
0 No control								
1 Content set point								
2 Vesible set point with outdoor ptemperature compensation								
3 Vesible set point with load ... compensation								
Humidity control								
0 No control								
1 Supply air humidity								
2 Supply air humidity control								
3 Room or ... air humidity control								

Table 4.6 Examples of tables defining the functions of automation, control and management with an impact on energy performance. Lighting control

AUTOMATIC CONTROL		DEFINITION OF CLASSES							
LIGHTING CONTROL		Residential				Non residential			
Occupancy control		D	C	B	A	D	C	B	A
0	Manual on/off switch								
1	Manual on/off switch + additional sweeping extinction signal								
2	Automatic detection Auto-On/Dimmed								
3	Automatic detection Auto-On/Auto-Off								
4	Automatic detection Manual on/Dimmed								
5	Automatic detection Manual on/Auto off								
Daylight control									
0	Manual								
1	Automatic								
BLIND CONTROL									
0	Manual operation								
1	Motorized operation with manual control								
2	Motorized operation with automatic control								
3	Combined light/blind/HVAC control (also mentioned above)								

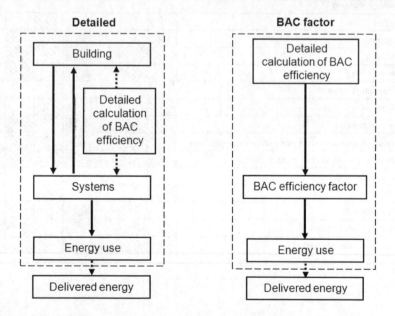

Figure 4.37 Flow charts of the detailed method and the method of the efficiency factors (EN 15232 Standard)

different boundary conditions, is suggested for making a rough estimate in the initial phase of the building project/provision and of the control system and energy management. However, in the opinion of the authors, it is the most important tool introduced by the standard. With reference to the most energy-intensive applications, the method based on efficiency factors makes it possible to assess, quickly and on a tabular basis, the impact of the automation control or management system chosen as a relative value based on the examined building energy needs. It also provides direct evidence of the economic value of the saving on operating costs that is achieved by adopting a given level of automation, control or management.

4.5.2.1 Detailed Calculation

The detailed calculation includes five different basic types:

1. Direct method, based on a simulation of the system in accordance with EN 13790 'Thermal performance of buildings – Calculation of energy use for space heating and cooling'.
2. Method based on mode of operation, considering sequentially the different states of each application (heating, lighting, etc.) and for each state determines the energy consumption, obtained by the sum of the total consumption of each application.
3. Method based on the time operation, which considers the duration of operation Total/Excluding/Partial of the devices, resulting in a savings ratio that characterizes the given function.
4. Method based on room temperature, which considers the influence of temperature control accuracy (low hysteresis) in energy savings' calculations.
5. A method of correction factors, used when the automatic control works in a combination of several factors, such as the effects of the presence of people, the temperature value and the time of operation.

Direct Method

The direct method is used to calculate the energy performance of a building based on an analytical simulation or a simulation on an hourly basis (EN ISO 13790). With the direct method it is possible to calculate the impact of a number of automation features, such as intermittent heating, varying the set-point temperature, heating or cooling, by pulling down the sun blinds and so on. Obviously, the method loses its effectiveness if the control would lead to corresponding changes in the time intervals of less than the simulation step.

Method Based on Mode of Operation

The method based on the operating mode is particularly useful in cases where the automatic control system allows it to act on the energy-intensive system in different modes. An example is an air conditioning system that allows two modes of ventilation (presence and absence of people), the heating may be normal or intermittent or otherwise. The approach to calculating the impact on consumption is to calculate the energy consumption in each mode and the total sum of the values thus obtained. The calculations are performed for each mode of operation, considering the state of the relative control system: for example, on/off fan. Of course, each mode corresponds to a given state of the control system.

Method Based on the Operation Time

The method based on the operation time can be used in cases where the control system has a direct impact on the period of operation (on/off) of a device (such as for the control of a fan or luminaire).

Consumption (E) over a given time interval (t) can be estimated using the following equation:

$$E = P \cdot t \cdot F_C$$

where P is the power of the controlled system and F_C is the coefficient representing the impact of the control system and is the ratio between the time when the controlled system is automatically switched off because not needed and the total period of time.

By extension, the method based on operation time can be used if the control system modulates the system rather than switches it on or off; in this case, it represents the equivalent operating time.

Method Based on Room Temperature

The method in question is devoted to cases where the control system has a direct impact on the room temperature. The principle is to consider the calculation of energy requirements according to EN ISO 13790 that will consider the impact of a temperature control system.

In general you should at least take account of the following:

- emission control (heating and cooling);
- intermittent emission and/or distribution control;
- optimization related to the coordination of various control devices;
- detection of errors in operating systems and techniques to provide support for the diagnosis;
- room control;
- intermittent heating control.

Consumption (E) over a given time interval (t) can be determined as:

$$E = L \cdot ((\theta_{SP} - \Delta\theta_C) - \theta_R) \cdot t$$

where:

- L is a transfer coefficient
- θ_{SP} is the set point
- $\Delta\theta_C$ is the impact of the actual control system (equal to 0 if the system is perfect, positive in case of heating and negative in case of cooling)
- θ_R is the reference temperature, for example the outdoor temperature

Method of Correction Coefficients

The method of correction coefficients can be useful when the control system acts on multiple factors such as time, temperature, etc.

Consumption (E) can be estimated as:

$$E = EPC \times C$$

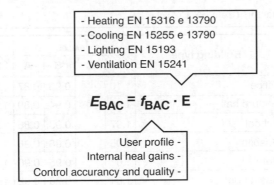

Figure 4.38   Conceptual framework of the application of factors in the BAC method (EN 15232)

where:

- EPC is the consumption in the reference case, for example, if the system is controlled ideally or if there is not a function of automation or management
- XC is the factor that represents the increase or decrease of energy consumption compared with consumption in the reference case. The values depend on the type of control but also depend greatly by the climate and the type of construction, etc.

Method of Factors: Boards and Formulae for Calculating

The determination of efficiency factors contained in the standard and therefore the effect of the adoption of automation, control and management functions for different types of buildings are the result of the comparison of annual consumption in a room of reference with that introduced in the same room under the same conditions (time of occupation, user profile, weather conditions, sun exposure) with the use of different levels of automation, control and management (A, B, C, D).

The method of efficiency factors is applicable to residential buildings as well as offices, reading rooms, schools and similar, hospitals, hotels, restaurants, malls and shops. The efficiency factors provided are different for different applications: electrical and thermal (heating, cooling). Please note that the energy required for artificial lighting was not taken into account in the calculation of BAC efficiency factors (Figure 4.38), therefore the impact of automation and control on the lighting system must be assessed separately using the EN 15193 standard.

The following tables (from the EN 15232) show directly the relationship between the consumption of room/building with Class A, B or D and the class C, (i.e. the reference class). Table 4.7 shows efficiency factors.

As a summary of what is reported in the tables, it is worth noting that compared with an environment with minimum C requirements, for example automation of cooling in an office, there are energy savings and therefore economic savings on quantifiable consumptions as reported in Tables 4.8 and 4.9. Even if the standard takes the class C as reference, practically there are many cases where the situation (existing or alternative basis) for comparison is the one corresponding to class D, without any automation, control or management. In this

Table 4.7 Efficiency factors (from EN 15232)

Building	Building type	$f_{BAC.EL}$				$f_{BAC.HC}$			
		D	C	B	A	D	C	B	A
Non residential	Offices	1.10		0.93	0.87	1.51		0.80	0.70
	Lecture hall	1.06		0.94	0.89	1.24		0.75	0.5[a]
	School	1.07		0.93	0.86	1.20		0.88	0.80
	Hospital	1.05		0.98	0.96	1.31		0.91	0.86
	Hotel	1.07	1	0.95	0.90	1.31	1	0.85	0.68
	Restaurant	1.04		0.96	0.92	1.23		0.77	0.68
	Wholesale and retail trade service buildings	1.08		0.95	0.91	1.56		0.73	0.6[*]
Residential	Single family houses Apartment block Other res. Building or similar	1.08		0.93	0.92	1.10		0.88	0.81

case the gap increases considerably. In the case of an office, actually, the gap in consumption between a building in Class A and one without automation, that is class D, results in almost 54%.

For ease of calculation, from the definition tables of the efficiency factors BAC contained in EN 15232, the coefficients reported in Table 4.10 were obtained, in order to derive the consumption corresponding to different solutions, directly from the consumption corresponding to the starting Class D.

With f_{BACi} represented by the coefficients shown in the standard EN 15232 and with k_n being the multiplying coefficients with respect to Class D.

Table 4.8 Estimated percentage of the savings achievable by adopting different automation systems with respect to Class C. Example of an office

Class	Saving
B	$(1 - 0.8/1)\ 100 = 20\%$
A	$(1 - 0.7/1)\ 100 = 30\%$

Table 4.9 Estimated percentage of the savings achievable by adopting different automation systems with respect to Class C. Example of a dwelling

Class	Saving
B	$(1 - 0.88/1)\ 100 = 12\%$
A	$(1 - 0.81/1)\ 100 = 19\%$

Table 4.10 BAC coefficients reworked by type of energy consumption

	Thermal energy	
k_c	f_{BACc}/f_{BACd}	0.909
k_b	f_{BACb}/f_{BACd}	0.800
K_a	f_{BACa}/f_{BACd}	0.736
	Electrical energy	
k_c	f_{BACc}/f_{BACd}	0.926
k_b	f_{BACb}/f_{BACd}	0.861
k_a	f_{BACa}/f_{BACd}	0.852

Please note that using the method of the factors of efficiency, at an early stage, does not exclude the possibility of subsequent detailed calculation for a more accurate estimate or adjustment of system functions to make them, with successive approximations, as effective as possible. A complete example of applying the method of the efficiency factors is shown in Table 4.11 and refers to a complex of offices with known electrical and thermal consumption (respectively row 6a-6b and row 1). The heating requirements of the office complex is 100 MWh/y, both for heating and for cooling. Adding to the heating requirements (line 1), losses of related systems (line 2), we obtain the total thermal energy used in the offices during the reference period. To determine the consumption that can be obtained in passing from a Class C system to a Class B, using the method of factors it is sufficient to multiply the total consumption of thermal energy (line 3) for the relative factor BAC (line 4), obtaining the total thermal consumption (row 5) with the advanced automation system at advanced efficiency (Class B) and, with this, the energy savings obtained equals 20%. The power consumption of the aux services required to operate the heating and cooling (line 6a) represent, together with illumination (line 6b), the total electricity consumption of the offices. Using the same

Table 4.11 Example: offices changing from Class C to Class B

No.	Description	Calculations	Unit	Heating	Cooling	Fan	Lighting
1	Themal energy need (Class C)		MWh/y	100	100		
2	System losses (Class C)		MWh/y	33	28		
3	Total (Class C)	Σ 1÷2	MWh/y	133	128		
4	BAC $f_{BAC.HC}$ (Class B)			0.80	0.80		
5	Therma energy consumption (Class B)	3 × 4	MWh/y	106	102		
6a	Auxiliary services		MWh/y	14	12	21	
6b	Lighting						34
7	BAC factor $f_{BAC.e}$ (Class B)			0.93	0.93	0.93	0.90*
9	Electrical energy consumption (Class B)		MWh/y	13	11	20	31

*Calculated according to the Standard EN 15193

calculation procedure, this time by refererring to the factor $f_{BAC.e}$ (line 7), the savings amount to 7.4%, compared with the system in Class C.

The same table can be obtained with any spreadsheet.

Further Readings

Angelo Baggini and Annalisa Marra, *Efficacia Energetica negli edifici. Il contributo della domotica e della building automation*, Editoriale Delfino, Milano, Italy, 2010.

Donald Wallace (2003-09-01). 'How to put SCADA on the internet, *Control Engineering*, http://www.controleng.com/article/CA321065.html.

EN 15232 2007 Energy performance of buildings – Impact of Building Automation, Controls and Building Management.

Robert Lemos SCADA system makers pushed toward security, *SecurityFocus*, (2006-07-26), http://www.securityfocus.com/news/11402.

J. Slay and M. Miller, Lessons learned from the Maroochy water breach, *Critical Infrastructure Protection*, **253**, Springer, Boston, 2007, pp. 73–82.

5

Power Quality Phenomena and Indicators

Andrei Cziker, Zbigniew Hanzelka and Irena Wasiak

It is increasingly being accepted that the impact of poor power quality (PQ) on effective electric power utilization is significant and disruptive; it is also being ever more clearly understood that technical and financial impacts, where relevant, are far greater than had previously been recognized.

The true economic value of PQ is linked to the effects that electromagnetic disturbances have on equipment and other loads on the system. In the industrial sector, for example, the economic value of disturbances is actually increasing due to the extensive detrimental effects they can cause in modern, automated plants in which sensitive equipment and devices are integral components of highly complex processes. For such processes a PQ disturbance can cause downtime that can be directly correlated with lost production and, therefore, lost revenue and profits. The consequences are not limited only to electrical systems in non-residential environments, but can harm domestic systems and appliances as well as mainly introducing electrical danger into the home [1].

On the other hand, a poor power quality may increase power and energy losses in different components of power systems (electric lines, power transformers, etc.) or customers' receivers (mainly induction motors, power electronics, capacitor banks for power factor correction, etc.). These supplementary losses will diminish the energy efficiency for both utility and final users, increasing the customers' energy bills.

The economic impacts of power quality are usually divided into three broad categories:

- *direct economic impacts*: e.g. loss of production; unrecoverable downtime and resources (e.g. raw material, labour, capital); process restart costs; spoilage of (semi-) finished production; equipment damage; direct costs associated with human health and safety; financial penalties incurred through non-fulfilment of contract; environmental financial penalties; utility costs associated with the interruption; supplementary energy losses in the electric supply system;

Electrical Energy Efficiency: Technologies and Applications, First Edition. Andreas Sumper and Angelo Baggini.
© 2012 John Wiley & Sons, Ltd. Published 2012 by John Wiley & Sons, Ltd.

- *indirect economic impacts*: e.g. the costs to an organization of revenue/income being post-poned; the financial cost of loss of market share; the cost of restoring brand equity;
- *social-economic impacts*: e.g. uncomfortable building temperatures as relating to reduction in efficient working/health and safety; personal injury or fear, also as related to reduction in efficiency and health and safety.

It is an important part of the end users' armoury to arrive at the correct balance between what the organization's poor PQ costs are, the investment required for any PQ solution, as well as to assess whether unmitigated poor PQ can be tolerated. The power sector and equipment manufacturers also need to assess how great an effort to place behind helping end users to reduce the impacts of poor PQ.

The methods available in the specialized literature for the economic evaluation of PQ disturbances refer mainly to voltage dips and short or long interruptions, but also to harmonics [2]. The effects of other disturbances such as voltage level and fluctuations and unbalance are seldom considered. Their consequences and economic impact are that the *indirect* and *social* impacts often fail to be recognized and go unaddressed.

5.1   RMS Voltage Level

The complex value of the voltage in various characteristic points of the electric network represents one of the main parameters of the operation state of power systems.

Slow voltage variations are usually quantified by calculation of the rms value of the supply voltage. In assessing this parameter, measurement has to take place over a relatively long period of time to avoid the instantaneous effect on the measurement caused by individual load switching (e.g. motor starting or inrush currents) and faults. Standard EN 50160: 2009 quantifies slow voltage variations using the 10-minute mean rms value and considering a week as the minimum measurement period; in particular, the 99th percentile of the 10-minute mean rms values over one week is considered as the site index.

The voltage variation limits in different European countries are presented in [3] and [4]: (i) for the Norwegian PQ directives, variations in the voltage rms value, measured as a mean value over one minute, shall be within an interval of $\pm 10\%$ of the nominal voltage; (ii) in France, for MV customers, the supply contracts contain the voltage variation limit $U_C \pm 5\%$ of the nominal voltage for 100% of the time, where U_C must be in the range of $\pm 5\%$ around the nominal voltage; (iii) the regulation in Hungary contains three different objectives: (a) 100% of the 10-minute rms voltage shall be between 85% and 110% of the nominal voltage; (b) 95% of the 10-minute rms voltage shall be between 92.5% and 107.5% of the nominal voltage; (c) 100% of the 1-minute rms voltage shall be less than 115% of the nominal voltage.

The regulation in Spain fixes the following limit: the 95th percentile of the 10-minute mean rms value over one week shall be between 93% and 107% of the declared disturbance description.

The supply voltage amplitude can have slow variations, especially due to voltage drops on lines and transformers, produced by the variation of electric load at consumers. The voltage variations can also be produced by changes in the electric configuration scheme of the grid, as well as by changes in the operation stage of reactive power sources.

The calculation of the voltage drops has been presented in [1, 2, 5, 6].

The slow voltage variations can be characterized by the relative voltage deviation from the nominal value at a certain point of the grid and at a certain moment:

$$\varepsilon_U \, [\%] = \frac{\Delta U}{U_N}.100 \, [\%] = \frac{U_S - U_N}{U_N} \cdot 100 \, [\%] \tag{5.1}$$

where U_S is the line voltage of the electric grid at a certain point and at a certain moment (the operating voltage) and U_N represents the nominal voltage. The ratio U_S/U_N is called the voltage level.

The permitted variation limits depend on the voltage level with which the consumer is supplied. According to today's standards, the relative deviations for the voltage in the PCC in the case of grids with nominal values up to 220 kV must not be more than 10%.

Equipment manufacturers have to mention the immunity levels to voltage variations for each category of load and, generally, these values are within the interval $(\pm 5 - \pm 10)\%$. For example, the relative admissible deviation is: (i) $\pm 5\%$ for electric motors; (ii) $\pm 10\%$ for welding machines and (iii) $\pm 5\%$ for electric lamps.

For a better characterization of the voltage variation from the nominal values, the following indices are defined:

- *The average value for the voltage relative deviation* from the nominal value in a T time interval [7]:

$$\varepsilon_{Umed} = \frac{1}{T} \int_0^T \varepsilon_U \cdot dt \tag{5.2}$$

The ε_{Umed} index is a measure for the mean voltage in the supply bus bars and gives indications on the accurate choice of the position of the transformer tap-changer.

- *The rms value of the voltage deviations* is given by the relation [7]:

$$\varepsilon_q^2 = \frac{1}{T} \int_0^T \varepsilon_U^2 \cdot dt \tag{5.3}$$

This index offers an evaluation of power quality from the point of view of slow voltage variations, as follows:
- $\varepsilon_q^2 \leq 10\%$ very good quality
- $10\% \leq \varepsilon_q^2 \leq 20\%$ good quality
- $20\% \leq \varepsilon_q^2 \leq 50\%$ poor quality
- $\varepsilon_q^2 \geq 100\%$ inappropriate quality.

5.1.1 Sources

Voltage varies at different points of a power system, the voltage value being influenced by different factors occurring in the generating, transmission and distribution processes. The main factor that determines the voltage variation is load variation. Another factor that influences voltage is the balance between the reactive power consumed and that generated [2].

5.1.2   Effects on Energy Efficiency

Voltage deviation from the nominal value has various consequences according to the load characteristics and the position of the connecting point of the consumers to the grid. It is generally admitted that the optimal supply voltage for every load is its nominal voltage; any deviation from this value influences the operation of various loads [8].

5.1.2.1   Lighting Systems

Incandescent lamps are very sensitive to changes in the supply voltage. These characteristics are of great practical and economic importance as the voltage deviation influences the absorbed power P_n, the luminous flux Φ, the luminous efficacy η, the average lifetime T and the colour temperature. Due to its reduced luminous efficiency, the usage of such lamps will be limited in the near future according to the Directive 2006/32/EC of the European Union [9].

Fluorescent lamps are less sensitive to voltage variations; thus, a voltage variation of 1% changes the lumen output of the lamp by an average 1.25%. These lamps can actually function even if the voltage is at 90% of the nominal value.

5.1.2.2   Electric Motors

In the case of asynchronous motors, the main power load in industrialized countries, a decrease of voltage will increase the rotor slip, will reduce the speed and, implicitly, the productivity of the driven equipment will decrease. The active torque of asynchronous motors is proportional to the square voltage and this is why, when the voltage drops, the starting conditions become more difficult, sometimes even leading to a full stop of the motor (the maximum torque of the motor C_M is below the resistant static torque C_s of the working machine). Thus, when the voltage drops by 10%, the active torque of the asynchronous motor decreases by 19%, the rotor slip increases by 27.5%, the rotor current increases by 14% and that of the stator increases by 10% [10]. The direct consequence is the increase of power losses, reducing the motor's energy efficiency; due to this supplementary heating of the motor, the insulating layer will wear about twice as quickly as in the case of a nominal voltage. With an increase in voltage of 10%, the active torque increases by 21%, the rotor slip drops by 20%, the current in the rotor decreases by 18% and that in the stator by 10%. Figure 5.1 presents the variation in the motor speed of rotation Ω according to the variation of supply voltage U (p is the number of pairs of poles in the machine) [7].

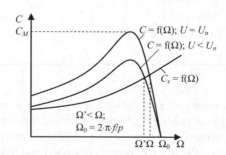

Figure 5.1   Variations in the characteristics of the asynchronous motor according to the power voltage

5.1.2.3 Electric Ovens

In the case of electric ovens with resistors, reducing the power voltage has negative effects on energy efficiency and the technological process, while increasing the supply voltage leads to the reduction of the heating elements' lifetimes. The lifetime of the heating elements of electric ovens with resistors depends on the operating temperature, material characteristics, the construction specifications of the element, etc. The reduction of the lifetime is determined by the acceleration of the oxidation processes of the resistive material and thus by the gradual alteration of the mechanical and electrical parameters.

In the case of three-phase electric arc furnaces, one of the main consumers of electricity in the industry, the deviation of the power quality indices from the contracted values determines – most of the time – an increase in the operating expenses. Thus, in the case of voltage value deviations from the contracted value, determining damage is done based on the extra amount of electric energy needed in order to bring the material to the temperature required by the technological process.

Changing the supply voltage determines a change in the duration of processing of the material. In the usual case of electric arc furnaces used to process steel, a reduction of supply voltage at the bus bars by 8% leads to an increase in the electricity consumption by about 7% and a decrease in productivity of about 6%.

In the case of electromagnetic induction ovens, which are supplied with an industrial frequency, a voltage reduction of 5% shows a productivity reduction of about 10% and a specific consumption increase of about 8%.

Modern installations used for metal processing in electromagnetic induction ovens are powered by means of a frequency converter, so that the reduction of supply voltage by 10% determines an increase in the supply voltage frequency by about 2% and the specific power consumption increases by about 5%.

5.1.2.4 Other Consumers

In the case of resistance welding equipment (spot, seam or butt welding), the variation in the supply voltage within an interval of ± 10%, in the case of regular types of steel, has consequences only on the duration of the technological process, without affecting the quality of the product. Voltage deviations of over 10% lead to the reject of the operation in most cases.

In the case of special types of steel, experimental results show the need to maintain the supply voltage within the interval $(0.95 \ldots 1.05)U_c$, where U_c is the contracted voltage, generally equal to the nominal voltage of the welding machine.

An increase of voltage on the supply rods determines the overcharge of the electric insulation and thus a reduction in its lifespan.

5.1.2.5 Cost

Establishing an optimal variation interval for the supply voltage of the consumers connected to a grid is both a technical and economic issue and its solutions should take into account both the expenses that are needed to reach a certain voltage quality benchmark, as well as the economic damage that can occur to the consumers and the grid placed upstream of the point where the voltage is regulated; this must be done according to the variation affecting the voltage in the electric network located before that specific point.

In most practical cases, the damage D, determined by the supply voltage deviation U from the rated voltage U_r of the loads can be expressed in the form of a second degree polynomial [7]:

$$D = a \cdot (U - U_r) + b \cdot (U - U_r)^2, \tag{5.4}$$

where the a and b values are specific to any technological process. Undoubtedly the relation, (5.4), can be applied only for a relatively narrow range of supply voltage variations. Reducing the supply voltage to below a critical value U_{cr1}, as well as increasing it above a value U_{cr2} can lead to significant damages.

The deviation in the supply voltage from the loads rated voltage leads, most of the time, to a reduction in the absorbed power and, subsequently, to a reduction in the productivity of the working equipment and to a reduction in the quality of the manufactured products.

5.1.3   Mitigation Methods

Maintaining the voltage within permitted values at the terminals of all the consumers can be accomplished only by combining centralized control activity (at a level of the whole power system) with local voltage controls. The central control maintains the voltage at a certain average level in the grid, while the local control systems continuously monitor the voltage levels at various points in the grid and brings them back to within established limits, if necessary.

Central control consists of regulating the voltage at the level of the generators' terminals by means of variations in the excitation current. This control can be achieved both automatically and manually. Thus, the automatic control maintains the voltages at an approximately constant value at the terminals of the generators, irrespective of the variation of the charge, while the manual control allows the increase or decrease of voltages at the generators' terminals according to the charge of the system.

Securing a certain value of the voltage with the help of central control is necessary but not sufficient to ensure admissible voltage at the connection point of all consumers. It is possible that the average level is acceptable, and yet the voltage at the connection point of the consumers is very different, some higher and some lower than the admissible values, hence the need for a local control in order to maintain the voltage values.

5.1.3.1   Voltage Control by Introducing Additional Voltage

Controlling the voltage from a point on the grid can be done by introducing additional voltage, for example by changing the transformation ratio. The devices used for the introduction of additional voltages in the electric grids are [2, 6, 7]:

Static Equipments with Adjustable Taps
- On-load or off-voltage ratio transformers and autotransformers. They are used for the direct control.
- Static boost and buck devices, transformers and autotransformers especially designed for additional voltage. They are used for indirect, longitudinal and transversal control.

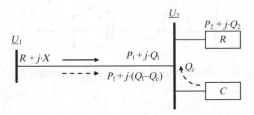

Figure 5.2 Line diagram of an electric network with a controllable installation C capable of absorbing or supplying reactive power

Static Equipments with Slow Control without Taps
- Autotransformers with a short-circuited coil
- Transformers and autotransformers with pre-magnetization
- Transformers with a control coil core

These devices can be used for consumers with high demands in terms of voltage quality.

Induction controllers – this category comprises widely spread direct or indirect controlling devices with on-load tap-changers.

5.1.3.2 Controlling Voltage by Changing the Reactive Power Flow

Figure 5.2 takes into account an electric network for energy transmission. If at the end of the line there is a controllable installation C, capable of absorbing or supplying reactive power, then, based on the relation:

$$\underline{U}_1 = U_2 + \frac{P_2 \cdot R + Q_2 \cdot X}{U_2} + j \cdot \frac{P_2 \cdot X - Q_2 \cdot R}{U_2} = U_2 + \Delta U + j \cdot \delta U \qquad (5.5)$$

it is possible to control the voltage U_2 at the end of the line for any load variation. The flow of the reactive power through the lines is accompanied by power losses, so that the choice of controlling method by modifying reactive power flow must be coordinated with the general issue of balancing reactive powers and energy losses.

The methods used to regulate the voltage by changing the reactive power flow are divided into two main categories (presented in Chapter 13):

1. rotating compensators: synchronous and asynchronous compensators, synchronous generators; and
2. static compensators.

5.1.3.3 Controlling Voltage by Changing the Impedance of the Grid

The inductive reactance of the lines can be reduced by a series connection of a capacitor bank, as presented in Figure 5.3. Reducing the inductive reactance of the grid greatly improves the voltage regime by decreasing voltage drops. If $X_L = \omega \cdot L$ is the inductive reactance of the

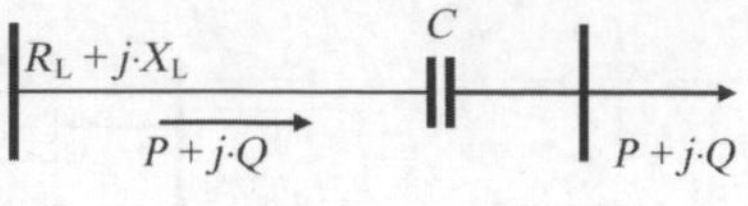

Figure 5.3 Compensating the inductive reactance of a line by placing a series capacitor bank on that line

line, then by placing a capacitor of capacitance C on the line, the resultant reactance of the line becomes:

$$X = X_L - X_c = \omega \cdot L - \frac{1}{\omega \cdot C} \qquad (5.6)$$

According to the value of the reactance X_C or of the ratio $\lambda = X_C/X_L$, also called the compensation degree, the following situations are possible:

- partial compensation $X_C < X_L$, $\lambda < 1$;
- total compensation $X_C = X_L$, $\lambda = 1$;
- overcompensation $X_C > X_L$, $\lambda > 1$.

The voltage on the series capacitors is $U_C = I \cdot X_C$ and does not go beyond 10% of the phase voltage of the grid. This fact allows the usage, in the case of longitudinal compensation, of a capacitor with a rated voltage much smaller than the nominal voltage of the line. For example, for 110 kV, 6–10 kV capacitors are used, and in the case of 35 kV, the required range is 1–3 kV. Longitudinal compensation is not considered as a means of controlling the voltage because it is difficult to control the value of the capacitance of the capacitors installed in series according to voltage variations. Occasionally, connecting capacitors in series in the lines can lead to resonance phenomena, to self-excitation of electrical machines, or sub-harmonic oscillations. There is also an increase in possibility of resonance phenomena on the higher harmonics.

5.2 Voltage Fluctuations

Voltage fluctuation is a series of the rms voltage value changes or a cyclic variation of the voltage waveform envelope (Figure 5.4) [2]. In the case of this disturbance we can use the terms: *voltage fluctuation waveform* (the voltage peak value envelope represented as a function of time – periodic or non-periodic, determinate (seldom) or random (more often)), the *amplitude of voltage changes* (difference of the maximum and minimum rms or peak voltage values occurring during the disturbance) and the *voltage changes rate* (the number of voltage changes per unit of time) or *frequency* (for periodic waveforms).

5.2.1 Disturbance Description

Two basic methods for voltage fluctuations evaluation can be identified. The *first method* consists of the quantitative assessment of the phenomenon, based on the variation of rms value over time. In this case voltage fluctuations are represented as points in the coordinate system: relative voltage change ($\Delta U(t)/U$)-number of voltage changes per unit of time (Figure 5.5).

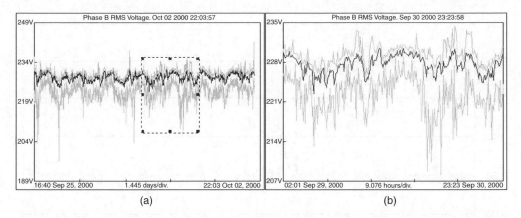

Figure 5.4   Example of (a) rms voltage fluctuation and (b) selected fragment of the waveform

Research on the visual perception of flicker caused by periodic voltage fluctuations resulted in evaluating the borderline curve of flicker severity (irritability) due to luminous flux changes.

The characteristics in Figure 5.5, prepared for different supply voltages of a light source, divide the plane into two parts. The area above the curve defines the unacceptable level of voltage fluctuation, whereas the area below the curve is relevant to acceptable levels of voltage fluctuation.

The *second method* consists of indirect measurement, i.e. measuring the phenomenon of flicker by means of a flickermeter. This instrument, where the input signal are voltage changes, instead of actual changes in the luminous flux, uses a model of the incandescent light source

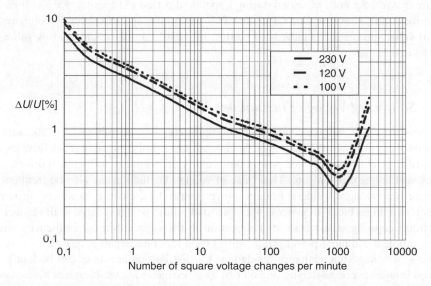

Figure 5.5   Flicker perception borderline characteristic for square-shaped, equidistant voltage changes applied to a 60 W incandescent lamp ($P_{st} = 1$)

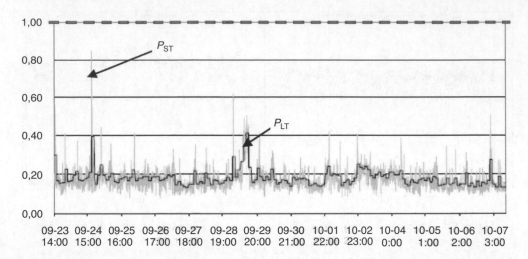

Figure 5.6 Example characteristics of P_{ST} and P_{LT} indices measured over seven days in an LV (low voltage) network

(60 W, 230 V tungsten bulb) and a model of the human reaction to light stimulus. The aim of the measurement is the evaluation of an observer's reaction – a discomfort or annoyance – to disturbance in visual perception. Two dimensionless numerical measures of voltage fluctuation are obtained at the instrument output: the short-term flicker severity index (P_{ST}) refreshed every 10 minutes and the long-term flicker severity index (P_{LT}) refreshed every 2 hours (Figure 5.6).

From at least seven-day measurements of these indices, particularly the P_{LT}, it is determined the value that has not been exceeded during a specified period of time, e.g. 95% or 99% of the measurement time (the so-called CP95 or CP99 percentile) and this value is compared with the limit value specified in standards or guides. In most European countries this value in LV networks equals one.

5.2.2   Sources of Voltage Fluctuations

The primary cause of voltage changes, including fluctuation, as follows from formula (5.5), is the time-variability of the load power. Such loads are mainly, but not exclusively, industrial loads with large individually rated power, particularly arc furnaces, rolling mill drives, mains' winders, welding equipment, etc. The cause of voltage fluctuation can also be frequent starts of electric motors, spot welders, boilers, power controllers, electric hammers, lifts, hoists, capacitor switching, etc. – in general, variable load whose power is large with respect to the short-circuit capacity at the point of connection to the supply. Similar influences are, e.g., X-ray equipment and large-power photocopiers used for commercial purposes.

Sources of voltage fluctuation in residential LV distribution networks can be loads whose operation implies cyclic switching on and off, like: refrigerators, washing machines, cookers, air conditioners, etc. Because of small individual powers their influence is limited only to the consumers/loads that are connected to the supply system not far from the terminals of fluctuating equipment, e.g. in one flat or house. Their adverse impact is, however, increased due to the simultaneous operation of many small loads, e.g. electric water heaters, during evening hours.

A particular source of voltage fluctuation can be mutually coupled data transmission and remote control signals (interlocks, protection systems, etc.) that occur in power networks, which are a source of interharmonics modulating the supply voltage waveform.

Voltage fluctuations occur also as the effect of switching processes in the power system (e.g. improper operation of transformer tap-changers) and operation of certain distributed energy sources, e.g. wind turbines.

A symmetrical generator with a constant load, excitation current and rotational speed, produces a constant voltage at its output terminals. If any of these quantities becomes disturbed, voltage fluctuations may occur. The rotational speed changes are dominant because the time constant of the generator excitation circuit effectively reduces the influence of the generator's flux variations on voltage changes.

In reciprocating engine-driven generators poor quality fuel or inadequate maintenance may lead to the engine's erratic ignition and, consequently, to changes in the output power.

In very large, low-speed engines output torque fluctuations occur even during normal operation. The frequency of the generator output power changes with the reciprocating engine pistons strokes as: $f_F = (Nn/25k)$ Hz, where: N is the number of cylinders; n is the generator rotational speed in revolutions per minute, coefficient k takes the value of 2 for two-stroke engines and 4 for four-stroke engines [11]. During the proper operation of a reciprocating engine, even in the case of low-speed generators, the change in the output power due to the engine pistons' strokes is sufficiently fast and fluctuations do not occur. If a misfire occurs in a cylinder(s) the frequency of voltage changes in Hz can be expressed as: $f_F = (n/25k)$. In many, commonly used generators the frequency of voltage changes coincides exactly with the most unfavourable area. For instance, a four-cylinder, 900-rpm engine produces fluctuations with frequency 7.5 Hz. At 1800 rpm the situation is better – the fluctuation frequency is 15 Hz, i.e. it is located within the area of lower sensitivity of the human eye to flicker. The basic way of preventing this effect is to maintain fuel quality control and proper engine maintenance [11].

Figure 5.7 shows the waveforms of the chosen phase voltage at the output of a rotating UPS supplying an air-conditioning system. A switching process (from power network supply to UPS) and the voltage fluctuations during steady-state operation are noticeable in both the signal waveform and spectrum.

5.2.3 Effects and Cost

5.2.3.1 Light Sources

A change in the supply voltage causes a change in the luminous flux of a light source, known as the flicker effect. It is a subjective impression of the luminous flux variation whose luminance or spectral distribution undergoes changes with time [12]. This phenomenon occurs in both incandescent and fluorescent light sources, though its mechanism and frequency range, as well as the limit values of disturbing components, are different. The rectifiers in electronic

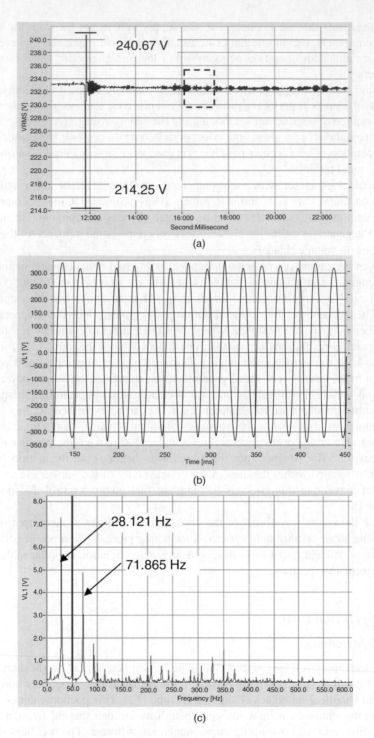

Figure 5.7 The waveforms of: (a) the chosen phase voltage rms value, (b) the instantaneous value at the output of a rotating UPS supplying an air-conditioning system and (c) the voltage spectrum

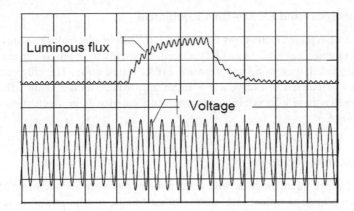

Figure 5.8 The effect of a voltage change on the luminous flux of an incandescent lamp [13]

starters transfer the instantaneous voltage value changes to the DC side, thereby changing the discharge conditions in a lamp and causing the flicker effect.

The luminous flux of an incandescent light source is proportional to the voltage according to $\Phi \sim U^{\gamma}$, where the exponent γ takes values in the interval 3.1–3.7 [12]. The exponent value for fluorescent lamps is: 1.5–1.8 [12].

Figure 5.8 shows an example of the change in luminous flux of an incandescent lamp in response to a temporary change in the supply voltage.

In low power incandescent lamps the filament temperature drops faster due to a smaller time constant, therefore such lamps are more prone to flickering (Figure 5.8).

Fluorescent lamps and present-day energy-efficient light sources are generally less sensitive to voltage fluctuation, though it is not a general rule. In this case the voltage peak value has a stronger influence on the luminous flux change than the rms value [13]. Certain solutions of electronic dimmers increase the flicker effect [14], though this rule also has its exceptions [15].

The flicker significantly impairs vision, causes general discomfort and fatigue, and deterioration in work quality. It considerably hampers reading and watching TV. Affecting the perception process and the human brain reactions, it can be the cause of work accidents and epileptic seizures. The psychophysical nature of the disturbance is complex [12].

5.2.3.2 Electric Machines

Voltage fluctuations at the terminals of an induction motor cause changes in torque and thereby in slip and, consequently, affect the production process. In the extreme case they may lead to excessive vibration, a reduction in mechanical strength and a shortening of the motor service life.

5.2.3.3 Static Converters

The supply voltage change in phase-controlled converters with DC side parameter control usually results in a reduction in the power factor and is a cause of non-characteristic harmonics and interharmonics generation. In the event of drive braking in an inverter mode, it can result in a commutation failure.

5.2.3.4 Electrolysers and Electro-Heat Equipment

In the presence of voltage fluctuations both the useful life of electrolyser equipment and the efficiency of technological process can be reduced. Elements of the high-current line become significantly degraded, increasing maintenance and/or repair costs. The efficiency of electro-heat equipment is reduced in every case – in an arc furnace as the result of a longer melt time, but generally these effects are noticeable only when voltage fluctuation amplitude is considerably large.

5.2.4 Mitigation Methods

The effects of voltage fluctuations depend first on their amplitude and their rate of occurrence. Whereas the amplitude is influenced, among other things, by the power system supplying fluctuating loads, the rate of changes depends upon the load type and its mode of operation. The rate of their occurrence is determined by the technological process. So far mitigation measures are focused on limiting the amplitude of voltage fluctuations; the technological process is influenced to a lesser extent. Example of these measures, in the case of an arc furnace, can be: a series reactor (also a controlled saturable reactor), proper functioning of the electrode control system, segregation and initial preparation of charge, admixing electrode material, etc. – methods well known to the steel-making process engineers. In the case of wind turbines it is a reduction of the number of switchings by keeping the turbine in standstill condition until the wind reaches a steady speed that is greater than the turbine start-up speed.

As follows from (5.5), the amplitude of voltage fluctuations – after modifications to the technological process (if possible) – can be limited in two ways:

1. *increasing the short-circuit power* (with respect to the load power) at the point of a fluctuating load coupling; practical means include: (a) connecting the load to higher nominal voltage system bus bars; (b) supplying this category of loads directly from a high voltage system through dedicated lines, supplying fluctuating loads and steady loads from either separate windings of a three-winding transformer or from separate two-winding transformers (separation of a fluctuating load); (c) supplying the fluctuating load from a transformer of a larger rated power or/and lower short-circuit voltage; (d) installing series capacitors, etc.;
2. *reducing reactive power changes* in the supply network by means of the so-called dynamic voltage compensator/stabilizer (Chapter 13).

A separate category of measures is the improvement of the loads' immunity to voltage fluctuations [16].

5.3 Voltage and Current Unbalance

Synchronous generators are the sources of a three-phase voltage in a power system. The voltages at the generator phase terminals are equal in magnitude and displaced from each other by $120°$. This vector system is called symmetrical. If individual phases of the system are equally loaded the currents also form a symmetrical vector system. In such conditions, if the components of the power system are linear and symmetrical, the voltages measured at load buses during normal system operation remain symmetrical.

5.3.1   Disturbance Description

In practice, it is not possible to obtain the full symmetry at all nodes of the power system. A condition in which the three-phase voltages in three-phase systems are not equal in magnitude and/or the displacement angles between them are different from $120°$ is defined as voltage unbalance. The analogous definition is applied to currents.

For the analysis of unbalance the method of symmetrical components is commonly applied. The method was introduced to calculations of electric power systems at the beginning of the 20th century. The main idea of it consists in the substitution of any three-phase unsymmetrical vector system of currents or voltages with the sum of three three-phase positive-, negative- and zero-sequence symmetrical systems. The equations for the respective symmetrical sets of voltage may be written as follows:

positive-sequence system (1) $\underline{U}_{1A} = \underline{U}_{1A} \quad \underline{U}_{1B} = a^2\underline{U}_{1A} \quad \underline{U}_{1C} = a\underline{U}_{1A}$ (5.7a)

negative-sequence system (2) $\underline{U}_{2A} = \underline{U}_{2A} \quad \underline{U}_{2B} = a\underline{U}_{2A} \quad \underline{U}_{2C} = a^2\underline{U}_{2A}$ (5.7b)

zero-sequence system (0) $\underline{U}_{0A} = \underline{U}_{0B} = \underline{U}_{0C},$ (5.7c)

where a is the rotational operator: $a = \exp(j2\pi/3) = -0.5 + j\sqrt{3}/2$. The respective phase voltage is the sum of relevant components:

$$\begin{aligned}
\underline{U}_A &= \underline{U}_{1A} + \underline{U}_{2A} + \underline{U}_{0A} \\
\underline{U}_B &= \underline{U}_{1B} + \underline{U}_{2B} + \underline{U}_{0B} = a^2\underline{U}_{1A} + a\underline{U}_{2A} + \underline{U}_{0A} \\
\underline{U}_C &= \underline{U}_{1C} + \underline{U}_{2C} + \underline{U}_{0C} = a\underline{U}_{1A} + a^2\underline{U}_{2A} + \underline{U}_{0A}.
\end{aligned}$$ (5.8)

The symmetrical component method, when applied to describe a three-phase circuit, allows the diagonalizing of the impedance matrix, which eliminates the couplings between phases and significantly simplifies the analysis of the circuit. A detailed description of the method is given in [2].

The unbalance factor K is commonly taken as a measure of unbalance. It is the ratio of the negative- and/or zero-sequence component to the positive-sequence component of voltage (current) of any phase:

$$K_{2U} = \frac{U_{2(1)}}{U_{1(1)}}100\% \quad K_{0U} = \frac{U_{0(1)}}{U_{1(1)}}100\% \quad \left(K_{21} = \frac{I_{2(1)}}{I_{1(1)}}100\% \quad K_{01} = \frac{I_{0(1)}}{I_{1(1)}}100\%\right).$$ (5.9)

The subscript (1) in the formulae above denotes that definitions are referred to the first harmonic.

The negative-sequence component generated by multi-phase loads is very important for the description of unbalance. It is transformed by a transformer irrespective of a windings connection, similarly to that in the case of the positive-sequence component. The zero-sequence component is normally present only in low-voltage (LV) networks, and the delta-connected transformer prevents it from transferring to the network of a higher voltage.

According to the definition, the amplitudes of the positive and negative-sequence voltage components must be known for determination of the unbalance factor. In [17–20] the relations are given that allow the calculation of the unbalance factor making use of measurements of rms values of line and phase voltages.

Most international standards and documents are in agreement on the definition of the unbalance phenomenon and its parameters. The compatibility level for LV and MV systems is 2%, and under special conditions 3% [18, 21–23].

5.3.2   Sources

Unbalanced operating conditions in an electric power system are mainly caused by the operation of unbalanced loads. Most low-voltage loads and certain medium-voltage ones, e.g. an electric traction, are single-phase appliances. The operation of such equipment in a three-phase system results in unbalance of load currents. Consequently, unsymmetrical voltage drops in individual phases of the supply system are produced, thus the voltage at nodes of the network becomes unbalanced.

Three-phase loads that may introduce unbalance to the power system are arc furnaces. The disturbance results from different impedances of high-current paths of the furnace and not equal phase loads, this being the effect of the physical nature of the melting process, i.e. variations of the arc impedance.

As arc furnaces are devices of relatively large power (tens or even hundreds MVA), the furnace load unbalance may result in significant voltage unbalance in the supply system.

The sources of unbalance can also be the three-phase components of the transmission system, in particular, overhead lines. Owing to different towers geometries, the conductors of individual phases are not simultaneously at the same location to each other and to earth. Following this, the line has different values of phase parameters, so the values of a voltage loss in individual phases are also different.

5.3.3   Effect and Cost

5.3.3.1   Power System

The negative-sequence and zero-sequence currents flowing in an electric power system result in technical and economical effects:

- additional losses of power and energy;
- additional heating, the consequence of which is the limitation of line transmission – capability for positive-sequence currents;
- voltage unbalance at nodes of the network that affects the operation of equipment.

Additional power and energy losses are the main effects of current unbalance in three-phase four-wire LV grids where the number of single-phase loads is significant. Under unbalance conditions the sum of the phase current vectors is not equal to zero and the resultant current flows through neutral wire. Active power losses are then expressed by:

$$\Delta P_{un} = \left(I_A^2 + I_B^2 + I_C^2\right) R_L + I_N^2 R_N, \tag{5.10}$$

where I_A, I_B, I_C are phase currents, I_N is the current flowing through neutral, R_L and R_N are phase and neutral wire resistances, respectively. Additional losses are proportional to the neutral current I_N.

As an example, let us consider the current unbalance caused by the difference in current amplitudes in one phase, phase C. Assuming, that $\underline{I}_A = I$, $\underline{I}_B = a^2\underline{I}_A = a^2I$ and $\underline{I}_C = ak\underline{I} = akI$, where $0 < k < 1$, it can be easily proved that the neutral current $I_N = (1-k)I$. The unsymmetrical current vector system can be substituted by symmetrical components:

$$\underline{I}_1 = \frac{1}{3}\left(\underline{I}_A + a\underline{I}_B + a^2\underline{I}_C\right) = \frac{1}{3}(2+k)\underline{I} \quad \underline{I}_2 = \frac{1}{3}\left(\underline{I}_A + a^2\underline{I}_B + a\underline{I}_C\right) = \frac{1}{3}(k-1)a^2I$$

$$\underline{I}_0 = \frac{1}{3}(\underline{I}_A + \underline{I}_B + \underline{I}_C) = \frac{1}{3}(k-1)aI. \tag{5.11}$$

The active power losses in the line wires can be calculated using sequence components according to (5.12):

$$\Delta P = 3I_1^2 R_P + 3I_2^2 R_P + 3I_0^2 R_P + 9I_0^2 R_N, \tag{5.12}$$

where R_p and R_N are the resistances of the phase and neutral wire, respectively. Taking into account (5.11) and assuming that $R_p = R$ one can obtain

$$\Delta P = I^2 R(2k^2 - 2k + 3) \tag{5.13}$$

To analyse the impact of unbalance, power losses are compared with the losses calculated for positive-sequence currents:

$$\Delta P_1 = I^2 R \left(\frac{1}{3}k^2 + \frac{4}{3}k + \frac{4}{3}\right) \tag{5.14}$$

Figure 5.9 shows the losses ΔP in relation to ΔP_1 as a function of ratio I_N/I_1 for different ratios of phase wire cross-section to neutral wire cross-section s_p/s_n. This simple example illustrates

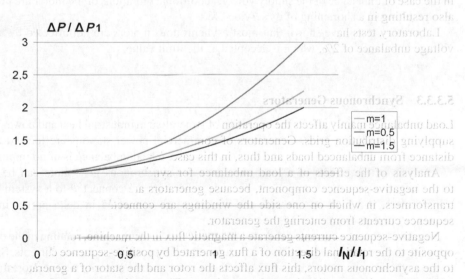

Figure 5.9 Increasing active power losses in three-phase four-wire line as a result of current unbalance, where $m = s_p/s_n$

the negative impact of unbalance on power and energy losses and thus on the efficiency of power delivery. Increasing neutral cross-section can be a means for the losses' reduction.

5.3.3.2 Asynchronous Motors

Asynchronous motors have their windings connected usually in a delta or star with an isolated star point. For this reason the operation of a motor is affected only by the positive-sequence and the negative-sequence component. The negative-sequence currents create a flux rotating in the direction opposite to the rotor. This flux causes:

- increased heating of the stator windings;
- additional losses of active power in the stator;
- additional torque operating in the opposite direction to the torque produced by the positive-sequence flux;
- inducting additional currents in the windings and rotor iron of a motor, and thereby additional power losses in the rotor.

The motor current under unbalance conditions can be several times higher than the rated current. Long-lasting unbalance can cause the motor insulation to deteriorate more quickly and its life shortened. Motors may be equipped with protections, which detect an overcurrent and can cause them to switch off.

Additional power losses due to the unbalance of a supply voltage reduce the maximum power of the motor to an extent that is dependent on the degree of unbalance, the type of the motor, and its construction.

A negative-sequence torque causes a reduction of the useful torque of the motor. Moreover, in the case of unbalance of the supply voltage, additional vibrations of the motor are produced, also resulting in a shortening of its service life.

Laboratory tests have shown that most asynchronous motors are not affected by a supply voltage unbalance of 2%, which is accepted as the limit value.

5.3.3.3 Synchronous Generators

Load unbalance mainly affects the operation of generators in industrial heat and power stations supplying distribution grids. Generators of commercial power stations are located at some distance from unbalanced loads and thus, in this case, a load unbalance is of no importance.

Analysis of the effects of a load unbalance for synchronous generators can be limited to the negative-sequence component, because generators are connected to a system through transformers, in which on one side the windings are connected in delta preventing zero-sequence currents from entering the generator.

Negative-sequence currents generate a magnetic flux in the machine, rotating in the direction opposite to the rotational direction of a flux generated by positive-sequence currents. Similarly to the asynchronous motors, this flux affects the rotor and the stator of a generator; it induces eddy currents, increases heating and power losses. The negative-sequence flux also produces additional mechanical forces acting on the rotor and the stator of a generator, which are hazardous to the strength of structural components.

The fundamental criterion for evaluating the permissible operation of a generator under unbalanced conditions is the additional heating of the rotor.

In general, unbalanced loads are not a great problem for the operation of synchronous generators; unbalance may cause more severe hazard during disturbances, for example, during unsymmetrical short-circuits.

5.3.3.4   Static Rectifiers

Converter equipment in most cases is supplied from a three-phase three-wire system, thus its operation can be affected only by the negative-sequence component of the voltage. It generates:

- an additional variable component of a rectified voltage (current) whose amplitude depends on the unbalance factor,
- harmonics that are non-characteristic for the given topology of a converter and interharmonics [24].

5.3.3.5   Other Loads

Unbalance can also affect the operation of other three-phase loads, changing the electric power, exploitative characteristics and their service life. Moreover, voltage unbalance associated with the change of voltage magnitude has an effect on the operation of single-phase loads. Some of them may be under the influence of a supply voltage that is too high or too low. Disturbances can also occur in the functioning of control systems, resulting in a disturbance and even an interruption of the operation of equipment.

5.3.4   Mitigation Methods

The unbalance of power system components is eliminated by a suitable design thereof. In the case of overhead lines, a transposition of conductors is applied for this purpose. The line is divided into sections, the number of which is divisible by 3, with three sections forming one transposition cycle. In each section the conductor of a given phase is routed at a different position to the other phases, which causes the line, taken as a whole, can be considered as a symmetrical one.

The methods for unbalance mitigation primarily concern balancing the load. The traditional approach to load balancing is to connect nominal loads evenly to each phase. Normally this is sufficient, so that severe unbalance of voltage does not appear very frequently. Where significant load unbalance is unavoidable, particularly in the case of large single-phase loads, special balancing equipment for disturbance compensation should be applied. The purpose of this is usually the elimination or limitation of the negative-sequence and zero-sequence components of the load currents. This process is called balancing.

5.3.4.1 Principle of Balancing

In three-phase medium-voltage (MV) systems, which usually operate as isolated or compensated ones, unbalanced loads are connected to a line voltage. In this case, there is no zero-sequence component of currents, so the balancing resolves itself into the elimination or limitation of the negative-sequence component. LV grids are four-wire earthed neutral systems, so there the negative-sequence as well as the zero-sequence component is present.

A balancing device (BD) is connected in parallel with the unbalanced load (UL) (Figure 5.10). This equipment causes the currents I_{Ak}, I_{Bk}, I_{Ck} to flow, which added to the load currents I_{Al}, I_{Bl}, I_{Cl} result in the symmetrical set of the source currents I_A, I_B, I_C.

There are a variety of systems for balancing load currents that differ in the number and type of elements and the art of connection. Comprehensive information on this can be found in [25]. BDs accomplish their tasks at a constant value of loads. For time-varying loads, such as electric traction or arc furnaces, a follow-up compensation is necessary. This kind of compensation is implemented through:

- *Shunt compensators* (static VAr compensators – SVC e.g. TCR, FC/TCR, STATCOM, Chapter 13). Connected to a node of an electric power network, they may be considered as a controlled parallel susceptance. Balancing is one of the tasks that can be accomplished by these devices. They are usually also applied for the compensation of reactive power or the limitation of voltage fluctuations and a light flicker phenomenon. The idea of compensation is to control currents or voltages in such a way as to minimize the negative- and zero-sequence components.
- *Series connected compensators* applied for mitigation of the supply voltage unbalance. The purpose of the system is to generate, for instance by means of a PWM converter, the boosting voltage $\underline{U}_b$ whose value and phase in relation to the source voltage $\underline{U}_s$/current $\underline{I}$ are regulated. The converter can be considered as a voltage source represented on the dq plane by two orthogonal components with regulated values:
  ○ component U_{bd} (in phase or in opposite phase in relation to a line current) decides the exchange of active power between a supply system and a compensator,
  ○ component U_{bq} (orthogonal to a line current) decides the value (also the character) of a series compensator reactive power.

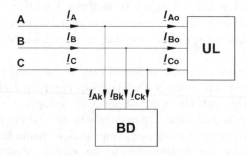

Figure 5.10 Diagram of unbalanced load with balancing device

5.4   Voltage and Current Distortion

The component parts of a power system are designed to work in a sinusoidal condition, where the curves of the voltage and electrical currents in the circuit have a sinusoidal waveform with the same frequency as the nominal frequency. In reality, nowadays in most of the nodes of the power system, the waveforms of currents and voltages present distortions from a sinusoidal curve [2, 7, 24].

The periodical deviation from the sinusoidal shape of the electric current or voltage wave-forms is called distortion or deformation.

The distortion is produced by the non-linearity of the components of the power system. The nonlinear characteristics of the user's loads are the most frequent source of distortions in the curve of the electrical current; distorted currents lead to distortions in the curve of the power supply voltage. With the development of command and control systems using power semiconductors, the sources of distorted currents became widely spread; their presence in the power system led to supplementary actions to identify disturbance sources, to assess the disturbance levels and effects, and to implement actions that could limit the damages caused by them, both in the electrical grid, as well as in the case of consumers connected to a harmonic polluted grid.

In order to assess the effects of the distortion of both electrical currents and voltages, special attention should be paid to establishing suitable indices that define the distortion level and their admissible values.

5.4.1   Disturbance Description

According to the theory elaborated by Fourier, in the case of periodic signals, the distorted curves can be regarded as an overlapping of a number of sinusoidal oscillations with frequencies that are an integral multiple of the fundamental frequency. In such conditions, an analysis of the distorted curves can be done in the frequency area, by evaluating the characteristic values of the component oscillations. This evaluation method is nowadays widely used and is the basis for all methods used today in measuring, analysing and evaluating various methods of limiting the distortion level.

Figure 5.11 presents a distorted waveform and the harmonic spectrum that compose it. Using the Fourier series, any periodic distorted signal can be decomposed under the form:

$$u(t) = U_0 + \sum_{k=1}^{\infty} U_k \cdot \sqrt{2} \cdot \sin(k \cdot \omega_1 \cdot t + \alpha_k), \tag{5.15}$$

where U_0 is the constant component of the signal, U_k is the r.m.s. value of the k-th harmonic, and α_k is the phase angle of the k-th harmonic from a reference axis.

Taking into account the fact that, in an electrical grid, the condition for a phenomenon to become periodical is not always being met, the use of (5.15) is allowed only on such intervals in which the electric quantities can be regarded as periodic. Based on this decomposition, a series of indices are used to define the spectral components of the distorted curve. Any index on the frequency domain cannot unequivocally characterize a distorted curve, which means

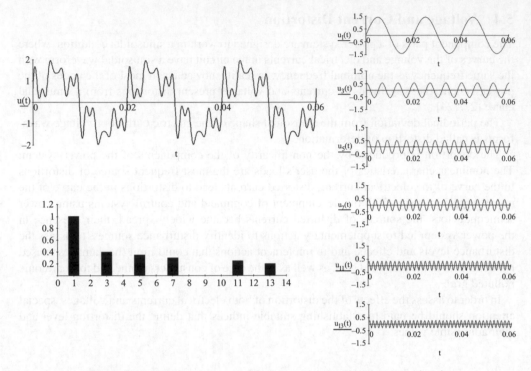

Figure 5.11   Distorted curve and the harmonic spectrum that comprises it

that several indices should be used. Table 5.1 presents the main indices used to characterize the distortion state [2, 3, 26, 27].

The harmonics standardization process can be considered as quite complete since this problem has already been studied for several decades. On the other hand, the same cannot be said about the interharmonics standardization process, which is work in progress, with knowledge and measured data still being accumulated.

Standards prescriptions define the limits on voltage supply, measurement method and instruments, mitigation process, etc. The main IEC standards dealing with harmonics and interharmonics are presented in [2]. The essential standards regarding the harmonics are EN 50160 [21], IEEE 519-1992 [28] and IEEE 1159-2009 [29].

5.4.2   Sources

The presence of distortion sources in the power system determines the creation and distribution of deformed curves of voltages or electrical current into the electrical grids. In practice, most common distortions have frequencies from just a few hertz to almost 10 kHz.

Table 5.2 presents some of the most important distortion sources for power systems, their ranks and the amplitude of harmonics that they generate [3, 26, 30].

Table 5.1 The main indices of the distortion state

Index	Calculation	Observations
R.m.s. value	$$U = \sqrt{\frac{1}{T} \cdot \int_0^T u^2 \cdot dt}$$	u defines the voltage variation in time, and T the integration interval.
Average value for half a period	$$U_{med1/2} = \frac{2}{T} \cdot \int_0^{T/2} u \cdot dt$$	This expresses the DC voltage that can be obtained after the rectification of an alternative signal during a period T.
Peak factor	$$k_v = \frac{u_M}{U}$$	u_M is the maximum value of the periodical non-sinusoidal curve. The peak factor can have the following values: • in the case of a sinusoidal curve, $k_v = \sqrt{2}$; • in the case of a sharp curve, $k_v > \sqrt{2}$; • in the case of a flattened curve, $k_v < \sqrt{2}$. The voltage curves characterized by a peak factor $k_v > \sqrt{2}$ can determine dangerous overloads for the electrical equipment insulation.
Form factor	$$k_f = \frac{U}{U_{med1/2}}$$	The form factor can have the following values: • for a sinusoidal curve, $k_f = 1.11$; • for a periodical curve more pointed than a sinusoidal form, $k_f > 1.11$; • for a periodical curve more flattened than a sinusoidal form, $k_f < 1.11$.
Harmonic level	$$\gamma_k = \frac{U_k}{U_1}$$	The harmonic level is an important index when evaluating the distortion level, the norms giving the maximal admissible values for the voltage curve within the nodes of the power system.
Distortion (harmonic) content	$$U_d = \sqrt{u^2 - U_1^2} \cong$$ $$\cong \sqrt{U_0^2 + U_2^2 + Y_3^2 + \ldots}$$	The distortion content is a measure of the thermal effect created by the harmonic components of the distorted signal.
Distortion factor	$$THD = \frac{U_d}{U_1}$$	The distortion factor, THD, is one of the indices used to evaluate the maximum level of distortion; its maximum allowed values in the nodes of power system are indicated by norms.

5.4.3 Effects and Cost

The greater importance of the nonlinear elements in the power systems, both in terms of installed capacity as well as in terms of types of equipment, leads to an increase in their harmonic pollution levels and to an increase of negative effects determined by the presence of harmonics in the electric grid. These effects can be classified into two main categories, according to the information presented in Table 5.3.

Table 5.2 Pollution sources for the energy systems

Distortion source	Rank and amplitude of harmonics they generate
• Controlled or semi-controlled single-phase rectifiers (full bridge) with resistive charge or filtered current, practically continuous when exiting the rectifier • AC power controller with resistive load • Home appliances.	• Odd rank harmonics • The amplitude of the harmonics decreases with the increase of their rank • The disappearance of certain frequencies for various values of the thyristor phase angle, in the case of semi-controlled or controlled rectifiers.
• Single-phase rectifiers (half-bridge) with a resistive charge or filtered current, basically continuous upon exit. • p-pulse rectifiers	• Odd and even rank of harmonics • The amplitude of the harmonics decreases when the rank increases. • $k = m \cdot p \pm 1$ ($m = 1, 2, 3 \ldots$) harmonics • the amplitude of harmonics decreases together with their rank, approximately according to the relation: $$I_k = \frac{I_1}{k^{1.2}},$$ where I_1 is the rms value of the fundamental and k is the harmonic rank • The disappearance of certain frequencies for various values of the thyristor phase angle in the case of semi-controlled converters.
• Saturated universal electrical motors • Fluorescent tubes	• Odd harmonics • The amplitude of the 3rd harmonic is below 15% of the fundamental one • The rapid decrease of the amplitude of the harmonics when their rank increases • The appearance of even harmonics during heating up.
• Automatic washing machines	• Odd harmonics • Even harmonics when rectifying only one half-wave • Decreasing amplitude when the rank of harmonic increases.

• Colour television sets	• Systems using the rectification for both half-waves
	  o Odd harmonics
	  o The amplitude of the 3rd current harmonic may reach 80% of the fundamental amplitude of the supply current
	  o The amplitude of the harmonic decreases with the increase of its rank
	• Systems using rectification for a single half-wave:
	  o Odd and even rank harmonics;
	  o The amplitude of the 2nd current harmonic is below 45% of the fundamental amplitude of the electric current;
	  o The amplitude of harmonics decreases with the increase of their rank.

Table 5.2 (*Continued*)

Distortion source	Rank and amplitude of harmonics they generate
• Electric ovens	• Odd and even rank harmonics • The amplitude of the 2nd current harmonic, about 5% from the fundamental amplitude of the electric current • The amplitude of the harmonics decreases with the increase of their rank.
• Static compensators for electric arc furnaces	• harmonics of rank 5, 7, 11, 13, . . . • the amplitude of the 5th current harmonic is below 20% from the fundamental electric current amplitude • the amplitude of the harmonics decreases as their rank increases.
• Electric locomotives with single-phase rectifiers	• Odd harmonics • The amplitude of the 3rd current harmonic is below 20% from the fundamental electric current amplitude • The amplitude of the harmonics decreases as their rank increases.

5.4.3.1   The Decrease of Energy Efficiency in Power Systems

The flow of harmonic currents through the elements of the electric grid produces an increase of active power losses in the conductive, magnetic and dielectric materials [2]. The supplementary power losses in the conductor material are generated by the harmonic currents that pass through the elements of the electric grid, as well as by the increased electric resistance of these elements (the latter is influenced by the skin and the proximity effects).

Assuming continuous component neglect, these losses can be calculated from the equation [6, 7]:

$$\Delta P = 3 \cdot \sum_{k=1}^{\infty} R_k \cdot I_k^2, \tag{5.16}$$

Table 5.3 Classification of negative effects produced by the harmonic distortion

Technical point of view	Economic point of view
• The system components are sensitive either to the harmonic currents (Joule losses, disturbances in the audio frequency), or to the deformed voltages (losses in the magnetic circuits and dielectric materials; over-voltages that, in certain cases, can be above the permitted levels). • The proper operation of some equipment is affected by the presence of voltage and/or current harmonics (control systems, equipment that are synchronized with the voltage of the grid, etc.).	• There are increased manufacturing expenses in order to limit the specific non-linearity of various equipment or in order to increase the immunity levels to disturbances (the equipment should comply with imposed immunity norms). • There is an increase in operation expenses for preventive or corrective maintenance. • There is an increase in electric energy production costs and generally an increase in investments in the power systems due to the over-sizing of the grid elements.

where I_k is the rms value of range k harmonic current; R_k is the electrical resistance of the element, calculated for harmonic frequency of rank k. An element crossed by alternating current presents an electrical resistance, $R_{a.c.}$, calculated as:

$$R_{a.c.} = R_{d.c.} \cdot (1 + r_s + r_p),$$ (5.17)

where $R_{d.c.}$ is electrical resistance in DC; r_s is the factor taking into account the skin effect; r_p is the factor taking into account the proximity effect. Assuming that the resistance R does not vary with frequency and the DC component is zero:

$$\Delta P = 3 \cdot R \cdot \sum_{k=1}^{\infty} I_k^2 = 3 \cdot R \cdot I_1^2 \cdot \left(1 + THD_I^2\right),$$ (5.18)

where THD_I is the total harmonic distortion of the current.

Equation (5.18) shows that the active losses in conducting elements can greatly increase if the system is functioning in non-sinusoidal condition rather than the sinusoidal condition.

The losses in the magnetic materials are determined by the hysteresis phenomenon and/or eddy currents. The losses in the dielectric materials are mainly located in the dielectric of the capacitors and in the insulation of overhead and underground electric lines. Losses are produced by the active component of the electric current and by the insulation resistance of the dielectric material and their value depends on the value of the loss angle on the k-th harmonic.

5.4.3.2 Skin Effect

The skin effect is the tendency of alternating current to flow on the outer surface of a conductor. This effect is more pronounced at high frequencies; in fact it is normally ignored because it has very little effect at power supply frequencies, while above approximately 350 Hz (the seventh harmonic and above), it becomes significant, causing additional loss and heating. The AC resistance to DC resistance ratio is dependent on r/δ where r is the conductor radius and δ is the current penetration thickness, which can be expressed as $\delta = (2\rho/\omega\mu)^{1/2}$, where μ is the magnetic permeability (H/m); ω the frequency (rad/s); and ρ the resistivity (Ω mm^2/m). It is evident that δ is dependent on the frequency; in particular it decreases as frequency increases [2].

5.4.3.3 Over-Voltages in the Power Grid

The increase of voltage in the nodes of electric grids or at the terminals of various equipment can be determined by the resonance on a voltage harmonic or by the increase of the potential of the neutral point in the case of star-connected transformers or three-phase receivers.

5.4.3.4 The Increase of the Neutral Point Potential in the Case of Star Connections

A well balanced load or a downward three-phase transformer connected to a balanced three-phase electric grid, with sinusoidal voltages on a fundamental frequency, has the neutral point potential equal to zero with reference to the ground if it presents a star connection. If the electric grid is affected by a non-sinusoidal periodical state, the equipment terminals will receive a harmonic voltage, which determines on the neutral point a potential with reference to the ground and which has a value that depends on the ratio between the harmonic impedance of the active phase and that of the neutral circuit.

5.4.3.5 Overcharging the Neutral Circuit of Three-Phase Grids

In the case of balanced electric grids with four conductors, the existence of harmonic currents determines the flow of a harmonic current through the neutral conductor, obtained by summing up the triplen harmonics. The harmonic current $3k$ is placed over that determined by a likely current unbalance; as a result, an overheating of the conductor is possible, especially since its cross-section – in regular constructions – is inferior to that corresponding to the conductors on the active phases. The problem of the overheating of the neutral circuit appears especially in low-voltage distribution networks, where a large percentage of the consumers are computers and lighting installations with gas and metallic vapour discharge; they are characterized by a high value of the 3rd harmonic (its value may reach 80%), so that in the neutral conductor high current will flow. As this conductor has no protective devices, the risk of overheating and fire might prove to be important.

5.4.3.6 Effects on Three-Phase Transformers

The operation of three-phase transformers in harmonic polluted electric grids can have the following consequences [7]:

- an increase of the active power losses in the conductor materials due to the increase of the winding's electric resistance for different current harmonics;
- an increase in the losses of magnetic materials in the presence of higher harmonics due mainly to the increase of the losses through eddy currents;
- the increase in the electric stress of insulations, determined both by the maximal value of the terminal voltage and its variation speed;
- additional mechanical stresses;
- the increase in the value of the distortion factor of the electric current in case of operation in the nonlinear portion of the of the magnetization curve (due to overcharge, a state that can be determined just by the harmonic pollution; in this case it is possible that a reduced voltage harmonic level generates a high level of current harmonics).

The main effect of an electric transformer functioning in a non-sinusoidal state is represented by the increase in temperature due to the supplementary losses in the windings and core. To avoid heating above the maximum temperature allowed by the manufacturer, a reduction of charge is required, i.e., applying *a depreciation factor of the nominal power*:

$$S = k_t \cdot S_N \qquad\qquad (5.19)$$

where S is the apparent power in the non-sinusoidal state and S_N is the nominal power of the transformer. The depreciation factor k_t can be determined using the relation:

$$k_t = \frac{1}{\sqrt{1 + 0.1 \cdot \sum\limits_{k=2}^{\infty} \left(\frac{I_k}{I_N}\right)^2 \cdot k^{1.6}}} \qquad (5.20)$$

in which I_N is the rated current of the transformer and I_k is the rms value of the rank k harmonic.

5.4.3.7   The Effects on the Operation of Rotating Machines

The main negative effects that appear in the case of electric rotating machines, determined by the harmonic pollution of the grid to which they are connected, are:

- an increase in temperature of the windings and magnetic core due to additional losses in the conductive and magnetic materials;
- changes in the torque of the electric machine, leading to a reduction in its efficiency;
- the presence of oscillations of the running torque of the shaft of the electric machine, thus contributing to the ageing of the material and extra vibrations;
- changes in the magnetic induction in the machine gap due to the higher rank harmonics;
- interactions between the magnetic flux determined by the fundamental harmonic and the magnetic flux determined by the superior harmonics.

A specific problem appears in the case of variable speed drives (VSD), where the electric motors are fed by means of frequency static converters. They (except those that contain PWM controlled inverters) create a highly distorted voltage that can lead to heavy thermal and mechanical wear in the motor. All these cases require an analysis of the practical possibilities that can reduce the distortions and the motor wear.

5.4.3.8   The Effects of the Non-Sinusoidal Periodical State on Electronic Equipment

Electronic equipment used in control systems is generally powered by a sinusoidal voltage, but they themselves could represent distortion sources for the grid to which they are connected, due to the specific way in which they change the controlled values (phase control, time control, etc.). When a non-sinusoidal voltage is applied to the terminals of these devices, the technical characteristics of the installation are altered and this could lead to negative consequences on the control and operation of the equipment.

Harmonic pollution affects electronic devices in several ways, the main ones being as follows:

- the possibility of multiple zero crossings of the voltage curve as a result of harmonic distortion represents a problem because a large number of electronic circuits are operated based on synchronization with the zero crossing of the grid voltage;
- the amplitude of the voltage curve and the value of its peak must be taken into account because some electronic sources use this information to charge the filter capacitor. However,

the presence of harmonics may determine the increase or reduction of the voltage in the grid; therefore, the voltage given by the source is modified, even if the rms value of the input voltage equals the rated value. The operation of the devices supplied by the source is affected starting from an increased sensitivity to voltage dips up to significant misoperation. In order to avoid such an effect, some computer manufacturers limit the peak values to $k_v = \sqrt{2} \pm 0.1$ and others demand that the total distortion factor should not be more than 5%;

- interharmonics and subharmonics can affect the operation of displays and television sets by modulating the amplitude of the fundamental frequency. In the case of such components, for levels above 0.5%, some periodic changes in the image may appear (for cathode tubes).

5.4.3.9   Effects on Capacitor Banks

In the case of three-phase electric grids, in which higher harmonics appear, the capacitor banks belonging to filters or installations used for power factor correction are subjected to the following overloads:

- extra heating due to dielectric losses;
- long-lasting overcharges due to the amplification of the pollution level;
- resonance phenomena (of voltage or current) that have an amplifying effect on already existing system harmonics: in the case of a series resonance, there is an increase in the value of the current on that particular harmonic and of voltages on the components of the resonant circuit; in the case of a parallel resonance, there are overcharges in the voltage in the system and high currents in the capacitors.

5.4.3.10   Economic Damages

The presence of harmonics in the electric grid causes damage due mainly to [7]:

- the increase in manufacturing costs in order to limit the non-linearity that is specific to various equipment or to increase the immunity level to distortions (by meeting the requirements of norms referring to immunity);
- the increase of operating expenses in the case of preventive or corrective maintenance;
- the increase of expense in the case of electric energy production and a general increase in investments in power systems due to over-sizing of the elements of the electric grid.

5.4.4   Mitigation Methods

Basically, the pollution of a power system becomes a problem if:

- the source of the harmonic currents is too strong;
- the part of the system that is traversed by the harmonics is too long (from an electric point of view), thus resulting either in an inadmissible distortion of the system voltage, or in unacceptable interferences with other systems (generally telecommunications interference);
- the frequency response of the system amplifies one or more harmonics.

Table 5.4 Solutions for the mitigation of the disturbed regime

Methods and means of limiting the disturbed state	Decreasing the generated harmonic currents	Rectifiers	Mounting AC line reactors at the converter input
			Increasing the value of the DC side reactors
			Using multi-pulse methods
		PWM invertors	Mounting AC line reactors on the input circuit
		Electric arc furnaces	Limiting reactors
			Decreasing short-circuit reactance
	Changing the frequency response of the power system	Increasing the power of the supply transformer	
		Mounting a series detuning reactor with the capacitor bank for power factor correction	
		Changing the placement of the capacitor bank for power factor correction	
		Splitting up the capacitor banks	
	Limiting the area where the harmonic currents can flow	Power supply using suitable insulating transformers	
		Using Y/Z transformers	
	Harmonic filters	Passive	With one step
			With several steps
		Active	Series
			Parallel
		Hybrid	

Taking into account these negative consequences of the presence of the disturbance, there is a need to initiate certain actions that can mitigate or remove this problem. The main options to achieve this are [2, 6, 7]:

• reducing the harmonic currents produced by the nonlinear loads;
• limiting the area crossed by the harmonic currents;
• changing the frequency response of the power system;
• mounting filters that could reduce the bidirectional flow of harmonic currents between the power system and the distribution installation of the consumer.

Nowadays there are many methods that are used. Table 5.4 gives an overview of these methods, giving examples for some of the disturbing receivers frequently encountered in industrial applications [7].

The classic solutions to reduce the distorted state are based on using passive components (inductors, capacitors and transformers) or on interventions on the configuration and/or the structure of the power supply system of the consumer in order to reduce the distortion level of the voltage at the heaviest charged point of common coupling. Developments in the technology of semiconductor devices – mostly achieved in the last decades – have made an important mark on the techniques and equipment used for reducing the distortions. Thus, the traditional methods have been completed by modern ways of removing harmonic pollution, based on devices and specialized highly performing semi-conductive circuits.

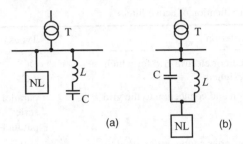

Figure 5.12   Suppressing filter (a) parallel; (b) series (NL – nonlinear load)

5.4.4.1   Passive Filters

Starting from the frequency analysis of the current distorted waveform, one of the most effective solutions to eliminate the harmonic currents is to use passive filters. Basically, a passive filter is composed of a number of *LC* series resonant circuits for the aimed harmonics, presenting a zero reactance for the rank *k* harmonics that are about to be filtered. These filters can be classified according to their application field [2, 27]:

- passive suppressing filters;
- passive absorber filters.

The first type of filters is used in already existing installations when a limit of the harmonic transfer due to nonlinear receivers should be obtained, Figure 5.12 [30]. Nonlinear consumers with a known steady state use passive absorber filters as one of the most effective means of limiting the harmonic transfer in the electric grid and of limiting the charge of the capacitors on the banks for power factor correction.

Absorber filters are formed of series circuits tuned on harmonic *k*, Figure 5.13. The filters are achieved in such a way that each of its resonant circuits has a capacitive character for the frequencies lower than the resonance frequency defined by the *k* harmonic and an inductive character for the frequencies higher than this value. Thus, for the fundamental frequency, each of the circuits generates reactive power that will be taken into account when sizing the power factor correction system. The presence of the absorber filter at the consumer power supply bus bars stops the harmonics and therefore the power voltage is free of distortions [7, 30].

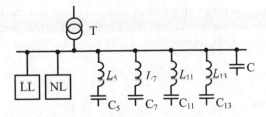

Figure 5.13   Connecting the absorber filter to the bars of the distortion consumer (LL – linear load, NL – nonlinear load)

Table 5.5 Classification of active filters

Classification criterion	Types of active filters
The type of circuit or electric grid for which the filter is designed	DC AC
Type of operation and connection to the grid	parallel series parallel–series
Static power converter configuration used to implement the filter	voltage source inverter current source inverter
Structure and technologies used for the implementation	active hybrid
Number of phases	single phase three phase

5.4.4.2 The Active Filter

An active filter basically represents a static power converter made in such a way that it synthesizes a current or voltage signal in the desired form; the signal is injected/applied in a certain point of an AC power grid, and simulates the desired impedance by meeting the specified values of the absolute value and argument.

The advantages of active filters over other conventional (passive) means used to reduce the harmonic pollution of electric grids are mainly:

- the high response speed;
- flexibility in defining and/or implementing the functions of the filter (modern active filters are capable of simultaneously executing several functions, which can easily be activated or deactivated and/or modified);
- elimination of the resonance problems of the compensating device – distribution grid assembly, specific to conventional solutions (passive filters, capacitor banks, transformers and specially designed transformers).

A classification of active filters is presented in Table 5.5.

According to the type of the synthesized signal (current or voltage), there are several ways of connecting the active filter to the grid – parallel or series, namely, active parallel filter (shunt) or series. In the context of the total power quality management concept, several designs of complex equipment that combine a structure of current source and one type of voltage source inverter have been designed and produced. These are called active parallel–series filters and they combine the functions and performances of both schemes. Table 5.6 shows the main types of active filters.

5.4.4.3 Hybrid Filters

This category of filters resulted from the combination of active and passive filters in order to obtain lower costs for more efficient structures. Figure 5.14 shows three hybrid filter structures and Table 5.7 presents a comparison of these filters [27].

Table 5.6 Types of active filters

Type of active filter	Basic diagram	Description
Parallel	nonlinear load i_s $Z_{ech,R}$ i_R i_{FA} G control circuit active parallel filter	The parallel active filter is a device that accomplishes the functions of a current source. The current signal synthesized by the filter i_{FA}, is injected into the connection point with the purpose of: • compensating the current harmonics produced by the consumer; • compensating the reactive power consumption of the receiver; • damping the effects produced by potential resonance phenomena that could occur due to the presence of a parallel passive filter connected to the same point; • balancing the current system absorbed in the connection point.
Series	nonlinear load u_{FA} T adapting transformer G Active series filter	The series active filter fulfils the function of a voltage source (the synthesized signal produced by the filter has been marked with u_{FA}), which is connected in series to the distribution grid using a matching transformer in order to: • reduce the distortions of voltage signals; • balance the voltage system; • reduce voltage dips and voltage fluctuations; • assure only short interruptions in power supply; • decrease the transitory high frequency phenomena that distort the voltage system; • reduce the flicker levels; • regulate the voltage rms.

(continued overleaf)

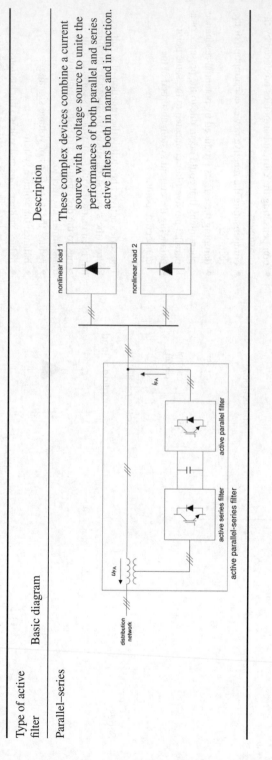

Table 5.6 (*Continued*)

Type of active filter	Basic diagram	Description
Parallel–series		These complex devices combine a current source with a voltage source to unite the performances of both parallel and series active filters both in name and in function.

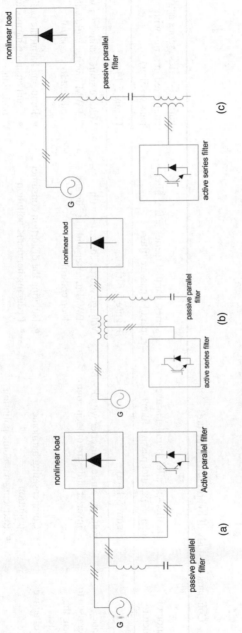

Figure 5.14 Hybrid filters: (a) parallel–parallel; (b) series–parallel; (c) parallel–series

Table 5.7 Comparative study of hybrid filter structures

	parallel–parallel (parallel active filter + parallel passive filter)	series–parallel (series active filter + parallel passive filter)	parallel–series (series active filter connected in series with a parallel passive filter)
Configuration of hybrid structure	Figure 5.14(a)	Figure 5.14(b)	Figure 5.14(c)
Type of static converter used to implement the active filter	Voltage source converter with PWM inverter and current loop control	Voltage source converter with PWM inverter without current loop control	Voltage source converter with PWM inverter with/without current loop control
Functions of active filter	• Compensating harmonic pollution • Reducing harmonic pollution	• Isolating the distortion consumer • Reducing harmonic pollution	• Reducing harmonic pollution • Compensating harmonic pollution
Advantages of the configuration	• Permits the use of every type of active parallel filter • Allows compensation for the reactive power consumption	• Permits the use of already existing passive filters • The active filter blocks the flow of harmonic currents	• Permits the use of already existing passive filters • Permits an easy and efficient protection of the active filter
Disadvantages of the configuration	Splitting the working frequencies range between the passive filter and the active one	• Difficulties in the overcharge protection of the active filter • Does not allow compensation for reactive power consumption	Does not allow compensation for reactive power consumption

| Observations | The two filters parallel connected complete each other and it is recommended that the range of harmonic currents be split in such a way that the active filter does not act over the frequency of the passive filter. The presence of the active filter considerably improves the efficiency of the compensation and allows the reduction in the number and size of the passive filters used but does not prevent their overcharge with harmonics propagated through the distribution grid, neither does it prevent the appearance of resonance between the passive filters and the grid. | The series active filter presented in this hybrid structure acts as an isolating element between the distribution grid and the load, thus avoiding on one hand the pollution of the grid by the customer and, on the other hand, overcharging the downstream passive filter due to harmonic currents in the grid. The disadvantage of this structure that the matching transformer of the series active filter is run through by a significant amount of current (the entire load current), hence the high costs and weight together with the difficulty in protecting the active filter in the case of a short circuit. | In this configuration, the active filter acts only on the harmonic currents, thus improving the effectiveness of the passive filter by (i) avoiding the amplification of harmonic voltages with a frequency above the anti-resonance frequency of the passive filter; (ii) a strong reduction of harmonic currents that circulate between the customer and the grid by lowering the impedance of the circuit. Since the active filter is not run through by the entire load current, it can be re-sized accordingly, thus resulting in reduced costs and weight (especially for the matching transformer). The structure is recommended for the reduction of harmonic pollution in high power grids and higher voltages (medium-voltage networks). |

References

[1] P. Caramia, G. Carpinelli and P. Verde, *Power Quality Indices in Liberalized Markets*, John Wiley & Sons Ltd, 2009.

[2] A. Baggini (Ed.), *Handbook of Power Quality*, John Wiley & Sons Ltd, 2008.

[3] Carmen Golovanov *et al*. *Modern Measurement Problems in Power Systems*, Tehnică Publishing House, Bucharest, 2002 (in Romanian).

[4] Al. Kusko and M. T. Thompson, *Power Quality in Electrical Systems*, McGraw-Hill, 2007.

[5] R. C. Dugan *et al*., *Electrical Power Systems Quality*, 2nd edn, McGraw-Hill, 2004.

[6] M. Eremia, Electric Power Systems. Electric Networks, Academiei Române Publishing House, Bucharest, 2006.

[7] N. Golovanov, P. Postolache and C. Toader, *Power Quality and Energy Efficiency*, AGIR Publishing House, Bucharest, 2007. (in Romanian).

[8] http://www.leonardo-energy.org/webfm_send/2703

[9] Directive 2006/32/EC on energy end-use efficiency and energy services (repealing Council Directive 93/76/EEC).

[10] M. Bercovici, A. Arie and Al. Poeată, *Electrical Networks. Electrical Design*, Tehnică Publishing House, Bucharest, 1974. (in Romanian).

[11] T. A. Short, *Electric Power Distribution Handbook*, CRC Press, 2004.

[12] Z. Hanzelka, A Bień, L. Flicker, *Power Quality Application Guide*, http://www.lpqi.org

[13] UIE: Guide to quality of electrical supply for industrial installations, part 5: Flicker, WG 2, Power Quality, UIE, 1999.

[14] T. A. Short, *Distribution Reliability and Power Quality*, Taylor & Francis, 2006.

[15] C.-S. Wang, Flicker-insensitive light dimmer for incandescent lamps, *IEEE Tran. Ind. Electron.*, **2**(55), 2008, 767–772.

[16] EN 61000-4-14: Voltage fluctuation immunity test.

[17] IEC 61000-2-1: Electromagnetic compatibility – Description of the environment – Electromagnetic environment for low-frequency conducted disturbances and signalling in public power supply systems.

[18] IEC 1000-2-12: Electromagnetic compatibility – Compatibility levels for low-frequency conducted disturbances and signalling in public medium-voltage power systems.

[19] IEC 61000-4-27: Electromagnetic compatibility – Testing and measurement techniques – Unbalance, immunity test.

[20] IEC 61000-4-30: Electromagnetic compatibility – Testing and measurement techniques – Power quality measurement methods.

[21] EN 50160-2009: Voltage characteristics of electricity supplied by public distribution system.

[22] IEC 61000-2-5: Electromagnetic compatibility – Classification of electromagnetic environments.

[23] UIE: UIE Guide to quality of electrical supply for industrial installations, part 4: Voltage unbalance, WG 2, Power Quality, UIE, 1996.

[24] Z. Hanzelka and A. Bien, *Harmonics – Interharmonics*, Leonardo Power Quality Application Guide, http//www.lpqi.org

[25] Z. Kowalski, *Asymmetry in Power Systems* (in Polish), PWN, Warsaw, 1987.

[26] A. Arie, E. Neguş, C. Golovanov and N. Golovanov, *Pollution Power Systems Operated Unsymmetrical Conditions*, Academiei Române Publishing House, 1994. (in Romanian).

[27] M. Chindriş *et al*., *Harmonic Mitigation of Industrial Distribution Networks*, Mediamira Publishing House, Cluj-Napoca, 1999. (in Romanian).

[28] IEEE 519-92: IEEE Recommended practices and requirements for harmonic control in electrical power systems.

[29] IEEE 1159: Recommended practice for monitoring electric power quality.

[30] S. Gheorghe *et al*., *Power Quality Monitoring*, Macarie Publishing House, Targoviste, 2001. (in Romanian).

Further Readings

J. Arrilaga, N. R. Watson and S. Chen, *Power System Quality Assessment*, John Wiley & Sons Ltd, 2000.

M. Bollen and I. Gu, *Signal Processing of Power Quality Disturbances*, John Wiley & Sons Ltd, 2006.

EN 61000-2-4: Compatibility levels in industrial plants for low-frequency conducted disturbances.

EN 1000-3-3: Limitation of voltage changes, voltage fluctuations and flicker in public low-voltage supply systems, for equipment with rated current ≤ 16 A per phase and not subject to conditional connection.

ER P28 Electricity Association Eng. Recommendation P28: Planning limits for voltage fluctuation caused by industrial, commercial and domestic equipment in the United Kingdom, 1989.

EURELECTRIC: Electromagnetic compatibility in European electricity supply networks, EMC & Harmonics, August 2001, Ref: 2001-2780-0020, EURELECTRIC.

A. Ghosh and G. Ledwich, *Power Quality Enhancement using Custom Power Devices*, Kluwer Academic Publishers, 2002.

IEC 50 (161): International Vocabulary (IEV) – Electromagnetic compatibility.

A. Johansson and M. Sandström, *Sensivity of the Human Visual System to Amplitude Modulated Light*, ISSN 1401-2928, http://www.arbetslivsinstitute.se

http://www.allaboutcircuits.com/vol_2/chpt_9/6.html

Z. Kowalski, *Voltage Fluctuations in Power Systems*, (in Polish), WNT, Warsaw, 1985.

C. M. Lefebvre, *Electric Power: Generation, Transmission and Efficiency*, Nova Science Publishers, New York, 2007.

D. Povh and M. Weinhold, Improvement of power quality by power electronic equipment, *CIGRE Session 2000*, **13**(14), Paper 36–06.

A. Robert J. Marquet, Assessing voltage quality with relation to harmonics, flicker and unbalance, *CIGRE*, 1992.

UIE: Guide to quality of electrical supply for industrial installations. Part 1: General introduction to electromagnetic compatibility (EMC), types of disturbances and relevant standards, WG 2, Power Quality, UIE, 1994.

UIE: Guide to quality of electrical supply for industrial installations, part 3: Harmonics, WG 2, Power Quality, UIE, 1995.

6

On Site Generation and Microgrids

Irena Wasiak and Zbigniew Hanzelka

Nowadays, the conventional electric power systems that rely on centralized power generation using fossil fuels face various challenges relating to secure energy supply with the required power quality, efficiency and environmental concern.

The fossil fuel resources exploited today are limited and capturing new ones may be expensive and technically difficult. Moreover, energy production from such sources causes the increasing emission of gases resulting in environmental pollution. Reducing emission means increased generation costs. Without expanding existing generation capacity it will be difficult to meet future energy needs.

Centralized energy generation needs bulk and complex power networks to transfer energy to distributed customers. Investment and maintenance costs are high and energy efficiency is reduced because of relatively high power losses in the networks. On the one hand the existing power system infrastructure (transmission networks) is becoming old and, on the other, there are environmental problems with constructing new components, mainly high voltage transmission lines (e.g. lack of space, public objections). The important issue on the distribution side of the system is maintaining the required quality of supply. Customer demands are still increasing, together with the increasing number of sensitive loads used by them. Utilities are under an obligation to provide customers with good quality energy in an effective and reliable manner. On the other hand, we can observe an increasing number of loads that are becoming the sources of electromagnetic disturbances.

In many European countries special support systems and mechanisms have been developed in recent years that, according to the EU policy [1], promote an increasing share of distributed generation (DG) in total energy production. Difficulties in the development of new high voltage transmission systems (so-called constraints) contribute to the development of distributed generation, which does not need a large power system to transfer electricity to consumers.

Electrical Energy Efficiency: Technologies and Applications, First Edition. Andreas Sumper and Angelo Baggini.
© 2012 John Wiley & Sons, Ltd. Published 2012 by John Wiley & Sons, Ltd.

In these conditions one can expect that the number of distributed energy resources (DERs) installed in the network will grow continuously.

DERs produce power on a customer's site or at a local distribution utility and supply electrical energy to the local distribution network. Installing DERs may delay the need to upgrade distribution infrastructure in the case of network operation near its capacity limits. Individual electricity consumers can use DERs as backup sources to supply energy when the power from the grid is unavailable. In normal operation conditions DERs may deliver energy to the hosting grid for the benefit of the owner. DERs owned and operated by power utilities are located at specific sites in the distribution grid to supply the local power or power and heat demand. Their capacity may be relatively high from a hundred kW to MW. DERs owned by independent power providers are scattered within the grid. They have a capacity from a few kW to hundreds of kW.

The integration of a considerable number of DERs into the grid may lead to a stressed power grid operation and cause difficulties in maintaining the required power quality (PQ). DERs may generate disturbances such as voltage variations, asymmetry or harmonics. The problems may be heightened by disturbing loads if they are installed in the grid. For a specific network, one can determine so-called 'hosting capacity' [2], which is the highest amount of DG that can be integrated without the PQ limits being violated. In many practical cases, the network hosting capacity limits wide penetration of DG, unless some measures are applied that will facilitate the integration process and assure the required power quality. Such measures include additional compensating devices and management and control systems.

6.1   Technologies of Distributed Energy Resources

The pressure of improving the overall efficiency of power delivery has forced the power utilities to reduce the losses, especially at the distribution level to make the system more efficient and reliable. The losses can be reduced in many ways. One technique is the use of DERs – energy sources and energy storage systems. Integration of DERs into an existing network can result in several benefits. These benefits, in addition to line loss reduction, also include reduced environmental impacts, peak shaving, increased overall energy efficiency, relieved transmission and distribution congestion, voltage support and deferred investments to upgrade existing generation, transmission and distribution systems. Furthermore, the electricity price in the spot market displays frequent fluctuations. DERs dispatched for the purpose of avoiding wholesale electricity purchases must be both dispatchable and flexible or be on-line at full capacity during the peak hours of highest electricity prices. Crucial factors in the applications of DERs for loss minimization are the size and location of DERs (see Section 6.3).

6.1.1   Energy Sources

Different types of DG technologies are in use today (e.g. Table 6.1). They can be classified into two categories depending on the prime fuel used. Micro-turbines, fuel cells and reciprocating engines are based on gas and are fully dispatchable. Photovoltaics, wind and hydro sources use renewable energy that is intermittent in nature, so they are non-dispatchable. Prime fuel local availability, investment and operating costs are the main factors influencing the selection of the source type (Table 6.1).

Table 6.1 Comparison of distributed energy sources [3]

	Combined cycle gas turbine	Gas turbine	Diesel engine	Micro-turbine	Gas engine	Fuel cell (high temperature)	Photovoltaic	Wind turbine	Small hydro power plant
Electrical efficiency [%]	35–55	25–42	35–45	27–42	25–43	40–60	6–19	25	
Overall efficiency [%]	73–90	65–87	65–90	90	70–92	90			
Typical power range [MW]	3–300	0.3–50	0.2–20	0.03–1	0.003–6	0.00–100	20 W– few MW	200 W– 5 MW	25 kW– 100 MW
Heat	Medium-grade steam or high temperature hot water	High-grade steam or hot gas	Low-pressure steam and medium temperature water	Hot gas or hot water	Low and medium temperature hot water	Steam or hot water			
	CHP technologies (see Section 6.2.1.1)								

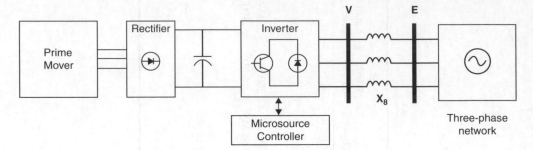

Figure 6.1   Typical structure of a microsource

Some of the DG technologies require a power electronics interface in order to convert the energy into grid-compatible AC power. The generic structure of the microsource that uses a power electronic inverter is shown in Figure 6.1.

For sources with AC output such as high speed micro-turbines the voltages are first rectified to produce a DC output and inverted to produce the desired AC voltage and frequency. The inverter is typically a voltage source converter and provides the necessary controls over the phase and magnitude of the bus voltage. Storage can be added at the DC bus to decouple the dynamics of the prime mover from the output. Sources such as fuel cells and PV panels produce DC directly and only an inverter is needed to produce AC output at the desired voltage and frequency. The power electronics interface can also contain protective functions for both the distributed energy system and the local electric power system, which allows paralleling and disconnection from the electric power system. A more detailed overview of DG technologies is reported in, for example, [4–6].

Some types of distributed generation can also provide combined heat and power by recovering some of the waste heat generated by the source – Combined Heat and Power (CHP).

6.1.1.1   CHP

Waste heat is produced in energy conversion processes. Through CHP generation (cogeneration) this heat is collected and used for spatial or process heating, which results in improving fuel-to-energy efficiency. Cogeneration means the simultaneous generation in one technological process of thermal and electrical energy. The overall efficiency of cogeneration is defined as the annual sum of electricity and useful heat production divided by the input fuel used to produce the sum of electricity and useful heat output. The efficiency is dependent on cogeneration technology, and the size of the unit (Table 6.1); an average value reaches 85%.

The comparison of cogeneration with separate production of electricity and heat is illustrated in Figure 6.2.

According to [3], the following efficiency is assumed for the comparison: for cogeneration an electrical efficiency of 35% and a thermal efficiency of 50% (based on a gas engine); for separate generation, a power plant with a typical efficiency of 43% and a boiler of 95% efficiency. Assuming that the same amount of electricity and heat is produced, the amount of fuel used in separate generation is 25% higher than in CHP systems, which means an energy efficiency improvement of 25%.

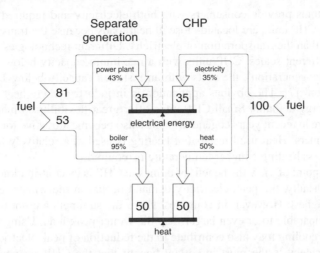

Figure 6.2 Comparison of a CHP system with separate electricity and heat production

A methodology for determining the efficiency of the cogeneration process is given in [7]. The amount of primary energy savings (PES) provided by cogeneration production can be calculated using the following:

$$PES = \left(1 - \frac{1}{\dfrac{\eta_{Q,CHP}}{\eta_{Q,Ref}} + \dfrac{\eta_{E,CHP}}{\eta_{E,Ref}}}\right) \cdot 100\%, \tag{6.1}$$

where $\eta_{Q,CHP}$ and $\eta_{E,CHP}$ are the thermal and electrical efficiency of the CHP calculated as the ratio of annual useful heat output and annual electricity, respectively, to the fuel input, and $\eta_{Q,Ref}$ and $\eta_{E,Ref}$ are the efficiency reference values for separate production of heat and electricity for which the cogeneration process is intended to substitute.

For the values from Figure 6.2, it yields:

$$PES = \left(1 - \frac{1}{\dfrac{0.5}{0.95} + \dfrac{0.35}{0.43}}\right) \cdot 100\% = 25.4\% \tag{6.2}$$

The efficiency reference values can be calculated taking into account the same fuel categories. Each CHP unit is compared with the best available and economically justifiable technology for the separate production of heat and electricity on the market in the year of construction of the cogeneration unit.

Cogeneration, which provides primary energy savings of at least 10% compared with the references for separate production of heat and electricity, is classified as high-efficiency cogeneration. Production from small scale and micro-cogeneration units providing primary energy savings may be regarded as high-efficiency co-generation.

The CHP systems provide consumers with both electricity and required heat simultaneously. Typically, CHP units are located close to heat loads because the transportation of heat is more difficult than the transportation of electricity. Different technologies of CHP systems are applied at different scales. CHP units with a maximum capacity below 50 kW are classified as 'micro-cogeneration', the units with an installed capacity below 1 MW are 'small scale cogeneration' [7]. The obvious application is in industry where heat demand usually accompanies energy demand. Small CHP units offer great flexibility in matching heat load demand, therefore in recent years cogeneration has also become attractive for commercial and residential customers. Heat used for spatial heating is used at a relatively low temperature, whereas for process heating a higher temperature is required.

From a utility point of view the benefit of domestic CHP is its contribution to the reduction of peak loads. Usually, the peak electricity demand occurs in the winter season when heat demand is the highest. However, in countries where the summers are hot the summer peak load may be comparable to or even larger than the winter peak load. Using CHP systems to provide building cooling may also contribute to the reduction of peak electricity loads [8].

Improved efficiency is the most important benefit that the CHP systems can offer. The development of this technology is supported and promoted by the European Commission [7] and one can expect that the share of electricity from cogeneration to overall power production will be increasing.

6.1.2 Energy Storage

Electric energy storage is one of the contemporary electrical engineering challenges. Its purposes are, *inter alia*, improvement of electric power generation effectiveness by levelling power demand, better management of transmission and distribution systems and improvement in electric power delivery. The key drivers for energy storage systems are the increasing popularity of distributed energy sources and their characteristic features: location, usually remote from public network, output power often not correlated with demand and varying over a wide range, low average power generating capacity utilization (0.2 wind generators, 0.1 photovoltaic sources), and still high energy cost, as well as, relatively low efficiency of centralized generation.

Depending on the unit power and required autonomy time, the energy storage issues can be divided into two groups concerning: power supply quality – chiefly consumer protection during voltage dips and short supply interruptions, the so-called *backup storage energy supply* (BSES), and energy management. The latter requires significantly larger power and longer autonomy times (Figure 6.3).

In utility applications, storage units should have large power capacity (to a few dozen MW) and storage time in a range from minutes to hours – depending on the task performed. In end-user applications, there is lower power capacity (from a few dozen to few hundred kW) but a larger storage time is needed.

The devices used in distribution networks can be also classified into three categories according to the energy discharge rate [9, 10]:

1. systems to supply energy within seconds or less to address ancillary services including PQ events compensation;

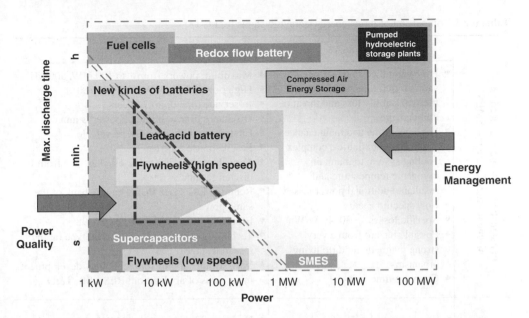

Figure 6.3 Classification of selected energy storage systems versus autonomy time and unit power (based on [6, 11])

2. systems to supply energy within minutes to address power shortages (uninterruptible power supply);
3. systems to supply energy for more than a few minutes to several hours for load levelling, peak shaving and energy management, including island operation.

There are many technologies that make energy storage possible [10]. The most commonly used are pumped hydro power plants due to the cost of storage and their effectiveness. Other technologies are nowadays used at a smaller scale, or in special cases such as flywheels, which are used when a large amount of power, but not necessary a large amount of energy, is needed (as in the emergency power supply to cover energy needs before a diesel generator becomes fully operational – no longer than 1 minute). The systems designated for long-term storage are pumped hydro, compressed air, hydrogen storage, batteries (including redox flow) and thermal storage. Flywheels, superconductive coils and supercapacitors (as well as some kinds of batteries) are generally meant to store lower amounts of energy; they are, however, capable of being 'charged'/'discharged' with large power that, combined with their good dynamics (like reaction time), makes them perfect for power quality regulation, for example. A major challenge to the market in the near future comes from applications such as electric vehicles or plug-in hybrids and grid-coupled PV systems with batteries, which will be used to suit the application and not primarily to support the grid. Tables 6.2 and 6.3 sum up the energy storage methods.

There are some technologies that are typical for their areas of application. In energy management applications, for a discharge time of up to 1.25 h, lead–acid batteries are the most

Table 6.2 Comparison of direct energy storage systems [6, 11–13]

	Disadvantages	Advantages
In magnetic field – superconducting magnetic energy storage – SMES	• High capital cost and stringent safety requirements, danger of thermal stablity loss and threat of coil damage • High operating and maintenance costs, large auxiliaries, complex cooling system to maintain cryogenic temperature and resultant additional power losses • Low specific energy • Specific losses: *c.* 30–40 W/Wh • Possible hazard from a very strong magnetic field (to living organisms) • Charging time: min–h	• Maximum powers from 1 to 100 MW, 500 MJ • High power density W h/kg: 30–100 • In-service time up to 30 years • Discharge time at nominal power: s min • Environmentally friendly, recyclable • Fast response time (ms) • Very short charging time, less than 90 s from full discharge state • Requires less space than a battery bank of the same power • Efficiency (charge/discharge): 85–98% • Technology readiness level – prototype (at the research stage) • Very strong support for technology development, main areas of application: RESs, SR, T&D*
In electric field – supercapacitors (ultracapacitors or double-layer capacitors) – SC	• Expensive technology • Low specific energy: 1–10 W h/kg • Low cell voltages – a series connection is needed to achieve higher voltages, voltage balancing required where a larger number of capacitors are connected in series • Requires advanced power electronics • Voltage reduction during discharge	• High specific power (10 000 W/kg), nominal power: 0.01–100 kW, 10 MJ • Less weight than batteries • Very slow self-discharge process • Efficiency (charge/discharge): 85–98%, specific losses: 0.026 W/W h • Capability for operation at very low temperatures, even (−40°C) • Simple charge methods – no full-charge detection is needed; no danger of overcharge, very fast charging/discharging time 0.3–30 s • Low impedance – enhances load handling when put in parallel with a battery • Virtually unlimited cycle life, maintenance-free (over 500 000 charging/discharging cycles) • Very strong support for technology development, main areas of application: RESs, SR*

*RESs – Renewable energy sources; SR – Support for RESs; G&SR – Generation and system reserve; T&D – Support for transmission and distribution; UPS – Uninterruptible power supply systems

economical [9]. Other technologies such as nickel–cadmium, sodium–sulphur or Li–Ion can offer a better performance in terms of cycle life or efficiency, however their costs are higher or they are currently under development.

Supercapacitors and fly-wheel energy storages are appropriate solutions for applications where large power capacity is required. In PQ applications ultracapacitors have attracted the great attention [14, 15].

Table 6.3 Comparison of indirect energy storage systems [6, 11–13]

		Disadvantages	Advantages
Mechanical	Pumped hydroelectric storage plants	• Geographic and geological constrains on the location of the installation • High capital cost • Long construction period • Dedicated for very large powers (0.1–1 GW)	• Mature technology – employed in the power sector for a long time • Relatively cheep operation • Discharge time at nominal power: h–days • Efficiency (charging/discharging): 65–80% • In-service time: 30–50 years • Main application areas: RESs, G&SR[*]
	Kinetic energy storage systems	• Low specific energy (2000–10 000 W h/kg) • High idling losses	• Mature technology – mass production • High specific power • Nominal power: 0.01–10 MW, 15 MJ • Discharge time at nominal power: s–min • Efficiency (charging/discharging): 90% • Specific losses low speed: 2000–10 000 rpm – 2.2 W/Wh; high speed: 10 000–1 000 000 rpm – 1.2 W/W h • In-service time: $c.$ 20 years • Very strong support for technology development, considerable application potential, main areas of application: RESs, SR[*]
	Compressed air energy storage (adiabatic CAES)	• Geographic and geologic constrains on the location of the installation • High capital cost • Long construction period • Dedicated for large installations • Long starting time, poor compatibility with distributed energy sources	• Mature technology • High specific energy and power • Nominal power: 0.1–1 GW • Discharge time at nominal power: hrs–days • Efficiency (charging/discharging): 60% • In-service time: 20–40 years • Strong support for technology development, area of application: RESs, G&SR[1]
Electrochemical	Fuel cells (varicus technologies)	• Expensive technology (expensive catalysts) • Difficult fuel production (hydrogen) • No current overload capability • Specific energy: $c.$ 11 000 W h/kg	• Mature technology (mass production) with very strong support for further development • Nominal power: up to 5 MW • Efficiency (charging/discharging): 40–55% • Capability for cogeneration (high temperature fuel cells) • Discharge time at nominal power: min–h • In-service time: 20–30 years • Main application areas: RESs, SR, G&SR, T&D and UPS

(continued overleaf)

Table 6.3 (*Continued*)

			Disadvantages	Advantages
Electrochemical	Battery – SB	Lead–acid battery	• Specific power: 1000 W/kg • Specific energy *c*. 10–100 W h/kg • Limited number of charging/discharging cycles • Limited operating voltages and currents • Requires maintenance and technical surveillance • Battery capacity dependence on temperature • High deep-discharge voltage • Expensive recycling • Efficiency: *c*. 80%	• Mature, readily available technology, extensive production and operating experience • Long in-service time: 10–20 years • Low price • Easy production • Modular construction, suitable size • Possible large power installations, up to 10 MW, over 600 MJ • Suitable voltage characteristic • Charging time: h, discharge time: min–h • Specific losses 0.023 W/W h • Efficiency (charging/discharging): *c*. 70–85% (concerns all battery storage technologies) • Main application areas: RESs, SR, G&SR, T&D and UPS
		Sodium–sulphur (Na-S) or NaNiCl battery	• Expensive technology, sodium polysulphides are highly aggressive – a container must be chromium and molybdenum plated (both are very expensive) • High operating temperature, necessary thermal insulation and temperature monitoring necessary, ceramic electrolyte is sensitive to thermal cycles	• Mature technology • Powers up to 8 MW, 60 MW h • Large power and energy densities • Efficiency: 70–80% • Very cheep electrode material • Applications: RESs, SR, G&SR, T&D and UPS
		Redox flow battery (vanadium) (VRB)	• Expensive technology at the development level • Difficult standardization	• Intended for large applications (1 MW – 1 h; 3 MW – 1.5 s) • High power and energy densities • Efficiency: 60–75% • Very strong technology development support, considerable application potential
		Zinc–bromine battery (Zn-Br)	• Expensive technology at the development level • High operating costs • Contains corrosive and toxic materials	• High power and energy densities • Intended for large applications • Efficiency: 70–80%

Table 6.3 *(Continued)*

			Disadvantages	Advantages
Electrochemical	Battery – SB	Li–Ion battery	• Technology at the development level, expensive and 'difficult' operation, instability of lithium in the presence of air causes hazard of battery failure, particularly in portable equipment • Short in-service time	• High power and energy densities • Efficiency: 90–95% • Minimum maintenance required • Small dimensions and weight
		Nickel-cadmium battery (Ni-Cd)	• Expensive technology • Contains toxic materials (cadmium) • Fast self-discharge (particularly at high temperatures) • Limited capability for charging in high temperatures • Application to low power systems	• Mature technology • High mechanical robustness • High energy density • Efficiency: 60–70% • Permissible high charging currents d • Long in-service time (up to 25 years) • Insignificant influence of ambient temperature on the battery capacity • Large number of deep discharges • Possible and recommended battery storage in discharged state
		Nickel metal hydride (Ni-M-H)	• Expensive technology • Sensitive to high temperatures, similarly as Ni-Cd batteries • Difficult recycling process	• Mature technology • Efficiency: 80–90% • High mechanical robustness • High energy density • Long in-service time • Smaller number of toxic compounds as compared with Ni-Cd batteries

*RESs – Renewable energy sources; SR – Support for RESs; G&SR – Generation and system reserve; T&D – Support for transmission and distribution; UPS – Uninterruptible power supply systems

6.2   Impact of DG on Power Losses in Distribution Networks

Active power losses in distribution networks depend on both active and reactive power flow:

$$\Delta P = 3I^2 R = 3\left(\frac{S}{\sqrt{3U}}\right)^2 R = \frac{P^2}{U^2}R + \frac{Q^2}{U^2}R \qquad (6.3)$$

The source connected to the node k of the network presented in Figure 6.4 introduces active power P_k and reactive power Q_k. Then, power flow in the network is changed. The active power in a $(\alpha - 1, \alpha)$ branch is equal to:

$$P_{\alpha-i,\alpha} = \sum_{i=\alpha}^{k-1} P_i - P_k + P_{k+1} \qquad (6.4)$$

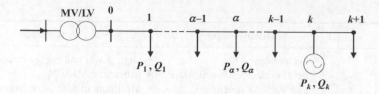

Figure 6.4 Connection of a source in an LV grid

and the reactive power to:

$$Q_{\alpha-1,\alpha} = \sum_{i=\alpha}^{k-1} Q_i \pm Q_k + Q_{k+1}. \tag{6.5}$$

The total losses in the network, after the source connection, are expressed as:

$$\Delta P_{T(s)} = \frac{1}{U_n^2} \cdot \sum_{\alpha=1}^{k+1} \left[R_{\alpha-1,\alpha} \left(P_{\alpha-1,\alpha}^2 + Q_{\alpha-1,\alpha}^2 \right) \right], \tag{6.6}$$

where

$$P_{\alpha-1,\alpha} = \sum_{i=\alpha}^{k-1} P_i - P_k + P_{k+1} \qquad \text{for} \quad \alpha \le k-1$$

$$P_{k-1,k} = P_{k+1} - P_k$$

$$P_{k,k+1} = P_{k+1}$$

and

$$Q_{\alpha-1,\alpha} = \sum_{i=\alpha}^{k-1} Q_i \pm Q_k + Q_{k+1} \qquad \text{for} \quad \alpha \le k-1$$

$$Q_{k-1,k} = Q_{k+1} \pm Q_k$$

$$Q_{k,k+1} = Q_{k+1}.$$

In general, a source can produce or consume the reactive power, however in most practical cases the source reactive power is equal to zero. Under such an assumption, connecting the source into the network will influence the active power component of the losses. The difference in active power losses of the feeder after introducing P_k active power to the node k will be:

$$\delta(\Delta P_T) = \Delta P_{T(s)} - \Delta P_T = \frac{1}{U_n^2} \cdot \sum_{\alpha=1}^{k+1} \left[R_{\alpha-1,\alpha} \cdot \left(P_{\alpha-1,\alpha}^2 - P_{0(\alpha-1,\alpha)}^2 \right) \right], \tag{6.7}$$

where $P_{0(\alpha-1,\alpha)}$ is the active power flowing in the branch $(\alpha - 1, \alpha)$ when $P_k = 0$. The total losses in the network will be reduced if $\delta(\Delta P_T) < 0$.

Usually, the energy source covers the local load demand, which is a relief for the supplying network. For most feeder branches the losses are reduced. However, if the source power is relatively large, the reduction in losses may not occur.

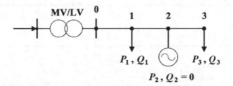

Figure 6.5 Example of a three-node feeder

Let us consider the simple three-node feeder like in Figure 6.5. After connecting the source to node 2, the active power flow in the feeder branches will be:

$$P_{12} = P_3 - P_2$$
$$P_{01} = P_1 + P_3 - P_2. \tag{6.8}$$

The difference in active power losses in the feeder will be:

$$\delta(\Delta P_T) = \frac{R_{01}}{U_n^2}\left[P_2^2 - 2P_2(P_1 + P_3)\right] + \frac{R_{12}}{U_n^2}\left(P_2^2 - 2P_2 P_3\right). \tag{6.9}$$

Assuming that R_{01} is equal to R_{12}, $\delta(\Delta P_T) < 0$ if

$$0 < P_2 < P_1 + 2P_3. \tag{6.10}$$

Changes in power losses as a function of power generated by the source are presented in Figure 6.6. Power losses ΔP_T are expressed in relation to the value ΔP_{T0} obtained for $P_2 = 0$. The source power P_2 is referred to the sum of load power ($P_1 + P_3$). The following data were assumed for the feeder: line nominal voltage $U_n = 0.4$ kV, wire cross-sections $s_{01} = s_{12} = s_{23} = 35$ mm^2, wire conductivity $\gamma = 34$ m/Ω mm^2, and feeder section lengths $l_{01} = l_{12} = l_{23} = 500$ m. The loads consume the constant reactive power with tg φ_1 = tg $\varphi_3 = 0.4$, the source reactive power is equal to zero.

It is clear from the figure that the range over which the power losses are reduced depends on the location of the local loads and their powers. The larger the load at the end of the feeder, the larger range of power generated by the source for which the losses in the feeder are reduced. If the total load power is located at node 1, then the reduction of losses will occur only if the source power does not exceed the load power. In this case, the network supplies only the reactive power. On the other hand, if the total load power is located at node 3, then the reduction of losses will occur for generated power from 0 to twice a total load power. For each individual case one can determine the source power that gives the minimum losses in the feeder.

Similar analysis can be performed for different feeder data and the influence of feeder section length or cross-section can be easily determined. For any individual load case and the source power it is also possible to select the optimal location of the source for which the feeder operation is the most effective. For larger distribution networks different optimization methods are applied that allow one to determine the optimal allocation of DG with the aim of minimizing power losses and keeping the voltage profile and reliability levels in the required range [16].

In practice, the connection of a single source may not cause a large reduction in losses, however in the case where many sources are connected in the microgrid this effect may be

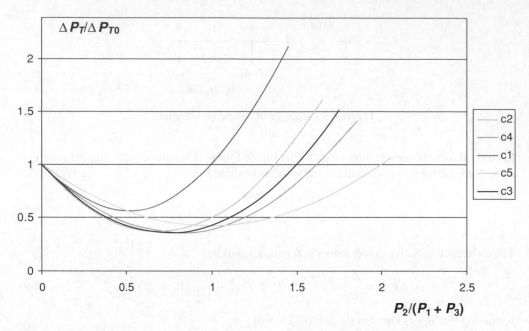

Figure 6.6 Changes of power losses in the example feeder from Figure 6.5 depending on the source power: $c1 - P_1 = 80$ kW, $P_3 = 0$ kW; $c2 - P_1 = 50$ kW, $P_3 = 30$ kW; $c3 - P_1 = 40$ kW, $P_3 = 40$ kW; $c4 - P_1 = 30$ kW, $P_3 = 50$ kW; $c5 - P_1 = 0$ kW, $P_3 = 80$ kW

significant. The results of studies are presented in [17] that show that for countries where peak demand occurs on hot summer days, PV generation contributes to the reduction of losses in some distribution networks by up to 40%. A similar effect can be obtained if peak demand occurs on a winter evening and CHP is applied in the network.

6.3 Microgrids

6.3.1 Concept

Together with a growing use of DERs, conventional distribution networks change their structure from passive to active. The active network in which the processes of energy generation, distribution, and use are executed in a controllable way forms an electrical power microgrid.

The microgrid integrates energy sources, including renewables, storage systems, controllable and uncontrollable loads. Energy sources produce power and heat mainly on the local load demand. The storage equipment supports the operation of variable power sources and is involved in energy balancing in order to optimize economical profits. Additional 'custom power' devices may be applied to improve supply quality and the system stability. The operation of a microgrid is monitored and supervised in real-time by the control system. This is a key issue in the microgrid concept, which decides on how smartly the microgrid can act. Real-time monitoring makes the network observable. The measured data are transferred to the management and decision centres. Smart prognostic and decision-making algorithms

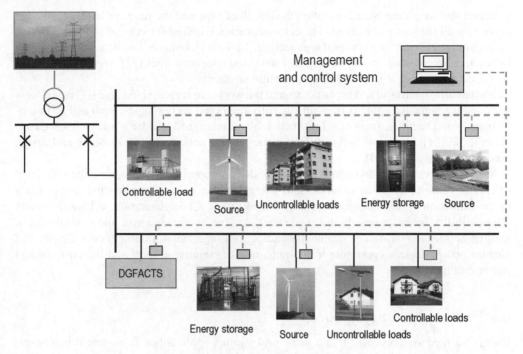

Controllable load

Source   Uncontrollable loads   Energy storage   Source

DGFACTS

Energy storage   Source   Uncontrollable loads   Controllable loads

Management and control system

Figure 6.7   Illustration of LV microgrid

enable them to predict the consequences of the existing situation and to make decisions that are optimal with respect to a given criterion.

So the characterized microgrid is a smart grid, which in more broad sense is defined by European Technology Platform as '*an electricity network that can intelligently integrate the actions of all users connected to it – generators, consumers and those that do both – in order to efficiently deliver sustainable, economic and secure electricity supplies*'. This definition is not constrained to any voltage level, however the microgrids refer rather to low voltage networks. The microgrid is illustrated in Figure 6.7.

Many research studies and demonstration activities regarding microgrid development are reported in the literature. Reference [18] gives an overview of ongoing projects in Europe, Japan, the USA and Canada.

The basic configuration of a microgrid is radial with several feeders supplied from a distribution substation [5]. These feeders could be part of the distribution system or a building's electrical system. A static switch usually connects the microgrid with the LV bus of the distribution transformer. Usually, the microsystem operates in connection with the electrical power grid, but autonomous operation is also possible as well as a transition between these two modes [19–21]. In grid-connected mode the excess power generated by the microgrid may be sold to the utility; in this case the microgrid will participate in the market operation or provide ancillary services [22]. The transition to island operation follows the loss of connection with the grid in fault conditions. During autonomous operation, which lasts until the connection with the network is restored, there is partial or full energy delivery to customers. Where there is intentional long-time island operation the length of time and the range of loads supplied is agreed by all the parties involved. The communication required for control and protection is an important element of microgrid architecture. Various techniques can be used such as the internet, radio communication, power-line carrier and telephone lines [23]. Work is continuing on the establishment of standard communication protocols.

The existing microgrid test networks around the world are reviewed in [21,23]. Test systems include representative sources from all currently important technologies, such as PV, micro-turbines, wind turbines, fuel cells. Information about microgrid standards and technologies is given in [20, 21]. Economic and regulatory issues are the concern of [8]. Planning and design aspects are presented in [4].

Microgrid technology and architecture provides an opportunity to improve the efficiency of energy consumption. The installation of energy sources close to the demand centres leads to the reduction of transmission and distribution losses. CHP generation technology offers the possibility for customers to use waste heat. An energy management and control system and the application of storage enable the microgrid to operate at minimum cost. Finally, the custom power devices contribute to the reduction of negative effects and the costs of bad power quality.

6.3.2   Energy Storage Applications

Energy storage systems have found many and various applications in electrical power microgrids. Storage technologies can provide solutions for the key issues that nowadays affect microgrid development, such as the intermittency of renewable energy sources (RES), demand fluctuation and power quality. They can help in better utilization of electrical energy for the benefit of both utilities and end-users. Utility applications include: the integration of RES with the grid, load levelling and ancillary services (see Section 6.3.5), whereas the end-user applications concern mainly peak shaving and uninterruptible power supply.

6.3.2.1   Integration of Renewable Energy Sources

Much work has been carried out on the integration of renewable energy sources, such as wind and solar systems, with the supply network. Daily power generation of such sources is stochastic and does not correspond to a typical demand curve. Usually, the highest generation is during off-peak hours. The application of a storage system enables one to capture part of this generation and transfer it to peak hours. The system stores the surplus energy generated during low-load periods and releases it during high-load periods. In this way local generation decreases peak power flows in the grid, which in consequence leads to the reduction of losses.

From the market point of view it means that energy is stored at times of low energy prices (low load) and injected to the grid at times of high prices. The benefit for the source owner can be in saving energy delivered from the grid to cover his demand. As regards big wind farms supplying energy to the grid, it also means economical profits. The technical and economical impact of this application is analysed in [24, 25].

Using a storage system as a power and energy buffer makes dispatching power generated by RES possible. Another purpose of the application is the mitigation of power output fluctuations, thus improving power quality. This function is important when there is a high level of RES penetration. Storage can lead to the increasing use of renewables, which in consequence contributes to a reduction in the use of fossil fuels and an increase in grid efficiency.

6.3.2.2   Load Levelling

In every electrical power grid power generation must match power consumption. The consumer's demand changes according to the daily load curve. This requires the grid to have a proper capacity to handle demand fluctuation, regardless of how it runs. Traditionally, electricity networks are dimensioned on peak power. Investigations showed that the energy transported through lines as a percentage of the total energy capacity needed for the peak demand is no more than 68% [26]. Energy storage systems can provide balancing service, reducing the stress on the network for peak loads and keeping the power transmitted from the network at lower level. It allows one to dimension the network for a lower value of the demand power.

In conditions when the network equipment is overloaded as a result of growing energy demand, energy storage systems can bring relief and may contribute to the increase in the network stability and reliability. From an economic point of view, it means the development of existing networks while postponing or omitting network reinforcement investments.

6.3.2.3   Ancillary Services

Ancillary service applications concern storage systems connected to the grid through coupling inverters and may include voltage control, Var support, and power quality improvement. Even though technically proven, this type of applications is not yet widely used but may be a promising opportunity for future microgrids.

The typical control strategy of a storage inverter assumes two objectives:

1. The generation or consumption of active power according to the energy management system (charging and discharging strategy)
2. Compensation of PQ phenomena caused by disturbing loads or fault events.

PQ control can include:

- compensation of disturbances introduced to the network by loads – voltage amplitude variations, harmonics, flicker, unbalance
- mitigation of voltage dips coming from the network.

The specific tasks performed by the unit depend on its operation mode.

In the current control mode, which is assumed for load compensation, the inverter must inject currents to the network in such a way that undesirable components of the load current e.g. harmonic, reactive and negative sequences are cancelled and the network current becomes a fundamental harmonic and a positive sequence. At the same time the inverter is expected to generate or consume the active power to meet the requirements of the management system. It is also possible to stabilize the voltage at the point of connection. To stabilize the voltage, the inverter must inject a reactive current of fundamental frequency and positive sequence, which gives an appropriate voltage drop on the reactances of the supply network.

6.3.2.4   Peak Shaving

In many countries a distinction is made in tariffs between peak and off-peak hours. As the energy delivered during peak hours is more expensive, end-users may be interested in lowering their consumption in those periods, which is possible using energy storage. The storage unit is charged during off-peak hours when energy is cheaper and is discharged during peak load periods supplying customer loads. For industrial users who are charged not only according to energy consumption but also according to the highest power demand, the reduction of peak power demand means decreasing demand charges. This application is profitable for customers with a high peak to low demand ratio, particularly where there is a large difference between peak and off-peak energy charge.

6.3.3   Management and Control

Apart from new architecture, the key issue in the microgrid concept is the management and control system. It is intended to coordinate and optimize the operation of various energy sources and loads and to achieve the integration of distributed resources in the electrical power network without redesigning the network itself [27].

The system may perform the following tasks [22, 28, 29]:

- optimize production of local sources and power exchanges with utility;
- ensure the proper operation of sources satisfying the operating limits;
- increase the system reliability and efficiency through using storage;
- optimize heat utilization;
- ensure the uninterruptible supply of sensitive loads.

The system should also be responsible for the smooth transfer of the microgrid from grid-connected to island operation and supervise its operation in island mode [28].

An optimal power sharing scheme is established by the minimization of the total costs of electricity production, while providing the required power to the load. Load profiles and RES generation are known from daily forecasts. Mathematically the problem is formulated as multiobjective optimization at given constraints: find the settings for the energy supply equipment such that the total cost of electricity production is a minimum. The objective function may include the following components:

- fuel consumption rate, operation and maintenance costs, start-up costs – for each controllable unit,
- costs of power exchange with the supplying network,
- environmental emission costs,
- penalty costs when violating boundary conditions.

Objective constrains may encompass:

- power balance,
- generation capacity limits, minimum up/down time limits, and the maximum number of starts and stops – for each source,

- load carrying capacity for lines and transformers,
- voltage limits in nodes of the grid.

Various forms of the objective function are reported in literature for different microgrid architectures. The environmental impact of generation emission is taken into account in [30–32]. Renewable sources are assumed to work at the maximum available power and they are considered as negative loads. In [31, 32] energy storage is taken into account; this is used for storing excess energy to support the operation of renewable sources. The state of charge of the storage is monitored and storage power is considered in power balance. In microgrids with heat demand, optimization involves both electricity and the production of heat [33–36]. If storage systems are included in the optimization, storage power is added as a new variable and the problem is formulated as minimization of overall energy costs over the scheduling time period. This is because the operation of storage in any time period affects its operation in the following periods. In such cases the charging and discharging pattern of the storage is controlled [34, 35, 37].

The microgrid concept also assumes demand-side control. This is an important feature that, in practice, has not so far been present in conventional networks. It needs smart metering and controllable load devices to be installed at customer sites. Controllable loads may be entered into the demand response control strategy executed and supervised through the energy management controller to reduce peak load and level the load profile [19]. Demand-side control may also be implemented by customers individually. As the customer is provided with information about tariffs and energy consumption, he or she can control the loads in order to minimize energy costs. In this way active demand is accomplished, by the interaction of consumers with utilities. Decreasing peak loads through demand-side management (DSM) contributes to more reliable and efficient power supply. Demand-side management is used to control both electrical and thermal consumption [4] and is beneficial for end-users and utility.

Control strategy may be centralized and decentralized [19, 22]. The centralized control scheme is hierarchical and consists of central management which optimizes the microgrid operation and local control of generators, loads or storages. The central controller provides set points to the local control. Its common structure includes modules with different functionalities. In [34, 37, 38] three modules are proposed: forecasting, economic optimization, and on-line control. In the first module load profiles and RES generation forecasts are produced for a day ahead. The second module determines set points for all controllable devices based on the forecasts. An on-line control module adjusts the operating points for sources to maintain the power balance and the required power quality during real-time operation. Local controllers follow the orders of the central controller and are designed to control the operating points of sources and loads. The measured data for local controllers are local voltages and currents.

The use of centralized control is beneficial for small-scale microgrids when the owners of microsources and loads have common goals [22].

The decentralized control (also called autonomous control – [19, 22, 30]) is used in microgrids that have different owners and operate in market conditions. This control would be appropriate if local microsources have other tasks besides supplying power to the grid, such as heat production, maintaining the voltage at a certain level or providing a backup supply for local critical loads [22]. In this case each unit participating in the market performs its action autonomously. A multi-agent system approach is proposed to solve the problem [18, 39].

So far, most approaches assume a hierarchical-type two-level management and control scheme: central energy management and local power and voltage control of individual units [40]. However multi-agent-based approaches are proposed for future smart grids [23,41].

A separate issue concerns microgrid control under islanding operation [40]. This includes maintaining suitable voltage and frequency levels for islanded loads. Energy sources have to be able to follow a load time course, so generation and load management become very important. To balance power generation and consumption various techniques can be used for both generation and load sides (load shedding).

Microgrid management and control is currently the subject of intensive research. It is a very challenging option in the search for energy savings. According to [35, 41], through optimized operation and management of microgrids that supply buildings, about 20% to 30% of building energy consumption can be saved. By considering the microgrid as an integrated energy system that provides customers with both electricity and heat, the optimal microgrid structure and technology of the sources may be identified. The objective is to meet customers' requirements with minimum costs. Research on designing microgrids using this approach has already been undertaken and some examples are presented in [17,33,35]. The development of the control strategy is the key issue in the practical transition towards microgrid-oriented system designs [4].

6.3.4   Power Quality and Reliability in Microgrids

Maintaining a high level of supply quality is one of the tasks of microgrid operation. High quality of power is required by customers and it contributes to the improvement of energy efficiency. Power quality control includes the stabilization of power produced by renewable sources and the mitigation of disturbances occurring at the grid, which deteriorate power quality. Traditionally, additional 'custom power' equipment has been applied for the mitigation of electromagnetic disturbances and improvement in power quality. Compensating devices such as DSTATCOM or APF systems play a similar role in distribution networks as FACTS systems do in transmission networks, therefore they are also called 'DGFACTS'.

Improving power quality by means of additional equipment is cost effective; therefore the decision to install a new device should be taken on the basis of detailed cost–benefit analysis. Customers' requirements are different; some of them may need the high level of power quality, while others do not. As power quality is controlled locally the level of power quality can be tailored more precisely to the requirements of end-users. According to [17] it might yield considerable benefits.

Networks with distributed generation provide the operator with new possibilities and solutions. Some of DER uses a power electronic interface to convert energy obtained by the source into AC power that is compatible with the grid. These converters can be used for accomplishing tasks similar to those performed by power electronic compensators. The additional functions of grid-coupling converters beyond their basic task, i.e. transferring active power/energy to the grid, are called ancillary services. The capability of DER to provide ancillary services is a unique option for microgrids to improve power quality and energy efficiency [21].

Figure 6.8 illustrates the possible operation of an inverter connected energy source providing ancillary services. In this case the system compensates for the reactive power and load unbalance. In 0.5 s the active power set value was decreased from 50 kW to 30 kW (load

(a)

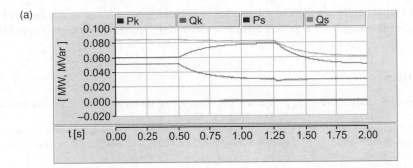

(b)

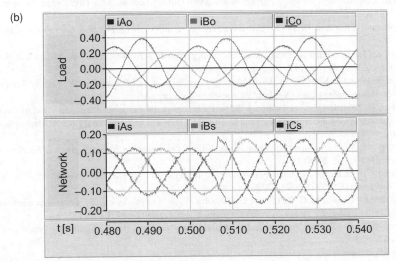

Figure 6.8 Operation of inverter connected energy source with ancillary services – results of simulation: (a) Pk, Qk are active and reactive powers of the source, Ps, Qs are active and reactive powers of the network; (b) iAo, iBo, iCo are load currents, iAs, iBs, iCs are network currents.

power constant), then in 1.25 s the load power was decreased by 30% proportionally in each phase (Figure 6.8(a)). The source reactive power changes as a result of load power changes (constant power factor) in order to keep the reactive power flowing to the grid constant. At the same time the system accomplishes the task of current compensation; load and network currents are presented in Figure 6.8(b).

Apart from various technical problems associated with DG integration, there are some non-technical issues that should be considered before the development of microgrids. They concern regulatory issues, pricing, decision priority, responsibility and incentives.

For example, taking advantage of the autonomous operation of the microgrid needs changes in the existing regulations, which require DES to trip in the case of any disturbance in the supplying network. Further, for DER with different owners the question is who will be responsible for controlling the operation of the microgrid. Or what are the incentives needed to get the DER owner interested in making his DER available for control or ancillary services?

The practical implementation of microgrids is a challenging task and needs further research and significant changes at the policy level in order to fully utilize the potential benefits.

References

[1]   Directive No 2001/77/WE of 27 September 2001. Directive on the Promotion of Electricity Produced from Renewable Energy Sources in the Internal Electricity Market.

[2]   M. H.-J. Bollen, Y. Yang and F. Hassan, Integration of distributed generation in the power system – A power quality approach, *Proc. 13th International Conference on Harmonics and Quality of Power*, Wollongong, Australia, 2008.

[3]   R. Smit, *Cogeneration, Power Quality Application Guide*, Section 8 – Distributed Generation, LPQI, www.lpqi.org

[4]   J. Driesen and F. Katiraei, Design for distributed energy resources, *IEEE Power & Energy Magazine*, May/June 2008.

[5]   H. Jiayi, J. Chuanwen and X. Rong, A review on distributed energy resources and MicroGrid, *Renewable & Sustainable Energy Reviews*, **12**, 2008, pp. 2472–2483.

[6]   Z. Styczynski and P. Komarnicki, *Distributed Energy Sources and Power Quality, Energy Storage Systems*, Leonardo Power Quality Initiative, http://www.leonardo-energy.org/

[7]   Directive No 2004/8/WE of 11 February 2004. Directive of the European Parliament and of the Council on the Promotion of Cogeneration Based on a Useful Heat Demand in the Internal Energy Market and amending Directive 94/42/EEC.

[8]   C. Marnay, H. Asano, S. Papathanassiou and G. Strbac, Policymaking for microgrids, *IEEE Power & Energy Magazine*, May/June, 2008.

[9]   A. Oudalov, T. Buehler and D. Chartouni, Utility scale applications of energy storage, *Proc. 2008 IEEE Energy 2030 Conference (ENERGY 2008)*, 2008.

[10] A. Oudalov, D. Chartouni, C. Ohler and G. Linhofer, Value analysis of energy storage applications in power systems, *Proc. 2006 IEEE Power Systems Conference and Exposition (PSCE'06)*, 2006.

[11] P. Tomczyk, *Energy Storage Systems* (in Polish), http://www.smartgrid.agh.edu.pl/index.php/materialy5

[12] http://www.piller.com

[13] http://www.amsuper.com

[14] S. C. Smith and P. K. Sen, Ultracapacitors and energy storage: Applications in electrical power system, *Proc. 40th North American Power Symposium (NAPS'08)*, 2008.

[15] H. Wei, W. Xin, G. Jiahuan, Z. Jianhua and Y. Jingyan, Discussion on application of super capacitor energy storage system in microgrid, *Proc. International Conference on Sustainable Power Generation and Supply (SUPERGEN'09)*, 2009.

[16] C. L. T. Borges and D. M. Falcão, Optimal distributed generation allocation for reliability, losses, and voltage improvement, *Electrical Power and Energy Systems*, **28**, 2006, pp. 413–420.

[17] C. Marnay, G. Venkataramanan, M. Stadler, A. Siddiqui, R. Firestone and B. Chandran, Optimal technology selection and operation of commercial-buildings microgrids, *IEEE Transaction on Power Systems*, **23**(3), 2008.

[18] N. Hatziargyriou, H. Asano, R. Iravani and C. Marnay, Microgrids. An overview of ongoing research, development and demonstration projects, *IEEE Power & Energy Magazine*, July/August, 2007.

[19] F. Katiraei, R. Iravani, N. Hatziargyriou and A. Dimenas, Microgrid management. Controls and operation aspects of microgrids, *IEEE Power & Energy Magazine*, May/June, 2008.

[20] B. Kroposki, T. Basso and R. DeBlasio, Microgrid standards and technologies, *IEEE Power and Energy Society General Meeting: Conversion and Delivery of Electrical Energy in the 21st Century*, 2008.

[21] B. Kroposki, R. Lasseter, T. Ise, S. Morozumi, S. Papathanassiou and N. Hatziargyriou, Making microgrids work, *IEEE Power & Energy Magazine*, May/June, 2008.

[22] R. Zamora and A. K. Srivastava, Controls for microgrids with storage: Review, challenges, and research needs, *Renewable and Sustainable Energy Reviews*, 14, 2010, pp. 2009–2018.

[23] N. W. A. Lidula and A. D. Rajapakse, Microgrids research: A review of experimental microgrids and test systems, *Renewable and Sustainable Energy Reviews*, **15**, 2011, pp. 186–202.

[24]　S. Faias, P. Santos, F. Matos, J. Sousa and R. Castro, Evaluation of energy storage devices for renewable energies integration: application to a Portuguese wind farm, *Proc. 5th International Conference on European Electricity Market (EEM 2008)*, Lisbon, Portugal, 2008.

[25]　K. Qian, C. Zhou, Z. Li and Y. Yuan, Benefits of energy storage in power systems with high level of intermittent generation, *Proc. 20th International Conference on Electricity Distribution – Part 1 (CIRED 2009)*, Prague, Czech Republic, 2009.

[26]　E. Veldman, M. Gibescu, J. G. Slootweg and W. L. Kling, Technical benefits of distributed storage and load management in distribution grids, *Proc. 2009 IEEE Bucharest Power Tech Conference*, Bucharest, Romania, 2009.

[27]　R. H. Lasseter and P. Piagi, Microgrid: A conceptual solution, *PESC'04*, Aachen, Germany, 20–25 June, 2004.

[28]　B. Awad, J. Wu and N. Jenkins, Control of distributed generation, *Elektrotechnik & Informationstechnik*, **125**(12), 2008.

[29]　A. Chuang and M. McGranaghan, Functions of a local controller to coordinate distributed resources in a smart grid, *IEEE Power and Energy Society General Meeting – Converse and Delivery of Electrical Energy in the 21st Century*, 2008.

[30]　E. Alvarez, A. C. Lopez, J. Gomez-Aleixandre and N. Abajo, On-line minimization of running costs, greenhouse gas emission and the impact of distributed generation using microgrids on the electrical system, *IEEE PES/IAS conference on Sustainable Alternative Energy*, 2009.

[31]　F. A. Mohamed and H. N. Koivo, On-line management of microgrid with battery storage using multiobjective optimization, *International Conference on Power Engineering, Energy and Electrical Drive POWERENG 2007*, 12–14 April, 2007.

[32]　H. Vahedi, R. Noroozian and S. H. Hosseini, Optimal management of microgrid using differential evolution approach, *7th International Conference on the European Energy Market (EEM)*, 23–25 June, 2010.

[33]　H. Asano and S. Bando, Economic Evaluation of Microgrids, *IEEE Power & Energy Society General Meeting*, 2008.

[34]　W. Gu, Z. Wu and X. Yuan, Microgrid economic optimal operation of the combined heat and power system with renewable energy, *IEEE Power Energy Soc. Gen. Meet.*, 25–29 July, 2010.

[35]　X. Guan, Z. Xu and Q.-S. Jia, Energy-efficient buildings facilitated by microgrid, *IEEE Trans. Smart Grid*, **2**(1), 2011.

[36]　C. A. Hernandez-Aramburo and T. C. Green, Fuel consumption minimization of a microgrid, *39th Industry Application Conference Annual Meeting*, 3–7 October, 2004.

[37]　S. Chakraborty and M. G. Simoes, PV-microgrid operational cost minimization by neural forecasting and heuristic optimization, *IEEE Industry Applications Society Annual Meeting*, 2008.

[38]　A. Borgetti *et al.* An energy resource scheduler implemented in the automatic management system of a microgrid test facility, *International Conference on Clean Electrical Power, ICCEP'07*, 21–23 May, 2007.

[39]　A. Dimeas and N. Hatziargyriou, Operation of a multi-agent system for microgrid control, *IEEE Transaction on Power Systems*, **20**(3), 2005.

[40]　J. A. Pecas Lopes, C. L. Moreira and A. G. Madureira, Defining control strategies for microgrids islanded operation, *IEEE Trans Power Syst.*, **21**(2), 2006, pp. 916–24.

[41]　C. M. Colson and M. H. Nehrir, A review of challenges to real-time power management of microgrids, *IEEE Power & Energy Society General Meeting*, 2009.

Further Readings

T. Ackermann, G. Andersson and L. Söder, Distributed generation: a definition, *Electric Power System Research*, **57**, 2001, pp. 159–204.

I. J. Balaguer, U. Supatti, Q. Lei, N.-S. Choi and F. Z. Peng, *Intelligent Control for Intentional Islanding Operation of Microgrids*, ICSET, 2008.

F. Barklund, N. Pogaku, M. Prodanovic, C. Hernandez Aramburo and T. C. Green, Energy management system with stability constraints for stand-alone autonomous microgrid, *Systems of Systems Engineering*, 2007.

S. Chowdhury, S. P. Chowdhury and P. Crossley, *Microgrids and Active Distribution Networks*, IET renewable energy series 6, London IET, 2009.

N. Jenkins, R. Allan, P. Crossley, D. Kirschen and G. Strbac, *Embedded Generation*, IEE Power and Energy series 13, IET, London, 2000.

R. Mieński, R. Pawełek, I. Wasiak, P. Gburczyk, C. Foote, G. Burt and P. Espie, Power quality improvement in LV network using distributed generation, *IEEE 11th International Conference on Harmonics and Quality of Power*, Lake Placid, USA, 12–15 September, 2004.

R. Mieński, R. Pawełek, I. Wasiak, P. Gburczyk, C. Foote, G. Burt and P. Espie, Voltage dip compensation in LV networks using distributed generation, *IEEE 11th International Conference on Harmonics and Quality of Power*, Lake Placid, USA, 12–15 September, 2004.

R. Pawelek, I. Wasiak, P. Gburczyk and R. Mienski, Study on operation of energy storage in electrical power microgrid – modelling and simulation, *Proc. 14th International Conference on Harmonics and Quality of Power, September 2010*, Bergamo, Italy.

S. C. Smith, P. K. Sen and B. Kroposki, Advancement of energy storage devices and applications in electrical power system, *Proc. 2008 IEEE Power and Energy Society General Meeting – Conversion and Delivery of Electrical Energy in the 21th Century*, 2008.

G. Venkataramanan and C. Marnay, A larger role for microgrids, *Energy Magazine*, May/June, 2008.

7

Electric Motors

Joris Lemmens and Wim Deprez

The energy efficiency of electric motor systems has strongly gained in importance and interest over the last decade. The underlying reason for this evolution can be found in the context of the challenges the energy sector is facing. A fundamental issue is the increasing consumption of energy coupled with the forecasted growth in the global economy. Unsustainable pressure on natural resources and the environment is inevitable if energy demand is not (at least partly) decoupled from economic growth. A global revolution concerning energy supply and use is needed. Amongst several technologies and scenarios, a far greater efficiency of 'Energy using Products' (EuPs) is a core requirement to reduce global energy consumption. It should be understood that increasing electric motor efficiency must be considered as part of this top category of measures.

Electric motor systems in general are widespread in industry, the service sector and buildings. In terms of energy consumption, electric motor systems account for roughly 40% of the total electricity demand of today. Approximately 65% of the electricity that is used in industry is consumed by electric motors [1], the majority (more than 96%) of the industrial motors being AC motors of which approximately 90% are induction machines (IM). For the enlarged European Union (EU-27), in the year 2006 the total electricity consumption was 3268 TWh, in industry 1143 TWh and for industrial motor systems in particular 742 TWh [2]. By 2020, the industrial electricity consumption in the EU-27 is estimated to approach 1432 TWh and the consumption of industrial electric motors 859 TWh if no action is taken. However, if the EU attains all the feasible energy savings in those motor systems (estimated at 31%), the result would be an annual saving of 270 TWh, equivalent to the total electricity consumption of Spain in 2000 [1]. Note that the total electricity consumption and potential savings of motor systems are even higher as motor systems in the service sector and in buildings should also be taken into account.

The European (EU-15) low voltage AC motor market in 2006 was estimated at 9 million units. In common applications such as heating and cooling facilities, elevators, escalators and transportation, induction motors are mostly used because they are robust, maintenance free,

---

Electrical Energy Efficiency: Technologies and Applications, First Edition. Andreas Sumper and Angelo Baggini.
© 2012 John Wiley & Sons, Ltd. Published 2012 by John Wiley & Sons, Ltd.

versatile and efficient and can be produced cost-effectively. This is why IMs are often called 'the workhorse of industry' and, since the introduction of variable speed drives (frequency converters or VSDs), even 'the race horse of industry'. In the last decades, most attention and effort concerning energy efficiency (technology and standardization) focused on IMs. However, the changing boundary conditions, i.e. environmental concern, increasing energy costs, material costs, mandatory minimum efficiency requirements, etc., mean that other motor types and technologies become an increasingly interesting alternative. It is expected that for low power and/or specific applications other motor types, for example permanent magnet synchronous motors (PMSM), are necessary to reach the highest requirements [3].

In this chapter on electric motor efficiency, it is not the intention to present an in-depth technical overview of state of the art motor technology but rather to identify the main points of attention concerning motor system efficiency. It is important to understand that electric motors transform electric energy rather than consume it. Most of the electric input power is converted into mechanical (shaft) power with inevitably, a certain amount of motor (and converter) losses that depend on the applied motor technology, materials, design, environmental conditions, etc. Investments in more energy-efficient drives to reduce these losses are led by economic/ecologic considerations and legislation. Electric motors belong to the so-called 'Energy using Products' or EuPs. The purchase and maintenance cost is only a fraction of the lifetime energy cost, especially for continuous duty applications. As an example, consider a 7.5 kW IM in full load, continuous duty operation with an efficiency of 89%, an expected lifetime of 25000 hours, a purchase cost of 70 €/kW and an energy cost of 10 c€/kWh. The purchase cost is only €525 and the energy cost €21 067, 2.5% and 97.5% respectively of the total cost. Keeping in mind the ever-increasing energy prices and the large amount of energy used in motor systems, investments and technological advances in more efficient electric drives are necessary from an economic and ecologic point of view.

This brief introduction should indicate the different points for attention in motor driven system efficiency and, consequently, the items addressed in the next sections of the chapter. There are two distinct aspects concerning energy efficiency of motor systems. On the one hand, there is the technological side concerning motor, converter and mechanical losses. On the other, there is the aspect of policy, legislation and standardization. The latter is required to enable and secure the market of efficient electric drives and also to ensure the implementation of high efficiency motors (HEMs) in products of, for instance, OEMs (Original Equipment Manufacturers).

The first section of this chapter focuses on the losses in electric motors. First the different loss components are described. Next, the influence of practical operating conditions on efficiency is addressed. The second section is devoted to standards on efficiency testing and classification. In the last section, the technology used in high efficiency motors is briefly discussed by looking at materials, design and manufacturing evolutions.

7.1 Losses in Electric Motors

In order to understand the meaning of energy efficiency and the improvement potential in electric motors and drives, a basic understanding of the loss components is required. There are numerous types of electric motors (Figure 7.1), but in industrial applications, the squirrel cage induction motor is the most common, as already stated in the introduction. Consequently,

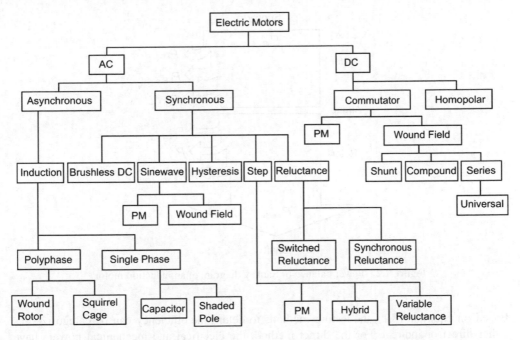

Figure 7.1 Classification of electric motors according to operating principle

the discussion about motor efficiency and losses over the last decades was mainly focused on IMs. Most of the IM loss components generally return in the other motor types, of course in a different order of magnitude and mutual distribution. Therefore, the treatment of losses in this chapter is confined to the IM.

7.1.1 Power Balance and Energy Efficiency

The losses in an IM consist of joule losses in the stator windings P_s and the rotor cage P_r, iron losses P_{FE} due to hysteresis effects and eddy currents in the laminated steel core, mechanical losses $P_{fr,w}$ due to friction in the bearings and windage losses of the cooling fan and additional or stray load losses P_{SLL}, which represent a number of losses not categorized under the previous items.

In Figure 7.2, the power balance or Sankey diagram for an IM in motor mode is shown, the thickness of the individual loss arrows is proportional to the specific power value concerned.

Theoretically, the determination of the efficiency for motor or generator mode is quite straightforward. It is the ratio between output mechanical power P_{mech} and input electric power P_{el}, usually expressed as a percentage (7.1). From the Sankey diagram, clearly the difference between the input and output power flow consists of the losses P_{loss} so the efficiency can also be expressed as a function of electric power and losses:

$$\eta_M = \frac{P_{mech}}{P_{el}} = 1 - \frac{P_{loss}}{P_{el}} \tag{7.1}$$

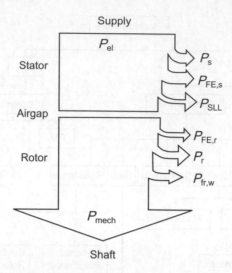

Figure 7.2   Power balance or Sankey diagram of an induction motor

Based on equation (7.1), the testing methods to determine efficiency can be categorized as either direct or indirect. For the direct method, the electrical and mechanical powers have to be measured. This necessitates an accurate power analyser, speed and torque measurement device. For accurate torque measurements, the range of the torque transducer should be adapted to the rated torque of the motor. Additionally, efficiency values obtained by this method depend on ambient and motor temperature, which is not desirable for a transparent and reproducible efficiency rating. Motor efficiency values obtained for different (ambient) temperature conditions cannot be compared using this approach. Moreover, it can be proven that the accuracy of this direct method is low, even when very precise equipment is used.

The other approach, the indirect method, is based on the separate determination of the individual loss components, the so-called 'segregation of losses'. It is applicable to several DOL (Direct On Line – thus for motors that are directly connected to the grid) motor types, but is mostly used for induction machines. With a load and zero-load test, it is possible to identify precisely the conventional losses, including stator and rotor joule loss, iron loss, and friction and windage loss. However, as early as 1912, it was shown that there exists a reasonable difference between the total conventional losses determined from the loss segregation method and the total losses found from a direct load test. This difference is covered by a loss component that is not included in the conventional losses. This additional power loss component, termed stray load loss, is caused by a number of parasitic effects due to the non-ideal nature of a practical machine (see further). It is impossible to measure this loss component directly, but there are several possibilities to account for them. The chosen approach for the determination of stray load losses is the most important difference between standards for efficiency measurement of IMs. The advantages of indirect over direct methods are mainly that the overall accuracy is higher and that they allow the correction of the different loss components to a specified ambient and reference motor temperature, and reduce measurement errors.

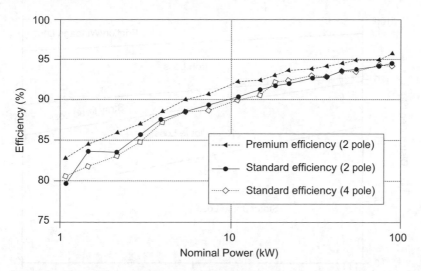

Figure 7.3 Rated load efficiency of cast iron frame induction motors for a two- and four-pole standard efficiency motor and a two-pole premium efficiency motor based on manufacturer data

7.1.2 Loss Components Classification

According to scaling laws, efficiency rises with increasing volume of active parts (core and conductors), thus frame size or power rating. This can be illustrated by plotting the rated efficiency versus power rating for a series of motors (Figure 7.3).

Figure 7.4 indicates the mutual distribution of loss components according to the rated motor power. The relative proportions of iron and rotor losses are almost independent of the frame size. On average, they each amount to about 20% of the overall losses. The introduction of high efficient IM designs with, for example, a copper rotor changes this general picture very much. This is due to the interrelations between these loss components as elucidated more in detail later on. For motors of lower power ratings, the ohmic losses in the stator windings are dominant. They amount to approximately 50% of the overall losses, whereas for large machines they are reduced to about 25%. The additional or stray load losses rise from a very low relative contribution to well above 10% for the MW range motors. A more or less similar contribution can be recorded for the part of the friction and windage losses.

From the nomenclature of the losses it is obvious that they all have their specific location within the IM. Such a classification results in only three possible areas: the windings (stator and rotor), the magnetic circuit (stator and rotor iron) and the bearings and air-gap (mechanical losses). The stray load losses are also located in one or more parts of the stator and/or rotor.

A classification of the electromagnetic losses, consisting only of winding and core losses, subdivides these loss components into fundamental, space harmonic and time harmonic losses. In general, time harmonic components have very little importance in constant speed (mains) applications when supposing good power quality. However, they do have to be considered when the IM is supplied from power electronic converters. The spatial harmonics of the stator and rotor MMF are related to practical (constructional) limitations, such as slotting, magnetic saturation and eccentricity. All harmonic components give rise to additional losses in stator and rotor core and windings.

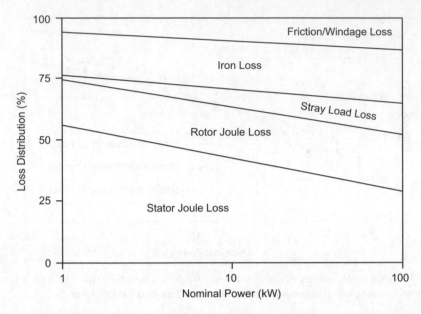

Figure 7.4   Distribution of loss components according to the rated power of the induction motor

Another possible classification is used as a basis for the loss segregation in several efficiency standards. This approach classifies the losses into two parts: constant and load dependent. Loss components assumed to be constant, are friction, windage and iron losses. The calculation of these losses based on motor geometry and material data is not trivial, especially the calculation of the iron losses. Steinmetz derived an expression (7.2) to approximate core losses:

$$P_{core} = P_{hyst} + P_{eddy} = k_{hyst} f B^n + k_{eddy} f^2 B^2 \tag{7.2}$$

where P_{core} is the core loss, P_{hyst} is the hysteresis loss, P_{eddy} is the eddy-current loss, f is the frequency of the sinusoidal magnetic field, B is the flux density and k_{hyst}, k_{eddy} and n are coefficients that depend on the lamination material, thickness, conductivity and other (geometry) factors. It can be used in analytic or FEM (finite-element modelling) approaches, but even in the case of the latter, the main problems with core loss determination still exist: the flux density is constantly changing in magnitude (up to local saturation) and/or direction and has harmonic components. Moreover, the degradation of magnetic properties due to mechanical manufacturing is not constant and is difficult to take into account in a model.

To determine these constant losses from tests, the so-called practical no-load test is mostly used. More about this test and the decomposition into the different components is discussed in Section 7.2.2, dealing with the different standards. It must be noted that the effect of slotting is also present at no-load, as in this mode space harmonics are present. Therefore, the iron losses in fact include additional losses, termed no-load stray losses.

The load dependent losses are the copper losses in stator windings (P_s), in rotor conductors (P_r) and stray load losses (P_{SLL}). The stator copper losses are calculated from the current corresponding to the load point considered and from the winding resistance. The winding resistance is measured, and corrected for a specified reference temperature according to the standard used.

The joule losses in the rotor conductors are also proportional to the load current of the machine. In a practical steady-state no-load situation, the actual speed is very close to the synchronous one. The rotor slip, being very small, gives rise to a torque to cover mechanical losses. The corresponding rotor current and frequency and their resulting rotor copper and core losses are very small. In fact, they are covered by the constant losses. When the machine is loaded, the slip increases in order to balance the load torque. The resulting rotor currents now start to produce increasing joule losses in the rotor. As the rotor frequency remains fairly low in the entire operating range of an IM, all the load dependent rotor losses can be considered as being joule losses. The rotor losses can be determined as the slip fraction of the air-gap power. For these losses also different correction methods according to standards, apply.

The stray load losses at least that part of harmonic losses not covered by the (considered as constant) iron losses, are the last, but most notorious load-dependent losses. They can be linked to three practical machine restrictions:

• Magnetic property limitations of electrical steel in the motor core, which leads to local saturation;
• The fact that a practical geometry is used, with slots and a discrete instead of perfectly sinusoidal winding giving rise to space harmonics and leakage flux;
• Industrial imperfections due to manufacturing, for example, the cross-bar currents that can occur due to the imperfect insulation of rotor cage bars.

They consist of supplementary losses occurring at load in the core and other metal pieces of the machine and eddy-current losses in both stator and rotor conductors caused by current depending leakage fields. They result in additional heating of the motor and a reduction in motor torque. These losses have been the subject of many research projects and publications investigating their origin, consequences, determination and simulation. A comprehensive overview of the origin, components, measurement methods and effects of additional losses can be found in several publications [4, 5].

7.1.3   Influence Factors

IMs, or other motor types, are only one of the components of an electric motor application. The overall system efficiency depends on several factors such as motor efficiency, temperature and sizing, motor control, mechanical transmission, maintenance practices, mechanical efficiency, losses in the supply system, supply quality, etc. Therefore, a short overview of the important factors influencing the system efficiency is given. It is almost impossible to give an exhaustive overview of points of interest as each application has particular requirements and boundary conditions. However, the most common and typical issues are addressed and some examples are given.

On the nameplate of a motor or in catalogues, mostly only the rated value of the efficiency, determined according to a certain standard, is given. This value and even the partial load values are determined based on fixed boundary (e.g. temperature, voltage) conditions. The rated values correspond to a steady-state situation, which means that under normal surrounding conditions, the motor can operate continuously (practically non-stop) at the specified load. In fact, the motor then operates at the thermal equilibrium for which it is designed, and the temperature of the different motor parts is under control. In practice these conditions are rarely met and as a consequence the resulting motor temperature will deviate from the standard test conditions. And in practice motors are often not operated at rated power, but at a lower power level or exceptionally at a higher. This over-rating (or under-rating) can have several reasons (starting torque, for example). Consequently, the practical efficiency will mostly deviate from the one listed on the nameplate. The most important 'deviating' practical motor operating conditions are [6]:

- Load conditions that differ from the rated output;
- Different operating temperatures;
- Unfavourable conditions due to bad maintenance;
- Deviating electrical supply conditions (power quality issues).

7.1.3.1 Temperature Effects

The different loss components are temperature dependent. The winding resistances rise by approximately 4% per 10K of temperature rise, and thus copper losses will increase proportionally. It can be shown that, depending on the motor size, the effect of this dependency on the rated-load efficiency can amount from 0.2 percentage points up to 2 percentage points. Windage losses vary inversely proportionally with temperature. With a relative humidity of 80% and a normal atmospheric pressure, the windage losses in the considered temperature range decrease approximately 4–5% for each 10K of temperature rise. Iron losses decrease with rising temperature, depending on sheet quality and magnetic flux density, they are by 4–8% lower at 100°C than at ambient temperature [6]. From Figure 7.5 it is clear that the temperature conditions affect the practical efficiency value.

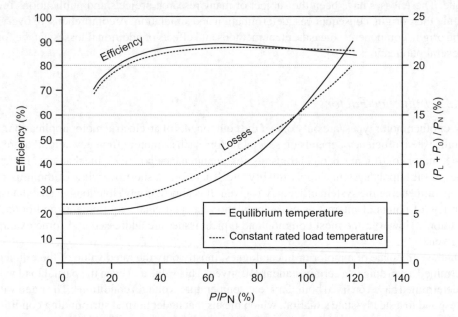

Figure 7.5 Influence of operating temperature on motor efficiency and total specific loss ($(P_L + P_0)/P_N$), P_L is the load dependent part of losses, P_0 are the constant losses, P_N is the rated power [7]

7.1.3.2 Partial Loading

The dotted efficiency characteristic of Figure 7.5 represents a typical efficiency characteristic, i.e. the load dependence of the motor efficiency as recorded according to an efficiency standard. From this figure it is also clear that a practical, continuous duty, partial-load efficiency differs from that determined by the standards, mainly due to temperature effects. There is an approach that can be used to estimate these effects. It is known that the load dependence of the efficiency (at constant reference temperature) can be mathematically defined [8]. This approach splits the losses at rated power (P_N) into a load independent P_0 and a load dependent component P_L. The efficiency can therefore be approximated by (7.3):

$$\eta_{(p)} = \frac{1}{1 + \dfrac{v_0}{p} + v_L \cdot p} \tag{7.3}$$

with $p = \frac{P}{P_N}$, $v_L = \frac{P_L}{P_N}$ and $v_0 = \frac{P_0}{P_N}$. From this, it can be found that the maximum efficiency (7.4) is located at the load point $p^* = \sqrt{v_0/v_L}$; this is the point for which the load dependent losses become equal to the constant losses:

$$\eta^* = \frac{1}{1 + 2\dfrac{\sqrt{P_0 P_L}}{P_N}}, \tag{7.4}$$

7.1.3.3 Maintenance

With these mechanisms in mind, it is possible to deduct some aspects and influences that bad maintenance practices may have on (induction) motors. For instance, when the system is operated in an environment involving a lot of dirt, process spills or other waste, it is important to clean the motor and especially its cooling fins and the cooling fan inlet. If this is not done regularly or properly, this may prove to be detrimental to cooling performance with increased overall losses and a decreased efficiency as a consequence. It is already noted that even if decreased efficiency would not be an issue, the higher temperature at thermal equilibrium of such cases can lead to early failure of the machine. Another example concerns the maintenance of bearings. If bearings are not greased/replaced in time, they cause higher friction losses.

7.1.3.4 Deviating Supply Conditions

Electrical supply conditions can have a significant influence on motor losses and efficiency. Many individual aspects concerning this issue are described in publications [9]. There are also separate standards in which even typical values for the distinct influences are given. For instance, IEC 60038 and EN 60150 are concerned with supply quality, EN 61000-2-4 for EMC noise immunity, IEC 60034-28 for voltage unbalance and according derating and EN 60204-1 for requirements on the electrical equipment of machines. In particular, because of the sensitivity of squirrel-cage IMs to voltage unbalance and/or harmonics, the latter describes the effects and include data on how they influence thermal conditions. These altered thermal conditions mostly lead to increased winding temperature and therefore, also the extent to which the continuous output power should be reduced in order to protect the winding insulation (i.e. derating), is stipulated.

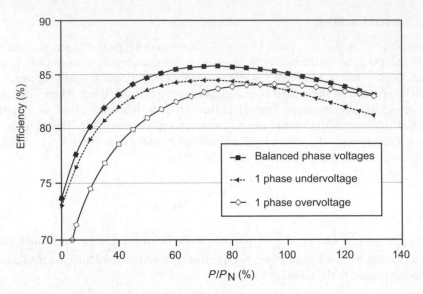

Figure 7.6 Efficiency characteristic according to IEEE112-B for a standard general purpose 7.5 kW induction motor. The unbalanced cases represent a voltage unbalance factor (VUF) of 4% by a one phase over or under voltage

As an example, Figure 7.6 gives the efficiency characteristics of an IM for the standard test conditions and for unbalanced voltage conditions. In a three-phase system, a voltage unbalance is the phenomenon in which the rms values of the (fundamental) voltages or the phase angles between consecutive phases are not equal. When an IM is fed by unbalanced voltages, an additional breaking torque is created, resulting in extra losses and possibly in motor failure [6]. Next to the lower motor efficiency when subjected to unbalanced supply conditions, the figure also shows a shifted efficiency curve with a maximum at higher or lower power compared with the balanced supply according to the present unbalance.

7.1.3.5 Converter Operation, Power Quality and Supply System

As has already been indicated above, harmonics cause additional motor losses. Several publications that are based on simulations and experiments are devoted to this subject. They indicate increased losses due to the harmonics associated with the use of frequency converters, but also increased noise and losses in variable speed induction motor drives with (IGBT) inverters, voltage reflections and associated grounding issues and even bearing damage [10–12].

The losses in the supply system components such as cables and transformers are also known as 'distribution losses'. These losses cannot be neglected as overloaded components may lead to increased losses. Moreover, these system components should be designed allowing for the deviating power factor in partial load conditions.

7.1.3.6 Mechanical Transmission

Some applications require a mechanical transmission. There are several possible solutions for such a transmission, i.e. gearboxes, V-belts, toothed-belts, cogwheels (helical or bevel gears), worm wheels, etc. Generally, belts have a high efficiency, i.e. 90% for V-belts to 97% for toothed V-belts. With cogged wheels an efficiency of 98% may be possible. The efficiency can also be a function of the loading of the gear. Worm wheels have very low and highly load-dependent efficiency.

7.2 Motor Efficiency Standards

7.2.1 Efficiency Classification Standards

The efficiency of motor-driven systems is of major importance, especially in the case of induction motors constituting the bulk energy use in industry. Even a modest increase in motor efficiency will yield considerable benefits in environmental, economic and strategic terms. Governments, with a responsibility to inform and to regulate, have been increasingly proactive in matters regarding device efficiency. There are numerous examples of national and international agreements, incentives and initiatives worldwide. For Europe, this is reflected in different ongoing programmes: 'The European Motor Challenge Program', the former classification scheme of industrial AC motors with the EFF1, EFF2, EFF3 labels of CEMEP (European Committee of Manufacturers of Electrical Machines and Power Electronics) voluntary agreement, the '4E Electric Motor Systems Annex' initiative and finally the new IEC 60034-30 IE-classification standard. This standard is the result of several initiatives and pleas to harmonize the different motor classifications and in fact efficiency standards worldwide. Figure 7.7 and Table 7.1 give a representation of these IE efficiency classes compared with the American and former European classifications. This IE-classification merges the main aspects

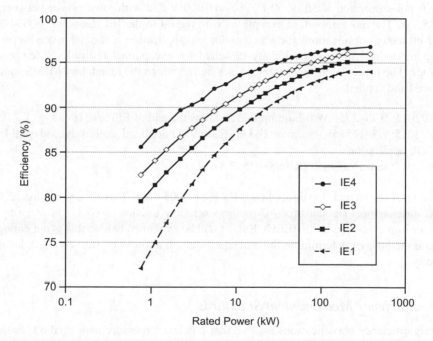

Figure 7.7 Representation of the new IE efficiency classes (four poles) defined in IEC Std 60034-30

Table 7.1 Harmonization of European and American classification by IEC 60034-30

Efficiency	IEC 60034-30	Europe (CEMEP)	United States
Super premium	IE4		
Premium	IE3		NEMA Premium
High	IE2	EFF1	Epact
Standard	IE1	EFF2	

of the American NEMA Premium/Epact and European CEMEP classes and therefore it is now valid for 50 as well as 60 Hz machines of 2, 4 and 6 poles. It is a worldwide, voluntary standard for AC motors between 0.75 and 375 kW. There are four classes (IE1 to IE4), leaving room for future classes. The IE4 class will also be valid for rated and partial load conditions of inverter fed motors such as PMSMs. In particular, this last feature is a progressive measure that may enhance the correct promotion of energy efficient drives.

This new standard IEC 60034-30 deals with the variety of different efficiency classes world-wide, whether voluntary or mandatory, which is a positive step. However, past experiences with these different approaches have taught us one thing: mandatory MEPS positively in-fluence the penetration of the superseding efficiency level (settled by a voluntary measure). Therefore, the European Commission translated this new IEC classification into a Directive (Dir. 2009/125/EG) in 2009.

The application range of this Directive is more limited compared to the IEC classification, but it introduces the mandatory aspect of minimum efficiency performance standards (MEPS) in Europe. It applies for three-phase, single-speed, continuous-duty squirrel cage motors of 2, 4 or 6 poles operated at 50 or 60 Hz, up to 1000 V and with rated power between 0.75 and 375 kW. The most important exceptions with regard to the IEC Standard 60034-30 are submersed motors, applications for which motor and application cannot be treated separately, braking motors and motors specifically designed for exceptional (altitude or temperature) conditions. The Directive foresees three phases during which the mandatory MEPS should be introduced and applied:

- 16/06/2011: 0.75–375 kW minimum IE2 and thus the end of EFF2/IE1;
- 01/01/2015: 7.5–375 kW minimum IE3 or IE2 in VSD-application, thus the end of EFF1/IE2 for DOL applications;
- 01/01/2017: same for small motors 0.75–7.5 kW.

The efficiency test standard to be applied is IEC Std. 60034-2-1. This revised edition (2007) of the IEC measurement standard for AC motors excluding traction motors is now harmonized with other important standards (IEEE). But, as will be explained below; this is not enough yet as several methods of determining the efficiency are still allowed by this standard, leading to ambiguity.

7.2.2   Efficiency Measurement Standards

Evidently efficiency classifications assume that efficiency measurement methods are estab-lished well beyond reproach and agreed upon in national and international standards. However, as with the efficiency classification standards, different efficiency measurement standards are used worldwide. A closer examination of standards reveals that there are major discrepancies between methods proposed by them [13–16], causing differences in the resulting efficiency values. This has serious consequences both in terms of issuing certificates and credibility of the declared efficiency values for decision making when purchasing motors. Motor system components are 'world widely' traded commodity goods. As they are subject to different local or national testing standards and performance and labelling requirements, large variations in the market penetration of high efficiency motors around the globe can be found.

Based on the criticism in several publications, questions from parties involved and other initiatives, the IEC started a process to update the controversial version of their motor efficiency standard. After a long process, the new version of the IEC Standard 60034-2-1 was published in September 2007. The method most used for efficiency determination described by this new standard is now aligned with the other important standard in this field, IEEE Standard 112-B. However, an additional method (EH-Y) for determining stray load losses is given. Given the numerous 'sub-methods' allowed by the respective standardization organizations, it cannot be stated that the standards for efficiency determination of (induction) motors are clear, unambiguous or harmonized. It should be noted however, that this standard is currently under revision again and, most likely, this contested method (EH-Y), will be removed.

Additionally, as described previously, the prescribed methods do not account for non-ideal operating conditions of a real application as opposed to the ideal test conditions described in the standard. For instance, standards are conspicuously silent on matters pertaining to unbalanced supply or poor power quality. Also, the practice of certifying efficiency on the basis of a single rated load efficiency value can be questioned.

7.2.2.1 General Procedure and Test Setup

In Section 7.1.1, two expressions (equation (7.1)) for the efficiency of IMs were introduced (direct and indirect) and the different loss components were indicated and discussed. The difference between the direct and indirect efficiency determination is introduced. The direct method is indicated as less accurate as it directly applies the measured mechanical and electric power. The most important standards are all based on the indirect method, based on the segregation or summation of losses. As already indicated, five loss components have to be determined using an indirect method: stator joule losses (P_s), rotor joule losses (P_r), iron losses (P_{FE}), friction and windage losses ($P_{fr,w}$) – these four components are mostly indicated as conventional losses P_{conv} – and the additional or stray load losses (P_{SLL}). These 'additional' losses can be estimated as the difference between the total measured losses (the difference between electric input power and mechanical output power) and the conventional losses.

The main difference between standards consists of the method to determine the stray load losses. The previously mentioned method to account for P_{SLL} is often termed the indirect 'Input–Output' method. Several publications indicate this method as most appropriate for an accurate determination of the stray load losses and IM efficiency [17–19]. Note that this method requires an accurate measurement of torque and speed, but it does not directly apply the measured mechanical power to obtain efficiency.

A second indirect method prescribed by some standards is based on a so-called fixed or variable allowance to estimate the stray load losses. In such methods, the mechanical power is not measured (no torque measurement needed) and the stray load losses are arbitrarily estimated to be equal to a certain percentage – the allowance – of the full-load input power (Table 7.4).

A third method recently included in the new version of the IEC60034-2-1 standard is the Eh-star (Eh-Y) method. In this method the stray load losses are determined from a separate test during which the IM is operated in no-load on a special (asymmetric) single-phase connection with an auxiliary resistance (R_{eh}). This method is very controversial, and most probably will be removed in a next revision of this standard.

It should be clear that the stray load loss estimation is the major difference between standards. Clearly, as there are several methods to determine the stray load losses, the outcome of an efficiency measurement depends mainly on the standard used. Nevertheless, the general measurement procedure and thus the number of required tests for each of the different standards are fairly similar. The general procedure and the test bench required for the determination of the five loss components and the efficiency are described now.

Figure 7.8 shows the schematic of a test setup that can be used for testing motors in VSD or DOL application. It is a facility with the capability of mechanically loading the motor and drive at the different operating points (torque and also speed in VSD application) as determined by the test protocol. To establish this loading, a controlled AC or DC drive may be used. The setup is also equipped with a torque and speed measurement device. A digital power analyser is used for the electrical measurements, i.e. voltage, current, frequency and power, which are

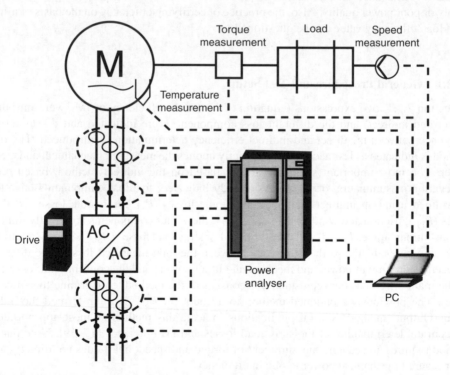

Figure 7.8 Schematic of the test setup that can be used for efficiency measurement of VSDs as well as for DOL motors; in the latter case the drive is by-passed

measured before and after the power electronic converter. This allows one to determine the motor and inverter efficiency separately. The measurements are controlled in such a way that they are recorded simultaneously. The accuracy of the electrical quantities at the fundamental frequency and the torque have a certified accuracy of ±0.2%. The range of the torque transducer should be adapted to the motor power and the accuracy of the speed measurement is ±1 rpm. The test setup should also be equipped with a temperature measurement device to keep track of the motor temperature. This temperature sensor should be located as near to the stator winding as possible, preferably in a slot or on the end-windings. As a more convenient alternative, the temperature sensor can also be mounted on the motor frame.

However, for VSD testing, given their nature, harmonics at the input and output of the converter are involved and should be considered when selecting and programming the measurement equipment. Therefore, the frequency range of the digital power analyser should be several kHz, preferably 200–300 kHz.

In general, an efficiency measurement procedure consists of three types of tests: (ambient) temperature and winding resistance determination, a load test (with or without torque measurement) in which a number of load torques are applied to the motor, and a no-load test in which the stator voltage is decreased gradually. Except for the method where the stray load loss determination is based on the Eh-Y procedure, all major standards prescribe a similar procedure for these three types of tests. For the Eh-Y test, an additional test is required.

From the no-load test, the iron and friction and windage losses – the no-load losses – can be determined separately. The stator joule losses are determined from the stator current measured in each point of the load test and the stator resistance value. Depending on the standard used, the losses are corrected to a reference ambient temperature based on winding resistance corrections. Next, the rotor joule losses can be calculated as the slip fraction of the air-gap power, which can be found by subtracting the iron and stator joule losses from the electrical input power.

Note that, for some standards, there is also a (temperature dependent) slip correction influencing the rotor joule loss to account for temperature deviations during the test. Consequently, if the stray load losses are determined according to 'input–output' method, they are also influenced by this correction. The different standards and methods each determine a specific procedure for the different phases of the measurements and also (temperature) corrections to be applied when processing the measurement results to achieve final efficiency values. Besides the different approach in determining the stray load losses, differences in these procedures and corrections ensure that these standards are not unambiguous.

7.2.2.2 Differences between Measurement Standards

As outlined above, and also underlined in several publications, the main difference between most efficiency determination methods for induction motors is to be found in the determination of the stray load losses. The value obtained for the efficiency of an induction machine depends on the method and standard used. Such differences can amount to up to several percentage points in the efficiency (Figure 7.9). In general, the differences between standards are not only at this technical level of determination of additional losses. There are other dissimilarities or peculiarities that can be distinguished. Some examples of these differences are in terminology, scope, titles, required instrumentation, accuracy, (number of) proposed methods, etc. In

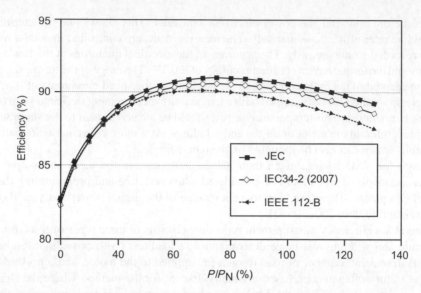

Figure 7.9 Efficiency of a four-pole 7.5 kW induction motor, measured according to three different standards. This results in efficiency differences of up to several percent

particular, the last two examples have a specific impact on the final result of the efficiency determination. Care should be taken when comparing or discussing different methods and their results, since the accuracy of this outcome should also be taken into account.

The two most important standards for the determination of induction machine efficiency are IEC Standard 60034-2 and the IEEE Standard 112. The IEC 60034-2 dates from 1972, but had updates/amendments in 1995 and 1996. The IEC 61972, deals only with squirrel cage induction motors, whereas the IEC60034-2 is for rotating machines in general, excluding traction motors; it was developed as a possible replacement of the IEC 60034-2 version of 1996, but was neither confirmed nor published. The version of IEC 61972 considered here, is the (first) edition of November 2002. Instead of this failed version, the IEC published a new version in September 2007, numbered IEC 60034-2-1. It is important to know that these standards offer different methods to determine efficiency. This is the same with the IEEE 112, divided in about eleven methods, named A, B, B1, C, ... of which B is the most important and best known method. The IEEE Standard 112 was introduced in 1964 and the latest revisions were made in 1996 and 2004, with only minor changes in the latest version. An overview of standards worldwide is given in Table 7.2.

In spite of this diversity of methods offered by the different standards, most manufacturers, test labs and others only use one or a few methods. The most significant methods are indicated in Table 7.3. For the IEEE 112 the most important method is method B, an Input–Output method requiring the measurement of mechanical torque and speed. The most commonly used method of the IEC 60034-2 (1996) is an indirect method: not requiring one to measure the mechanical torque. The stray load losses in this method are determined based on a fixed allowance (5% of the input power). In fact, there are four important methods: Input–Output, fixed allowance, variable allowance (Table 7.4) and Eh-Y. These differences in approach to determining the stray load losses cause the efficiency results to differ.

Table 7.2 Overview of methods for efficiency determination of induction motors according to recent (versions of) standards

	Status	Year	Significant methods	Remarks
IEEE 112	Valid	2004	Method B	Input–Output method, requiring mechanical torque and speed measurement. It provides several other methods.
IEC 60034-2 Ed. 3	Outdated	1996	Fixed allowance for P_{SLL}	Indirect method. Mechanical torque is not required. It provides some other methods, such as the reverse rotation test (RRT).
IEC 61972	Failed	2002	2	Developed as possible replacement of the 34-2, provides two (main) methods: • Input–Output similar to 112B • Indirect w/ variable allowance.
IEC 60034-2 Ed. 4	Valid	2007	3	Recently published, in fact a compromise. Holds the same methods as the IEC 61972 with an additional (new) Eh-Y test to determine P_{SLL}, but, e.g., still holds the RRT.
CSA C390-98	Valid	2005	1	Canadian standard. Very similar to IEEE112B, is used in the context of local minimum efficiency performance standards.
AS 1359.102	Valid	1997	3	Australian standard. Provides the choice between three methods: an indirect method similar to IEC 60034-2 Amd. 2 (1996), a calorimetric method (very accurate but time consuming and expensive) and an Input–Output method as in IEC 61972. It is expected that it will adopt the new version of the IEC 60034-2 (2007).
JEC-2137-2000	Valid	2000	2	Japanese standard. Originally introduced as JEC-37 on February 23rd, 1934 and revised several times. On March 27th, 2000 it was revised again and renamed as JEC-2137. It proposes several methods; e.g. Input–Output method and fixed allowance, ...
GB/T 1032-2005	Valid	2006	Fixed allowance for P_{SLL}	Chinese standard 'Test procedures for three-phase induction motors' issued on September 19th 2005 and implemented on June 1st 2006. It is similar to the old IEC 60034-2 (1996), proposing an indirect method using a fixed allowance of 0.5% of the input power for the determination of the stray load losses.

Table 7.3 Overview indicating, for each of the four main indirect methods, if it is included in the standardized test method and additional remarks

P_{SLL} method	IEEE 112	IEC 60034-2 (1996)	IEC 60034-2 (2007)	IEC 61972	Main differences between IEC versions and the IEEE version
Input–Output	Yes Methods B & B1	No	Yes Load test with torque measurement	Yes Method 1	Main differences between IEC versions and the IEEE version: • Nomenclature: IEC uses residual and stray load losses ⟷ IEEE uses stray load and corrected stray load losses; • Minimum correlation factor for linear regression analysis indicating the quality and repeatability of the test differs: IEEE specifies 0.9, IEC 0.95.
Fixed allowance	No	Yes	No	No	It is assumed that the total value of the stray load losses at full load is equal to 0.5% of the rated input power ($P_{SLL} = 0.5\% \ P_{el}$). Observe the difference with the variable allowances.
Variable allowance	Yes Methods E1, F1 and E1/F1	No	Yes Assigned allowance	Yes Method 2	Now the allowance depends on the motor rating (P_N), but the assigned values are different for IEC and IEEE standards, Table 7.4. Moreover, for IEC it is a percentage of P_{el} whereas for IEEE of P_N.
Eh-Y circuit	No	No	Yes	No	

Table 7.4 Assigned allowance for stray load losses according to IEEE 112 and IEC 61972 and IEC 60034-2 (2007)

	IEEE		IEC
P_N [kW]	Allowance [% of P_N]	P_N [kW]	Allowance [% of P_{el}]
-90	1.8	<1	2.5
91–375	1.5	1–10000	$2.5 - 0.5\log(P_N)$
376–1850	1.2	>10000	0.5
>1850	0.9		

7.2.3 Future Standard for Variable Speed Drives

Many modern applications require accurate control of speed or torque, which can be facilitated with the use of power electronic converters. Moreover, other motor types such as permanent magnet synchronous motors or switched reluctance motors inherently operated with power electronic converters are increasingly filling up niches in the market. They are even becoming economic alternatives to induction motors in VSDs.

Although IEC is preparing a 'Guide for the selection and application of energy efficient motors including variable-speed applications' labelled IEC Std 60034-31, to date, there is no internationally accepted test protocol that allows the determination of drive system efficiency at different load points. However, for proper design, measurement and classification of the energy performance of drive systems, the classic motor efficiency standards cannot be used as they concern direct-on-line application only. Several international initiatives try to fill this lacuna. For instance, the new IE4 efficiency limits are already formulated for an entire torque–speed range. Additionally, IEC is working on a new standard for the determination of efficiency of VSDs, labelled 60034-2-3: 'Rotating electrical machines: Specific test methods for determining losses and efficiency of converter-fed AC motors'.

There are three main issues concerning the assessment of VSD system efficiency. First, there is the problem of the determination of the efficiency for different load points. Compared with direct-on-line efficiency, there are much more possible operating points (speed, torque combinations) and the method of loss segregation as it is installed in most efficiency determination standards is not readily applicable. Secondly, the nature of VSDs means that there are multiple 'degrees of freedom' that influence the system's efficiency. In fact, there is a mutual influence of motor, converter, control algorithm and parameter settings on the respective losses in motor and converter. For which setup, boundary conditions and settings should the efficiency be determined to allow an objective comparison between different motors and drives? And thirdly, there is not yet any harmonized system for the visualization or classification of the VSD efficiency over the entire operating range. However, in the context of (electrical) drive trains for vehicles a useful concept has already been in use for years, namely the so-called efficiency maps or iso efficiency contours [20]. They are in fact the loci of equal efficiency values plotted as functions of speed and torque on the abscissa and ordinate respectively (Figure 7.10).

In fact, the IEC's efficiency classification already suggests the similar approach of using efficiency maps for the efficiency limits of IE4 motors if the motors are rated for a certain speed–torque range [21]. However, no specific method for the determination of the system efficiency is yet agreed. It is only mentioned that the direct efficiency should be measured

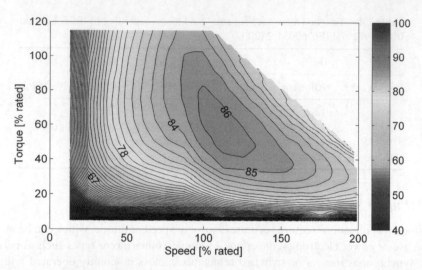

Figure 7.10 Efficiency map of an IE1, 11 kW, 400 V, four-pole induction motor + converter

over the motor's winding connection and the mechanical output (shaft power). Thus this does not take the mutual interaction between motor and converter, or the efficiency of the converter itself into account. Even more importantly, if no extra precautions and conditions are imposed, this will not guarantee an accurate or reproducible efficiency determination. This would be in contrast to the discussions concerning direct-on-line induction motor efficiency determination and the associated motivations for using the segregation of loss method instead of the direct methods. Moreover, in recent publications introducing and promoting the new IEC efficiency classification for IMs and discussing VSD efficiency, a similar approach using a matrix of test points (pairs) for speed and torque is mentioned [22]. There, a suggestion is made to limit the number of testing points according to two common application types: constant torque applications such as conveyors and quadratic torque applications such as pumps.

7.3 High Efficiency Motor Technology

In order to improve motor-driven system efficiency on a large scale there are two main areas on which to work. First, there is the standardization, classification and legislation part that stimulates the implementation of high efficiency motor systems by imposing voluntary or mandatory minimum efficiency limits based on (inter)national classification and measurement standards (Section 7.2.1). Second, technologic improvements to reduce losses on a system level are necessary. The highest savings can be obtained by matching the motor output to the speed and torque demand of the load by using a variable-speed drive, for example in pumping applications. In a next step, the consumption of the electric motor itself should be reduced by replacing it with a so-called high efficiency motor (HEM). Regarding these HEMs, the question arises as to in which aspects they differ from standard efficiency motors. What is HEM-technology? The answer is difficult to formulate, as there are many possible approaches to improving overall motor efficiency (Figure 7.11) with better materials, design and manufacturing technologies.

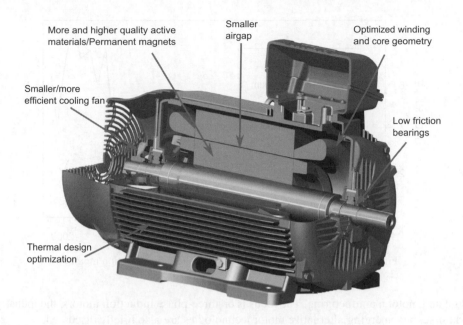

Figure 7.11 Cut-away view of a high efficiency induction motor showing a number of approaches to improving the efficiency. Reproduced by permission of WEG

Motor losses must be reduced while keeping the mechanical output at a constant level. Secondary motor performance characteristics such as starting and breakdown torque, power factor and locked rotor current should also remain within an acceptable range. To accomplish this task, an engineer must thoroughly understand the origin of losses in a motor and identify the main influencing factors, discussed in the first section of this chapter.

The traditional approach to reducing the main losses is simply to add more and higher-quality active materials to an existing motor design. Adding core or conductor material by increasing the main dimensions will respectively lower the flux and current density in the motor, yielding lower iron and joule losses for the same power-output rating. On the other hand, using higher quality materials, such as a high-grade lamination steel or copper instead of aluminium rotor conductors, will give lower losses for the same flux or current density. However, this is a very simplistic approach. Designing an HEM would not be such an advanced problem if the cost of a motor was not a significant design parameter. In the current economic reality, motor purchase cost is of primary concern to customers, especially original equipment manufacturers (OEMs) who have little interest in running costs. The design of a HEM comes down to finding an optimal balance between efficiency and product cost. This is why a well-designed HEM is a fundamentally different motor, built not only with better (and/or more) active materials, but also with optimized geometries and components in order to reduce the material cost considerably. The final factor influencing motor efficiency is the manufacturing and construction process. Improved techniques should have fewer imperfections and allow for smaller tolerances. In the next sections, a comprehensive overview of technological issues concerning motor efficiency, divided in three main categories, is given: motor materials, motor

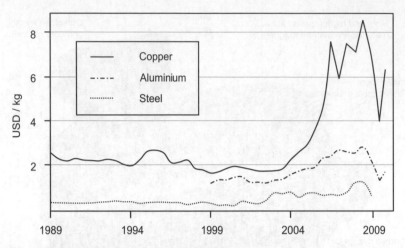

Figure 7.12   Price history of raw materials used in induction motors

design and motor manufacturing. The focus is on three-phase induction motors, the industry's workhorse, but emerging alternative motor technologies are also briefly discussed.

7.3.1   Motor Materials

In the last decades, high efficiency induction motors used standard stator and rotor designs with additional active material by increasing the stack diameter or length. More magnetic steel in the core, copper in the stator windings and aluminium in the rotor cage decreased the losses considerably. Before 2004, copper, steel and aluminium prices were relatively stable and attractive. However, in the last five to ten years quite a turnaround occurred in high efficiency motor development, caused by raw material price swings as a result of global economic changes. Note that Figure 7.12 shows only raw material prices, and finished lamination steel for example is approximately a factor of two more expensive.

Table 7.5 illustrates this evolution by giving a comparison between the material costs of a small 0.75 kW induction motor in the year 2009 versus that in 2000 [23]. The raw material price swings have compromise the cost-effectiveness and have put a hold on HEM development. The key to cost-effective construction of HEMs today is by a close cooperation between material suppliers and motor designers in order to improve the material quality while keeping control of the total cost by optimizing motor-design. Less is more: less material with higher quality combined with a better motor design. Developments and state of the art of core materials, permanent magnets and conductor materials are discussed in the following sections.

7.3.1.1   Core Material

The magnetic circuit formed by the ferromagnetic material of the stator and rotor core is the most expensive part of an induction motor and provides the flux with a low reluctance magnetic path. Compared with paramagnetic and diamagnetic materials, the permeability of

Table 7.5 Comparison of the material cost of a 750 W induction motor in the years 1999 and 2009

Induction motor 750 W	Quantity [kg]	1999 [US$/kg]	1999 [US$]	2009 [US$/kg]	2009 [US$]
Steel laminations	14	0.9	12.6	1.8	25.2
Copper conductors	1.7	4.5	7.65	9	15.3
Aluminium rotor	0.5	2.2	1.1	4.4	2.2
Total material cost			**21.35**		**42.7**

ferromagnetic materials such as construction steel or advanced electrical steel alloys is 100 to even 50 000 times higher. Hence, the magnetizing current and affiliated stator copper losses, required to drive the flux through the circuit will be reduced. It is possible to build a motor with construction steel core material, but the efficiency of such a motor would be insufficient due to the high core losses and the relatively low permeability. Because of the magnetostriction in the core material when subjected to an alternating electric field, loss is generated proportional to the surface of the hysteresis loop. This loss is, according to the Steinmetz equation (7.2), is approximately proportional to the frequency and the flux density B (Tesla) squared. Soft magnetic materials used in motor cores have a narrow hysteresis-loop with a coercive field strength of less than 1000 A/m, yielding smaller hysteresis losses. The eddy-current losses are caused by induced circular currents in the magnetic core and are proportional to the square of frequency and the square of flux density. To reduce these phenomena, a low-conductivity soft magnetic steel with isolated laminations is used (Figure 7.13).

Iron loss and copper loss are related quantities. The mechanical torque produced by a motor is proportional to the active rotor current and the flux. Hence, increasing the non-saturated magnetic flux density with a better highly permeable core material will allow the motor to operate at a higher flux density, reducing active current and copper loss but increasing iron loss. An optimal balance between flux (iron loss) and active torque producing current (copper loss) can be found by design optimizations.

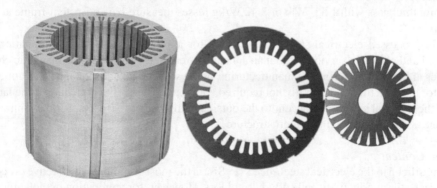

Figure 7.13 Electrical steel laminations are stacked to form the stator and rotor core

Transformers have a fixed flux orientation in the core and therefore use grain-oriented soft-magnetic materials. In motor cores however, the flux direction is variable, so non-oriented soft-magnetic materials are used. They can be divided further according to material composition, ranging from standard silicon steel found in most motors today to more exotic and expensive alloys:

• Silicon steel
• Nickel-iron and cobalt-iron alloy
• Soft-magnetic composites (powder core)
• Amorphous and nano-crystalline magnetic material

To summarize, high-grade core materials have the following properties:

• High magnetic flux density (high permeability): reduce stator copper loss;
• Low conductivity: reduce eddy-current loss;
• Soft-magnetic, low coercivity: reduce hysteresis loss.

In the last 20 years, considerable advancements have been made to improve electrical steel quality for high-efficiency motors [24]. Next, a number of evolutions and core material specifications are given.

Semi-processed and Fully-processed Electrical Lamination Steels

Semi-processed electrical steels are delivered in a cold rolled, un-annealed condition, and must be annealed after punching the laminations to develop optimum magnetic properties [25]. After stamping, the laminations are typically annealed at temperatures between 790 and 845°C for about one hour in a decarburizing atmosphere to recrystallize the microstructure. The objectives of the annealing treatment include:

• Eliminating punching stresses;
• Promoting and controlling grain growth;
• Further reducing impurities, particularly carbon, nitrogen and sulphur.

For example, the non-oriented semi-processed silicon steel with ASTM-code 47S155 has a 0.47 mm thickness with 1.55 W/lb or 3.42 W/kg losses measured in an Epstein-frame at 1.5 T and 60 Hz.

Fully processed electrical steels are intended for applications where the laminations are punched and placed in service without an annealing treatment. The desirable magnetic characteristics are produced during the manufacturing of the steel, so an additional heat treatment by the rotor manufacturer is generally not required. Obviously during the punching of the laminations, the material structure adjacent to the outside cutting edges deforms, but the laminations are used with this sub-optimal microstructure.

Silicon Content

Adding silicon to the electrical steel sheets ($FeSi_3$) is the most common and effective procedure to decrease the conductivity and eddy-current loss. However, for application in high efficiency motors, the Si-content is limited because it lowers the non-saturated magnetic flux density (permeability) and increases mechanical hardness, which makes the production more expensive due to increased tooling costs. Low conductivity and high permeability were often regarded as irreconcilable, but work has been done in this area [26] with the development of cobalt and nickel iron alloys.

Interlaminar Insulation
Insulating coatings on the laminations reduce iron loss by restricting the eddy currents to flow in an individual lamination of the core. A wide range of coatings can be classified by the types C0, C1, C2, C3, C4 and C5, with different insulation resistance, heat resistance, punchability and weldability [27]. Most common for small motors are oxide coatings (C0, C1), formed naturally or during the annealing process of punched laminations. However, the interlaminar insulation of the oxide layer is not consistent, which makes it unsuitable for use in high efficiency motors. Basic organic coatings (C3) are more consistent but are not able to withstand annealing temperatures on semi-processed steels. In high efficiency motors, inorganic (C5) and hybrid coatings (C4/C5/C6) are used that have excellent insulation consistency, thermal resistance and punchability [28].

Lamination Thickness
Thinner laminations have less eddy current losses but increase the costs considerably, both for steel suppliers and motor manufacturers. Below 0.5 mm the sheet material becomes difficult to handle and process. Special manufacturing and stacking techniques are required to integrate ultra-thin (0.05–0.2 mm) laminations and insulation layers with package densities (stacking factor) up to 98% [29].

Cobalt-iron and Nickel-iron Alloys
High-grade soft-magnetic materials based on cobalt–iron or nickel–iron alloys have superior properties and offer considerable advantages for high efficiency motors compared with standard silicon–steel [30]. Cobalt–iron has very high magnetic flux density and permeability, which allows lower magnetizing currents (copper losses) for the same core size or higher power densities for equal magnetizing current. Nickel–iron alloys have lower magnetic flux density but also lower losses due to the low coercivity force and high electrical resistivity. Table 7.6 gives a comparison between a typical silicon–iron and high grade nickel and cobalt alloys [29].

Grain Size
The grain size of magnetic steel is of crucial importance for minimizing the iron losses. Smaller grains on the laminations edges due to mechanical processing (punching) will increase hysteresis losses because the friction between magnetic domain walls will be larger. Furthermore, eddy current loss is proportional to grain size. An optimum exists at 100–150 μm [26]. Typical silicon steel has a grain size of 30 μm before and 50 μm after stress relief annealing, still much smaller than the optimal value. The growth of grains is limited by small inclusions that interact with the grain boundaries. Lowering the amount of small inclusions and coarsening them by adding aluminium and rare-earth elements to molten silicon–steel will improve grain-growth during stress-relief annealing considerably (up to 70 μm).

Table 7.6 Comparison between strip material specifications of silicon–iron, cobalt–iron and nickel–iron

	Lamination thickness	Coercivity force	Electrical resistivity	Induction at $H = 16$ A/cm	Saturation polarization	Loss at 50 Hz (1 T)
	[mm]	[A/cm]	[$\mu\,\Omega$m]	[T]	[T]	[W/kg]
Silicon–iron (47F165)	0.47	0.2	0.40	1.45	2.03	1.17
Cobalt–iron	0.2	0.4	0.40	2.25	2.35	0.6
Nickel–iron	0.2	0.04	0.45	1.55	1.55	0.23

Future Developments

Soft magnetic composites (SMC) are being developed as an alternative core material for the classic lamination sheets [31]. SMC is produced by bonding iron powder under pressure with an epoxy or other organic resin in a particular shape. The bonding material provides great strength and insulation properties. SMCs have average to good magnetic properties (permeability) coupled with low eddy-current losses, especially at high frequencies, due to the insulation between the grains provided by the bonding resin. This makes SMC an interesting material to use in high efficiency motors for inverter operation (high-frequency harmonics). The main advantage however is the freedom that SMC gives to the motor designer to make complex core forms with a very smooth surface finish as the electrical resistance is isotropic (unlike laminations), imposing no constraints on the design.

Amorphous magnetic materials (AMM) offer future potential to further reduce iron losses in motor cores. AMM is produced by quenching liquid alloy compositions (Fe, Co, Ni, Si, etc.) rapidly in a very thin layer in order to avoid the normal equilibrium crystalline structure and retain an amorphous disordered structure similar to glass [32]. The advantage of this disordered atomic structure is a very low iron loss due to the high resistance, typically less than 0,2 W/kg at 50 Hz and 1.5 T (2–4 W/kg for silicon steel). The saturation level and permeability is only average because of the need to add glass-forming elements (Si), so slightly larger cores must be used with AMM. The biggest prohibitive factor is still the cost. AMM has the potential to be produced economically in the future, but the brittleness of the extremely thin gauges (0.04 mm max) make AMM difficult to process in terms of cutting out lamination shapes and obtaining a good stacking factor.

7.3.1.2 Rotor Conductor Material

Joule losses ($R\,I^2$) in the rotor cage account for the second largest portion of loss in an induction motor. Squirrel cage material conductivity, cross-section and length determine the resistance and also the joule losses in the rotor. The choice of highly conductive materials in mass-production motors is limited by cost, excluding exotic materials and leaving the relatively cheap aluminium and the highly conductive copper as the only two feasible options. The rotor cages of induction motors in the small to medium power range are generally pressure die-cast from aluminium because this allows cost effective mass production and fairly good

Figure 7.14 High efficiency induction motor with a die-cast copper rotor. Reproduced by permission of Siemens

conductivity and efficiency. The alternative, to use copper for the rotor cage, has gained interest in the last decade, driven by the need to improve motor efficiency according to the MEPS. The 70% higher conductivity of copper can reduce joule losses, leading to a lower operating temperature, again decreasing conductor resistance. Apart from the efficiency benefit, the lower motor temperature allows for a smaller fan with less air friction and will increase motor lifetime according to the empirical law stating that a decrease in 10° operating temperature will extend motor lifetime by a factor of two.

Copper rotors are not a new technology, in fact they are the standard in large power induction motors with fabricated rotor cages built from solid copper bars welded to the end rings. For mass-produced small to medium range induction motors, fabricated copper rotors are not economically feasible but copper die-casting was problematic due to the high melting point of copper. For a long time, this has been a barrier for copper rotor breakthrough, but work has been done in this area by a number of major motor manufacturers to overcome this problem [33]. Today, the technology has reached a mature status able to improve motor efficiency considerably (Figure 7.14). Unfortunately, another barrier was created due to the large increase in the cost of copper the last five years, which has put a serious hold on the cost effectiveness of copper rotors. This is why the market penetration of die-cast copper rotor motors is limited today. Copper rotor technology has considerable and proven energy-saving potential, but the discussion about the economics can still be considered open and the outcome will depend mainly on the future copper price evolution [34].

Copper Die-casting Problems and Solutions

The rotor lamination stack and die inserts are placed in the casting machine and liquid copper is injected under pressure to fill the cavities of the rotor bars and end rings. The higher melting temperature of copper (1083°C) compared with aluminium (660°C) gives rise to a number of difficulties with the die-casting process [33]:

- Heat checking: a thermal fatigue phenomenon caused by the cyclic heating and cooling of the die material. The die surface is in contact with the hot molten copper and is at a higher temperature than the inner portion of the die. This results in an unequal expansion and a large tensile stress on the die surface leading to cracking and premature die failure. The higher die temperatures can also lead to decarburization and softening of the steel dies. Replacing ordinary die steel with high performance alloys (nickel, tungsten or molybdenum) and using electrical resistance pre-heaters for the dies appeared to be a good solution to these problems.

- Risk of overheating, welding or annealing of rotor laminations, which can compromise material properties.
- Oxidation of the copper melt. For small numbers of castings a shot-by-shot induction melting is appropriate to minimize oxygen exposure. For larger batches, induction melting in an enclosed furnace under a nitrogen atmosphere is necessary.
- Porosity or incomplete casting gives rise to unbalanced rotors and higher stray-load losses. Large pores in the rotor bars locally reduce the cross-section and cause an uneven current distribution in the rotor cage. With 3D fluid flow simulations, optimal casting machine parameters and melt temperature can be calculated to reduce these air inclusions during casting.

Direct Replacement

The most straightforward approach is to leave the stator and rotor lamination stack design unchanged and substitute the aluminium for copper. In [33] this approach was found to give a 1.2% efficiency gain in an 11 kW motor. Other motor characteristics are also influenced by the lower rotor resistance. The torque–speed curve will shift according to Figure 7.15(b), yielding a slight increase in operating speed, lower starting torque, breakdown torque at higher speed and a rise in start-up current. The approximately three times higher mass density of copper compared with aluminium will also considerably increase rotor inertia in a direct replacement. These changes can cause problems for dynamic applications and a non-conformity with motor performance parameters defined in standards. It also should be remarked that copper rotor motors are slightly more vulnerable to voltage unbalance [6]. These negative influences can be reduced by appropriate motor design.

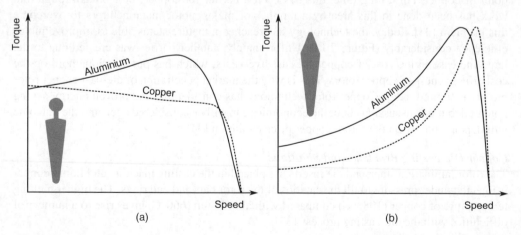

Figure 7.15 Influence of rotor slot geometry on the torque–speed characteristic of aluminium and copper rotor induction motors

Replacement with Rotor Slot Geometry Optimization

The direct replacement of aluminium by copper already significantly improves efficiency, but an even greater enhancement is possible when this substitution is combined with a redesign of the motor, including choice of core material, active part dimensions and rotor slot geometry. The cross-section of the copper bars can be designed to be smaller, so less volume is required, lowering material cost and inertia. Lower starting torque and increased inrush current are expected with a higher rotor conductivity, but by adapting the rotor bar shape, the designer can change motor behaviour to counteract this theory. The key is to make optimal use of the frequency-driven diffusion effects (skin-effects) in the rotor cage [35]. Figure 7.15(right) shows a tapered bar shape used typically in aluminium rotors, maximizing conductor surface with a uniform rotor tooth width to avoid saturation. A double bar slot shape in Figure 7.15(a) will increase apparent rotor resistance at start-up because the electrical frequency seen by the rotor is around the line frequency, forcing the currents to the upper starting bar with a reduced section. This results in an increased start-up torque compared with a conventional slot shape. At running conditions the rotor frequency is low and the currents will flow mainly in the low resistance running bar. It must be remarked that the stray load losses are also influenced by this slot shape as the space and time harmonic frequencies will experience a higher apparent resistance, meaning higher losses.

7.3.1.3 Permanent Magnets

Replacing the rotor squirrel cage of an induction motor with permanent magnets can offer a significant improvement in efficiency. According to the stator winding structure and supply voltage shape (trapezoidal or sinusoidal) this is called a brushless DC (BLDC) or a permanent magnet synchronous motor (PMSM) respectively. The three most common permanent magnet materials are:

- NdFeB: neodymium iron boron (rare earth);
- SmCo: samarium cobalt (rare earth);
- Ferrite (ceramic).

Table 7.7 gives a comparison between the material properties. It is clear that NdFeB is by far superior to the 'classical' ceramic magnets in terms of energy density. In [23] it is stated that the available power of a motor varies approximately with the square of the energy product. This translates in more compact and lighter motor designs with NdFeB-magnets.

Table 7.7 Comparison of magnet materials

	Energy product	Coercivity	Remanence	Mass density
	[kJ/m^3]	[kA/m]	[1]	[g/cm^3]
Neodymium iron boron	200–440	750–2000	1–1.4	7.5
Samarium cobalt	120–200	600–2000	0.8–1.1	8.4
Ferrite	10–40	100–300	0.2–0.4	4.5

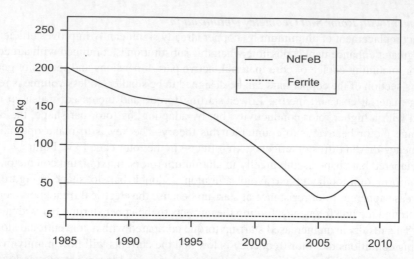

Figure 7.16   Historic price of rare-earth (NdFeB) and ferrite permanent magnet material

Permanent magnets come in many forms, depending on the manufacturing technique:

- Sintered/fully dense: Maximum attainable energy density, limited to simple geometries, brittle
- Compression bonded: Mixture of magnetic powder bonded with a resin, simple geometries, lower energy density, brittle
- Injection moulded: Complex geometric shapes, not brittle, lowest energy density.

Until recent years it was assumed that the use of high energy density (rare earth) magnets in mass-produced motors was economically not feasible. Until the beginning of the 21st century, conductor and core materials were relatively cheap and rare earth magnets expensive (Figures 7.12 and 7.16).

Looking at the changes in the price of materials in the last decade, it is clear that rare earth magnet costs have dropped significantly and copper and steel costs have moved the other way. The drop in the price of NdFeB is due to the proven abundant resources of China. Of course this market-dependency on China as a rare earth magnet supplier also yields risks for future magnet pricing as it is predicted that the market share of permanent magnet motors will continue to increase, for example in electric vehicles. The fact that permanent magnet motors are smaller for the same output power also reduces the need for conductor and core material. The dramatic cost reduction and improvements in magnetic and thermal properties of permanent magnet technology has made it an interesting alternative to induction motors. A comparison between pre-1999 and 2009 material costs of a permanent magnet motor is given in Table 7.8 [23].

7.3.2   Motor Design

The motor efficiency requirements imposed by legislation will increase substantially in time. The straightforward approach of adding active material to meet these efficiency requirements is no longer sufficient or even possible without switching to a larger frame size (shaft height).

Table 7.8 Comparison of the material cost of a 750 W permanent magnet motor in the years 1999 and 2009

PM motor 750 W	Quantity	1999		2009	
	[kg]	[US$/kg]	[US$]	[US$/kg]	[US$]
Steel laminations	6	0.9	5.4	1.8	10.8
Copper conductors	1.1	4.5	4.95	9	9.9
NdFeB magnets	0.3	100	30	50	15
Total material cost			**40.35**		**35.7**

Furthermore, motor efficiency is not the only design target that has to be met. Cost and performance specifications are equally important. High efficiency motor design can therefore be considered as a constrained optimization problem of numerous geometry and material parameters [36].

7.3.2.1 Stator, Rotor and Air-gap Geometry Optimization

In the past, the optimization of motor geometry, if executed at all, used to be a trial and error procedure relying mostly on designer experience and rules of thumb. Today, powerful finite-element modelling (FEM) software packages are used as a design tool for the optimization (Figure 7.17).

With FEM-tools it is possible to simulate the electric, magnetic, mechanical and thermal behaviour of a motor in the design stage. Calculated quantities such as flux and current

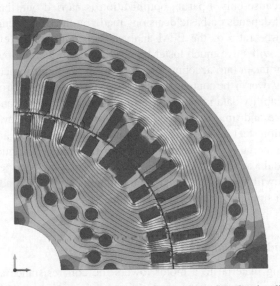

Figure 7.17 Finite-element model showing the flux lines and the flux density distribution of an induction motor

densities, torque, temperature distribution and losses are the input for an iterative algorithm, which searches for the optimal geometry. The objective of the optimization is to improve the efficiency while decreasing the amount of active material and maintaining motor performance figures within an acceptable range:

- *Core length and diameter*: Output torque is proportional to stack length and the square of the diameter, while iron losses are a function of flux density. Today's optimized HEM-designs yield lower losses with equal or even less volume.
- *Rotor cage geometry*: The shape of the conducting rotor bars in an induction motor determines rotor slip losses. A loss reduction of more than a percent is possible without compromising starting performance or material costs [37]. Skewing of the rotor bars is also often applied to decrease ripple torque caused by space and time harmonics.
- *Air-gap length*: About 70% of the magnetizing current is required to drive the flux through the air-gap of the motor; the remaining 30% is due to the lamination steel permeability. Hence, a smaller air-gap can make a considerable reduction in magnetizing current. Of course, this imposes higher requirements on geometrical tolerances and manufacturing techniques.
- *Stator slot geometry*: This is a crucial optimization exercise between lowering current density in the stator conductors and preventing teeth saturation. Furthermore, stray losses and ripple torque due to slot space harmonics, caused by the chopping of the flux as it crosses the air-gap, can be reduced by adequate slot geometry design.

Typically only a limited number of parameters are optimized with an algorithm. All other dimensions and parameters are constrained to a fixed value because the computing time needed to perform an optimization is proportional to n^2, with n the number of design variables. A stator slot geometry optimization alone, for example, already has 16 variables [38]. Even with current and future computing performance, an optimization including all design variables will remain a utopia. Because only a partial optimization is carried out, the maximal attainable motor efficiency still depends on basic decisions made by the design engineer. The accuracy versus calculation time ratio of the FEM analysis depends on the level of detail. A 2D-model, for example, will solve much faster than a full 3D-model but parasitic effects of the end windings are not taken into account, unless equivalent circuit impedances are used. The biggest challenge however in the context of accuracy is the modelling of iron losses. Lamination steel loss and permeability data is needed for an extended frequency and induction range to account for the space and time harmonic losses in a motor core. However, the conditions under which lamination steel samples are tested with the Epstein frame are distinctly different from those that the material is subjected to in a motor [39]. The catalogue data is mostly only available for sinusoidal excitation at certain frequencies and a fixed temperature. Furthermore, manufacturing steps will alter the magnetic behaviour locally, which adds even more to the discrepancy between simulations and measurements.

7.3.2.2   Stator Winding Layout

Adding more winding material to a motor lowers the coil resistance and thereby also the stator copper loss. The increase of the copper wire cross-section or the additional windings will, however, create space problems in the stator slots of traditional radial motors. In older high-efficiency motor designs, the slot width and/or height were increased to provide space for the additional copper. Accordingly, the length and diameter of the core must be increased to keep the flux density and core reluctance the same and to prevent saturation in the teeth. This is of course not an issue in axial flux motors, since windings can be added without affecting the magnetic circuit. Keeping in mind the lamination steel cost, a more interesting approach is to maintain the slot geometry, but to increase the slot fill factor with more advanced winding techniques and wire material. Thinner winding insulation coatings and slot isolation are used to improve the slot fill factor. End winding length is another crucial factor in winding designs because, for small motors, up to 40% of the winding is inactive but loss-producing end winding. In a conventional slotted radial motor, containing a distributed winding, applying a multilayer, chorded winding can reduce end winding length considerably. An interesting evolution in this area is a new construction method for radial motors called segmented core or cut core design. This technique uses individual pole stampings of lamination steel to be assembled into a full radial core. The core segments are wound individually with a concentrated winding, yielding very short end turns and a higher fill factor, which leads to more compact and efficient motors. Up to now, this technique is mainly used in low power motors, since the manufacturing cost is still prohibitive [23].

Next to the previously discussed ohmic losses, the stator winding topology also directly influences harmonic losses (stray losses). Space harmonics are due to the discrete division of windings in slots and time harmonics are caused by non-sinusoidal supply conditions. This leads to non-sinusoidal flux causing lower torque and provoking additional core losses and joule losses in the conductors due to diffusion effects. Full-pitch, non-distributed windings (coil width equal to one pole pitch) generate a significant amount of 7th and 5th harmonics, which are the most harmful to motor performance. By employing multilayer chorded (coil pitch < pole pitch) and distributed windings, these harmonics can be suppressed.

7.3.2.3   Cooling and Thermal Design

The thermal design of an air-cooled high efficiency motor is as equally important as the electrical or mechanical design. Thermal analysis of cooling by conduction, radiation and convection is typically performed with FEM-tools and equivalent thermal resistance networks in which the losses (heat sources) are distributed at the appropriate nodes. Stator and rotor conduction losses are a function of the resistance, which is proportional to the temperature. In [40] the efficiency of a 3.7 kW induction motor was increased with 0.25% by a coil temperature reduction of 10°C. This can be achieved by a reduction of losses in the motor with the previously discussed measures. Next step is better heat conduction from the active parts to the housing components from which the heat is dissipated in the surrounding air. A good thermal contact between the core outside diameter and the housing is crucial, requiring a precise machining of both parts. Another possibility is potting: filling the air cavities inside the frame with an electrically isolating but thermally conductive resin. The housing material and the cooling fin dimensions are also of importance, e.g. an aluminium frame has higher conductivity than cast iron. The next step is the optimization of fan performance. Once again, a trade-off between temperature rise and windage loss can be made, but is must be noted that small HEMs run so cool that a fan is not even needed. For larger motors in applications with a fixed direction of rotation, it could be interesting to replace the standard bidirectional fan with a higher efficiency unidirectional fan.

7.3.2.4   Bearings

Bearing friction represents a relatively small loss segment, but it is an easy way of gaining a few tenths of percent points in efficiency. Bearing manufacturers provide high efficiency groove-ball bearings that can save 30–50% compared with standard bearings [41]. For a continuously running 7.5 kW four-pole motor this is about 100 kWh per year. The savings are due to the following innovations.

- Non-petroleum greases on a lithium or synthetic basis offer lower losses, longer maintenance intervals and are able to withstand higher temperatures. Maintenance practice is also of crucial importance; over- or under-greased bearings have higher friction leading to premature failure.
- Low friction ceramic ball bearings are another interesting innovation; they run cooler, have a prolonged lifetime and further reduce losses.
- Enclosures of general-purpose motors should meet IP55 requirements as a minimum. Sealed bearings are used but have additional friction, especially in small motors. On the other hand, they allow longer service intervals because the grease is kept in place. Non-contact, low friction seals are the next evolution.
- Friction is proportional to bearing size. For some applications bearing size can be reduced, but shaft loading (for example with belts transmissions) will be limited. In high-efficiency motors the opposite drive end bearing is lightly loaded and can be chosen to be smaller than the drive-end-bearing [24].

7.3.2.5   Alternative Motor Technology

Induction motors (IM) are still the industry's workhorse and represent nearly the entire installed base of constant speed and variable speed general purpose motors [42]. The IM is mature technology, cheap, robust and can reach high efficiencies. Levels exceeding IE3 are possible when material, design and manufacturing are optimized. However, the present and future market for motors places ever increasing value on operating efficiency, power density, low cost and reliability [43]. Since current induction motor technology is reaching the physical boundaries of what is possible in terms of efficiency improvement at reasonable cost, other technologies like permanent magnet (PMSM) or switched reluctance motors (SRM) are considered (Figure 7.18).

SRMs are an alternative in variable speed traction operation, for example in electric or hybrid cars, with a simple and robust rotor structure and a high efficiency. Each motor phase consists of two opposite stator poles carrying concentrated excitation windings. The SRM requires one converter unit per phase. When a motor phase is supplied with DC current, the energized stator pole pair attempts to pull the closest rotor pole pair into alignment, which in turn minimizes the reluctance of the magnetic path. The operating principle is inherently accompanied by vibration, acoustic noise and a relatively high torque ripple. Furthermore, smooth operation of an SRM requires a complex controller and accurate position feedback to

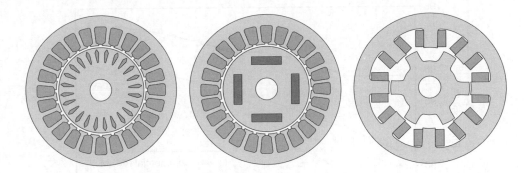

Figure 7.18 The three main AC motor drive technologies, from left to right: induction motor, permanent magnet synchronous motor and switched reluctance motor

profile the current waveforms [44]. This is why SRM technology will probably not replace the IM in standard frame general purpose motors, but is rather to be used in specific variable speed applications in which its benefits can be used to the full advantage.

Another more promising technology that is gaining ground in standard frame general purpose motors is the permanent magnet motor. According to the stator current waveform one can distinguish the brushless DC (BLDC) or permanent magnet synchronous motor (PMSM) with respectively rectangular or sinusoidal current waveforms. Note that the stator of a PMSM has an identical layout to that of an induction motor, with a three-phase distributed or concentrated winding.

PMSMs have a number of substantial benefits over the IM.

- *Higher efficiency*: The permanent magnets in the rotor supply the flux. Practically no magnetizing current flows in the stator winding and no currents are induced in the rotor, reducing the corresponding copper losses. Furthermore, due to the synchronous operation, the flux variation causing the iron losses in the rotor core is, apart from harmonics, zero.
- *Higher power density*: The reduction of losses and the corresponding lower heat generation allows one to increase the stator current keeping the same operating temperature with higher torque output. It is possible to deliver the same output in smaller frames, significantly reducing the weight and volume of the PMSM.
- *Wider speed range with constant torque*: PM motors are more efficient than induction motors throughout the speed range, especially at lower speeds. Therefore, the cooling system efficiency reduction at lower speed is compensated for by the loss reduction, thus allowing constant torque in the entire speed range.
- *Maintenance*: The bearings run cooler, prolonging the lifetime and the lubrication intervals.
- *Inertia*: Reduced rotor weight is an advantage in dynamic applications.

PM motor technology is not new, but was up till now mostly used in servo motors or in special-purpose traction motors in low speed, high torque applications. The improved performance to cost ratio of rare earth (NdFeB) permanent magnet materials has reached a level where the increased efficiency and power density offsets the cost of the magnets [43]. At present, a

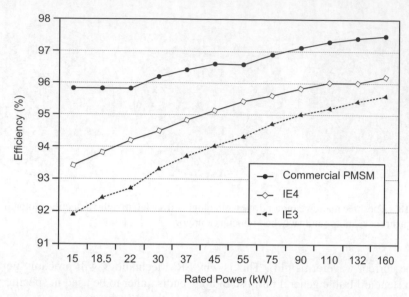

Figure 7.19 Efficiency of a commercially available PMSM range (two pole) compared with the IE3 and IE4 MEPS levels

number of major motor manufacturers offer this technology in a complete range of standard frame general purpose motors for variable speed application, as an alternative for the IM. The efficiency gains are (Figure 7.19) exceeding the current premium (IE3) efficiency level by 2% and comfortably reaching the IE4-level.

A drawback of a PMSM is that it cannot be directly connected to the grid, even for constant-speed applications. It needs an inverter with field-oriented control (FOC) that adds to the losses and cost in a constant speed application. A line start permanent magnet synchronous motor (LSPMSM) is in this case a good solution. It is in fact a hybrid IM/PMSM with permanent magnets in the rotor combined with a squirrel cage at the rotor surface that provides the starting and accelerating torque when connected to the grid. When synchronism with the grid frequency is reached, the motor operates with zero slip, invariant of load, and practically no losses are induced in the cage. The efficiency is 1–2% better than premium efficiency induction motors, but power density is about equal so they come in the same standard frame sizes. This technology is readily available in the catalogues of major manufacturers. It is specifically aimed at a low to medium power range as a direct replacement for induction motors in constant-speed applications. Variable speed (V/f or scalar) control is also possible, but efficiency will be decreased due to the losses induced by switching harmonics in the rotor cage.

7.3.3   Motor Manufacturing

Next to material and design innovations, motor designers are now also turning to the production plant in their search for higher motor efficiency. It is widely recognized that manufacturing and construction techniques have a large influence on the efficiency of the final product.

A motor can be well designed with high quality materials, but the manufacturing must be on the same high level to achieve a quality end-product. This requires a close cooperation between manufacturing engineers and design engineers, a system of quality control during the production stages and investments in plant, tooling and training of manufacturing personnel. Limiting the production tolerances and thereby the efficiency bands is crucial. In the past, designers often added supplementary material to have enough 'buffer' to account for efficiency deviations due to production. This is not an option any more, given the current cost-efficiency requirements. In recent years, a number of loss increasing manufacturing processes have been identified and improved that enabled designers to reduce the amount of material previously added to allow for these variations.

7.3.3.1 Core Manufacturing

Magnetic steel suppliers guarantee the losses and permeability of their products by Epstein frame test data on unmanufactured steel strips, but the properties will deteriorate significantly as the material is transformed into a motor core. 'Manufacturing iron loss' is due to the mechanical and thermal stress to which lamination steels are subjected during motor core manufacturing. Iron losses can almost double, but with an adequate manufacturing process, the influence can be minimized.

- A first possible source of stress is the cutting operation on the thin metal sheets. The microstructure of the steel is affected (smaller grains), which results in a local decrease of permeability and an increase of hysteresis losses of 5–30% in a region of 1–6 mm from the cutting edge [45]. A number of different cutting techniques are used, each with a different influence on the iron loss [46]. Laser cutting is commonly used for small production batches of prototypes or special machines. It is reported to be the worst technique because it subjects the material edges to very high temperatures. Punching is the most common technique and leads to high stress levels in the lamination material. However, when the punching tools are kept in good condition by regrinding on time to prevent burrs and excessive stress in combination with stress relief annealing, the influence can be reduced to an absolute minimum. The best technique, however, is spark erosion with a very low mechanical and thermal impact on the material.
- The core laminations stack is held together under axial pressure by a variety of methods including welding, bolting, cleating or semi-pierce interlocking [47]. Compressive stress should be minimized and welds, cleats or bolts should be positioned in strategic places of low flux density.
- The stator core is often subjected to a large radial force due to the shrink-fit in the housing, which must be reduced to a minimum.

7.3.3.2 Stator Winding

For the stator winding, high slot fills and short tight end windings are important for reducing ohmic losses, but are more difficult to install and insulate. Automatic winding machines often produce random wound coils, which results in a relatively low slot fill and loose end windings. Machines able to produce high density layer wound coils for a specific slot geometry are necessary. Another possibility to increase slot fill is by reducing the wire insulation volume, which on a 0.8 mm conductor represents 12% of the cross-sectional area [28]. This requires the development of materials with an equal insulation value for a smaller thickness.

7.3.3.3 Rotor Cage Die-casting

Aluminium or copper squirrel cage rotor die-casting is a critical process, in which the filling of the slots and end rings and the purity of the material (air or dirt inclusions) are crucial factors influencing the rotor losses. Furthermore, the cast conductor bars form bridges between the laminations, shorting out the insulation layers. This gives rise to additional stray load and eddy current losses. By making use of the different contraction rates of steel and aluminium/copper, quenching the rotor in cold water after die-casting will cause the two to separate [28]. A better way, however, is to apply an insulating layer to the rotor slots before casting.

7.3.3.4 Geometrical Tolerances

High-efficiency motors are typically designed with a smaller air-gap. This puts higher demands on geometrical tolerances when manufacturing and assembling the parts to prevent vibrations or even collision due to eccentricity. To ensure air-gap concentricity it is common practice to turn the motor to size from the motor shaft centres [28]. The risk exists that laminations are burred together to form a conducting surface over the rotor laminations if the condition of the tooling is insufficient. High frequency rotor surface losses due to the skin effect of space and time harmonics will be doubled if no precautions are taken. For smaller motors, the best solution is to increase the accuracy of the punching operation, so laminations can be punched to size directly. For larger motors this is not practical, and burrs must be avoided or removed. The tolerances on the machining of the frame and bearing end shield will also be tighter to ensure concentricity between the housing inner-surface and the bearing houses.

References

[1] H. De Keulenaer, R. Belmans and E. Blaustein, *Energy Efficient Motor Driven Systems. Can Save Europe 200 Billion kWh of Electricity Consumption and over 100 Million Tonnes of Greenhouse gas Emissions a Year?*, European Copper Institute, Brussels, Belgium, 2004.

[2] P. Bertoldi and B. Atanasiu, *Electricity Consumption and Efficiency Trends in the Enlarged European Union – Status report 2006*, European Commission, DG TREN, Joint Research Centre and Institute for Environment and Sustainability, 2007, p. 66.

[3] A. T. De Almeida, F. J. T. E. Ferreira and J. A. C. Fong, Standards for efficiency of electric motors, *Industrial Electronics Magazine, IEEE*, **17**, 2011, p. 8.

[4] A. A. Jimoh, R. D. Findlay and M. Poloujadoff, Stray losses in induction machines: Part I, Definition, origin and measurement, *IEEE Transactions on Power Apparatus and Systems*, **PAS-104**, 1985, pp. 1500–1505.

[5] A. A. Jimoh, R. D. Findlay and M. Poloujadoff, Stray losses in induction machines: Part II, Calculation and reduction, *IEEE Transactions on Power Apparatus and Systems*, **PAS-104**, 1985, pp. 1506–1512.

[6] W. Deprez, *Energy Efficiency of Induction Machines: A Critical Assessment*, KU Leuven, 2008.

[7] H. Auinger, Efficiency of electric motors under practical conditions, *Power Engineering Journal* see also *Power Engineer*, **15**, 2001, pp. 163–167.

[8] H. Auinger, Determination and designation of the efficiency of electrical machines, *Power Engineering Journal* see also *Power Engineer,* **13** 1999, pp. 15–23.

[9]   P. Pillay, P. Hofmann and M. Manyage, Derating of induction motors operating with a combination of unbalanced voltages and over or undervoltages, *IEEE Transaction on Energy Conversion*, **17**, 2002, pp. 485–491.

[10] S. Van Haute, A. Malfait, R. Reekmans and R. Belmans, Losses, audible noise and overvoltages in induction motor drives, *III Power Electronics Specialists Conference*, Atlanta, Georgia, 1995.

[11] F. J. T. E. Ferreira, P. Pereirinha and A. T. de Almeida, Study on the bearing currents activity in cage induction motors using finite-element method, *i iICEM 2006, XVII International Conference on Electrical Machines*, Chania, Crete, Greece, 2006.

[12] B. Bolsens, J. Knockaert, J. V. d. Keybus and R. Belmans, Common-mode poles and eigenmodes for various grounding configurations in motor applications with variable speed drives, *16th International Conference on Electrical Machines (ICEM 2004)*, Cracow, Poland, 2004.

[13] A. Boglietti, A. Cavagnino, M. Lazzari and M. Pastorelli, International standards for the induction motor efficiency evaluation: a critical analysis of the stray-load loss determination, *IEEE Transactions on Industry Applications*, **40**, 2004, pp. 1294–1301,.

[14] W. Deprez, Understanding the pitfalls between efficiency measurement standards and the practical performance of motor driven systems (invited paper), *Journades Europees d'Energia a l'ITSEIB* Barcelona, Spain, 2007.

[15] B. Renier, K. Hameyer and R. Belmans, Comparison of standards for determning efficiency of three phase induction motors, *IEEE Transactions on Energy Conversion(IF 0.360)*, **14**(3), 1999, pp. 512–517.

[16] A. Nagornyy, A. K. Wallace and A. V. Jouanne, Stray load loss efficiency connections, *Industry Applications Magazine, IEEE*, **10**, 2004, pp. 62–69.

[17] P. Van Roy and R. Belmans, Assessment of efficiency of low voltage, three phase motors, *Energy Efficiency in Motor Driven Systems (EEMODS'02)*, Treviso, Italy, 2002.

[18] A. T. de Almeida, *European Ecodesign Directive on Energy – Using Products (EuPs) – Project: Lot 11 Motors*, 2006.

[19] P. Angers, Comparison of existing standard methods of determining energy efficiency for three-phase cage induction motors, *EEMODS'09: Energy Efficiency in Motor Driven Systems*, Nantes, France, 2009.

[20] W. Deprez, J. Lemmens, D. Vanhooydonck, W. Symens, K. Stockman, S. Dereyne and J. Driesen, Iso efficiency contours as a concept to characterize variable speed drive efficiency, *XIX International Conference on Electrical Machines (ICEM)*, 2010, pp. 1–6.

[21] M. Doppelbauer, Guide for the selection and application of energy-efficient motors, *EEMODS '09: Energy Efficiency in Motor Driven Systems*, Nantes, France, 2009.

[22] A. I. de Almeida, C. U. Brunner, P. Angers and M. Doppelbauer, Motors with adjustable speed drives: testing protocol and efficiency standard, *EEMODS '09: Energy Efficiency in Motor Driven Systems*, Nantes, France, 2009.

[23] J. Petro, Cost-effective construction of high efficiency motors, *EEMODS*, Nantes, 2009.

[24] J. Malinowski, J. McCormick and K. Dunn, Advances in construction techniques of AC induction motors: preparation for super-premium efficiency levels, *IEEE Transactions on Industry Applications*, **40**, 2004, pp. 1665–1670.

[25] Copper Development Association, Copper Motor Rotor Project, 2010.

[26] M. Takashima, N. Morito, A. Honda and C. Maeda, Nonoriented electrical steel sheet with low iron loss for high-efficiency motor cores, *IEEE Transactions on Magnetics*, **35**, 1999, pp. 557–561.

[27] M. Lindenmo, A. Coombs and D. Snell, Advantages, properties and types of coatings on non-oriented electrical steels, *Journal of Magnetism and Magnetic Materials*, 2000, pp. 79–82.

[28] D. Walters, Energy efficient motors saving money or costing the earth? Part I, *Power Engineering Journal*, **13**, 1999, pp. 25–30.

[29] Vacuumschmelze, Application notes: VAC alloys for motor and generator applications, 2010.

[30] E. Krol and R. Rossa, Modern magnetic materials in permanent magnet synchronous motors, *Electrical Machines (ICEM), 2010 XIX International Conference on*, pp. 1–3.

[31] G. Cvetkovski and L. Petkovska, Performance improvement of PM synchronous motor by using soft magnetic composite material, *IEEE Transactions on Magnetics*, **44**, 2008, pp. 3812–3815.

[32] L. A. Johnson, E. P. Cornell, D. J. Bailey and S. M. Hegyi, Application of low loss amorphous metals in motors and transformers, *IEEE Transactions on Power Apparatus and Systems*, **PAS-101**, 1982, pp. 2109–2114.

[33] *Technology Transfer Report – The Die-Cast Copper Motor Rotor*, Copper Development Association and International Copper Association, April 2004.

[34] A. Boglietti, A. Cavagnino, L. Feraris and M. Lazzari, Energy-efficient motors, *Industrial Electronics Magazine, IEEE*, **2**, 2008, pp. 32–37.

[35] J. L. Kirtley, J. G. Cowie, E. F. Brush, D. T. Peters and R. Kimmich, Improving induction motor efficiency with die-cast copper rotor cages, *Power Engineering Society General Meeting, 2007. IEEE*, 2007, pp. 1–6.

[36] J. F. Fuchsloch, W. R. Finley and R. W. Walter, The next generation motor, *Industry Applications Magazine, IEEE*, **14**, 2008, pp. 37–43.

[37] D. G. Walters, I. J. Williams and D. C. Jackson, The case for a new generation of high efficiency motors–some problems and solutions, *Electrical Machines and Drives, 1995. Seventh International Conference on (Conf. Publ. No. 412)*, 1995, pp. 26–31.

[38] A. M. Knight and C. I. McClay, The design of high-efficiency line-start motors, *IEEE Transactions on Industry Applications*, **36**, 2000, pp. 1555–1562,

[39] A. Boglietti, A. Cavagnino, D. M. Ionel, M. Popescu, D. A. Staton and S. Vaschetto, A general model to predict iron losses in PWM inverter-fed induction motors, *IEEE Transactions on Industry Applications*, **46**, 2010, pp. 1882–1890.

[40] M. K. Yoon, C. S. Jeon and S. K. Kauh, Efficiency increase of an induction motor by improving cooling performance, *IEEE Transactions on Energy Conversion*, **17**, 2002, pp. 1–6.

[41] M. Janssens, SKF Energy Efficient deep groove ball bearings for higher driveline efficiency, *EEMODS*, Nantes, 2009.

[42] X. Feng, L. Liu, J. Kang and Y. Zhang, Super premium efficient line start-up permanent magnet synchronous motor, 2010 *XIX International Conference on Electrical Machines (ICEM)*, pp. 1–6.

[43] M. Melfi, S. Evon and R. McElveen, Induction versus permanent magnet motors, *Industry Applications Magazine, IEEE*, **15**, 2009, pp. 28–35.

[44] Z. Q. Zhu and D. Howe, Electrical machines and drives for electric, hybrid and fuel cell vehicles, *Proceedings of the IEEE*, **95**, 2007, pp. 746–765.

[45] L. Vandenbossche, S. Jacobs, F. Henrotte and K. Hameyer, Impact of cut edges on magnetization curves and iron losses in e-machines for automotive traction, *25th World Battery, Hybrid and Fuel Cell Electric Vehicle Symposium & Exhibition*, Shenzhen, China, 2010.

[46] W. M. Arshad, T. Ryckebusch, F. Magnussen, H. Lendenmann, B. Eriksson, J. Soulard and B. Malmros, Incorporating lamination processing and component manufacturing in electrical machine design tools, *Industry Applications Conference, 2007. 42nd IAS Annual Meeting. Conference Record of the 2007 IEEE*, 2007, pp. 94–102.

[47] D. Miyagi, K. Miki, M. Nakano and N. Takahashi, Influence of compressive stress on magnetic properties of laminated electrical steel sheets, *IEEE Transactions on Magnetics*, **46**, 2010, pp. 318–321.

8

Lighting

Mircea Chindris and Antoni Sudria-Andreu

Nowadays, lighting significantly contributes to energy consumption and greenhouse gas emissions. The electricity used to operate lighting systems corresponds to an important fraction of total electricity consumed worldwide: 5–15% of the national electricity consumption in the industrialized countries, and even higher than 80% in some developing countries. In 2003, the use of artificial light was estimated to result in the consumption of approximately 650 Mtoe of primary energy, representing 8.9% of total global primary energy consumption [1]; accordingly, lighting-related CO_2 emissions were estimated at 1900 Mt, equivalent to approximately 8% of world emissions. Even so, these data are not accurate for the present day but they do serve to indicate the size of the problem.

On the other hand, there is a widespread use of obsolete lighting technologies, especially in the residential sector. According to [2], two thirds of all lamps currently installed in the European Union are energy inefficient (in EU homes, about 85% of lamps are energy inefficient); in the UK, there are estimated to be some 375 million incandescent lamps used in homes, about 60% of the total [3].

At the same time, the energy saving potential in the lighting sector is high even with the existing technologies. Indeed, according to literature, the average luminous efficacy has scarcely increased from about 18 lm/W in the 1960s to roughly 50 lm/W in 2005; however, this is not uniform across all lighting applications, ranging from only a little above 20 lm/W in the residential sector to slightly above 50 lm/W in the commercial sector and to about 80 lm/W in the industrial sector. As other lighting technologies are emerging on the market, lighting applications represent a good objective for demand-side energy efficiency initiatives targeting large-scale implementation of efficient lighting technologies. These initiatives can offer very appropriate solutions, with significant benefits for all those involved, as shown below [4]:

- *Customers*: energy savings, reduced bills, mitigation of impact of higher tariffs;
- *Utilities*: peak load reduction, reduced capital needs, reduced costs of supplying electricity;
- *Governments*: reduced fiscal deficits, reduced public expenditures, improved energy security;
- *Environment*: reduced local pollution and reduction in greenhouse gas emissions.

Electrical Energy Efficiency: Technologies and Applications, First Edition. Andreas Sumper and Angelo Baggini.
© 2012 John Wiley & Sons, Ltd. Published 2012 by John Wiley & Sons, Ltd.

In lighting, the energy efficiency is defined as optimization of energy consumption, with no sacrifice in lighting quality. Although the amount of electricity consumed by lighting is an important issue, to reduce the human effectiveness of the illumination on the basis of energy efficiency would be a serious retrograde step for human performance and, in the long term, counter-productive. Therefore, the problem that the lighting specialists must address is how to provide illumination at best practice standards while, at the same time, using the minimum amount of electricity necessary [5]. The right solution will be a combination of several items from the following: accurate choice of illumination level, appropriate selection of lamps and luminaries, and thoughtful design and choice of a fitting control system. Furthermore, the integration of the lighting system in the environment or space that is being lit must be also considered.

8.1 Energy and Lighting Systems

8.1.1 Energy Consumption in Lighting Systems

A very comprehensive analysis of energy consumption in lighting systems is presented in [6]. According to this report, more than 33 billion lamps operated worldwide in 2005, consuming annually around 2650 TWh of energy (19% of global electricity consumption, or slightly more than the total electricity consumption of OECD Europe for all purposes); over half of this electricity consumption is in IEA member countries, but their share is declining: by 2030, non-OECD countries are expected to account for more than 60% of global lighting electricity demand. The annual cost of this service including energy, lighting equipment and labour is US\$ 360 billion, which is roughly 1% of the global GDP (electricity accounts for some two thirds of this).

The total lighting-related CO2 emissions were estimated to be 1900 million tons (equivalent to 70% of the emissions from the world's light passenger vehicles), namely about 7% of the total global CO_2 emissions from the consumption and flaring of fossil fuels.

The largest proportion of electricity used in lighting is consumed by the indoor illumination of tertiary-sector buildings; on average, lighting accounts for 45% of tertiary-sector electricity consumption and 14% of residential consumption in OECD countries (in non-OECD countries these shares are usually higher; globally, they are 31% for residential and 43% for commercial). Outdoor stationary lighting (mainly including street, roadway, parking and architectural lighting) uses less than one-tenth of total lighting electricity consumption (about 8%), while an estimate of 490 TWh/an of final electricity was consumed by industrial lighting (amounting to about 18% of total lighting electricity consumption and just over 8.7% of total electricity consumption in the industrial sector).

On the other hand, according to [7], the global electricity consumption for lighting is distributed as follows: approximately 28% to the residential sector, 48% to the service sector, 16% to the industrial sector, and 8% to street and other lighting. Other information regarding the energy consumption in EU-27 is presented in [8] and [9]. In 2007, lighting represented 10% of the final electricity consumption with the most important weight in the residential sector (at 10.5% it represents the third main consumer after electricity for heating and cold appliances), tertiary-offices (21.6%) and street lighting (4.7%); in industry it is generally not more than 5% [10].

With reference to the USA, lighting systems consume approximately 20% of the electricity generated [11] or about 30% of the total energy consumption in the US [12]; in Japan, about 14% of generated electric energy is consumed for lighting [13]. Demand for artificial light is strongly linked to per-capita GDP, but there are important variations between economies; [6] presents an estimation for per-capita consumption of electric light in 2005, expressed in megalumen-hours: North America: 101 Mlmh; Europe: 42 Mlmh; Japan/Korea: 72 Mlmh; Australia/New Zealand: 62 Mlmh; China: 32 Mlmh; Former Soviet Union: 32 Mlmh; Rest of the world: 8 Mlmh (India, for instance, uses only 3 Mlmh).

Over the last decade, global demand for artificial light grew at an average rate of 2.4% per annum, slower in IEA countries (1.8%) than in the rest of the world (3.6%). With the current economic and energy efficiency trends, it is projected that global demand for artificial light will be 80% higher by 2030 and will still be unevenly distributed. If this comes to pass and the rate of improvement of lighting technologies does not increase sufficiently, global lighting electricity demand will reach 4250 TWh. Furthermore, without additional energy efficiency policy measures, lighting-related annual CO_2 emissions will rise to almost 3 Gt by 2030.

8.1.2 Energy Efficiency in Lighting Systems

According to [14], energy efficiency in lighting systems is defined as the optimization of energy consumption, with no sacrifice in lighting quality; its implementation supposes the thoughtful design and selection of appropriate lamp(s), luminaire(s) and control system, and the informed choices of the illumination level required (the integration and awareness of the environment or space that is being lit must be also considered). The aim is to achieve a lighting system enabling tasks to be performed efficiently and effectively whilst providing a good balance of cost and energy consumption. Every visual task has different lighting requirements and the following elements must be taken into account to assure an appropriate visual perception: (i) level of lighting; (ii) luminance in the field of vision; (iii) absence of irritating reflection (glare); and (iv) colour rendering. As a result, the most suitable lighting technology for each circumstance varies and the best practice in lighting depends highly upon the application.

On the other hand, for economic evaluation of different lighting solutions, a life cycle cost analysis has to be made as, usually, only the initial (investment) costs are taken into account [15]. People are not aware of the variable costs, which include energy, lamp replacement, cleaning and reparation costs. The energy costs of a lighting installation during its entire life cycle are very often the largest part of the whole life cycle costs. It is essential that in future lighting design practice, maintenance schedules and life cycle costs will become as natural as, for example, illuminance calculations already are.

Fortunately, there is a significant potential to improve energy efficiency of old and new lighting installations already with the existing technologies. The following three basic steps to improving the efficiency of lighting systems are indicated in [11]:

1. Identify necessary light quantity and quality to perform visual task.
2. Increase light source efficiency if occupancy is frequent.
3. Optimize lighting controls if occupancy is infrequent.

Step 1, identifying the proper lighting quantity and quality is essential to any illuminated space, whilst steps 2 and 3 are options that can be explored individually or together; they can both be implemented, but often the two options are economically mutually exclusive.

The first step is very important especially in lighting retrofits: in these cases, it is often overlooked because most energy managers try to keep the same illumination level as in the old system, even if this is over-illuminated and/or contains many sources of glare. Unfortunately, although levels recommended by international standards have continuously declined, nowadays there are still many excessively illuminated spaces in use. Energy managers can obtain remarkable savings by simply redesigning a lighting system in order to achieve proper illumination levels.

The second step addresses the efficiency of lighting system components, including in this term the lamp, the ballast and the luminaire. From this point of view, increasing the efficiency simply means a better global luminous efficacy (getting more lumens per watt out of the lighting system). This goal may be reached by increasing the source efficacy, replacing the magnetic ballasts (by more efficient versions or by electronic ballasts) or improving the fixture efficiency. Increasing the efficacy of the light source is the most popular choice because energy savings can more or less be guaranteed if the new system consumes fewer watts than the old one. Nevertheless, this energy criterion should be compatible with other lighting design criteria: in many applications, the optical features (colour temperature, colour rendering index, light distribution curve, etc.) are frequently the leading criteria instead of the lamp efficacy in choosing lamp types (lamp efficacy may become a secondary consideration).

Finally, the third step aims to implement a variety of control techniques in order to reduce the energy consumed while the system is operating (by daylight, for instance) or to turn artificial lighting systems off when they are not needed. Nevertheless, it is important to underline that influencing user's behaviour can also bring important savings if the use of control systems is too costly.

On the other hand, when dealing with energy efficiency in lighting installations, the concern in other topics such as directing light to where it is needed or using reflective room surfaces is also important.

In conclusion, efficient and effective lighting systems will [14]:

- provide a high level of visual comfort;
- make use of natural light;
- provide the best light for the task;
- provide controls for flexibility;
- have low-energy requirements.

The biggest energy saving and lighting quality opportunities are found in older, over lit buildings that use inefficient technologies or in locations where utility costs are very high and where lighting is uncontrolled and left on all night. Apart from the owners' economic motivation, the implementation of efficient lighting systems must be supported by policy measures addressing the following objectives [6]:

- phasing out or substantially reducing the use of low-efficiency lamps and control gear;
- encouraging the adoption of high-efficiency luminaires and discouraging the use of their low-efficiency counterparts;

- encouraging or requiring the use of appropriate lighting controls;
- ensuring lighting systems are designed to provide appropriate lighting levels according to national or international norms;
- stimulating better lighting design practice to encourage task lighting, individual user control of lighting needs and dynamic integration with daylight rather than uniform artificial illumination;
- encouraging greater and more intelligent use of daylight in the built environment, resulting in energy, health and productivity benefits;
- reducing light pollution, especially for outdoor applications;
- stimulating the development and early adoption of new, more efficient lighting technologies;
- overcoming market barriers to efficient lighting and negating the overemphasis on first costs in favour of life cycle costs;
- protecting consumers from poor-quality lighting components, such as low-quality compact fluorescent lamps (CFLs) and linear fluorescent lamps (LFLs) with a lifespan, light output or efficacy that does not meet declared and/or minimum values.

According to the European Lamp Companies Federation, more than 50% of all lamp technologies installed in Europe are still not the most energy efficient [2]; therefore, the potential for improvements and savings (of energy, costs and CO_2 emissions) for Europe is significant. The majority of these savings (between 75% and 80%) can be achieved in the area of professional lighting (in streets and offices for example); for that reason, the public sector has an important role to play in setting an example and influencing the market place through green procurement. 'Green' procurement is a key element in controlling consumption of energy in Europe. Europe's local authorities spend 14–16% of EU GDP on public procurement each year; this money can be used wisely to help save energy consumption through purchasing energy efficient technologies, such as modern lamps or ballasts. Although in general this equipment is initially more expensive, based on their 'total cost of ownership', savings can be made through operational costs in electricity, maintenance and disposal.

8.2 Regulations

Different lighting guidelines, with a significant influence on the amount of light provided and, hence, on the amount of lighting energy required, have been issued at the national or international levels. The recommendations in the guidelines have evolved considerably and frequently over the last 80 years and yet remain divergent from one jurisdiction to another, with large associated implications for lighting energy demand [6]. The main European standards and directives are as follows.

1. EN 12665: Light and lighting. Basic terms and criteria for specifying lighting requirements
2. EN 12464-1: Light and lighting. Lighting of indoor work places
3. EN 13201 series: Road lighting
4. EN 13032 series: Light and lighting. Measurement and presentation of photometric data of lamps and luminaries
5. EN 60598 series: Luminaires

6. EN 61000-3-2: Electromagnetic compatibility. Limits. Limits for harmonic current emissions (equipment input current ≤ 16A per phase)
7. prEN 15193: Energy performance of buildings. Energy requirements for lighting
8. Directive 2006/32/EC on energy end-use efficiency and energy services (repealing Council Directive 93/76/EEC)
9. Directive 98/11/EC on Energy labelling of household lamps
10. EuP Directive 2000/55/EC on energy efficiency requirements for ballasts for fluorescent lighting
11. Directive 2002/91/EC on the energy performance of buildings
12. Directive 2004/108/EEC on Electromagnetic Compatibility (EMC).

8.3 Technological Advances in Lighting Systems

A lighting system is made up of lamps, luminaires and the control gear (which controls switching, ignition and regulation system). Designing an appropriate lighting system for each specific task takes into consideration a range of performance characteristics and, beyond the quantity and quality of light, the choice of technology is influenced by considerations of economy, durability and aesthetics. Over the last decades the lighting equipment industry has made considerable advances in the energy efficiency of equipment, including new lamp technologies, improved optical performance of luminaries and high frequency, low-energy electronic control gear for discharge lamps.

8.3.1 Efficient Light Sources

The lamp is the first component to consider in the lighting design process; the choice of lamp determines the light quantity, colour rendering index, colour temperature, as well as other different technical and economic characteristics of the whole lighting system. According to [6], the average luminous efficacy of lighting systems has improved from about 18 lm/W in 1960 to roughly 48 lm/W in 2005; the rate of improvement appears to have been relatively constant from 1960 to 1985, at about 2.8% per year, but since 1985 onwards it slowed to 1.3% per year.

Most artificial light is nowadays generated by three processes: (i) incandescence (the glow from hot solids); (ii) direct emission in gas discharges; (iii) luminescence (conversion of ultraviolet into visible radiation). The last decades have witnessed a new technology emerging, based on light emission from semiconductor devices (SSL or LED).

Figure 8.1 presents the existing types of electric lamps while their main characteristics are listed in Table 8.1. In practice, various lamp technologies are used for different applications depending, in part, on the quality of the light they generate; this covers the amount and colour characteristics of the light produced, the speed of ignition and the time taken to reach full output, and the ease of control.

Incandescent lamps represents the oldest electric lighting technology; they are also the least efficacious (only 5% of the electricity is converted into light, 95% of the power taken up is lost as heat) and have the shortest life. Alternatively, incandescent lamps have a relatively high quality light (especially excellent colour rendering), being preferred by many consumers because of their familiarity, and low purchase price (however, if life cycle cost analyses

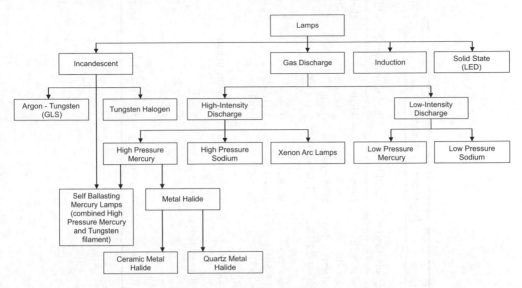

Figure 8.1   Categories of light sources

are used, incandescent lamps are usually more expensive than other lighting sources with higher efficacies); these bulbs are quiet, dimmable, and turn on instantly. Because of their low efficiency, many governments have adopted regulations in order to replace inefficient incandescent lamps with more efficient alternatives.

Linear fluorescent lamps are the most common gas discharge lamps and have been used in commercial and industrial settings since their commercialization in 1937 due to lower operating costs; nowadays they provide the bulk of global lighting.

Fluorescent tubes have much higher efficacy levels and much longer operating life than incandescent lamps. With a high colour rendering index (CRI), these lamps can also be designed to provide a large diversity of colour temperatures, ranging from 2700 K (as found with an incandescent lamp) to 7500 K (daylight). Fluorescent lamp technology has made significant advance in performance in the last two decades, and the old T12 lamps, operated by inefficient magnetic ballasts, have been replaced by new versions (T8 and T5).

The T8 lamps were developed in the early 1980s and rapidly became the best solution (along with the efficient magnetic or high frequency electronic ballasts that drive them) for fluorescent upgrades for several years. Recent advances in T8 lamps are improvements in colour rendering (due to its tri-phosphor coating), a longer life (20% longer than standard T8 lamps), and a rated lumen maintenance of 0.94.

The T8 lamp with an electronic ballast combination provides a light level comparable to the T12 with an electromagnetic ballast system, and has the benefit of consuming up to 40% less electrical energy; also, the T8 lamp life is about 50% greater than the T12 lamp. Most T8 lamps can operate on conventional electromagnetic ballast, and can be used to directly replace T12 lamps; however, it is worth noticing that not all electromagnetic ballasts and T8 could harmoniously work together, and a trial of a few selected luminaire should be conducted before large-scale replacement [16].

Table 8.1 The main characteristics of existing light sources

Characteristics	Incandescent		Low intensity discharge		High intensity discharge			Solid state
	Light globes	Quartz halogen	Fluorescent tube	Compact fluorescent	Mercury vapour	Metal halide	Sodium vapour	LED
Efficacy* [lm/W]	Low (8–17)	Low (20–30)	Moderate to high (60–100)	Moderate to high (50–65)	Low to high (15–70)	High (60–100)	High (75–160)	High (200)
Lamp life [h]	Shortest (less than 2000)	Short to moderate (750–12 000)	Moderate to long (7500–24 000)	Moderate to long (10 000–20 000)	Moderate to long (6000–24 000)	Moderate (1500–15 000)	Long (14 000–24 000)	
Colour rendering	Excellent (100)	Excellent (100)	Medium to good (50–98)	Medium to good (50–80)	Poor (15–50)	Medium to good (60–90)	Poor (17–25)	Good (>80)
Normal wattage range** [W]	Up to 1500	Up to 1500	8–220	4–40	40–1000	70–2000	70–1000	
Installation cost	Low	Low	Low	Low	Moderate	Moderate to high	Moderate to high	
Running costs	Highest	Highest	Moderate to low	Moderate to low	High to moderate	Moderate to low	Low	
Replacement costs	Low	Medium	Low	Medium	Low	High	High	
Relight time	Immediate	Immediate	Immediate	Immediate –3 sec	3–10 min	10–20 min	Less than 1 min	Immediate

* Includes power consumption of control gear or ballast
** Lamp only

The *high-performance T8 lamps* are part of a dedicated lamp/ballast system that can save about 19% energy over standard T8 systems with the same light output and double the lamp life (on program-start ballasts); supplementary, they exhibit high lumen maintenance (95%) and high CRI (86).

The T5 lamps come in two distinct and different families: *standard* (high-efficiency) and *high output* (HO); all these lamps are powered only by electronic ballasts. They are designed to peak in their lumen rating at 35°C as opposed to 25°C for T12 and T8 lamps; this characteristic provides higher light output in confined applications where there is little or no air circulation. Standard T5 lamps are 12–18% more efficient than T8 lamps and 10–15% more efficient than the T5HO. T5s employ rare-earth phosphors with CRI greater than 80 and lamp lumen maintenance rated at 95%.

High output T5 linear lamps are physically the same size as standard T5 lamps, but provide higher lumen output. T5HO lamps generate from 1.5 to 2 times the light output of the standard T5 and nearly twice the light output (188%) of T8 and T12 systems with the same number of lamps; unfortunately, they can be up to 8% less efficient than standard T8 systems.

Normally, T5 cannot replace T8, as this lamp requires its own high frequency ballast that is different from that of a conventional T8. Also, the T5 tube is slightly shorter in length than the T8, and the holding end-caps on the luminaire have to be spaced differently. In conclusion, a retrofit of existing lighting of T8 or T12 to use the new T5 will require the replacement of the ballast and probably the entire luminaire as well; the result is a higher cost of retrofit. To avoid retrofitting the luminaire when switching from T8 or T12 on electromagnetic ballast to T5 on electronic ballast, the Plug&Enhance (PnE) technology, introducing a quasi-electronic ballast (QEB) in the tube replacement, is nowadays available [16]. The QEB is an electronic device that is attached as an end cap or inside a fitting, and works with the original electromagnetic ballast to light up the fluorescent tube. With the QEB, a T5 can be fitted directly in place of the T8 or T12, and the overall efficiency after replacement is similar to a simple T5 on electronic ballast arrangement. A short circuit component is required to replace the original starter for some systems.

Compact Fluorescent Lamps (CFLs) usually consist of 2, 4 or 6 small fluorescent tubes that are mounted in a base attached to a ballast (for ballast-integrated models), or are plug-in tubes (for the non-integrated alternative). Integrated lamps use either a screw-in base or bayonet cap in the same way as standard incandescent lamps; more recent models are available in a variety of screw-in diameters and represent one of the most efficient solutions available today for improving energy efficiency in residential lighting. In fact, CFLs consume 20 to 25% of the energy used by incandescent light bulbs to provide the same level of light (about 25% of electricity consumed is converted to visible light compared with just 5% for a conventional incandescent lamp). However, the perceived ratio is nearer 30%; the reason for this is probably because the spectral distributions of the two lamp types are very different in that the incandescent lamp has smooth distribution, similar to a black body, while a CFL has a distribution with many peaks and troughs [5].

They were first commercialized in the 1980s and improvements to CFL technologies have been occurring every year since they became commercially offered. Products available today have a reduced environmental impact (use mercury as an amalgam), provide higher efficacies, instant starting, reduced lamp flicker, quiet operation, smaller size and lighter weight. They also come in a broader series of colour temperatures ranging from the same temperature as incandescent lamps up to much higher values nearer to daylight. Life expectancy is from

6000 hours (mainly marketed for the residential sector) to beyond 15 000 hours and it is no longer affected by switching (the current standards for accreditation require over 3000 switching cycles per 8000 hours of tested life; moreover, some manufacturers produce heavy duty CFLs with up to 500 000 switching cycles capability and 15 000 hours life). CRI typical values (80 to 85) are high enough for most applications and prices have also fallen substantially over the preceding decade.

Dimmable CFLs are now available (screw-base dimmable CFLs were introduced in 1996) and the next generation (also known as Super CFL) will increase their performances: fully dimmable (smooth dimming down to 10% light output, with no colour shift), the capacity for the lamp to restart at any light level setting, high power factor, higher efficacy (a minimum of 70 lm/W), etc.

As a drawback, the current THD from electronically ballasted compact fluorescent lamps can exceed 30%, though nowadays low-harmonic units are also available. Compact fluorescent lamps with magnetic ballast typically produce THD of 15–25%, which is acceptable in most applications. Technological improvements and cost reductions in compact fluorescent lamp electronic ballasts make them economically viable, providing instant starting, three-lamp capabilities, reduced flickers, hum, size and weight, and efficacy increase of 20%. Even if the electronic compact fluorescent lamp system costs several times more than the comparable incandescent lamp, the life-cycle cost is usually worthy of consideration.

High Intensity Discharge (HID) lamps produce light by discharging an electric arc through a tube filled with gases at a greater pressure than fluorescent lamps. Originally developed for outdoor and industrial applications, they are also used in office, retail and other indoor applications. Most of these lamps have relatively high efficacies and long life expectancy (from 5000 to more than 24 000 hours); unfortunately, they require time to warm up and should not be turned on and off for short intervals. There are three popular types of HID sources (listed in order of increasing efficacy): mercury vapour, metal halide and high-pressure sodium.

High-pressure mercury gas discharge lamps are based on the oldest HID technology and can be considered obsolete as they are relatively inefficient (30–60 lm/W), provide poor CRI and have the highest lumen depreciation rate of all HIDs.

Metal halide lamps produce light by passing an electric arc through a high-pressure mixture of gases (argon, mercury, and a variety of metal halides) and are among the most energy-efficient sources of white light available today (up to 100 lm/W). With nearly twice the efficacy of mercury vapour lamps, metal halide lamps are commonly used in industrial facilities, sports arenas and other spaces where good colour rendition is required (colour rendering is very good, up to 96). The high-wattage pulse-start versions (175 to 1000 W), operated by electronic ballasts, are rapidly replacing standard metal halide lamps, providing a quicker re-strike time (3–5 minutes) versus previous models (8–15 minutes). The implemented technological improvements result in higher lamp efficacy (up to 110 lm/W), enhanced lumen maintenance (up to 80 %), and consistent lamp-to-lamp colour (within 100 K).

In the last years, ceramic metal halide lamps (CMH or CDM), operated by electronic ballasts, also became commercially available; in this case, a polycrystalline alumina (PCA) arc tube is used. These systems have the following advantages: 10–20% higher lumen output (i.e. a better efficacy), the best colour stability, high CRI (83–95), limited colour shift (from ± 75 K to ± 200 K), excellent lamp-to-lamp colour consistency, and good lumen maintenance (0.70–0.80).

High Pressure Sodium (HPS) lamps produce golden (yellow-white) light and represent an economical choice for most outdoor and some industrial applications where good colour rendition is not required (the luminous efficacy is about 100–150 lm/W, but colour rendering is low, at about 25). Nowadays, they are losing position to white-light sources, such as metal halide or fluorescent; to face this situation, several improvements have been introduced in new-generation HPS lamps such as the elimination of end-of-life cycling (characteristic of standard high-pressure sodium lamps), reduced or zero mercury content, etc.

Induction lamps were introduced in the early 1990s and work on the basis of electromagnetic induction, i.e. they use an electromagnetic field (produced by a high-frequency electronic generator) to induce a plasma gas discharge into a tube or bulb that has a phosphor coating. Therefore, these lamps have no electrodes and will operate 5–8 times longer than fluorescent and metal halide systems and about four times longer than HPS systems; they have good luminous efficacy (80 lm/W) and colour rendering (80–90). Additionally, induction lamps come on relatively quickly and have short re-strike time compared with HID lamps. As the long life (60 000–100 000 hours) is the primary advantage of these systems, a good payback can be provided where maintenance labour cost is high.

Solid state lighting (SSL) represents a lighting technology involving light-emitting diodes (LEDs) and organic light-emitting diodes (OLEDs), semiconductor devices that emit light when a current passes through them. Solid state lighting is making substantial progress and initially has experienced rapid gains in niches of the illumination market (for instance task lighting, signal purposes, portable lighting, or the use of LEDs with photovoltaic); however, they have recently become available for general illumination, both indoor and outdoor. Manufacturers are developing new types of LED bulbs for general lighting and anticipate rapid cost reductions.

With significant advantages (long lifetime, colour-mixing possibilities, spectrum, design flexibility and small size, easy control and dimming, no mercury content, etc.) it is expected that LEDs will radically transform lighting practices and the market in the near future. The energy performance of LEDs is continuing to improve considerably; according to [2], the maximum luminous efficacy of phosphor converted cool-white LEDs is expected to be around 200 lm/W by 2015 (from 120 lm/W today), while the luminous efficacy of warm white LEDs is expected to be above 140 lm/W (from 80 to 100 lm/W at present).

OLEDs are another promising technology that ultimately could produce devices that are significantly less expensive than LEDs; although OLEDs are some way behind LEDs in performance improvements, they have already reached efficacies of 32–64 lm/W at 1000 cd/m^2 [3, 17].

As efficacy increases, the thermal management, one of the major problems with LEDs, is simplified, and the system costs can be reduced. Indeed, to a large extent, the general performances depend on the luminaire, which should keep the LED chip cool by dissipating the heat efficiently. In other words, the design of a luminaire, especially the provision of heat sinks, plays an important role in determining the life expectancy of the LEDs.

8.3.2 Efficient Ballasts

All discharge lamps have a negative impedance characteristic, and a supplementary device (called ballast) should be used for proper operation; it supplies a high voltage to initiate the discharge arc and then limits the current to levels that allow the discharge arc to be stabilized during normal operation (it may also include capacitors to correct the power factor). There are two broad categories of ballasts: electromagnetic (also known as ferromagnetic or core and coil ballasts) and electronic (also called high-frequency or solid-state ballasts). Electromagnetic ballasts comprise a magnetic core of several laminated steel plates wrapped with copper windings and are the most popular. Although they enable step or continuous lamp-dimming capability (however, not below 20% for the last versions), electromagnetic ballasts have large size and weight, low efficiency, and sensibility to voltage changes [18]. Electronic ballasts can overcome all these drawbacks offering simultaneously more possibilities for lighting control; they use electronic reactors to allow lamps to be operated at much higher frequencies (over 20 kHz, where the lamp efficacy can rise by 10–15%). Other benefits of lamps operated by electronic ballasts are [6]: an increased mean lumen output (may be up to 20%, depending on lamp type), longer lamp life (up to 30%), smaller size and lower weight; much less lamp flicker, less lamp noise, improved lumen maintenance, better starting and operational control of the lamp, a power factor of one without the need for a power factor correction capacitor, more accurate lamp and circuit control that enables full dimming capability or a seamless integration into building energy management systems (BEMS), etc.

The latest products offer additional functions: they may incorporate photocells, automatically dimming the output of a fluorescent lamp according to the daylight availability (saving significant amounts of energy); some ballasts include circuits that automatically detect and adapt to the input voltage, allowing a single ballast to be used on multiple voltages; the use of intelligent circuitry to optimize lamp starting and restarting, which allows fluorescent lamp life to be increased (this function is important if the lamp is being activated frequently through, say, the use of energy-saving occupancy sensors).

In conclusion, electronic ballasts are energy efficient, produce less heat load for the air conditioning system, and eliminate flicker and hum. Products are available for compact fluorescent and full-size lamps, connecting up to four lamps at a time (3–lamp and 4-lamp ballasts reduce material and energy costs, as compared with 2-lamp ballasts, because less units will be required). Fewer types of ballast can also increase the overall system efficiency as the gear loss of 4-lamp electronic ballast is typically less than the sum of the gear loss of four 1-lamp electronic ballasts. However, as the cable between the ballast and the lamps will be well screened for electrical safety as well as to restrict the electromagnetic interference to an acceptable limit, the multiple lamps on single ballast arrangement are usually applied to lamps in the same luminaire [16].

As an integral component of the lighting system, ballasts need power in order to function; ballast power losses range from a few percent to as much as 40% of the total lighting system consumption, depending on the efficiency of the ballast adopted (the latest generation of high-efficiency magnetic ballasts reduce losses to about 12% while electronic ballasts are typically 30% more energy efficient). In order to steadily phase out inefficient ballasts, the European Union adopted the Directive 2000/55/EC on energy efficiency requirements for ballasts for fluorescent lighting that came into effect on 21 May 2002 (the purpose of this directive is to improve the efficiency of the lighting systems by limiting the ballast losses); it requires manufacturers to mark ballasts, indicating their efficiency, and it bans the sale of inefficient ballasts throughout the EU. Similar norms have been introduced in the USA [19].

For this purpose, CELMA (Federation of National Manufacturers Associations for Luminaires and Electrotechnical Components for Luminaires in the EU) developed a classification system that takes both the lamp and the ballast into account; ballasts are classified according to their Energy Efficiency Index (EEI) in the following classes [15]:

- Class D: magnetic ballasts with very high losses
- Class C: magnetic ballasts with moderate losses
- Class B2: magnetic ballasts with low losses
- Class B1: magnetic ballasts with very low losses
- Class A3: electronic ballasts
- Class A2: electronic ballasts with reduced losses
- Class A1: dimmable electronic ballasts.

Dimmable ballasts are considered A1 if they fulfil the following requirements:

- At 100% light output setting, the ballast fulfils at least the demands belonging to A3.
- At 25% light output setting, the total input power is equal to or less than 50% of the power at the 100% light output setting.
- The ballast must be able to reduce the light output to 10% or less of the maximum light output.

The sale of Class D ballasts has been banned since 21 May 2002 and Class C ballasts since 21 November 2005. A date for the phasing out the less efficient Class B ballasts (Class B2) may be set if EU sales of Class A ballasts do not increase sufficiently [20].

8.3.3   Efficient Luminaries

A luminaire (fixture) is a unit consisting of the lamps, ballasts, reflectors, lenses or louvers and housing. The main function is to focus or spread light emanating from the lamp(s) without creating glare. The efficiency of a fixture is the percentage of lamp lumens produced that actually exit the luminaire in the intended direction; it varies greatly among different fixture and lamp configurations.

Improperly designed or maintained luminaires can cause lamp or ballast overheating and reduce product lifetime, while those with a low luminaire maintenance factor (LMF) will require an over-dimensioning because the LMF depreciation factor must be taken into account when installations are calculated according to the existing standards. On the other hand, in order to obtain a high utilization factor (UF) it is important to direct the light to the intended surface or area (the higher the UF is, the lower is the energy needed to operate the lamp). Several technical solutions for an increase in LMF, but also in luminaire utilization factor are:

- a higher degree of ingress protection (IP rating);
- use of new materials (for instance glass with self-cleaning activated by UV rays);
- use of internal top reflectors or reflective top shielding covers;
- appropriate use of transparent protector material with reduced diffuse optic properties and high optic transmittance;
- multi-facet reflector technology for highly asymmetric light distribution;
- the change from a painted reflector to an aluminized reflector.

8.4   Energy Efficiency in Indoor Lighting Systems

Although the use of modern, energy efficient lighting technologies has been increasing over the last several years, particularly in the commercial sector, a large proportion of total lighting energy is used by inefficient and outdated technologies, e.g. incandescent lamps, low-efficacy fluorescent lamps with low-efficiency ballasts, etc. As an example, in 2007, in the EU-27 the incandescent lamps (GLS) still held the dominant position with 767 million units sold and some 54% (2.6 billion units or 13.1 lamps/household) of the existing stock; one third (33%) of the sold non-directional GLS lamps are 60 W and 31.6% are 40 W [21].

8.4.1   Policy Actions to Support Energy Efficiency

Recognizing that the incandescent technology is very inefficient and that higher-quality and lower-priced new lamps are becoming increasingly available and popular, a number of countries have enacted since early 2007 legislation or regulations to phase out incandescent lamps and other low-efficient lighting sources within their jurisdictions [17]. The intention of the regulations already adopted or under preparation is to encourage the usage of higher efficiency lamps (most notably CFLs) in place of standard GLS lamps and thereby eliminate a key source of energy waste. The timing of the phase-out varies between different countries; in most cases, and for all the larger markets, a phased approach is being adopted where certain parts of the GLS market are prohibited from sale earlier than others. The start dates for when the phase-out regulations begin to enter into force vary from as early as the end of 2008 in the case of Australia to the beginning of 2012 for the United States, Canada and Korea; end dates for first tier requirements are between 2009 and 2014 for those that are already known. In the EU, the adopted regulations operate phased stages beginning in September 2009 and concluding in 2012. In addition to the regulatory approaches being adopted, many softer options are being implemented or are under development, including large-scale market transformation programmes, utility energy efficiency schemes, retailer initiatives and fiscal/financial incentives. Many of these schemes are of sufficient scale to have a major impact on local GLS and CFL markets.

Alternatively, the EU Regulation also imposes minimum requirements on discharge lamps, ballasts and luminaires, resulting in minimum efficiency and quality requirements on remaining products and in phasing-out as follows [21]:

- Linear T12 and T10 halo-phosphate lamps will be banned starting 2012 with the exception of lamps for special purposes.
- Minimum performance requirements for T8 and T5 linear lamps will be imposed and the T8 halo-phosphate lamps will be banned starting 2010.
- From 2012 new luminaires must be sold with electronic ballasts and from 2017 magnetic ballasts will not be permitted even for replacement in existing luminaires.
- Requirements on minimum lumen maintenance levels have been introduced.
- From 2017 (8 years after the Regulation takes effect) all fluorescent lamps must be designed to work with an electronic ballast.

The criterion to be used to determine if a light source will be allowed on the EU market is its energy label class (ranging from A to G, with A being the most energy efficient class

Table 8.2 Energy label classes corresponding to different ranges of the energy efficiency index

Energy label class	Energy efficiency index (E_1)
B	< 0.60
C	$=$ or > 0.60 and < 0.80
D	$=$ or > 0.80 and < 0.95
E	$=$ or > 0.95 and < 1.10
F	$=$ or > 1.10 and < 1.30
G	$=$ or > 1.30

and G the least). A lamp belongs to the energy label class A if its wattage (W) fulfils the condition [3]:

$$W \leq 0.15\sqrt{L} + 0.0097L \tag{8.1}$$

for a CFL without an integrated ballast, or

$$W \leq 0.24\sqrt{L} + 0.0103L \tag{8.2}$$

for other lamp types; in (8.1) and (8.2), L is the lamp lumen output. If the lamp cannot meet the requirements for energy label class A, an energy efficiency index is calculated. The energy efficiency index (E_1) is the ratio of the actual wattage (W) and the reference wattage (W_R) of the lamp, given by (8.3):

$$W_R = 0.88\sqrt{L} + 0.049L. \tag{8.3}$$

Table 8.2 shows the energy label class assigned to different ranges of the energy efficiency index.

The available alternatives to replace incandescent lamps are:

- *Conventional low-voltage halogen lamps*: available until 2016 (can reach C-class efficiency and can live up to 4000 hours);
- *Halogen lamps with xenon gas filling (C-class)*: use about 25% less energy for the same light output compared with the best conventional incandescent lamps;
- *Halogen lamps with infrared coating (B-class)*: use about 45% less energy for the same light output compared with the best conventional incandescent lamps and can live up to 3000 hours;
- *Compact fluorescent lamps (CFLs)*: use 65–80% less energy for the same light output compared with the best conventional incandescent lamps;
- *Light-emitting diodes (LEDs)*: a fast emerging technology having the luminous efficacy in the same range as CFLs (moreover, they do not contain mercury and live longer).

A comparative analysis of these technologies is presented in Table 8.3. Owing to the above regulations, EU citizens are expected to save close to 40 TWh and reduce CO_2 emission by about 15 million tons per year; about 5–10 billion Euros are expected to be reinvested in the EU economy [21].

Table 8.3 Efficiency of lamp technologies compared with conventional incandescent lamps (E-class)

Lamp technology	Energy savings	Energy class
Incandescent lamps	–	E, F, G
Conventional halogens (mains voltage 230 V)	0–5%	D, E, F
Conventional halogens (low voltage 12 V)	25%	C
Halogens with xenon gas fillings (mains voltage 230 V)	25%	C
Halogens with infrared coating	45–50%	B (lower end)
CFLs with bulb-shaped cover and low light output, LEDs	65%	B (higher end)
CFLs with bare tubs and high light output, LEDs	80%	A

Despite these advantages, the adoption of energy efficient lighting has to overcome several barriers [22]:

- *Economic Barriers*: The initial cost of new energy-efficient appliances is high; therefore, many household owners may be unable or unwilling to spend money for this purpose because of the slow rate of return on the investment despite rising energy costs. On the other hand, in businesses and industries, the main driver for the adoption of energy-efficiency measures are higher energy costs; unfortunately, sometimes the investment in retrofitting projects may have a long payback time (two or more years), especially for small and medium-sized businesses. This represents a barrier to adoption of the best solutions because the beneficiaries are not sure if they will be in business long enough to recover their costs. In addition, the overall economic slump also creates a disincentive to invest in these appliances.
- *Institutional Barriers*: Usually, any new construction code-changing process takes to much time to be passed into law, implemented and then enforced. This aspect also represents an important barrier to the adoption of energy-efficient lighting. For older buildings, only the high cost of retrofitting may represent an economic barrier.
- *Consumer Barriers*: From the consumer point of view, a lack of knowledge or awareness about the availability and potential of energy efficient lighting may represent a significant barrier to the adoption of new solutions (this is especially true for residential costumers). In the commercial and industrial sectors, despite a greater incentive from the point of view of the cost to benefit ratio, a large number of retailers and businesses still use inefficient solutions. Other consumer barriers include poor aesthetics because the user may not like or may not be used to certain colour characteristics of the new lighting product. Unacceptable performance is also a barrier, for example, people have sometimes reported flickering or buzzing sounds in their CFLs.
- *Manufacturing Barriers*: Different failures in the manufacturing process, negatively affecting the final product performances (life time, CRI, lumen output, etc.), also represent a barrier to the availability and adoption of efficient lighting products. Furthermore, emerging technologies, such as white LEDs are still at the developmental stage and have not yet attained a high level of overall performance.
- Conversely, numerous commercial buildings are leased and so tenants may be unmotivated to invest in a building they do not own, while the landlords are unmotivated because they usually make their tenants responsible for the utility bills.

8.4.2   Retrofit or Redesign?

When assessing the opportunities for improvement to an existing lighting system, the first step is to measure how effectively the existing light levels and characteristics serve their function. Next, the basic option facing the facility manager is whether to retrofit or redesign. In a retrofit, new lamps and ballasts are installed in existing luminaires, while the existing controls can be replaced (improved); in a redesign, the fixtures themselves may be replaced or moved.

8.4.2.1   (a) Redesign

If the building's primary spaces have been designated to new purposes for which the existing lighting system provides insufficient lighting conditions, the space may benefit from a redesign. Generally, in order to achieve an energy-efficient solution, the following design criteria have to be implemented and fulfilled [16]:

- light sources of high luminous efficacies;
- lamp control-gears of low energy losses;
- luminaires of high light output ratios;
- room surfaces of high reflectance;
- optimum mounting height.

However, the energy-efficiency criteria interact with other lighting effect criteria, and appropriate trade-offs may be necessary. For instance, from the energy efficiency point of view, it is recommended light sources are chosen with high luminous efficacies; nevertheless, such energy criterion must be compatible with other lighting design criteria, i.e. colour rendering. Moreover, in many applications, the optical features are frequently the lead criteria in choosing lamp types and lamp efficacy may become a secondary consideration.

A careful selection of luminaires results in a reasonable illuminance, minimum direct glare, reflected glare and veiling reflections; as a result, both the tasks of visibility and productivity can be improved. Some important factors in this procedure are the Utilization Factor (UF), luminaire dirt depreciation (LDD), room surface dirt depreciation (RDD), lamp lumen depreciation (LLD), and lamp failure factor (LFF).

8.4.2.2   (b) Retrofit

There are many opportunities for cost-effective retrofits to an existing lighting system and it is possible to simultaneously increase lighting levels and use less energy if the most efficient technologies and practices are implemented. The business benefits of an appropriate lighting system should also be considered. From an economical point of view, retrofitting the existing lighting system of less energy efficient equipment into new energy efficient equipment reduces its operating costs, but can also improve the light quality, reduce operating costs, increase the working productivity, improve the visual environment, etc. [16].

On the other hand, a basic choice will be whether to replace the existing lighting system all at once in a planned upgrade or replace individual components as they fail. Apparently, replacing individual components appears to be the easiest path forward as it avoids the upfront cost of equipment and installation labour and the potential disruption of a renovation. Based on these arguments, most companies replace lamps when someone notices a lamp is burned out [11]; however, a planned upgrade presents several major advantages [11, 19]:

- good lighting performance, uniformity and space appearance by switching from a type of light source to another all at once, avoiding confusion resulting from maintaining luminaries with different light output;
- higher energy savings and greater lighting quality resulting from re-evaluating the existing lighting system and upgrading it to current best practices;
- bulk purchasing may yield savings;
- reduced time and labour costs.

The rule of thumb suggests that group re-lamping must be done at 50–70% of the lamp's rated life as the lamp's operating interval depends on site-specific factors i.e. quality of power supply, environmental conditions, number of On–Off commutations, user behaviour, etc. As a result, the same light source may have different operating characteristics and lives from one location to another and must be replaced at different times. Therefore, it is important for the energy manager to maintain records on lamp and ballast replacements and determine the most appropriate re-lamping interval; this also helps keep track of maintenance costs, labour needs and budgets.

Evaluation of various retrofit options can be made in terms of their payback periods. Although there are numerous potential combinations of lamps, ballasts and lighting systems, a few retrofits are very common; they are presented below.

1. Lamp Changes
In this case, the most used solutions are:

- Incandescent lamps to CFLs: it represents one of the most efficient solutions available today for improving energy efficiency in residential lighting. More details helping to identify suitable replacements for both standard and soft output incandescent lamps can be found in [3];
- Incandescent lamps to halogen incandescent lamps (C and B energy class);
- Incandescent lamps to LED lamps (the efficacy of LEDs has been improving rapidly and has doubled roughly every two years since the 1960s);
- Standard halogen lamps to white-LED lamps;
- Fluorescent lamps to fluorescent lamps (T12 to T8 or T8 to high-output T5 lamps; the T5/T8 lamp with electronic ballast combination is practically the most popular choice);
- Fluorescent lamp to LEDs.

2. Ballast Type Changes
The existing solutions are:

- Magnetic ballast to magnetic ballast with low or very low losses;
- Magnetic ballast to electronic ballast.

3. Luminaire Changes

When it is decided to go for retrofitting, reflectors replacement can be a very simple and cost effective option; in this case, the luminaire efficiency is increased because less light is trapped and wasted within the luminaire fitting. The effectiveness of the reflector depends on its geometrical shape, coating material and the efficiency of the luminaire. Reflector replacement can usually reduce the number of fluorescent lamps in three- and four-lamp luminaires for the same light output levels (usually, the remaining lamps may need to be relocated in order to maximize light output and uniformity). In combination with higher output fluorescent lamps and electronic ballasts, light output may be increased significantly with suitable reflectors. Old and degraded luminaires that cannot be rectified by cleaning alone are generally excellent reflector retrofit candidates.

Conversely, shielding devices are usually installed in luminaires to control uncomfortable glares but inefficient shielding devices will degrade even the most efficient lamps and ballasts. Therefore, balancing visual comfort (glare control) and luminaire efficiency should be the key to achieving success in reflector retrofit.

Sometimes replacing rather than retrofitting the luminaires can be more economical and cost effective, especially where there is a major change in the lighting system requirements: new luminaires can optimize efficiency, visual performance, technology compatibility and aesthetics, etc. [16].

4. Lighting Control Upgrades

Taking into account the complexity of this problem, the lighting control is presented in detail in the next section.

8.4.3   Lighting Controls

Nowadays, the inappropriate operation of lighting systems is causing a large proportion of electric light to be delivered to spaces where no one is present, or for which there is already adequate daylight. Energy consumption can be reduced by suitably controlling the number of operating hours and/or the level of lumen output; lighting controls offer the ability for systems to adjust their characteristics to the existing specific conditions.

Lighting controls may be manual or automatic, or a combination between the two; they must also be 'user friendly'. There are several control technology upgrades for lighting systems, ranging from simple (for instance, installing manual switches in proper locations) to more sophisticated (installing occupancy sensors). The choice of lighting controls has a large impact on total lighting energy use; according to [6], using advanced automatic controls will save an important fraction of energy (20–35% is typical) and can be highly cost-effective. Energy used for electric lighting depends upon the degree to which the lighting controls reduce or turn off the lights in response to [16]:

- the availability of daylight;
- changes in visual tasks;
- occupancy schedules;
- cleaning practices.

Improvements in the technology and continuously decreasing prices of lighting controls, combined with economic and environmental considerations, are leading to the increased use of advanced lighting controls, particularly in the commercial and public sectors, where lighting represents an important share of total energy costs. At the same time, automatic controls should be designed to avoid annoying the occupants; for instance, abrupt changes in the illuminance could be irritating in critical task areas and may create labour or security problems. As a result, often the automatic controls do not provide the saving expected because of user dissatisfaction or sabotage (they are not sufficiently user friendly) [5].

Nowadays, automatic lighting controls are used in only a small fraction of the cases where it would be currently economic to install them; consequently, lighting control offers important energy-saving opportunities. Currently, various control strategies are implemented, their suitability depending on the specific applications and patterns of energy usage. All of them can be framed into one of the following two major types:

- On/Off control;
- level control.

For the first type, controls offer only the ability to turn on and off the existing lighting system; by using the level control, the luminous flux emitted by lamps may be adjusted continuously or in steps (usually, two or three). All control strategies can be implemented either manually or automatically.

8.4.3.1   (a) Manual Control

Manual control is the simplest form of control, providing mainly on/off functions; in this case, the energy conservation device is represented by the standard manual, single-pole switch. The efficiency of manual control mainly depends on switch location because if switches are far from room exits or are difficult to find, occupants are more likely to leave lights on when walking out a room (occupants do not want to walk in darkness to find exits) [11]. An interesting opportunity for retrofitting consists in conveniently dividing the existing supply circuits and installing supplementary switches everywhere needed in order to obtain local lighting systems.

Solid-state dimmers for incandescent lamps level control are also available but in the future their importance will diminish (incandescent lamps will be banned). Fortunately, some ballast producers nowadays offer the opportunity to use these dimmers in combination with energy-efficient CFL lamps, thus facilitating the upgrading of incandescent systems.

8.4.3.2   (b) Automatic Control

Automatic lighting controls are also known as *responsive illumination*; despite representing a mature technology, they are being adopted only slowly in buildings and thus still offer a huge unrealized potential for cost-effective energy savings [6].

(b1) Automatic on/off Control
Energy consumption can be reduced by controlling the number of operating hours namely by turning off the light when it is not needed; all spaces that are not used continuously should

Table 8.4 Estimated savings from occupancy sensors [11]

Application	Energy savings [%]
Offices (Private)	25–50
Offices (Open spaces)	20–25
Rest rooms	30–75
Corridors	30–40
Storage areas	45–65
Meeting rooms	45–65
Conference rooms	45–65
Warehouses	50–75

have automatic switching, allowing lights to be turned off when the space is not in use. The automatic on/off control can use occupancy detection or timers.

Nowadays, *occupancy (or motion) sensors* are accepted as an effective energy-saving device (Table 8.4); there are three types of occupancy sensor, all of which are based on the detection of motion: passive infrared *PIR* (detects the motion of the heat sources), ultrasonic *US* (detects objects moving in the space) and hybrids (called also Dual-Technology (*DT*) sensors, they have both ultrasonic and passive infrared detectors; this technology avoids false turning on produced by wind-blown curtains or papers, for instance. Another dual technology control incorporates a microphone sensor, which detects small sounds, such as the turning of pages, even though an occupant would not show any appreciable movement in the room).

When the sensor detects motion, it activates a control device that turns on the lighting system; if no motion is detected within a specified period, the lights are turned off until motion is sensed again. With most sensors, sensitivity (the ability to detect motion) and the time delay (difference in time between when sensor detects no motion and lights go off) are adjustable. Occupancy sensors can be combined with dimming controls or stepped switching so that illuminance levels might be lowered to low ambient levels when the occupants leave a space [6, 11].

Occupancy sensors are available as wall-mounted or ceiling-mounted units; the first version may be used for small spaces that do not have obstacles to signal detection; the latter is recommended for larger spaces, irregularly shaped rooms and those with partitions. It is important to underline that energy savings may not be realized if the sensors are improperly installed or are disabled by frustrated occupants. For instance, *PIR* detectors must have a direct line of sight to the occupants to detect motion; on the other hand, *US* sensors do not require a direct line of sight to occupants, but wind-blown curtains or papers can trigger the sensor incorrectly.

Timers or *Time Clocks*, available in electronic or mechanical technologies, switch lights on and off at pre-set times; therefore, they can be used to control light systems when their operation is based on a fixed (imposed) schedule. However, regular check-ups are needed to ensure that the time clock is controlling the system properly, or to change setting according to different seasons (normally, the timer should be re-set at least twice annually to suit the sun summer/winter set/rise times). After a power loss, electronic timers without battery backups can get off schedule-cycling on and off at the wrong times.

Occupancy sensors and timer switches are occasionally configured to a manual on/automatic-off arrangement, by which the occupants turn lights on in their work space at the beginning of the day, or when needed [16]; timers are frequently used in combination with very low level emergency-egress lighting to ensure that users can always find their way safely around a space.

(b2) Automatic Level Control

In this case, electric-lighting levels are automatically adjusted in response to the detected illuminance level to maintain a pre-set value. Automatic level controls are a combination of photosensors and electronic ballasts with dimming functions: sensors continuously measure the illuminance level, while ballasts automatically adjust the lumen output of artificial light sources. Such an automatic dimming system has the following advantages:

- It compensates for the wasted power due to lamp lumen depreciation and luminaire dirt depreciation.
- It adjusts the lighting levels in accordance to the activity levels.
- It responds to variations in daylight availability (see the next subsection).

Generally, electronic ballasts with dimming functions operate linear fluorescent lamps; many of these products start the lamps at any dimmer setting, and do not have to be ramped up to full-light output before they dim. Most dimmable ballasts now have separate low-voltage control leads that can be grouped together to create control zones, which are independent of the power zones. The dimmer modules used in the high-power lighting systems are suitable for interfacing with time clocks, photocells or computers and may assure dimming ranges from 100 to 1%. Several manufacturers offer dimming ballasts for higher-wattage rapid-start compact fluorescent lamps; the lowest dimming limit is 5% and the dimming range varies with the manufacturer. They are designed to accept the AC phase-control signals from incandescent wall-box dimmer controls in order to facilitate upgrading an older incandescent system to an energy-efficient CFL system (no new wiring required).

The control generally uses DC 0–10 V signals, even if dimming ballasts designed to accept AC control signals are also available on the market [11]. Different control standards provide the interface platform, in particular for equipment and devices of the same manufacturer; in order to guarantee the exchangeability of dimmable electronic ballast from different manufacturers, the international standard DALI (Digital Addressable Lighting Interface) is now available. It was developed to overcome the problems associated with the analogue 1–10 V control interface and provides a simple (digital) way of communication between intelligent components in a local system; by using DALI, each luminaire of a lighting system could be individually addressed and programmed (as a result, different groups of luminaries may be programmed to fulfil various tasks) [23].

The impact of electronic ballasts on power quality sometimes can be harmful; fortunately, most of the models available on the market respect the restrictions imposed by [24], having a current total harmonic distortion (THD) of less than 15% throughout the dimming range. Normally, dimmable lighting systems are expensive and may not be applicable for every installation. It is recommended that dimmers should be used only where it is anticipated that lighting level control is needed.

Compared with on–off controls, level controls generally increase energy savings, better align lighting with human needs, and extend lamp life; they are also useful for spaces that have more artificial (electric) lighting than is currently needed, and have the added benefit of dimming lights further when natural light from outside is available. Such systems can also be used to dim lights for other reasons, such as for presentations. Dimming armatures by as much as 50% may be barely noticeable to building occupants, unless they are involved in tasks requiring visual acuity [11].

Micro-processor control can also be applied to a lighting system, from a stand-alone system in a single space to the entire system in a building. It is usually connected with controlling devices such as timers, photosensors, etc. to provide the desired lighting group/system performance. It offers great flexibility to lighting control as typical control functions usually include the following:

- automatic compensation for lamp lumen depreciation and luminaire dirt depreciation;
- fine tuning of lighting level to suit actual requirements;
- scheduling of lighting operations to minimize the operating hours;
- automatic daylight compensation control.

8.4.4 Daylighting

The sustainable development concept has revived the interest for daylighting, i.e. for the use of daylight as a primary source of illumination in a space; in fact, daylight is a flicker-free source, generally with the widest spectral power distribution and highest comfort levels. By daylighting a space, both psychological and energy efficiency benefits can be obtained [25]:

- Psychologically, the presence of controlled daylight improves the overall attitude and well-being of the occupants (increases in worker productivity and mood, and decreases in absenteeism and errors are commonly stated advantages).
- From an energy efficiency point of view, daylighting can offer great energy savings due to reduced electric lighting loads. Reported savings in lighting energy consumption from daylight design and harvesting vary between 15% and 80% [6]. Because preponderant edifice occupancy patterns are high for non-domestic buildings during the day, there is a greater potential to save energy by using daylighting in this zone (it is often the most energy-efficient lighting option, particularly for warehouses that have large roof areas with large open spaces).

On the other hand, poorly introduced daylight can negate some of these benefits by direct sunlight introducing disabling glare or distracting veiling reflections, or solar gains causing uncomfortable thermal conditions.

Over the last decades there has been a steady increase in interest in the use of daylight in architecture, especially in European countries. Depending on cloud conditions and building location, daylight luminance levels may be highly variable. However, they are almost always more than high enough to provide necessary minimum internal illuminance levels, on condition that the design allows enough daylight to enter the building and be distributed appropriately. Various studies have proved that the area close to the window receives enough daylight for all illumination needs to be provided on some occasions, only by daylight; deeper into the room, a mixture of artificial light and daylight is needed, and deeper still almost all light is needed from artificial light. Energy savings are between 60% and 70% in the day-lit area, 30% and 40% in the mixed-light area and 5% and 20% in the artificial-light area [6].

The previous data indicating the energy saving potential give an idea about the value of having the daylight-responsive control systems and the extra savings that can be achieved by increasing the availability of daylight deeper into the floor plan. Unfortunately, daylight is a dynamic source of lighting, i.e. the illuminance from the sky is not constant, and the variations in daylight can be quite large depending on season, location or latitude, and cloudiness. These variations must be compensated by the control system, in order to maintain a constant and uniform illuminance in the task area, avoiding the occupants' discomfort.

There are at least two dimensions to daylight-responsive controls [11, 26]: the control of the daylight input to the space, and the control of the electric lighting output. The first is critical for providing adequate quantity and quality of daylight in interior spaces (simultaneously, with proper fenestration solar control, solar gains during cooling load periods can be mitigated and solar gains during heating load periods can be beneficial, reducing both the overall cooling and heating requirements of a space); the second saves energy and improves the overall distribution of light when and where daylight is insufficient.

The daylighting control employs strategically located photosensors and uses continuous dimming techniques that allow users to adjust lighting levels over a wide range of lighting output and offer far more flexibility than step-dimming controls. The fluorescent lighting is dimmed to maintain a required band of light level when there is sufficient daylight present in the space. As continuous dimming follows the daylight pattern very closely, it is often more acceptable to occupants, and can produce higher energy savings, particularly in areas with highly variable cloud cover. Continuous dimming also responds to changes in light output due to dirt depreciation on luminaires and lamps, and lamp lumen depreciation due to lamp aging. It is achievable using either analogue or digital ballasts. As classic control systems present some difficulties to adjust their performances to the rapid changes in daylight and to occupants' preferences, new analysis methods and controllers based on Artificial Intelligence techniques (for instance, Artificial Neural Networks and Fuzzy Control) have been proposed [26–28].

Usually, the room is divided into several zones, according to the existing fenestration and working tasks; the correct placement of sensors is critical to balancing lighting quality and energy saving (various type of automatic sensors should be located in a manner such that the portion of the lighting zone being controlled experiences fairly uniform daylight illuminance levels). Electric lighting systems should be designed to be compatible with the daylighting system in respect of luminance ratios, controls and colour rendition. This co-ordination helps to enhance the daylight quality and improve user acceptance of the energy saving features.

8.5 Energy Efficiency in Outdoor Lighting Systems

Globally, an estimated 218 TWh of final electricity was consumed by outdoor stationary lighting in 2005, amounting to about 8% of total lighting electricity consumption; the average luminous efficacy of the used light sources was 74 lm/W, higher than in any other sector except industry [6]. Road and street lighting represents a large amount of energy consumption each year as light intensity is often excessive when traffic density is low; therefore, energy conservation for large-scale illumination tasks such as street lighting is gaining considerable importance. By correct investments and utilizing the existing technology it is possible to reduce today's energy consumption in this sector by as much as approximately 60% (for Europe as a whole, this stands for about 36 TWh a year); in addition, a significant saving on maintenance costs can be achieved [29].

8.5.1 Efficient Lamps and Luminaires

At present, high intensity discharge (HID) lamps are the popular choice as light sources in road lighting; the HID lamp family includes high-pressure sodium (HPS), low-pressure sodium (LPS), metal halide (MH), and mercury vapour lamps [30]. According to [6], almost 62% of total outdoor light is provided by high- and low-pressure sodium lamps, 30% by mercury vapour lamps and 6% by metal halide lamps; the remaining 2% is mostly provided by halogen and incandescent lamps. HPS lamps are the most widespread type of light source used for outdoor applications, due to two important economic considerations: low initial and operating costs. Despite these economic advantages, MH lighting systems are gaining continuously wider acceptance for outdoor applications in large part because they have become more competitive on reducing system costs, both initial and operating. In addition, MHLs are more efficacious for night-time applications because, for the same wattage, MH spectra are better tuned to the spectral sensitivity of the human retina at mesopic light levels [31].

Mercury vapour street lamps are outdated and use far more energy than HPS or MH devices. These light sources are extremely inefficient and use obsolete technology, which results in a very high cost of ownership through high energy use; if all the inefficient mercury vapour street lighting in Europe was upgraded to the latest technology, Europe would achieve 1 billion Euros in running cost savings (based on 2006 energy prices) [32]. This is equivalent to:

- 4 million tons of CO_2 savings;
- 14 million barrels of oil per year;
- the equivalent consumption of 200 million trees.

In recent years, LED array illumination has received attention as an energy-reducing light source. LED road illumination requires only about 30–50 % of the electric power needed for HID lighting, while its lifecycle can be more than three times longer [33]; the amount of time needed to exchange defective fixtures could be reduced, and it is expected that an LED system would be comparatively maintenance free. Other LED advantages address the free content of mercury, lead or other known disposal risks and instant operation, without run-up time or re-strike delay. Further, while MH and HPS technologies continue to improve incrementally, LED technology is constantly improving very fast in terms of luminous efficacy, colour quality, optical design, thermal management, and cost [33]. In such a background, and as a result of the significant improvements to luminescent efficiency in recent years, LED lighting can be expected to fully replace previously used light sources within our lifetimes.

However, current LED product quality can vary significantly between manufacturers, so due diligence is required in their proper selection and use. LED performance is highly sensitive to thermal and electrical design, weaknesses that can lead to rapid lumen depreciation or premature failure. Further, long-term performance data do not exist given the early stage of the technology's development.

Energy efficiency of outdoor lighting systems cannot be obtained by simple selection of more efficient lamps alone; indeed, it also encompasses the optical efficiency of the luminaire, and how well the luminaire delivers light to the target area without casting light in unintended directions. The goal is to provide the necessary illuminance in the target area, with appropriate lighting quality, and only efficient luminaires along with the lamp of high efficacy may achieve the optimum efficiency. Luminaires differ in their optical precision. Photometric reports for outdoor area luminaires typically state downward fixture efficiency, and further differentiate downward lumens as 'streetside' and 'houseside' [34]. Mirror-optic luminaires with a high output ratio and bat-wing light distribution can save more energy; simultaneously, the luminaire should ensure that discomfort glare and veiling reflections are minimized. System layout and fixing of the luminaires also play a major role in achieving energy efficiency. In conclusion, fixing the luminaires at optimum height and usage of mirror optic luminaries leads to energy efficiency.

LED luminaires use different optics than MH or HPS lamps because each LED is, in effect, an individual point source. Effective luminaire design exploiting the directional nature of LED light emission can translate to lower optical losses, higher luminaire efficacy, more precise cut-off of backlight and up-light, and more uniform distribution of light across the target area. Better surface illuminance uniformity and higher levels of vertical illuminance are possible with LEDs and close-coupled optics, compared with HID luminaires. Their special design offers the following advantages:

- a reduction in luminous intensity in the 70° to 90° vertical angles to avoid glare and light trespass;
- zero to little intensity emitted between 90° and 100°, the angles that contribute the most to sky glow;
- a higher reduction in light between 100° and 180° (zenith), which also contribute to sky glow.

Several comparative data between luminaires containing HID lamps and LED luminaires presented in [34] highlight the following potential benefits offered by well-designed LED luminaires for outdoor area lighting:

- the uniformity was improved by more than a factor of two;
- a better control of the illuminance;
- improved uniformity ratio.

Moreover, since HID lamps are high-intensity near-point sources, the optical design for these luminaires causes the area directly below the luminaire to have a much higher illuminance than areas farther away from the luminaire; this over-lighting represents wasted energy, and may decrease visibility since it forces adaptation of the eye when looking from brighter to darker areas. In contrast, the smaller, multiple point-source and directional characteristics of LEDs can allow better control of the illuminance distribution.

The EU Regulation concerning the eco-design requirements includes the following issues referring to the outdoor lighting [21]:

- Minimum performance requirements are introduced for high intensity discharge (HID) lamps, consisting in phasing out of high-pressure mercury (HPM) lamps following an agreed schedule, the largest wattages being phased out first.
- Requirements on minimum lumen maintenance levels are introduced.
- 90% of the high-pressure sodium (HPS) lamps should have a lifetime of more than 16 000 hours.
- Metal halogen lamps should have a minimum life time of 12 000 h for 80% (frosted) and 90% (clear) reduction on lumen output.
- Requirements of directional light sources for street lighting luminaires (not only HID) are established in order to reduce light pollution.
- New minimum performance requirements are introduced for all HID lamps to minimize mercury content.

The following recommendations are of interest in the case of output lighting systems retrofit:

- Replacement of mercury vapour lamps with higher-efficiency alternatives
- Replacement of probe-start metal halide lamps with higher-efficiency options (e.g. pulse-start metal halide lamps)
- Installation of metal halide lamps for colour critical applications when higher illumination levels are required
- Installation of high pressure sodium vapour lamps for applications where colour rendering is not critical
- Ensure high-pressure sodium and metal halide lamps permitted on the market have adequate efficacy levels and are matched to good ballasts and lighting controls
- Higher luminaire efficiency
- The photometric performance of all luminaires should be measured and available according to standard test procedures. If the latter are not yet developed for specific luminaire or lamp combinations they need to be established
- Outdoor luminaires should generally be of a cut-off variety to avoid light trespass and minimize light pollution unless there is a strong aesthetic argument to the contrary
- Better control systems

Finally, some aspects regarding power quality are of interest in outdoor lighting systems. On the one hand, all high-pressure discharge lamps are non-linear loads generating a significant amount of current harmonics (important levels of current total harmonic distortion (THD) were observed) [35]; on the other hand, all lamps are single-phase loads producing the unbalance of the three-phase supply system. These electromagnetic disturbances produce supplementary power losses in all line and neutral conductors [36, 37]; for large outdoor lighting systems, the amount of these additional losses may become unacceptable, and measures to mitigate the generated disturbances must be implemented. Intolerable voltage drops can also appear for too long supply lines.

8.5.2   Outdoor Lighting Controls

Currently, the outdoor lighting control systems range from very simple to the most modern applications. The amount of light necessary on the road depends on legal requirements, tarmac, traffic volume, type of road, speed limit and surroundings. In the case of one of these parameters changes during the night (e.g. traffic volume, speed limit), the luminous flux can be decreased by reducing power (voltage) of the lamps or switching lamps off at all. The reduction in voltage is however limited, since the usual physical voltage drop, which is a function of the length of cable, is the limiting factor for lamps to come on at the end of an installation's cable. In order to estimate savings, the structural condition of the affected installation has to be analysed individually.

8.5.2.1   High-intensity Discharge Lamp Dimming

Generally, HID lamp dimming can result in energy savings, peak demand reduction and greater flexibility in multi-use spaces. For the first purpose, dimming can be achieved either manually (via input from a switch) or automatically (via input from a control device such as timers, occupancy sensors, photocells, etc.); for the second one, dimming can be scheduled using a time-programmable controller during times of peak demand. It is important to point out that shutting off and restarting the HID lamps does not represent a practical solution taking into account their characteristics (long warm up time and significant shorter lamp life if the burn time per start decreases below 10 hours); on the contrary, if the lamps are dimmed in response to a signal from an occupancy sensor or time-programmable controller, significant energy savings can occur during these periods, but the lamps will be able to achieve full light output quickly when the space becomes occupied again. HID lamps can be dimmed using step-level or continuous-dimming systems [38].

(a) *Step-level dimming* enables wattage reduction, usually from 100% to a step between 100% and 50% of rated power; accordingly, this solution is also known as a two-level dimming system. Depending on the lamp type and wattage, in a bi-level dimming system, the Low level may be 15–40% of light output and 30–60% of wattage (with possible energy savings as high as 40–70%). Tri-level dimming systems operating at three fixed light levels are also possible, offering a greater degree of flexibility to address multiple uses of the space. This dimming method usually employs constant-voltage autotransformer magnetic ballast with one or two additional capacitors, depending on whether the ballast provides bi- or tri-level dimming. Relay switching of the capacitors results in additional impedance, which reduces the lamp current and the wattage. The capacitor circuit configuration may be a parallel or series connection.

The solution is less expensive than continuous dimming systems and allows individual luminaire control (it is suitable for retrofit). In addition, fixtures are available with a dedicated occupancy sensor and dimming ballast, appropriate for direct fixture replacement. Ideal applications for step-dimming include spaces that may be unoccupied for long periods of time but still need to be lighted, such as parking lots, warehouses, supermarkets and malls. High pressure sodium lamps are typically used for parking lots and warehouses, while metal halide lamps are generally used for supermarkets and malls. It is worth mentioning that when HID lamps are dimmed below 50% of rated power, they may experience degradation in service life (by 90%), efficacy, colour and lumen maintenance, or they may even extinguish. As a result, NEMA recommends that the maximum dimming level is 50% rated lamp wattage for both metal halide and high pressure sodium lamps.

(b) Continuous (line-voltage) dimming enables a smooth, continuous reduction of lamp wattage; this technique is used anywhere it is advantageous to adapt the lighting system to a wide range of light levels. The control system may be one of the following three types:

1. Variable voltage transformer: reduces the primary voltage supplied to the ballasts, reducing light output and electrical input (enabling a reduction in rated power down to 50%).
2. Variable reactor: changes lamp current without affecting the voltage (enabling a reduction in rated power down to 50%).
3. Electronic control circuits: change the waveform of current and voltage input to the ballasts enabling a reduction in rated power down to 50%.

All these control solutions use electromagnetic ballasts with large size and weight, low efficiency, and sensibility to voltage changes. Electronic ballasts can overcome these drawbacks, supplementary offering more possibilities for lighting control. There are three main dimming methods for HPS lamps with electronic ballasts [18]: (i) variation of operating voltage of inverter; (ii) variation of operating frequency of inverter; and (iii) variation of pulse-width of inverter output signal (PWM). Variation of the voltage or frequency is easy to implement, but the colour of light will be affected; PWM is more complicated, but the colour of light output will be affected only minimally. No matter which ballasts are applied, the dimming range is limited for HPS lamps, and colour rending starts to shift even at 60% of rated power.

Several manufacturers offer solid state electronic ballasts for MH lamps, claiming that these ballasts provide better performance in a smaller package, have a high power factor, save more energy, generate less heat, and have lower maintenance costs compared with electromagnetic ballasts. These ballasts are more commonly available for lamps below 150 watts.

For all cases, the light output will be reduced further than the wattage reduction. In general, light output reductions are about 1.2–1.5 times the power reduction for metal halide lighting systems, and about 1.1–1.4 times the power reduction in high-pressure sodium lighting systems.

8.5.2.2 LED Dimming

LEDs offer many advantages for control and operation. This feature, combined with the ready ability to control each LED in an array individually via a microcontroller, may offer potential benefits in terms of controlling light levels (dimming) and colour appearance. Dimming drivers can dim LEDs over the full range from 100% to 0% by reducing the forward current, pulse width modulation (PWM) via digital control, or more sophisticated methods. Dimming does not result in a loss of efficiency as the LEDs are still operated at the same voltage and current as during full light output. In addition, lamp life is not affected by dimming, as is sometimes the case with frequently dimmed HID lighting. Therefore, dimming LEDs may lengthen the useful life of LEDs, because dimming can reduce operating temperatures inside the light source [39]. As LED driver and control technology continues to evolve, this is expected to be an area of great innovation in lighting. Dimming, colour control, and integration with occupancy and photoelectric controls offer potential for increased energy efficiency and user satisfaction.

8.5.2.3 Digital Lighting Control Systems

Modern digital lighting control aims to integrate the lamp dimming capabilities into more complex management systems consisting of three main elements [40]:

1. control unit in the luminaire;
2. CPU and GSM module in each of the switchboards of the installation;
3. remote data processor for management and controlling the individual installations.

In this case, modern technologies for supervision of the traffic will optimize the lighting: it will still need to supply for the worst case scenario but should also be able to automatically adapt fully to the current needs. If the light levels of a lighting system are adjusted according to predefined parameters such as time, traffic density, weather conditions, etc, the lighting system can be called *adaptive* or *dynamic*; such a system can be *intelligent* when light levels are adjusted in real time according to real needs. An intelligent road lighting control system is defined as a modern lighting control system based on the technology of computer science, communication, automation, and power electronics, which can automatically collect system information, analyse, deduce and estimate the collected information, and realize the optimum lighting control effect by changing the lamp light output in real time according to real needs [18]. Its main purpose is to save electricity and maintenance costs without negative effects on traffic safety. From the lighting point of view, this means providing the optimum luminance level based on the prevailing traffic and weather conditions.

These systems are capable of operating each luminaire individually from an installation. Moreover, they inform online about the state of each luminaire and its individual components including detailed fault detection. The savings are claimed to be up to 30% for electricity and 40% for maintenance. Investment costs for such systems, especially when retrofitting, are high and life cycle costs have to be calculated individually to estimate the payback time [40].

Communication is the key issue in an intelligent road lighting control system. As the control centre is far from the remote control unit and lamps, different communication solutions must be used. Wireless communications, such as radio and GSM/GPRS, are possible answers with high transmission rate and high capability for large-scale long-distance; however, they have relatively high operation costs. Wired communication, such as optic fibre, telephone lines or even power line carriers, are also potential solutions, depending on the budget. The use of optic fibre is very expensive but it has high capability and transmission rate; the telephone line is an economical solution, but has a low capacity and low transmission rate, while power line communication is economical due to the existing wires, but it is easily affected by interference. The choice of communication systems depends on the specific applications, requirements, and budget. There is only a right solution for the right application, whereas there is no single solution that is better than others in all cases [18].

In conclusion, the obvious advantage of using adaptive lighting, with built in intelligence, is reduced energy consumption and reporting from the lamp of the current status. This gives better control of the installation and ensures that the equipment actually delivers what is required to the customer, leading to an improvement in the quality of the delivered product, 'road lighting' [29]. Better control also gives increased predictability and secondarily it lowers maintenance and running costs, by being able to achieve better planning and better implementation of error corrections in the installations. Adaptive lighting introduces demands for better energy measurements in the installations allowing the automatic regulation of both electrical parameters, burning hours and light levels. If the same system implements the measurement function and the control function then the communication expenses will be reduced, it will minimize the number of components needed in the system so reducing costs and simplifying the operation and maintenance.

8.6 Maintenance of Lighting Systems

For all lighting systems, the general performances decease over time. This degradation can be the result of lamp lumen depreciation, lamp and ballast failure, dirt accumulation on lamps and luminaires, luminaire surface deterioration, room surface dirt depreciation, and many other causes; in combination, these factors commonly reduce the light output by 20–60% [18]. It is evident that all lighting installations need to be maintained to perform at maximum efficiency.

Designing lighting systems to compensate for lack of system maintenance can waste lighting energy since more light is provided initially than is necessary; assessing the future maintenance condition of the installation in the design step always results in a significant add-on margin. In some extreme cases, the margin can be as high as 40–50%, which means that the installation is over designed by 40–50% at its initial operating period [16]; however, the energy consumption may be reduced by using appropriate control systems. Taking into account this aspect, but also the labour cost, the maintenance schedule should be reviewed frequently during the initial operating period of an installation so that an optimum frequency for maintenance can be established.

The optimum interval for bulk re-lamping and the cleaning depends on the following factors [16]:

- *Type of premises*: governs the operational needs of the lighting installation (whether the requirement of maintaining the illumination level is stringent or loose, whether there are specific personnel to be responsible for the utilities management or not, etc.).
- *Location of premises*: affects the rate of dirt accumulation in luminaires.
- *Usage rate of a particular space and the existing control*: affects how long will it take for the illumination of a lamp to depreciate to an unacceptable level.
- *Type of luminaire*: affects the ease of accumulation of dirt, the loss in Utilization Factor resulted, and the labour efforts for cleaning.
- *Type of lamp*: governs the characteristic of the lumen depreciation as well as the nominal average lamp life of the installation.
- *Electricity and labour costs*: key factors in the economic analysis of the maintenance schedule (according to [40], total costs of a typical street lighting installation over a period of 25 years consist of 85% maintenance and electricity and only 15% investment costs).

From a maintenance point of view, it may not be cost effective to replace the lamp at a time when the lamp burns out, because it is labour consuming and the burn out of lamps occurs more or less in a random manner; it is therefore more cost effective to carry out group replacement at the economic lamp life. In this issue, intelligent road lighting control systems have great potential in the optimization of lighting maintenance in the following respects:

- Based on the information of burning hours and lamp types, it is possible to predict the end of lamp life, which can be used in planning the group replacement schedule.
- The control centre can detect lamp fault instantly (consequently, work time, response time, and inspection frequency can be reduced).
- With luminance meters or light output sensors, light levels can be adjusted to provide the proper amount of light.

References

[1] N. Lefèvre *et al.*, *Barriers to Technology Diffusion: The Case of compact Fluorescent Lamps*, International Energy Agency, COM/ENV/EPOC/IEA/ SLT(2006)10, http://www.oecd.org/env/cc/aixg.
[2] European Lamp Companies Federation, *The European Lamp Industry's Strategy for Domestic Lighting. Frequently Asked Questions & Answers on Energy Efficient Lamps*, 2007, http://www.elcfed.org
[3] B. Jacob, Lamps for improving the energy efficiency of domestic lighting, *Lighting Res. Technol.*, **41**, 2009, pp. 219–228.
[4] D. R. Limaye *et al.*, *Large-Scale Residential Energy Efficiency Programs Based on CFLs*, Project P114361 (2009), http://www.esmap.org
[5] D. L. Loe, Energy efficiency in lighting – considerations and possibilities, *Lighting Res. Technol.*, **41**, 2009, pp. 209–218.
[6] International Energy Agency, *Light's Labour's Lost. Policies for Energy-efficient Lighting. In support of the G8 Plan of Action*, http://www.iea.org/textbase/npsum/lll.pdf
[7] R. Van Heur, *Power Quality and Utilisation Guide*, Energy Efficiency, 2007, http://www.leonardo-energy.org
[8] B. Atanasiu and P. Bertoldi, Analysis of the electricity end-use in EU-27 households, *Proceedings of the 5th International Conference EEDAL'09*, Berlin, Germany, pp. 189–201.
[9] P. Bertoldi, Energy efficiency in lighting & project of EU quality charter for LED lighting, *CELMA/ELC LED Forum*, Frankfurt, 2010.
[10] N. Anglani *et al.*, Energy efficiency technologies for industry and tertiary sectors, the European experience and perspective for the future, *IEEE Energy2030* Atlanta, GA, USA, 2008.
[11] W. C. Turner and S. Doty (eds), *Energy Management Handbook*, 6th edn, The Fairmont Press, 2007.
[12] R. Mehta *et al.*, *LEDs – A competitive solution for general lighting applications, IEEE Energy2030*, Atlanta, GA, USA, 2008.
[13] T. Yoshihisa, Energy saving lighting efficiency technologies, *Quarterly Review*, **32**, 2009, pp. 59–71.
[14] National Framework for Energy Efficiency (Australia), *The Basics of Efficient Lighting. A Reference Manual for Training in Efficient Lighting Principles*, 1st edn, 2009, http://www.energyrating.gov.au/wp-content/uploads/2011/02/2009-ref-manual-lighting.pdf
[15] EN200212 CELMA Ballast Guide, http://www.eu-greenlight.org/pdf/BallastGuideEN200212.pdf
[16] EMSD Hong Kong, *Guidelines on Energy Efficiency of Lighting Installations*, 2007, http://www.emsd.gov.hk
[17] P. Waide, *Phase out of Incandescent Lamps. Implications for International Supply and Demand for Regulatory Compliant Lamps*, IEA, Energy Efficiency Series, 2010, http://www.iea.org
[18] L. Guo *et al.*, *Intelligent Road Lighting Control Systems*, Helsinki University of Technology, Department of Electronics, Lighting Unit, Report 50, Espoo, Finland 2008, http://www.lib.tkk.fi/Diss/2008/isbn9789512296200/article2.pdf.
[19] C. DiLouie, *Fluorescent Magnetic T12 Ballast Phaseout: It's Time to Upgrade Existing Lighting and Control Systems*, 2010, http://www.aboutlightingcontrols.org/education/papers/2010/2010_lighting-upgrades.shtml
[20] Energy Saving Trust, *Energy Efficient Lighting-Guidance for Installers and Specifiers, CE61 (2006)*, http://www.est.org.uk/housingbuildings

[21] P. Bertoldi and B. Atanasiu, *Electricity Consumption and Efficiency Trends in European Union – Status Report 2009*, http://www.jrc.ec.europa.eu

[22] O. Soumonni, *Lighting Energy Efficiency Potential in Georgia: a Technology and Policy Assessment*, Georgia Institute of Technology, 2008, http://stip.gatech.edu/wp-content/uploads/2010/05/intern08-diran-report.pdf

[23] Zumtobel Staff, *The Lighting Handbook*, 1st edn, 2004, http://www.zumtobelstaff.com.

[24] EN 61000-3-2 Electromagnetic Compatibility. Limits. Limits for Harmonic Current Emissions (equipment input current $\leq$ 16 A per phase).

[25] Architectural Energy Corporation, Daylighting metric development using daylight autonomy calculations in the sensor placement optimization tool, 2006, http://www.archenergy.com/SPOT/SPOT_Daylight%20Autonomy%20Report.pdf

[26] A. Cziker, M. Chindris and A. Miron, Implementation of fuzzy logic in daylighting control, *11th International Conference on Intelligent Engineering Systems INES2007*, 2007, Budapest, Hungary, pp. 195–200.

[27] A. Cziker, M. Chindris and A. Miron, Fuzzy controller for a shaded daylighting system, *11th International Conference on Optimization of Electrical and Electronic Equipment OPTIM'08*, 2008, Braşov, Romania, paper 586.

[28] S. L.Wong *et al.*, Artificial neural networks for energy analysis of office buildings with daylighting, *Applied Energy*, **87**, 2010, pp. 551–557.

[29] Intelligent Energy Europe, *Guide for Energy Efficient Street Lighting Installations*, 2007, http://www.e-streetlight.com/Documents/Homepage/0_3%20Guide_For%20EE%20Street%20Lighting.pdf

[30] M. Chindris *et al.*, A new policy in public lighting of Cluj-Napoca, *4th European Conference on Energy-Efficient Lighting*, 1997, Copenhagen, Denmark, 1, pp. 229–232.

[31] M. S. Rea *et al.*, Several views of metal halide and high-pressure sodium lighting for outdoor applications, *Lighting Res. Technol.*, **41**, 2009, pp. 297–320.

[32] BeneKit, *Storybook Efficient White Light – Think Inside the Box*, 2007, http://www.lighting.philips.com/microsite/benekit/index.php.

[33] Y. Aoyama and T. Yachi, An LED module array system designed for streetlight use, *IEEE Energy2030*, Atlanta, GA, USA, 2008.

[34] EERE Information Center. Building Technologies Program, *LED Application Series: Outdoor Area Lighting*, 2008, http://www.netl.doe.gov/ssl

[35] F. Pop and M. Chindris, Analysis of the Harmonic Distortion of the Public Lighting Network, *8th European Lighting Conference*, Amsterdam, 1997.

[36] M. Chindris *et al.*, Neutral currents in large public lighting networks, *International Conference on Renewable Energies and Power Quality (ICREPQ'04)*, 2004, Barcelona, Spain (paper 210 on CD).

[37] M. Chindris, A. Cziker and A. Miron, Computation of power losses in public lighting networks, *4th International Conference ILUMINAT2007*, 2007, Cluj-Napoca, Romania, pp. 8.1–8.6.

[38] C. DiLouie, *HID Lamp Dimming*, 2004, ecmweb.com/mag/electric_hid_lamp_dimming

[39] C. DiLouie, Controlling LED lighting systems: introducing the LED driver, *LEDs Magazine*, December 2004, http://www.ledsmagazine.com/features/1/12/6

[40] Intelligent Energy Europe, Energy Efficiency in Streetlighting and Transport Infrastructure. Reference material from COMPETENCE, http://www.transportlearning.net/competence/docs/Competence_reference%20material_urbandesign_en.pdf

Further Readings

P. Bertoldi *et al. The European GreenLight Programme 2000–2008 – Evaluation and Outlook*, 2010, http://www.jrc.ec.europa.eu

M. Bross and A. Pirgov, Do new types of energy saving lamps change the markets?, *Proceedings of the 5th International Conference EEDAL'09*, Berlin, Germany, pp. 612 621.

Bureau of Energy Efficiency, *Lighting System – Guide Book*, http://www.em-ea.org/Guide%20Books/book-3/Chapter%203.8%20Lighting%20System.pdf

Department of transportation, *Minnesota Roadway Lighting Design Manual*, 2003, www.dot.state.mn.us/trafficeng/lighting/lightingdesignmanual.pdf

Electronic Ballasts Market Forces and Demand Characteristics, 5th edn, Darnell Group, January 2009.

S. Fassbinder, *How Efficient are Compact Fluorescent Lamps?*, 2008, http://www.leonardo-energy.org

W. L. O. Fritz and M. T. E. Kahn, Energy efficient lighting and energy management, *Journal of Energy in Southern Africa*, **17**(4), 2006, pp. 33–38.

R. N. Helms and M.C. Belcher, *Lighting for Energy Efficient Luminous Environments*, Prentice Hall, 1991.

HERA, *LED – The Future of Energy Efficient Lighting*, 2010, http://www.hera-online.de

HOLOPHANE, *Lighting Guide. The Fundamentals of Lighting*, 2010, http://www.holophane.com

K. Johnsen and R. Watkins, *Daylight in Buildings*, ECBCS Annex 29/SHC Task 21 Project Summary Report, 2010, http://www.ecbcs.org

N. Khan and N. Abas, Comparative study of energy saving light sources, *Renewable and Sustainable Energy Reviews*, **15**, 2011, pp. 296–309.

Lighting Industry Federation Limited: Lamp Guide 2001, UK, http://www.lif.co.uk

G. Logan, *Lighting Efficiency Standards in the Energy Independence and Security Act of 2007: Are Incandescent Light Bulbs Banned?*, CRS Report for Congress, 2008, http://www.policyarchive.org/handle/10207/bitstreams/18981.pdf

MEMO/09/368 (Brussels 2009), FAQ: phasing out conventional incandescent bulbs, http://www.energy.eu/DG-TREN-releases/MEMO-09-368_EN.pdf

Sustainability Victoria, *Energy Efficiency Best Practice Guide. Lighting*, http://www.sustainability.vic.gov.au

P. Van Tichelen *et al.*, *Preparatory Studies for Eco-design Requirements of EuPs*, Final Report, Lot 8, Office lighting, 2007/ETE/R/115, http://www.eup4light.net/assets/pdffiles/Final/VITO_EuP_Office_Lighting_Projectreport.pdf

P. Van Tichelen *et al.*, *Preparatory Studies for Eco-design Requirements of EuPs*, Final Report, Lot 9, Public street lighting, 2007/ETE/R/021, http://www.eup4light.net/assets/pdffiles/Final/VITOEuPStreetLighting Final.pdf

Washington State University Cooperative Extension Energy Program, *Daylight Dimming Controls*, 2003, http://www.energy.wsu.edu/ftp-ep/pubs/building/light/dimmers_light.pdf

9

Electrical Drives and Power Electronics

Daniel Montesinos-Miracle, Joan Bergas-Jané and Edris Pouresmaeil

Power electronics plays a key role in energy efficiency in electrical systems. Power electronics is the enabling technology for renewable energy sources, such as photovoltaic, wind energy and fuel cell systems, and others, where the injected energy is subjected to strong regulation in order not to perturb the grid.

In distribution, power electronics is becoming more and more present. The inclusion of HVDC systems and STATCOMs are two examples of power electronics based devices that are present in the distribution system.

Also in energy storage systems, e.g. batteries, flywheels and SMES, power electronics is the technology that makes the system run. In addition, power electronics is present in each piece of electrical or electronic device or equipment, whether it is portable or not.

The role of power electronics is to control energy flow, while transforming the energy. Of course, this must act according to electrical specifications, and also with high efficiency. Nowadays, power electronic and control devices can meet these two objectives at the same time.

Electrical drives are one of the major energy-consuming devices used in industry. Transforming electrical energy to mechanical energy is a necessary process in most industrial processes, e.g. manufacturing, pumps, fans and compressors in HVACs.

Motor driven systems use 30% of the electrical energy in Europe and make up 60% of the industrial use of electric energy [1]. These figures show how important electrical drives are in terms of energy savings. A large number of studies and initiatives are currently running with this objective. The Motor Challenge Programme (MCP) is an initiative promoted by the European Commission to help industries to improve energy efficiency in motor driven systems; it has had great success in the last years.

In motor driven applications, energy is the main cost of the system, more than the initial cost and maintenance, so a reduction in energy consumption is a must both for energy and financial savings.

Electrical Energy Efficiency: Technologies and Applications, First Edition. Andreas Sumper and Angelo Baggini.
© 2012 John Wiley & Sons, Ltd. Published 2012 by John Wiley & Sons, Ltd.

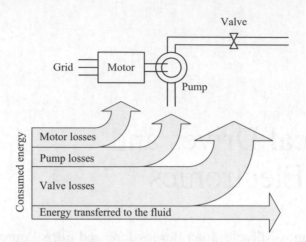

Figure 9.1   Energy losses in a traditional pumping system

Figures 9.1 and 9.2 illustrate an example of energy saving in the electrical drives of a pumping system. Traditional systems (Figure 9.1) use valves to control flow rate. At first, the pump transfers the maximum energy to the fluid using a direct connection of the motor to the grid, and then, to adjust the flow rate, a valve is used to dissipate the excess energy. This is not an efficient system.

In order to increase efficiency, it seems more reasonably to transfer only the necessary amount of energy to the fluid, as shown in Figure 9.2. However, pumps and fans can do this only using a varying rotational speed for the pump or the fan. This can only be done by modifying the rotational speed of the motor. Because the grid has fixed frequency and amplitude, a new interface element must be inserted between the grid and the motor; this provides the needed frequency and voltage amplitude for the motor. This element is the VSD.

A variable speed drive is a power electronics device that allows energy control of the grid to the motor and, consequently, to the fluid. The VSD can adjust the frequency and the voltage

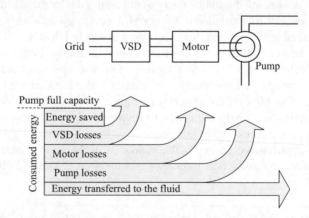

Figure 9.2   Energy savings in a pumping system using a variable speed drive (VSD)

amplitude needed for the system in real-time, transferring only the necessary amount of energy to the fluid.

Available variable speed drives are more than just systems to change the frequency and amplitude of the voltage applied to the motor. They are used as protection devices for the motor; they act as a soft starter, reducing stress on input lines and transformers, and avoiding an oversized design.

In addition, VSDs can act as decentralized control systems because normally they include some automation features, including analogue or digital communications buses.

Adding a new element can increase the cost of the system, and payback analysis must be used in this type of application. The payback is strongly dependent on duty cycle and energy cost, and it can vary from 8 months to 2 years. Even in the worst case of 2 years, this technology can give significant operational cost savings and, of course, a reduction in CO_2 emissions.

The induction motor has been the workhorse for industrial applications for many years. The advantages of induction motors over other types of motors are well known. The first reason is the cost, because of the simple design, simple manufacturing process and the low cost of materials used. Another reason is that it does not need any control devices, and it can be directly connected to the grid, giving the solution as shown in Figure 9.1.

Affordable power electronics and variable speed drives are a 'novelty' of the last 20 years. And, in the last 20 years new motor technologies with power electronics have appeared.

Permanent magnet synchronous motors (PMSM) are synchronous motors with magnets in the rotor instead of windings as traditional synchronous motors. The presence of these permanent magnets gives some advantages over induction motors. PMSM are more efficient than induction motors because they do not require magnetizing current [2]. Because of the lack of windings in the rotor, there are no rotor losses, so thermal design of the rotor is much simpler than that of an induction motor. In addition, the rotor can be made with lower inertia, so increasing the ratio of torque to inertia. This is the reason that this motor is the workhorse in such high dynamic applications as servos and robotics.

However, there are some disadvantages. The relative high cost of the permanent magnets, and the need for position sensors and power electronics. The PMSM cannot be directly connected to the grid because of instability at relative low operational speeds, below the nominal speed [3].

Increasing the energy efficiency of the variable speed drives can be achieved in different ways. The first is the control method used. Different control methods can lead to different energy efficiencies with the same performances, so the most efficient control technique must be used. However, high energy efficient control solutions demand a much higher computational cost.

From the converter point of view, there are three main topics that influence system efficiency: first, converter topology, second, modulation technique and, third, semiconductor devices.

The topology of the converter can influence drive efficiency. Some topologies are able to reduce current harmonics in the motor, so reducing losses and torque ripple, and also reducing stress on semiconductor devices, or allowing for lower voltage semiconductors, which normally are more efficient. Multilevel converters and matrix converters are key topologies

Generating a high quality sinusoidal voltage to be applied to the motor is the task of the modulator. Different modulation techniques exist and, depending on the modulation technique used, the efficiency of the motor and the converter can be improved.

In power electronic devices, the switching devices are semiconductors that act as on–off switches. Operation in the linear region of the semiconductor is not allowed because of high losses. Even accounting for the lower losses of a semiconductor working at saturation, during conduction and switching there are some losses that must be taking into account. These losses cause the semiconductor to heat to destruction. In this case, as in the motor, the limit is a thermal limit, and a heat removal system must be used. In this sense, semiconductor research focuses on reducing conduction and switching losses, but also on increasing the working temperature of semiconductor devices, reducing the heat removal system energy consumption, and also increasing the drive efficiency.

The most used material in semiconductors is silicon, and different technologies for silicon semiconductor exist. New materials, such as SiC (silicon carbide), GaN (gallium nitride) and even diamond, which have lower losses and a higher working temperature capability, are appearing and these will be the future of semiconductor devices.

Electrical drives usually work in motor mode. However sometimes, and depending on the application, they can work in braking mode. The braking is done electrically: it means that the braking energy is transferred back from the mechanical load to the motor, and converted back into electrical energy.

Variable speed drives must handle this energy. Usually, this energy is dissipated into resistors, but new technologies allow this energy to be stored in supercapacitors or batteries, or fed back into the grid, or even used by another drive that is working in motor mode.

This chapter focuses on all these improvements in drives.

9.1 Control Methods for Induction Motors and PMSM

There are several control methods for induction and PMSM, but only three are considered here because of its importance in industrial applications. The V/f or scalar control is based on steady-state electrical equations of the motor. This method is mainly used in applications that do not require high dynamic performance, such as, HVAC, pumps and fans, because it directly controls the torque developed by the motor.

For applications with high dynamics such as servos and robotics, vector control or DTC methods are more useful. These methods are based on state space motor equations, and they offer torque control. These high performances come at a cost: they need higher computation and sensing effort.

9.1.1 V/f Control

The V/f control is based on the equivalent per phase model of an induction motor, as shown in Figure 9.3, operating at steady state.

In the equivalent circuit of Figure 9.3, the torque can be expressed as:

$$T_{em} = \frac{P_m}{\omega_m} = 3I_r'\frac{1-s}{s}R_r'. \tag{9.1}$$

The motor equations are

$$u_s = R_s i_s + j\omega_s \lambda_s, \tag{9.2}$$

where u_s is the applied voltage, R_s is the stator resistance, i_s is stator current, ω_s is the stator voltage frequency and λ_s is the stator flux linkage.

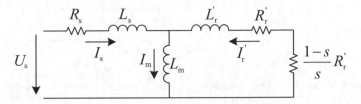

Figure 9.3   Per phase equivalent circuit of an induction motor

From (9.2), the normalized stator magnitude can be calculated as

$$u_s = \sqrt{(R_s i_s)^2 + (\omega_s \lambda_s)^2} \tag{9.3}$$

and, neglecting the stator resistance and for a constant stator flux linkage, equation (9.3) can be expressed as

$$\frac{u_s}{\omega_s} = \lambda_s = ct. \tag{9.4}$$

In the V/f control, this flux linkage λ_s is supposed constant on the whole speed range up to the base speed (ω_{sb}).

The applied voltage versus the applied frequency is a linear relation, and can be seen in Figure 9.4. Above base or rated speed, the voltage is then maintained constant while the frequency increases because the voltage capability of the converter is limited.

At low speed, the resistive term in (9.3) cannot be neglected and, to maintain a constant flux, some extra voltage must be applied to the motor as shown in Figure 9.4. This extra voltage depends on the current level (load) of the motor, as shown in (9.3).

Figure 9.5 shows a block diagram of the V/f control scheme. The control algorithm computes the voltage amplitude, proportional to the desired frequency. Some resistive drop compensation can be added at this point, but current measurements must be made. The voltage magnitude and frequency are used in the modulator to synthesize the switching signals for the voltage source inverter.

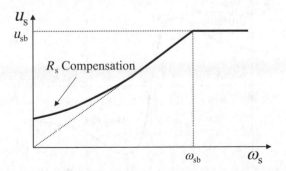

Figure 9.4   Voltage vs. frequency in V/f control method. The resistive drop compensation can also be seen

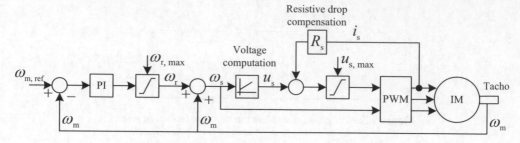

Figure 9.5 Block diagram of the V/f control method

The direct connection of a motor to the grid is a particular case of V/f control. But, by adding a VSD, this particular case can be extended to the whole speed range, adding features as soft starting, protection, and speed regulation.

These features, added to the easy implementation of this control method, have led to the great success of the V/f control method and it is now one of the most used control method for induction motors.

In the case of the induction motor, because it is an asynchronous machine, the mechanical and stator voltage speeds are not the same and depend on load. The slip ratio, s, is the relation between mechanical speed and stator voltage speed, and can be defined as

$$s = \frac{\omega_s - \omega_m}{\omega_s} = \frac{\omega_r}{\omega_s}, \tag{9.5}$$

where ω_m is the mechanical speed. Figure 9.6 shows the torque-speed characteristic of an induction motor. To assure the operation in the stable region of this curve, limiting the maximum slip, a PI controller can be added in an external loop as shown in Figure 9.5. This controller assures not only the stable operation and exact desired speed, but can act as current protection for the motor because rotor speed and current (or torque) are related, as shown in equation (9.3) and Figure 9.6.

As shown in Figure 9.5, a voltage limiter must be used because of the limited voltage of the converter, as will be seen in next sections, but this method needs a speed sensor.

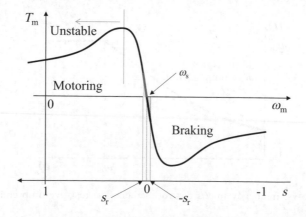

Figure 9.6 Torque vs. speed characteristic of an induction motor under V/f control

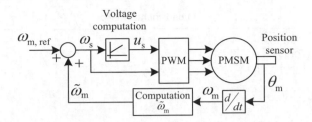

Figure 9.7 Stabilization loop with speed measurement for PMSM

In the case of PMSM, the V/f control strategy cannot be applied directly because of instability at low speeds [4]. The use of PMSM with V/f control requires the use of a squirrel cage in the rotor, combined with permanent magnets [5], or the measurement of the rotor mechanical speed in order to synchronize stator currents and rotor position, as seen in Figure 9.7, but this solution increases the cost and reduces reliability because of the speed sensor.

In [3], a sensorless stable V/f control is proposed. The method, instead of using speed perturbations to stabilize the motor, uses the power perturbations. The power can be computed using motor phase current measurements. This method shows a good response at nominal speed with nominal torque step changes, but shows a bad performance at low speeds.

9.1.1.1 Pumping, Fans and HVAC Applications

Continuous running applications such as pumps, fans and HVAC applications are at the focus of energy saving. Sixty-three percent of energy used in motors is in compressors, ventilation and pumping [1]. The energy cost of these applications is around 75% of the total cost, including investment and maintenance. The Motor Challenge Programme of the European Commission demonstrates that 30% of the energy use in pumping, fans and HVAC applications can be saved using VSD.

Typical installations are designed for the worst-case load conditions (maximum flow), and they only occur 5% of the time. Flow control is normally achieved using an output control valve. However, as can be observed in Figure 9.8, the energy savings using a VSD can be high. As shown in Figure 9.8, pumps and fans are quadratic torque loads with rotating speed.

From (9.1), it can be obtained that the $T_{em} \propto V^2/f^2$, but, as said, the load characteristic is $T_{pump} \propto f^2$. Obviously, at the constant speed operating point, the two torques are equal. Then, the VSD must be programmed to follow a $V \propto f^2$ law, instead of $V \propto f$.

As can be seen in Figure 9.9, in valve control applications, the desired flow rate is obtained by increasing the head of the pump or fan (operating point O_1). This is because the pump or fan characteristics cannot be modified (operation at fixed rotational speed), and the flow rate can only be modified by creating a fictive head (or losses) with the valve. The transferred power to the fluid ($Head \times Flow\ rate$) is the dashed area below point O_1.

By using a VSD, the characteristic of the pump or fan can be moved to the desired flow rate by modifying the rotational speed from n_1 to n_2 (operating point O_2). Now, the transferred power to the fluid is just the power needed, increasing the efficiency of the VSD-Motor-Pump (or fan)-tube system.

Figure 9.9 also shows that as low as the flow rate is, the energy savings are higher.

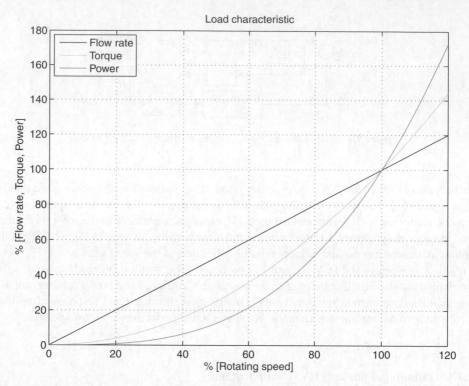

Figure 9.8 Load characteristics of a fan or pumping system. It must be noted that at 60% of rated speed, the torque is 35%, and the power is 20% of rated

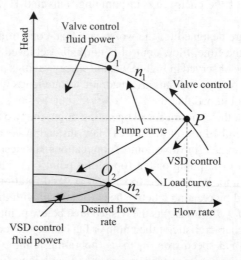

Figure 9.9 Operation of a pumping of fan system using valve control or VSD control

9.1.2 Vector Control

Vector control is based on the instantaneous decomposition of the electrical equations of the motor in order to obtain a relationship between the stator currents and the rotor flux with the torque produced, as in a DC motor. The decomposition is based on the well known Park transformation [6], and the result is the decomposition of the flux-producing current and the torque-producing current, as in a separate commutator motor, where the flux current corresponds to the excitation current and the torque-producing current corresponds to the armature current.

The principles of vector control apply equally to synchronous and asynchronous motors. The main difference is that in the case of PMSM, the rotor flux is produced by permanent magnets, which yield to an increase of efficiency of PMSM in front of an induction or asynchronous motor because there is no need to create rotor flux from stator currents.

From the Park transformation, the transformed equations are:

$$u_{qs} = R_s i_{qs} + p\lambda_{qs} + \omega_s \lambda_{ds}$$

$$u_{ds} = R_s i_{ds} + p\lambda_{ds} - \omega_s \lambda_{qs}$$

$$u_{qr} = 0 = R_r i_{qr} + p\lambda_{qr} + (\omega_s - \omega_r)\lambda_{dr}$$

$$u_{dr} = 0 = R_r i_{dr} + p\lambda_{dr} - (\omega_s - \omega_r)\lambda_{qr}$$

$$\lambda_{qs} = L_s i_{qs} + L_m i_{qr}$$

$$\lambda_{ds} = L_s i_{ds} + L_m i_{dr}$$

$$\lambda_{qr} = L_m i_{qs} + L_r i_{qr}$$

$$\lambda_{dr} = L_m i_{ds} + L_r i_{dr}$$

$$T_{em} = \frac{3PL_m}{22L_r}\left(\lambda_{dr} i_{qs} - \lambda_{qr} i_{ds}\right), \tag{9.6}$$

where u_{qs} and u_{ds} are the stator applied voltages, R_s and R_r are the stator and rotor resistance respectively, i_{qs} and i_{ds} are the stator currents, p is the derivative operator, λ_{qs} and λ_{ds} are the stator flux linkages, ω_s is the stator voltages' and currents' frequency, u_{qr} and u_{dr} are the rotor voltages, which are equal to zero because the rotor windings are supposed short-circuited, i_{qr} and i_{dr} are the rotor currents, λ_{qr} and λ_{dr} are the rotor flux linkages, ω_r is the rotor voltages' and currents' frequency, L_s, L_m and L_r are the stator, magnetizing and rotor inductance respectively, T_{em} is the electromagnetic produced torque, and P is the number of poles.

The advantage of vector control is that, with proper selection of the angle in the Park transformation, equation (9.6) can be simplified. Choosing the angle oriented with rotor d axis flux, and then controlling the rotor q axis flux to be zero ($\lambda_{1qr} = 0$), the torque equation becomes

$$T_{em} = \frac{3pL_m}{22L_r}\lambda_{dr} i_{qs}, \tag{9.7}$$

which is similar to the torque equation of a DC motor. Moreover, if at the same time $p\lambda_{dr} = 0$, then λ_{dr} and i_{qs} are orthogonal, producing the maximum possible torque [7].

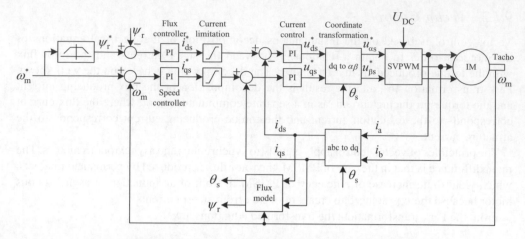

Figure 9.10   Block diagram of the vector control scheme

However, to maintain these conditions, the motor needs an accurate and dedicated control method.

In the case of PMSM, the rotor flux λ_{dr} is created by permanent magnets and has a fixed value. To control the torque it is necessary only to control i_{qs}, but in the case of an induction motor, the flux λ_{dr} must be created from stator currents. From the rotor voltage equation in (9.5), the relation between the rotor flux λ_{dr} and the stator current i_{ds} can be obtained as

$$\lambda_{dr} = i_{ds} \frac{L_m}{\tau_r p + 1}, \tag{9.8}$$

where $\tau_r = L_r/R_r$ is the rotor time constant. As mentioned in (9.7), the relation is of first order and must be computed in the controller in order to estimate the rotor flux λ_{dr} from the stator current i_{ds}, which can easily be measured.

Figure 9.10 shows that the torque and the flux can be independently controlled. In addition, in this control strategy, current limitation is directly implemented in the control scheme. It allows an inherent protection of the motor and converter. This limitation can be adapted depending on the motor working conditions, such as temperature, allowing the use of the motor in different ranges.

9.1.3   DTC

Direct Torque Control (DTC) was introduced in 1986 by Takahashi and Noguchi [8]. Meanwhile, Depenbrock developed Direct Self-Control (DSC) [9], a control scheme similar to DTC. It was this last author who developed the first VSD commercially available in 1995, based on DTC by ABB.

In DTC, the torque and the stator flux are controlled simultaneously by inverter voltage space-vector selection using a look-up table.

The main advantage of DTC compared with vector control is its simplicity. This simplicity allows DTC to implement a speed sensorless control system.

The main drawbacks of DTC are the high torque ripple and the variable switching frequency caused by the hysteretic controllers.

The DTC is based on the fact that the stator flux vector can be known from the voltage applied to the stator as

$$\Psi_{dqs} = \int \left(u_{dqs} - R_s i_{dqs} \right) dt. \tag{9.9}$$

On the other hand, the torque produced can be computed knowing if this stator flux vector and stator current vector are known:

$$T_{em} = -\mathrm{Im} \left(\Psi_{dqs} i^{\bullet}_{dqs} \right) = - \left\| \Psi_{dqs} \right\| \left\| i_{dqs} \right\| \mathrm{Im} \left(e^{j(\theta_s - \theta_1)} \right), \tag{9.10}$$

where θ_s is the angle of the stator flux vector and θ_1 is the angle of the stator current vector.

The DTC, supposing that the stator flux vector is known, determines which voltage vector must be applied to the motor depending on whether the flux must be increased or not, and also whether the torque must be increased or not.

Figure 9.11 shows, for a given stator flux vector, that each of the eight stator voltage vectors determine a new stator flux vector.

At the instant of time in Figure 9.11, the magnitude of the flux must be reduced in order maintain the stator flux vector within the flux hysteresis band. Then, only stator voltage vectors V_3 and V_4 can be used.

The decision to choose V_3 or V_4 is taken in the knowledge of whether the torque must be increased or reduced. To make this decision, another hysteresis comparator is used. In the case presented in Figure 9.11, V_3 increases the torque, and V_4 decreases the torque. The two zero vectors, V_0 and V_7, can be used when the torque needed is zero.

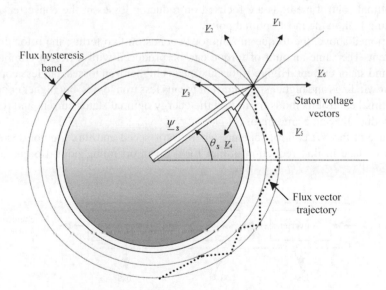

Figure 9.11   Flux trajectory in DTC control

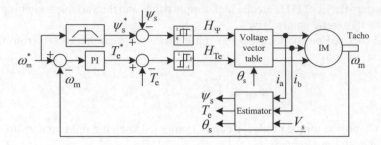

Figure 9.12   Block diagram of the DTC control method

From this methodology, a switching table can be obtained for all cases that determine which voltage vector is to be applied to the stator at each instant in time [8].

As shown in Figure 9.12, the DTC is a sensorless torque controller technique. The speed closed loop system requires a speed sensor but, as in case of vector control, speed sensorless techniques can be used with DTC.

9.2   Energy Optimal Control Methods

Figure 9.13 shows that in a variable speed drive there are three main sources of losses: the converter, the motor and the mechanical transmission system. Also cabling and connections can be considered in this loss count.

The motor control method can only deal with converter (inverter) and motor losses.

Transmission losses, and losses generated in the grid because of the rectifier, are not related to the motor flux level and current, so the control method cannot help to reduce these losses. Energy optimal control methods are focused on reducing losses in the converter and motor, and operating both in the most optimal point.

As mentioned above, the torque in a motor depends on two terms: the rotor flux and the stator current. The same amount of torque can be produced with different combinations of rotor flux and stator current. Higher rotor flux means higher iron losses, but less copper losses in the stator windings and higher stator current means less iron losses, but higher copper losses. Energy optimal control methods try to find the energy optimal stator current and rotor flux in order to produce the same amount of torque.

In the case of the V/f control scheme, where rotor speed and flux are not decoupled, the main disadvantage in flux reduction is that for a given operating point (torque and speed)

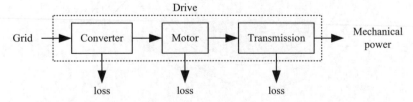

Figure 9.13   Energy process in a drive showing the losses of each element

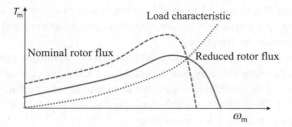

Figure 9.14 Torque-speed characteristics for a given operating point with nominal rotor flux and reduced rotor flux

the stator frequency must be increased, and then the maximum torque is reduced, making the system more sensitive to load perturbations, as shows in Figure 9.14.

In the case of vector control and DTC where the torque and flux are decoupled, the flux can be independently optimized, giving better performance, but the problem is the same (Figure 9.15). What is the optimal flux for a desired torque? Selecting the desired flux level is a balance between efficiency and not reducing the performance of the drive.

Calculating the optimal flux level in terms of efficiency is a complex task that must consider all the losses in the system. Computing all the losses in the VSD is an impossible task in real systems, because some losses are difficult to predict or not all the necessary measurements are available to compute them. Adding extra measurements to estimate the losses better will increase the cost of the system and also will reduce reliability of the control method and VSD.

9.2.1 Converter Losses

Converter losses are due to non-ideal behaviour of switching devices. They can be divided into conduction losses and switching losses.

Conduction losses are due to the resistive behaviour of power devices (transistors and diodes) when they are conducting. These losses depend on circulating current, the modulation index, the power factor of the converter and transistor parameters, which are strongly affected by many parameters such as temperature and voltage. These parameters are difficult to estimate and datasheet values are normally used.

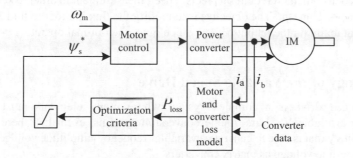

Figure 9.15 Block diagram of the optimization process

Switching losses are due to non-ideal switching behaviour of the power devices. Also, these losses depend on many parameters such as current, voltage temperature and device parameters, and are linearly dependent on frequency.

In order to reduce switching losses, adequate modulation technique must be used. As will be explained later, choosing the adequate modulation technique can reduce the number of commutations, reducing the switching losses, but not reducing the performance of the drive.

In the case of DTC, predicting the converter losses is a difficult task because conduction intervals and switching frequency are not constant.

9.2.2   Motor Losses

Motor losses are divided into copper losses and iron losses. Also, mechanical losses such as friction and windage can easily be taken into account.

Copper losses are due to the Joule effect in the stator and rotor windings (in the case of induction motors). Copper losses depend on the current and winding resistor, which depends on temperature and frequency due to skin effect.

Core losses include eddy currents and hysteresis, and both are due to magnetic flux variation in iron.

9.2.3   Energy Optimal Control Strategies

Many optimal control strategies for energy efficiency are present in the literature [7]. The most important methods are grouped and briefly explained here: simple state control, model-based optimization method, and search control methods. All these methods can be applied to V/f, vector and DTC control methods.

The simplest method is called simple state control. The objective of this method is to maintain a constant power factor operation point. A closed loop system measures cos (φ) and adjusts the flux to maintain a constant power factor.

Model-based methods are based on a motor model. The loss model can be solved analytically or numerically. In both cases, the currents are measured and then the model is solved, providing the optimal operating point.

Solving the model numerically allows for the inclusion of non-linearities, but it requires much time. The analytical solution can directly give the optimal flux.

The search control method requires a precise speed measurement in order to keep the output power of the motor constant. The objective is to modify the flux in order to find the minimum input power but maintain the output power, which gives the optimal operating point.

9.3   Topology of the Variable Speed Drive

A variable speed drive is a complex system where many elements must be present. Commercially available VSD are double conversion systems, even then there exists some converter topology that is able to generate a variable frequency-amplitude voltage from a fixed frequency-amplitude voltage as matrix converters.

Figure 9.16 shows the block diagram of the VSD. When operating in motor mode, the power from the grid is transformed into mechanical power. Normally, the motor can work

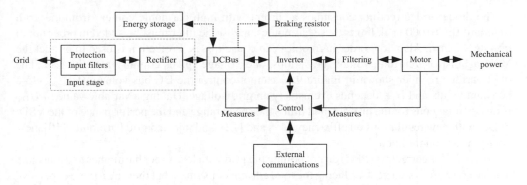

Figure 9.16   Block diagram of a VSD

in regenerative braking, but not the rectifier stage, so this braking energy must be dissipated in a braking resistor, or in more advanced applications, and can be stored in batteries or supercapacitors to be used in the future. Also, a rectifier can be designed to also inject energy to the grid from the motor.

Some measures must exist for control purposes for protection (currents, position/speed, DC bus voltage, grid voltage, etc.). All these measures are used by the control device to generate the appropriate signals to the inverter and other systems to control all the VSD functions. An external communications module is imperative in an actual VSD, not only for VSD configuration and parameterization, but also in a distributed control system, to interact with other control and supervision elements.

9.3.1   Input Stage

As shown in Figure 9.16, the input stage is formed by a protection and filter block and the rectifier. The rectifier is used to convert AC fixed frequency and magnitude voltage from AC grid into the DC voltage need by the inverter.

Figure 9.17 shows the diode bridge rectifier as a cheap and simple solution. It needs no control and line-commutated diodes offer good efficiency.

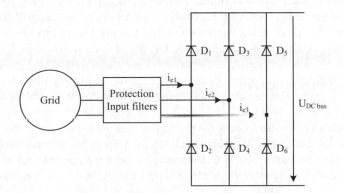

Figure 9.17   Input filter and rectifier on a VSD

The diode bridge rectifier consumes a current with high harmonic content from the grid. Standard IEC61000 [10], limits these harmonics, and some filter must be added in order not to pollute the grid. Also, low-order harmonics can cause losses. Protection is used to protect the VSD against overcurrent, voltage deviations, and voltage spikes coming from the grid. The diode bridge rectifier shown in Figure 9.17 cannot control the DC bus voltage ($U_{DC\,bus}$) to a constant value, and U_{DC} depends exclusively on grid voltage. Having a variable value of U_{DC} is disadvantageous for the inverter, and limits the speed range and the performance of the VSD.

But with new regulations on efficiency, [11] and [12], and injected grid harmonics [10], new solutions are on the shelf.

Power factor correctors (PFC) do not pollute the grid with low-order harmonics and consume a unity power factor current, reducing the total current consumed, but they are a more expensive solution, need more components and efficiency can be reduced. The main advantages are that the filter can be reduced, only high frequency harmonics must be filtered, and also that the PFC is able to maintain a constant U_{DC}.

When the motor is in generating mode, the diode bridge rectifier and PFC are unable to feed back this energy to the grid, and it must be dissipated in the braking resistors or stored in batteries or supercapacitors. The stored energy can be used in the future, increasing the total efficiency of the VSD.

But an active rectifier can be used as an input front-end. The active rectifier allows for a bidirectional power flow, and energy in the generating mode can be back injected to the grid, increasing the efficiency of the system. The topology of the active rectifier is the same as for the inverter, and can be controlled to consume sinusoidal current from the grid and give a constant power factor. The complexity and the cost of the system are high because of the large number of expensive elements and complex control that must run in a microprocessor.

9.3.2   DC Bus

The DC bus is an energy buffer formed mainly of an electrolytic capacitor to supply power transients to the inverter and maintain a constant DC voltage.

When connecting the VSD to the grid, a pre-charge system for DC bus voltage must be used in order to prevent in-rush currents. Frequently, for low-power VSD NTC resistors are used, but for high-power VSD, a resistor and a by-pass switch is desirable.

The bulk energy in the DC bus is stored in electrolytic capacitors. The C rate of this bus depends on the power of the converter. Electrolytic capacitors are a source of losses because they have a high ESR (and hence, high losses) and are not high temperature components (100°C). Care must be taken when designing the DC bus in order not to degrade the life of the bus with high temperature.

The high ESL of electrolytic capacitors can cause high dV/dt in transistors, destroying them by over-voltage. To reduce these voltage spikes in transistors, MKP capacitors are connected close to the transistor.

When the motor works in a regenerative mode, the mechanical energy at the rotor shaft is transferred to the DC bus by the inverter, increasing the voltage in the capacitors. If the input stage is a diode bridge rectifier or a PFC, this energy cannot be injected back to the grid and is dissipated in the braking resistor. Figure 9.18 shows the DC bus capacitor, braking chopper and resistor.

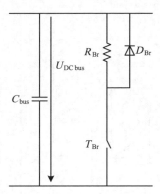

Figure 9.18   DC bus capacitors and braking chopper and resistor

This braking resistor must be sized to dissipate the predicted braking energy. To increase efficiency, this resistor is not always connected to the DC bus. A control system monitors the DC bus voltage. When this voltage reaches some predefined level, the chopper is used to dissipate the excess energy in the braking resistor.

However, this is not an efficient way to use electrical energy. Recently, in some applications where regenerative braking is common, such as in lifts, methods for energy storage with super-capacitors are becoming popular and effective [13]. Using supercapacitors to store the braking energy not only allows for a more efficient way of using energy, but it decreases the power of the rectifying stage. With a storage system, the input rectifier must supply only the average power and system losses, and not the peak power. It can reduce the cost of the system, input filter, protection, and cabling.

9.3.3   The Inverter

In general, when talking about VSD, the inverter is the power electronic element responsible for delivering the appropriated power to the motor [14]. For high voltages (400–800 V), high-power (4 kW–MW), the power electronic converters arena is dominated by the use of power semiconductors, which are mainly used in switching mode applications. This leads to some basic principles and operation modes that apply to all power electronics circuitries.

There are two different types of inverters for VSD [15]. Those that deliver a variable amplitude and frequency voltage, so called Voltage Source Inverters (VSI); and those that deliver a variable amplitude and frequency current, so called Current Source Inverters (CSI). In the medium to low-power drives market, VSI is the most common inverter, due to its simplicity and reliability; in the high-power drives market, both inverters share a part of the market, with a tendency for VSI to increase this share.

The converter output characteristics depend on the converter structure, the type of electronic devices and the method of control. Besides the connection circuit, they impose the propri-eties of the output voltage and, consequently, the global impact of the power converter on load supplied.

Silicon-based (Si) static switches are, by far, the prevailing switches in VSD inverters [16]. Insulated gate bipolar transistors (IGBTs) or metal-oxide field-effect transistors (MOSFETs)

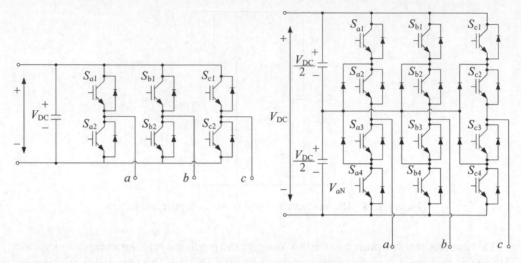

Figure 9.19   Schematics of a three-phase two-level inverter, and a three-phase three-level inverter

are the most usual type of power semiconductor switches; IGBTs in the medium to high-power applications and MOSFETs in the low-power ones. As these switches are the main source of losses in an inverter (conduction and commutation losses), there are great efforts to switch from the Si-based devices to the more energy efficient wide bandgap (WBG) based devices, with silicon carbide (SiC) and gallium nitride (GaN) being at the forefront [17] and [18].

The topology of the power converter in a drive plays an important role in energy efficiency considerations. The most spread inverter topology is the two-level inverter, but multilevel topologies, for high efficiency or high voltage drives, have to be considered. Figure 9.19 shows the three-level and the multilevel NPC topology.

Finally, the way in which switches are turned on and off, i.e. which modulation technique is used, has a great impact on the behaviour of the drive system, affecting losses in the inverter itself, due to commutations, losses in the motor due to harmonic currents, or to a weakened magnetic field, caused by a reduced voltage feeding.

9.4   New Trends on Power Semiconductors

Power electronics uses semiconductors to control energy flow between the source and the load. There are different semiconductor materials, but silicon-based semiconductors are the most used [19]. New materials such as SiC (silicon carbide) and GaN (gallium nitride) are being investigated for use as semiconductors in power electronics.

Semiconductors in power electronics are used only in saturation, i.e. they are only used in an on and off state, but not in the linear operating region. Figure 9.20 shows the switching process of a semiconductor transistor. In the figure there are two main sources of losses: conduction losses and switching losses.

Operation of the converter at high switching frequencies results in a size reduction of the passive components at the expense of increased switching losses. However, Wide Band Gap

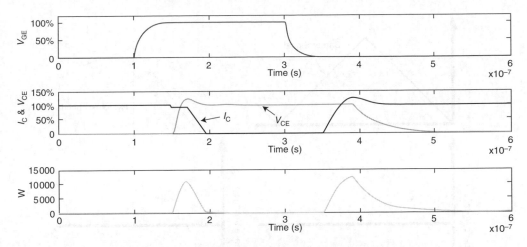

Figure 9.20   Behaviour of an IGBT (turn-on and turn-off)

(WBG, such as SiC and GaN) power devices have the potential to operate at high switching frequencies without the penalty of significant loss because of their fast switching times and their ability to work at high temperatures compared with similar Si devices [20]. Table 9.1 shows different properties of different materials.

9.4.1   Modulation Techniques

The inverter responsible for converting the DC voltage, from the DC bus, to variable amplitude and frequency AC voltages. This conversion from DC to AC can be realized using different techniques, which are usually known as modulation techniques.

In general, all the modulation techniques consist of turning the switches of the inverter on and off at a high frequency in such a way that the average value of the inverter output voltage equals that of the voltage reference. The longer the switch is on compared with the off time, the higher the average output voltage is.

There are several techniques, each with its own advantages and disadvantages, so for each application one should select the one that best fits the demands of the application. Next, a brief description of the main modulation techniques is presented, focusing on the differences between them, in order to make a selection of the modulation technique clear to the end user.

Table 9.1   Different properties of Si, SiC and GaN materials [21]

Property	Si	4H-SiC	GaN
Bandgap, Eg (eV)	1.1	3.26	3.45
Dielectric constant, ε_r	11.9	10.1	9
Breakdown field (kV/cm)	300	2200	2000
Thermal conductivity (W/cm K)	1.5	4.9	1.3

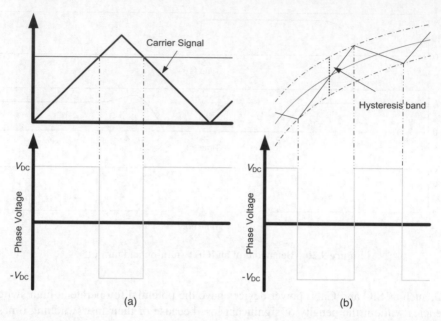

Figure 9.21 Schematics of: (a) a carrier-based modulation and (b) hysteresis modulation

The different modulation techniques can be classified into two large groups, according to the constancy of the switching frequency:

1. The first group, presenting a constant switching frequency, is usually known as Carrier-Based PWM methods (CBPWM). Figure 9.21 shows an example of a CBPWM where the main signals and characteristics of these modulation techniques arise. In these modulation techniques, the current spectrum presents as harmonic at the switching frequency and at its multiples [22], making it easy to filter such harmonic components. The higher the switching frequency, the higher the order of the current harmonics. Considering that in electrical motors, current harmonics implies torque vibration, in applications where mechanical resonances are of great importance, this is the kind of modulation that is preferred, thanks to its ability to move torque frequencies away from the resonances ones.

2. The second group, presenting a variable switching frequency, is usually known as hysteresis modulations (when they are implemented in a hardware manner), or dead-beat control (when they are implemented in a software manner). Compared with the previous group, they deliver a superior performance in terms of the bandwidth of the close loop, but presenting a worst current spectrum.

Apart from the aforementioned difference, modulation techniques differ in other important aspects, the next being the most important as discussed below.

- *Modulation Index (MI)*: in some sense, the modulation index is a figure of merit of the use made by the modulation technique of the DC bus voltage. That is, one inverter controlled by two different modulation techniques, each one with its own MI, can obtain different output voltages from the same DC bus. This is the modulation method with the greater MI, the one that obtain the maximum output voltage. From the energy efficiency side, this has a great impact on the efficiency of an induction motor operated at a fixed speed and close to its nominal speed. In these situations, the maximum output voltage of the inverter tends to be below the nominal voltage of the motor, feeding the motor with a reduced voltage, which finally produces additional losses.

- *Commutation losses*: each time that a switch of an inverter is turned on or off, a certain amount of energy is lost. Some modulation techniques make better use of these commutations (optimize the commutation sequence, obtaining the same voltage distortion) than others with higher commutations, implying reduced commutation losses and thus higher efficiencies.

The selection of a modulation technique for a particular application is often a balance between these competing advantages and disadvantages. Next, the principal modulation techniques are introduced, ordered in a chronological order, and with a brief introduction to each one.

9.4.2   Review of Different Modulation Methods

9.4.2.1   The Square-Wave Modulation

The square-wave modulation was the first method used and a simpler one (better known today as the *six-steps* modulation). Of all the various modulation techniques, this is the one that obtains the maximum fundamental output voltage, but giving, at the same time, the maximum current harmonic distortion.

Owing to this high current THD (Total Harmonic Distortion), this method is currently rarely used, and its use is restricted to applications for controlling the speed of induction motors in the areas of maximum speed and maximum torque, where the inverter has to deliver the maximum possible voltage. In this case, current harmonics are minimized thanks to the effective impedance of the phase motor, which increases linearly with the voltage frequency, and it reaches its maximum at the rated speed.

Nevertheless it is important to calculate the maximum output voltage of the *six-step* modulation, because the modulation index of a modulation method, is defined as the ratio between the maximum voltage of this modulation method and the maximum voltage of the *six-step* method [23]. As is obvious, the modulation index of the *six-step* modulation, by definition, is one.

The *six-step* modulation consists of turning on and off each leg of the inverter at the frequency of the reference output voltage. In Figure 9.22, the three leg voltages in respect to the virtual mid-point of the inverter are shown. Considering a balanced load (as is usually the case in a three-phase induction motor), it shows the voltage between the neutral point of the load and the virtual mid-point of the inverter [22].

The line-to-line voltages are shown in Figure 9.23 along with the line-to-neutral voltage. The presence of the *quasi-square* waveform can be observed in the line-to-line voltages, which gives the name to this kind of modulation; and in the line-to-neutral waveform, the presence of six 'steps' can be observed, which gives the name to this other kind of modulation. From Figure 9.23, it is possible to calculate the line-to-line voltage using the Fourier series (Table 9.3).

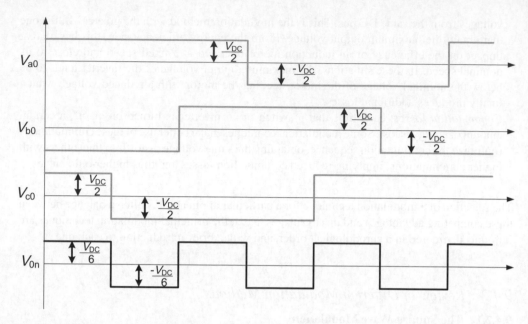

Figure 9.22   Phase to virtual mid-point of the DC bus voltages in the six-step modulation and neutral to virtual mid-point of the inverter voltage

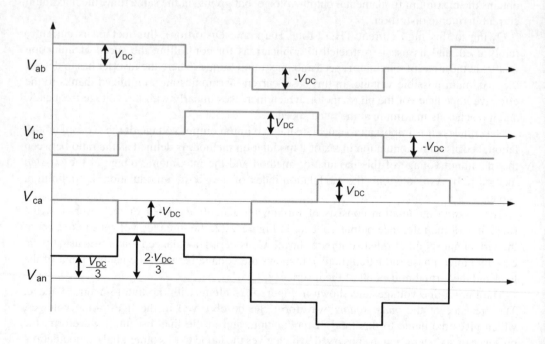

Figure 9.23   Line-to-line voltage and line-to-neutral voltage in a six-step modulation

9.4.2.2 Sinusoidal Carrier-Based PWM (Figure 9.24)

The sinusoidal carrier-base Pulse Width Modulation (SPWM) is one the most widely used modulation technique [24]. It was first introduced by Schönung in 1964, and it consists in the comparison of a carrier signal (usually a triangular one) with a modulated signal (in this case a sinusoidal control signal). The upper switch of an inverter leg is turned on each time the modulated signal is greater than that of the carrier, and on the other hand, when the modulated signal is below that of the carrier, it is the lower switch that is turned on, and the upper switch is turned off [15].

Although this modulation technique has been presented in its single-phase case, it is easy extended to its three-phase implementation. It is worth noting that this kind of modulation technique is easily implemented in a digital manner, thanks to what are known as PWM outputs, which are implemented in micro-controllers or digital signal processors (DSP) that are intended for digital motor control.

The maximum output voltage of the PWM modulation is presented in Table 9.3. Along with the maximum output voltage and the maximum modulation index, Table 9.3 shows the ratio between the maximum output voltage and the three-phase input voltage, assuming that the inverter is preceded by a three-phase diode rectifier, and that the DC bus gets charged with the peak voltage of the input three-phase voltage (which would be the idealized value when the DC link capacitors tend to infinity). This value of 0.86 means that, if a three-phase induction motor that was specified to work directly connected to the mains, was fed by an inverter controlled by the SPMW, this motor could never be working at its rated speed or

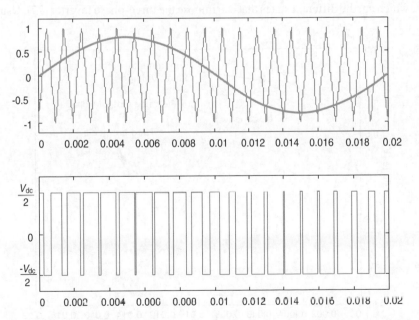

Figure 9.24 Sinusoidal carrier-based PWM. In the upper figure the carrier signal) and the modulated signal are depicted. In this case, the modulated signal consists of a 50 Hz sinusoidal. In the bottom figure, the voltage of the phase to mid-point of the inverter is shown

otherwise it would work in an undervoltage condition, which always ends with an increment in the losses of the motor, and a drop in its efficiency.

9.4.2.3   Third Harmonic Injection PWM (THIPWM) (Figure 9.25)

In 1997, Buja and Indri presented the THIPWM to overcome this inconvenience of the SPWM method [25]. This modulation technique consists in adding an adequate third harmonic waveform to each of the reference three-phase voltages, so allowing an increase in the fundamental maximum output voltage by a factor of $\sqrt{3}/2$, which is nearly 15.47% higher than the SPWM.

The THIPWM has been object of an intensive research, and different authors have proposed different *appropriate* third harmonic waveform. Some have proposed a maximum peak value of 1/6 and others peak values for the sinusoidal modulating signal [24], while others have proposed discontinuous zero-sequence components, not always sinusoidal [26].

9.4.2.4   Space-Vector PWM (SVPWM)

With the generalization of the Park strategy to control the torque and the speed of the induction motor, the carrier-based modulating techniques were no longer suitable for control of the inverter. Instead, a complex voltage vector for control of the inverter arose, the space-vector PWM.

This method, in a naturally manner, includes in its algorithm the discretization in eight unique vectors, which are the different states that can impose the three-phase inverter [27]. Usually, the

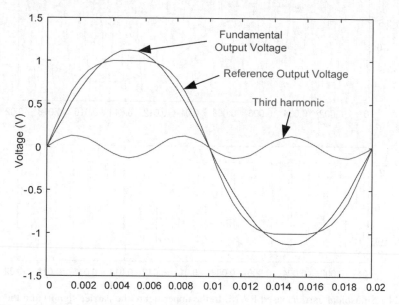

Figure 9.25   THIPWM: The fundamental output voltage, reference output voltage and third harmonic waveform

voltages resulting from these eight states are expressed in Park's variables. The highly coupled nature of the inverter loads, such as induction and synchronous machines, has led to the use of *artificial* variables rather than actual (phase) variables [22]. The mathematical expression which allows one to obtain from the actual variables the Clarke variables is as follows:

$$
\begin{pmatrix} v_0 \\ v_\alpha \\ v_\beta \end{pmatrix} = \sqrt{\frac{2}{3}} \begin{pmatrix} {}^1\!/\sqrt{2} & {}^1\!/\sqrt{2} & {}^1\!/\sqrt{2} \\ 1 & -{}^1\!/_2 & -{}^1\!/_2 \\ 0 & \sqrt{3}/_2 & -\sqrt{3}/_2 \end{pmatrix} \begin{pmatrix} v_r \\ v_s \\ v_t \end{pmatrix}.
\tag{9.11}
$$

In Figure 9.26 a brief explanation of the Clarke transformation is presented. Representing in Cartesian coordinate system the phase variables (v_r, v_s, v_t), of a symmetrical balanced three-phase voltage, all the points are located in the homopolar plane. That is, at any instant $v_r + v_s + v_t = 0$. Now, one can define a new coordinate orthogonal frame, with two of the axes located in the homopolar plane $(\alpha - \beta)$ and the third axis perpendicular to this same plane (0).

The SVPWM modulation consists of synthesizing a reference voltage, expressed in Clarke variables $\underline{V}^{\bullet} = (V_\alpha, V_\beta)$, by the combination of any of the eight possible states of the inverters, as depicted in Figure 9.27 and 9.28. Table 9.2 summarized the eight possible vectors for a two-level inverter.

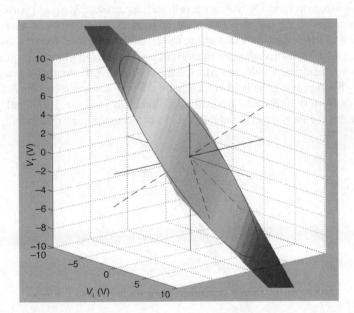

Figure 9.26 Interpretation of the Park's transformation. The Cartesian coordinate system, formed by the three-phase vectors, and the homopolar plane (that is, all the points of the space where $v_r + v_s + v_t = 0$), where the $\alpha - \beta$ axes are located

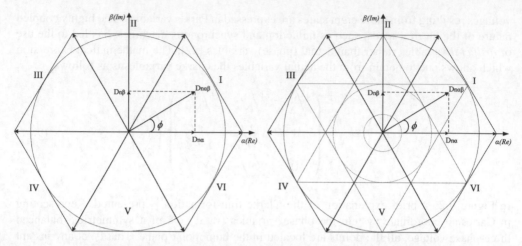

Figure 9.27   A two-level space-vector PWM and a three-level SVPWM

9.4.2.5 Multilevel Inverter Topologies

Recently, multilevel power converters have received a great deal of attention in numerous high-power medium-voltage industrial applications [28–33]. A multilevel converter uses a series of power semiconductor switches to perform the power conversion by synthesizing the AC output terminal voltage from several DC voltage levels and, as a result, staircase waveforms can be generated. Compared with standard two-level converters, multilevel converters offer great advantages such as lower harmonic distortion, lower voltage stress on loads, a lower common-mode voltage, and less electromagnetic interference. By reducing filtering requirements, they not only improve the efficiency of converters, but also increase the load power and, hence, the load efficiency by improving the load voltage with a lower harmonic content.

Multilevel converters have been basically developed to increase a nominal power in the converter. The higher number of voltage levels in these topologies results in higher quality

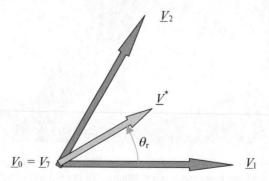

Figure 9.28   Synthesizing of a reference voltage by the two adjacent vectors and the two homopolar ones

Table 9.2 The eight different states of a three-phase inverter, and the vectors resulting from these states. In the fifth column it expressed in terms of the phase voltages, and in the last column in the $\alpha-\beta-0$ frame

	S_1	S_2	S_3	$\underline{V}_{rst} = (V_r, V_s, V_t)$	$\underline{V}_{park} = (V_\alpha, V_\beta, V_0)$
$\underline{V}_0$	-1	-1	-1	$\left(\dfrac{-V_{DC}}{2}, \dfrac{-V_{DC}}{2}, \dfrac{-V_{DC}}{2}\right)$	$\left(0, 0, \dfrac{-\sqrt{3}V_{DC}}{2}\right)$
$\underline{V}_1$	1	-1	-1	$\left(\dfrac{V_{DC}}{2}, \dfrac{-V_{DC}}{2}, \dfrac{-V_{DC}}{2}\right)$	$\left(\dfrac{\sqrt{2}V_{DC}}{\sqrt{3}}, 0, \dfrac{-\sqrt{3}V_{DC}}{6}\right)$
$\underline{V}_2$	1	1	-1	$\left(\dfrac{V_{DC}}{2}, \dfrac{V_{DC}}{2}, \dfrac{-V_{DC}}{2}\right)$	$\left(\dfrac{V_{DC}}{\sqrt{6}}, \dfrac{V_{DC}}{\sqrt{2}}, \dfrac{\sqrt{3}V_{DC}}{6}\right)$
$\underline{V}_3$	-1	1	-1	$\left(\dfrac{-V_{DC}}{2}, \dfrac{V_{DC}}{2}, \dfrac{-V_{DC}}{2}\right)$	$\left(\dfrac{-V_{DC}}{\sqrt{6}}, \dfrac{V_{DC}}{\sqrt{2}}, \dfrac{-\sqrt{3}V_{DC}}{6}\right)$
$\underline{V}_4$	-1	1	1	$\left(\dfrac{-V_{DC}}{2}, \dfrac{V_{DC}}{2}, \dfrac{V_{DC}}{2}\right)$	$\left(\dfrac{-\sqrt{2}V_{DC}}{\sqrt{3}}, 0, \dfrac{\sqrt{3}V_{DC}}{6}\right)$
$\underline{V}_5$	-1	-1	1	$\left(\dfrac{-V_{DC}}{2}, \dfrac{-V_{DC}}{2}, \dfrac{V_{DC}}{2}\right)$	$\left(\dfrac{-V_{DC}}{\sqrt{6}}, \dfrac{-V_{DC}}{\sqrt{2}}, \dfrac{-\sqrt{3}V_{DC}}{6}\right)$
$\underline{V}_6$	1	-1	1	$\left(\dfrac{V_{DC}}{2}, \dfrac{-V_{DC}}{2}, \dfrac{V_{DC}}{2}\right)$	$\left(\dfrac{V_{DC}}{\sqrt{6}}, \dfrac{-V_{DC}}{\sqrt{2}}, \dfrac{-\sqrt{3}V_{DC}}{6}\right)$
$\underline{V}_7$	1	1	1	$\left(\dfrac{V_{DC}}{2}, \dfrac{V_{DC}}{2}, \dfrac{V_{DC}}{2}\right)$	$\left(0, 0, \dfrac{\sqrt{3}V_{DC}}{2}\right)$

Table 9.3 Summary of the four modulation techniques. For each technique, the following results are presented: the maximum output voltage (with respect to the voltage of the DC bus), the maximum modulation index and, finally, the ratio between the output voltage and the three-phase input voltage, assuming that the inverter is preceded by a three-phase diode rectifier, and that the DC bus is charged with the peak voltage of the input three-phase voltage

Modulation technique	Maximum output voltage	$\dfrac{V_{eff}\text{out}}{V_{eff}\text{in}}$	Modulation index
Square-wave	$\dfrac{\sqrt{6}}{\pi} \cdot V_{DC} = 0.779 \cdot V_{DC}$	1.1	1
PWM	$\dfrac{\sqrt{3}}{2 \cdot \sqrt{2}} \cdot V_{DC} = 0.612 \cdot V_{DC}$	0.86	0.785
PWM + Third harmonic injection (THIPWM)	$\dfrac{1}{\sqrt{2}} \cdot V_{DC} = 0.707 \cdot V_{DC}$	1	0.907
Space-vector PWM	$\dfrac{1}{\sqrt{2}} \cdot V_{DC} = 0.707 \cdot V_{DC}$	1	0.907

output voltages. The concept of multilevel converters was introduced in 1975, and the term 'multilevel' first meant 'three-level' [34] but now refers to converters with more than a two-level output voltage. Multilevel topologies have been developed by increasing the number of semiconductor switches or the number of power converter modules (i.e., multiple converter modules). The trend toward a greater number of voltage levels is necessary because of the benefits of higher voltage ratings with a very low total harmonic distortion. By increasing the number of voltage levels, the converter's fundamental output voltage can be produced with a lower harmonic content, and it will significantly improve the quality of the output voltage and eventually approach a desired sinusoidal waveform.

The conventional two-level converter can produce high-quality outputs for low-power applications by using a high switching frequency. However, for medium- and high-power applications, the maximum switching frequency is limited by the switching devices due to the high switching losses. In this case, multilevel converters can be used to lower the switching frequency, and a high-quality output waveform can be produced. The superior features of multilevel converters over two-level converters can be briefly summarized as follows [28, 35, 36].

- They can generate output voltages with very low THD. Multilevel output PWM voltage can reduce the inverter switches blocking voltage and the dv/dt stress on the load such as a motor. The lower voltage stress on a load can reduce the number of Electro-Magnetic Compatibility (EMC) problems.
- They can produce a lower common-mode voltage. Thus, the lifetime of a motor connected to a multilevel motor drive can be increased due to the reduced stress on the bearings of the motor.
- By generating a staircase voltage waveform, they can produce lower converter input current distortion. The lower current ripple can reduce the size of a capacitor filter in a DC link.
- They are capable of operating at both a fundamental switching frequency and a high switching frequency PWM. In high-power applications, a lower switching frequency can reduce the switching loss, resulting in an efficiency increase.
- However, the trade-off for such an increased performance in multilevel converters is that they require a greater number of power switching devices. The number of semiconductor switches together with their related gate drive circuits can increase the overall system cost and the control complexity. In addition, several DC voltage source are required, which are usually provided by capacitors. Balancing the voltages of these capacitors according to an operating point is a difficult challenge.
- Despite these drawbacks, multilevel converters have turned out to be a very good alternative for high-power applications, since the cost of the control for these cases is a small portion of the whole system cost. Furthermore, as prices of power semiconductors and DSPs continue to decrease, the use of multilevel topologies is expected to extend to low-power application (those of less than 10 kW) as well. Fast power devices (CMOS transistors), which can operate at very high switching frequencies, can be used for low voltages. Therefore, the values of the reactive components will undergo significant reduction. Furthermore, new power devices are expected to appear in the near future, and these may also extend the application of multilevel topologies.

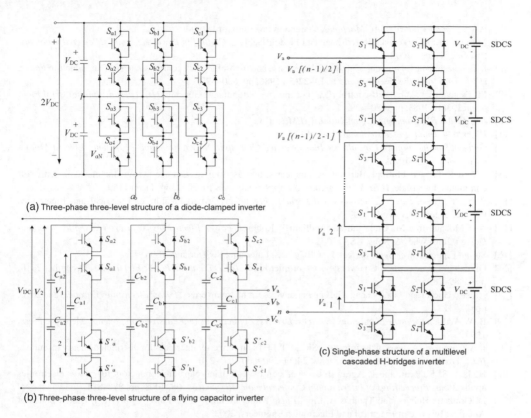

(a) Three-phase three-level structure of a diode-clamped inverter

(b) Three-phase three-level structure of a flying capacitor inverter

(c) Single-phase structure of a multilevel cascaded H-bridges inverter

Figure 9.29 Three different structures of multilevel converters

Multilevel topologies are classified into three categories (Figure 9.29):

1. Diode clamped (neutral-clamped) converter
2. Flying capacitors (capacitor clamped) converter
3. Cascaded H-bridges converter with separate DC sources.

References

[1] European Commission. (February 2003) *The Motor Challenge Programme*, http://re.jrc.ec.europa.eu/energyefficiency/motorchallenge/index.htm
[2] G. R. Slemon, Electrical machines for drives, *Power Electronics and Variable Frequency Drives. Technology and Applications*, IEEE Press, 1997.
[3] Daniel Montesinos-Miracle, P. D. Chandana Perera, Samuel Galceran-Arellano and Frede Blaabjerg. Sensorless V/f control of permanent magnet synchronous motors, *Motion Control*, In-tech, 2010.
[4] R. S. Colby and W. Novotny, An efficieny-optimizing permanent-magnet synchronous motor drive, *IEEE Trans. Ind. Applicat.*, **24**(3), 1988, pp. 462–469.
[5] Thomas M. Jahns, Variable frequency permanent magnet AC machine drives, *Power Electronics and Variable Frequency Drives*, IEEE Press, 1997.

[6] P. Vas, *Vector Control of AC Machines*, Clarendon Press, Oxford, 1990.

[7] Marian P. Kazmierkowski, R. Krishnan and Frede Blaabjerg, *Control in Power electronics. Selected problems*, Elsevier Science, 2002.

[8] I. Takahashi and T. Noguchi, A new quick response and high-efficiency control strategy of an induction motor, *IEEE trans. on Industry Applications*, **IA-22**(5), 1986, pp. 820–827.

[9] M. Depenbrock, Direc self-control (DSC) of inverter-fed induction machine, *IEEE Trans. on Industrial Electronics*, **3**(4), 1988, pp. 420–429.

[10] *IEC 61000 Electromagnetic compatibility (EMC)*, IEC.

[11] *IEC60034 Rotating electrical machines*, IEC.

[12] *General Specification for Consultants, Industrial and Municipal: NEMA Premium® Efficiency Electric Motors (600 V or Less)*, NEMA.

[13] Alfred Rufer and Philippe Barrade, A supercapacitor-based energy-storage system for elevators with soft commutated interface, *IEEE Trans on Industry Applications*, **38**(5), 2002, pp. 1151–1159.

[14] Bimal K. Bose, *Power Electronics and Variable Frequency Drives*, Institute of Electrical and Electronics Engineers, Inc., New York, 1997.

[15] Ned Mohan, Tore M. Undeland and William P. Robbins, *Power Electronics: Converters, Applications and Design*. Jonh Wiley & Sons, USA, 2003.

[16] Stefan Linder, *Power Semiconductors*, CRC Press, Lausanne, Switzerland, 2006.

[17] Qingchun Zhang *et al.*, SiC power devices for microgrids, *IEEE Trans. on Power Electronics*, December 2010, pp. 2889–2896.

[18] Alberto Guerra and Jason Zhang, GaN power devices for micro inverters, *Power Electronics Europe*, June 2010, pp. 28–31.

[19] B. W. Williams, *Power Electronics: Devices, Drivers, Applications and Passive Components*, Metropol Press, 1992.

[20] J. A. Carr *et al.*, Assessing the impact of SiC MOSFETs on converter interfaces for distributed energy resources, *IEEE Transactions on Power Electronics*, **24**(1), 2009, pp. 260–270.

[21] H. Jain, S. Rajawat, and P. Agrawal, Comparision of wide band gap semiconductors for power electronics applications, *Proceedings of International Conference on Microwave – 08*, 2008, pp. 878–881.

[22] D. Grahame Holmes and Thomas A. Lipo, *Pulse Width Modulation for Power Converters*, United States of America: Institute of Electrical and Electronics Engineers, 2003.

[23] J. Holtz, Pulsewidth modulation – a survey, *IEEE Transactions on Industrial Electronics*, December, 1992, pp. 410–420.

[24] Edison R. C. Da Silva, Euzeli C. Dos Santos and Cursion B. Jacobina, Pulsewidth modulation strategies: nonsinusoidal carrier-based PWM and space vector modulation techniques, *Industrial Electronics Magazine*, June 2011, pp. 37–45.

[25] G. Buja and G. Indri, Improvement of pulse width modulation techniques, *Archiv fur Elektrotechnik*, **57**(5), 1977, pp. 281–289.

[26] A. Trzynadlowski, Nonsinusoidal modulating functions for three-phase inverters, *IEEE Trans. Power Electronic*, **4**(3), 1989, pp. 331–339.

[27] H. W. Van Der Broeck, H. Skudelny and G. V. Stanke, Analysis and realization of a pulsewidth modulator base on voltage space vectors, *IEEE Transactions on Industry Applications*, **24**(1), 1988, pp. 142–150.

[28] W. Bin, *High Power Converters and AC drives*. John Wiley & Sons, Inc, New Jersey, 2006.

[29] L. Garcia Franquelo *et al.*, The age of multilevel convertes arrives, *IEEE Industrial Electronics Magazine*, **2**, 2008, pp. 28–39.

[30] P. W. Hammond, A new approach to enhance power quality for medium voltage AC drives, *IEEE Trans. on Industry Applications*, **33**, 1997, pp. 202–208.

[31] L. Jih-Sheng and P. Fang Zheng, Multilevel converters–A new breed of power converters, *IEEE Trans. on Industry Applications*, **32**, 1996, pp. 509–517.

[32] J. Rodriguez, S. Bernet, W. Bin, J. O. Pontt and S. Kouro, Multilevel voltage-source converter topologies for industrial medium-voltage drives, *IEEE Trans. on Industrial Electronics*, **54**, 2007, pp. 2930–2945.

[33] J. Rodriguez, L. Jih-Sheng and P. Fang Zheng, Multilevel inverters: a survey of topologies, controls and applications, *IEEE Trans. on Industrial Electronics*, **49**, 2002, pp. 724–738.

[34] A. Nabae, I. Takahashi and H. Akagi, A new neutral-point clamped PWM inverter, *IEEE Trans. of Industrial Application*, **IA-17**(Sept./Oct.), 1981, pp. 518–523.

[35] M. H. Rashid, *Power Electronics Handbook*, Academic Press.
[36] R. Teichmann and S. Bernet, A comparison of three-level converters versus two-level converters for low voltage drives, tracion and utility applications, *IEEE Trans. on Industry Applications*, **41**, 2005, pp. 855–865,

Further Readings

F. Abrahamsen, Energy optimal control of induction motor drives, Ph.D. Aalborg University, Denmark, 2000.

M. Alonso-Abella, E. Lorenzo and F. Chenlo, PV water pumping systems based on standard frequency converters, *Prog. Photovolt: Res. Appl.*, **11**, 2003, pp. 179–191.

Angelo Baggini, *Handbook of Power Quality*, John Wiley & Sons Ltd, West Sussex, 2008.

R. H. Baker and L. H. Bannister, *Electric Power Converter*, US Patent: 3 867 643, February 1975.

Vítezslav Benda, John Gowar and Duncan A. Grant, *Power Semiconductor Devices: Theory and Applications*, John Wiley & Sons Ltd, West Sussex, 1999.

J. Biela, M. Schweizer, S. Wafffer and J. W. Kolar, SiC versus Si—Evaluation of potentials for performance improvement of inverter and DC–DC converter systems by SiC power semiconductors, *IEEE Transactions on Industrial Electronics*, **58**(7), July 2011, pp. 2872–2882.

F. Blaschke, The principle of field orientation as applied to the new Transvector –closed loop control systems for rotating field machines, *Siemens Review*, 1972, pp. 162–165.

G. H. Bode and D. G. Holmes, Load independent hysteresis current control of a three level single phase inverter with constant switching frequency, *PESC. Power Electronic Specialists Conference 2001*, 2001, pp. 14–19.

K. Corzine and Y. Familiant, A new cascaded multilevel H-bridge drive, *IEEE Transactions on Power Electronics*, **17**(1), 2002, pp. 125–131.

F. Z. Peng and J. S. Lai, Multilevel cascade voltage-source inverter with separate DC source, US patent: 5 642 275, June 24, 1997.

10

Industrial Heating Processes

Mircea Chindris and Andreas Sumper

The industrial sector uses about one-third of the total final energy consumed annually in the United States [1] and 28% in the EU [2]; for the latter, the two most energy-intensive users are the iron and steel industry, consuming 20% of industrial energy use, and the chemical industry, with a share of 16.3%. If we are talking about the total electricity consumption in the EU, industry takes 41%.

Process heating is essential in most industrial sectors, including those dealing with products made from metal, plastic, rubber, concrete, glass and ceramics; the energy sources involved in this process are quite varied, as shown in Figure 10.1 [3]. In the figure, 'other' fuels usually refers to opportunity fuels (frequently, waste products such as sawdust, refinery gas, petroleum coke, etc.) and to by-product fuel sources, such as wood chips, biogases or black liquor. In many industries, they account for a large portion of the energy use.

According to Figure 10.1, there are many options for industrial processes involving heating, melting, annealing, drying, distilling, separating, coating, drying, etc. The old existing solutions use direct fuel burning (oil, gas, by-products or waste products), hot air, steam, water, etc. On the other hand, the last several decades have witnessed the rapid expansion of different electric-based process heating systems (usually called *electrotechnologies* or *electroheating technologies*) that assure the necessary heat by transforming the electrical energy into thermal energy. The practical implementation of these systems is based on direct, indirect or hybrid heating methods.

Direct heating methods generate heat directly within the work piece, based on one of the following techniques: passing an electrical current through the material; inducing an electrical current (eddy current) into the material; or exciting atoms and/or molecules within the material with electromagnetic radiation (e.g., microwaves). *Indirect heating methods* use one of these three methods to heat a special element; the latter transfers the heat to the work piece by either conduction, convection, radiation, or a combination of these, depending on the work temperature and the type of heating equipment. *Hybrid systems* use a combination of process

Electrical Energy Efficiency: Technologies and Applications, First Edition. Andreas Sumper and Angelo Baggini.
© 2012 John Wiley & Sons, Ltd. Published 2012 by John Wiley & Sons, Ltd.

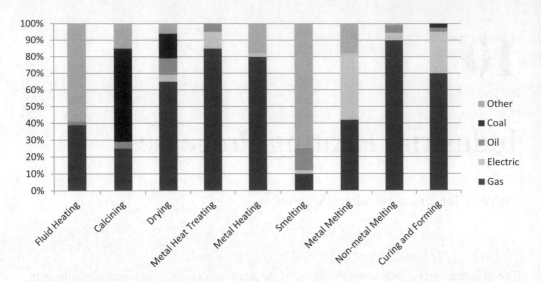

Figure 10.1   Energy sources for key industrial process heating operations (based on [3])

heating systems based on different energy sources or different heating methods based on the same energy source.

The rapid development of electrotechnologies in the last decades is based on the following aspects [2, 3, and 4]:

- *Huge progress in power electronics.* The widespread industrial implementation of power electronics was based on their high performances and reduced prices, largely contributing to the successful application of electricity in heating processes. The main contribution was the possibility of changing the frequency of industrial distribution networks; indeed, the enhancement of heating process efficiency requires that power can be supplied to the application at an appropriate (depending on the heating method) controlled frequency;
- *Accurate modelling.* New, very powerful and accurate modelling programs have been developed. Based on the increasing calculation power of modern computers, they allow Maxwell's equations to be solved point by point in space and are so accurate that the impasse is more in obtaining the correct knowledge of the material parameters (and their variation during the heating process);
- *Manufacturing of advanced materials.* New advanced materials for heat sources or enclosures, for instance, allowed the improvement of the working parameters (temperature, pressure, type of atmosphere, etc.) and the system's efficiency, or even introducing new technologies (laser processing, microwave processing, plasma processing, ultraviolet processing, etc.);
- *Finiteness of energy resources.* In the 1970s there was a growing awareness of public opinion of the finiteness of available energy resources. Coinciding with the abrupt increase in fossil fuel price, this encouraged the implementation of new technologies based on alternative (renewable) energy sources;

- *Acid rains.* Generation of electricity from coal and the use of heating technologies based on direct fuel burning have generated important emissions of sulphur dioxide and nitrogen oxides. The resultant acid deposition produced important environment pollution problems;
- *Climate change.* The principal reason for the environmental problems is inappropriate human activities, i.e. the burning of ever-larger quantities of oil, gas, and coal, the cutting down of forests, certain farming methods, etc. Since many countries have agreed to implement the Kyoto Protocol for reducing greenhouse gas emissions, the way in which energy is looked upon in industry has thoroughly changed. The Kyoto Protocol links energy efficiency with costs, so that the industrial world has started to realize that in the near future CO_2 emission costs will have to be added to the traditional energy bill. Any electrotechnology will now be accepted only if it proves its merits in energy efficiency, thus reducing greenhouse gas emissions;
- *Economic challenges.* Globalization, the migration of manufacturing to countries with much lower remuneration and increasing energy prices are forcing manufacturing industries in different developed countries to become more efficient, produce superior quality products, and remain competitive in order to survive. A possible solution consists in replacing the old heating technologies by electric-based process heating systems;
- *Economic policies.* At the moment, many companies focus on productivity-related issues. While productivity and output are clearly important, significant energy cost savings are also achievable in industrial utility systems, including process heating systems, and these opportunities are often overlooked. Taking into account these aspects, but also the international policy regarding sustainable development, a plant manager's responsibilities include minimizing the amount of energy used in the plant, maximizing product quality, maintaining the highest possible rate of production and reducing, as much as possible, the environmental impact of the operation.

Many techniques are nowadays used for transferring the supplied electrical energy into thermal energy to heat up the material: resistance heating (direct, indirect), infrared radiation, induction, dielectric heating, etc.; all of these methods have both advantages and drawbacks, so the choice often has to be made on a case-by-case basis. In some cases, electric-based technologies are chosen for their unique technical capabilities, as the application cannot be used economically without an electric-based system; in other cases, the relative price of natural gas (or other fossil fuels) and electricity is the deciding factor.

Often, the electroheating technologies compete with concurrent technologies using fossil fuels, reducing on industry's investment and operating costs, energy costs and primary energy requirements. For some industrial applications, electric-based technologies are the most commonly used; in others, these are only used in certain niche applications. It is important to mention that in all cases, efficiency and performance must be considered together, always based on a life cycle cost analysis.

Regarding the whole industrial process, the implementation of electroheating fulfils the following expectations [4, 5]:

- *It reduces the energy required to produce goods (a better energy efficiency).*
  Heating a manufactured component by producing heat inside the material (as many electrotechnologies do), rather than by heating it from the outside in (typical of classical heating technologies), requires less energy. Moreover, in many cases, the existing heating equipment

must be kept fired continuously during the whole production process, increasing the fuel consumption.

• *It reduces greenhouse gases.*
By reducing the total energy consumption, electroheating processes also assure a decrease in greenhouse gas emissions. The implementation of some special technologies, heating in controlled atmosphere enclosures (for instance, induction and radio-frequency techniques) further reduces the environmental impact.

• *It improves product quality.*
The quality of numerous finished products fundamentally depends on the performances of the heating process. In many applications, electrotechnologies have the best ability to achieve a certain product quality under constraints (for example, high throughput and low response time); more importantly, they can be applied to simultaneously improve product quality and value while saving energy.

• *It increases the speed of production.*
Classical heating technologies are very time-consuming processes; indeed, a product normally requires a quite long time to reach the imposed temperature due to material thickness, its thermal conductivity, and surface properties. On the other hand, the temperature limits and temperature gradient should be accurately imposed in order to avoid product damage or undesirable changes. Some electric technologies (direct resistance heating, induction or dielectric heating) can speed up the rate of material heating without exceeding surface temperature limits; the efficiency of these processes is normally much higher than that of conventional ovens, heating tunnels and radiant heaters. An improved rate of heat transfer and higher efficiency translates into a faster production rate and lower energy costs.

• *Electric-based process heating systems are controllable.*
In most equipment, a quite simple control system used during the heating procedure maintains the proper values for power input and other electrical or technological parameters.

• *It improves the work environment for employees.*
Since electrotechnologies do not use fossil fuels, the work environment is usually improved by lowering temperatures and eliminating combustion products from the shop floor.

10.1 General Aspects Regarding Electroheating in Industry

Electroheating technologies comprise high-power heating processes that are powered through electrical energy and cover a large percentage of industrial electricity consumption, ranging from 20 to 40% within the EU. Figure 10.2 [6] represents the percentages of different electroheating processes in industry (in this chart, 'Others' include techniques as ultraviolet, plasma, air knife, electric arc, etc.).

As was previously indicated, the industrial applications of electroheating technologies lead to energy savings and product quality improvements versus other processes. They have the following main characteristics [7, 8]:

• Electroheating technologies use electric currents or electromagnetic fields to heat a large variety of materials (metals, ceramics, natural fibres, polymers, foodstuffs, etc.).
• Currently they are associated with many industrial processes involving heating, melting, annealing, drying, distilling, separating, coating, drying, etc.

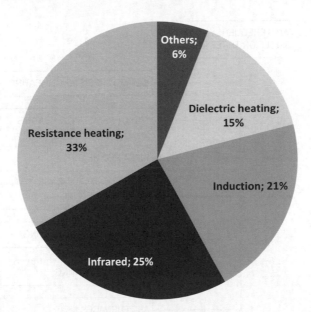

Figure 10.2   The percentages of different electroheating processes (based on [6])

- Most of the electroheating installations can be very accurately controlled and/or process materials in controlled atmosphere enclosures. These attributes guarantee a better quality product, less material and energy wasted, and reduced operation time, that is energy savings, reduced costs, and reduced CO_2 emissions.
- Sometimes, the electroheating technologies are the unique solution for some particular technological processes (for instance, high-temperature heating and melting or localized hardening of metal parts) and in the production of new materials. They also support the development of new technologies such as nanoelectronics and optoelectronics.
- In high-temperature applications, the electroheating installations are generally more energy efficient than their alternatives (furnaces based on direct fuel burning). The optimal efficiency of an electric furnace can reach up to 95% process efficiency, whilst the equivalent for a gas furnace is only 40–80%.
- Normally, the electric-based heating systems do not emit, at the location of use, gases (NOx, SOx, CO_2 or other gaseous products of heating processes), dust or hazardous salts and metals, so improving the working environment.
- They are the only solution for technological processes involving very high temperatures.

Figure 10.3 shows a basic classification of electroheating technologies currently used in industry [7,9]. An overview of application potential in industry is presented in Table 10.1 [5]. The table shows the electrotechnologies that may be applicable to specific industry groups depending on the actual technological processes in the plant; the intersections are labelled with a qualitative potential for application (High, Medium or Low).

The industrial practice clearly confirmed that, apart from technological benefits, electroheating processes can be very advantageous from the point of view of energy consumption and CO_2-emissions. Figure 10.4 [7] presents a global comparison of different drying technologies

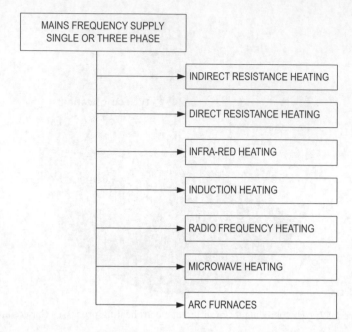

Figure 10.3 Classification of the principal industrial electroheating processes

and highlights that in this area the dielectric heating (microwave in this case) represents the best choice compared with old technologies (oil heated and gas fired) but also compared with another electrical solution (resistance heating).

Figure 10.5 illustrates an additional significant example; here, the energy consumption and CO_2-emissions for melting aluminium are compared for two technologies: a large gas-fired furnace and a large electric channel induction furnace (CIF).

In Figure 10.5, the analysed quantities refer to the following aspects:

- *Primary energy*: represents the energy required to heat the furnace (for the gas-fired furnace) and the energy required for generating the necessary amount of electric energy at a power plant (for the CIF), respectively;
- *Final energy*: the energy required by the melting process;
- *CO_2 emission:* the figures presented relate to the primary energy used.

For each bar, the upper section represents the energy equivalent for the melt losses, while the lower section is the energy used in the melting process. After a comparative analysis of the data presented, the following conclusions can be made:

- If the energy equivalent for the substitution of the melt losses is taken into account, the electric furnace consumes less primary energy than the gas-fired furnace for melting an equal amount of aluminium.
- The final energy is also lower for CIF.
- The channel induction furnace emits less CO_2.
- The relative difference between primary and final energy is the largest for the CIF (this ratio is typical for generation of electric power).

Table 10.1 Application potential of electrotechnologies in industry (based on [5])

Electrotechnology	Food	Textile product	Clothing	Leather allied	Wood products	Paper	Printing & related	Petroleum & coal products	Chemical manufacturing	Plastic & rubber	Non-metallic mineral products	Primary metal manufacturing	Fabricated metal products	Machinery manufacturing	Computer & electronic product manufacturing	Electric equipment, appliances & components	Transportation equipment manufacturing	Furniture and related product manufacturing
Infrared	M	H	M	M	M	H	H	M	M	H	H	M	H	H	H	H	H	M
Induction	L	M	L	L	L	L	L	L	M	M	M	H	H	H	M	H	H	M
Radio-frequency	M	H	L	L	M	M	L	L	L	H	L	L	H	H	L	H	L	M
Microwave	H	L	L	L	L	L	L	L	M	M	M	M	L	L	L	L	H	L
Direct Resistance	M	L	L	L	L	L	L	L	L	L	M	H	H	M	M	M	M	M
Indirect Resistance	H	M	M	M	M	M	M	M	M	M	M	M	H	M	M	M	M	M

Legend: Application potential – **H** = High; **M** = Medium; **L** = Low

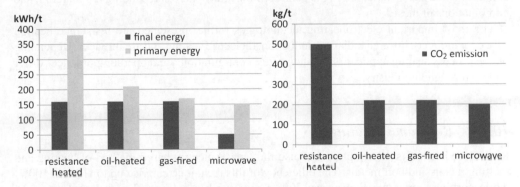

Figure 10.4 Final and primary energy and CO_2-emission in different drying technologies (based on [7])

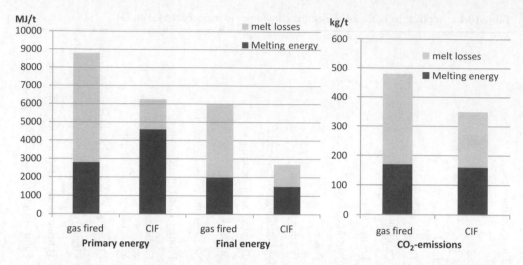

Figure 10.5   Energy requirements and CO_2-emission for melting of aluminium (based on [7])

10.2   Main Electroheating Technologies

Different electric-based process heating systems assure the necessary heat by transforming the electrical energy into thermal energy transmitted to the material to be treated. In the case of direct heating methods (e.g., direct resistance, dielectric heating, induction), the heat is generated within the material; for indirect methods (e.g., indirect resistance, indirect induction, infrared, electric arc, laser, etc.) energy is transferred from a heat source to the material by conduction, convection, radiation, or a combination of these techniques.

In most processes, an enclosure is needed to isolate the heating process and the environment from each other. The enclosure reduces thermal losses and assures the containment of radiation (e.g., microwave or infrared), the confinement of combustion gases and volatiles, the containment of the material itself, the control of the atmosphere surrounding the material, and combinations of these.

The most important electroheating technologies further presented are resistance heating (direct or indirect), infrared radiation, induction, dielectric heating and arc furnaces. Depending on the process heating application, system sizes, configurations, and operating practices differ widely throughout industry.

10.2.1   Resistance Heating

Resistance heating is the simplest (but also the oldest) electric-based method of heating and melting metals and non-metals. The efficiency of this technique can rise up to close to 100%, while working temperatures can exceed 2000°C, so it can be used for both high-temperature and low-temperature applications. With its controllability and rapid heat-up qualities, resistance heating is used in many applications from melting metals to heating food products. There are two basic types of this technology: indirect and direct resistance heating.

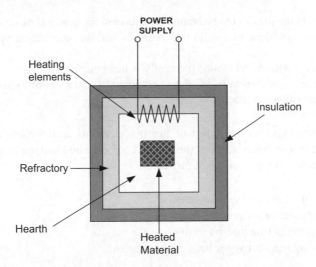

Figure 10.6   Outline of indirect resistance heating ([7])

10.2.1.1   Indirect Resistance Heating

1. Principle

Indirect resistance heating involves passing line-frequency current through a set of heating elements made of a high resistance material such as graphite, silicon carbide, or nickel chrome. When current flows throughout this element, heat is generated by the Joule effect and then transferred to the work piece via conduction, convection, and/or radiation. Heating is usually done in a furnace, with a lining and interior that varies depending on the target material; the outline of a system of indirect resistance heating is shown in Figure 10.6.

The process material temperatures can range from ambient to 1700°C or more (with an inert atmosphere), depending on the application and type of heating elements. Resistance heating systems that rely on convection as the primary heat transfer method are mainly used for temperatures below 650°C, while those that employ radiation are used for higher temperatures, sometimes in vacuum furnaces.

Usually, this type of heating is typically performed in a well-insulated enclosure; characteristic furnace linings are ceramic, brick and fibre batting, whilst furnace interiors can be air, inert gas, or a vacuum (at high temperatures an inert atmosphere is required around the work piece in order to prevent corrosion of the surface of the material). This solution minimizes thermal losses and provides a high heating efficiency, typically in the 80% range.

2. Types of Systems and Applications

Indirect resistance heating can be implemented in several ways [4, 5, 8]:

• custom heating solutions, using a wide variety of encased heaters, in which the resistive element is enclosed in an insulator (the coat is needed to isolate the heating element and the working environment from each other). Usually, this solution is adopted for low-power applications;

- direct contact with the material to be heated (in this case the heater is placed in the liquid that needs to be heated or close to a solid that requires heating, e.g. immersion-element water heater);
- by heating an intermediate substance (typically water or air);
- as the heat source in a thermally insulated enclosure (as a rule, in high-power industrial applications – resistance furnaces).

For furnace applications, various types of heating elements and enclosures may be used, depending on the temperatures needed, the product to be heated and the process used. There are four general categories of resistance furnace:

- normal or controlled atmosphere furnaces;
- batch process or continuous-process furnaces;
- batch feeding furnaces (manual, by motor, etc.);
- continuous feeding (e.g. conveyor belt, etc.) furnaces.

Indirect resistance heating is used in a wide variety of applications, including:

- high-temperature heating and melting in metal and glass industries, including metal preheating or reheating for forging and sintering (melting can be combined with refining processes, which demand the increase of temperature to remove impurities and/or gases from the melt);
- low-temperature heating and melting (for non-metallic liquids and solids);
- heat treatments, extensively used in metals production, and in the tempering and annealing of glass and ceramics products;
- calcining (familiar applications include construction materials, such as cement and wallboard, the recovery of lime in the craft process of the pulp and paper industry, the production of anodes from petroleum coke for aluminium smelting, and the removal of excess water from raw materials for the manufacture of specialty optical materials and glasses, powders, grains, etc.);
- agglomeration and sintering (commonly used in the manufacturing of advanced ceramics and the production of specialty metals);
- cooking, baking and roasting of food products;
- drying and curing processes (application of coatings to metallic and non-metallic materials, including ceramics and glass; in stone, clay, and glass industries, where the moisture content of raw materials, such as sand, must be reduced; in the food processing, textile manufacture, and chemical industry, in general);
- liquids (water, paraffin, acids, caustic solutions, etc.), air and gas heating;
- steam generation.

Most of the above-mentioned applications can be also performed by a wide variety of fuel-based process heating systems or steam-based systems. In many cases, resistance heating is chosen because of its simplicity and efficiency; on the other hand, there are also many hybrid applications, including 'boosting' in fuel-fired furnaces to increase production capacity.

3. Advantages and Limitations

Indirect resistance heating has several advantages justifying its extended use. Some of them are as follows:

- Although an old technique, it is simple and has a high flexibility of application, assuring an easily replacement of fuel oil or gas burners.
- It can be conveniently controlled and automated.
- The maintenance costs are low.
- Heating installations do not generate smoke, dust or combustion gases, improving the working conditions and reducing the environment pollution.
- The equipment is compact and may be adopted for a wide variety of heating technological processes (including those requiring special atmospheres or vacuum), generally improving the quality of the product and reducing energy consumption.

On the other hand, indirect resistance heating has some limitations and problems to be solved:

- The working temperature is limited by the melting point of the refractory material (for instance, fire clay and silica: 1700°C; chrome iron: 2000°C; zirconium: 2400–2500°C; graphite: 3000°C, etc.) and the maximum operating temperature for the electrical heating elements used (nickel: 1000°C; graphite: 1800°C, molybdenum: 2500°C; wolfram: 2800°C, etc.; it must be underlined that their service life can be severely reduced if the working temperature goes beyond the maximum allowable limit). Several types of heating elements are described in Table 10.2;
- Efficiency is strongly influenced by the heat transfer rate between the heating elements and load and also by the quality of the insulation; sometimes the operating cost may be high.

4. Typical Performance

Typical performances of indirect heating technologies are as follows:

- Conversion of electrical energy: higher than 95%.
- Power density: about 15–25 kW/m^2 of surface of furnace wall for installations equipped with metallic resistances and up to 70 kW/m^2 of surface for furnaces equipped with non-metallic resistances.

5. Application Considerations

For selecting the suitable heating technique and accurate design of the equipment, the subsequent elements have to be considered:

- the desired heating application (heating and melting, preheating, heat treating, baking, etc.) and load characteristics: nature (metal or non-metal, solid or fluid, etc.), geometrical dimensions, and working temperature (it decisively influences the heat transfer method: for instance, over 500°C the radiation is predominant);

Table 10.2 Heating element types ([5])

Heating element family	Material	Maximum temperature of element [°C]	Application–Remarks
Iron-Nickel-Chrome Alloys	Fe-20Ni-25Cr	900	Widely used because of their availability and low cost. Wide range of temperatures. Used in oxidizing atmospheres.
	Fe-45Ni-23Cr	1050	
	Fe-65Ni-15Cr	1100	
	80Ni-2-Cr	1150–1200	
Iron-Chrome-Aluminium Alloys	Fe-22Cr-14, 5Al	1280	Higher temperatures than Ni-Cr at about same cost. Embrittlement on first heating. Used in oxidizing atmospheres
	Kantal AF	1400	
Non-metallic Alloys	SiC (silicon carbide)	1600	Bars are brittle – susceptible to thermal and mechanical shocks. Used in oxidizing or reducing atmospheres
	$Cr_2O_3La_2O_3$ (lanthanum chromite)	1800–1900	
	Graphite	2500	Low cost of bars. Used in neutral or reducing atmospheres, or under vacuum.
Noble metals	Molybdenum	2300	Wire or plates. Very high cost. Only in neutral or reducing atmospheres, or under vacuum.
	Tungsten	2500	
	Tantalum	2500	

- family (according to desired temperature) and shape (related to furnace contour and size) of heating elements;
- furnace operation method (continuous/discontinuous) and maximum allowable thermal losses through walls and openings.

10.2.1.2   Direct Resistance Heating

1. Principle

Direct resistance heating involves passing an electric current directly through the product to be heated, causing an increase in temperature; based on this, direct resistance heating is often referred to as 'conduction heating'. Direct resistance heating is an example of the Joule law or effect at work: the resistance of the work piece to the current passing through generates

3x400/230V

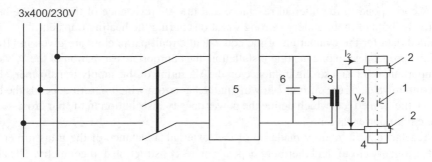

Figure 10.7 Outline of a direct resistance heating installation: 1 – heated part; 2 – contacts; 3 –power transformer; 4 – flexible jumper leads; 5 – symmetrizing system; 6 – power factor correction capacitor

heat. The material to be treated must have a reasonable electrical conductivity, but metals with higher resistivity, such as steel, create more resistance and more heat, which makes the process more efficient.

As shown in Figure 10.7, a direct resistance heating installation basically consists of a single-phase power transformer fed at constant voltage, of fixed bus bars and flexible jumper leads and of the contact assemblies whereby the current is carried into and out of the bar to be heated. Consumable or non-consumable electrodes are used to physically make contact with the product being heated; different solutions are nowadays implemented: fixed (clamp types) or mobile connectors (roll type) for metal heating applications and submerged electrodes for metal melting or liquid heating. The connector and/or electrode material has to be compatible with the processed material.

The temperature is controlled by adjusting the current, which can be either alternating or direct. Low-frequency current (DC or 50 Hz) heats the part throughout; high frequency current tends to heat only the surface of the work piece.

The equivalent circuit of the installation can be represented by the simplified series circuit shown in Figure 10.8 [10, 11].

In this circuit, the external reactance X_e takes into account the series reactance of the supply transformer (referred to the secondary terminals) and the external reactance of the high-current circuit; R_e represents the resistance of the same elements, while R_c is the contacts' resistance.

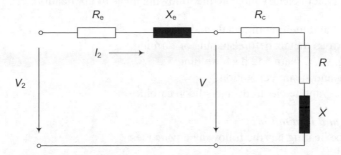

Figure 10.8 Simplified equivalent circuit of a direct resistance heating installation (V_2 – transformer secondary voltage; I_2 – heating current)

X and R correspond to the internal reactance and the AC resistance of the bar to be heated, respectively; both of them undergo strong variations during the heating transient.

Detailed data for the evaluation of the equivalent circuit parameters are given in [10] and [12]. For a proper design of a heating installation, the appropriate estimation of the X_e value is very important as it has a strong influence on the VA rating of the supply transformer, also on the maximum value of the current I flowing in the bar (from which in turn depends the life of contacts), the skin effect (which define the power density distribution in the bar cross-section) and the circuit power factor.

The reactance X_e is mainly made up of the 'external' reactance of the transformer high-current secondary circuit and therefore depends on its construction and geometry. The above-mentioned references present a series of theoretical procedures for the calculation of X_e; they consider various cross-sections and geometrical layouts of conductors. The consequence of proximity and skin effects may be taken into account or neglected (the current density in the heated bar is considered to be uniform).

Appropriate procedures also exist for the internal impedance (resistance and reactance) of the heated element; in this case, the following elements are considered: the nature of the material, its geometrical form and dimensions and the frequency of the heating current (DC or AC). It is important to mention that the dynamic values of these parameters are current- and temperature-dependent. However, modern computing software allows a detailed numerical analysis of the heating process, providing an in-depth knowledge of all these dependencies at each instant of the heating transient; for this purpose, the variations of the electrical and thermal material's physical characteristics with temperature and the local magnetic field intensity (for magnetic permeability) are taken into account.

2. Types of Systems and Applications
Nowadays, direct resistance heating is quite extensively used in industry [5]:

- metal heaters (billet heating for further forming operations, such as extrusion, heat treating, etc.) and resistance welders (spot welders, seam welders, etc.);
- non-metal heaters (heating and/or melting glass, silicon carbide, salt baths, etc.);
- food cookers and sterilizers;
- steam generators (e.g. high voltage electrode boilers, humidity generators in building HVAC systems).

These types of system clearly indicate the following areas of application:

- heat treatment and melting of metals and glass;
- heating of ferrous metals before shaping or forming;
- metal joining: spot, seam and flash welding;
- water heating and steam generation;
- production of graphite electrode and silicon carbide.

3. Advantages and Limitations
Direct resistance heating has the following advantages:

- It has high efficiency, due to two important elements: (i) only the work piece is heated, and (ii) the radiation and convection losses from the piece surface are very small.

- There is a reduced heating time (rapid rate of heating) due to the high-power density (up to 105 kW/m^2) existing in the working material, whilst the final temperature, as well as temperature differences in the cross-section of the heated product can be very precisely controlled.
- The technology allows heating of only parts of a complex piece.
- The process can start very quickly (seconds) and does not generate combustion products.
- There is a lower equipment space requirement and moderate capital investment.

At the same time, the following limitations must be considered:

- The contact surfaces must be clean and scale free for good electrical connection, whilst the number of contacts and the contact pressure must be suitably selected in order to avoid local melting.
- The piece configuration must provide realistic high resistance to current flow but uniform cross-section is required for uniform heating.
- The work piece must be long and slender (i.e. length-to-diameter at least 6:1), direct heating being better suited to smaller cross-sections (i.e. < 3 cm diameter).
- Large systems (e.g. glass melting) may be limited to moderate production rates by power supply ratings.
- The power factor can be quite low (0.3–0.95).

4. Typical Performance
- electrical energy conversion: above 95%;
- overall process efficiency: typically 75– 95%;
- energy consumption: 200–350 kWh/t.

5. Application Considerations
For correct design of the equipment, the following elements have to be taken into consideration:

- the shape, size and homogeneity of material and its electrical resistivity;
- connection resistances (local overheating);
- thermal losses through the surfaces (radiation, convection) and connections (conduction);
- voltage and power supply (DC, AC, capacity required);
- skin effect (dissipated power varies according to operating frequency and depth of penetration of the current).

10.2.2   Infrared Heating

10.2.2.1   Principle

Infrared radiation heating is practically a variant of indirect resistive heating; it uses radiation emitted by electrical resistors, usually made of nickel-chromium or tungsten, heated to relatively high temperatures – Figure 10.9 [4, 10]).

The used infrared region (wavelength λ: 0.78–10 μm) is commonly divided into three ranges: near-infrared or short-infrared ($\lambda = 0.78$–2 μm), intermediate-infrared or medium-infrared ($\lambda = 2$–4 μm) and far-infrared or long-infrared ($\lambda > 4$ μm). Although the electric infrared

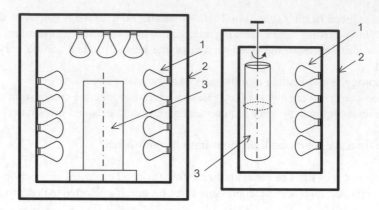

Figure 10.9 Outline of infrared heating: 1 – IR emitters (including the reflector system); 2 – furnace wall; 3 – heated part

is most often used in applications in which only the surface of an object needs to be heated by infrared, it can also be used in bulk heating applications.

As Figure 10.9 shows, electric infrared heating systems typically consist of an emitter, a reflector system, and controls (because operation time can be as little as seconds and an exact temperature of the work piece to be heated is usually required, accurate control is critical). Most electric infrared applications also have a material handling system and a ventilation system.

10.2.2.2   Types of Systems and Applications

Infrared energy is usually used when the object being heated is in line-of-sight of the emitters and/or reflector; however, some infrared systems can cure coatings that are not in line-of-sight (for example, curing a coating on the inside of a pipe using infrared focused on the outside of the pipe: while the curing is being accomplished by conductivity, it is using infrared processing).

Emitters used in infrared heating systems are solid resistors emitting energy in the infrared regions mentioned above; emitter technologies and performances correspond to each of the three infrared radiation spectrum bands, as presented in Table 10.3 [5, 8, 13, 14].

Because of the shape of the curve giving the spectral distribution of the radiated energy versus temperature and wavelength, the following rules must be respected in order to place the radiated energy within the desired band of wavelengths:

- for short and medium infrared radiation, the emitter temperature must be much higher than that defining the upper limit of the band;
- for long infrared radiation, this temperature must be close to the limit of the lower emission band, or even in the upper part of the medium infrared band.

Electric infrared processing systems are used by many manufacturing sectors for heating, drying, curing, thermal-bonding, sintering and sterilizing applications. They are often combined

Table 10.3 Infrared sources

Emitter	Near infrared		Intermediate infrared		Far infrared
	Tungsten filaments under vacuum		Nickel-chromium wire in quartz or silica	Steel-clad elements	Elements embedded in Pyrex or ceramic
	Glass or quartz lamp	Quartz tube with reflector			
Power	150 W	12–14 cm: 500W	30–250 cm: 6000 W	600–6000 W	150–1000 W
	250 W	27–28 cm: 1000 W			
	375 W	64–70 cm: 5000 W			
Operating temperature	2000°C	2200°C	Quartz tubes: 1050°C	750°C	Pyrex panels: 350°C
			Silica panels: 650°C		Ceramic elements: 300–700°C
Maximum product temperature	300°C	600°C	Quartz tubes: 500°C	400°C	Pyrex panels: 250°C
			Silica panels: 450°C		Ceramic elements: 500°C

with hot air to remove vapours and to help prevent premature surface hardening. Examples of applications include:

1. Drying and polymerization of coatings on various supports
   o drying of paint, varnish, adhesives, inks or vitreous enamels on metal, wood, glass, paper and textiles;
   o coatings on leather or hides and on paper;
   o polymerization of resins on textiles and of different plastifying agents.
2. Dehydration and partial drying
   o papers, cardboards, textiles (including water paints and inks);
   o drying of plastic granules and washed sheets;
   o preheating of plastic pipes before bending and curing of powder coatings.
3. Miscellaneous heating
   o drying of washed metallic parts, preheating of metals before shot blasting or welding and heat treatment (annealing, hardening, etc.);
   o baking dehydration of breads, biscuits and cookies;
   o reheating of food products, roasting of meat;
   o pasteurization and sterilization of milk and fruit juice;
   o drying of wood panels and tobacco;
   o heat shrinking of plastic wrapping.

10.2.2.3   Advantages and Limitations

In the above mentioned applications, infrared radiation may represent a highly efficient and attractive means of heating. Its main advantages are:

- a high energy efficiency (energy is transferred from the emitter to the surface of the work piece without contact and without any significant direct absorption by the environment, according to the laws of optics);
- a low thermal inertia of the heating system and high temperature rise due to the possibility of high-power densities (up to 30 W/cm^2);
- short start-up periods and reduced heating time (rapid processing);
- the energy radiated can be concentrated, focused, directed or reflected, allowing a convenient control of heating process;
- compact equipment (it is several times smaller in size than a hot air oven due to quite low processing temperatures).

The limitation of this electrotechnology results from its operation principle:

- It is best suited to products in layers or sheets (for bulk heating, the heat takes time to travel from the surface through the material).
- It is difficult to treat complex shaped pieces or reflective coatings.
- The maintenance of IR emitters is higher in dirty environments.

10.2.2.4   Typical Performance

- The power densities that can be exchanged are much higher than with convection.
- There is low to moderate capital cost depending on application.

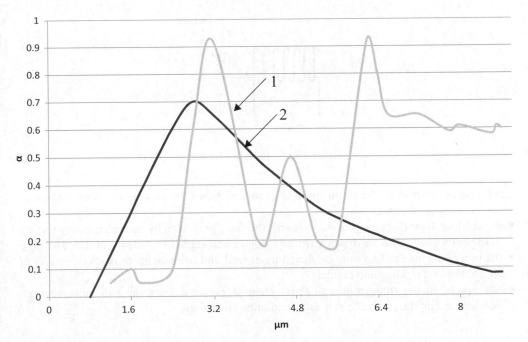

Figure 10.10 Optimum use of the absorption spectrum of the product to be heated (1) and of the emission spectrum of source (2)

- There is low maintenance, primarily emitter cleaning and replacement.
- There is high overall efficiency compared to alternative heating processes.

10.2.2.5 Application Considerations

As previous stated, in infrared radiation the heating energy is transferred from the emitter to the surface of an object according to the laws of optics. Absorption of the radiation by the piece to be heated is gradual and takes place at a certain depth from the surface; it is also a selective phenomenon that depends greatly on the incident wavelength and on material thickness. On the other hand, reflection depends mostly on the surface condition and the wavelength of the incident radiation.

It results that, for a high efficiency, the work piece to be heated must have a reasonable absorption to infrared and a reduced reflection. For this, it is desirable that the wavelengths emitted by heating sources be located within the product's maximum absorption range, Figure 10.10 [4, 10].

When analysing the possibility of selecting the infrared heating, the following factors must be taken into account:

- the heating operation to be performed (drying, heating, curing, etc.) and the required emission wavelength, based on the absorption factor of the product to be treated – the latter depends

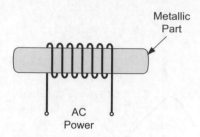

Figure 10.11   Induction heating

on type of material or the formulation (e.g. solvent-based vs. powder coating), thickness, surface conditions, etc.;
- the product shape (for complex shaped parts, the energy received by each elementary surface of the body varies with the distance from the source and angle of incidence of the radiation);
- the processing method (continuous/batch treatment) and oven characteristics (e.g. risks of shock between product and emitters);
- the type of emitter (temperature of active element, power density, etc.), distance between emitters and product to be heated, and ventilation conditions.

10.2.3   Induction Heating

1. Principle

Induction heating consists of applying an alternating magnetic field created by an inductance coil (inductor) to an electrically conducting object, Figure 10.11. The variable (oscillating) magnetic field produces an electric current (called an induced current) that flows through this body and heats it by the Joule effect. In addition to the heat induced by eddy currents, magnetic materials also produce heat through the hysteresis effect: magnetics naturally offer resistance to the rapidly alternating electrical fields, and this causes enough friction to provide a secondary source of heat (however, this secondary heating process has a reduced weight and disappears at the temperature at which the material loses its magnetic properties, the Curie point).

Induction heating uses the same principle as a power transformer [15]. The primary circuit (the inductor) creates a variable magnetic field, while the heated metal piece, through which the short circuit current flows, represents the secondary circuit. As alternating currents are concentrated on the outside of a conductor (the *skin effect*), the currents induced in the material to be heated are the largest on the outside and diminish towards the centre. The skin effect is characterized by its so-called *penetration depth*, δ, defined as the thickness of the layer, measured from the outside, in which 87% of the power is developed (Figure 10.12 [16]). The penetration depth, δ, results from Maxwell's equations:

$$\delta = 503 \cdot \sqrt{\frac{\rho}{\mu_r \cdot f}} \ [\text{m}] \tag{10.1}$$

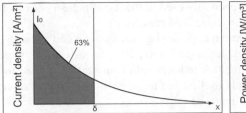
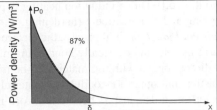

Figure 10.12   Penetration depth ([16])

where ρ is the material resistivity (Ω m); μ_r is the magnetic permeability of the material to be heated; and f is the supply frequency (Hz).

Both the frequency of magnetic field and the characteristics of the material to be heated (ρ, μ_r) influence the penetration depth. The frequency dependence offers a convenient possibility to control the penetration depth and, in this way, other parameters of the heating system: energy efficiency, power factor, mixing and melting rates, etc. Table 10.4 [13] gives several values of interest with regard to the influence of the material and magnetic field frequency on the depth of penetration.

The heating intensity diminishes as the distance from the surface increases, so small or thin pieces generally heat up more quickly than large thick parts, especially if the larger parts need to be heated all the way through.

2. Types of Systems and Applications

Induction heating is accomplished by placing an alternating current-carrying coil around or in close proximity to the material. When talking about the heating of processed material, there are two variants of this electrotechnology:

1. *Direct induction:* this occurs when the material is heated directly by the alternating magnetic field (eddy currents induced within the piece flow against the electrical resistivity of the metal, generating precise and localized heat without any direct contact between the piece

Table 10.4   Penetration depth ([13])

Material		Steel 20°C	Steel 20°C	Copper 20°C	Copper 900°C	Graphite 20°C
ρ [$\mu\Omega$ m]		0.16	0.16	0.017	0.086	10
μ_r		40	100	1	1	1
δ [mm]	50 [Hz]	4.50	2.85	9.31	20.87	225.08
	100 [Hz]	3.18	2.01	6.58	14.76	159 15
	1 [kHz]	1.01	0.64	2.08	4.67	50.33
	10 [kHz]	0.32	0.20	0.66	1.48	15.92
	100 [kHz]	0.10	0.06	0.21	0.47	5.03
	1 [MHz]	0.03	0.03	0.07	0.15	1.59

and the coil). Direct induction heating is primarily used in the metal industry for melting, heating and heat treatment (hardening, tempering, and annealing).

2. *Indirect induction:* in this case, the electromagnetic field generated by a coil induces eddy currents into an electrically conducting material (also referred to as susceptor), which is in direct contact with the material to be treated. Indirect induction heating is used to heat plastics and other nonconductive materials by first heating a conductive metal susceptor that afterwards transfers heat to the work-material.

The inductor and the heated part behave as an inductive load and are compensated with capacitors, while a frequency converter feeds the coil with a single-phase current at the desired frequency. Other components of an induction installation are: the cooling system (for frequency converter and inductor), a transport system (if required by the heating process) and the necessary control system. From the power supply point of view, the following variants can exist:

- 50 Hz installations: the compensated load is directly connected to the industrial distribution network;
- frequency converters with thyristors (rated power up to 10 MW), working in the frequency range 0.1–10 kHz with an efficiency of 90–97%;
- frequency converters with transistors (rated power up to 500 kW), with frequency range up to 500 kHz and efficiency 75–90%;
- frequency converters with vacuum tubes (rated power up to 1.2 MW): frequency range up to 3 MHz and efficiency 55–70%.

The applications of induction are usually associated with the melting of metals, heating of metals for forging, brazing and welding, and all sorts of surface treatments; the heating of non-metallic materials are also possible (however, the induction systems involved in this kind of applications require high-frequency magnetic fields).

Direct Heating of Metals (Metallurgical Applications)
- Smelting and melting of steel, cast iron, aluminium, copper, zinc, lead, magnesium, precious metals and alloys;
- heating prior to forming/forging (slabs, sheets, tubes, bars, etc.) and selective heating of parts and bonding of metals and non-metallic bodies;
- heat treatments (hardening of gears and annealing of tubes, welds, wire, sheets, etc.) or heating prior to surface treatments of metal elements (cleaning, stripping, drying, galvanizing, tinning, enamelling, organic coatings, etc.);
- welding and brazing.

Indirect Heating of Materials in Metal Containers
- Heating of dies and press platens in the plastics industry or of chemical reactors in the manufacture of resins, paints and inks;
- heating of vats in the food industry;
- melting and crystallization of glass, refractory oxides and nuclear waste.

3. Advantages and Limitations

Owing to its operation principle, induction heating has a number of essential advantages, such as:

- It has a very quick response (practically instant) and a high efficiency (the overall efficiency is 70–75% on average, and as much as 90%; in addition, there is no energy loss during idling time), for faster production rate and very high temperatures.
- It allows heating very locally (no risk in heating undesired components) and controlled depth of penetration, at extremely high heating speeds (due to the high-power density: 50–50 000 kW/m^2).
- Remarkable purity is possible in the absence of any physical contact between the energy source and the object to be heated and/or by working under vacuum or inert atmospheres.
- There is a reduction of oxidation losses and an absence of decarburization.
- There is a substantial improvement in working conditions (less heat and noise released to the local area) and environmental conditions (elimination of combustion products).
- The power input to the piece can be accurately controlled by the shape of the coils, the intensity of the field and the time applied.
- It uses bath stirring in the melting processes (this favours the homogeneity of alloys and rapid melting).
- Good repeatability (often used where repetitive operations are performed: once an induction system is calibrated for a part, work-pieces can be loaded and unloaded automatically).

The next limitations must be considered:

- A large investment is needed that must be considered and compared with alternative heating techniques.
- It is not well suited to irregularly-shaped parts during forging.
- It is necessary to change the inductor and sometimes compensate for pieces that are non-repetitive in shape.
- The power factor of the inductor and the load usually lies around 0.05–0.6.

4. Typical Performance

- The efficiency depends on the operating parameters: geometry of the inductors, distance between inductor and material, nature of materials to be treated and the properties of the inductor's conductors, etc.;
- It has a specific power output up to several MW.

5. Application Considerations

- According to the material properties, the direct (metallic bodies) or indirect (non-metallic materials) heating is selected but the latter has only specific areas of application,
- The size and shape of the piece must be analysed and, sometimes (for long and slender work-pieces), the direct resistance heating can be more efficient.
- The configuration of the inductors must be as close as possible to that of the materials (in order to reduce the air gap between the inductor and the load, so improving the power factor and efficiency).

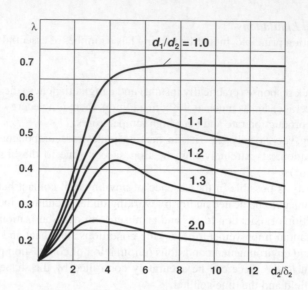

Figure 10.13 Variation in power factor λ for an induction heating system: d_1 – interior diameter of the inductor; d_2 – exterior diameter of the part; δ_2 – penetration depth in the part

- Great attention must be paid to the parameters of the material to be heated (magnetic permeability, electric resistivity and thermal conductivity), and to their changes with temperature.
- The choice of the frequency of the variable magnetic field is important because it decisively influences the energy efficiency and the power factor, Figure 10.13 [4, 10].

10.2.4 Dielectric Heating

Dielectric heating is used for heating materials that are poor electrical conductors (but also poor conductors of heat) and corresponds to a term covering two similar approaches: *radio-frequency heating* (RF heating) and *microwave heating* (MW heating). In both techniques, the material to be heated is exposed to an electromagnetic field that is continuously reversing direction (alternating) at a very high frequency.

When a non-electrically-conducting material is placed in a high frequency electric field, the electron and proton charges within the molecules of the material try to align themselves with the applied field. This results in a rapid agitation of the molecules, which is converted into heat within the material as a result of molecular interaction (e.g. friction); this is known as heating by dielectric hysteresis or, in short, dielectric heating. The heat produced depends mainly on the dielectric properties of the material to be treated. In RF heating the frequency is between 1 and 300 MHz; for MW heating, the frequency lies between 300 and 30 000 MHz, typical 2.45 GHz. To avoid conflict with communications equipment, several frequency bands have been set aside for industrial dielectric processing.

The essential advantage of dielectric heating resides in the generation of heat directly within the material to be heated. In comparison with other more conventional heating techniques (hot air, infrared) in which the material is heated via the outer surface, dielectric heating is much more rapid (as materials are poor conductors of heat, the transfer of the heat by conduction

Table 10.5 Dielectric properties

Material	Temperature [°C]	30 [MHz]		2500 [MHz]	
		ε'	ε''	ε'	ε''
Water	−12	3.8	0.7	3.2	0.003
	+25	78	0.4	77	13
	+85	58	0.3	56	3
Salt solution 0.1	+25	76	480	76	20
0.5 molar	+25	75	2400	68	54
Alumina ceramic	+25	8.9	0.0013	8.9	0.009
Quartz glass	+25	3.78	<0.001	3.78	<0.001
Nylon66	+25	3.2	0.072	3.02	0.041
Polyethylene	+25	2.25	<0.0004	2.25	0.0007
Teflon	+25			2.05	<0.0005
PVC	+20	2.86	0.029	2.85	0.016

from the outer surface takes time) and more efficient (savings from 15 to 40% in primary energy can be obtained). However, a forced-air system can be added to a dielectric heating system; the moving air removes moisture from the heating chamber and the drying processes are accelerated.

Different materials react differently to alternating electromagnetic fields; therefore, not all materials are equally suitable for dielectric heating. For instance, water heats up very fast as it absorb microwaves very easily; rubber is another good absorber. The ease with which a dielectric material can be heated is represented by what is known as the *loss factor*, reflecting the two phenomena playing a role in the dielectric heating of a material (the polarization and the molecular friction), and expressed by:

$$\varepsilon'' = \varepsilon' \tan \delta, \tag{10.2}$$

where ε' is the *relative permittivity* or *dielectric constant* of the material and δ is the *loss angle*.

The higher the loss factor, the more energy can be absorbed in the material. For a given material, the loss factor is variable, depending on temperature, moisture content (adding salt or carbon to a material increases its loss factor) and frequency. Other elements, such as the orientation of the electrical field, can also have an effect. When objects consist of materials with different loss factors, interesting applications are possible (e.g. pasteurizing pharmaceuticals and foodstuffs within their packages without burning the packaging material). Table 10.5 gives an overview of the dielectric properties of some common materials [17].

10.2.4.1 Radio-frequency Heating

1. Principle

In an RF system the field is generated between two conducting plates (electrodes), to which an alternating voltage is applied; the material to be heated is placed between these plates (see Figure 10.14 [5]).

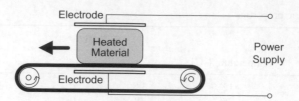

Figure 10.14   Radio-frequency heating ([5])

2. Types of Systems and Application

Generally, as shown in Figure 10.14, an RF heating system includes the power supply, the applicator (or operating space) containing the product that is to be heated and a material handling system (sometimes a ventilation system may also be added). The following types of systems, based on different criteria, are used in industrial applications [5]:

- fixed or variable frequency operation;
- batch or continuous heating (i.e. conveyor) systems;
- conveyor systems (characterized by the electrode configuration) – flat plate types for thick objects, stray field types for thin webs, and staggered electrode types for thick sheets;
- tubular heating systems (for liquids).

The RF generators are either controlled frequency oscillators with a power amplifier (also called '50-ohm' or 'fixed impedance'), or a power oscillator in which the load to be heated is part of the resonant circuit (also known as 'free-running' oscillators); the 50-ohm generators are used most prevalently in industrial processes. As stated above, only authorized frequencies may be used for open systems; for furnaces that are closed and suitably shielded, the frequencies used may range from 1 to 300 MHz.

Applicators can be constructed in different ways, depending on the specific characteristics of the product or the process; from an electrical point of view, every configuration is a capacitor with the material to be heated acting as its dielectric (for this reason, the radio-frequency heating is also referred to as capacitive heating). In radio frequency installations several configurations are possible, depending on the application, Figure 10.15 [17].

1. *'Stray field' electrodes*: the electrodes, produced as tubes or rods, are located on one side of the product to be heated, successive electrodes having a reversed polarity. This configuration is used for products or layers very thin (up to 10 mm), flat and having a large surface area.
2. *'Staggered-through' electrodes (garland arrangement):* in this configuration, the product to be treated is located between two rows of electrodes. The solution usually offers a more homogeneous field, while the field strength can be regulated easily by varying the electrode distance.
3. *Plate electrodes*: the electrodes are parallel plates located on either side of the treated product (forming a flat capacitor); this arrangement is mostly used when the product is thick or complex in shape.

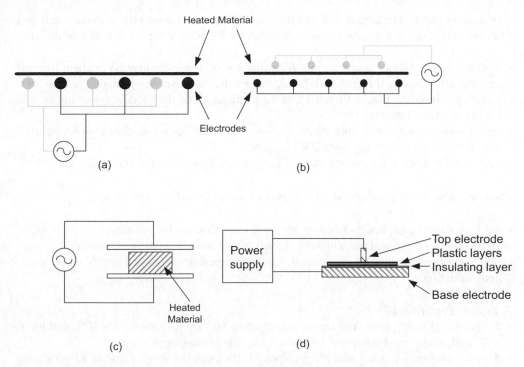

Figure 10.15 Main types of applicators for radio-frequency heating: (a) 'stray field'; (b) staggered through'; (c) flat plate; (d) welding electrodes (based on [17])

4. *Welding electrodes*: the configuration is used in such applications as thermoplastic welding and consists of a plate-like electrode and an electrode that is especially designed for the application (required pressure is exerted on the top electrode).

The main applications of capacitive heating are:

- the heating of any dielectric material, so long as it has a convenient loss factor (glues, plastics, resins, etc.);
- evaporation of water in dielectric and fairly regular in shape materials (paper, cardboard, board, textiles, wood, etc.), or drying of water-based coatings, inks and adhesives in paper manufacturing and converting;
- the welding and sealing of plastics;
- post-baking drying and moisture control of biscuits, crackers, cereals and other food products;
- heat treating, de-infestation and pasteurizing of bagged materials.

3. Advantages and Limitations
Based on the general advantages of the dielectric heating (the material is heated directly and instantly, there is the possibility of applying high-power densities throughout the process,

significantly reduced treatment time and floor space, increased productivity, instant start and stop, etc.) the following specific positive attributes for RF heating applications can be derived:

- high efficiency, mainly in applications requiring a good water-elimination gradient (overall process efficiency in the range 50–70%), due to the selective water heating – moisture-levelling within the material treated and no overheating of the product (resulting in less energy used and better quality);
- power consumption with humid materials alone (advantageous for separate or spaced parts), with possible power density up to 200 kW_{RF}/m^2;
- heat transfer more or less independent of temperature and volume of ventilating air.

The main limitations in the industrial implementation of capacitive heating are:

- the high capital cost, but applications are more cost-effective for products with high added value or when combined use with less expensive processes (e.g. infrared, hot air);
- only authorized frequencies can be used, requiring expensive protection against electromagnetic radiation.

4. Typical Performance
- Efficiency of supply converter (rated output up to 900 kW): typically 55–70%, but up to 80% with newer solid-state, high frequency amplifier technology;
- floating-frequency systems offer a higher overall efficiency, but a higher level of RF-shielding is required;
- tube service life: 5000–10 000 hours.

5. Application Considerations
When selecting this technology or designing the equipment, the following aspects have to be considered:

- The materials to be treated must be dielectric and the chosen frequency (floating or fixed, according to the material reaction) has to correspond to a high loss factor (otherwise the electric field is totally reflected at the piece surface and does not penetrate).
- The treatment (heating, drying, gluing, etc.) may be performed in batches or continuously.
- The parts to be treated should preferably be regular or flat in shape, their characteristics imposing the electrode and conveyer types, power density and total power to be installed.
- The possibility of combining the high-frequency technique with another one, such as infrared or hot air should seriously be considered in order to reduce capital costs.

10.2.4.2 Microwave Heating

1. Principle
In microwave ovens, material is heated by means of microwaves, packages of energy travelling through space; the higher the frequency of this wave, the higher its energy. The frequency typically used in industry is 2.45 GHz. Microwaves have a higher power density than radio-frequency waves and usually heat material faster.

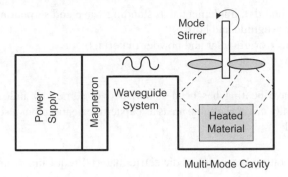

Figure 10.16   Microwave heating (based on [5])

The electromagnetic waves are generated in a wave generator (magnetron), which is situated outside of the heating chamber, Figure 10.16 [5]. As microwaves have a wavelength that is comparable to the dimensions of the installation, the energy cannot be transported via standard conductors and discrete networks. In this case, the waves are sent to the heating chamber through a tube, called the waveguide, produced as metal (copper or aluminium) pipe, mostly with a rectangular cross-section; the dimensions are dependent on the frequency. Waveguides can be both straight and curved, but the inside surface must be smooth and clean. The waves are then absorbed by the material placed inside the heating chamber.

2. Types of Systems and Applications

The MW heating can be implemented either as Batch Systems or Continuous Systems, with the following alternatives:

- single-mode (the product runs though a folded rectangular wave guide), Figure 10.17(a) [17];
- multi-mode (the product to be heated is placed in a large heating cavity or oven), Figures 10.16 and 10.17(b);

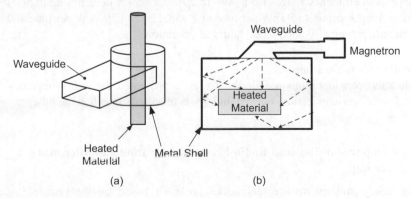

(a)   (b)

Figure 10.17   Microwaves installations: (a) monomode applicator; (b) multimode applicator (based on [17])

- tubular types (for liquids), implemented as standing wave and serpentine or coiled tube;
- radiating slotted waveguide and horn;
- rotary tube systems (for granular and powder materials);
- slotted waveguide (for webs).

Nowadays, microwave heating has been established in some key industries; the available applications practically cover the same areas as capacitive heating, but also some specific uses may be encountered:

- heating and evaporating water in any dielectric material requiring drying (even of compli-cated shapes);
- preheating and vulcanizing rubber products;
- heating and tempering frozen meat and other food products;
- drying and hardening coatings on carpets, textiles, paper, plastics, and electronics;
- sintering ceramics;
- production of plasma in chemical processes.

3. Advantages and Limitations
MW heating has the same advantages as radio-frequency technology. The limitations are also quite similar, but we can add the following:

- Only low power units are available and that may require multiple power sources or may restrict the quantity of heated material.
- Undesirable heating effects can appear in some applications, depending on the processed material (runaway temperature rise, burning).
- It may be difficult to treat large areas uniformly.

4. Typical Performance
Nowadays, the typical performances of MW heating equipment are:

- overall process efficiency ranges from 50% to 70% for power densities up to 500 kW/m^2;
- generator unit power: up to 30 kW per tube at 2.45 GHz, and 100 kW per tube at 915 MHz;
- power tube life: typically 5000–8000 hours of operation.

5. Application Considerations
The application considerations are similar to those presented for radio-frequency heating. However, for a given application, one technology is usually better than the other.

10.2.4.3 Comparisons Between Radio-Frequency Heating and Microwave Heating

Radio frequency and microwave techniques are both based on the principles described above. The operating frequency is, however, different and the two electrotechnologies differ notably in the behaviour of the various materials treated and the nature of the components used [5]:

- Power outputs available from RF sources (tubes or solid-state amplifiers) are higher than for microwave sources, thus allowing a scale reduction in costs (RF: up to 900 kW_{RF}; MW: up to 75 to 100 kW_{MW}). Overall, the capital cost for RF equipment is about half as much as for MW equipment.
- There is no RF power dissipation when there is no load (unlike MW).
- RF is better suited to large (thick), flat materials (uniform power and applicator type) while irregularly shaped products are more easily treated in MW multimode cavities.
- There is a wider choice of RF frequencies to adapt to different situations.
- MW heating is more suited to materials with low dielectric-loss factor and lends itself better to the application of high-power densities without creating a breakdown (e.g. arcing).

10.2.5   Arc Furnaces

1. Principle

Electric arc furnaces are process heating systems that heat materials by means of an electric arc created when a current (AC or DC) passes through an ionized gas between two electrodes. Such arcs are quite powerful and allow temperatures of up to 4000°C to be reached; these temperatures are high enough to melt steel, iron scrap or other materials. The electric arcs are used in furnaces as radiant heat sources or as submerged arcs, and the charge is usually heated both by current passing through the charge and by the radiant energy from the arc.

2. Types of Systems and Applications [3]

There are three types of electric arc heating systems, namely direct arc (contact) furnaces, indirect arc furnaces and submerged arc furnaces. In the first case, the furnace consists of a water-cooled refractory-lined vessel, covered by a retractable roof through which graphite or carbon electrodes (typically there are three electrodes) protrude into the furnace. As the electric arc strikes from an electrode to the metal charge, the distance between the electrode and the melt surface must be adjusted, in order to assure a constant arc length (the electrode wear is compensated by a positioning system that lower the electrodes into the furnace during operation).

The indirect arc furnaces have a horizontal barrel-shaped steel shell, lined with refractory, the arc being drawn between two carbon electrodes positioned above the load. Because the heat is transferred by radiation from the arc to the metal being melted, excessive heating of the refractory above the melt level may occur.

In the case of submerged arc furnaces the electrodes are deep in the furnace and the reaction takes place at the tip of the electrodes.

The main application of direct arc furnaces is in processes for melting of metals, largely iron and steel from scrap steel and iron as raw materials; applications for smaller arc furnaces include the melting of iron, steel and refractory metals. Direct arc furnaces used in foundries are usually for producing iron for casting operations, including the continuous casting for flat products such as steel plates.

Indirect arc furnaces are common in the production of copper alloys (these units are generally much smaller than direct arc furnaces), while submerged arc furnaces are used in smelting processes to produce materials such as silicon alloys, ferromanganese, calcium carbide, and ferronickel.

3. Advantages and Limitations
The most important advantages of arc furnaces are:

- They have large production capabilities (more than 400 tons in industrial-scale processes that make steel from scrap steel), allowing very high operation temperatures (up to 4000°C).
- Direct arc furnaces are less expensive (in terms of money per ton of steel capacity) than basic oxygen furnaces.
- They have higher efficiency than basic oxygen furnaces or integrated blast furnaces.
- The final product has a very good quality even when starting from scrap steel and scrap iron.

The limitations refer mainly to the excessive wear of the refractory and the negative impact on power quality in the point of common coupling (mainly flicker, voltage variation and harmonic pollution).

4. Typical Performance
- Energy consumption ranges between 400 and 500 kWh per short ton, depending on furnace production capacity (about 1/3–1/10 from the energy required by basic oxygen furnaces or integrated blast furnaces).
- Production capacities are from less than 10 tons (in foundries that melt iron and steel for castings) to more than 400 tons (in industrial-scale processes that make steel from scrap steel).

5. Application Considerations
Electric arc furnaces used for steelmaking are usually employed where there is a plentiful and inexpensive supply of electric power.

10.3   Specific Aspects Regarding the Increase of Energy Efficiency in Industrial Heating Processes

In practice, the increase of energy efficiency in industrial heating processes may be obtained in the following ways:

- the replacement of traditional technologies (direct fuel burning hot air, steam, water, etc.) by existing electrotechnologies, selecting the most appropriate technology for the required application;
- increasing the efficiency of the existing electroheating equipment;
- ensuring suitable operation and maintenance of the industrial electroheating equipment.

10.3.1 Replacement of Traditional Heating Technologies

The main advantages (from different points of view) of existing electrotechnologies and practical examples referring to the substitution of old technologies are further presented [5,7].

10.3.1.1 Resistance and Infrared Heating

Resistance heating is the simplest and oldest electrotechnology, used for both high-temperature and low-temperature industrial applications. With its excellent efficiency, good controllability, and rapid heat-up qualities (especially for direct heating), resistance heating can substitute the fuel-based process heating equipment in many applications, from melting metals to heating food products.

1. Energy Efficiency/Energy Savings
Modern indirect resistance heating equipment uses well-insulated enclosures, based on new advanced materials, consequently minimizing the thermal losses; as a result, this technology is more energy-efficient than the existing alternatives, especially at higher temperatures (for instance, the optimal efficiency of an electric oven can reach up to 95%, while those of a gas furnace can amount to 40–80%; in fact, the real heating efficiency of a gas furnace averages 15–20%).

Direct resistance heating and infrared heating also have good energy efficiency characteristics. With the advent of new power sources, materials and controls, both technologies are used in numerous manufacturing sectors. In many applications, electric infrared systems are used in conjunction with conventional direct-fired process heaters; frequently, the infrared system pre-dries the product, and then the process is finished in a conventional oven. For example, auto body production lines use infrared to rapidly set the paint on the body, and then the car goes into a convection oven to complete the curing process (the rapid setting of the coating on the body eliminates dust damage). An additional benefit of a hybrid system is the potential to increase throughput by increasing line speed.

2. Other Benefits
Environmental

- There is no output of harmful combustion products (NOx, SOx, COx), hazardous salts or metals.
- As the heating process is very precisely controlled, there is less heat transferred to the environment.

Technical

- Resistance equipment can be operated in a flexible way and less space is required (especially for direct and infrared heating).
- Precise temperature control and operation in controlled atmosphere are also possible, with benefits regarding the quality of final products.
- Melting in a resistance furnace can decrease dross or material loss.

Financial

- Investments depend on the size and type of the installation and are usually comparable to those required by fossil-fired systems.
- Return on investment is about 2 years (for high temperature indirect furnaces and direct heating equipment) or less (for infrared heating systems).
- There are reduced operation costs as little maintenance is required.

10.3.1.2 Dielectric Heating

The concept of using radio waves to heat material was known in the late 19th century, but industrial applications depended on the techniques for generating high-power at high frequency. Radio frequency generators were developed in the 1930s, while the microwave processing technology development was a result of research on radar systems during World War II; however, interest in microwaves considerably increased only in the 1980s as a way to raise productivity and reduce costs.

There are currently many successful applications of radio-frequency and microwave processing in a variety of industries, including food, rubber, pharmaceutical, polymers, plastics and textiles.

1. Energy Efficiency/Energy Savings

The conventional way of heating dielectric materials is to apply heat to the surface, which is further transferred to the interior by means of thermal conduction (a very slow process because these materials are also poor conductors of heat). As electromagnetic fields penetrate the material, dielectric heating occurs inside the piece and the thermal conduction is of minor importance during the heating process (this results in very fast heating).

As a result, heating processes can be shortened from hours to minutes and, partly due to this gain in time, dielectric heating offers large energy savings. Another reason for the low energy consumption is that only the work pieces are heated (no energy is lost in heating the walls or other parts of the oven or the air inside); supplementary, in drying or polymerization processes, energy is absorbed in a selective way, i.e. only the moisture and adhesives inside the work piece absorb supplied energy.

In general, dielectric heating can reach an efficiency of 65–75% (however, in the majority of applications the efficiency lies between 50 and 70%); for comparison, the efficiency of conventional heating methods is about 35–50%. Therefore energy savings of 15–40% compared with conventional methods are common; other sources indicate savings of between 25 and 50%. The high energy efficiency of dielectric heating systems is the main reason why manufacturers install them as replacements for conventional systems (e.g. hot air, steam).

2. Other Benefits

Environmental

- Dielectric heating is environment-friendly (no medium is required to transfer heat to the product and no fuels are burned at the production site).
- The electromagnetic waves are kept inside the heating chamber, so the technology does not put workers' health at risk and does not create other problems in adjacent areas.

Technical
- Dielectric heating assures high production rates (due to reduced processing time and because the heating process can be started up and shut down quickly, enabling just-in-time production).
- The course of the temperature during heating can be controlled very accurately, improving the quality of final products.
- Sometimes dielectric heating can represent a unique method of manufacturing different products.
- The heating chamber of a dielectric system is small and simple, compared with other systems, resulting in easier and faster maintenance.

Financial
- Because the quality of the products is very high, very few products are rejected.
- Dielectric heating systems are smaller than comparable conventional systems (space savings of 50–90% are possible).

10.3.2   *Selection of the Most Suitable Electrotechnology*

As stated above, nowadays many electric-based technologies can be used for various industrial heating processes. However, each of these technologies has its one characteristic and performance; that is why, in order to benefit fully from all the existing advantages, a suitable electrotechnology must be selected for an existing industrial heating process.

The first example was presented in Figure 10.4; the data offered there highlight that for drying processes, dielectric heating (in this particular example, microwave heating) is more suitable than resistance heating from both energy consumption and CO_2 emissions points of view.

Another example is direct resistance welding. In this case, an electric current is sent through the point where the metallic pieces touch and the resistance to the current flow generates heat that melts the metal; when the current is interrupted, the metal cools down and the two objects are fused. In a factory, the conventional machine, using an alternating current, was replaced by a new device; equipped with modern power electronics, the latter enables the system to produce a controlled direct current. This action assured significant energy savings (64% in annual energy consumption) and better product quality, Table 10.6 presents several comparative results (for a production of 1 320 000 drums per year). The cost of replacing an old unit is paid back in less than 2 years; a new, additional unit is paid back in less than 1 year.

Table 10.6   Energy consumption of welding systems ([7])

Parameter	Old system	New system	Difference
Power demand [kW]	250	90	160
Annual energy consumption [kWh]	390 323	140 620	249 700
Energy cost [€]	45 503	16 393	29 110

10.3.3   Increasing the Efficiency of the Existing Electroheating Equipment

A continuous maintenance process is required in order to maintain the good efficiency properties of any electric-based process heating system; on the other hand, the industrial implementation of new power electronic devices, advanced materials and new control techniques allows an improvement in the initial characteristics of existing electroheating equipment. However, it is important to mention that in general electroheating equipment requires less maintenance activities (and costs) than conventional heating technologies.

Several actions, adapted from [2, 4, 8] are further presented.

10.3.3.1   Resistance Heating

For the resistance-based process heating systems, the following actions are of interest:

- taking a continuous survey of the furnace envelope in order to notice as soon as possible any supplementary heat losses source (for instance, wall cracks);
- maintenance actions must include the cleaning of heating elements (clean resistive heating elements can improve heat transfer and process efficiency);
- for high-power direct resistance heating systems, the implementation of solutions aiming to reduce the unbalance (for instance, Steinmetz circuitry) can assure the required levels of electromagnetic disturbances in the point of common coupling;
- a permanent improvement of the control system can have both technological and energy saving benefits; good control systems allow precise application of heat at the proper temperature for the correct amount of time.

10.3.3.2   Infrared Heating

In this case, the subsequent measures can be valuable:

- The existing system must be used only for suitable materials (the absorption spectrum of the product to be heated must match the emission spectrum of the IR source).
- Maintenance actions have to include the cleaning of heating elements, but also of reflectors and end caps and additional reflectors can be installed in the oven in order to reduce energy losses and to re-radiate stray infrared energy back to the product.
- The oven shape must be reconsidered for every piece to be heated as the IR banks or panels are straightforwardly moveable (proper emitter positioning with respect to the product clearly improves energy efficiency).
- Zoning the IR system is also very energy efficient as it assures the consumption of the electricity only in a constrained area. This action is necessary especially when pieces having diverse dimensions are heated in the same oven; zoning can be configured horizontally or vertically, and can be specifically profiled for the product, due to the controllability of electric infrared energy (unfortunately, a more sophisticated control system will be required).

- A high-quality control system is also important: in addition to providing for zoning, an effective control system can also provide for a variable control system instead of simple on/off control, precisely delivering the required amount of radiant energy to the product, even if product size, shape, or colour, etc. might vary.

According to [3], better efficiencies (lower cost/part) – from 10% to 30% in existing ovens – have been demonstrated with the employment of these recommendations.

10.3.3.3 Induction Heating

Taking into account the flexibility of induction heating systems, possible actions differ according to the operation areas:

Melting
- When medium or high frequency is required, solid state power supplies are more efficient then rotative generators (they also allow operation with a variable frequency and require less maintenance).
- Continuous surveying of the furnace refractory is important (this reduces heat losses and also avoid an inductor breakdown).
- Bus bars system must be as short as possible and conductors must be selected according to the operation frequency.
- For highly conductive metals such as aluminium, copper alloys and magnesium, channel furnaces are more efficient than crucible furnaces.
- If the production line and schedule allow, two furnaces can share the same power supply by taking advantage of an optimized melting program.
- Melting without a cover on the crucible can account for approximately a 30% energy loss.
- For high-power mains frequency furnaces, the implementation of solutions aiming to reduce the unbalance (for instance, Steinmetz circuitry) is compulsory.

Heating and Heat Treating
- Usually, the heating and heat treating systems require high frequency; as previously mentioned, the use of solid state power supplies is strongly suggested.
- A dual-frequency design also has positive effects; in this case, a low-frequency design is used during the initial stage of the heating when the bar retains its magnetic properties, and a higher frequency is used in the next stage, when the bar becomes non-magnetic. However, this solution can be avoided by using a self-adaptive frequency converter (the supply works near the resonance frequency of the circuit represented by the heating system and the capacitor bank for power factor correction and follows the parameter modification during the heating process).
- For coreless systems, the use of flux concentrators is recommended. As these passive devices provide a contained pathway for the magnetic fields, stray magnetic losses are reduced.

- For good efficiency, the configuration of the inductor must be as close as possible to that of the heated materials (in many industrial applications, the same coil is used to produce a number of different products; actually, using coils designed specifically for a product will improve efficiency by up to 50%).
- Any existing gap (for inspection or work access) needs to be shielded to reduce heat loss.

10.3.3.4   Dielectric Heating

Dielectric heating is also widespread in modern industry. The following action must be considered:

Microwave Heating
- Systems involving water evaporation require a frequent visual inspection to avoid a dangerous water deposition on the heating chamber.
- The maintenance actions must include cleanliness of the wave guides and the operating condition of all motors and drives associated with the process.
- The shielding against electromagnetic radiation must be also periodically established.
- Whenever the characteristics of the material to be heated change (e.g. a change in width, depth or weight), the system must be re-evaluated in order to maintain its efficiency and all necessary modifications must be implemented.
- Power tube life is limited (as indicated before) so that the ageing generators must be replaced according to the vendor's recommendations (this action will assure the rated energy efficiency and will reduce system down time).

Radio-Frequency Heating
- The same maintenance actions previously mentioned for microwave heating are necessary (here, water deposition on the applicator system must be avoided).
- A good control system allows for better product quality and energy efficiency.
- The implementation of a hybrid radio-frequency/convection heating system must almost always be studied: the efficiency of a convection dryer drops significantly as the moisture level in the material decreases. At this point, radio frequency is more efficient for removing the moisture.

10.3.3.5   Arc Furnaces

The number of arc furnaces is lower than that of other electric-based process heating systems. However, due to the rated power of the existing units and their influence on the power supply, the experts pay special attention to the accurate operation of these systems. Some such actions are as follows:

- The injection of an inert gas (e.g., argon) in the bottom of the arc furnace can increase the heat transfer in the melt and the interaction between slag and metal (increasing liquid metal).
- By using ultra-high-power transformers, the furnace operation can be converted to ultra-high-power, increasing productivity and reducing energy losses.

- By introducing heat recovery systems, the waste heat of the furnace is used to preheat the scrap charge, so that the required energy for the technological process decreases.
- New advanced materials can be used for furnace insulation (for instance, ceramic low thermal mass materials instead of conventional ceramic fibre).
- The implementation of a hybrid heating system can ensure some advantages: for example, using a fuel-based system in the first part of the heat cycle saves energy by increasing heat transfer and reducing heat losses.
- Post-combustion of flue gases optimizes the benefits of oxygen and fuel injection: the carbon monoxide in the flue gas is oxidized to carbon dioxide, while the combustion heat of the gases helps heat the steel in the arc furnace ladle.
- Advances in power electronics allow the use of variable speed drives on flue gas fans (this reduces heat loss as well as the electricity consumption of the driving system itself).
- The maintenance must pay a special attention to the electrodes' positioning system as the accuracy of the electrodes' positions during the melting process decisively influences the energy efficiency and the operation costs (electrodes wearing).
- The negative impact on power quality at the point of common coupling (mainly flicker, voltage variation and harmonic pollution) must be compensated by appropriate means (harmonic filtering, active power factor compensation, etc.).

References

[1] U.S. Department of Energy. Energy Efficiency and Renewable Energy, *Energy Technology Solutions: Public-Private Partnerships Transforming Industry*, 2007, http://www1.eere.energy.gov/industry/bestpractices/pdfs/itp_successes.pdf

[2] K. Van Reusel *et al.*, Up to date arguments for selling electrotechnologies in Europe or how to use the political framework as evolved from the Kyoto agreement, *4th World Congress on Microwave and Radio Frequency Applications*, Austin, TX, 2004, http://www.esat.kuleuven.be/electa/publications/fulltexts/pub_1326.pdf.

[3] U.S. Department of Energy. Energy Efficiency and Renewable Energy, *Improving Process Heating System Performance: A Sourcebook for Industry*, 2nd edn, 2007, http://www1.eere.energy.gov/industry/bestpractices/pdfs/process_heating_sourcebook2.pdf

[4] M. Ungureanu, M. Chindris and I. Lungu, *End Use of Electricity*, EDP, Bucharest, 2000.

[5] CEA Technologies Inc. (CEATI), *Electrotechnologies. Energy Efficiency Reference Guide For Small to Medium Industries*, 2007, http://ebookbrowse.com/electrotechnologies-energy-efficiency-reference-guide-ceati-pdf-d177429937

[6] K. Van Reusel and R. Belmans, Technology bound and context bound motives for the industrial use of dielectric heating, *40th Annual Microwave Symposium Proceedings*, Boston, MA, 2006, pp. 15–18.

[7] EURELECTRIC, *Electricity for More Efficiency, Electric Technologies and their Energy Savings Potential*, 2004, www.uie.org/webfm_send/5

[8] N. Golovanov and I. Sora (eds), *Electrothermal Conversion and Electrotechnologies, Vol. I, Electrothermal Conversion*, Editura Tehnica, Bucharest, 1997.

[9] A. C. Metaxas, *Foundations of Electroheat*, John Wiley & Sons Ltd, 1996.

[10] D. Comsa, *Industrial Electroheating Installations*, Editura Tehnica, Bucharest, 1986.

[11] S. Lupi *et al.*, Characteristics of installations for direct resistance heating of ferromagnetic bars of square cross-section, *International Scientific Colloquium Modelling for Electromagnetic Processing*, Hanover, 2008, pp. 43–49.

[12] D. I. Romanov, *Direct Resistance Heating of Metals*, Mashinostrenie, Moskow, 1981.

[13] J. Callebaut, Leonardo Energy Power Quality & Utilisation Guide. Section 7, Energy Efficiency. Infrared Heating, 2007, http://www.leonardo-energy.org

[14] M. Orfeuil, *Electric Process Heating. Technologies. Equipment. Applications*, Battelle Press, 1987.

[15] GH-IA-Induction-Heating-Guide, www.inductionatmospheres.com

[16] J. Callebaut, Leonardo Energy Power Quality & Utilisation Guide. Section 7, Energy Efficiency. Induction Heating, 2007, http://www.leonardo-energy.org
[17] J. Callebaut, Leonardo Energy Power Quality & Utilisation Guide. Section 7, Energy Efficiency. Dielectric Heating, 2007, http://www.leonardo-energy.org

Further Readings

N. Anglani *et al.*, Energy Efficiency Technologies for Industry and Tertiary Sectors: the European Experience and Perspective for the Future, *IEEE Energy2030*, Atlanta, GA, 2008.
R. Belmans, Energy-demand management, AIE Conference Building a *Sustainable European Energy Market: Impact And Strategies For The Electrical Industry*, Brussel, 2004, http://www.esat.kuleuven.be/electa/publications/search.php
M. Chindris *et al.*, *Energy Management. Applications*, Casa Cartii de Stiinta, Cluj-Napoca, 2004&2009.
International Energy *Agency World Energy Outlook 2006*, ISBN 92-64-10989-7, Paris, 2006.

11

Heat, Ventilation and Air Conditioning (HVAC)

Roberto Villafáfila-Robles and Jaume Salom

Heating, ventilation and air conditioning (HVAC) systems are responsible for meeting the requirements of an indoor environment, with regard to air temperature, humidity and air quality by heating or cooling the spaces. HVAC systems represent an important part of the energy needs in industry and building sectors as it has been shown that such systems consume most of the energy in buildings, requiring almost 50% of the total energy demand [1].

HVAC systems are part of an energy chain for conditioning a space using different energy sources and converting them into thermal energy to meet the required level of comfort (see Figure 11.1). HVAC systems use fuels and electricity; fossil fuels are the most common source for heating whereas electricity is the almost the only source for cooling. Apart from cooling, electricity is also used to move fluids to enable thermal energy to reach the required spaces.

Fans and pumps used in thermal generation and transportation processes are the main electric energy consumers in an HVAC system. These are driven by electric motors and their energy efficiency is related to the use of power electronics-based drives, a topic that is principally covered in Chapter 9. Apart from optimizing the consumption of these electrical devices, there is a great potential for energy savings in order to decrease the energy consumption of HVAC systems as a whole, as well as in heat and cold generators. This can be achieved by reducing the thermal energy demand by passive methods and the use of renewable energy sources to support the production of both heat and cold. This chapter deals mainly with energy saving measures to reduce the thermal energy demand in HVAC systems and, as a consequence, achieve higher energy efficiency ratios.

Heat transfer is the basis of the performance of an HVAC system in order to heat or cool a space. This transference is performed in a set of heat exchangers (evaporators and condensers) that can be on their own or a part of a cool or heat generation device. Heat exchange processes are studied using thermodynamics.

Electrical Energy Efficiency: Technologies and Applications, First Edition. Andreas Sumper and Angelo Baggini.
© 2012 John Wiley & Sons, Ltd. Published 2012 by John Wiley & Sons, Ltd.

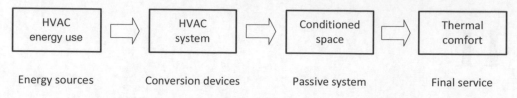

Figure 11.1 The map of energy flow to deliver thermal comfort [2]

In this chapter, first the basic concepts of thermodynamics with regard to HVAC systems are reviewed, with a special focus on cooling. HVAC systems are used in a variety of activities and places, so the thermal needs differ from one application to another. Secondly, the parameters used for defining indoor conditions are described.

HVAC systems consist of a set of equipment, a distribution network and terminals for delivering thermal energy to condition a space either collectively or individually. The interaction of these during their performance makes an assessment of the energy balance a complex task. Thirdly, the most common HVAC systems are described and the energy conversions involved are analysed.

Lastly, energy efficiency measures are listed for each link of the energy flow presented in Figure 11.1. Most of such measures are focused on decreasing energy demand, that is, thermal demand, which leads to a general reduction in the consumption of energy.

11.1 Basic Concepts

It is well known that heat flows naturally from a hot area to a cold one. Such heat transfer does not require the use of energy. So heat exchangers, evaporators and condensers are simple devices that normally consist of two fluids at different temperatures that flow through two different circuits. The hotter fluid transfers its heat to the colder fluid. After such a transfer, the hot fluid loses temperature and the cold one increases its temperature.

However, the reverse process it is not possible by itself: it needs the use of a thermal engine called a refrigerator. Thermal engines are, generally speaking, devices that have a cyclic operation, removing heat from one region and injecting it into another. A fluid (called a refrigerant) is normally used as the medium to transfer such heat during its cycle.

Refrigerators work as shown in Figure 11.2(a) for a vapour-compression process. The fluid (refrigerant) enters the compressor as vapour and is compressed up to condenser pressure, achieving a high temperature. When it flows through the condenser, it cools and condenses as it delivers its heat to the hot environment. It then reaches the expansion valve where both its pressure and temperature fall drastically due to the throttle effect. At this low temperature, it enters the evaporator and evaporates, absorbing heat from the space to be refrigerated. The cycle is completed when the fluid leaves the evaporator and flows into the compressor again.

The objective of a refrigerator is to eliminate heat, Q_L, from a refrigerated space. In order to achieve this, it requires external work, W_{net}. The efficiency of a refrigerator is expressed in terms of the coefficient of performance (COP), which is the ratio of useful heat movement to

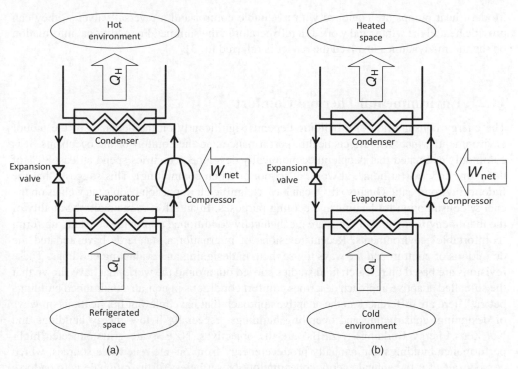

Figure 11.2   (a) Refrigerator and (b) heat pump

work input (equation (11.1)). The work input corresponds to the consumption of the electrically powered pumps.

$$COP_{\text{refrigeration}} = \frac{Q_L}{W_{\text{net}}}. \qquad (11.1)$$

The work capability of refrigerators is expressed in terms of $COP_{\text{refrigeration}}$ to avoid confusion, since it can have a value greater than unity as the heat removed from the refrigerated space can be greater than the work input.

The heat pump is another device that uses the same cycle as a refrigerator and can transfer heat from an area with low temperature to another area with high temperature (see Figure 11.2(b)). Nevertheless, the aim of a heat pump is to keep a space heated. This goal is achieved by absorbing heat from a cold environment and delivering that heat to the required space.

The work capacity of a heat pump is also expressed by the coefficient of performance according to equation (11.2). It can also have values greater than unity. A typical value of COP_{heating} is between 2 and 3. The COP_{heating} falls below 1 if the temperature of the cold environment is too low. In such a case, the heat pump changes its operation mode to electric heating.

$$COP_{\text{heating}} = \frac{Q_H}{W_{\text{net}}}. \qquad (11.2)$$

Modern heat pumps are equipped with a suitable control and a reversal valve, so they can provide heat in the winter and work as a refrigerator in the summer. For additional information on the thermodynamics involved the reader is referred to [3].

11.2   Environmental Thermal Comfort

The energy consumption of a building depends significantly on the demands of the indoor environment, which also affects health, performance and the comfort of the occupants. It is commonly estimated that people in economically developed countries spend at least 80% of their time indoors (at home, at work, at school or when commuting). This suggests that the Indoor Environmental Quality (IEQ) can have a significant impact on people, as well as on the energy consumption for heating and cooling purposes. In an effort to maintain the quality of the indoor environment, buildings are mechanically conditioned to provide constant, uniform, 'comfortable' environments. Recent revisions of international standards have updated the definitions of comfort and the ways to use them in designing and evaluating buildings. These revisions are based on research field studies carried out around the world that have shown that the so-called adaptive approach describes comfort conditions in non-air-conditioned buildings better. There are differences in the adaptive approach that have relevant implications on ways of designing, constructing and operating buildings, especially if low-energy buildings and Net Zero Energy Buildings (NZEBs) are the objectives. NZEBs are grid-connected high-performance buildings that annually produce energy from on-site renewable sources, which compensates for the annual energy consumption. Nevertheless, if the rationale is to endorse the design of environmentally friendly buildings that promote sustainable development, then reasonably strict requirements on energy efficiency should be satisfied while providing high levels of IEQ.

IEQ should address four aspects: indoor air quality, thermal environment, lighting and acoustics. All of these are related and have their effects on health, comfort and performance. 'Health' is understood very broadly as the state of complete physical, mental and social well-being and not merely the absence of disease. 'Comfort' expresses satisfaction with the environment. 'Performance', or productivity, is related to the ability to perform demanding tasks. Health, comfort and productivity can be influenced by physiological, behavioural factors or social and organizational variables, not only by IEQ conditions. Although light and noise are important constituents of IEQ, they will not be treated in depth here.

The most important variables that affect thermal comfort are the air temperature, the mean radiant temperature, the relative air velocity, the relative humidity (or the water content in the air), the type of activity undergone by people (which determines the heat production in the body) and their clothing (which offers thermal resistance to the heat transfer between the human body and the environment). Research conducted mainly during the 1970s by Fanger [4] in well-controlled environments has been partially taken and reorganized into international standards where thermal comfort is defined as 'that condition of mind which expresses satisfaction with the thermal environment and is assessed by subjective evaluation'.

Thermal conditions affect health. This is illustrated by the mortality rates during extremely hot weather conditions. In the case of less extreme conditions, elevated temperatures have been associated with the increased prevalence of symptoms typical of sick-building syndrome

(SBS) and several studies indicate that it is beneficial to keep the temperatures in buildings at the lower level of thermal comfort.

Thermal conditions can affect the performance or work through several mechanisms. An unpleasant sensation of being too hot or too cold (thermal discomfort) can distract people from their work and disturb their feelings of wellbeing. This may lead to reduced concentration and decreased motivation to work. The consequence of such a state is usually a reduction of productivity. Seppanen *et al.* (2006) [5] showed that performance increased with temperatures up to 20–23°C and decreased with temperature above 23–24°C. Maximum performance was predicted to occur at a temperature of 21.6°C. The data were obtained in office environments, factories, field laboratories and school classrooms. Nevertheless, other studies and surveys show that people accept a wider temperature and humidity swing if they can influence their own environment, especially with operable windows [6]. Recent projects such as the ThermoCo Project [7] analysed user satisfaction in buildings with regards to the upper temperature limits of the adaptive comfort model, showing the need for further scale surveys, particularly in hot (European) climates.

IAQ (Indoor Air Quality) is an important parameter that characterizes the indoor environment and is strongly related to the health of a building's occupants. The quality of indoor air can be defined as the level of the contaminants/pollutants in indoor air. The operation and the performance of the ventilation system directly determine the IAQ of an indoor space. In addition, the air flow and paths inside the building have an important influence on the thermal comfort of the occupants, especially during the summer. It is known that slight air velocities up to 1.5 m/s improve thermal comfort for moderate activities.

So the impact of thermal comfort expectations and the impact of ventilation on the energy performance of a building is crucial. When the outdoor fresh air supply is increased, the cooling and heating energy requirements are also increased when the cooling or heating system is in operation. On the other hand, there are energy conservation strategies, such as free cooling operation and the use of natural ventilation in buildings, which becomes much easier when indoor temperature comforts are applied as a consequence of an adaptive approach.

The results of recent research works have shown that when variable indoor temperatures comfort standards based on adaptive theory are used, remarkable energy savings may occur [8]. Savings of up to 18% or up to 34% in residential buildings in a Mediterranean climate may be achievable.

The most recent standards to specify the combination of indoor thermal environmental factors and personal factors that will produce thermal environmental conditions that are acceptable to a majority of the occupants are ANSI/ASHRAE Standard 55-2010 and EN-15251-2007. Both standards propose that acceptable temperature ranges actually depend on the type of system used to provide comfort. The European standard EN-15251 specifies the parameters for design addressing indoor air quality, lighting and acoustics, not just the thermal environment.

Standard AHSRAE 62.1 is used for determining the minimum ventilation requirements for high-performance buildings. The minimum air exchange ratio of fresh air is defined as a function of the occupation density and the surface of the conditioned area. Table 11.1 illustrates as an example of such ratios.

The required ventilation will be the result of the addition of the two requirements. The standard allows changes in the quantity of fresh air needed depending on the level of occupancy or other operational conditions.

Table 11.1 Air-exchange ratio requirements in buildings

Type	Requirement per person [l/s person]	Requirement per surface [l/s m^2]
Offices	2.5	0.3
Residential	2.5	0.3
Classrooms	3.8	0.3
Restaurants	3.8	0.9

According to EN-15251, several criteria are used for the definition of the fresh air ratios. Two of them are aligned with the method described in ASHRAE 62.1 and considers both the ventilation needed to dilute the pollutants emitted by the occupants and the building. The standard classifies the buildings into three categories (I, II, III), depending of the percentage of dissatisfied persons. Table 11.2 gives the recommended values for the three categories.

With regard to the control of the ventilation systems, in periods when the building is not occupied by people, the ventilation can be reduced to minimum values (0.1–0.2 l/s m^2 in non-residential buildings, 0.05–0.1 l/s m^2 in residential buildings).

With regard to thermal comfort, which drives the design of heating and cooling systems, EN-15251 distinguishes the buildings that are provided with mechanical cooling systems from those without. For buildings provided with mechanical air-conditioning systems, the criteria for determining thermal comfort are based on the method of Fanger and on the PMV–PPD (Predicted Mean Vote–Percentage of People Dissatisfied) indexes. ASHRAE 55 made a similar distinction and recommends the ranges for typical applications to be PPD < 10 and PMV (−0.5 < PMV < 0.5). The values of the indexes are shown in Table 11.3.

EN 15251 allows the use of an adaptive model for buildings without mechanical cooling. In the definition section, 'buildings without mechanical cooling' are defined in the standard as 'buildings that do not have any mechanical cooling and rely on other techniques to reduce high indoor temperature during the warm season like moderately-sized windows, adequate sun shielding, use of building mass, natural ventilation, night time ventilation etc. for preventing overheating'. Mechanical cooling is defined as 'cooling of the indoor environment by mechanical means used to provide cooling of supply air, fan coil units, cooled surfaces, etc.'. The adaptive model can be applied to spaces equipped with operable windows that open to the outdoors and that can be readily opened and adjusted by the occupants of the spaces. There shall be no mechanical cooling in operation in the space. Mechanical ventilation with

Table 11.2 Recommended values for obtaining fresh air according to EN-15251

Category	Air flow by person [l/s person]	Air flow per surface [l/s m^2]		
		Very low polluted building	Low polluted building	Polluted building
I	10	0.5	1	2
II	7	0.35	0.7	1.4
III	4	0.2	0.4	0.8

Table 11.3 Values of PPD and PMV indexes for thermal comfort

Category	PPD	PMV
ASHARE 55	<10	−0.5 < PMV < 0.5
EN-15251-I	<6	−0.2 < PMV < 0.2
EN-15251-II	<10	−0.5 < PMV < 0.5
EN-15251-III	<15	−0.7 < PMV < 0.7

unconditioned air (in summer) may be used, but opening and closing of windows shall be of primary importance as a means of regulating thermal conditions in the space. There may, in addition, be other low-energy methods of personally controlling the indoor environment such as fans, shutters, night ventilation, etc.

ASHRAE Standard 55 makes a similar distinction but does not have exactly the same wording, allowing the application of an adaptive model (based on outdoor monthly average temperatures), in 'occupant-controlled naturally conditioned spaces' defined as 'those spaces where the thermal conditions of the space are regulated primarily by the occupants through opening and closing of windows'.

In general, the application of the adaptive model indicates that indoor thermal comfort is achieved with a wider range of temperatures than by implementation of the Fanger model. The temperature limits shown in Figure 11.3 for EN 15251 should be used for dimensioning

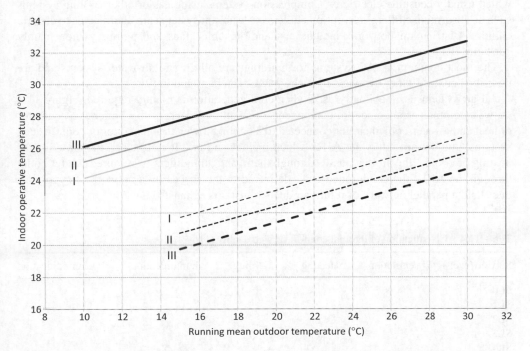

Figure 11.3 Acceptable indoor operative temperature ranges (cooling season) for buildings without a mechanical cooling system as a function of outdoor air daily running mean temperature (from EN 15251)

the passive means of preventing overheating in summer and are function of the mean ambient temperature. When the adaptive temperature limits (upper limits) cannot be guaranteed by passive means, the design criteria for buildings with mechanical cooling should be used. In [5] it is shown that using some of the indexes proposed by EN 15251 and their intended use (start with its adaptive variant and, if comfort conditions for the chosen category cannot be met, switch to the Fanger variant) implies the presence of discontinuities in the procedure. This is because with common assumptions on metabolic rate and clothing certain conditions will be above the comfort range for Fanger and below the range for adaptive.

11.3 HVAC Systems

The service of HVAC systems covers air circulation, control of temperature and the control of humidity. The needs of ventilation and air conditioning vary widely depending on the requirements of the indoor environment. Consequently, there is a variety of HVAC systems. Depending on their capability to provide such services, HVAC can be classified into five service levels (Table 11.4).

HVAC systems contain a large variety of components that include equipment for generating heat and cool, air-handling units (AHU), distribution networks and terminals. The more services that are demanded, the more complex the system will be.

Heating technologies comprise boilers, furnaces and unit heaters, which typically use fossil fuels. Cooling technologies include chillers, cooling towers and air conditioning equipment, which usually consume electricity. Compression systems using gas or other cooling systems such as absorption chillers, adsorption chillers or desiccant systems are alternatives to electric chillers. Distribution networks include air-handling units, fans and pumps, which mainly consume electricity.

The most common HVAC systems for buildings are all-air and all-water systems and are described next. A wider overview of such systems can be found in [9].

All-air systems are habitually used when cool generation is mainly required. Figure 11.4 shows a common arrangement of components of an all-air system with a boiler as a generator of heat and a water-cooled chiller connected to a cooling tower as a generator of cool. In such systems, outdoor air enters the AHU, which includes a fan, filters to clean the air, coils for heating or cooling the air that passes through them, and humidifiers (if required). After going through the AHU, the conditioned air is delivered to the spaces by a network of supply air ducts and a parallel network of return air ducts transports exhausted air.

Table 11.4 Classification of thermal comfort service quality [1]

Service level	Ventilation	Heating	Cooling	Humidification	Dehumidification
SL0	√				
SL1	√	√			
SL2	√		√		
SL3	√	√	√		
SL4	√	√	√	√	
SL5	√	√	√	√	√

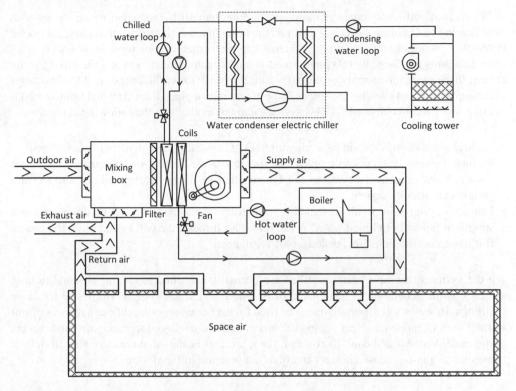

Figure 11.4   Elements of typical all-air HVAC system with chilled and hot water generation

The majority of all-air systems have a single duct system that provides either heat or cool to the conditioned spaces. HVAC single duct systems can be divided into the following:

- Single zone systems are the simplest all-air system as they deliver a constant volume of air with the same temperature to the entire facility. They are commonly used, for example, in department stores, factory spaces and auditoriums.
- Variable air volume (VAV) systems permit one to control the temperature individually in separate zones. They are widely used in large buildings.
- Reheating is achieved by adding coils to both former systems to allow conditioning individually in each zone. They are used in spaces with precise humidity requirements such as museums and some industrial processes.
- Multizone systems have a dedicated duct directly connected with the AHU that supplies conditioned air independently to each zone. They are almost obsolete except in small buildings with few zones and short duct distances.

Apart from widespread single duct systems, there are dual duct systems that have separate ducts for heating and cooling purposes. Both distribution networks run in parallel and the air is mixed in mixing boxes in each zone. They are used in buildings with strict requirements for temperature and humidity. HVAC dual duct systems can be divided into constant air volume (CAV) systems and variable air volume (VAV) systems.

Water, in its different stages, is more appropriate than air for transporting heat energy in both heating and cooling systems. However, the humidity, which affects air quality, cannot be controlled. Water distribution piping systems can be formed by one pipe, by connecting all terminals units in a loop; by two pipes, one to supply each terminal unit and the other for the return; three pipes, one supply of heated water, the other for chilled water and the last for a common return; and even by four pipes, supply and return pipes from terminal units for both heating and cooling. All-water HVAC systems can be divided into the following categories:

- Natural convection is the simplest all-water HVAC system. It uses hydronic convectors.
- Radiant heating systems can employ electric resistance heating or combustion as a heat source. Low temperature systems can be installed in floors, while medium temperature systems are panel mounted.
- Fan-coils systems are small air-handling units for a single space with a coil connected to a supply of hot and/or chilled water. A fan blows air through the coil to condition the space. If it has an outside inlet, it is called a unit ventilator.

All the systems that have been described are normally classified as central air-conditioning systems as the generation of heat and cool is centralized. Such systems are useful for large buildings. However, smaller systems can be used for the same purposes, although there is a limit to their area of influence. Heat pumps are an example of packaged terminal air-conditioners that can deliver heat and cool. In contrast, there are unit heaters that are only able to deliver heat, such as fan-forced unit heaters and high temperature infrared heaters.

11.3.1   Energy Conversion

HVAC systems consist of a huge number of interconnected elements and subsystems that make a global assessment of the energy conversion procedures complex. The representation of the energy conversion processes by HVAC loops [1, 10] facilitates the identification of such processes by assessing the energy involved in each transformation and then the energy balance.

Figures 11.5 and 11.6 represent the all-air HVAC system shown in Figure 11.4 through HVAC loops. Figure 11.5 (a) depicts the thermal energy chain in heating mode via three loops (if boiler is used):

- Combustion loop: heat is generated in the boiler by burning fuel.
- Hot water loop: water is heated in the boiler and is cooled in the coil; water is driven by pumps.
- Air loop: heat is delivered to the conditioned space; air is driven by fans.

Figure 11.5(b) also illustrates the energy chain in heating mode via four loops (if a heat pump is used). In this case they are as follows:

- Heat extraction loop: heat is extracted from the environment to the evaporator. Air is driven by fans.
- Refrigerant loop: the refrigerant works on a vapour-compression cycle extracting heat from the evaporator and transfer it to condensing water. The refrigerant is driven by a compressor.

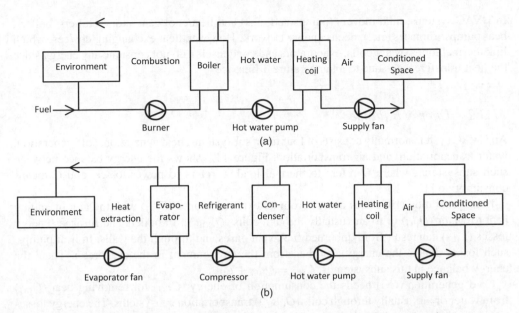

Figure 11.5   HVAC thermal chain for heating: (a) hot water boiler; (b) air to water heat pump

- Hot water loop: water is heated in the condenser and is cooled in the heating coil. Water is driven by pumps.
- Heat extraction loop: heat from the heating coil is delivered to the conditioned space. The air is driven by fans.

Figure 11.6 shows the cooling mode via five loops:

- Air loop: heat is extracted from the conditioned space to the chilled water. Fans drive air through the ducts and then it passes through cooling coils to produce heat transfer.
- Chilled water loop: chilled water is heated in the cooling coils and it is cooled in the chiller evaporator. The chilled water is driven by pumps.
- Refrigerant loop: the refrigerant works on a vapour-compression cycle extracting heat from the evaporator and transferring it to the condensing water. The refrigerant is driven by a compressor.
- Condensing water loop: heat from the condenser is transported to the cooling tower. The water is driven by pumps.
- Heat rejection loop: heat is ejected to the environment in the cooling tower.

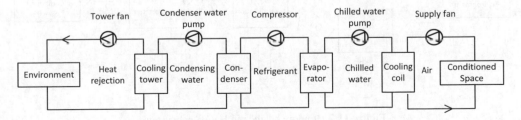

Figure 11.6   HVAC thermal chain for cooling

In HVAC systems, fluid movers (pumps and fans) and heat/cool generators (chillers, boilers, heat pumps, furnaces, etc.), needs energy to work. However, heat exchanging devices, which link thermal loops (water coils, evaporators and condensers), do not consume any energy since the heat transference is due to a temperature difference.

11.3.2   Energy Balance

An HVAC system normally consists of four main subsystems: heat generation, cold generation, water transportation, and air transportation. Figure 11.7 shows the energy balance between such subsystems, where Q refers to thermal load, L refers to power losses, and C means consumption [1].

Heat generation (HG) uses energy from an external source (C_{HG}) with the aim of adding heat to water (Q_{HG}) or to air (usually through coils (Q_{COIL})). In boilers and furnaces energy losses (L_{HG}) can result from unburned fuel, vent gases and through the walls. In heat pumps, such losses denote the energy absorbed from the environment in the evaporator. Then, the energy balance in this subsystem is: $C_{HG} = Q_{HG} + L_{HG}$.

Cold generation (CG) needs the consumption of energy (C_{CG}) for removing heat (Q_{CG}) from water or air (usually through coils (Q_{COIL})) transportation subsystems. The energy losses (L_{CG}) are the energy rejected to the environment by the cooling subsystem, including pumps and fans that are used in refrigerant compression, air-cooled devices and water-cooled devices. Then, the energy balance in this subsystem is: $C_{CG} = Q_{CG} + L_{CG}$.

A water transportation (WT) subsystem uses energy from an external source (C_{WT}) for feeding pumps that drive water from the generation equipment to the water coils through pipes. Heat exchange between generation subsystems and water is known as primary load (Q_{PRI}). If it is positive it means heat generation and if it is negative it means cold generation. The energy losses (L_{WT}) are caused by inefficiencies in the pumping process, water escape and thermal losses in the water distribution network. Then, the energy balance for this subsystem is: $Q_{PRI} = Q_{COIL} + L_{WT} - C_{WT}$.

An air transportation (AT) subsystem uses energy from an external source (C_{AT}) for feeding fans that drive conditioned air (Q_{SP}) from the water coils to the conditioned space through air

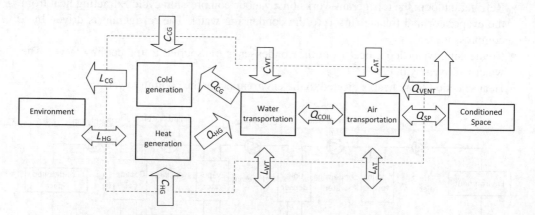

Figure 11.7   Energy balance in HVAC subsystems

ducts. Fans are also used for ventilation that considers the energy contained in return air and exhaust air (Q_{VEN}). The energy losses (L_{AT}) are produced by inefficiencies in fans, air leaks or infiltrations, and thermal losses in the air distribution network. Then, the energy balance for this subsystem is: $Q_{COIL} = Q_{SP} + Q_{VEN} - C_{AT} + L_{AT}$.

11.3.3 Energy Efficiency

The final energy consumption of an HVAC system depends on the building in which it is placed, the environmental parameters considered at the design stage, the equipment installed, the leaks and infiltrations in fluid transportation subsystems, and the operating controls established according to building requirements, among other factors.

Therefore, the energy efficiency of such a system varies widely depending on the service it provides. Moreover, it is rather difficult to establish one energy efficiency index in order to compare its performance with other systems. The comparison of HVAC systems is easily made by means of relative energy efficiency.

The relative energy efficiency of the air conditioning systems mentioned in this chapter is shown in Figure 11.8. It can be seen that CAV dual duct HVAC systems are comparatively less efficient than the rest of the HVAC systems. On the other hand, individual units, heat pumps

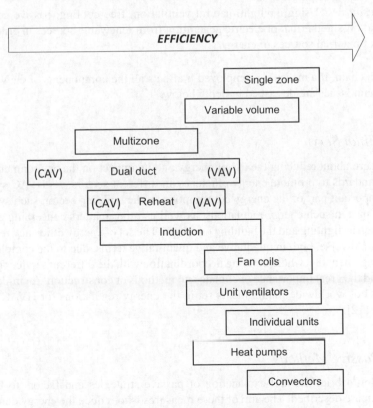

Figure 11.8 Relative energy efficiency of HVAC systems [11]

and convectors are comparatively the most efficient HVAC systems. The most appropriate system for each application can be consulted in [9].

11.4   Energy Measures in HVAC Systems

The energy efficiency in HVAC systems can be handled at each of the stages of the HVAC energy chain (Figure 11.1). It can be improved by focusing on the specific technical components in each component of the chain:

- *Final service*: Establishing limits on both ventilation rates and indoor comfort conditions through HVAC energy policies in order to encourage a rational use of energy.
- *Passive system*: Designing envelopes using high insulation, good air tightness and heat recovery ventilation systems in order to meet low-energy standards needs.
- *Conversion device*: Reducing consumption of thermal generators and fluid movers whose losses are due to wasted energy or equipment inefficiencies. Fuels are the most common source for heating while electricity is almost the unique source for cooling. Moreover, electric energy is used in transport subsystems and auxiliary equipment such as valve actuators. The use of solar thermal systems and an integrated building design approach can save energy by reducing heating requirements.
- *Energy resources*: Using daylight, natural ventilation, free cooling, passive cooling, heat pumps and on-site thermal or electric generation from renewable sources in order to reduce the need of external sources of energy.

As there is a chain, if a measure is employed it affects all the components of the whole chain. Thus, the main issues are described in-depth below.

11.4.1   Final Service

Global concern about reducing the use of energy and its impact on the environment has led to laws and standards to promote energy efficiency in different sectors. As HVAC systems represent an important part of the energy consumption in the building sector, such systems have a dedicated part in such energy regulations, as well as other matters concerning energy consumption such as lighting and the building envelope. The set of energy efficiency requirements for HVAC are diverse, both in qualitative and quantitative terms, due to the complexity of the energy management of a whole building in coordination with the different services (e.g. HVAC systems), and the great importance of climate and methods of construction. An in-depth review can be found of worldwide requirements set in the energy regulations for HVAC systems in buildings in [12].

11.4.2   Passive Methods

In this section a brief list and explanation of passive strategies and factors to be taken in consideration are described. The aim of these measures is to reduce the energy demand of the buildings and increase the user satisfaction in regard to their requirements of IEQ.

11.4.2.1 Local Climatic Conditions

In order to be able to exploit the local climate, it is essential that one analyses the climate type within which the site is located. The first task should be to collate the relevant climatic data (temperature, humidity, solar radiation, wind, etc.) that will inform the appropriate strategic design. The geographical location where the building is placed is the primary condition, but local conditions can differ significantly and will have implications for design.

11.4.2.2 Site Planning

The site location of the building can be influenced by some external factors related to the surroundings or the urban environment. These factors can slightly modify the local conditions, creating a microclimate. The microclimate is affected mainly by local topography, urbanization and vegetation. The use of vegetation as a bioclimatic measure has different effects: on the one hand, it improves the quality of air and conditions around the building in an urban environment that is beneficial to human health and comfort, on the other hand, from an energy perspective, the strategic placement of deciduous trees creates shade depending on the weather station and could also help to protect from the wind. Mainly important in the case of urban buildings, the features of the urban context could have a great influence regarding direct access to solar radiation, the shadows by surrounding buildings, the implications of the wind and air movement and the increase of air temperature referred to as the 'heat island' effect. In the case of urban sites, some aspects to be considered should be: the shape of the buildings (*L, H. I*, etc.), the height, the width of the streets and the existence of open spaces or backyards.

11.4.2.3 Orientation

The orientation of the building with respect to the sun's path and wind direction is an aspect that will have a direct impact on the design of the building and of the envelope solution applied to the different orientations of the façades. The optimum situation in terms of solar orientation, for a linear building, is to be oriented west–east axis, where the west and east façades have minimal openings. Façades with good access to solar radiation can improve daylighting and reduce heating needs in winter due to the solar gains through the windows. Appropriate shading protection systems in hot climates or during the summer season will help to reduce cooling loads for south façades without losing the benefits of solar radiation at other periods of the year. In addition, the orientation of the building is important as it enables one to take advantage of the main directions of the wind and maximize the effects of natural ventilation by night if we want to cool the building in a hot period. On the other hand, orientation and façade design can help protect from cold winds in the winter.

11.4.2.4 Building Plan Section

Building plan section and form is influential in determining the potential interactions between the building and the environment. One of the key aspects is the shape of the building, which determines its compacity. The compacity relates the volume to the surface of the building that is external. A compact form minimizes heat losses to the exterior through the building envelope but makes it harder to profit from natural resources such as solar radiation or wind. The internal planning and distribution of spaces depending on the planned use (hall, offices, bedrooms, living rooms, service areas, etc.) will have implications for the energy and comfort performance. For example, in the case of a dwelling, internal planning that allows air circulation between opposing façades has great benefits with regard to the possibility of natural crossed ventilation. Buffer spaces, which are spaces that are used occasionally, can act as a thermal buffers. These semi-conditioned (tempered) buffer zones, such as atria, winter garden, accessible double façades, attic, etc. increase the building's energy performance.

11.4.2.5 Envelope

With regard to the external envelope of the building, there are several strategies that should be applied in order to reduce heating and cooling loads. Thermal insulation is a primary way of avoiding heat losses. Building regulations recommend maximum U-values for the walls and a maximum glass ratio (also known as WWR – Window to Wall Ratio) in order to minimize fabric heat losses. The U-value is a standard measure of the rate of heat loss conducted through a building component. In addition to the level of insulation or the percentage of windows, other important aspects for the opaque part of an external wall must be taken into account. Some of these aspects are the type of construction and the position of the insulation layer in a wall which affects the thermal mass of the building and its dynamic thermal performance. Vertical walls should be carefully treated, as well as roofs and floors, especially if they are in contact with unconditioned spaces or with the ground. Thermal bridges also increase the heat losses. Although it is impossible to build a house without thermal bridges, proper wall construction can reduce these to a minimum. The properties of the external surface of the wall with respect to its solar absorbance or reflectivity will affect the amount of solar radiation absorbed or rejected by the wall. Improving of the tightness of the building using, for example, better windows, helps to control undesired air infiltration from outside. Another key aspect is the properties of the windows or transparent elements in the façades. The WWR should be optimized with regard to the location of the building, its typology (residential, office, school, etc.) and the shading devices. They are two key figures that defines the thermal behaviour of a transparent element: one is the U-value and the other is the SHGC (Solar Heat Gain Coefficient) (also known as the g-value). The U-value in windows is larger than in the opaque elements. The U-value should be minimized to some extent if the reduction of heat loss is the driving strategy or if there is a high WWR. The SHGC is the ratio that measures the heat gain entering an indoor space through a window with respect to the incident solar radiation on the window's surface. When solar protection or better isolation properties are needed a variety of windows with different SHGC and U-values are available: sun protection, simple/double/triple glazing windows, glazing systems filled with gas, low-emissivity windows, etc.

11.4.2.6 Solar Protection

The installation of a high percentage of transparent elements in façades that are oriented to the south, which is beneficial during the winter months, can lead to a situation of significant solar gain, which could result of a decrease of thermal comfort in the summer. In order to control the radiation that is directly responsible for solar gains, shading devices should be considered. The shading element can be fixed or mobile and should be designed according to the orientation of

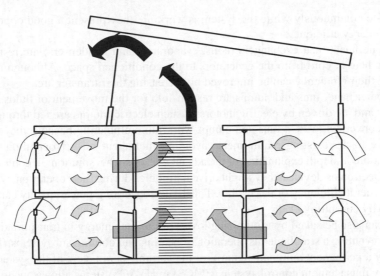

Figure 11.9 Example natural ventilation in summer at night

the opening and taking into account the direction of incidence of the solar radiation at different times of the year and during different times of day. The installation of solar protection should be properly sized so that the incident solar radiation is not compromised in the winter months.

11.4.2.7   Natural Ventilation

Ventilation is a way of providing the renewal rate required in terms of fresh air to ensure adequate indoor air quality. All the spaces of a building should be adequately ventilated in order to remove the pollutants that are generated during normal use of the spaces. Natural ventilation is a means of providing enough fresh air so it is an alternative to mechanical ventilation systems, which need electricity to operate. Additionally, natural ventilation can increase the thermal comfort of occupants, providing air movement to the indoor spaces and can be used to dissipate heat from a building without using air-conditioning. One of the strategies used in the summer or intermediate seasons is night ventilation. Night ventilation consists in leaving the air to circulate through the building by night (or even during the day) when the temperature of the outside air is below the required inside temperature. So the fresh air renews the heated air inside and also cools the mass elements that are part of the building (see Figure 11.9). Ground ventilation is a strategy that supplies air to the building from the outside through an air-hearth heat exchanger buried several metres in the ground. This technique allows one to pre-heat the air in winter or to cool it in summer due to the stable temperatures that occur two or three metres below the external ground surface.

11.4.3   Conversion Device

An HVAC system can improve its efficiency if efficient equipment is used at each of the four subsystems, specifically, air transportation, water transportation, cold generation and heat generation. The main elements of such subsystems are heat exchangers, fans and pumps. As these operate continuously while the system is working, they present a good opportunity for improving energy efficiency.

On the one hand, heat exchangers (coils, evaporators and condensers) are responsible of transferring heat or cold from the generators to the conditioned space. Although they do not use energy, their efficiency can be improved by increasing their transfer area.

On the other hand, fans and pumps are responsible for the movement of fluids (air, water, refrigerant) and are driven by electric motors. If such electric motors are fed through variable frequency converters (VFC), fans and pumps work more efficiently because they adapt their performance to the existing requirements avoiding inefficient on–off operation. Chapter 9 provides a more in-depth explanation of the advantages of such a solution. In contrast, as VFC are power electronics devices, they might drive harmonics during their operation, which might affect the general efficiency of the facility. This issue concerning power quality can be further understood by consulting [13].

Energy can also be saved by minimizing losses of thermal energy in transportation subsystems and preventing a simultaneous operation of heating and cooling subsystems. This can be achieved by a permanent automatic control of the variable parameters of the system. The use of an energy management control system (EMCS) in HVAC systems allows one to efficiently manage the energy in such systems as a whole. For example, an HVAC system in a large building covers different zones that are likely to have different thermal requirements. In order to provide an independent supply or extraction of the heat in each area an EMCS is fundamental.

At the core of an EMCS is a computer-based program where the data obtained from the sensors are processed for monitoring, controlling and optimizing the HVAC system, while achieving comfort and efficiency at the same time. The extra investment put into the devices and systems described earlier has a reduced payback due to the energy and operation savings obtained by its use.

The EMCS is not limited to HVAC applications and consists of a set of hardware devices and software connected via a communication network. The hardware includes:

- Sensors for obtaining the principal variable parameters of the system, that is, temperature, humidity, pressure, flow of air and water.
- Actuators that perform the action set by the control devices; such actuators normally are electric and pneumatic driven devices.
- Control devices that control the performance of actuators according to the data obtained from the sensors that are processed via the algorithms implemented in such devices. Such control is achieved by direct digital controllers, such as programmable logic controllers (PLC). Auxiliary devices such as relays might be required to interact with the actuators.
- Energy measuring devices allow one to obtain the energy consumption whilst the systems are working.

The software comprises a supervisory control and data acquisition (SCADA) system and data storage. The energy consumed by the whole HVAC system is monitored by the SCADA system. This allows one to supervise the operating conditions of the system and assess if the control strategies are performing well. Moreover, coordination with other systems of the building has to be considered, as for example with the fire-fighting system in order to stop air circulation in case of fire to facilitate fire extinction. A comprehensive description on the control of HVAC systems can be consulted in [14].

The communication link between hardware devices and software can be made according to different standard communication protocols (BACnet, LonWorks, Modbus, DeviceNet, Ethernet, TCP/IP, etc.), which can be either wireless or wired (which includes twisted pair cable, Power Line Carrier-PLC-, and fibre optic). Gateways are also usually needed in order to guarantee the interoperability, that is, the exchange of data between all the devices.

11.4.4   Energy Sources

The energy demand of an HVAC system is related to the building in which it works. A building is a very complex energy system where a large number of physical phenomena interact in a dynamic way. The building performance of a building is dependent on the site where it is placed, represented mainly but not only by the climatic conditions and the type of activity that might be carried on inside. Then, an Integrated Energy Design strategy is used for reducing the energy consumption.

The appropriate approach to designing buildings that provide high levels of comfort to its occupants and have the minimum impact in terms of the use of non-renewable resources is to use Integrated Energy Design. Many initiatives have been devoted to improving building codes [15,16], to developing guidelines for the industry [17], to improving design tools [18,19] and to increasing awareness in the final user who is seeking a building with low-energy consumption or even a zero energy building [17].

The Integrated Energy Design approach, by definition, is an inclusive approach that seeks the participation during the design process of all the design team members as well as the owner, project manager, maintenance representative, commissioning professional, etc. The other key point is to consider the building as a whole complex system where most of the aspects have a strong relationship. There are three key principles in Integrated Energy Design of low-energy buildings: 1) energy saving measures, 2) energy efficient systems and 3) renewable energy systems. The success of any energy building/project depends largely on how these three issues are addressed according to building type, use, location, climate environmental conditions (including availability for renewable resources), local energy infrastructure and financial resources, to name a few. Figure 11.10 illustrates the three key principles of Integrated Energy Design (IED): Passive Approaches (PA), Energy Efficient Systems (EES) and Renewable

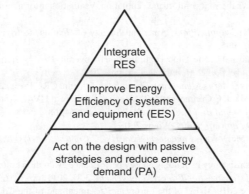

Figure 11.10   Principles of the Integrated Energy Design

Energy Systems (RES). The first principle in the IED process focuses on reducing the amount of energy needed through passive approaches. Given the inherent needs of artificial lighting and possible heating and/or cooling, the second principle aims at implementing energy-efficient systems. The renewable energy systems are needed to offset in large measure the energy demand required for lighting, heating and cooling (the third principle).

The HVAC designer will be aware of any passive heating and cooling strategies that can be promoted during the early stage of the design. By the same token, every kilowatt obtained from renewable energy to provide pre-heating and/or pre-cooling will have a major impact on HVAC equipment sizing. Through integrated design, the HVAC engineer will have the capacity to influence some key features of the building envelope such that passive strategies will be put forward. The cost benefit of renewable energy will also be taken into account by the HVAC engineer during the course of the design. The integration of renewable energy is often seen as an optional design feature, but when included in a whole building simulation analysis it has the potential to offset a significant portion of the heating and cooling loads. Therefore, the HVAC design of a low or nearly zero energy building will be part of an integrated design approach with an emphasis on the role of passive strategies and the integration of renewable as a way to reduce the peak loads and consequently the size of the HVAC equipment.

References

[1] L. Perez-Lombard, J. Ortiz and I. R. Maestre, The map of energy flow in HVAC systems, *Applied Energy*, **88**(12), 2011, pp. 5020–5031.

[2] L. Pagliano and P. Zangheri, Comfort models and cooling of buildings in the Mediterranean zone, *Advances in Building Energy Research*, **4**, 2010, pp. 167–200.

[3] M. J. Moran and H. N. Shapiro, *Fundamentals of Engineering Thermodynamics*, 5th Edition, John Wiley & Sons Ltd, 2006.

[4] P. O. Fanger, *Thermal Comfort: Analysis and Applications in Environemntal Engineering*, Mc Graw-Hill, New York, 1970.

[5] I. Sartori, J. Candanedo, S. Geier, R. Lollini, F. Garde, A. Athienitis and L. Pagliano, Comfort and energy efficiency recommendations for net zero energy, buildings, *Eurosun 2010 Conference*, March 2010.

[6] *Transsolar Climate Engineering*, High Comfort Low Impact, FMO Publishers, Germany, 2009.

[7] ThermoCo Project,Thermal comfort in buildings with low-energy cooling – Establishing an annex for EPBD-related CEN-standards for buildings with high energy efficiency and good indoor environment, Intelligent Energy – Europe (IEE), www.thermoco.org, 2009.

[8] M. Santamouris. Adaptive thermal comfort and ventilation, Ventilation information paper No. 12, *INIVE EEIG*, 2006.

[9] *ASHRAE Hanbook: HVAC Applications*, American Society of Heating, Refrigerating and Air-Conditioning Engineers, Inc., Atlanta, 1995.

[10] L. Lu, W. Cai, Y. Soh Chai and L. Xie, Global optimization for overall HVAC systems. Part I problem formulation and analysis, *Energy Conversion & Management*, August 2004.

[11] W. C. Turner and S. Doty, *Energy Management Handbook*, 6th edn, CRC Press, 2007.

[12] L. Perez-Lombard, J. Ortiz, J. F. Coronel and I. R. Maestre, A review of HVAC systems requirements in building energy regulations, *Energy and Buildings*, **43**(2–3), 2011, pp. 255–268.

[13] A. Baggini, *Handbook of Power Quality*, Wiley, 2008.

[14] R. Montgomery and R. McDowall, *Fundamentals of HVAC Control Systems*, ASHRAE, Elsevier Science, IP edn, 2008.

[15] E. Cubi and J. Salom, Consistency in building energy performance evaluation systems: a review and discussion, *International Journal of Construction Project Management*, **3**(3), 2011, pp. 1–16.

[16] EPBD (recast). Directive 2010/31/EU of the European Parliament and of the Council of 19 May 2010 on the energy performance of buildings.

[17] IEA SHC Task 40, ECBCS Annex 52: *Towards Net Zero Energy Solar Buildings*, http://www.iea-shc.org/task40/index.html

[18] EnergyPlus Energy Simulation Software, http://apps1.eere.energy.gov/buildings/energyplus

[19] *TRNSYS, The Transient Energy Simulation Tool*, www.trnsys.com

Further Readings

ASHRAE Handbook: Systems and Equipment, American Society of Heating, Refrigerating and Air-Conditioning Engineers, Inc., Atlanta, 1992.

IEE, Intelligent Energy Europe programme, http://ec.europa.eu/energy/intelligent

EU-Project, *CommonCense*, http://www.commoncense.info

D. Mumovic and M. Santamouris, *A Handbook of Sustainable Building Design & Engineering, An integrated approach to Energy, Health and Operational Performance*, Earthscan, 2009.

O. Seppänen, and W. J. Fisk, Some quantitative relations between indoor environmental quality and work performance or health, *Proceedings of Indoor Air 2005*, Beijing, China, 2005, pp. 40–53.

12

Data Centres

Angelo Baggini and Franco Bua

A data centre can consume up to 100–200 times as much electricity as standard office premises. With such a large energy consumption, they are prime targets for energy efficient design measures that can save money and reduce electricity use.

However, the reliability and the high power density capacity required by a data centre puts many design criteria far above energy efficiency.

Designing an energy efficient data centre therefore requires attention, skill and investment, but if done correctly it can provide substantial benefits. It is important to develop a holistic strategy and management approach to the data centre. This approach should grant the desired reliability, economic, utilization and environmental benefits.

The three main factors to consider when designing a data centre are:

1. reliability, a feature that is often guaranteed by the redundancy of equipment;
2. scalability, which can be achieved by using modular components that allow one to adjust to situations that may change over time and to avoid unnecessary over sizing;
3. the choice of high efficiency components.

These features must be taken into account in the selection of each component, whether it is IT equipment, components of the power supply infrastructure, or part of the HVAC system.

12.1 Standards

A variety of activities on the topic of Green Data Centres are currently ongoing in standardization organizations. An overview of the main initiatives in this field is presented here.

With regard to general data centre standardization, ANSI/TIA-942:2005 represented the first activity in the field of 'telecommunications infrastructure', providing a number of key definitions but with some content in relation to energy efficiency. Work is now being undertaken

Electrical Energy Efficiency: Technologies and Applications, First Edition. Andreas Sumper and Angelo Baggini.
© 2012 John Wiley & Sons, Ltd. Published 2012 by John Wiley & Sons, Ltd.

in the USA by BICSI to produce a Data Centre standard that builds on and expands the content of ANSI/TIA-942.

ISO and IEC are working on this subject with ISO-IEC-JTC 1. This joint working group identified the energy efficiency of Data Centres as a significant topic in the industry and has established a Study Group on Energy Efficiency of Data Centres (EEDC) to investigate market requirements for standardization, initiate dialogues with relevant consortia and to identify possible work items for JTC 1.

The telecommunication world has published with ETSI a series of documents entitled 'Broadband Deployment – Energy Efficiency and Key Performance Indicators' of which ETSI TS 105174-2-2 specifically addresses operators' data centres.

CENELEC has set up a specific working group to discuss the potential standardization actions in relation to energy efficiency within data centres. This working group (CLC BTWG 132-2) has reviewed the development and, where completed, outcomes of work relating to data centre energy efficiency undertaken by the European Commission, standards bodies external to CENELEC and recognized international fora. The BTWG 132-2 recommended to CEN, CENELEC and ETSI that they establish a joint CEN-CLC-ETSI group to manage and coordinate European activity in this field.

More specifically within CENELEC, TC 215 has initiated work on the EN 50600 series of standards, which addresses facilities and infrastructures to support effective operation of telecommunications cabling and equipment within data centres. Amongst other issues, these standards will specify:

- relevant measurement methods of parameters that may be used to determine energy efficiency;
- infrastructures necessary to enable the measurement of those parameters.

With reference to de facto standardization, the primary industry forum is Green Grid™ an open industry consortium of end users, policy-makers, technology providers, facility architects, and utility companies that has the aim of improving the resource efficiency of data centres. Also in this field the European Commission DG-JRC has instituted the Code of Conduct on Data Centres, which is a voluntary scheme targeted at improving the energy efficiency of data centres and which comprises an important document concerning best practices.

12.2   Consumption Profile

One of the basic operations to achieve high energy efficiency for a data centre is the measurement or estimation of consumption of the equipment that compose it.

In the case of a new data centre it is advisable to estimate its energy consumption in the design stage, so that the choice of equipment is as correct as possible. Very often the data centre is placed inside a building that is not completely dedicated to it; therefore it is not so easy to detect the energy consumption of a data centre if energy meters are not installed for this purpose.

In addition to total energy consumption, it is also important to understand and monitor how energy is distributed within the data centre, by the various energy flows and correlate them with areas or a single piece of equipment.

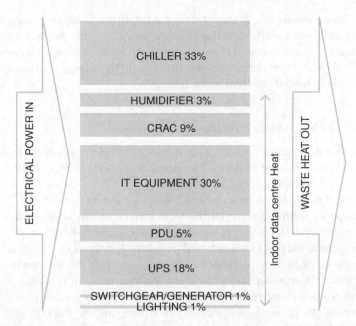

Figure 12.1   Typical data centre energy balance (Elaboration of data from Green Grid)

In this way, it is possible to identify not only the areas or equipment with the highest energy consumption, but also those whose functions are unnecessary and can be switched off or put on standby.

This is an assessment that can be time-consuming, given the large number of measurements needed, but it is rather critical, and fundamental in order to build the energy map of the data centre.

A typical data centre energy balance can be divided into three main parts: power, cooling and IT equipment. (See Figures 12.1 and 12.2.) The useful work is represented by data processing

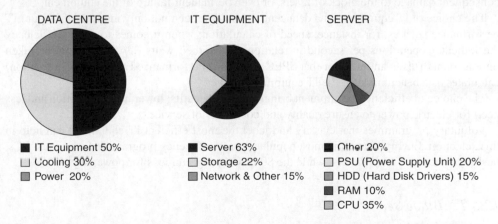

Figure 12.2   Typical values for single data centre components (Elaboration of data from ENEA)

in IT devices whose consumption is about half of the total energy used by a data centre, which must also cover the needs for HVAC and ancillary systems (e.g. fire safety, intrusion, etc.).

The energy used by the IT system is converted into heat, which must be dissipated; increasing the efficiency of these devices implies a direct decrease in the energy consumption as well as a reduction in cooling (about one third of the total).

12.2.1   Energy Performance Index

Evaluating the energy efficiency of a data centre is a complex operation. In principle, the efficiency should be calculated based on the useful work done, but it is difficult to have a standard measure of the useful work for a data centre.

Internationally, a performance indicator has been proposed by Green Grid™: PUE (Power Usage Effectiveness) and its inverse DCIE (Data Center Infrastructure Efficiency). These indexes were initially based on the measurement of electrical power (kW); the tendency is now to shift from these indexes to others based on the measurement of energy (kWh). In early 2010 four categories of PUE were defined: zero, one, two and three depending on whether you refer to power or energy, depending on where the measurement is made.

The PUE zero, the first to be defined, is the ratio between the total electric power consumption by a data centre and the power consumption by IT equipment; the theoretical ideal would be 1, but values up to 1.5 are considered good.

Existing data centres have average values of around 2.5; this indicates that there is a significant degree of improvement and suggests the opportunity for investment in energy efficiency measures.

12.3   IT Infrastructure and Equipment

IT equipment can be considered as the end user of a data centre. Their function, however, can only be ensured by the presence of an adequate supply system and cooling infrastructure. The supply system has to ensure electrical continuity and quality, while the cooling infrastructure has to dissipate the heat produced by the IT equipment and to avoid overheating, with consequent damage to the block of assets, or even permanent failure of the equipment.

The choice of IT equipment in a data centre should be seen not only in terms of its pure IT performance such as, for instance, speed of calculation; when it comes to energy efficiency the indicator operations per second in relation to absorbed watts (*Ops/W*) should be taken into account. This is an indicator that SPEC (Standard Performance Evaluation Corporation) calculates as a benchmark for the IT equipment market.

The choice of efficient IT equipment brings indirect benefits: lower heat production and less need for electric power to ensure quality and continuity of service.

Voluntary programmes that classify and label the most efficient IT equipment can help in this selection. Energy Star is the most popular energy efficiency programme for servers, PCs, desktops, monitors and printers, while the 80 PLUS programme is for power supplies.

12.3.1   Blade Server

This is a particular type of server that includes a high concentration of computer components (CPU, RAM, storage, etc.) that share the auxiliary components (power supplies and fans).

This type of server can be either introduced in new racks, or in place of the classic-type ones. For example 14 blade servers occupy the space of seven classic. This solution thus allows the elimination of classic-types of units and replaces them with blade-types, which can be considered high density and high performance computing components. Increasing the power of computing in a smaller space will result in large increase of heat per unit area, up to 28 kW per rack, so special cooling systems need to be provided. This type of server provides an overall saving compared with the classic setup of about 10%, with the same computing power.

12.3.2 Storage

The choice of technology storage systems, solid state storage (SSD Solid State Drives) technology versus traditional HDDs (Hard Disk Drives) can give advantages both in terms of efficiency and speed of data selection.

The SSD have lower power consumption and produce less heat. The speed of data selection in a storage system is measured in input and output operations per second (IOPS Input/Output Operations Per Second). In the case of HDDs, with a rotation speed of 15 000 rpm, 300 IOPS are possible, while SSDs can reach up to 30 000 IOPS per unit. In order to have the same speed input/output data of an SSD 100 HDDs are needed.

According to Sun/Oracle estimates, the ratio IOPS/watt for the SSD is around 10 000, while in the case of the HDD it is about 20. (Figure 12.3) At the same HDD IOPS consumption is about 500 times.

SSDs diffusion is still hindered by high costs compared with traditional HDDs, even if this difference should decrease in the future.

12.3.3 Network Equipment

The new generations of network equipment pack more throughput per unit of power; there are active energy management measures that can also be applied to reduce energy usage as network demand varies. Such measures include idle state logic, gate count optimization, memory access algorithms and Input/Output buffer reduction.

Peak data transmission rates continue to increase, requiring dramatically more power; increasing energy is required to transmit small amounts of data over time. Ethernet network

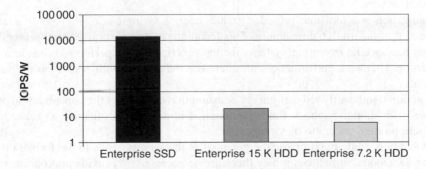

Figure 12.3   Comparison of IOPS/W ration for different storage devices

energy efficiency can be substantially improved by quickly switching the speed of the network links to the amount of data that is currently transmitted.

12.3.4 Consolidation

Where dealing with an existing data centre the possibility of replacing old machines with new ones should be considered.

The calculus performance of processors and memories evolves so fast – according to Intel estimates – that every four years it is possible to completely replace all machines with the latest models and reap the benefits in terms of energy consumption.

For example the consolidation of nine 2005 racks with a 2009 single rack can allow about a 10 times reduction in power demand.

Server consolidation can be achieved in several ways. Two typical approaches are:

1. combining applications on to a single server and a single operating system instance and
2. virtualizing application workloads in virtual machines and hosting them on a hypervisor.

Consolidation can also be applied to physical space. A data centre with lower supply fan power and more efficient cooling system performance, in fact, can be achieved when equipment with similar heat load densities and temperature requirements are grouped together. Isolating equipment by environmental requirements of temperature and humidity allows cooling systems to be controlled to the least energy-intensive set points for each location.

This concept can be expanded to data facilities in general. Consolidating underused data centre spaces in a centralized location can ease the utilization of data centre efficiency measures by condensing the implementation to one location, rather than several.

12.3.5 Virtualization

Virtualization is a form of consolidation and can be applied not only to servers but also to other components or applications, such as desktop, storage, application, and network virtualization.

Server virtualization is a technology that allows for the consolidation of computing workloads that historically required a dedicated physical server per workload. Organizations can achieve energy savings by reducing physical servers while maintaining an ample level of computing capacity on a single virtualized server or, perhaps, a condensed array of virtualized servers.

Virtualization is a tool that provides greater energy savings, allowing one to maximize the potential of the available machines. This technique allows several 'virtual servers', each of which has specific operating systems for the functions to be performed, inside a physical machine. Hardware performances are therefore maximized with respect to the software employed.

This action significantly reduces energy consumption as it allows the consolidation of some machines with shorter payback times, although it is difficult to quantify as related to the configuration of the system to be virtualized.

Virtualization, while maximizing the potential of the servers, leads to the formation of hot areas inside a room. Sometimes it may therefore be necessary to provide appropriate cooling systems on servers that contain the virtual machines.

Some of the main benefits of adopting virtualization include:

- greater flexibility;
- better management processes;
- reduction in the number of machines;
- reduction of operating costs (power and cooling);
- more space available to the data centre.

12.3.6 Software

The choice of data processing software can also have its effects on energy efficiency. In some cases, software that perform the same operations require a different percentage of CPU utilization. Often, the operations within the data centre are repetitive and software that requires fewer calculations to the CPU, and therefore less energy to perform the same functions, can give interesting results.

It is therefore advisable to check the load of the processors, memory and other server resources for different applications in the evaluation, before making the choice, and take it into account, especially when there is no large difference in terms of computational results.

12.4 Facility Infrastructure

12.4.1 Electrical Infrastructure

IT equipment needs a continuous and high-quality power supply. Very often, however, the electrical supply system is of poor quality and can damage equipment or interfere with its functioning.

The main equipment of the power supply system are as follows:

- UPS (Uninterruptible Power Supply);
- PDU (Power Distribution Unit);
- PSU (Power Supply Unit);
- cables.

12.4.1.1 UPS (Uninterruptible Power System)

The choice of UPS must ensure not only the maximum user protection from power disturbances and discontinuities, but also high efficiency.

A static UPS consists of rectifier, inverter and battery. The rectifier converts the AC current to DC for recharging the batteries, while the inverter converts the DC to AC supplying the end user. Double transformation of the energy from the network applies.

Static UPS can have three different architectures:

1. *double conversion*: in normal mode of operation, the load is continuously supplied by the converter/inverter combination in a double conversion technique, i.e. AC/DC and DC/AC;
2. *line interactive*: in a normal mode of operation, the load is supplied with conditioned power via a parallel connection of the AC input and the UPS inverter. The inverter is operational

to provide output voltage conditioning and/or battery charging. The output frequency is dependent upon the AC input frequency;

3. *passive standby*: in a normal mode of operation, the load is supplied by the AC input primary power via the UPS switch. Additional devices may be incorporated to provide power conditioning as voltage stabilizer systems. The output frequency is dependent on the AC input frequency. When the AC input supply voltage is out of UPS preset tolerances, the UPS enters a stored energy mode of operation, when the inverter is activated and the load transferred to the inverter directly or via the UPS switch (which may be electronic or electromechanical).

The most appropriate mode of operation to evaluate the energy efficiency of a static UPS is a double conversion (also known as online), which guarantees complete protection against power disturbances.

Each UPS is sized for the load that it has to feed and its efficiency is generally a maximum for load values close to the maximum.

For the latest generation of UPS, although the efficiency is more constant as load decreases, the tendency is to use modular groups that can adapt to the conditions and work load required in order to ensure the highest efficiency.

The energy lost in conversion turns into heat, which must be extracted by the HVAC system; this means that greater efficiency means less cooling required and lower operating costs.

Dynamic or flywheel UPS have the same function as static, and are generally installed without batteries; in this case, however, in order to provide continuity of supply for more than 10–15 seconds or so they must be coupled with gensets or the protection will cover transient and short duration disturbances only.

This type of UPS is formed by a flywheel, which rotates at high speed accumulating kinetic energy that is converted into electricity when a voltage disturbance occurs. When the flywheel UPS is equipped with batteries their useful life will increase as the flywheel reduces the number of charge/discharge cycles to which they are subjected.

Dynamic UPS are characterized by high efficiency, around 96–97%. They also have the advantage of being able to operate in environmental conditions that are less restrictive than those required by a static UPS, with the possible advantages of reduced consumption for air conditioning.

When selecting the type of UPS, the network characteristics of the construction site must be taken into account: few disturbances allows a UPS to work in by-pass mode with high efficiency; if the frequency of power disturbances is high and their duration is limited to few seconds, the use of a rotary UPS could be considered.

The choice should always be made taking into consideration the specific characteristics of each installation and carrying out a life cycle cost analysis, which takes into account the variables such as the cost of cooling, maintenance, etc.

12.4.1.2   PDU (Power Distribution Unit)

PDUs (Power Distribution Units) have the function to distribute power to the various end users in the data centre. The losses introduced are low, but they should not be neglected as they ultimately turn into heat, which has to be removed.

In the design phase it is very important to choose their position properly as it influence the power cables routes; in fact power cables should preferably be placed in an orderly manner to facilitate maintenance and to avoid obstructing the passage of the cooling air. They are generally mounted at the rear of the cabinet, and fixed to the floor or the ceiling.

It is advisable to choose PDUs equipped with instruments that can measure the instantaneous power consumption of the connected equipment, which allows for the monitoring of the system.

12.4.1.3   PSU (Power Supply Unit)

Most data centre equipment uses internal or rack mounted alternating AC/DC power supplies. Historically, a typical rack server's power supply converted AC power to DC power at efficiencies of around 60–70%. Today, through the use of higher-quality components and advanced engineering, it is possible to find power supplies with efficiencies up to 95%.

Using higher efficiency power supplies will directly lower a data centre's power bills and indirectly reduce cooling system cost and rack overheating issues.

There are also several certification programmes currently in place that have standardized the efficiencies of power supplies; for example the voluntary programme 80PLUS, labels power supplies with high levels of efficiency, and can help in the selection process.

12.4.1.4   Lighting

In designing the electrical system for the data centre, lighting should also be considered. Data centre spaces are not uniformly occupied and, therefore, do not require full illumination during all hours of the year.

Careful selection of an efficient lighting layout, lamps and ballasts will reduce not only the lighting electrical usage but also the load on the cooling system.

Therefore, it is recommended that one use the most efficient lighting technologies, such as LEDs, combined with an automatic control system zone based on occupancy sensors.

12.4.2   HVAC Infrastructure

The design of the HVAC system must be based on the actual load of IT and electrical equipment, trying to avoid as much as possible unnecessary over sizing and ensuring high reliability.

The main parameters to consider are:

- the total heat to be dissipated developed by all the equipment;
- the spatial distribution of thermal power to be dissipated with the identification of 'hot spots';
- the type of cooling system to be used;
- the operation temperature of IT equipment.

The designer's work can be supported by appropriate simulation software of thermal loads and air conditioning systems. Computational Fluid Dynamics (CFD) may be very useful for studying the movement of air flows.

In the case where a data centre has a very significantly different specific heat load per unit area (kW/m^2) or per unit of rack (kW/rack), different cooling systems can be used according to the needs of individual areas and/or individual systems.

The main technical characteristics of a high efficiency HVAC system are:

- continuous monitoring of temperature and humidity;
- optimized control system based on the actual requests of IT equipment;
- high efficiency air conditioning system components, among which it is worth mentioning:
  - high efficiency motors, pumps and fans with variable speed drives;
  - use the free-cooling system;
  - high EER (Energy Efficiency Ratio) chillers.

12.4.2.1   Cooling Best Practices

Low Density Rooms
The cooling of IT equipment is guaranteed by streams of cold air from the raised floor or ceiling. The cold air is supplied by air conditioning units CRAC (Computer Room Air Conditioning), located on either side of the room, or by centralized systems.

In this case it is of fundamental importance to minimize or eliminate mixing between the cooling air supplied to equipment and the hot air rejected from the equipment. In order to separate the two flows, hot air aisles are created, alternating them with cold air aisles, where air intake vents are placed. To ensure that no mixing of hot and cold air occurs, the hot aisles or the cold ones can be completely closed; panels can be installed to properly delineate the flows.

In addition to a complete separation of the aisles, a series of ducts can be designed with air intake vents to return hot air back to the CRAC. This cooling system is no longer sufficient when the thermal power dissipated per cabinet exceeds 3–4 kW (Figure 12.4).

High-Density Rooms
In the rooms where the installed capacity per cabinet is greater than 5–6 kW the use of dedicated cooling systems, placed near the heat source, has to be evaluated.

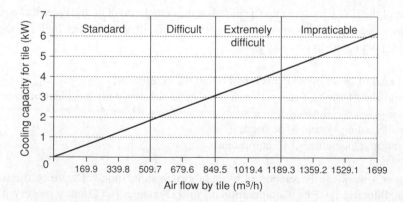

Figure 12.4   Air flow as a function of the cooling capacity

The main technological solutions available as power density increases are as follows:

- cooling systems for aisles: they are air conditioning units that are installed between two racks, integrating with the row; they can be mounted in the vicinity of high-density cabinets to compensate for the cooling that is not guaranteed by the classical configuration;
- cooling systems per unit (or rack): the cooling system (compressor and heat exchangers) is contained in the rack itself, which is hermetically sealed or placed separately on the rack. The heat released in the external environment can be removed or transferred to the HVAC system;
- direct cooling systems on the component within the server. A fluid, usually cool water, passes through the appropriate channels in direct contact with the CPU and electronic components, removing the heat. The main advantage of this system is its high thermal power dissipation; the disadvantages are the danger of loss of fluid in the vicinity of IT equipment and the higher cost.

These systems can be integrated with a traditional low density cooling system or replaced completely.

New IT systems, almost always coupled with virtualization, lead to the formation of 'hot spots'; this problem can be overcome by the installation near the load of a dedicated air conditioning units.

In these cases the presence of a raised floor for air distribution is not required; this usually implies a higher ceiling with respect to standard buildings.

The main advantages of a high-density rack are: lower operating costs, due to the dedicated cooling, and an increase in the space available in the data centre.

Free Cooling

Free Cooling (FC) allows one in certain conditions to use external air to directly cool the data centre, without using the HVAC system, except for dehumidification.

There are two types of FC:

1. *direct*: where you enter outside air at a low temperature inside the room; air must be properly filtered and humidified or dried to make it suitable for internal requirements;
2. *indirect*: where cooling is achieved by the single passage of the coolant in the cooling tower, without the intervention of the refrigeration compressor.

In both types, although external conditions are not such as to allow a total FC, there is still a further range of temperatures over which you can perform a free pre-cooling, reducing the compressors work required to maintain the set temperature.

The FC and the pre-cooling FC operating hours vary depending on the climate of the place of construction and the internal temperature settings.

Taking the average temperature in different months of the construction site of the data centre, the hours of operation in free-cooling mode, the operating costs and payback time can be estimated.

The FC, especially the indirect FC, can be also introduced as a retrofit in existing data centres, by adding some components in the system, without radically changing the HVAC system configuration.

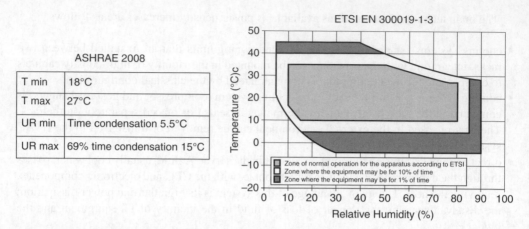

Figure 12.5 Reference values for temperature and relative humidity in data centres defined by ASHRAE and by ETSI

Temperature and Humidity Set Points

The choice of the temperature and humidity set points inside the room are two factors that can significantly affect the efficiency of a data centre.

The energy consumption of HVAC system, which generally accounts for roughly 30% of the total consumption, depends, in fact, on these values.

A rise in temperature may on the one hand allow a reduction in the consumption of the HVAC system and increase the hours of operation per year of the free-cooling system if one is present; on the other hand it reduces the time available for intervention in case of a downshift of the HVAC system and may increase the consumption of IT equipment for the operation of internal fans, which depends on the temperature set point in the bios of the machines.

The temperature set point optimization can be an energy saving opportunity but it requires careful analysis.

The reference values for temperature and relative humidity in the data centres defined by ASHRAE (American Society of Heating, Refrigerating and Air Conditioning) and by ETSI (European Telecomunications Standards Institute) in its standard EN 300019-1-3, are summarized in Figure 12.5.

12.5   DG and CHP for Data Centres

The use of DG and CHP in data centres result in cost savings to the facility operator in the form of:

- reduced energy-related costs and enhanced economic competitiveness – from reduced fuel and electricity purchases, resulting in lower operating costs;
- increased reliability and decreased risk from outages – as a result of reliable on-site power supply; and
- increased ability to meet facility expansion timelines – by avoiding the need for utility infrastructure upgrades.

Because of their very high electricity consumption, data centres have high power costs. Installation of CHP systems with absorption cooling can often reduce energy costs by producing power more cheaply on-site than can be purchased from the utility supplier. In addition, waste heat from the power generation can drive absorption chillers that displace electric air conditioning loads.

Data centres require both high-quality and extremely reliable power. Of all customer types, data centres, telecommunication facilities, and other mission-critical computer systems have the highest costs associated with poor power quality.

Data centres almost always have UPS systems to condition power and eliminate power disturbances. Battery backup is generally used to provide a short-term outage ride-through of a few minutes to an hour. Longer-term outages are typically handled with standby diesel generators.

On-site power generation, whether it is an engine, fuel cell, microturbine or other prime mover, supports the need for reliable power by protecting against long-term outages beyond what the UPS and battery systems can provide. DG/CHP systems that operate continuously provide additional reliability compared with emergency backup generators that must be started up during a utility outage. Backup generators typically take 10–30 seconds to pick up load in the case of an outage.

Developing on-site power sources gives data centre operators increased flexibility in both the expansion and design of new facilities. Upgrading older, smaller data centres with new equipment can result in a large increase in power demand to the facility that the utility might not be able to meet in the near term. Incorporating continuous prime power DG/CHP options can facilitate expansion and facility development on a more rapid schedule than can sometimes be possible by relying solely on the existing utility grid. Minimizing external power demand also reduces additional utility infrastructure requirements and associated costs that might be required for new or expanded facilities.

12.6   Organizing for Energy Efficiency[1]

To engineer the migration from a traditional energy-consuming data centre to a modern energy efficient data centre properly requires an organizational alignment that facilitates such a migration.

Figure 12.6 illustrates an IT organizational structure that integrates the expertise of personnel who understand both IT systems and physical infrastructure systems. The new organizational wrinkle involves the integration of an IT facilities arm into the rest of the IT organization.

This organizational alignment presents several advantages. For years, IT and facilities departments have operated as separate entities and evolved separate cultures and even used separate languages.

As a result, most data centre design/build or upgrade projects are painful, lengthy, and costly.

This new IT facilities group is a separate group from the traditional 'building' facilities group. The IT facilities group acts as a liaison between IT and the facilities building group, but is under the direct control of IT.

---

[1] Green Grid, White Paper no. 1 *Guidelines for Energy-Efficient Data Centres*.

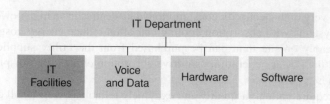

Figure 12.6   IT organizational structure proposed by GreenGrid

The IT facilities group addresses data centre issues specific to hardware planning, electrical deployment, heat removal, and physical data centre monitoring.

This organizational alignment allows a data centre team to rapidly deploy an energy efficient data centre upgrade policy that addresses both IT systems and physical infrastructure systems.

Further Readings

M. Bramucci, D. Di Santo and D. Forni, Linee guida per la progettazione di datacenter ad alta efficienza, FIRE, 2011.

European Commission Joint Research Centre, *2011 Best Practices for EU Code of Conduct on Data Centres*.

The Green Grid. Library and Tools, http://www.thegreengrid.org

PG&E, *High Performance Data Centres: A Design Guidelines Sourcebook*, 2006.

US DoE, *FEMP Best Practices Guide for Energy-Efficient Data Center Design*.

13

Reactive Power Compensation

Zbigniew Hanzelka, Waldemar Szpyra, Andrei Cziker
and Krzysztof Piątek

Reactive power is essential to the operation of AC electrical equipment, but its generation and transmission affects the operation of the power system. The adverse effects of reactive power generation and transmission include the increase in active power and energy losses, increased voltage drops, reduced capacity of network components and, consequently, higher costs of electricity supply. Power losses in a symmetrically loaded network component with the resistance R (e.g. a line or transformer) can be calculated from the formula:

$$\Delta P = 3I^2 R = \frac{S^2}{V^2} R = \frac{P^2}{V^2 \cos^2 \varphi} R = \frac{P^2}{V^2} R + \frac{Q^2}{V^2} R = \Delta P_a + \Delta P_r, \tag{13.1}$$

where I is the current flowing through the network component, R is the resistance of network component, S is the apparent power flowing through the network component, $\cos \varphi$ is the power factor, V is the phase to phase voltage, P and Q are the active and reactive power flowing through the network component respectively, ΔP_a and ΔP_r are the power losses due to active (a) and reactive (r) power flow respectively and φ is the the angle (in degrees) between current and voltage. The ratio of ΔP_r to ΔP_a vs. the power factor $\cos \varphi$ is shown in Figure 13.1.

The voltage drop ΔV across a symmetrically loaded network component with the resistance R and reactance X can be calculated from the formula:

$$\Delta V = \frac{PR + QX}{V} = \frac{PR}{V} \left(1 + \frac{QX}{PR} \right) = \frac{PR}{V} \left(1 + \tan \varphi \frac{X}{R} \right), \tag{13.2}$$

where X is the the reactance of a network component and $\tan \varphi$ is the reactive to active power ratio.

Equation (13.2) shows that the voltage drop depends on both the ratio of the reactive power to active power $\tan \varphi$ and on the ratio of reactance to the resistance X/R of a network component.

Electrical Energy Efficiency: Technologies and Applications, First Edition. Andreas Sumper and Angelo Baggini.
© 2012 John Wiley & Sons, Ltd. Published 2012 by John Wiley & Sons, Ltd.

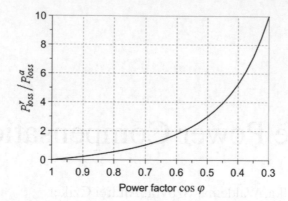

Figure 13.1 The ratio of active power losses caused by the reactive power flow to active power losses caused by the active power flow vs. the power factor

For overhead power lines the ratio X/R is $0.5 \div 3$, and for transformers it can be up to 20. The relative increase in the voltage drop as a function of power factor at different values of the X/R ratio is shown in Figure 13.2. The voltage drop value at power factor $\cos \varphi = 1$ was adopted as a benchmark.

Power network components are sized to the maximum rms current or the maximum apparent power that can be transmitted through them. The presence of reactive current or reactive power reduces the component's capacity for active power. For example, if the element capacity for the apparent power equals 1 p.u. the increase of reactive power from 0.6 to 0.8 p.u. reduces its active power capacity from 0.8 to 0.6 p.u. (see Figure 13.3).

Active power transmission at a low power factor requires higher ratings of electrical equipment (e.g. an increase in wire cross-section or the rated power of transformers), and consequently an increase in investment. The increase in investment as a function of power factor is illustrated in Figure 13.4.

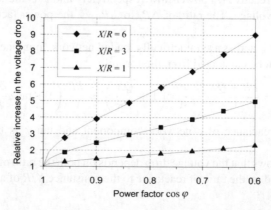

Figure 13.2 Relative increase in the voltage drop as a function of power factor at different values of the X/R ratio

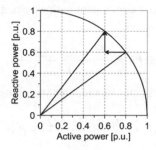

Figure 13.3 Limiting load for the active power due to the increased load reactive power

13.1 Reactive Power Compensation in an Electric Utility Network

Reactive power compensation in a power system is particularly important for both operational and developmental planning since the installation of capacitors near to electricity consumers contributes to a reduction of power and energy losses, and a reduction in voltage drop. It increases network capacity for active power, improves voltage regulation and power quality and, in the case of designing a new network, it also reduces investment cost.

The task of reactive power compensation in a power system is usually solved separately for transmission and distribution networks. In the case of transmission networks the primary reason for installing reactive sources is to improve voltage stability, thus increasing the security of the energy supply. In the case of distribution networks the compensation objectives are different for the electric utility network and for industrial networks. The aim of reactive power compensation in an electric utility network is to reduce distribution costs and improve power quality, while in industrial networks it is aimed at reducing the costs of energy supply.

In this chapter we will discuss the issues related to the effectiveness of the installation of capacitor banks in distribution networks.

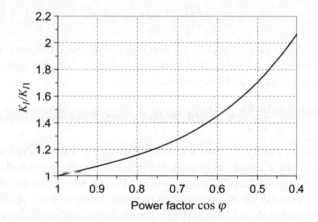

Figure 13.4 Relative investment increase vs. power factor: K_I is capital expenditures for $\cos \varphi < 1$, K_{I1} is capital expenditures for $\cos \varphi = 1$

Power losses ΔP_K in the network component with a capacitive load (a capacitor) connected can be calculated from:

$$\Delta P_K = \frac{P^2 + (Q - Q_K)^2}{V^2} R. \tag{13.3}$$

The difference between the losses prior to and after switching on the capacitor is:

$$\Delta P_R = \Delta P - \Delta P_K = \frac{Q^2 - (Q - Q_K)^2}{V^2} R = \frac{2 Q Q_K - Q_K^2}{V^2} R. \tag{13.4}$$

As seen from (13.4) the effect of reactive power compensation, i.e. the power loss reduction, depends on both: the compensator power Q_K and the reactive power Q flowing through the element before the compensation. With this dependence it is also apparent that the greatest reduction in power loss is obtained for the $Q_K = Q$ (active power losses reach a minimum value equal to the losses derived from the transmitted active power).

Since the load of the network changes over time, the amount by which the energy losses will decrease due to compensation requires knowledge of the load reactive power variation over time. If the voltage is assumed constant in time, the amount by which energy losses will be reduced during a time period T as a result of reactive power compensation with a fixed value of Q_K can be calculated according to:

$$\Delta W_R = \int_{t=0}^{t=T} \Delta P_R(t)\, dt = \frac{R}{V^2} \int_{t=0}^{t=T} \left(2 Q_K Q(t) - Q_K^2 \right) dt, \tag{13.5}$$

where $\Delta P_R(t)$ is the instantaneous value of the active power losses reduction, $Q(t)$ is the instantaneous reactive power flowing through the network component, V is the voltage and T is duration of the time period (usually taken to be $T \approx 8760$ h). It should be noted that $\int_{t=0}^{t=T} Q(t) dt$ is equal to the amount of reactive energy W_r that has flown through the element R during time T. Thus (13.5) for the reduction of active energy losses takes the form:

$$\Delta W_R = \frac{R}{V^2} \left(2 W_r Q_K - Q_K^2 T \right). \tag{13.6}$$

By differentiating (13.6) with respect to Q_K and comparing the differential to zero the compensator's rated power at which the largest reduction of active energy losses in the circuit is achieved during the period T can be determined:

$$\frac{\partial \Delta W_R}{\partial Q_K} = \frac{2R}{V^2} \left(W_r - Q_K T \right) \Rightarrow Q_K^{\max} = \frac{W_r}{T}. \tag{13.7}$$

The amount of reactive energy W_r can be measured using a reactive energy meter.

Power distribution grids consist of high, medium and low voltage networks. A high voltage network operates in a closed configuration while MV and LV networks, although they are often built as closed structures, always operate in open configurations. A simplified diagram of a medium voltage network fed from a HV/MV transformer station is shown in Figure 13.5.

In that network reactive power compensation capacitors can be connected to the MV busbar in the distribution station supplying the network, or in transformer substations to the low voltage windings of MV/LV transformers. The subject of these considerations is the

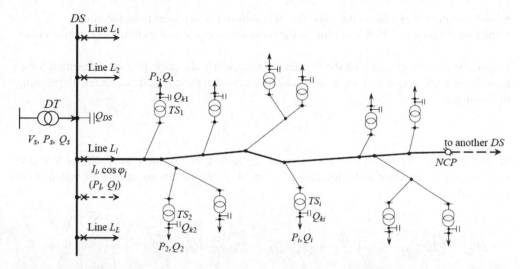

Figure 13.5 Simplified diagram of a medium voltage network: DS – HV/MV distribution station; DT – HV/MV distribution transformer; P_1, Q_1 active and reactive power supplying the line L_1; TS_i – i-th MV/LV substation; P_i, Q_i – active and reactive power load of the i-th substation; Q_{ki} – rated power of the capacitor to be installed in the i-th transformer substation; Q_{DS} – rated power of the capacitor to be installed in the distribution station; P_S and Q_S are the active and reactive power supplying distribution station; NCP – point of networks division (network cut point), other symbols are explained in the text

distribution network as shown in Figure 13.5, in which the compensation capacitors are to be connected to the low voltage windings of MV/LV transformers.

In order to determine the effect of capacitors installation on power and energy losses in the network it is necessary to know the network components' loading. Unfortunately, because of the large number of network components and the lack of measurements, information about the network components' loading is not available. We therefore employ a simplified method for calculating the power and energy losses in networks with such configurations. Several simplified methods for calculating the losses in distribution networks are presented in [1–7]. In one of these methods, each individual circuit L_l of the network shown in Figure 13.5 can be represented by the equivalent circuit with the load distributed evenly along the line, as shown in Figure 13.6.

The use of these simplified methods requires the following assumptions:

- all MV/LV transformers supplied by the circuit are charged in proportion to their rated power;

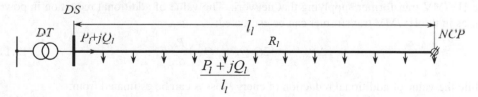

Figure 13.6 Circuit with evenly distributed load: l_l – length (from distribution station to NCP) of main line of circuit l [m]; R_l – resistance of main line of circuit l [Ω]

- load variation over time is the same for all transformers supplied from the circuit;
- the power factor of MV/LV loading is the same and is equal to the circuit load power factor.

Power losses in the circuit with the load distributed evenly along the line are three times lower than those in the circuit with the same load concentrated at the end and can be calculated using the relation:

$$\Delta P = \frac{P_l^2 + Q_l^2}{3V^2} R_l. \tag{13.8}$$

Taking into account the (13.8), the value of the reduction in power losses in an MV network single circuit after installing capacitors in MV/LV substations can be estimated using the equation:

$$\Delta P_{R_l} = \left(2Q_l Q_{K_l} - Q_{K_l}^2\right) \frac{R_l}{3V^2} + \left\{2Q_l(Q_{K_l} - Q_{M_l}) - (Q_{K_l} - Q_{M_l})^2\right\} \frac{\sum\limits_{i \in NTS_l} P_{K_l}}{\left(\sum\limits_{i \in NTS_l} S_{n_i}\right)^2}, \tag{13.9}$$

while the value of energy losses reduction can be estimated from the formula:

$$\Delta W_{R_l} = \left(2W_{r_l} Q_{K_l} - Q_{K_l}^2 T\right) \frac{R_l}{3V^2} + \left\{2W_{r_l}(Q_{K_l} - Q_{M_l}) - (Q_{K_l} - Q_{M_l})^2 T\right\} \frac{\sum\limits_{i \in NTS_l} P_{k_i}}{\left(\sum\limits_{i \in NTS_l} S_{n_i}\right)^2}, \tag{13.10}$$

where NTS_l is a set of substations supplied by circuit l, Q_{K_l} is the sum of powers of capacitors installed in substations supplied by line l, Q_{M_l} is the sum of the magnetizing power of MV/LV transformers, P_{k_i} is the rated load ('full-load') loss of i-th MV/LV transformer and S_{n_i} is the rated power of i-th MV/LV transformer.

Equations (13.9) and (13.10) are derived under the assumption that the power of capacitors installed in substations is proportional to the rated power of transformers in these substations. If the amount of reactive energy received from the substation is known, the output capacitor should be determined from (13.7).

Reactive power compensation in MV network will also reduce power and energy losses in the HV/MV transformer supplying that network. The value of additional reduction in power losses in the HV/MV transformer can be estimated:

$$\Delta P_{R_{DT}} = \left(2Q_s Q_{K_l} - Q_{K_l}\right) \frac{P_{k_{DT}}}{S_{n_{DT}}^2}, \tag{13.11}$$

while the value of additional reduction of energy losses can be estimated from:

$$\Delta W_{R_{DT}} = \left(2W_{r_{DT}} Q_{K_l} - T Q_{K_l}^2\right) \frac{P_{k_{DT}}}{S_{n_{DT}}^2}, \tag{13.12}$$

where Q_s is the reactive power received from the distribution transformer, P_{kDT} is the rated load loss of the distribution transformer and S_{nDT} is the rated power of the distribution transformer.

Equations (13.11) and (13.12) also allow one to calculate the amount of loss reduction when a capacitor is connected only at the switchgear in distribution stations supplying a medium voltage network. In this case, the total power of capacitors installed in substations Q_{K_l} should be replaced with the power of the capacitor installed in this station Q_{DS}.

13.1.1   Economic Efficiency of Reactive Power Compensation

Annual savings resulting from the installation of capacitors can be calculated on the basis of a reduction in power loss at peak load ΔP_{R_s} and in energy loss during a year ΔW_{R_y}. These savings can be calculated from:

$$O_y = \Delta P_{R_s} k_P + \Delta W_{R_y} k_A, \tag{13.13}$$

where k_P is the unit cost of power losses and k_A is the unit cost of energy losses. In order to assess the economic efficiency of the installation of capacitors it is necessary to know the capacitors' prices and the installation and operation costs. To simplify calculations the capacitor price can be approximated by a linear function (see Figure 13.7):

$$C_p = k_v Q + k_f, \tag{13.14}$$

where C_p is the capacitor price, Q is the capacitor rated power, k_f is a constant, independent of the power, capacitor cost component and k_v is the variable, depending on the power, capacitor cost component.

To the price of the capacitor must be added the cost of installation k_i. The cost of purchase and installation of capacitors of the total rated power Q_{K_l} installed in NTS_l substations supplied by a circuit l is:

$$K_{l_l} = k_v Q_{K_l} + NTS_l \left(k_f + k_i \right), \tag{13.15}$$

where k_i is the cost of capacitor installation and NTS_l is the number of capacitors installed in substations supplied by circuit l.

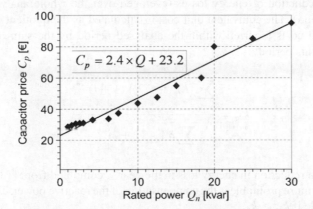

Figure 13.7   Price of MKPg type capacitors as a function of rated power

As a measure of economic efficiency the net present value ratio (*NPVR*) can be used, calculated for n years of the capacitors' operation:

$$NPVR = \frac{\sum_{y=0}^{n} \left(CI_y - CO_y\right) (1+i)^{-y}}{\sum_{y=0}^{n} K_{I_y} (1+i)^{-y}}, \tag{13.16}$$

where K_{I_y} is the capital expenditure incurred in year y, CI_y is the cash inflow in year y, CO_y is the cash outflow in year y, n is the calculation period (the period of construction + expected period of operation) and i is the discount rate (interest rate) expressed as a decimal fraction.

In (13.16) the investment K_{I_y} is substituted by the cost of the capacitors' purchase and installation calculated from (13.15), for the cash inflow CI_y are inserted the savings on reduction of power and energy losses calculated from (13.9), (13.10), (13.11) and (13.12); the expenditures CO_y are substituted by the capacitors' operating costs. It is usually assumed that operating costs are proportional to the investment value, thus the expenditure CI_y will be:

$$CO_y = K_{I_y} k_{ec}, \tag{13.17}$$

where k_{ec} is the fixed operating costs factor. Assuming that the purchase and installation of capacitors occurred at the end of year $y = 0$, and the savings resulting from the reduction of losses are the same in subsequent years, the formula for $NPVR_l$ of the circuit l takes the form:

$$NPVR_l = \frac{O_{y_l} SD - K_{I_l} (1 + k_{ec} SD)}{K_{I_l}} = \frac{O_{y_l} SD}{K_{I_l}} - (1 + k_{ec} SD), \tag{13.18}$$

where SD is the sum of discount factors:

$$SD = \sum_{y=1}^{n} (1+i)^{-y} = \frac{(1+i)^n - 1}{i (1+i)^n}. \tag{13.19}$$

The higher the *NPVR* value, the more profitable is the compensation of reactive power.

Another measure of the effectiveness of reactive power compensation can be the equivalent unit cost of the reduction of energy losses (averaged over the whole analysed period using the discount method). The equivalent unit cost is calculated as the quotient of the sum of the discounted annual costs incurred within the analysed period by the sum of the discounted annual effects in this period:

$$k_{eq} = \frac{\sum_{y=0}^{n} \left(K_{I_y} + CO_y\right) (1+i)^{-y}}{\sum_{y=0}^{n} \delta W_{R_y} (1+i)^{-y}}, \tag{13.20}$$

where ΔW_{R_y} is the reduction in energy loss in the year y, computed from (13.10) and (13.12). The lower k_{eq}, the more profitable is the compensation of the reactive power. The compensation becomes infeasible if $k_{eq} \geq k_A$.

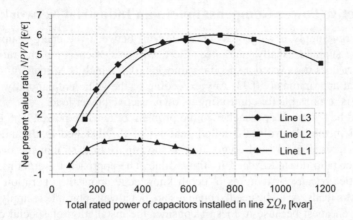

Figure 13.8 The dependence of the net present value ratio dependence on the total rated power of capacitors installed in lines

Figure 13.8 illustrates the dependence of the Net Present Value Ratio (NPVR) on the total rated power of the capacitors installed in three circuits powered by a 25 MVA transformer installed in a 110/15 kV distribution station. Figure 13.9 shows the equivalent unit cost of energy losses reduction on the total rated power of the installed capacitors.

Analysis of the above results shows that installation of reactive power compensation capacitors in MV/LV stations can be an effective method for energy loss reduction in distribution networks. It is, however, not always economically feasible, e.g. in the line L1 its cost-effectiveness is low. The optimum power of capacitors installed in a given circuit is the power at which the highest *NPVR* is achieved.

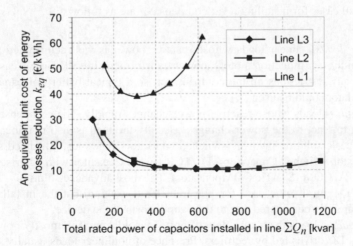

Figure 13.9 The dependence of the equivalent unit cost of energy loss reduction on the total rated power of capacitors installed in lines

13.2   Reactive Power Compensation in an Industrial Network

Electrical energy supply is subject to an agreement between the energy supplier and the consumer. The supplier usually guarantees the tariff costs of energy if the ratio of reactive to active power does not exceed a specific value, e.g. 0.4, which is equivalent to keeping the power factor at approximately 0.93. Any exceeding of the value is financially penalized, so the consumer is responsible for controlling its own reactive power load.

There could be two main approaches to the problem of reactive power. The first approach using 'natural methods' involves detailed analysis of the installed loads and a change in its working pattern, controls and connection to the supply line, so that the overall reactive to active power ratio doesn't exceed its limit value. The approach is sometimes difficult to apply in practice because it requires in-depth knowledge about the operation and working conditions of the loads. Also the needed changes are not always possible to apply. The method is usually the cheapest because it doesn't consider the installation of special compensation equipment. However the analysis could suggest changing a load to one that is better suited to the operation. An example of natural method application is to consider the controlled rectifier, whose nominal DC voltage $V_{DC,N}$ is significantly lower that the maximal one $V_{DC,max}$. The rectifier output DC voltage V_{DC} is approximately equal to: $V_{DC} \approx 1.35\,V \cos\alpha$, where V is phase to phase supply voltage and α is the firing angle. The maximal DC voltage is therefore equal to: $V_{DC,max} \approx 1.35\,V$ and depends only on supply voltage. Neglecting higher harmonic components, the power factor (PF) of the rectifier is equal to: $PF = \cos\alpha$, which yields the power factor $PF = V_{DC}/(1.35\,U)$. In the case of a large difference between $V_{DC,max}$ and $V_{DC,N}$ the control angle is large and consequently the power factor, PF, is low. Decreasing the supply voltage of the rectifier, e.g. by exchanging the supply transformer for one with a lower secondary side nominal voltage, improves the power factor in this case.

The second approach is reactive power compensation – in other words on-site generation of inductive reactive power by means of dedicated compensators. The method is very effective, provided that the natural method was used to properly design the whole system, the compensator could be used to keep the power factor below the limit.

Some typical cases for installing a var compensator are as follows:

- *linear static loads* – linear loads whose reactive power is well determined and does not changes rapidly, e.g. induction motors, power transformers. It is the easiest case, and it does not impose any restrictions on the compensator, so a typical battery of capacitors with a typical regulator could be used.
- *dynamic loads* – loads whose reactive power changes rapidly. The compensator has to be fast enough to react to the power changes. Usually the load is also nonlinear, e.g. an arc furnace. A basic solution involves switched capacitors. The most advanced solutions are: TSC (Thyristor Switched Capacitors) FC/TCR (Fixed Capacitor with Thyristor Controlled Reactor), STATCOM (STATic COMpensator). The most advanced solution – Active Power Filter (APF) – can also be used for power factor correction, but the installation of APF should be carefully considered due to the overall costs of the device.
- *nonlinear loads* – loads that cause flow of non-sinusoidal current, mostly power electronic devices, e.g. DC drives fed by rectifiers. Presence of nonlinear loads usually also indicates a non-sinusoidal supply voltage due to voltage drop. There is the possibility of resonance, so special care should be taken during when designing the compensator. If the power does

not vary rapidly, a battery of capacitors could be equipped with a series reactor that changes the spectral properties of the battery and introduces series resonance. In this case, the compensator becomes a passive filter. The filter could be further extended to compensate for different harmonic components. The most advanced solution involves installation of an APF, which also provides harmonic filtration.

When considering the location of the compensator, three cases should be mentioned:

1. *Individual compensation:* loads are compensated for separately by a capacitor of proper power.
2. *Centralized compensation* of all loads in one point. The compensator is installed in the main supply station close to the point of common coupling (PCC) and settlement point. The compensator is usually of varying reactive power so it can compensate varying load to the desired power factor.
3. *Group compensation* of the group of loads, applied when there is a wide internal network i.e. several substations connected to the main supply station. Compensators are installed in the substations in order to avoid reactive power flowing over the internal network. Note that settlement of energy is done in the main station, so the power factor at the station should be kept below the limit.

A typical case arises when an energy consumer is fed only from one supply line. There is only one point of energy settlement and the point is taken into consideration. A more complex case arises when there are several feeding lines i.e. several settlement points when energy is calculated conforming the tariff. The desired reactive power, i.e. power factor (or tangent equivalently), should be kept in all of the points.

13.2.1 Linear Loads

Reactive power balance rule:

$$Q_{des} = Q_{nat} - Q_{comp}, \tag{13.21}$$

where Q_{des} is the desired amount of reactive power in PCC; usually it depends on active load power and is derived as a percentage of the latter, e.g. in most cases the feeder guarantees the cost of energy if the ratio of active to reactive power is less than 40%, that is if $\tan \varphi < 0.4$, which is equivalent to a power factor of approximately 0.93; Q_{nat} is the natural or nominal reactive power of the load; Q_{comp} is the reactive power of the compensator, the amount of reactive power that is needed to keep the desired power Q_{des}. The compensator acts as a source of (inductive) reactive power. Assuming that compensation does not change active power flow, i.e. losses are neglected and compensation is achieved by means of power capacitors, the balance rule could be transformed into:

$$Q_{comp} = P_{load} \left(\tan \varphi_{nat} - \tan \varphi_{des} \right), \tag{13.22}$$

where P_{load} is the the load active power; $\tan \varphi_{nat}$ is the natural, i.e. non-compensated, power $\tan \varphi$ value; $\tan \varphi_{des}$ is the value that should be achieved by means of the compensator.

The value of the natural reactive power Q_{nat} and $\tan \varphi_{nat}$ consequently could be obtained by means of two approaches:

1. *Analytical calculation*: knowing the parameters of the device and its working pattern, reactive power could be calculated. This is usually done during the design stage by the design engineer. The method could be inaccurate because equipment rarely works with its nominal conditions and its working pattern is also difficult to determine during the design stage because, for example, the actual production size is unknown. For centralized compensation there is also simultaneity of working loads to be taken into consideration. Consequently the analytical calculation gives only an approximate value of the natural reactive power.

2. *Measurement*: performed by an analyser capable of determining active and reactive power in a sufficient period of time, for example a week, with a sufficient averaging window, for example, 10 minutes. The main advantage of this method is that the measured powers are the actual ones consumed by the load. Provided that the measurement period is long enough to cover the whole technological process, detailed information about value and changes of reactive power could be obtained. Measurement can also be useful to check the correctness or to adjust compensators designed on the basis of analytical calculation. The method obviously can not be used at the design stage. The main drawback is that the measurement could only be made when the company is working under normal conditions, so var compensation would have to be designed by the other method.

13.2.1.1   Example 1 – Centralized Compensation

Let us consider a 400 V switchboard that is fed from the transformer with rated power $S_N = 400\,\text{kVA}$ and short-circuit voltage $e_\% = 4.5\%$. The switchboard supplies loads with total active power ranging from $P = 190\,\text{kW}$ to 208 kW and a power factor varying from 0.62 to 0.68. Determine the maximal and minimal power of the compensator in order to compensate reactive power to the power factor value of 0.92.

For the purpose of this calculation the power system short-circuit capacity, resistances of components and compensator active power losses can be neglected. Assume the supply voltage is non-distorted.

As the active power varies so the power factor and the load reactive power are variable. Since there is no information on the correlation between the power factor and the active power, we shall assume extreme cases and calculate both the maximal and minimal compensating power for the given Q/P value. The maximal power is drawn at the maximal active power and the maximal difference between $\tan \varphi$ values, i.e. at the minimal power factor of 0.62. Correspondingly, the minimal power occurs for the minimal active power and maximal power factor 0.68. Therefore, we obtain

$$Q_{k,\min} = 190(1.078 - 0.426) = 123.93\,\text{kvar}, \quad Q_{k,\max} = 208(1.265 - 0.426) = 174.61\,\text{kvar},$$

where $Q_{k,\min}$ and $Q_{k,\max}$ are minimal and maximal compensator powers respectively. Thus the compensator reactive power has a certain constant value of $Q_{const} = 123.93$ kvar, whereas the regulation range is $Q_{reg} = 50.61$ kvar.

13.2.2 Group Compensation

Group compensation is useful when a consumer maintains an internal network that covers, for example, a wide area. In this case there are several substation connected to the main supply station by means of overhead lines or cables. In order to avoid reactive power flow between the main station and substations, compensation of local loads is deployed in the local substations. Therefore the purpose of the compensator is not only keeping reactive power below the limit, but also the minimization of internal network losses. Consequently it is essential to take into consideration the line resistances during the evaluation of local var compensators.

Lets assume that there exists i local substations supplied from the main station. Each substation feeds its own loads with active power P_i and reactive power Q_i. The total reactive power Q_K needed to compensate the loads so that a specified value of $\tan \varphi_{\mathrm{des}}$ is kept in the main station is equal to:

$$Q_K = P_{\mathrm{tot}} \left(\tan \varphi_{\mathrm{tot}} - \tan \varphi_{\mathrm{des}} \right), \tag{13.23}$$

where P_{tot} is the sum of the local load powers:

$$P_{\mathrm{tot}} = \sum_i P_i, \tag{13.24}$$

where P_i is the active power of loads in substation i; $\tan \varphi_{\mathrm{tot}}$ is the total natural reactive to active powers ratio derived from:

$$\tan \varphi_{\mathrm{tot}} = \frac{Q_{\mathrm{tot}}}{P_{\mathrm{tot}}} = \frac{\sum_i Q_i}{\sum_i P_i}, \tag{13.25}$$

where Q_{tot} is the total natural reactive power of the loads equal to the sum of the reactive powers in substations; Q_i is the natural reactive power of local loads in substation i.

The next step is to distribute the total reactive power Q_K into the local substations according to conduction losses. Local var compensator in substation i should have power equal to:

$$Q_{k,i} = Q_i - (Q_{\mathrm{tot}} - Q_K) \frac{R}{R_i}, \tag{13.26}$$

where Q_i is the natural reactive power, Q_{tot} is the total natural reactive power, Q_K is total compensation power, R_i is the resistance of i-th substation connection to the main station, R is the total equivalent resistance, which can be derived from:

$$\frac{1}{R} = \sum_i \frac{1}{R_i}. \tag{13.27}$$

Such a selection of compensator powers could cause power tangents in the local substation to be different from the desired value. However, the reactive to active power ratio is kept at the desired level in the main supply station. The difference depends on line resistances, i.e. conduction power losses.

13.2.2.1 Example 2 – Group Compensation

Let us consider a main supply station that supplies four local substation (LSS). The parameters are shown in the Table 13.1 where P_n is the nominal active power, Q_n is the nominal (natural) reactive power. The system is schematically shown in Figure 13.10.

Design var compensation so that the power factor does not exceed $PF_{des} = 0.93$ (equivalent to $\tan \varphi_{des} = 0.4$). Assume that the internal supply lines are made of copper with unit conductance $\gamma = 55$ [$m\Omega/mm^2$]. Inductance and capacitance of the lines could be omitted. The problem could be solved by means of two approaches:

1. group compensation with compensation of each substation to $PF_{des} = 0.93$ ($\tan \varphi_{des} = 0.4$), i.e. equally distributed compensation;
2. group compensation with a desired power factor in MSS equal to $PF_{des} = 0.93$ ($\tan \varphi_{des} = 0.4$), i.e. loss minimization compensation.

Table 13.1 Group compensation (Example 2)

	Parameters			
	LSS 1	LSS 2	LSS 3	LSS 4
P_n [kW]	30	35	45	70
Q_n [kvar]	30	46.5	42	75
Line length, l [m]	150	30	120	60
Cross-section, s [mm^2]	16	25	35	35
Equally distributed compensation – tangent and reactive power				
$\tan \varphi_i$	1	1.328	0.933	1.071
PF_i	0.707	0.601	0.731	0.628
$Q_{k,i}$ [kvar]	18	32.5	24	47
Power losses in equally distributed compensation case				
R_i [$m\Omega$]	170.45	21.82	62.34	31.17
$\Delta P_{p,i}$ [W]	958.81	167.05	788.96	954.55
$\Delta P_{q,i}$ [W]	153.41	26.73	126.23	152.73
Loss minimization compensation – tangent and reactive power				
$Q_{k,i}$	25.77	13.44	30.43	51.86
$\tan \varphi_i$	0.414	0.944	0.257	0.331
PF_i	0.924	0.727	0.968	0.949
Power losses in loss minimization compensation case				
$\Delta P_{p,i}$ [W]	958.81	167.05	788.96	954.55
$\Delta P_{q,i}$ [W]	19.07	149.02	52.16	104.31
Power losses in centralized compensation case				
$\Delta P_{p,i}$ [W]	958.81	167.05	788.96	954.55
$\Delta P_{q,i}$ [W]	958.81	294.85	687.27	1095.78

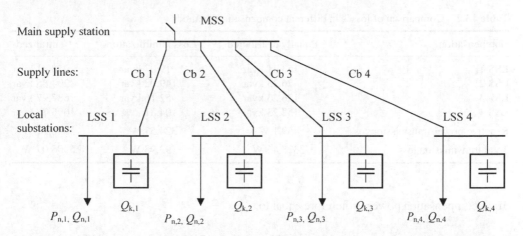

Figure 13.10   Supply system diagram for Example 2 – group compensation in local substations

The first approach, group compensation with equally distributed var compensation does not take into consideration losses, however the compensation is deployed in local substations. The objective is to compensate local reactive power so that in each substation the power factor is equal to the desired value $PF_{des} = 0.93$ (the tangent $\tan \varphi_{des} = 0.4$). Consequently, the power factor in the main supply station would also be equal to $PF_{des} = 0.93$. The computations are similar to the those from the central case, but compensation is derived separately for each substation. The natural reactive to active power ratios and powers of compensators are equal to:

$$\tan \varphi_i = Q_i/P_i \quad Q_{k,i} = P_i \,(\tan \varphi_i - \tan \varphi_{des}).$$

Now the total losses could be considered. Let us consider two kind of losses separately: the power loss due to the active power transition $\Delta P_{p,i}$, and the power loss due to the reactive power transition $\Delta P_{q,i}$ in substation i which are equal to:

$$\Delta P_{p,i} = \frac{P_i^2}{V^2}R_i \quad \Delta P_{q,i} = \frac{(Q_i - Q_{ki})}{V^2}R_i,$$

where V is the supply voltage. The cable resistance R_i is equal to $R_i = l_i/(s\gamma)$, where l_i is the length of the cable, s is the cross-section and γ is the unit conductance of the cable. The results are shown in the Table 13.2. Total losses are:

$$\Delta P_p = \sum_i \Delta P_{p,i} = 2869.36\,[\text{W}] \quad \Delta P_q = \sum_i \Delta P_{q,i} = 459.10\,[\text{W}]$$

$$\Delta P = \Delta P_q + \Delta P_p = 3328.46\,[\text{W}].$$

The losses due to the reactive power transition are 13.79% of the total loss ΔP.

The second case, loss minimization compensation allows for further loss reduction. The total active and reactive power and power tangent is equal to:

$$P_{tot} = 30 + 35 + 45 + 70 = 180\,\text{kW} \quad Q_{tot} = 30 + 46.5 + 42 + 75 = 193.5\,\text{kvar}$$

$$\tan \varphi_{tot} = \frac{193.5}{180} = 1.075.$$

Table 13.2 Comparison of losses in different compensation methods

Compensation	Equally distributed	Loss minimization	Centralized
LSS 1	153.41 kvar	19.07 kvar	958.81 kvar
LSS 2	26.73 kvar	149.02 kvar	294.85 kvar
LSS 3	126.23 kvar	52.16 kvar	687.27 kvar
LSS 4	152.73 kvar	104.31 kvar	1095.78 kvar
Reactive power transmission loss	459.1 W	324.57 W	3036.71 W
Total transmission loss	3328.46 W	3193.93 W	5906.07 W

Total compensation power is therefore equal to:

$$Q_K = 180 \cdot (1.075 - 0.4) = 121.5 \text{ kvar}.$$

The total equivalent resistance could be derived from:

$$\frac{1}{R} = \frac{1}{170} + \frac{1}{114} + \frac{1}{62} + \frac{1}{15}, \quad R = 10.26 \,[\text{m}\Omega].$$

Consequently the local compensator powers and local power factors are presented in Table 13.1.

Assuming that the cable cross-sections are the same as in the group compensation case, losses could be computed, Table 13.1. Total losses are:

$$\Delta P_p = \sum_i \Delta P_{p,i} = 2869.36 \,[\text{W}] \quad \Delta P_q = \sum_i \Delta P_{q,i} = 324.57 \,[\text{W}]$$

$$\Delta P = \Delta P_q + \Delta P_p = 3193.93 \,[\text{W}].$$

Losses due to reactive power transition are 10.16% of the total loss ΔP.

For a better comparison let us consider centralized compensation. This approach takes into consideration only the main station and does not involve any loss calculation. There is only one compensator and its power is Q_K. However, the natural reactive power $Q_{n,i}$ should be transmitted to local substations, which causes additional power loss. The results are shown in Table 13.1. Total losses are:

$$\Delta P_p = \sum_i \Delta P_{p,i} = 2869.35 \,[\text{W}] \quad \Delta P_q = \sum_i \Delta P_{q,i} = 3036.71 \,[\text{W}]$$

$$\Delta P = \Delta P_q + \Delta P_p = 5906.07 \,[\text{W}].$$

The loss due to reactive power transition is 51.42% of the total loss ΔP. Note that transmission of reactive and active power in this case results in current flow that could overload supply lines. The main results are summarized in Table 13.2. It could be noted that the total reactive power needed for compensation is the same in every case. The difference lies in the placement of compensators.

13.2.3 Nonlinear Loads

Nonlinear loads cause the flow of non-sinusoidal currents, that is, currents that can be decomposed into a fundamental harmonic component and higher harmonic components. The case should be considered separately because a capacitor changes the spectral properties of the system. Consequently some of the higher harmonics could be amplified, causing compensator overcurrent and voltage distortion in the PCC.

Let us consider a simple system with a nonlinear load, a compensator and a transformer. The system could be represented as in Figure 13.11, where: $I_{(n)}$ is the n-th harmonic source current; X_{Tr} is the transformer reactance calculated for the fundamental harmonic; X_c is the compensator reactance for the fundamental harmonic; n is the relative frequency (with respect to the fundamental, so $f = nf_{(1)}$; integer n is equivalent to harmonic order). In this case the equivalent impedance (actually the reactance, assuming the resistance is neglected) is:

$$Z_z(n) = \frac{-nX_{Tr}X_c}{n^2X_{Tr} - X_c}. \tag{13.28}$$

The resonance conditions are related to the zeros of the numerator or denominator. For positive values of circuit components only the denominator can become zero. This occurs for: $n^2X_{Tr} - X_c$ and is the condition of parallel resonance; the resonance frequency n_r is: $n_r = \sqrt{X_C/X_{Tr}}$. Any harmonic current component will be causing voltage drop across the system impedance, i.e. transformer reactance is equal to $V_{(n)} = I_{(n)}Z_z(n)$, for $n = n_r$ the voltage drop is infinite. Noting that $THD = (\sqrt{\sum V_{(n)}^2/V_N}) \cdot 100\%$, where V_N is nominal rms value of the phase voltage. The harmonic component of frequency near n_r will significantly distort the supply voltage in the PCC.

In order to deal with the problem, a series reactor could be added. The reactor creates conditions for series resonance. The equivalent circuit for this case is shown in Figure 13.12, where X_d is the reactance of the series reactor. The equivalent impedance of this circuit is:

$$Z_z(n) = \frac{nX_{tr}(n^2X_d - X_c)}{n^2(X_{tr} + X_d) - X_c}. \tag{13.29}$$

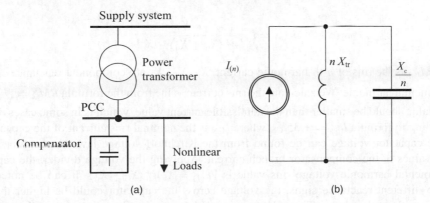

(a)   (b)

Figure 13.11 (a) A simple system composed of a nonlinear load, a power transformer and a compensator. (b) Equivalent circuit for the system

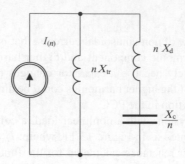

Figure 13.12   Equivalent circuit for the compensator with a choke

Two options are possible when the numerator or denominator equals zero. The nominator becomes zero if the relationship: $n^2 X_d - X_c = 0$ is satisfied, which is the condition of series resonance. If the resonance frequency n_s is known (e.g. chosen during the project process) the reactor reactance can be calculated from $X_d = X_C/n_s^2$. For the denominator, we obtain the relationship $n^2 (X_{tr} + X_d) - X_c = 0$, which is the condition of parallel resonance. The resonance frequency is:

$$n_r = \sqrt{\frac{X_c}{X_{tr} + X_d}}. \tag{13.30}$$

It should be noted that the parallel resonance frequency has been shifted towards the lower frequencies compared with that of the circuit without the series reactor. Therefore, when choosing the reactor the series resonance frequency should be selected to match the lowest harmonic present. In the other case, the reactor will shift the parallel resonance frequency towards the lower harmonic in the current spectrum, which could result in current resonance. This may increase the voltage distortion compared with that of the circuit without the reactor.

The correct selection of the compensator components, i.e. the capacitors and reactor, requires determining their operating conditions. The current in the compensator branch I_f can be found by employing Kirchhoff current law according to Figure 13.12.

$$I_{f(n)} = I_{(n)} \frac{n^2 X_{tr}}{n^2 (X_{tr} + X_d) - X_c}, \tag{13.31}$$

where $I_{(n)}$ is the rms of n-th harmonic current, X_d, X_{Tr}, X_c are component reactances for the fundamental harmonic. We calculate the rms current value from the formula $RMS = \sqrt{\sum I_{(n)}^2}$. This value should be smaller than the admissible current value, which is in some cases derived according to formula $I_{rms} = 1.35 I_{CN}$ where I_{CN} is the nominal rms current of the capacitor.

The capacitor voltage can be found from the Kirchhoff voltage law, or using the calculated values of the compensator branch current. Applying the voltage divider, the capacitor fundamental harmonic voltage rms value is $V_{C(1)} = X_C V / (X_C - X_d)$. It can be noted that, due to different reactance signs, the voltage across the capacitor could be higher than the voltage without the choke. The voltage is even higher due to distorted current flow through the capacitor. The rms value derived from $RMS = \sqrt{\sum V_{(n)}^2}$ should be less than the maximal permissible continuous voltage, determined as: $V_{C,rms} < 1.1\, V_{CN}$ where V_{CN} is the nominal capacitor voltage. The limit values are chosen according to [5].

The peak capacitor voltage condition should also be checked. The composition of the sinusoidal waves could yield a high peak magnitude, which could exceed the capacitor breakdown capacity. The worst case is determined from the formula $MAX = \sum V_{(n)}$ and the value should be less than the maximal permissible peak voltage, which is $V_{C,max} < 1.2\, V_{CN}$.

13.2.3.1 Example 3 – Var Compensation in the Presence of Nonlinear Loads

Let us go back to Example 1, where a group of loads of varying active and reactive powers was fed from the transformer with rated power $S_n = 400$ kVA and short-circuit voltage $e_\% = 4.5\%$. Assume that load current is distorted; the maximal values of the harmonic currents are given in Table 13.3.

Let us check whether the voltage distortion is maintained within the allowed limits and calculate the series reactor impedance, if needed.

Since the load current is distorted, there is a possibility that resonance may occur. Moreover, neglecting the resistance implies that attenuation introduced by certain elements will not be taken into account. It should be noted that harmonic currents values are small compared with the 400 kVA transformer rated current. In this case, the equivalent circuit in Figure 13.12 could be applied. The reactance values can be determined from simple formulae:

$$X_{c,min} = \frac{V^2}{Q_{k,max}} = \frac{400}{147.61 \cdot 10^3} = 0.916\,\Omega \quad X_{c,max} = \frac{V^2}{Q_{k,max}} = \frac{400}{123.93 \cdot 10^3} = 1.291\,\Omega$$

$$X_{Tr} = \frac{e_\%}{100}\frac{V^2}{S_{Tr}} = 0.045\frac{400}{400 \cdot 10^3} = 18\,m\Omega.$$

Table 13.3 Var compensation in the presence of nonlinear loads (Example 3)

	Maximal values of harmonic currents			
Harmonic order n	5	7	11	13
$I_{(n)}$ [A]	10	7.14	5.13	4.1
	Voltage harmonics computation			
$Z_{(n)}$ [Ω]	0.177	3.365	0.144	0.1
$U_{(n)}$ [V]	1.768	24.03	0.738	0.414
	Maximal voltage distortion THD, which will occur for the minimal compensator power			
$I_{(n)}$	10	7.14	5.13	4.1
$Z_{(n)}$	0.448	0.046	0.117	0.144
$U_{(n)}$	4.48	0.327	0.6	0.592
	Elements current and voltage rise – worst-case calculation results			
$I_{(n)}$	10	7.14	5.13	4.1
$I_{f(n)}$	59.81	4.54	2.1	1.57
$X_{C(n)}$	0.183	0.131	0.083	0.07
$U_{C(n)}$	10.94	0.59	0.17	0.11

As the compensator reactive power varies, so does its reactance and, consequently, the resonance frequency varies. Substituting the above values, we determine the maximal and minimal resonance frequency $n_{r,max} = 8.47$ for $Q_{r,min}$, $X_{r,max}$ and $n_{r,min} = 7.13$ for $Q_{r,max}$, $X_{r,min}$. Since the 7th harmonic is present in the load current, a resonance can occur when the capacitor bank is loaded with maximal power. This will result in an increase in the 7th harmonic current and, consequently, significant voltage distortion at the point of common coupling (at the switchboard bus-bars).

In order to check whether the capacitor bank operation is possible under such conditions, the maximal total harmonic voltage distortion, THD, should be calculated. The resonance will obviously occur at frequency n_r equal to 7.13, i.e. for the capacitor maximal reactive power. Knowing the maximal harmonic current values, the voltage drops and hence the voltage THD can be calculated.

Table 13.3 shows the values calculated for each harmonic. The calculated value of total harmonic distortion, THD, will be:

$$THD = \frac{\sqrt{1.768^2 + 24.03^2 + 0.708^2 + 0.414^2}}{230} \cdot 100\% = 10.48\%.$$

This value exceeds the limit of 8% set out in standard EN 50160 [1] with regard to low voltage. Such a high voltage distortion occurs only at the maximal capacitor bank power, therefore either the power needs to be reduced or a series reactor should be connected.

The reactor should create a series resonance for the lowest harmonic component in the current, i.e. for the 5th harmonic. A typical mistake is to choose the 7th harmonic because the parallel resonance occurs for the 7th harmonics. However, this will shift the parallel resonance frequency towards the 5th harmonic, causing voltage distortion to be far higher than the permissible 8%. Therefore the reactor will be chosen for the frequency n_s to be equal to the 5th harmonic. Then we obtain:

$$X_d = \frac{X_c}{n_s^2} = \frac{0.919}{5^2} 36.76 \, m\Omega.$$

The parallel resonance frequency will vary within the range $n_{r,max} = 4.86$ $n_{r,min} = 4.09$. It is thus sufficiently far from the current harmonics (namely the 5th harmonic).

Now we calculate the maximal voltage distortion, THD, which will occur for the minimal compensator power, i.e. at the resonant frequency 4.86. The calculations are in Table 13.3.

Maximal total voltage distortion THD is:

$$THD = \frac{\sqrt{4.48^2 + 0.327^2 + 0.6^2 + 0.592^2}}{230} \cdot 100\% = 1.99\%,$$

which is far below the admissible value. The next step is to check for the proper working condition of the elements. We calculate the fundamental harmonic current from the formula:

$$I_{f(1)} = \frac{V}{\sqrt{3} \, (X_c - X_d)} = 262.53 \, A$$

The worst-case calculation results are presented in Table 13.3.

The rms current value we calculate from $RMS = \sqrt{\sum I_{(n)}^2}$. This yields

$$RMS = \sqrt{262.53^2 + 59.81^2 + 4.54^2 + 2.1^2 + 1.57^2} = 269.31 \, A.$$

This value should be smaller than the admissible current value I_{rms}. The capacitor voltage increase can be found from the Kirchhof voltage law, or using the calculated values of the compensator branch current. Thus we obtain results as in Table 13.3. From the voltage divider the capacitor fundamental harmonic voltage rms value is: $V_{C(1)} = X_C V / (X_C - X_d) =$ 416.67V, which yields the rms value: $RMS = \sqrt{240.56^2 + 10.94^2 + 0.59^2 + 0.17^2 + 0.11^2} =$ 240.81 V. This value should be less than the maximal permissible continuous voltage. The peak capacitor voltage is determined from the formula $MAX = V_{(n)}$. This yields $MAX =$ 240.56 + 10.94 + 0.590.17 + 0.11 = 252.37 V. This value should be less than the maximal permissible peak voltage of the capacitor.

13.3 Var Compensation

The use of reactive power follow-up compensators is one of the most effective ways of improving the power factor and/or the voltage stabilization factor at the point of their connection.

Figure 13.13 shows the classification of various technical solutions. They are mostly three-phase systems of considerable rated power, designed for the compensation/voltage stabilization of a selected load or group of loads characterized by a fast variation of the fundamental harmonic reactive power.

13.3.1 A Synchronous Condenser

A synchronous machine is a traditional source of the fundamental harmonic reactive power, and is inductive or capacitive in nature. It can also be the source of a mechanical moment, when

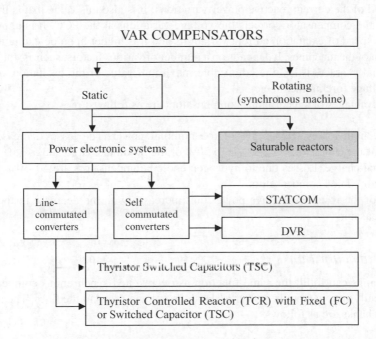

Figure 13.13   The classification of reactive power follow-up compensators/dynamic voltage stabilizers

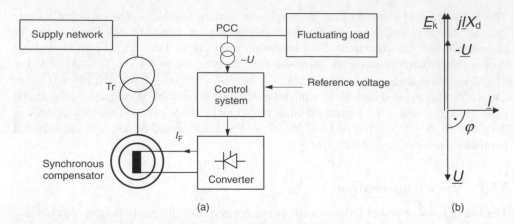

Figure 13.14 (a) A synchronous machine operating in the circuit of a dynamic voltage stabilizer and (b) a phasor diagram

it operates as both a condenser and a synchronous motor at the same time. The high dynamics of the compensation process as well as the continuous – not discrete – reactive power variation result from the fact that a controlled converter unit has been installed in the machine excitation circuit and that it works in a closed control circuit (Figure 13.14). The regulation of reactive power also makes it possible to stabilize the voltage at the point of machine connection.

What is assumed as the criterion of the operation quality in the case of a synchronous machine – in contrast to static compensators – is not the response time (in dynamic states) but the speed of the current/reactive power variations. It is also this value that is the measure of how fast the compensator corrects the voltage variations at the PCC. It is approximately equal to the speed of excitation current variations and it is higher upon de-excitation than on re-excitation of the machines. It depends on a number of various factors, such as: the machine's constructional characteristics, the admissible maximum value of the excitation voltage, the assumed voltage forcing factor, etc.

The disadvantages of synchronous compensators are as follows:

- Relatively high losses, 1.25–5.5% of the nominal power. The losses increase with the decrease of the charge, being 2.9–8% to 50% of the nominal charge and 5–14% to 25% of the nominal charge. Losses due to hydrogen cooled compensators are 10–30% lower than those for air cooled compensators.
- The cost of the installed reactive power unit increases with the decrease in the power of the compensator.

13.3.2   Capacitor Banks

Capacitor banks can fulfill the same function as overexcited synchronous compensators, so they can locally supply reactive power and reduce voltage drops in the lines. The advantages of capacitor banks are as follows:

- The cost for the installed reactive power unit varies very little with the total installed power of the installation.

- They can be mounted sequentially, so they do not need initial high investment.
- The power losses in the capacitors are 0.25–0.30% of the nominal power.
- The operating expenses are reduced more than in the case of synchronous machines.
- The price for the reactive energy produced in the capacitor banks is 3–4 times smaller than that for those produced in the other static compensation installation.

Capacitor banks have the inconvenience that they do not allow continuous control of the voltage, they do not absorb the capacitive power of the lines, their power decreases with the square of the voltage at their terminals.

13.3.3 Power Electronic Compensators/Stabilizers

Generally speaking, static compensators are circuits including reactors and/or thyristor controlled capacitors, connected to a power node. They can be treated as regulated parallel susceptance. They are also used to compensate reactive power, symmetrize currents/voltages, or limit voltage fluctuations. The most common solutions in this group include: thyristor switched capacitors (TSC), a fixed capacitor bank circuit (FC) and a circuit with an inductive current fundamental harmonic control switch (TCR), FC/TCR and TSC/TCR circuits as well as the STATCOM circuit.

13.3.3.1 Thyristor Switched Capacitors (TSC)

In such a circuit delta-connected capacitor banks are divided into sections; each section is switched on (or switched off) separately by means of AC thyristor switches (Figure 13.15(a)). The synchronization of the switch switch-on moment together with the initial battery charging makes it possible to avoid overcurrents and overvoltages, which appear in a non-synchronous capacitor connection (Figure 13.15(b)). Response time at symmetrical operation does not exceed 20 ms.

In addition, other configurations that could be called 'economical' are used; in these cases, one of the semi-conduction switch thyristors is replaced by a diode.

13.3.3.2 Fixed Capacitors/Thyristor Controlled Reactors (FC/TCR)

This solution is an example of an indirect compensation method, where, depending on the needs arising from the functions of a voltage stabilizer or a reactive power compensator, the value of the sum of two current components is regulated (Figure 13.16):

- a current fundamental harmonic i_{FC} of a capacitor, functioning almost always as a harmonic filter/filters;
- a current fundamental harmonic of a reactor i_{TCR}, regulated by phase control of an AC thyristor switch (T).

Figure 13.16(a) shows a one-phase schematic diagram of such an installation, whereas Figure 13.16(b) shows waveforms illustrating the working principle and reactor current waveforms for various values of the control angle α (in relation to voltage zero-crossing points).

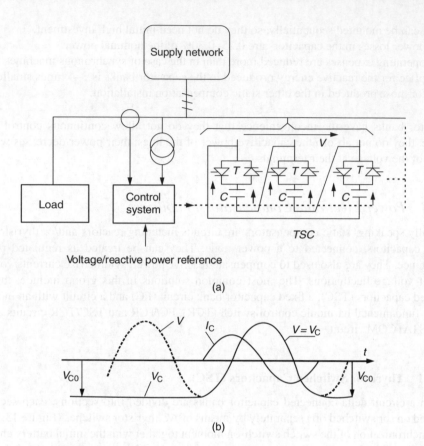

Figure 13.15 (a) Schematic diagram of a TSC compensator; (b) current and voltage waveforms during capacitor switching (V_{C0} – voltage initial condition of the capacitor)

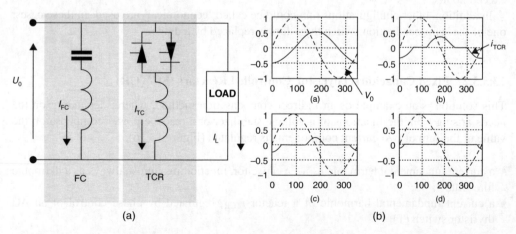

Figure 13.16 (a) Conceptual diagram of a one-phase FC/TCR compensator; (b) TCR current waveforms

In such a compensator, the reactor current (fundamental harmonic) is regulated by means of a phase-controlled thyristor switch, which consists of two antiparallel-connected thyristors. A capacitor bank, constituting the source of a capacitive reactive current, is parallel-connected to the reactor branch.

The control angle, together with the value of the reactor current fundamental harmonic and compensator current fundamental harmonic (as well as the value and sign of the voltage drop on the system impedance), can change in each supply voltage half-period, assuming any value from the range $(0.5\pi, \pi)$. What follows is the variation of reactor reactive power, from the maximum value to zero, respectively; also, the value (and sometimes also the nature – inductive or capacitive) of compensator resultant current changes.

The increase of the angle α results in the decrease of the fundamental harmonic in the reactor current, which means that its equivalent inductive reactance for this harmonic increases, while its input fundamental harmonic reactive power decreases.

If the current in the reactor branch equals zero ($\alpha = \pi$), the compensator returns the reactive power to the supply network, while its current is capacitive in nature. When the thyristors are fully controlled ($\alpha = \pi/2$), and the reactor power exceeds the capacitor power, the compensator takes reactive power, while its current is inductive in nature. The regulation of the compensator current within the range from I_{Cmax} to I_{Lmax} is continuous in nature.

Phase control of a thyristor switch results in the generation of higher harmonics of odd orders, and when retardation angles for both thyristors are not equal, even harmonics are generated as well [3]. 'Triple' harmonics (the orders of which are multiples of three) are eliminated from the compensator phase currents by connecting the reactor branches in a delta (Figure 13.17). These three-phase reactor configurations, together with parallel harmonic filters function for the supply network as equivalent susceptances switched on phase-to-phase (Figure 13.18). Their values change as a result of changing the control angles (α_{12}, α_{23}, α_{31}, where 1, 2, 3 $-$ 1, 2, 3 – ordinal numbers of phases), steplessly and independently of each

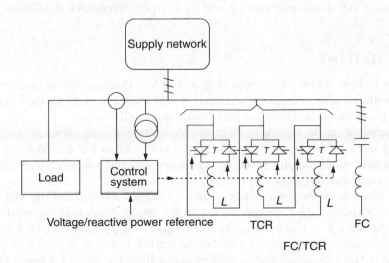

Figure 13.17   A schematic diagram of a three-phase FC/TCR compensator

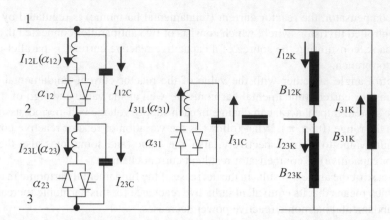

Figure 13.18 An FC/TCR compensator as a symmetrization device

other. Also, equivalent reactances of a transformer of a large short-circuit voltage (known as a *thyristor controlled transformer* circuit – TCT) can function as a reactor.

The static compensator circuit in question is particularly successful for the compensation of loads, the working conditions of which are asymmetrical and change quickly with time, for example for the compensation of an arc furnace.

In order to limit odd harmonics, two six-pulse circuits supplied by a three-winding Yyd transformer can be used (twelve-pulse circuit) or higher harmonics filters can be used as well.

In *Thyristor Switched Capacitors/Thyristor Controlled Reactors* (TSC/TCR) compensators, apart from TCR, thyristor switched capacitor steps are also used. The circuit has a significantly expanded regulation range in the area of capacitive currents. This solution, used almost exclusively in voltage stabilization circuits, combines the advantages of TSC and FC/TCR circuits. It makes it possible to minimize the value of harmonics of currents generated during TCR operation and to minimize active power losses in a compensation installation.

13.3.3.3 STATCOM

The circuit includes a converter – most frequently a VSI (Voltage Source Inverter). Its connection conditions of semiconducting elements (PWM modulation) decide the value and also the nature (inductive or capacitive) of the compensator reactive power.

A compensator can be regarded as a controlled voltage source, connected to the supply network by means of a reactor (Figure 13.19(a)). In order to control the reactive power flow the regulation of voltage amplitude is used. Such a compensator has possibilities analogous to these of a synchronous machine (but it works much faster).

For the phase angle between converter output voltages and the network voltage at the point of connection of the compensator $\delta = 0$, the circuit draws only the reactive power: capacitive when $V_{VSC} > V_{bus}$ or inductive when $V_{VSC} < V_{bus}$ (Figure 13.19(b)). In this way, the compensator is either a source or a load of the reactive power.

The STATCOM-type compensators applied in distribution systems for power quality improvement are called DSTATCOMs (Distribution STATCOMs). They can perform complex

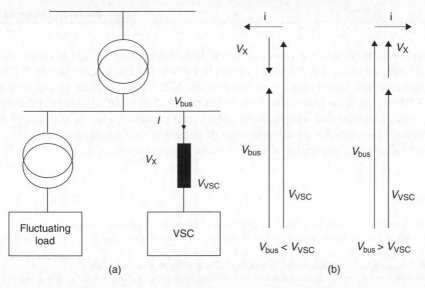

(a)   (b)

Figure 13.19   (a) Schematic diagram of a STATCOM compensator, connected to the supply network; (b) voltage and current phasor diagrams for different relations between V_{bus} and V_{VSC}, where X_k is the reactor or converter transformer reactance, V_{VSC} is the phase voltage at converter terminals, V_{bus} is the voltage drop on converter input reactance ($= X_k I$), I is the compensator current fundamental harmonic

Table 13.4   Comparison of selected compensators

	Synchronous machine	SVC		STATCOM
		FC/TCR	TSC (with TCR if it is needed)	
Compensation accuracy	Good	Very good	Good, very good with a TCR	Exceptionally good
Reactive power variability	Capacitive/ inductive	Capacitive/indirect inductive	Capacitive/Indirect inductive	Capacitive/Inductive
Control	Continuous	Continuous	Discrete/Continuous with TCR	Continuous
Response time	Relatively slow	1–2 periods	1–2 periods	Very short, depending on the control circuit and switching frequency
Harmonics	No	Large	No/Present in the case of a TCR	Small, depending on switching frequency
Losses	Medium	Low; increase together with the increase of the inductive power	Low; increase together with the increase of capacitive power	Very low; increase together with the increase of switching frequency
Symmetrization ability	Limited	Good	Limited/Good	Very good in the case of three one-phase units; limited in a three-phase configuration
Cost	High	Medium	Medium	From low to medium

tasks simultaneously: reactive power compensation, load balancing and harmonic filtering. The basic element of the DSTATCOM system is a PWM voltage inverter in which fully controlled semiconductor devices are used, for example IGBT, IGCT. A new quality to designing STATCOM circuits is the use of multi-level converters. Their main advantage is the reduction of harmonics emission and the possibility of a direct connection to the network of an ever higher voltage because of a series connection of bridges or semi-conductors.

Table 13.4 gives a comparison of a few selected compensators.

References

[1] EN 50160: Voltage characteristics of electricity supplied by public distribution systems.
[2] A. Ghosh and G. Ledwich, Power *Quality Enhancement Using Custom Power Devices*, Kluwer Academic Publishers, 2002.
[3] Angelo Baggini (ede.), *Handbook of Power Quality*, John Wiley & Sons Ltd, 2008.
[4] Z. Hanzelka and A. Bień, *Flicker. Leonardo Power Quality Application Guide*, http://www.lpqi.org
[5] IEEE Standard for Shunt Power Capacitors, Std. 18-2002.
[6] J. Kulczycki (ed.), *Electric Energy Losses in Distribution Networks* (in Polish), PTPiREE Publishers, Poznań, 2009.
[7] W. Szpyra, Simplified model of medium voltage distribution networks for calculation power loss and voltage drop (in Polish), *Proceedings of the Scientific and Technical Conference 'Optimization in Power Engineering'*, Jachranka, October 7–8 1999, pp. 125–134.

Further Readings

W. Szpyra, W. Tylek and A. Kot, Determination of power and energy loss in an extensive low voltage power network, *Proceedings of the Institute of Electrical Power Engineering*, Wrocław University of Technology, Serial: Conferences No.34, The IV Conference 'Electric Power Networks in Industry and Power Sector. Networks 2000', 2, 2000, pp. 403–410.

Index

active filter 156–9, 160–61
AHU-air handling unit 95–101, 103–106,
    342–3
amorphous 24, 28, 34, 39, 47–8, 50–52, 60,
    212, 214, 227
ampacity 9, 19
ancillary services 170, 179–81, 184–5
applicators 320–21
auxiliary losses 23–4, 43

backup storage energy supply 170
balancing 131, 143–4, 157, 172, 178, 181,
    247, 252, 290, 398
ballasts 232–5, 237–9, 240–42, 245–7, 250,
    252, 255, 257, 261, 365
batteries 6, 99, 105, 171–2, 175, 263, 266,
    277–8, 363–4
blade server 360–61
building automation 71, 81, 111, 113–14,
    124
buildings 2–4, 6, 46, 71, 77, 82, 113–16,
    119, 121–2, 124, 184, 186–7, 189,
    230, 232, 234, 244, 248, 251, 261–2,
    335, 338–44, 348–9, 353–5, 367

capacitive heating 320–22, 324
capacitor 12, 125, 131–2, 134, 138, 150,
    152–6, 171–2, 187–8, 240, 256, 266,
    277–9, 285, 290–91, 293, 307, 316,
    320, 331, 373–81, 387–96, 398
CHP 7, 163–4, 168–70, 178, 180, 368–9

classification standards 199, 201
CO_2 emissions 1, 229–31, 233, 265,
    299–300, 302, 329
cogeneration 2, 168–73, 186–7
cold-rolled 47–8, 51
compact fluorescent lamps 25, 233, 237–8,
    243, 250, 261
compensation 88, 101, 103, 117, 132, 143–4,
    160–61, 170, 181, 185, 187, 251, 267,
    333, 373–6, 378–86, 389, 391–3, 396,
    398
compressed air 171
conduction 10, 221, 266, 275–6, 280, 295,
    302–3, 306, 309, 318, 328, 383, 393
conductor resistance 12, 18, 215
control and management systems 71
consolidation 362
convection 10–11, 221, 295, 302–3, 308–9,
    312, 327, 332, 344
COP-coefficient of performance 336–7
copper rotor 193, 215–16, 228
core material 28, 39, 42, 46–7, 50–52, 64,
    210, 211–12, 214, 217–18
cross-sectional area 15–16, 18–19, 24, 49,
    226
current rating 12, 16, 19

data center 360, 370
daylighting 251–2, 260, 262, 349
DCIE 360
DCS 76–7, 85

Electrical Energy Efficiency: Technologies and Applications, First Edition. Andreas Sumper and Angelo Baggini.
© 2012 John Wiley & Sons, Ltd. Published 2012 by John Wiley & Sons, Ltd.

T0201282

AN INTRODUCTION TO BIOREACTOR HYDRODYNAMICS AND GAS-LIQUID MASS TRANSFER

AN INTRODUCTION TO BIOREACTOR HYDRODYNAMICS AND GAS-LIQUID MASS TRANSFER

Enes Kadic and Theodore J. Heindel
Department of Mechanical Engineering
Iowa State University
Ames, Iowa

Published by John Wiley & Sons, Inc., Hoboken, New Jersey
Published simultaneously in Canada

For general information on our other products and services or for technical support, please contact our
Customer Care Department within the United States at (800) 762-2974, outside the United States at
(317) 572-3993 or fax (317) 572-4002.

Wiley also publishes its books in a variety of electronic formats. Some content that appears in print
may not be available in electronic formats. For more information about Wiley products, visit our web
site at www.wiley.com.

Library of Congress Cataloging-in-Publication Data:

Kadic, Enes, 1973-
  An introduction to bioreactor hydrodynamics and gas-liquid mass transfer / Enes Kadic, Theodore J.
Heindel.
    pages cm
  Includes bibliographical references and index.
    ISBN 978-1-118-10401-9 (hardback)
1. Bioreactors. 2. Bioreactors–Fluid dynamics. I. Heindel, Theodore J., 1964- II. Title.
  TP248.25.B55K33 2014
  620.1′064–dc23

                                                               2013042733

Printed in the United States of America

10 9 8 7 6 5 4 3 2 1

EK would like to thank his parents, MK and EK, for their continued support and love. TJH would like to thank MLH, HMH, and OLH for their love and support.

CONTENTS

1 Introduction

The biological production of renewable fuels, chemicals, medicines, and proteins is not possible without a properly functioning bioreactor. Bioreactors are expected to meet several basic requirements and create conditions favorable to the biological matter such that the desired production is maximized. The basic requirements may include minimal damage to the biological matter, maximum bioreactor volume utilization, maximum gas–liquid mass transfer, and/or maximum mass transfer from the liquid to the biological species (Bliem and Katinger, 1988a). Even though gas–liquid mass transfer is often the limiting reaction process, the biological species may incur additional limitations. For example, biological species can be very sensitive to shear while others may not grow well in laminar flow conditions but thrive in very turbulent conditions (Bliem and Katinger, 1988a; Hoffmann et al., 2008). In other words, the bioreactor has to accommodate very specific environmental conditions, and the operator has to be mindful of those when choosing bioreactor design and operating conditions.

Once the broadness of the problem is absorbed, it becomes clear that one bioreactor design or design ideology is insufficient to meet the operational requirements for all bioreactor operations (Bliem and Katinger, 1988a). Therefore, each bioreactor design tries to produce a very specific set of conditions applicable to a certain cell or bacteria line. In order to help with this decision process, this book provides a survey of relevant gas–liquid and gas–liquid–solid bioreactors; defines the respective bioreactor pros, cons, hydrodynamic considerations, and gas–liquid mass transfer correlations; and identifies research needs and figures of merit that have yet to be addressed. Since a large portion of the bioreactor designs have been ported over from the chemical and petrochemical industries, a significant portion of the basic bioreactor knowledge has originated from those areas. Hence, bioreactors will often be referred to as simple reactors in order to signal that some of the research used for the discussion and conclusion have been adapted from nonbiological research areas.

The remainder of this book is organized as follows. All bioreactors have common modes of operation, which are described in Chapter 2. General gas–liquid mass transfer considerations are then summarized in Chapter 3. Various hydrodynamic and gas–liquid mass transfer measure techniques are then outlined in Chapter 4, followed by a summary of multiphase flow modeling

An Introduction to Bioreactor Hydrodynamics and Gas-Liquid Mass Transfer, First Edition.
Enes Kadic and Theodore J. Heindel.
© 2014 John Wiley & Sons, Inc. Published 2014 by John Wiley & Sons, Inc.

methods in Chapter 5. Chapters 6–8 then cover the three common bioreactor types, including stirred-tank bioreactors, bubble column bioreactors, and airlift bioreactors, respectively. Chapters 9 and 10 then cover less common bioreactor types, including fixed bed bioreactors and novel bioreactor designs. Some general figures of merit are then described in Chapter 11, followed by general conclusions.

2 Modes of Operation

Batch, semibatch, and continuous modes of operation are classified by the flow rates in and out of the system. Virtually all bioreactor types are capable of operating in one of these modes, depending on hardware configuration. This section will review the different modes by presenting some general information, operating procedures, and advantages and disadvantages. Discussion of operational modes for specific bioreactors can be found in the respective chapters.

2.1 BATCH BIOREACTORS

The batch bioreactor is the oldest and most used bioreactor in industry (Bellgardt, 2000b; Branyik et al., 2005). Its historical and most familiar use is in the production of alcoholic beverages (beer, wine, whiskey, etc.) and bread. Batch bioreactors combine all the necessary ingredients and then operate until the desired product concentration is reached at which point the product is extracted. In well-known processes where the final product is relatively cheap, product concentration can be correlated with time, leading to some process automation, lower capital needs, and lower operational costs (Bellgardt, 2000b). Batch bioreactor systems are also useful in modeling environmental issues (Fogler, 2005).

Biological application and experience have led to a differentiation based on substrate input or sterilization frequency. The simplest and least applicable variant is the batch cultivation system (Bellgardt, 2000b). Bioreactor sterilization is undertaken prior to the start of the process, followed by the medium being fed into the bioreactor creating a high substrate concentration (Bellgardt, 2000b; Williams, 2002). Inoculated microorganisms are introduced into the batch bioreactor at a low concentration to allow proper growth, which is practically uncontrollable until the process is finished. Ideally, the product is extracted once a satisfactory concentration is achieved, but the product in the batch cultivation system is also extracted if a necessary ingredient has been exhausted (Bellgardt, 2000b). Finally, the bioreactor is cleaned, and the process starts over again with bioreactor sterilization.

The need for more control over the biological process created the fed-batch (also known as the semibatch) cultivation system, which is the most widely used batch

An Introduction to Bioreactor Hydrodynamics and Gas-Liquid Mass Transfer, First Edition.
Enes Kadic and Theodore J. Heindel.
© 2014 John Wiley & Sons, Inc. Published 2014 by John Wiley & Sons, Inc.

bioreactor. This deviation is a variable volume process that introduces additives at specific time intervals, gradually creating a more responsive and friendly growth environment (Bellgardt, 2000b). In other words, the bacteria receives the right amount and type of nutrients at the appropriate growth stage, creating a more efficient and controllable process. The final result is a product that can be adjusted or extracted when it achieves the desired properties.

The fed-batch and batch cultivation systems share the same cleaning and sterilization process in which the bioreactor operation is stopped and the bioreactor is emptied. This stoppage creates considerable costs and operational downtime. The repeated or cyclic system, which can be applied to both batch and fed-batch cultivation systems, may be installed in order to maximize the productivity. The cyclic cultivation system does not enter the cleaning and sterilization process, but rather empties a portion of the bioreactor while preserving part of the batch for the next cycle. Another method to increase productivity is cell retention techniques such as fluidized beds, membranes, or external separators. These options allow multiple cycles without cleaning and sterilization, which is initiated only if it is deemed that mutation risks exceed tolerable levels (Bellgardt, 2000b).

Variations of the batch bioreactor try to limit problems or expand batch bioreactor applications, but some systematic advantages and disadvantages exist. For the most part, batch bioreactors have lower fixed costs due to the simple concept, design, and process control (Bellgardt, 2000b; Donati and Paludetto, 1999; Williams, 2002); however, variable costs are generally higher for several reasons. First, cleaning and sterilization often add significant downtime and labor costs (Donati and Paludetto, 1999; Williams, 2002). These costs, however, can be limited in the cyclic cultivation system. Second, batch bioreactors have heat recovery difficulties leading to high environmental impact and energy consumption (Donati and Paludetto, 1999; Schumacher, 2000; Williams, 2002). Third, the additive nature of fed-batch and cyclic cultivation systems force the operator to prepare several subcultures for inoculation, which adds further variable cost pressures (Williams, 2002). Finally, batch bioreactors are not steady-state processes. The biological matter grows uncontrollably, leading to a changing environment that can bring about safety issues, runaway growth, or unexpected products when mutations occur (Westerterp and Molga, 2006).

Runaway reactions are unlikely in biological systems, but the variable environment can create conditions that change the competitive situation favoring a different bacterial species than the initially dominant one (Hoffmann et al., 2008). Batch bioreactors have limited, albeit relatively simple, process control that can lead to inconsistent or unwanted products, especially in a batch cultivation system. This problem can get even more pronounced in operations with a high potential contact amid pathogenic microorganisms or toxins, adding to variable costs if more stringent cleaning and sterilization procedures are needed (Williams, 2002).

The fed-batch cultivation system makes process control more challenging by creating a variable volume process. Any control mechanisms, therefore, require much more labor or capital (Bellgardt, 2000b; Donati and Paludetto, 1999; Simon et al., 2006; Williams, 2002). According to Simon et al. (2006), a fed-batch system can have thousands of control variables requiring a modern and powerful supervisory control and data acquisition system, programmable logic controllers, trained personnel, and an 8-year upgrade cycle, all of which eliminate or limit upgradability of older systems or construction of larger batch bioreactor systems (Heijnen and Lukszo, 2006; Simon et al., 2006). The complexity limits practical batch bioreactor application beyond a certain size, while other bioreactor modes enjoy economies of scale for much larger operations (Donati and Paludetto, 1999; Heijnen and Lukszo, 2006; Simon et al., 2006; Williams, 2002).

Some of the batch system costs can be offset by its flexibility. Batch bioreactors are able to produce the desired product consistently. They are also capable of producing several types of products with the same equipment or making the same type of product with different equipment. Significant product modifications can also be implemented online (Donati and Paludetto, 1999; Heijnen and Lukszo, 2006). These traits offer flexibility and competitive advantages to batch bioreactor operations; however, many problems and complications are encountered when these bioreactor schemes are used for multiple separation processes, which is often the case in industry (Barakat and Sorensen, 2008).

Most batch bioreactors operate in a changing external environment especially with respect to product and ecological demands (Heijnen and Lukszo, 2006). Researchers are able to take batch bioreactors and investigate reactions, both chemical and biological, for which data are unavailable or have never been documented, while limiting contamination and experimental or dangerous risks (Donati and Paludetto, 1999). These research bioreactors should be used for scaling purposes with care since most reactions and biological growth are affected by hydrodynamics, which are a function of bioreactor scale and type.

Ultimately, batch bioreactors contain biological matter that tends to mutate. Growth periods, therefore, need to be kept short and controlled to prevent these microbial mutations, which could produce inconsistent or undesirable products (Williams, 2002). Some fermentation processes, however, are characterized by biological matter that mutates very little allowing for long reaction times (Donati and Paludetto, 1999). Either way, a positive side effect of the controlled growth period is a higher conversion level (Williams, 2002).

A specific batch bioreactor application depends on multiple internal and external factors; however, general rules of thumb and process-specific improvements can be employed to make a smarter and more profitable selection. Batch bioreactor selectivity is based on the following factors: economic balance, production scale, reaction times, production flexibility, and the nature of the process and product (Donati and Paludetto, 1999). Typically, batch bioreactors are used for smaller operations, specialty products, long growth periods (bioreactor of choice by elimination), operations in which flexibility is vital, unsteady processes, and experimental development (Donati and Paludetto, 1999; Simon et al., 2006; Williams, 2002).

Batch bioreactor operation can be made more efficient by implementing several simple managerial procedures. First, a disturbance strategy should be developed by which personnel are trained to respond and actively scan for problems in the process leading to "lines of defense" that limit contamination and loss of product (Westerterp and Molga, 2006). These "lines of defense" should include an operating condition within which personnel and management are comfortable, an early warning system, and a reaction procedure to accidents and malfunctions including proper training and equipment (Westerterp and Molga, 2006). Second, a decision support framework (DSF) should be developed so that all personnel and management are familiar with operating costs, benefits, objectives, etc. The DSF will make production more efficient and profitable; it provides a clear outline of benefits and costs associated with general and specific options. General models, such as ANSI/ISA88 or ANSI/ISA95, are available and can be applied to all batch bioreactors (Heijnen and Lukszo, 2006). Finally, two improvement strategies can be implemented to make batch reactions more efficient. The "cook book" or "recipe" approach has been shown to improve yields in batch process operations. The user is able to adjust the biological reaction online as needed and is able to draw on extensive experience and/or knowledge to have better process control and product quality and consistency. The second strategy, production schedule optimization, has been proven effective in situations where products are made with different equipment, or equipment is used to make different products by optimizing capacity utilization (Schumacher, 2000).

2.2   CONTINUOUS BIOREACTORS

Continuous bioreactors have several intrinsic properties that differentiate them from batch bioreactors. The largest distinction is that substrate and product continuously flow in and out of the bioreactor, which does not allow for cleaning or sterilization processes and extracts product regardless of identity or quality (Bellgardt, 2000b). If output does not meet specifications, the resulting product has to be either discarded or separated and recycled back into the bioreactor. Either option creates a negative economic impact by increasing (i) initial investment due to the necessary installation of a recycling system and (ii) variable costs due to the discarded product and the associated inputs (Williams, 2002). Product properties are controlled by substrate residence time which, by design, can only be controlled by material flow rate and bioreactor geometry. In order to ensure a homogeneous product, the process is assumed to be in steady state and conditions within the bioreactor are typically assumed to be independent of time (Williams, 2002). Therefore, continuous bioreactors are agitated mechanically and/or by gas injection. Substrate input is not used for agitation so as to decouple it from bioreactor hydrodynamics. In order to make the steady-state conditions easier to achieve and maintain, most continuous bioreactors are run in a constant volume setting, which induces uniform volumetric substrate and product flow rates. Efficiency is enhanced using cell retention techniques such as fluidized beds, membrane bioreactors, or cell recycle (Bellgardt, 2000b). The semicontinuous bioreactor, a hybrid between the batch and continuous bioreactor, is run in batch mode during start-up. Once necessary conditions are achieved, this bioreactor is operated continuously unless the product has not achieved the necessary properties, in which case the bioreactor is operated in batch mode until the desired specifications are met (Williams, 2002).

Any continuous bioreactor discussion is ultimately related to batch systems which are seen as a proven technology with processes designed around their capabilities and properties. In addition, operators have more experience and are more comfortable dealing with batch disturbances (Branyik et al., 2005). Continuous bioreactors, however, offer many advantages such as control, production, and the potential for optimization (Williams, 2002). Control can be achieved with several schemes. Substrate and cell concentrations or bioreactor conditions can be modified online to influence bacterial growth rates which, in turn, modify bioreactor and reaction dynamics (Bellgardt, 2000b; Gonzalez et al., 1998; Ramaswamy et al., 2005; Reddy and Chidambaram, 1995; Sokol and Migiro, 1996). These changes provide an indirect influence over product type and properties (Ramaswamy et al., 2005).

The continuous bioreactor offers more operational control and flexibility over batch bioreactors. Practical production is made simpler and more profitable due to the possibility for automation and the lack of cleaning and sterilization processes (Bellgardt, 2000b). The added equipment cost is offset with operational savings while control schemes, which are limited to a few hundred variables, are often much simpler than the corresponding batch bioreactor systems (Simon et al., 2006; Williams, 2002).

In addition to the control advantages, the steady-state operation of continuous bioreactors allows for the production of a consistent and economically attainable product quality (Williams, 2002). It also rectifies a major downside in the batch system—the bacterial concentration is very low in the initial stages while growth rates are sluggish at maturity, both leading to decreased productivity. The continuous system allows high bacterial concentrations throughout the process, boosting production capacity and consistency (Bellgardt, 2000b). Combined with the lack of cleaning and sterilization processes, continuous bioreactors maximize production time while providing lower labor and variable costs and maximum capital utilization (Bellgardt, 2000b; Williams, 2002).

Finally, continuous bioreactors allow for optimization. Operators are able to vary inputs and bioreactor conditions online, creating an optimized environment for the growth stage and age of the cell culture, resulting in a more consistent output (Bellgardt, 2000b; Williams, 2002). In some series bioreactor applications, process augmentation is necessary to prevent washout or inconsistent product (in case of an upstream disturbance and varying substrate concentration). These disturbances and variances are often experienced when a batch bioreactor or other limiting cycles are used to prepare a substrate for the continuous process (Ramaswamy et al., 2005).

Disturbance management depends greatly on the ability to investigate the cause of the disturbance. Batch reaction systems have complicated procedures to identify the cause of disturbances with the operator often left guessing (Schumacher, 2000). The continuous process provides a clear plan of attack. The effect of changing one variable while keeping all others constant allows for a clear relationship and better understanding (Williams, 2002). Researchers are able to form global correlations or models aiding development, adoption, or ongoing operations.

Much higher production capacity, efficiency, capital utilization, and lower variable costs for continuous bioreactors may make operations appear efficient, but continuous systems have a number of disadvantages. Biological growth typically requires a batch start-up period to achieve conditions within the bioreactor to promote optimal production (Bisang, 1997; Gonzalez et al., 1998; Williams, 2002). In addition, bacterial matter, especially in recycle and retention processes, age and mutate, which negatively impacts their productivity and efficiency, and may even lead to the production of unwanted products (Branyik et al., 2005; Domingues and Teixeira, 2000). Because of this, a continuous system may be unable to produce the prescribed product quality or it may require more equipment and control systems (Williams, 2002). Long-lasting processes also have contamination problems and biological growth on the walls, requiring the bioreactor to be controlled more closely; otherwise, faster growing cultures overtake the desired ones (Bellgardt, 2000b; Williams, 2002).

Controlling continuous production is also made more difficult because it is usually nonlinear (Ramaswamy et al., 2005; Reddy and Chidambaram, 1995). Several control mechanisms exist, but most make the mistake of varying substrate or bacterial concentrations based on indirect data, such as temperature or pH, and assuming a first-order response (Bellgardt, 2000b; Ramaswamy et al., 2005). This type of control mechanism usually results in a system that is too slow and has stability problems (Reddy and Chidambaram, 1995). Nonlinear controllers are available, but have been shown to be unreliable and complicated and usually report excessive variations (Gonzalez et al., 1998; Reddy and Chidambaram, 1995). Operators may also be limited to a certain gas or liquid input range. Bacteria require a certain amount of gas for proper growth, and a low gas flow rate could suffocate the bacteria or create a hostile environment decreasing bioreactor productivity. On the other hand, too much gas may create suboptimal conditions and have a negative impact on bacterial growth or bioreactor hydrodynamics (Bellgardt, 2000b).

The flexibility, reliability, efficiency, and cost reductions which the continuous systems are supposed to deliver can be offset by control issues, making the pure continuous bioreactor ineffective and even more expensive to operate than batch systems (Williams, 2002). In addition, some fermentation processes, such as beer production, are not able to easily use continuous systems since they change basic product properties, such as taste, color, and/or odor (Branyik et al., 2005).

Several scenarios exist for which continuous bioreactors are feasible and preferred, such as high volume production, use of mutation-resistant bacteria culture, wastewater treatment, and processes that do not require a sterile environment. Currently, continuous bioreactors are used for the production of vinegar, baker's yeast, alcohols, and solvents, and for wastewater treatment (Bellgardt, 2000b; Simon et al., 2006; Williams, 2002). Proper control systems also allow continuous principles to be used in other production systems. These requirements include that the process does not necessitate a sterile environment or that it can be easily controlled through alternative means such as flocculation, substrate and cell concentrations, or the use of secondary or genetically altered cell cultures (Bellgardt, 2000b; Bisang, 1997; Domingues and Teixeira, 2000; Gonzalez et al., 1998; Ramaswamy et al., 2005; Reddy and Chidambaram, 1995; Simon et al., 2006; Sokol and Migiro, 1996; Williams, 2002). The most practical option for the use of the continuous system would be with the semicontinuous bioreactor. This variant would require a smaller initial investment, offer productivity and optimization possibilities, and have less control problems than a continuous bioreactor (Williams, 2002). It would allow for start-up, residence time variations, more flexibility, and some positive batch bioreactor traits.

2.3   SUMMARY

The batch bioreactor is the oldest and most widely used bioreactor. The advantages of using a batch bioreactor are lower fixed cost, flexibility, and operational simplicity. Operational costs can be lowered if the process is well known. These bioreactors are preferred for use in smaller operations, specialty products, long growth periods, operations in which flexibility is vital, unsteady-state processes, and experimental development. Continuous bioreactors, on the other hand, provide for more production optimization, automation, and capital utilization. If the continuous process is used properly, the larger initial investment can be easily offset by lower variable costs and faster production times. Therefore, this process is preferred for high volume production, processes using mutation-resistant bacteria, and processes that do not require a sterile media. The production of alcohols, solvents, vinegar, or baker's yeast, and wastewater treatment are a few examples that take advantages of the continuous process.

Although the choice may seem simple, the experience has been otherwise. The continuous process has had a hard time breaking into industries that have traditionally used batch bioreactors. Some operators have found the switch to a continuous process to be laborious. Unknown challenges provide a situation which could be critical to the project's survival. For example, even though the continuous process offers huge time savings in the production of beer, the batch bioreactor remains the most popular. More risk loving and high volume operators, such as producers of industrial alcohols and solvents and wastewater treatment plants, find the continuous system to be a huge economic advantage. The choice should be made with an understanding of risk capacity for the particular application. The analysis should also include an economic sensitivity analysis and a foresight into the equipment availability. The process selection may ultimately limit the type of bioreactor used or its operational conditions.

3 Gas–Liquid Mass Transfer Models

Mass transfer operations in biological systems depend on a myriad of intermediate and parallel processes. Reactors for gas–liquid applications, which account for about 25% of all chemical reactions (Tatterson, 1994), fulfill two needs: dispersion and absorption (Oldshue, 1983). Dispersion requires that the entire reactor volume be used to mix the gas into the liquid. This step, however, is usually easily achieved or is not the critical system constraint (Oldshue, 1983). The low solubility of most gases limits the gas absorption to the point that gas–liquid mass transfer becomes the rate-limiting step for the overall reaction (Bouaifi et al., 2001; Fogler, 2005; Garcia-Ochoa and Gomez, 1998; Linek et al., 1996a; Moo-Young and Blanch, 1981; Ogut and Hatch, 1988; Oldshue, 1983; Vazquez et al., 1997). This limitation is even more severe in systems using very low solubility gases, such as carbon monoxide found in synthesis gas, some of which are very important in industrial applications (Moo-Young and Blanch, 1981). Thus, the easiest way to increase the productivity of gas–liquid bioreactors is to increase the gas–liquid mass transfer rate (Kapic, 2005).

3.1 GAS-LIQUID TRANSPORT PATHWAYS

The transfer of gas from the gas phase to a microorganism suspended in a bioreactor must take place along a certain pathway. Figure 3.1 schematically describes the general transport route and includes eight resistances to gas mass transfer that may exist between the gas bubble and the microorganism (Chisti, 1989); these resistances include (i) in the gas film inside the bubble, (ii) at the gas–liquid interface, (iii) in the liquid film at the gas–liquid interface, (iv) in the bulk liquid, (v) in the liquid film surrounding the cell, (vi) at the liquid–cell interface, (vii) the internal cellular resistance, and (viii) at the site of the biochemical reaction.

It should be noted that all resistances are purely physical except for the last resistance, and that not all mass transfer resistances may be significant for a given system. Many of the resistances may be neglected in most bioreactors except for those around the gas–liquid interface (Chisti, 1989; Moo-Young and Blanch, 1981). Thus, the transport problem is greatly simplified to a gas–liquid interfacial

An Introduction to Bioreactor Hydrodynamics and Gas-Liquid Mass Transfer, First Edition.
Enes Kadic and Theodore J. Heindel.
© 2014 John Wiley & Sons, Inc. Published 2014 by John Wiley & Sons, Inc.

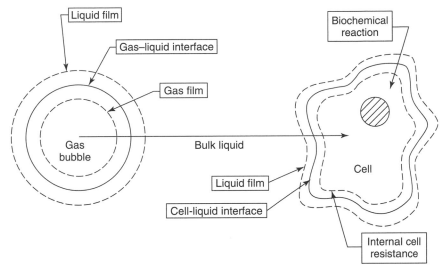

Figure 3.1   Mass transfer resistances encountered in gas–liquid dispersions containing active cells. Adapted from Chisti (1989).

mass transfer problem around the bubble. At this location, two transfer coefficients may be considered at the gas–liquid interface. The liquid-phase mass transfer coefficient is represented by k_L, whereas the gas-phase mass transfer coefficient is identified by k_G. Since the gas-phase mass transfer resistance is typically much smaller than the liquid phase, $k_G \gg k_L$ and gas–liquid mass transfer is controlled by k_L (Chisti, 1989); this value is modulated by the specific (gas–liquid) interfacial area a. The driving force for mass transfer is the gas concentration gradient between the gas phase C^* and the dissolved gas C. The mass transfer rate is then determined by

$$\frac{dC}{dt} = k_L a (C^* - C) \tag{3.1}$$

The volumetric gas–liquid mass transfer coefficient $k_L a$ is typically used when determining the mass transfer coefficient because it is difficult to measure k_L or a independently. Variations in the volumetric gas–liquid mass transfer coefficient during operation are often thought to be a direct result of changes in the interfacial area (Hoffmann et al., 2007; Stenberg and Andersson, 1988b), which would imply that homogeneous (bubbly) operation is more desirable than heterogeneous flow (Bouaifi et al., 2001). However, according to Linek et al. (2005b), concise conclusions are often troublesome because liquid-phase mass transfer is calculated using the gas–liquid mass transfer coefficient (k_L) and the specific interfacial area (a). Any measurement errors in either variable cause false conclusions or improper use of mass transfer models.

3.2 BASIC MASS TRANSFER MODELS

Many mass transfer models exist, but most of them depend on three assumptions and are simplified versions of actual mass transfer mechanisms, many of which occur simultaneously. The first assumption is that the different phases and the phase interface offer resistance to mass transfer in series, in a similar manner to heat transfer resistances. The second assumption maintains that mass transfer is controlled by the phase equilibrium near the interface, which changes more quickly than the bulk phase equilibrium (Azbel, 1981). In other words, mass transfer occurs at the microscale level (van Elk et al., 2007). Finally, gases are assumed to be single component. Multiple component problems are more complicated because each individual gas component making up the mixture has to be considered for the limiting gas–liquid mass transfer step. The complexity grows further once the relationships between each gas component and, for example, the bacteria in a bioreactor are considered.

Single-component gases are preferred for research purposes because direct and mechanistic relationships can be drawn. However, air has been used in many oxygen mass transfer studies because it is easily accessible; in this case, air is typically assumed to be composed of two components: oxygen and an inert gas comprised primarily of nitrogen. Nevertheless, its utilization in gas–liquid processes is often seen as inferior relative to other gases such as pure oxygen, carbon monoxide, carbon dioxide, or hydrogen (Bliem and Katinger, 1988b; Worden and Bredwell, 1998). For experimental purposes, air is often treated as a single-component gas with the realization that the driving force could change due to nontransferred gases (Worden and Bredwell, 1998). Multiple component gases, however, are often encountered in industrial settings mainly due to biological or chemical reactions. Therefore, mass transfer models and correlations are typically used to provide mass transfer estimates and possibly production bounds.

The oldest and most simplistic mass transfer model, which is often presented in undergraduate chemical reaction engineering textbooks (e.g., Fogler (2005)), is the film model originally presented by Nernst in 1904. The interface is assumed to be infinitesimally thin and its resistance is usually ignored. The liquid phase has a constant and definite boundary layer, or film, of thickness δ_{eff} which limits mass transfer (Azbel, 1981). Since these assumptions are made about the boundary layer characteristics, mass transfer is concluded to occur in a steady-state environment. Other limiting factors are the molecular diffusivity D and the driving force represented by the concentration gradient. The film model equation, therefore, predicts mass flux J as

$$J = \frac{-D}{\delta_{\text{eff}}}(C^* - C) \tag{3.2}$$

Molecular diffusivity and film thickness are combined into the liquid-phase mass transfer coefficient k_{L} such that

$$J = k_{\text{L}}(C^* - C) \tag{3.3}$$

This model has several limitations. The film model assumes that mass transfer is controlled by the liquid-phase film, which is often not the case because the interface characteristics can be the limiting factor (Linek et al., 2005a). The liquid film thickness and diffusivity may not be constant over the bubble surface or swarm of bubbles. Experiments also indicate that mass transfer does not have a linear dependence on diffusivity. Azbel (1981) indicates that others have shown that turbulence can have such a significant effect on mass transfer such that eddy turbulence becomes the controlling mechanism in which diffusivity does not play a role. In most instances, however, eddy turbulence and diffusivity combine to play a significant role in mass transfer (Azbel, 1981).

The border diffusion layer model was introduced as an amendment to the film model to present a more realistic description. It accounts for an undefined film thickness, turbulence effects, and the role of molecular diffusion. When the flow is turbulent, the flow around the bubble is split into four sections: the main turbulent stream, the turbulent boundary layer, the viscous sublayer, and the diffusion sublayer. Eddy turbulence accounts for mass transfer in the main turbulent stream and the turbulent boundary layer. The viscous sublayer limits eddy turbulence effects so that the flow is laminar and mass transfer is controlled by both molecular diffusion and eddy turbulence. Microscale eddy turbulence is assumed to be dominant in the viscous sublayer. Mass transfer in the diffusion sublayer is controlled almost completely by molecular diffusion (Azbel, 1981).

This model can be used as a rough estimate. It is still plagued by the steady-state assumption which is oftentimes an inadequate description of mass transfer. The diffusion sublayer δ is a function of viscosity (ν), diffusivity (D), and viscous sublayer thickness (δ_0), but realistic measurements are difficult, if not impossible, to obtain. An empirical correlation has been suggested for δ (Azbel, 1981):

$$\delta = \left(\frac{D}{\nu}\right)^{1/n} \delta_0 \tag{3.4}$$

The power n is an experimental variable with a value of about 2 for gas–liquid mass transfer. The liquid-phase mass transfer coefficient can therefore be related to diffusivity as (Azbel, 1981; Moo-Young and Blanch, 1981)

$$k_L \approx D^{1/n} \tag{3.5}$$

The Higbie penetration model for mass transfer compensates for transient behavior. It assumes that mass transfer occurs during brief phase contacts that do not allow enough time for steady-state conditions. In other words, the phases collide but do not have a definitive and continuous interface with respect to time. The mass transfer is prompted by turbulence that refreshes the interface, and the refresh rate is the limiting step in mass transfer. Eddies approach the surface at which point mass transfer by molecular diffusion is initiated and is described by Azbel (1981):

$$\frac{\partial C}{\partial t} = -D \frac{\partial^2 C}{\partial y^2} \tag{3.6}$$

Equation (3.6) assumes that the gas phase is not reacted during the mass transfer process at the interface. Fast reactions, however, predominantly occur very close to the interface and have to be accounted for by van Elk et al. (2007)

$$\frac{\partial C}{\partial t} = -D\frac{\partial^2 C}{\partial y^2} - r_C \tag{3.7}$$

where r_C represents the reaction rate of the gas phase at the liquid interface. In this case, fast reactions are assumed to occur when $Ha > 2$, where Ha is the Hatta number defined as the ratio of species absorption with and without reactions. The reactive absorption provides the advantage of separating the investigation of the interfacial area from the mass transfer coefficient (Hoffmann et al., 2007).

All turbulent eddies spend approximately the same amount of time at the interface defined by τ. This interface contact time is assumed to be proportional to the amount of time it takes the bubble to rise one bubble diameter d_B or

$$\tau = \frac{d_B}{U_B} \tag{3.8}$$

where U_B is the bubble rise velocity. The Higbie penetration model predicts (Azbel, 1981; Moo-Young and Blanch, 1981)

$$k_L = 2\left(\frac{D}{\pi\tau}\right)^{1/2} \tag{3.9}$$

The assumption that each turbulent eddy spends the same amount of time at the interface is unrealistic. Modified penetration models, such as Danckwerts' surface renewal theory model, allow each eddy to have an independent, variable interface contact time based on a statistical probability function. It uses a fractional renewal s to account for the rate of surface renewal, in which case:

$$k_L = (Ds)^{1/2} \tag{3.10}$$

The surface renewal theory model still shares the same power as the two previous models (Azbel, 1981). The difficulty with the surface renewal model is that the unknown variable s depends on operational conditions and reactor geometry. Hence, it would be difficult to use this model as a scale-up guide.

Later research concluded that reality lies somewhere between the film and penetration models. These film–penetration models, such as those of Hanratty and Toor (Azbel, 1981; van Elk et al., 2007), concluded that the original film model is accurate for highly diffusive gases, large interface contact times, or a small boundary layer thickness (relative to a penetration depth). If a penetration model is used during these situations, predicted mass transfer will be too high (van Elk et al., 2007). Other penetration-type or hybrid models allow for further modifications making their predictions more accurate or applicable for certain conditions. A common ground has been found in the exponential dependence of the diffusivity. Gas–liquid applications show the power n ranging from 0.5 to 1. If water is used as the liquid, the power ranges from 0.65 to 0.985 (Azbel, 1981).

Current research falls into one of two schools of thought: Calderbank's slip velocity model and Lamont and Scott's eddy turbulence model (Linek et al., 2004; Linek et al., 2005b). Even though both models are penetration-type models, they make very different assumptions. The slip velocity model assumes different behavior for small and large bubbles. It also assumes a significant difference between average velocities for the two phases. The slip velocity and the surface mobility control mass transfer and, in terms of penetration theory, surface renewal.

Small bubbles with a diameter $d_B < 1$ mm act as rigid spheres with an immobile surface and slipless interface (Linek et al., 2005a; Scargiali et al., 2007). The surface and its phase interface limit mass transfer. In this case, turbulence indirectly affects mass transfer by influencing the terminal bubble rise velocity and, therefore, the residence time (Poorte and Biesheuvel, 2002). Larger bubbles, those with $d_B > 2.5$ mm, have a mobile, ellipsoidal surface experiencing much larger drag forces than rigid spherical bubbles with the same total volume (Scargiali et al., 2007). Therefore, mass transfer in large bubbles is limited by eddy turbulence. In this case, other penetration models can properly describe mass transfer behavior. The transition area is highly variable. Bubble surface mobility in this region is dependent on liquid properties and surfactants (Linek et al., 2005a).

The eddy turbulence model, or simply eddy model, assumes that the small-scale eddies control surface renewal and, subsequently, mass transfer. This model acknowledges a scale dependence. Macroscale movements, those represented by the Reynolds number, Re, are assumed to have a small impact on surface renewal, where the Reynolds number is defined as

$$Re = \frac{Ud_B}{\nu} \tag{3.11}$$

where U is the velocity, d_B is the (hydraulic) bubble diameter, and ν is the fluid kinematic viscosity. The eddy model postulates that small bubbles would not recognize macroscale motions. Small-scale effects are needed to incur a relative impact. Eddies are assumed to pound the bubble surface and cause surface renewal regardless of bubble size and interface properties (Linek et al., 2004; Linek et al., 2005a; Linek et al., 2005b).

Some of these models predict opposing results. For example, the slip velocity model predicts mass transfer to decrease with increasing turbulence once the bubble surface becomes rigid. The eddy turbulence model predicts the opposite. These theories represent effects that occur concurrently. The strength of each mode depends on the bioreactor type and process. For example, complex reactions, such as polymerization or cellular cultures, are highly sensitive to micromixing (Nauman and Buffman, 1983). Mass transfer during these highly complex processes is affected more by micromixing than by macromixing (Hoffmann et al., 2008). These models and their predictions will be revisited in the appropriate bioreactor sections. Furthermore, the models do not account for surfactant effects on the interface, nor its influence on mass transfer behavior (Vazquez et al., 1997).

3.3   SUMMARY

The fundamental gas–liquid mass transfer models lack the ability to obtain and process all necessary information and factors integral to bioreactor operation. Gas–liquid systems are simply too complex. Therefore, a theoretical equation, which is widely applicable, does not exist (Garcia-Ochoa and Gomez, 2004). Empirical correlations have been developed to simplify analysis and design and have become exclusive in the literature and practice (Kawase and Moo-Young, 1988). Model parameters are chosen that are thought to influence the operation, and their powers and constants are fitted to the experimental data. The correlations are used for design and scale-up while theoretical mass transfer models are used to explain the influence of various operational inputs.

4 Experimental Measurement Techniques

A significant difficulty in characterizing and quantifying gas–liquid, liquid–solid, and gas–liquid–solid mixtures commonly found in bioreactor flows is that the systems are typically opaque (e.g., even an air–water system becomes opaque at fairly low volumetric gas fractions); this necessitates the use of specially designed invasive measurement probes or noninvasive techniques when determining internal flow and transport characteristics. Many of these probes or techniques were developed for a particular type of gas–liquid flow or bioreactor. This chapter first introduces experimental techniques to gauge bioreactor hydrodynamics and then summarizes gas–liquid mass transfer measurement techniques used in bioreactors.

4.1 MEASURING BIOREACTOR HYDRODYNAMIC CHARACTERISTICS

Hydrodynamics in gas–liquid systems have been studied extensively in the past due to their wide range of applications. Characteristics of interest include flow regimes, local pressure drop, gas residence time, axial diffusion coefficients, bubble size, bubble rise velocity, gas holdup, and power consumption. This section will summarize various experimental techniques to quantify some of these characteristics.

In general, the experimental techniques used in gas–liquid flows can be classified as intrusive where invasive probes are used to record local measurements, or noninvasive where data are recorded without altering the flow conditions. Measurement techniques can further be classified as time-average or transient techniques as well as local, regional, or global techniques. Depending on the bioreactor of interest, one technique may be more applicable than another. For example, global time-average gas holdup may be all that is desired in an internal-loop airlift bioreactor investigation, whereas local transient conditions may be required to validate turbulence modeling simulations in bubble columns. Several reviews of one or more experimental methods used in multiphase systems have appeared in the literature, and these can be applied in bioreactor characterization; additional details can be found in several of these publications (Azzopardi et al., 2011; Beck and

---

Williams, 1996; Boyer et al., 2002; Ceccio and George, 1996; Chaouki et al., 1996, 1997; Cheremisinoff, 1986; Heindel, 2011; Hewitt, 1978, 1982; Kumar et al., 1997; Mudde, 2010a; Powell, 2008; Prasser, 2008; Vatanakul et al., 2004; Vial et al., 2003; Williams and Beck, 1995).

4.1.1 Flow Regime Measurements

Flow regime identification is dependent on the geometry of the bioreactor. For example, flow regimes in bubble columns will be different from those identified in stirred-tank bioreactors. In some cases, the experimental techniques used to identify flow regimes are system independent, while in other cases, the technique was developed for a particular geometry. The specific flow regime definitions in common bioreactor types are described in detail in their respective chapters.

Many of the early flow regime studies were based on visual observations in optically accessible flow systems. In cases were optical access is problematic, techniques have been developed to analyze local process measurements such as local pressure fluctuations to correlate various measures with the observed flow regimes. For example, Boyer et al. (2002) indicate many different process measurements can be analyzed to identify bioreactor flow regimes, including measures from wall pressure transducers, microelectrodes imbedded in the wall, conductivity probes, acoustic receivers, optical probes, optical transmittance probes, temperature probes, hot film anemometry, and electrochemical probes. Many of these techniques are invasive to the bioreactor. The fluctuating signals from these probes have been analyzed with various signal processing techniques including classical statistical or spectral techniques, as well as newer fractal, chaotic, or time–frequency analysis techniques. Boyer et al. (2002) and Vial et al. (2000) provide an excellent overview of the flow regime identification methods based on pressure signal fluctuation analysis. Tomographic techniques have also been used to identify flow regimes in multiphase systems (Chaouki et al., 1996, 1997).

4.1.2 Local Pressure Drop

The pressure drop between two locations in a bioreactor is an important hydrodynamic characteristic because the actual measure is needed to size pumps and compressors. Average and transient pressure drop measures can also be analyzed to quantify phase holdup or identify flow regime.

Local pressure drop can be recorded through a variety of methods. Manometers were initially installed along multiphase flow columns to measure pressure signals (Hills, 1976; Kara et al., 1982; Merchuk and Stein, 1981; Zahradnik et al., 1997). More recently, pressure transducers have been used (Letzel et al., 1999; Lin et al., 1998; Luo et al., 1997; Su and Heindel, 2003, 2004, 2005a; Tang and Heindel, 2004; Tang and Heindel, 2005a; Tang and Heindel, 2005b; Tang and Heindel, 2006a; Ueyama et al., 1989), and they are usually flush mounted to the bioreactor wall so that the disturbance to the flow caused by the pressure transducers is minimized.

Local pressure drop measurements with flush-mounted sensors are an easy way to determine gas holdup in the region in which the pressure drop was measured

(Tang and Heindel, 2006a), is noninvasive, and does not interrupt bioreactor operation. With the price drop of piezoelectric pressure transducers and the development of computer data acquisition technology, this method is a convenient low cost measurement technique and is applicable to systems at high temperature and pressure (Letzel et al., 1999; Lin et al., 1998). This technique does not require a transparent fluid or containment vessel, nor does it have requirements on liquid electrical properties. Bioreactor pressure drop can be analyzed to measure the overall average gas holdup in a multiphase region between the two pressure measurement locations, as well as the global average gas holdup. Thus, it can be used to probe the axial gas holdup variation in a column (Hol and Heindel, 2005). Compared to radiation attenuation methods (e.g., γ-ray or X-ray tomography), the pressure difference method is much safer. Furthermore, in addition to estimating gas holdup, pressure signals can also be used to determine flow regime transition (Ruthiya et al., 2005; Vial et al., 2000) and average bubble size (Chilekar et al., 2005) in bubble columns. When a solid phase is present, the pressure difference method can be used to measure gas holdup if the liquid–solid slurry behaves as a pseudo-homogeneous mixture or if the solid concentration as a function of height is known (Kumar et al., 1997; Tang and Heindel, 2005b; Tang and Heindel, 2006a, 2007).

4.1.3 Mixing or Residence Time

Mixing time is defined as the time required to achieve a specified quality of mixing or homogeneity after the addition of some materials (Merchuk, 1985; Weiland and Onken, 1981). Mixing time is very important to several processes, including fermentation. If, for example, an additive does not disperse adequately or takes too long to mix, a local high concentration may be observed, which could lead to cell damage in that region. In batch processes, mixing time is typically associated with the time it takes to fully disperse an additive in the entire tank after a localized input. In continuous processes, mixing time and residence time are coupled, where residence time is the time a particular fluid element (or slug) remains in a given region of the bioreactor. For example, in an airlift bioreactor, residence time is typically associated with the length of time a fluid element remains in the riser and is used extensively as a key parameter in airlift bioreactor modeling.

As indicated by Rodgers et al. (2011), there are several experimental methods that have been used to determine mixing or residence time, including dye injection, pH shift, tracer monitoring, flow followers, and tomography imaging. For example, neutrally buoyant tracer particles were tracked by Kawase and Moo-Young (1986b) to determine the circulation time in an airlift bioreactor. The time it took the tracer particles to complete one loop was determined through multiple measures to calculate the average circulation time.

4.1.4 Axial Diffusion Coefficient

The axial diffusion (or dispersion) coefficient is a measure of mixing in a vertical bioreactor like a bubble column or airlift bioreactor. It has also been used to quantify backmixing in bubble columns. A simple method to measure the axial diffusion coefficient is to add a small amount of salt tracer to the liquid surface, and then measure the local fluid conductivity at a particular distance from the column surface as a function of time (Ohki and Inoue, 1970). Lorenz et al. (2005) extended this idea using a thermal pulse. In this method, a pulse of the same liquid, but at a higher or lower temperature than the bulk fluid, is injected into the system. Thermocouples placed along the column record the local fluid temperature as a function of time to track the thermal pulse. The axial dispersion coefficient D_L is then determined using a 1D dispersion model of the form (Ohki and Inoue, 1970):

$$T = \left(1 + 2\sum_{n=1}^{\infty}\left[\left(\cos\frac{n\pi}{H}z\right)\exp\left\{-\left(\frac{n\pi}{H}\right)^2 D_L t\right\}\right]\right)(T_0 - T_\infty)e^{\alpha t} + T_\infty \quad (4.1)$$

where H is the bioreactor height, z is the axial location within the reactor where the temperature is recorded, T_0 is the initial temperature, T_∞ is the final temperature, α and n are the constants, and t is the time.

4.1.5   Gas–Liquid Interfacial Area

The gas–liquid interfacial area (a) is a fundamental parameter in designing bioreactors because the knowledge of this parameter is required to calculate individual gas–liquid mass transfer rates (Vasquez et al., 2000). The interfacial area is a challenge to quantify because it is influenced by the bioreactor geometry and operating conditions, as well as the physical and chemical properties of the gas–liquid system. In some cases, the interfacial area is estimated by assuming a uniform bubble diameter d_B and measuring the overall gas holdup ϵ. In this case, the gas–liquid interfacial area is estimated from Chisti (1989):

$$a = \frac{6\epsilon}{d_B} \quad (4.2)$$

If the bubble size is not uniform, the bubble size distribution can be measured using a variety of measurement tools such as optical probes, multipoint needle probes, or hot-film anemometers, and the Sauter mean diameter (d_{SM}) can be substituted for the bubble diameter (Azzopardi et al., 2011) (see also Eq. (4.3)).

The interfacial area can be measured in specific systems using chemical reactions in which the absorption rate kinetics are a known function of the gas–liquid interfacial area. For example, Vasquez et al. (2000) compared three different chemical methods: (i) Danckwerts' method using the absorption of CO_2 in sodium or potassium carbonate buffer solutions, (ii) the sodium sulfite method involving the oxidation of sulfite ions, and (iii) the sodium dithionite method involving the oxidation of dithionite ions. All three methods were shown to produce similar interfacial area measurements.

4.1.6 Bubble Size and Velocity

Bubble size and bubble velocity measurement techniques have been reviewed by Saxena et al. (1988) and Boyer et al. (2002); the interested reader is directed to these sources for detailed descriptions of the available hardware and data analysis procedures.

It is common to report the average bubble diameter as the Sauter mean diameter d_{SM}, which is defined as the diameter of a bubble equivalent to the volume-to-surface area ratio of the entire dispersion (Saxena et al., 1988). Assuming all bubbles are spheres,

$$d_{SM} = \frac{6\epsilon}{a} = \frac{\sum d_B^3}{\sum d_B^2} \tag{4.3}$$

where ϵ is the total gas holdup, a is the total interfacial area, and d_B is the individual bubble diameter. Bubble size can be determined through visual observations and image analysis techniques.

If the time difference between successive frames of the same bubble is known and the bubble displacement can be measured, then bubble rise velocity can also be measured. However, visual methods are limited to systems with optical access, so observations are limited to regions near the wall even under moderate gassing rates in gas–liquid systems. The wall and liquid must also be transparent.

In addition to optical methods, bubble size can be determined using optical probes and electrical conductivity (resistivity) probes (Saxena et al., 1988). For example, Magaud et al. (2001) used dual optical probes to determine the local instantaneous presence of the liquid or gas in a bubble column. With this information, local bubble chord length and bubble rise velocity can be determined. One advantage of optical probes is that its operation is independent of the electrical properties of the medium surrounding the probe (Saxena et al., 1988). Electrical conductivity or resistivity probes can be configured as needle probes, which have been used to determine mean bubble chord length, bubble size, and bubble rise velocity.

Bubble size in opaque systems has been determined by Heindel and coworkers using flash X-ray radiography (Garner and Heindel, 2000; Heindel, 1999, 2000, 2002; Heindel and Garner, 1999). In this process, an intense burst of radiation is produced for a fraction of a second to provide stop-motion X-ray projections of bubble motion. Image analysis is then completed to determine bubble size and bubble size distribution. A significant drawback of this technique was that the flow was limited to quasi-2D flows and the bubble number density had to be small enough to distinguish individual bubbles.

Ultrasound Doppler velocimetry (UDV) has also been used to measure bubble velocity by measuring the frequency shift between the emitted ultrasound beam and the echo reflected from the gas–liquid interface (Vial et al., 2003).

Many researchers have used two or more measurement methods to characterize multiple aspects of gas–liquid flows. For example, Broder and Sommerfeld (2003) used a specially designed particle image velocimetry (PIV) and particle tracking velocimetry (PTV) techniques to simultaneously measure the bubble size as well as bubble and liquid velocities. Their experimental equipment was mounted on a traversing system that allowed them to follow rising bubbles with a high speed camera system. Knowing the bubble size and bubble and liquid velocities, and using sophisticated data analysis procedures, they could further determine bubble–bubble collision rates, coalescence rates, and coalescence efficiencies. Another example is provided by Kiambi et al. (2011) who used a bioptical probe and hot-film anemometry to measure local gas holdup, bubble velocity, bubble size, and liquid velocity in an external airlift reactor. They were able to provide radial distributions of the various measures of interest.

4.1.7   Global and Local Liquid Velocity

In fermentation processes like those found in airlift reactors, the difference in riser and downcomer gas holdup creates a hydrostatic pressure difference between the bottom of the riser and the bottom of the downcomer, which in turn acts as the driving force for liquid circulation. A mean circulation velocity U_c is defined as (Blenke, 1979)

$$U_c = \frac{x_c}{t_c} \tag{4.4}$$

where x_c is the circulation path length and t_c is the average circulation time for one complete circulation. However, liquid circulation velocity is not commonly used as a characteristic parameter for gas–liquid fermentation processes. The superficial liquid velocity in the riser (U_{Lr}) or downcomer (U_{Ld}) is more commonly used as they are more meaningful and allow for direct comparison of liquid circulation rates in reactors of varying sizes. The superficial liquid velocity is different from the true linear velocity because the liquid flow occupies only a portion of the flow channel—the space occupied by rising gas bubbles reduces the local cross-sectional area available for liquid flow.

The superficial liquid velocity cannot be directly measured and is usually determined from the knowledge of the linear liquid velocity (V_L) and gas holdup. In airlift reactors, U_L and V_L have both riser and downcomer components, yet the riser superficial liquid velocity (U_{Lr}) is the parameter of greatest interest and the one commonly reported in the literature.

The determination of riser and downcomer U_L is often accomplished using a tracer technique or specially calibrated flow meters and mathematical relationships to convert the measurable V_L to U_L. The tracer techniques commonly used to determine V_L are based on determining the time it takes for a given tracer to travel a set distance. For example, a potassium chloride salt tracer and conductivity electrodes are commonly used to measure the time it takes an injection of the salt solution to travel past two fixed locations from which V_L is calculated (Bello et al.,

1984; Chisti, 1989; Jones, 2007; Van't Riet and Tramper, 1991). Knowing V_L, the superficial liquid velocity is determined from Chisti (1989)

$$U_L = (1 - \epsilon)V_L \tag{4.5}$$

While the superficial liquid velocity is a function of riser and downcomer gas holdup, it also influences these holdups. Hence, for a given airlift reactor geometry, the superficial liquid velocity is a function of gas holdup. The superficial liquid velocity can only be changed in this reactor through geometry modifications (which will also affect gas holdup values) or through the use of a throttling device (Popovic and Robinson, 1988).

Local instantaneous liquid velocity measurements in bioreactors that can quantify turbulence statistics are challenging using conventional laser-based techniques because optical access is critical for effective signal acquisition. Laser Doppler anemometry (LDA) and PIV have been used to determine local liquid velocities within multiphase flows. Reviews of LDA and PIV with applications to multiphase flows have appeared in the literature (Boyer et al., 2002; Chaouki et al., 1997; Cheremisinoff, 1986).

Liquid velocity may also be determined using hot film anemometry (Boyer et al., 2002; Magaud et al., 2001). One advantage of this technique is that it is fairly inexpensive, accurate, and relatively easy to implement. However, proper implementation requires a uniform temperature in the bioreactor as well as a limited solid content, and the measurement technique is invasive.

Radioactive particle tracking using neutrally buoyant γ-ray-emitting particles has been used to determine liquid velocity within various bioreactors (Devanathan et al., 1990; Dudukovic, 2000; Khopkar et al., 2005; Luo and Al-Dahhan, 2008). In this technique, a single neutrally buoyant particle that has been tagged to emit γ-rays is inserted into a bioreactor, and γ-ray detectors are located at several locations around the periphery of the bioreactor. With proper signal calibration, the location of the tracer particle is then determined at any instant in time. By following the particle over long time intervals (several hours), mean and fluctuating liquid velocity components can be determined within the multiphase flow.

X-ray particle tracking velocimetry (XPTV) is an X-ray imaging technique where several X-ray absorbing objects (particles) are simultaneously tracked as a function of time (Seeger et al., 2001a). By tracking neutrally buoyant X-ray absorbing particles, Seeger and coworkers (Kertzscher et al., 2004; Seeger et al., 2001b; Seeger et al., 2003) were able to record the 3D liquid velocity field in a slurry bubble column.

4.1.8   Gas Holdup

Gas holdup (or gas fraction or void fraction) is defined as the volumetric fraction occupied by the gas phase in the total volume of a two- or three-phase mixture. It is one of the most important parameters characterizing gas–liquid and gas–liquid–solid hydrodynamics, because it not only gives the volume fraction of the gas phase but also is needed to estimate the interfacial area and thus the mass transfer rate between the gas and liquid phases (Shah et al., 1982).

Gas holdup can be measured by numerous invasive or noninvasive techniques, which have been reviewed by Kumar et al. (1997) and Boyer et al. (2002), and include changes in total bed expansion upon gassing, pressure drop measurements, dynamic gas disengagement (DGD), and tomographic techniques.

4.1.8.1 Bed Expansion. One of the simplest methods to measure global gas holdup, ϵ, is to measure the bed expansion upon gassing. Assuming that the containment vessel has a constant cross-sectional area:

$$\epsilon = \frac{H - H_0}{H} \tag{4.6}$$

where H is the liquid height at a given gas flow rate and H_0 corresponds to the initial liquid height. This assumes that the bulk liquid velocity is zero. The liquid expansion height is very easy to identify at low gas flow rates, but this identification is more challenging at high gas flow rates due to fluctuations at the free surface caused by bubble disengagement. In industry, this level may be identified with an electronic float to continuously monitor the gas holdup.

4.1.8.2 Pressure Drop Measurements. Pressure drop measurements are one of the most widely used techniques for measuring gas holdup. This method has been used in semibatch bubble columns (Letzel et al., 1999; Lin et al., 1998; Luo et al., 1997; Su and Heindel, 2003, 2004, 2005a; Su et al., 2006; Tang and Heindel, 2005a; Tang and Heindel, 2005b; Tang and Heindel, 2006a; Ueyama et al., 1989; Zahrad-nik et al., 1997), as well as airlift reactors (Al-Masry, 2001; Hills, 1976; Merchuk and Stein, 1981), and concurrent bubble columns (Kara et al., 1982; Tang and Hein-del, 2004, 2005b; Tang and Heindel, 2006a), where there is a net upward liquid flow. With this method, gas holdup is measured using the time-average static pressure drop along the column. The resulting gas holdup is an average value (both temporal and spatial) over the volume of the dispersion between the corresponding pressure taps. In semibatch bubble column operations, Kara et al. (1982) and Tang and Heindel (2005b) showed that the gas holdup values obtained via the pressure difference method matched well (within $\pm 3\%$) with those obtained via direct gas holdup measurement (i.e., estimating gas holdup by measuring the mixture or liquid level before and after DGD).

Assuming 1D isothermal flow, steady-state, constant cross-sectional area, negligible mass transfer between the gas and liquid phases, and constant properties in a cross section, Merchuk and Stein (1981) used a separated flow model of Wallis (1969) for vertical gas–liquid cocurrent flows to determine gas holdup in gas–liquid bubble columns and airlift reactors:

$$\epsilon = \left(1 + \frac{1}{\rho_L g} \frac{dp}{dz} \right) + \frac{4\tau_w}{\rho_L D_c g} + \frac{U_L^2}{g} \frac{1}{(1-\epsilon)^2} \frac{d\epsilon}{dz} \tag{4.7}$$

where ϵ and p are the local gas holdup and pressure at position z, respectively, ρ_L is the liquid density, g is the acceleration due to gravity, D_c is the column inner diameter, U_L is the superficial liquid velocity, and τ_w is the wall shear stress. Hills (1976) obtained a similar expression assuming a pseudo-homogeneous two-phase mixture.

The first term on the right-hand side of Eq. (4.7) accounts for the hydrostatic head, the second term describes wall shear effects, and the third term represents fluid acceleration due to void changes. The contribution of the acceleration term is typically $\sim 1\%$ of the total gas holdup (Merchuk and Stein, 1981). Hills (1976) has shown that in the worst case in a study with superficial liquid and gas velocities as high as 2.7 and 3.5 m/s, respectively, the acceleration term amounted to less than 10% of the total gas holdup. As a result, the acceleration term is usually neglected in practice (Al-Masry, 2001; Hills, 1976; Merchuk and Stein, 1981; Tang and Heindel, 2004; Zahradnik et al., 1997). Without the acceleration term, Eq. (4.7) becomes

$$\epsilon = \left(1 + \frac{1}{\rho_L g}\frac{dp}{dz}\right) + \frac{4\tau_w}{\rho_L D_c g} \tag{4.8}$$

To obtain the average gas holdup $\bar{\epsilon}$ in a column section between two locations separated by a distance $\Delta z = z_2 - z_1(> 0)$, average both sides of Eq. (4.8) from z_1 to z_2:

$$\frac{1}{\Delta z}\int_{z_1}^{z_2}\epsilon\, dz = \frac{1}{\Delta z}\int_{z_1}^{z_2}\left(1 + \frac{1}{\rho_L g}\frac{dp}{dz}\right)dz + \frac{1}{\Delta z}\int_{z_1}^{z_2}\frac{4\tau_w}{\rho_L D_c g}\, dz \tag{4.9}$$

Thus,

$$\bar{\epsilon}_I = \bar{\epsilon} = \left(1 - \frac{1}{\rho_L g}\frac{\Delta p}{\Delta z}\right) + \frac{4\bar{\tau}_w}{\rho_L D_c g} \tag{4.10}$$

where $\Delta p = p_1 - p_2(> 0)$ with p_1 and p_2 the pressures at locations z_1 and z_2, respectively, and $\bar{\tau}_w$ represents the average wall shear stress in the same column section. Tang and Heindel (2006a) denoted the gas holdup measurement based on Eq. (4.10) as Method I ($\bar{\epsilon}_I$) where it totally accounts for the wall shear stress effects and provides accurate gas holdup values based on the assumptions above.

The wall shear term in Eq. (4.10) is usually neglected for semibatch bubble columns (Su and Heindel, 2003, 2004, 2005a; Su et al., 2006; Ueyama et al., 1989; Zahradnik et al., 1997). For cocurrent bubble columns and airlift reactors, this term is small at low superficial liquid velocities (e.g., $U_L \sim 1\,\text{cm/s}$ in air–water systems). When the wall shear term is negligible, Eq. (4.10) can be simplified to

$$\bar{\epsilon}_{II} = 1 - \frac{1}{\rho_L g}\frac{\Delta p}{\Delta z} \tag{4.11}$$

The gas holdup measurement based on Eq. (4.11) completely neglects the effects of wall shear stress and has been identified by Tang and Heindel (2006a) as Method II ($\bar{\epsilon}_{II}$).

The wall shear term in Eq. (4.10) increases significantly with increasing superficial liquid (U_L) and gas (U_g) velocities and can amount to $\sim 20\%$ of the total gas holdup (Hills, 1976; Merchuk and Stein, 1981). This is because the wall shear stress $\bar{\tau}_w$ increases significantly with U_L and U_g (Liu, 1997; Magaud et al., 2001; Wallis, 1969). When the liquid phase is highly viscous, the wall shear term can be significant even at superficial liquid velocities on the order of $\sim 2-10\,\text{cm/s}$ (Al-Masry, 2001). Hence, it is necessary to include the wall shear effect in the total gas holdup value for most cocurrent or viscous flow bioreactors.

Calculation of the wall shear term in Eq. (4.10) requires estimation of the two-phase wall shear stress $\bar{\tau}_w$, which is a complex function of gas holdup, superficial gas and liquid velocity, liquid-phase rheological properties, and wall roughness. The models for $\bar{\tau}_w$ in gas–liquid two-phase flows are limited, and most are not general and cannot be extended beyond their restricted conditions (Gharat and Joshi, 1992). The two-phase wall shear stress is even more difficult to estimate when the liquid phase is non-Newtonian (Al-Masry, 2001). Even when a model for $\bar{\tau}_w$ is known, the model is usually a highly nonlinear function of gas holdup (Beyerlein et al., 1985; Herringe and Davis, 1978; Merchuk and Stein, 1981), and one has to solve a nonlinear version of Eq. (4.10) to obtain the gas holdup. This is inconvenient, especially when a large number of data points are acquired.

Tang and Heindel (2006a) have shown that Eq. (4.11) can be modified for concurrent multiphase flow systems to estimate gas holdup based on differential pressure measurements, with

$$\bar{\epsilon}_{III} = 1 - \frac{\Delta p}{\Delta p_{0,U_L}} \tag{4.12}$$

where $\Delta p_{0,U_L}$ is the pressure difference between z_1 and z_2 (the same locations corresponding to Δp) when $U_g = 0$ ($\bar{\epsilon} = 0$) and U_L is the same superficial liquid velocity at which Δp is measured. Equation (4.12) becomes Eq. (4.11) when $U_L = 0$; Tang and Heindel (2006a) define this as Method III ($\bar{\epsilon}_{III}$). As described by Tang and Heindel (2006a), Eq. (4.12) considers an estimation of the wall shear stress effect without modeling the two-phase wall shear stress or solving a nonlinear form of Eq. (4.10). The procedure is as simple as Eq. (4.11) but provides more accurate gas holdup values in cocurrent systems. Hence, using Eq. (4.12), more accurate gas holdup measurements in cocurrent multiphase systems can be made with only pressure measurements, and the calculation is as simple as that required by Eq. (4.11). Furthermore, no knowledge of wall shear stress is required for Eq. (4.12), which is not the case for Eq. (4.10). Tang and Heindel (2006a) have shown that gas holdup in a cocurrent air–water–fiber bubble column was simple and accurate with Eq. (4.12), while error could be significant for selected operational conditions with Eq. (4.11).

4.1.8.3 Dynamic Gas Disengagement (DGD).

DGD abruptly stops the aeration process in a gas–liquid or gas–liquid–solid bioreactor and then records the liquid level or pressure at different locations as a function of time (Daly et al., 1992; Deshpande et al., 1995; Fransolet et al., 2005; Patel et al., 1989; Schumpe and

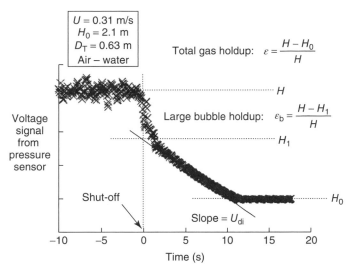

Figure 4.1 Sample data from a dynamic gas disengagement experiment (Krishna and Ellenberger, 1996).

Grund, 1986; Sriram and Mann, 1977). As summarized by Boyer et al. (2002), DGD can be used to record global gas and solid holdup, as well as estimate the holdup make-up according to different bubble size classes.

Krishna and coworkers (Ellenberger and Krishna, 1994; Krishna et al., 1997; Krishna and Ellenberger, 1996; Krishna et al., 2000; Krishna et al., 1999; Vermeer and Krishna, 1981) have employed the DGD technique extensively to determine dense and dilute phase gas holdup values. As shown in Figure 4.1, Krishna and Ellenberger (1996) followed the fluid-level decline as a function of time and iden-tified the dilute phase with the large fast-rising bubbles that disengage first and the dense phase with the small bubbles that disengage after the large bubbles. The demarcation between the dilute and dense phase bubble region was the change in the rate of fluid-level decline. The gas holdup for each phase, as well as the total gas holdup, can be determined from these data. Lee et al. (1999) determined that the disengagement of large bubbles had a significant influence on small bubbles, but the overall gas holdup was similar when measured by differential pressure drop, dispersion height, and PIV.

In general, the DGD technique is fairly straightforward and easy and inexpensive to implement, but has limited applications beyond the laboratory.

4.1.8.4 Tomographic Techniques. Tomography refers to the cross-sectional imaging of a system from either transmission or reflection data collected by illuminating the systems from many different directions (Kak and Slaney, 1988). A variety of tomographic techniques have been developed to determine local time-average phase fractions within an imaging volume. Current tomographic techniques for phase fraction determination include electrical impedance tomography (Ceccio and George, 1996; George et al., 2000; Tortora et al., 2006), electrical resistance tomography (Fransolet et al., 2005; Toye et al., 2005), electrical capacitance tomography (Du et al., 2006; Gamio et al., 2005; Ismail et al., 2005; Makkawi and Wright, 2002; Makkawi and Wright, 2004; Marashdeh et al., 2008; Pugsley et al., 2003; Warsito and Fan, 2003a, 2003b, 2005), ultrasonic computed tomography (Utomo et al., 2001; Vatanakul et al., 2004; Zheng and Zhang, 2004), gamma densitometry tomography (Dudukovic, 2000; George et al., 2001; Jin et al., 2005; Kumar and Dudukovic, 1996; Kumar et al., 1995; Mudde et al., 2005; Rados et al., 2005; Shaikh and Al-Dahhan, 2005; Tortora et al., 2006; Yin et al., 2002), X-ray computed tomography (Drake and Heindel, 2011, 2012; Ford et al., 2008; Franka and Heindel, 2009; Hubers et al., 2005; Kantzas, 1994; Marchot et al., 2001; Mudde, 2010a, 2010b; Prasser et al., 2005; Schmit and Eldridge, 2004; Schmit et al., 2004), positron emission tomography (Dechsiri et al., 2005), neutron transmission tomography (Harvel et al., 1999; Prasser, 2008), and magnetic resonance imaging (Muller et al., 2008; Powell, 2008; Rees et al., 2006).

As summarized by Marashdeh et al. (2008), tomographic systems are generally classified into soft field or hard field measurement systems. In soft field methods such as electrical capacitance tomography, a change in the measured property (e.g., capacitance) in one location changes the recorded field throughout the entire domain, resulting in a very complex reconstruction process that could produce multiple solutions. Typically for soft field methods, iterative and optimization techniques are utilized to find the most likely reconstruction. In hard field methods such as X-ray tomography, the field lines of the measured property (e.g., X-ray attenuation) remain straight and they are not influenced by property changes outside the line of sight. This makes the reconstruction easier, but, because of the detection systems and source strength, data acquisition is typically slow.

Of the ionizing radiation (hard field) techniques, X-ray imaging is safest because the sources only emit X-rays when they are powered and their energy can be controlled by varying the input voltage (Chaouki et al., 1997; Toye et al., 1996). Toye et al. (1996) also note that X-rays are preferred over γ-rays because the X-rays provide better spatial resolution due to improvements in the X-ray detector technology in recent years. X-ray tubes also provide a smaller spot size when compared to γ-ray sources of equivalent strength, which also provides improved spatial resolution.

Thatte et al. (2004) used a 67 μCi ^{137}Cs source (γ-ray) to measure gas holdup in a transparent, flat-bottomed, 0.57-m-diameter cylindrical tank equipped with a pitched blade downflow turbine or a disk turbine. For both impellers, the average gas holdup was obtained by integrating the local gas holdup and it matched well with the results obtained by visual observations. The reproducibility of the measurements was within ±10%. Khopkar et al. (2005) used a ^{137}Cs source with seven NaI detectors to measure gas holdup in a flat-bottomed, 0.2-m-diameter cylindrical tank with a shaft that extended to the vessel bottom. The total scan time was a little over 3 h. Khopkar et al. (2005) noted that CT results were sensitive with respect to the convergence criterion used during data processing. High energy γ-rays, unlike

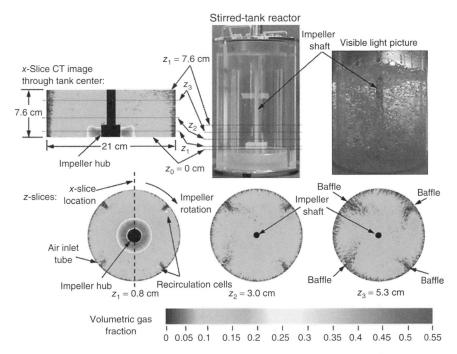

Figure 4.2   Gas holdup in a 21-cm-diameter stirred-tank reactor obtained using X-ray computed tomography imaging.

X-rays, also work on larger tanks because the γ-rays are strong enough to pass through substantial thicknesses of metals, overcoming the reactor wall thickness. For example, Veera et al. (2001) used a ^{137}Cs source to measure gas holdup in a three-phase, 4.9-m-diameter stirred-tank reactor equipped with two impellers.

Ford et al. (2008) used X-ray computed tomography to determine the local time-average gas holdup in a 21-cm-diameter stirred-tank reactor equipped with a Ruston-type turbine. The high resolution of the X-ray system allowed fine details such as recirculation cells behind the baffles to be visualized. An example of such imaging is shown in Figure 4.2, where X-ray CTs are used to determine the local time-average phase distribution anywhere within the imaging volume of the STR. The visible light picture shows a stop-motion image of the gas dispersion but there are so many bubbles that internal details are obscured. The details provided by the CT imaging show a high gas content in the impeller region and recirculation cells behind two of the baffles.

4.1.9   Liquid Holdup

Liquid holdup in a bioreactor can be measured by recording the residence time distribution (RTD) of a tracer that is injected into the bioreactor (Boyer et al., 2002). By following the tracer, the liquid mixing can also be characterized. The most common RTD method involves injecting a small amount of salt tracer and then measuring the liquid conductivity at two fixed points. The time it takes for the conductivity to have a step change between the two locations is then related to the liquid velocity and holdup. Radioactive particle tracking has also been used as a tracer to determine liquid holdup (Devanathan et al., 1990).

4.1.10 Power Measurements

Stirred-tank bioreactors mechanically agitate the gas–liquid dispersion, and the resulting power draw is an important parameter in these bioreactors. The measured power draw is used to quantify two dimensionless numbers in air-sparged stirred-tank bioreactors, the ungassed and gassed power numbers.

The ungassed power number (N_{po}) represents the ratio of the pressure differences producing flow to the inertial forces of the liquid dispersion and it is analogous to a friction factor or drag coefficient. N_{po} is usually based on the power input by the impeller for agitated vessels and takes the form:

$$N_{po} = \frac{P_o}{\rho N^3 D_i^5} \tag{4.13}$$

where P_o is the impeller power input into the liquid without sparged gas, ρ is the fluid density, N is the impeller speed, and D_i is the impeller diameter. The gassed power number N_{pg} is a dimensionless parameter that provides a measure of the power requirements for the impeller operation in a gas–liquid dispersion. The gassed power number represents the ratio of the pressure differences producing flow to the inertial forces of a gas–liquid dispersion. When gas is introduced into ungassed agitated vessels, the mixing power will drop and is related to a gassed power number defined by

$$N_{pg} = \frac{P_g}{\rho N^3 D_i^5} \tag{4.14}$$

where P_g is the impeller power input into the liquid when gas is sparged into the vessel.

A thorough review of power consumption in stirred-tank reactors has been provided by Ascanio et al. (2004). There are four main techniques for measuring power consumption in stirred-tank bioreactors, including electric measurements, calorimetric measurements, torque measurements, and strain measurement. Although calorimetric and strain techniques can be very accurate, particularly in a laboratory setting, the setup and required controls can be complicated and they are generally not utilized in an industrial setting. Torque measurements through dynamometers or torquemeters can provide accurate measurements of the power imparted to the fluid, provided losses due to no-load conditions are accounted for accurately. In an industrial scale, however, the operating torques can be very large, making these measurement systems impractical.

In an industrial setting, electrical measurement of the power consumption is the simplest and most commonly utilized technique. A wattmeter can be used to determine the total power draw by the motor. This measurement includes the power lost in the motor, gearbox (if any), and seal (if any), as well as the power imparted to the fluid to create the mixing process. Provided the losses can be accurately quantified, typically by running the system in air and then assuming the losses are constant, electrical measurements can provide the needed information and are commonly monitored in process control applications.

4.2   GAS–LIQUID MASS TRANSFER[1]

In the fermentation industry, usually two fermentation types are distinguished: (i) aerobic fermentation and (ii) anaerobic fermentation as discussed in detail by Bell-gardt (2000c). Aerobic fermentations require an oxygen supply, which is normally acquired from the surrounding environmental air. Anaerobic fermentations are con-ducted in an oxygen-free environment and may utilize carbon monoxide (CO) as the sole carbon source from a gas mixture. In both fermentation types, a general premise is that microorganisms take one component and biologically convert it to another component. Therefore, dissolved O_2 and CO mass transfer rates are very important to determine bioreactor performance. The gas–liquid mass transfer rate is typically measured by recording the dissolved gas content as a function of time. Many methods are available to measure dissolved oxygen content (Gogate and Pan-dit, 1999b; Linek et al., 1987; Sobotka et al., 1982; Stenberg and Andersson, 1988a; Tobajas and Garcia-Calvo, 2000; Van't Riet, 1979; Wilkin et al., 2001) and they are discussed in Section 4.2.1. Details of a specific method to measure dissolved CO content as a function of time are described in Section 4.2.2 because it is important to synthesis gas fermentation processes (Brown, 2005; Henstra et al., 2007; Kapic et al., 2006; Riggs and Heindel, 2006; Ungerman and Heindel, 2007; Zhu et al., 2008, 2009) and no dissolved CO sensors are currently available.

4.2.1   Dissolved Oxygen Measurement Techniques

There are several techniques used to determine the dissolved oxygen content in a fluid. In practice, five general methods exist: chemical, volumetric, tubing, optodes, and the electrochemical electrode (Carroll, 1991; van Dam-Mieras et al., 1992). This section will discuss these methods and some of their limitations and uses; the emphasis, however, will be on electrochemical electrodes as they are the most common dissolved O_2 sensors.

4.2.1.1   Chemical Method.   In the chemical method, a sample is taken from the reactor and the dissolved oxygen concentration is determined off-line using a titri-metric method. The use of chemical methods for systems that have rapidly changing dissolved oxygen content is limited because these methods are laborious, slow, and prone to error if not done correctly.

---

[1]Material in this section is based on the information summarized by Samuel T. Jones in "Gas-Liquid Mass Transfer in an External Airlift Loop Reactor for Syngas Fermentation," PhD Dissertation, Depart-ment of Mechanical Engineering, Iowa State University, 2007. Used with permission.

The most widely used chemical method is the Winkler method (iodometric method) developed by Lajos Winkler in 1888 (Anonymous, 2005), and is considered to be the most reliable and precise titrimetric procedure for dissolved oxygen analysis. This method involves several steps. First, adding a divalent manganese solution followed by a strong alkali to a sample in a gas tight container; this causes the dissolved oxygen to oxidize an equivalent amount of manganese ions to hydroxide. Second, an acid is added to convert the hydroxide to iodine. Third, the solution is titrated with a thiosulfate solution in the presence of a starch indicator to determine the number of iodine molecules in solution. The number of measured iodine molecules is proportional to the number of dissolved oxygen molecules in the original sample as shown by

$$1 \, \text{mol} \, O_2 \rightarrow 4 \, \text{mol} \, Mn(OH)_3 \rightarrow 2 \, \text{mol} \, I_2 \qquad (4.15)$$

As with any analytical method, the success of the Winkler method is highly dependent on how the sample is collected and prepared. Care must be taken during all steps of the analysis to ensure that oxygen is neither introduced nor lost from the sample. Furthermore, care must be taken to ensure that the sample is free of contaminants because they may oxidize the iodide or reduce the iodine, which are challenges commonly encountered with fermentation broths. Wilkin et al. (2001) stated that the Winkler method is the most accurate and precise of all methods for determining dissolved oxygen concentrations, and that it is also the most challenging technique to master and the most time consuming.

Other chemical methods such as the NADH oxidation and phenylhydrazine oxidation have been employed to determine dissolved oxygen content (van Dam-Mieras et al., 1992), but are not frequently used.

4.2.1.2 Volumetric Method. The volumetric method is simple and robust in principle, but rather inaccurate in practice. This method relies on the conversion of dissolved oxygen to carbon dioxide which is then driven out of solution. As the carbon dioxide is driven out of solution, it is collected and its volume is determined at a known pressure and temperature. Then, using the ideal gas law and an elemental balance for the oxygen to carbon dioxide reaction, the oxygen concentration is determined. While simple in theory and nearly unaffected by other compounds that might be in the sample, this method, similar to the chemical method, is slow and lacks the sensitivity needed for dynamic biological applications (van Dam-Mieras et al., 1992).

4.2.1.3 Tubing Method. The tubing method consists of using a very small-diameter thin-walled tube of semi-permeable material that is immersed in a fermentation broth (Turner and White, 1999; van Dam-Mieras et al., 1992). A slow stream of oxygen-free carrier gas is pumped through the immersed tube and allowed to absorb oxygen from the fermentation broth by diffusion. The oxygen concentration in the exit gas stream is then measured using a gas analyzer or electrode. This method is strongly influenced by the tubing type, length, diameter, carrier gas flow rate, wall thickness, temperature, and the mixing characteristics within the reactor vessel. Owing to the many factors that may influence the operation of this method, extensive calibration is required. The tubing method is also very slow and has been shown to have response times of 2–10 min (Turner and White, 1999). However, despite the long response times and the need for extensive calibration, this method can be very accurate, robust, and can withstand repeated sterilization cycles.

4.2.1.4 Optode Method. A photometric transducer or optode can be used to measure gaseous and dissolved oxygen concentrations (Koeneke et al., 1999). Many types of optodes exist, of these the fluorescence quenching optode is most widely used for oxygen measurements (Turner and White, 1999).

Optodes for oxygen sensing are constructed using an immobilized fluorophore (a special dye) attached directly to the end of an optic fiber. When excited by a reference light wave, the fluorophore will emit another light wave having a different wavelength with an intensity that depends on the quencher concentration. Thus, when the quencher is oxygen, the intensity of the emitted light is proportional to the dissolved oxygen concentration.

These sensors can be used in very harsh environments, do not consume oxygen, are very small, are very sensitive to oxygen concentration changes, and are not prone to response time issues common to other methods (Glazer et al., 2004; Koeneke et al., 1999; Kohls and Scheper, 2000; Terasaka et al., 1998; Turner and White, 1999). However, a few drawbacks such as ambient light interactions and photobleaching are issues that must be addressed before their use is widespread.

4.2.1.5 Electrochemical Electrode Method. Membrane-coated dissolved oxygen electrodes were developed in the 1950s and have become one of the most important process instruments for aerobic fermentations (van Dam-Mieras et al., 1992). Normally, the membrane used with these electrodes is only gas permeable and impermeable to most ions such as those used in the electrolyte solution, thus these electrodes do not disturb the biological process. For this reason, and the fact that dissolved oxygen electrodes are relatively easy to use, they are very popular and widely used in industry. Today nearly all oxygen electrodes can be classified as either polarographic or galvanic.

Polarographic or galvanic electrodes are based on the reduction of oxygen at the cathode, which is negatively polarized with respect to the anode. While these electrodes are similar in construction and operation, the main difference between the two is the source of the needed polarization voltage. Polarographic electrodes are typically charged with a negative voltage of 0.75 V by an external source, while galvanic electrodes utilize a negative 0.75-V potential created by the use of dissimilar metals.

It is important to note that both the polarographic and galvanic electrodes measure the oxygen tension of the medium in which they are placed (Doran, 2013). So when an electrode is placed in a liquid, it does not measure dissolved oxygen, but rather the dissolved oxygen partial pressure, which is proportional to oxygen

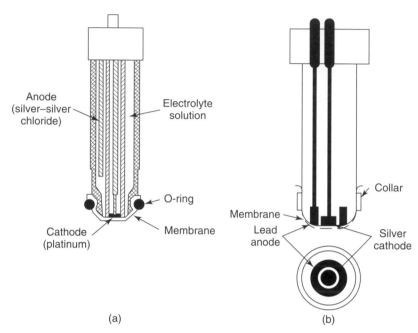

Figure 4.3 Schematics showing the typically construction of (a) polarographic and (b) galvanic electrodes. Adapted from Linek (1988).

tension in the fluid. It is necessary to know the oxygen solubility, pressure, and temperature of the fluid medium in order to determine the exact dissolved oxygen concentration.

Polarographic Electrodes. Polarographic electrodes usually contain a platinum or gold cathode, a silver/silver chloride anode, and a potassium chloride electrolyte. Figure 4.3a shows a schematic representation of a polarographic electrode. When the anode of the electrode is polarized by an external power supply, the following reactions take place at the surface of the electrode (Linek et al., 1985; Turner and White, 1999; van Dam-Mieras et al., 1992):

$$\text{cathode} : \quad O_2 + 2H_2O + e^- \rightarrow H_2O_2 + 2OH^-$$

$$H_2O_2 + 2e^- \rightarrow 2OH^-$$

$$\text{anode} : \quad Ag + Cl^- \rightarrow AgCl + e^-$$

$$\text{overall} : \quad 4Ag + O_2 + 2H_2O + 4Cl^- \rightarrow 4AgCl + 4OH^- \tag{4.16}$$

The potassium chloride electrolyte solution between the membrane and probe tip provides the chloride ions needed for the above reactions. Since chloride ions are consumed over time with this type of probe, it is necessary to periodically replace

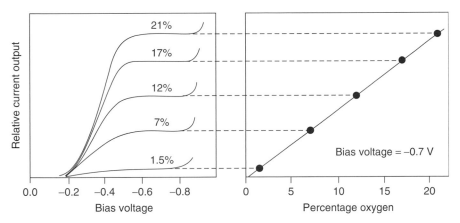

Figure 4.4 Typical polarographic electrode polarogram. Adapted from Lee and Tsao (1979).

the electrolyte solution. Owing to the reactions that take place at the electrode surface, a voltage-dependent current is created that can be related to the oxygen partial pressure as shown in the polarogram (current vs voltage diagram) in Figure 4.4. The rate at which the current-producing reaction takes place at the electrode surface in the plateau region shown in Figure 4.4 is limited by the diffusion rate of dissolved oxygen through the membrane and electrolyte as schematically represented in Figure 4.5 (Linek, 1988; Linek et al., 1985; van Dam-Mieras et al., 1992). Since these reactions are very quick, the diffusion rate is a function of the bulk fluid oxygen concentration. As shown in Figure 4.4, when the correct polarization voltage is selected for a particular electrode, the current output is linear with respect to dissolved oxygen concentration. Care must be taken to ensure that the voltage is not too high to prevent the formation of hydrogen peroxide due to water electrolysis as this will increase the current generation. On the other hand, if the voltage is too low, the current response will be nonlinear. Care must also be taken to ensure that the reaction at the electrode is sufficiently fast to prevent the built up of hydrogen peroxide that may promote hydrogen peroxide diffusion from the electrode tip. If hydrogen peroxide diffuses away, the electrode reaction stoichiometry will be altered. Likewise, it has been shown that the accumulation of OH$^-$ ions also retards the probe reaction rates (Linek et al., 1985). Thus, it can be concluded that a careful balance must be achieved to ensure proper electrode operation; however, on a positive note, this balance is relatively easy to achieve and maintain in practice.

Galvanic Probes. In contrast to the polarographic electrode, a galvanic probe utilizes an anode of zinc, lead, or cadmium and a cathode of silver or gold, where a silver cathode and lead anode are the most common (Linek et al., 1985). Figure 4.3b shows a schematic representation of a typical galvanic probe. The electrochemical reactions that take place at the probe surface are as follows (Linek et al., 1985; Turner and White, 1999; van Dam-Mieras et al., 1992):

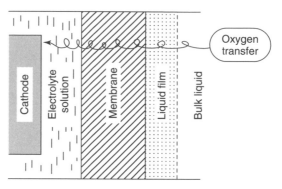

Figure 4.5   The typical oxygen transport path encountered at an electrode tip.

$$\text{cathode} : \quad O_2 + 2H_2O + 4e^- \rightarrow 4OH^-$$

$$\text{anode} : \quad Pb \rightarrow Pb^{2+} + 2e^-$$

$$\text{overall} : \quad 2Pb + O_2 + 2H_2O \rightarrow 2Pb(OH)_2 \qquad (4.17)$$

Similar to the polarographic probe, the galvanic probe is constrained by the rate-limiting step of oxygen diffusion across the probe membrane. Thus, the current output of the probe is linearly related to the dissolved oxygen concentration in the bulk fluid.

Electrochemical Electrode Time Constant. Despite their fundamental differences, both electrochemical electrodes presented earlier operate on the same basic principles, where the electrode behavior can be predicted using a simplified electrode model with the following assumptions (Lee and Tsao, 1979):

1. The cathode is well polished and the membrane is placed over the cathode surface to minimize the thickness of the electrolyte layer, allowing the electrolyte layer to be neglected in the mathematical model.
2. The liquid around the probe is well mixed so that the oxygen partial pressure at the membrane surface is the same as in the bulk fluid.
3. The electrochemical reaction at the surface of the electrode is much faster than oxygen diffusion through the membrane.
4. Oxygen diffusion occurs only in one direction, perpendicular to the probe.

These assumptions led to the development of the so-called one-layer model (Aiba et al., 1968; Lee and Tsao, 1979). A schematic representation of the one-layer model is shown in Figure 4.6a where oxygen diffusion to the electrode surface is only a function of the membrane layer. Under steady-state conditions for the

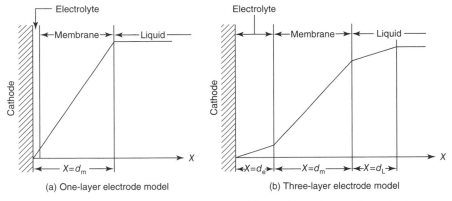

Figure 4.6   One- and three-layer electrode models used to estimate electrode time constants. Adapted from Linek (1988).

above-mentioned simplifications, Fick's first law describes oxygen diffusion from the bulk fluid to the membrane surface, showing that the electrode current output is linearly related to partial pressure of the bulk liquid oxygen. However, in application, the above-mentioned oversimplifications can rarely be used.

Although the one-layer model is an oversimplification of actual conditions, its application to the case where the oxygen partial pressure is allowed to change with time illustrates how electrode properties affect transient dissolved oxygen measurements. Fick's second law is needed to describe the unsteady-state diffusion in the membrane, and shows that the diffusion coefficient of the membrane directly determines how fast an electrode will respond to a step change in the oxygen partial pressure (Aiba et al., 1968; Lee and Tsao, 1979; Sobotka et al., 1982). Lee and Tsao (1979) showed mathematically that the electrode response time, for the one-layer model, depends on the electrode time constant defined as

$$\tau_e = \frac{\pi^2 D_m}{d_m^2} \qquad (4.18)$$

where D_m is the membrane diffusion coefficient and d_m is the membrane thickness. A large τ_e results in a fast probe response, which means that either the membrane is very thin or it has a high D_m. However, a small τ_e indicates that the membrane is impermeable to oxygen or that the membrane is too thick. Since electrode stability relies on membrane-controlled diffusion, a compromise between electrode response and stability is required.

As stated earlier, the one-layer model is an oversimplification of actual conditions typically observed, and hence, a three-layer model is typically employed. This model accounts for the effects of the electrolyte and the stagnant boundary layer as shown in Figure 4.6b (Aiba et al., 1968; Lee and Tsao, 1979; Sobotka et al., 1982). While the three-layer model is more suited to quantifying the electrode response to transient conditions, it only provides the foundation for determining the electrode response constant due to the many factors, as listed in the literature, that may affect it. Electrode design aspects such as membrane type, membrane thickness, cathode surface area, electrolyte, and electrode style all profoundly affect the behavior of the electrode response to oxygen partial pressure. Likewise, bulk fluid properties such as fluid type, viscosity, temperature, total pressure, oxygen partial pressure, fluid velocity, and solid loading can also affect electrode dynamics.

Electrochemical Electrode Response Time (τ_e). Owing to the complexity involved in estimating the probe time constant, most investigators opt to measure the electrode response time to a step change in the oxygen partial pressure. Typically, the electrode response time is defined as the time it takes the electrode to indicate 63% of the total change in dissolved oxygen concentration (Doran, 2013; Sobotka et al., 1982; Tribe et al., 1995; Vardar and Lilly, 1982). There are several experimental procedures described in the literature for obtaining τ_e when the probe is exposed to a stepwise concentration change (Linek et al., 1985) and these procedures are summarized below.

Procedure 1. The electrode is placed at the exit of a three-way valve and the interchange of fluids having different oxygen concentrations takes place when the valve is turned.

Procedure 2. The electrode is placed in a tube and the concentration change is produced by starting and stopping the flow of liquid saturated with air. While the flow of liquid is stopped, the concentration of oxygen in the liquid near the electrode decreases due to the chemical reaction at the electrode. The decrease in concentration will continue until nearly all the oxygen near the electrode is consumed. When this near-steady-state condition is reached, liquid flow is restarted causing a jump in the oxygen concentration near the electrode surface. This method is limited for use with electrodes that have a large cathode (i.e., ones that consume oxygen rapidly).

Procedure 3. The electrode is transferred between two vessels having liquids of different oxygen concentrations that are well mixed and thermostatically controlled (this may be the most popular).

Procedure 4. The electrode is transferred from air to a sulfite solution by inclining a vessel such that the probe, initially in air, is immersed in the sulfite solution.

Procedure 5. The electrode is rapidly transferred from a pure nitrogen environment to a vessel containing a liquid saturated with air. The liquid and hydrodynamic conditions in the test vessel should be the same as those in which the electrode will be used after calibration.

Procedure 6. The electrode is placed in a closed vessel containing a liquid saturated with oxygen and a stirrer. The stepwise concentration change is then facilitated by introducing a compound that immediately consumes all of the dissolved oxygen.

Regardless of the procedure used to find τ_e, care must be taken to ensure that the hydrodynamic conditions around the electrode during the response time test closely resemble those of the process in which the probe will be used, and that the step change is as rapid as possible.

To achieve reasonably accurate overall mass transfer values, a τ_e much smaller than $1/k_L a$ is recommended (Tribe et al., 1995; Van't Riet and Tramper, 1991) as problems occur when this is not the case. In practice, there are three gas–liquid mass transfer conditions of interest (Gaddis, 1999):

1. $\tau_e \ll 1/k_L a$. In this range, the response time of the electrode is much smaller than the dynamic oxygen concentration change in the reactor and the electrode is suitable for monitoring changes in oxygen concentration with a small error.
2. $\tau_e \approx 1/k_L a$. In this range, the response time is of the same order of magnitude as the reactor response time and considerable errors may be encountered when calculating the overall mass transfer coefficient. However, since this case is commonly encountered, models have been developed to account for this error.
3. $\tau_e \gg 1/k_L a$. In this range, the response time is much larger than that of the reactor and the use of electrodes to monitor changes in oxygen concentration is not recommended.

Electrochemical Electrode Response Models. Most oxygen-measuring electrodes used in biological processes have response times that range from 3 to 100 s (Gaddis, 1999; Van't Riet and Tramper, 1991), which may result in the need to correct oxygen concentration data depending on the reactor dynamics. Many models have been developed to correct for τ_e and are discussed in more detail in the literature (Chang et al., 1989; Chisti, 1989; Freitas and Teixeira, 2001; Gaddis, 1999; Kim and Chang, 1989; Lee and Luk, 1983; Lee and Tsao, 1979; Linek, 1988; Linek et al., 1981; Linek et al., 1991a; Linek et al., 1979, 1984, 1989; Linek et al., 1992; Linek et al., 1985; Lopez et al., 2006; Ruchti et al., 1981; Sobotka et al., 1982; Tobajas and Garcia-Calvo, 2000; Tribe et al., 1995; Van't Riet and Tramper, 1991; Vardar and Lilly, 1982). Lee and Luk (1983) and Sobotka et al. (1982) provide a good review of these model corrections. A selection of these models is presented below starting with the simplest and finishing with a few of the more popular complex models.

MODELS THAT NEGLECT THE ELECTRODE DYNAMIC RESPONSE. Van't Riet (1979) and Gaddis (1999) suggest that if $\tau_e < 3$ s, the overall mass transfer coefficient can be accurately measured without model correction. Hence, assuming ideal mixing and insignificant gas-phase concentration changes, the overall mass transfer coefficient may be calculated from

$$\frac{C^* - C_L}{C^* - C_0} = \exp(-k_L a \cdot t) \tag{4.19}$$

where C^* is the gas–liquid interface equilibrium molar concentration, C_L is the liquid-phase molar concentration, and C_0 is the steady-state molar concentration at

$t = 0$. Van't Riet (1979) cautioned that considerable corrections have to be made to coefficients calculated using this method if the gas residence time in the reactor is much greater than $1/k_L a$. These corrections were reported to greatly reduce the accuracy of Eq. (4.19). Linek et al. (1991b, 1987) also reported that the use of this model to relate experimental data to overall mass transfer coefficients would lead to an underestimation of $k_L a$ for air systems in which nitrogen transport is neglected, and that this model is really only sufficient for steady-state signals and marginally acceptable in the extreme case when oxygen concentration changes are much slower than τ_e.

MODELS CONSIDERING MEMBRANE DIFFUSION. The following model has been used when assuming that the electrode response is a first-order lag function, the liquid and gas phases are perfectly mixed, there is negligible nitrogen diffusion, and the interfacial area and oxygen concentration in the gas phase are constant (Blazej et al., 2004a; Chisti, 1989; Freitas and Teixeira, 2001; Fuchs et al., 1971):

$$\frac{C^* - C_L}{C^* - C_0} = \frac{(e^{-k_L a \cdot t} - k_L a \cdot \tau_e \cdot e^{-t/\tau_e})}{(1 - k_L a \cdot \tau_e)} \qquad (4.20)$$

In general, τ_e represents all the diffusional properties of the measurement system in the model (Sobotka et al., 1982). As with the previous model, the adequacy of this model depends on the ratio of τ_e and $1/k_L a$. When τ_e is much less than $1/k_L a$, Eq. (4.20) reduces to Eq. (4.19), and the resulting error associated with neglecting τ_e has been reported to be small (Gaddis, 1999; Merchuk et al., 1990; Nakanoh and Yoshida, 1983). This model is again subject to the same errors and limitations as the previous model, especially if nitrogen transport is neglected.

Jones and Heindel (2007) compared the $k_L a$ values calculated from dissolved oxygen measurements using the models described in Eqs (4.19) and (4.20). As shown in Figure 4.7a, when $\tau_e \ll 1/k_L a$, there is no real difference in how Eq. (4.19) and Eq. (4.20) represent the experimental data. When $\tau_e < 1/k_L a$ (Figure 4.7b), Eqs (4.19) and (4.20) begin to show a difference in how they model the experimental data. When $\tau_e \approx 1/k_L a$ (Figure 4.7c), Eq. (4.20) provides a better model of the experimental data. Table 4.1 summarizes the calculated $k_L a$ values using these two models for these three test conditions.

Linek et al. (1985) have suggested a different approach using a very sophisticated model in which the electrode time constant plays a major role. Rather than solving the model explicitly to determine $k_L a$, they suggested that since the electrode signal is most distorted during the initial response, one could find $k_L a$ by removing the distorted portion of the signal and using the remaining response and Eq. (4.19) to find $k_L a$. The proper application of this technique is discussed in detail in the literature (Linek, 1972; Linek et al., 1985).

If the system being studied can be assumed to have a perfectly mixed liquid phase and a constant oxygen concentration in the gas phase, Tobajas and

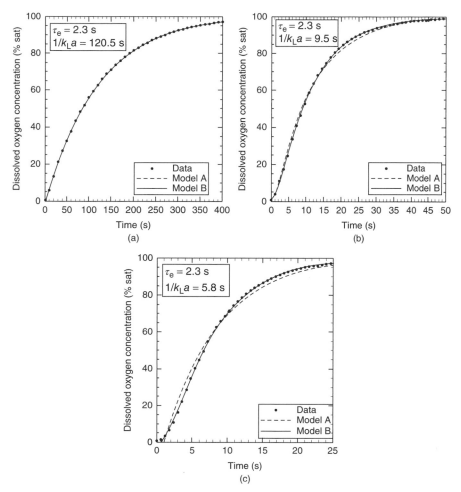

Figure 4.7 Comparison of the experimental data to Eq. (4.19) (Model A) and Eq. (4.20) (Model B) for (a) $\tau_e \ll 1/k_L a$, (b) $\tau_e < 1/k_L a$, and (c) $\tau_e \approx 1/k_L a$. Adapted from Jones and Heindel (2007).

Garcia-Calvo (2000) suggested that the following model be used to determine $k_L a$:

$$\frac{C^* - C_L}{C^* - C_0} = \frac{1}{1 - m}\left[1 - \exp\left(\frac{-m \cdot t}{\tau_e}\right) - m\left(1 - \exp\left(\frac{-t}{\tau_e}\right)\right)\right] \qquad (4.21)$$

where m is defined as

$$m = \frac{k_L a \cdot \tau_e}{1 - \epsilon} \qquad (4.22)$$

and ϵ is the total gas holdup.

TABLE 4.1   Equations (4.19) and (4.20) $k_L a$ Estimates for the Three Experimental Conditions where τ_e Ranges from $\tau_e \ll 1/k_L a$ to $\tau_e \approx 1/k_L a$

	Time Constant (τ_e) (s)	$1/k_L a$ (s)	$k_L a$ from Eq. (4.19) (s^{-1})	$k_L a$ from Eq. (4.20) (s^{-1})	Difference (%)
$\tau_e \ll 1/k_L a$ (Figure 4.7a)	2.3	120.5	0.0083	0.0083	0.1
$\tau_e < 1/k_L a$ (Figure 4.7b)	2.3	9.5	0.0919	0.1052	12.6
$\tau_e \approx 1/k_L a$ (Figure 4.7c)	2.3	5.8	0.1311	0.1711	23.4

Adapted from Jones and Heindel (2007).

MODELS CONSIDERING MEMBRANE DIFFUSION AND TIME DELAY. Lopez et al. (2006) and Vardar and Lilly (1982) suggested that when the electrode dynamic response was first order with a time delay, the following model can be used to correct the dissolved oxygen concentration data:

$$C_L(t - \tau_d) = C_E(t) + \tau_e \frac{dC_E(t)}{dt} \qquad (4.23)$$

where $C_E(t)$ is the recorded electrode concentration at time t and τ_d is the dead time. The dead time represents the time from the beginning of the concentration step change to the beginning of the change in the electrode signal. Once the concentration data (C_L) is corrected for the electrode dynamic response, Eq. (4.19) can be used to determine $k_L a$.

Sobotka et al. (1982), however, suggested that the following complex relationship be used to find $k_L a$ using the electrode data:

$$\frac{C^* - C_L}{C^* - C_0} = \frac{1}{(1 - k_L a \cdot \tau_e)} \left[\frac{1}{(1 - k_L a \cdot \tau_d)} e^{-k_L a \cdot t} - \frac{k_L a \cdot \tau_e^2}{(\tau_e - \tau_d)} e^{-t/\tau_e} + \right.$$
$$\left. \left((1 - k_L a \cdot \tau_e) - \frac{1}{(1 - k_L a \cdot \tau_d)} + \frac{k_L a \cdot \tau_e^2}{(\tau_e - \tau_d)} \right) e^{-t/\tau_d} \right] \qquad (4.24)$$

When $\tau_d \ll \tau_e$, Eq. (4.24) reduces to Eq. (4.20), and when $\tau_d \ll \tau_e \ll 1/k_L a$, Eq. (4.24) reduces to Eq. (4.19).

MODELS CONSIDERING MEMBRANE AND LIQUID FILM DIFFUSION. Models considering membrane and liquid film diffusion are quite complex as they are of second order in nature, and the solution to these models require numerical analysis or a method of moments due to their complexity (Sobotka et al., 1982). Linek et al. (1985), Ruchti et al. (1981), and Dang et al. (1977) suggested that while these models are more complex and involved, their solutions are much superior to any first-order model. However, due to their complexity, they are typically not used and the reader is referred to the literature for more information concerning these models.

MODELS CONSIDERING A MEMBRANE DIFFUSION MODEL. Sobotka et al. (1982) claimed that the empirical time delay models previously described do not properly consider the physical nature of the electrode response. Instead, they insisted that models based on Fick's second law are superior and encouraged their use to more accurately model system dynamics. Sobotka et al. (1982), in their review, presented many of the different diffusional models that have been developed and discussed their usefulness.

Summary of Electrochemical Electrode Response Models. As has been shown by a review of just a few of the models presented in the literature, the use of electro-chemical electrodes to accurately determine $k_L a$ can be complicated due to internal instrument dynamics as well as system dynamics. Hence, as implied by Tribe et al. (1995) and others (Keitel and Onken, 1981; Lee and Luk, 1983; Lee and Tsao, 1979; Linek et al., 1985; Sobotka et al., 1982), the proper selection of an elec-trode and method for evaluating its signal will greatly impact the accuracy of the experimental results.

4.2.2 Dissolved Carbon Monoxide Measurements

The determination of dissolved carbon monoxide concentrations and CO–liquid mass transfer rates is important to synthesis gas fermentation, where carbonaceous fuels, such as biomass, are gasified into flammable gas mixtures, sometimes known as synthesis gas or syngas, consisting of carbon monoxide (CO), carbon dioxide (CO_2), hydrogen (H_2), methane (CH_4), nitrogen (N_2), and smaller quantities of higher hydrocarbons and contaminants (Bridgwater, 1995). Syngas fermentation utilizes CO and H_2 as growth substrates to anaerobically produce a variety of fuels and chemicals, including methane, acetic acid, butyric acid, ethanol, and butanol (Bredwell et al., 1999). One potential bottleneck to the commercialization of syngas fermentation is the mass transfer limitations within gas–liquid bioreactors (Bredwell et al., 1999; Worden et al., 1997); this is a result of the low solubility of the major syngas components (CO and H_2) in the aqueous fermentation broth when it contains a high cell concentration. If the cell concentration is too low, the system yield will be low and mass transfer will be kinetically limited (Vega et al., 1989). To improve the CO–liquid mass transfer rates, actual dissolved CO concentrations and mass transfer rates must be determined, but there are no dissolved CO measurement probes like those used for dissolved O_2.

Dissolved CO concentrations can be determined using a myoglobin-protein assay as described by Kundu et al. (2003). This method was used by Riggs and Heindel (2006), Kapic et al. (2006), and Ungerman and Heindel (2007) to determine pure CO concentrations in water to assess CO–water mass transfer rates for various operating conditions. A summary of the bioassay measurement technique is provided next and further details can be found in Jones (2007).

4.2.2.1 Bioassay Overview. In the bioassay, a liquid sample is taken from the bioreactor and the dissolved carbon monoxide concentration is determined off-line using a protein-binding method. The use of the bioassay is limited, much like the chemical methods for determining dissolved oxygen concentrations, because the method is laborious, slow, and prone to error if done incorrectly.

The method involves several steps that include, first, preparing a myoglobin protein solution that is free of dissolved oxygen and carbon monoxide. Second, the myoglobin protein solution is added to the withdrawn liquid sample in a gas tight container; this causes the dissolved carbon monoxide to bind to the myoglobin. Third, the change in the absorbance spectrum (400–700 nm) of the sample is measured after the addition of the protein solution. The change in the absorbance spectrum is proportional to the number of dissolved carbon monoxide molecules in the original sample.

As with any analytical method, the success of the bioassay is highly dependent on how the samples are collected and prepared. Care must be taken during all of the steps of the analysis to ensure that oxygen is not allowed to bind with myoglobin and that all measurements are carried out very carefully. Furthermore, attention to sample acquisition must be taken to ensure that small gas bubbles are not entrained in the samples when they are collected as this leads to errors (Kapic, 2005). Although difficult to use, the bioassay technique, once mastered, may be successfully used to accurately measure dissolved carbon monoxide concentrations.

4.2.2.2   Needed Materials. Jones (2007) used various materials to complete the dissolved CO measurements using the bioassay technique. An Ocean Optics ChemUSB2-VIS-NIR spectrophotometer was used to measure the absorbance spectrum in the liquid sample. Figure 4.8 displays several syringes and a sample cuvette that were used in the measurements. Samples were prepared and scanned in 1.5-ml polystyrene disposable cuvettes that have a 10-mm path length. These cuvettes are usable for wavelengths ranging from 340 to 800 nm and have polystyrene caps to reduce contamination. Syringes used for liquid sample collection were gastight high performance 10-μl syringes from Hamilton (model 1701); the needles were cemented into this type of syringe by the manufacturer to minimize oxygen contamination. Several other syringes shown in Figure 4.8 are also used in the bioassay.

Myoglobin used in the dissolved carbon monoxide concentration measurements by Jones (2007) was purchased from Sigma-Aldrich (product number M1882) and derived from horse heart. The myoglobin comes as an essentially salt-free lyophilized powder of at least 90% pure that must be stored at minus 20 °C. One gram of the myoglobin powder is dissolved in approximately 25 ml of 0.1 M potassium phosphate at pH 7.0 buffer solution prepared by adding 3.3 g of dibasic potassium phosphate powder and 11.0 g of monobasic potassium phosphate powder to 1 l of deionized water. The final pH of the buffer solution is adjusted to 7.0 with either potassium hydroxide or o-phosphoric acid. To increase the shelf life of the myoglobin solution, the solution is run through a dialysis separation process for 24 h and then spun down in a centrifuge to remove impurities. The solution is then separated into 1-ml containers and frozen until needed.

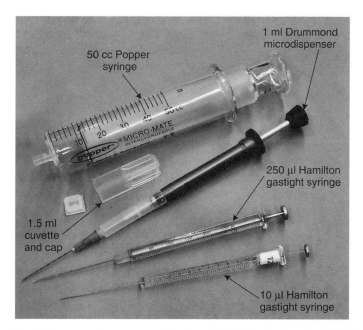

Figure 4.8   Syringes and cuvette used by Jones (2007) in the dissolved CO bioassay.

4.2.2.3 Liquid Sample Collection. Liquid samples were collected by Jones (2007) using 10-μl syringes; he determined that it is best if several syringes are first numbered sequentially and then inserted into the bioreactor through a septum (see, e.g., Figure 4.9). Care must be taken to ensure the syringe tips are all located in close proximity. Prior to introducing carbon monoxide into the bioreactor, a single sample is taken with syringe number 0 to measure the carbon monoxide concentration at time $t = 0$ (assumed to be zero and the sample confirms this). Once the carbon monoxide is introduced into the bioreactor, transient samples are withdrawn at specified time intervals depending on the operating conditions. Once the bioreactor is saturated with CO, three additional samples are taken to determine the steady-state concentration.

4.2.2.4 Identifying the Concentrated Myoglobin Solution Concentration. Prior to testing, the myoglobin solution must be thawed and the myoglobin concentration determined to calibrate the test solution. The goal in preparing the test solution is to obtain a peak absorption value near Abs = 1.5 for a saturated oxygen sample (Figure 4.10). The saturated oxygen peak occurs at 409 nm for myoglobin.

The concentration is determined by putting 1 ml of buffer solution into a cuvette and adding 1 μl of myoglobin protein. The absorbance is measured and more protein is added in 1 μl increments until the peak absorbance is near 1.5.

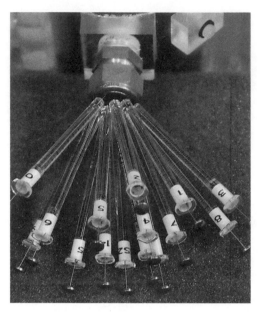

Figure 4.9   Sample syringes inserted into a bioreactor for liquid sample collection.

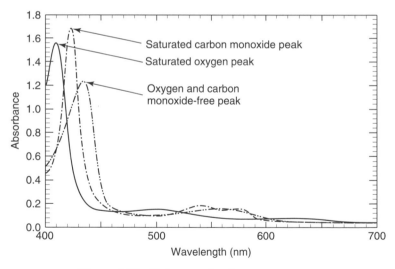

Figure 4.10   Reference absorbance spectrums.

Once the peak absorbance reaches Abs $\approx$ 1.5, the myoglobin concentration (C_p) and the dilution ratio (DR) are determined from

$$C_p = \frac{\text{Abs}}{\lambda \cdot \epsilon_m} \qquad (4.25)$$

$$DR = \frac{\text{microliters of myoglobin solution}}{\text{milliliters of buffer solution}} \qquad (4.26)$$

where Abs is the absorption value, λ is the path length of the cuvette, and ϵ_m is the extinction coefficient. For horse heart myoglobin, ϵ_m is reported to be 188 $\mu M/\text{cm}$ (Antonini and Brunori, 1971).

4.2.2.5 Sample Preparation for Analysis. The test solution used to analyze the liquid samples is prepared just prior to use because the myoglobin solution is temperature sensitive and its exposure to room temperature should be minimized. The test solution is prepared by pipetting 1 ml of buffer solution for every sample being analyzed into the 50-ml syringe. Then, using the previously calculated dilution ratio (Eq. (4.26)), an appropriate amount of myoglobin solution is added to the test solution, and the mixture is gently agitated. A 1 ml sample is set aside in a 1.5-ml cuvette and then placed in the spectrophotometer and scanned to record the "oxy" spectrum, corresponding to the test solution with dissolved oxygen. Finally, a small amount of sodium dithionite (Na_2SO_4) is added to the test solution to neutralize all the dissolved oxygen and the oxygen bonded to the myoglobin.

All measurements are initiated with 1 ml of the test solution being added to empty 1.5-ml cuvettes. A "deoxy" spectrum is determined by scanning a cuvette containing only the test solution; this spectrum corresponds to a sample containing no carbon monoxide. In a similar manner, a "saturated CO" spectrum is found by saturating the solution in one cuvette with an excess amount of carbon monoxide to ensure that all the myoglobin is bound to a carbon monoxide molecule. The resulting spectrum corresponds to the maximum amount of dissolved carbon monoxide that can be detected without increasing the myoglobin concentration in the test solution.

The liquid samples are analyzed after the three reference spectrums have been determined. After 1 ml of test solution has been placed in the cuvette, a $10\,\mu l$ liquid sample is injected into the cuvette and the cuvette is capped, gently agitated, and then scanned. This process is repeated for each of the acquired liquid samples. The resulting spectra will follow the trend shown in Figure 4.11, where an increase in the amount of dissolved carbon monoxide results in the peaks of the spectra initially shifting down and to the left and then up and to the left. Errors may occur in these measurements because of gas bubbles becoming entrained in the liquid sample when it is drawn; thus, care must be taken to ensure that the syringes are clean and properly located in the sample port.

4.2.2.6 Determining the Dissolved CO Concentration. Dissolved CO concentration in the liquid sample is determined by loading the spectra data files into a data analysis program (i.e., JMP 6.0) to fit the liquid sample spectra between the "deoxy" and "saturated" spectra; effectively interpolating between CO-free and CO-saturated samples. The software package uses a least squares fitting routine that outputs a percent similarity to each of the reference spectra. These output data

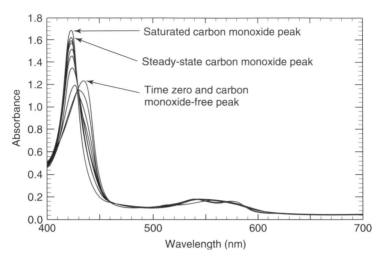

Figure 4.11   Absorbance spectra progression from carbon monoxide free state to a carbon monoxide saturated state.

are then used to determine the carbon monoxide concentration (C_{CO}) in the given liquid sample as a percent of the steady-state concentration using

$$C_{CO} = (C_p)(SS) \left(\frac{Vol_T}{Vol_S} \right) \tag{4.27}$$

where C_p is the myoglobin concentration in the test solution, SS is the percentage of the steady-state concentration exported from the data analysis program, Vol_T is the total liquid volume in the cuvette, and Vol_S is the sample volume of the dissolved carbon monoxide liquid. The CO concentration as a function of time is then used to determine the CO–liquid mass transfer rate using the models in Eq (4.19) or (4.20).

4.2.3   Determining Volumetric Gas–Liquid Mass Transfer Coefficient, $k_L a$

For two-phase gas–liquid systems, it has been shown that the gas transfer rate (GTR) can be described by

$$GTR = \frac{dC_L}{dt} = k_L a(C^* - C_L) = a \cdot J \tag{4.28}$$

where C^* and C_L are the equilibrium gas concentration at the gas–liquid interface and the dissolved gas concentration in the liquid phase, respectively. In bioreactors, dC_L/dt, C^*, and C_L can all be measured directly. However, as stated earlier, k_L and a are not so easily measured, so it is common to report the product of $k_L a$. This product is commonly called the *overall volumetric mass transfer coefficient* and has units of s^{-1}.

The most widely used methods for determining $k_L a$ in bioreactors have been summarized in the literature (Chisti, 1989; Gogate and Pandit, 1999b; Sobotka et al., 1982; Van't Riet and Tramper, 1991); they will be presented later. It is important to note that in using Eq. (4.28), it is assumed that the gas and liquid phases are well mixed so that $k_L a$ can be assumed constant over the entire gas–liquid system. These assumptions, however, are not always applicable to the system being evaluated, and further modeling of the gas–liquid system may be needed.

4.2.3.1 Gas Balance Method. The gas balance method can only be used in a gas-consuming system. Typically, this method is applied to a fermentation run where all the variables except $k_L a$ are measured. The gas concentration and the entering and exiting gas stream flow rates are monitored using a gas analyzer and mass flow meters. Using this information, the gas transfer rate (GTR) can be calculated from (Van't Riet and Tramper, 1991)

$$ \text{GTR} = \frac{F_i \cdot C_i - F_o \cdot C_o}{\text{Vol}_L} \tag{4.29} $$

where F is the respective gas flow rate, C is the respective gas concentration, and Vol_L is the liquid volume. Once the GTR is known, $k_L a$ can be calculated using Eq. (4.28).

The gas balance method is claimed to be the most reliable method for determining $k_L a$ (Doran, 2013; Poughon et al., 2003). However, this method requires the precise measurement of the gas inlet and outlet concentrations and flow rates. Since the difference between inlet and outlet conditions is typically very small, the accuracy of this method is determined in large part by the accuracy of the instrumentation (Poughon et al., 2003; van Dam-Mieras et al., 1992). Because of this, very precise instruments are required and the instrumentation cost for this method is often high. Hence, this method is usually justified only when expensive gas-monitoring equipment is also needed for process control.

This method is also limited by the underlying assumption that the gas phase is constant throughout the bioreactor. For large systems where the gas concentration may vary widely from inlet to outlet, gas-phase modeling is required to accurately estimate GTR and $k_L a$ (Van't Riet and Tramper, 1991).

4.2.3.2 Dynamic Method. The dynamic method involves measuring the dissolved gas concentration as a function of time for a step change in the inlet gas concentration. Similar to the gas balance method, this method can be applied to an actual fermentation or it can be applied to systems containing no microorganisms. Owing to its versatility and ease of use, this method is widely used and discussed in the literature (Blanch and Clark, 1997; Chisti, 1989; Doran, 2013; Sobotka et al., 1982; Van't Riet and Tramper, 1991; van Dam-Mieras et al., 1992). As a result, many variations of this method exist of which a selected few are discussed in more detail.

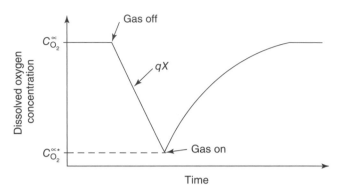

Figure 4.12   Typical dissolved oxygen concentration variation with time for the biological dynamic method. Adapted from Blanch and Clark (1997).

Biological Dynamic Method. The biological dynamic method is applied to actual fermentations using a step change in inlet gas concentration, where the change in dissolved gas concentration in the bulk fluid is recorded. The step change is initiated in one of several ways that will be discussed in more detail later. This method consists of three primary steps. First, the system is brought to an initial steady-state condition. Second, the inlet gas step change is initiated and the change in dissolved gas concentration is recorded. Typically, the dissolved gas concentration is reduced to a point just above the critical gas concentration needed to prevent cell death and/or an irreversible change in cell behavior (Blanch and Clark, 1997). Third, after a period of time, the inlet gas concentration is returned to its original state and the change in gas concentration is recorded as the system moves back to the original steady-state condition. Figure 4.12 illustrates the typical dissolved gas concentration profile obtained using the dynamic method.

The system mass balance for the dynamic method is as follows:

$$\frac{dC_L}{dt} = k_L a(C^* - C_L) - qX \tag{4.30}$$

where qX is the microbial gas consumption rate. If the gas phase disengages quickly from the liquid, then the transport term disappears in the above-mentioned relationship and it reduces to

$$\frac{dC_L}{dt} = -qX \tag{4.31}$$

Equation (4.31) can be used to find qX assuming that the microbial uptake of the gas is unaffected by stopping aeration. The volumetric mass transfer rate $k_L a$ is calculated using the overall system mass balance and does not require previous knowledge of qX, as qX can be replaced in Eq. (4.30) with the following expression (Doran, 2013):

$$qX = k_L a(C^* - C_\infty) \tag{4.32}$$

where C_∞ is the dissolved gas concentration in the liquid at steady state. Equation (4.30) then reduces to the following and can be solved directly for $k_L a$:

$$\frac{dC_L}{dt} = k_L a(C_\infty - C_L) \tag{4.33}$$

The instruments used in obtaining the liquid–gas concentration data for this method depend on the required fermentation gas. For processes that utilize oxygen, typically an oxygen electrode is used, although in rare situations another dissolved oxygen-measuring technique may be used. If an oxygen electrode is used, care must be taken to properly account for the probe dynamics as previously discussed (Chisti, 1989; Van't Riet and Tramper, 1991). For processes that utilize other gases such as hydrogen or carbon monoxide, specialized measuring instruments or techniques must be employed. For example, dissolved carbon monoxide concentration data can be obtained using the bioassay technique discussed in Section 4.2.2.

Blanch and Clark (1997) reported that the biological dynamic method is commonly used in both large- and small-scale equipment, primarily due to the fact that sterilizable oxygen probes permit the finding of $k_L a$ during fermentation without significantly upsetting the system.

Nonbiological Dynamic Method. This method is similar to the biological dynamic method in that it employs the use of an inlet gas concentration step change, though it differs from the previous method as the system either has microorganisms that have been terminated, had cell respiration blocked, or does not have any microorganisms present (van Dam-Mieras et al., 1992). This method is commonly used for systems that are void of microorganisms (Abashar et al., 1998; Sobotka et al., 1982; Van't Riet, 1979).

The nonbiological dynamic method begins by first removing the dissolved gas being studied from the vessel by (i) aerating the system with an inert gas such as nitrogen, (ii) using a vacuum to cause the dissolved gas to come out of solution, or (iii) adding a chemical compound to consume the dissolved gas. Once the dissolved gas has been removed, the system is then aerated and the change in gas concentration is recorded until steady state is reached.

Without cell respiration, the overall mass balance for the biological dynamic method simplifies from Eq. (4.30) to Eq. (4.28). The volumetric mass transfer coefficient $k_L a$ is then evaluated by integrating Eq. (4.28) and plotting $\ln[(C^* - C_L)/(C^* - C_o)]$ as a function of time, where $k_L a$ is the slope of the resulting line, or by curve fitting the data with a nonlinear regression software package.

This method is reported to offer accurate results if the system being studied does not vary significantly from the actual system containing respiring microorganisms (van Dam-Mieras et al., 1992). However, the accuracy of the results obtained using this method was reported to depend on the procedure used to initiate the concentration step change and electrode dynamics (if a dissolved oxygen electrode is used) (Linek et al., 1993; Linek et al., 1989; Linek and Sinkule, 1990; Linek et al., 1987; Van't Riet and Tramper, 1991).

Variations of the Inlet Step Change.   While variations for each of the dynamic methods have been reported in the literature, the variation of greatest importance seems to be in how to initiate the change in the dissolved gas concentration. The remainder of this section will review the most popular techniques used to initiate a step change in the inlet gas concentration.

GAS OFF/ON OR START-UP.   The gas off/on technique is used primarily for fermentations that have actively respiring cells. In such fermentation systems, the dynamic method is applied by turning the gas flow off and allowing the cells to deplete the dissolved gas until the critical gas concentration is reached and then the gas is turned back on (e.g., Figure 4.12).

One of the main advantages of using this technique is that the gas–liquid mass transfer is not affected by alternating the gas species, which has been reported to affect the calculation of $k_L a$ values (Linek et al., 1981). Another advantage is the low cost associated with this technique as it requires no additional equipment. However, this method has a couple of limitations that must also be realized. First, this method must be done quite rapidly and with extreme care to ensure that cell respiration is not affected by the change in dissolved gas concentration. Second, when the gas is turned off and then on again, the system hydrodynamics may be altered. Consideration must also be given to the time needed to once again reach steady-state hydrodynamic behavior because if the start-up time approaches or exceeds the length of the experiment, then the method cannot be used for calculating $k_L a$ (Gogate and Pandit, 1999b). For example, in the extreme case when vessels are very large or have a height greater than 1 m, the time to reestablish steady-state gas holdup conditions may be larger than the characteristic $k_L a$, resulting in inaccurate $k_L a$ estimates that are not representative of normal operation (Van't Riet and Tramper, 1991).

GASSING OUT OR GASSING IN.   Since dynamic methods are usually quite sensitive to the starting conditions of the experiment, a gas switching technique is used to eliminate hydrodynamic changes. The gassing-out technique is one of the most widely used techniques for the dynamic method when a simulated fermentation broth is used. This technique, as the name implies, begins by aerating with one gas and then switching at $t = 0$ to a second gas. For example, in an air–water system, the system may first be aerated with air until the water is completely saturated, and then aerated with nitrogen to replace the oxygen in solution (Figure 4.13). A wide variety of gas pairs have been used in the application of this technique in the literature, though air–nitrogen is the most common.

Van't Riet and Tramper (1991) reported that when deoxygenation with nitrogen was followed by an aeration switch, the average gas-phase residence time (τ_g) must

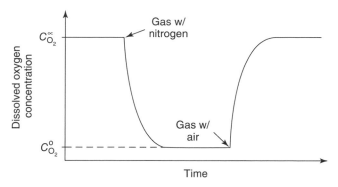

Figure 4.13 Typical dissolved oxygen concentration variation with time for the nonbiological dynamic method. Adapted from Blanch and Clark (1997).

be considered as the gas-phase concentration was no longer constant over the entire test. Van't Riet and Tramper defined τ_g as

$$\tau_g = \frac{H}{U_g} \left(\frac{\epsilon}{1 - \epsilon} \right) \tag{4.34}$$

where H is the unaerated liquid height in the vessel, U_g is the superficial gas velocity, and ϵ is the gas holdup. If τ_g is the same order of magnitude as $1/k_L a$, then the assumption of a constant gas-phase concentration used to derive Eq. (4.30) is no longer valid (Sobotka et al., 1982). Models to correct for this behavior have been proposed by Dunn and Einsele (1975) and Dang et al. (1977). These models have been reported to be useful only over a narrow range of conditions (Van't Riet and Tramper, 1991). Linek et al. (1981) reported that interphase nitrogen transport may significantly influence $k_L a$ estimations. Linek et al. (1993) indicated that errors in $k_L a$ estimation due to nitrogen transport can be as high as 25% for large $k_L a$ values. However, they also indicated that for low $k_L a$ values commonly encountered, the error due to nitrogen transport may be negligible. Stenberg and Andersson (1988a) found that the change from nitrogen to air had a small but significant effect on $k_L a$ measurements, but this error was smaller than other observed experimental errors.

Lopez et al. (2006) and Chang et al. (1989) suggested that a gas pair of air and oxygen-enriched air be used to improve this technique by eliminating the need for pure nitrogen. Lopez et al. (2006) showed that $k_L a$ values measured with this technique closely matched those obtained for the gas off/on technique. Kim and Chang (1989) indicated that the difference in inlet oxygen concentrations for this technique must be at least 20% in order to minimize errors.

PRESSURE STEP. Another widely used form of the dynamic method is the pressure step technique where the gas concentration is changed by suddenly increasing or decreasing the system pressure. The system pressure is typically changed by a small amount, for example, from 15 to 20 kPa, by the addition of gas into the reactor head space. The sudden pressure change is believed to instantaneously change the gas concentration in the gas phase throughout the vessel and be independent of system hydrodynamics.

The use of the pressure step technique was found by Blazej et al. (2004a) and Linek et al. (1991b) to be more accurate (by up to 60% for systems with non-coalescing liquids) in determining $k_L a$ values than the gassing-out technique. This increase in accuracy was attributed to model shortcomings related to the washing out of one gas by another for systems with non-coalescing liquids. The $k_L a$ values for the gassing-out method and pressure step method were found to be similar under some operating conditions. Linek et al. (1989, 1994) also reported that there was no difference in $k_L a$ values when gassing with air or pure oxygen, indicating that nitrogen transport was not a factor. They also reported that experimental results for this technique only match those from the gas off/on method for small values of $k_L a$.

NONIDEAL PRESSURE STEP. The nonideal pressure step technique is slightly different from the pressure step technique. The difference in the two is that in the pressure step technique, the pressure change is considered instantaneous or ideal. In the nonideal pressure step technique, however, the pressure step is actually achieved by throttling the exit gas stream to cause a pressure buildup, where the time lag for the pressure step depends on the gas flow rate and the vessel size. Linek et al. (1993) compared this technique with the pressure step technique and reported that the results from the two techniques agreed very well.

CONCENTRATION STEP. The concentration step technique is a rarely used technique that deoxygenates the liquid phase by the addition of a small amount of a chemical compound like sulfite without interrupting aeration. This technique should not be confused with the chemical sorption methods as only a small amount of the compound is added with the intent of causing a dissolved gas concentration step change. For this method to work properly, the system must be very well mixed to ensure uniform dissolved gas concentrations. Also, care must be taken to ensure that the chemical compound being added does not alter the hydrodynamics or enhance mass transfer rates.

Dynamic Method Drawbacks. The dynamic methods are affected by several factors:

1. These methods assume that both the gas and liquid phases are well mixed. However, if either one of these phases is anything other than well mixed, which is often the case, especially for large or tall vessels, the $k_L a$ measurement accuracy decreases (Blanch and Clark, 1997).
2. Since air is commonly used for experimental purposes, the effect of simultaneous oxygen and nitrogen transport may affect the accuracy of experimentally determined $k_L a$ values (Gogate and Pandit, 1999b; Letzel et al., 1999; Linek et al., 1981; Stenberg and Andersson, 1988a).

3. Changing from one steady state to another, where the gas-phase residence times are significant, will cause the $k_L a$ estimate to be inaccurate. This is especially true when the time to move from one steady-state condition to another is of the same order of magnitude as $1/k_L a$ (Gogate and Pandit, 1999b; Van't Riet, 1979).

4. The rapid change in dissolved oxygen concentrations with time may lead to oxygen electrode outputs that are not directly related to the instantaneous oxygen concentration unless the output is conditioned to adjust for electrode dynamics (Van't Riet, 1979). Tribe et al. (1995) showed that neglecting electrode probe response time, while using any of the dynamic measurement methods, would cause errors in $k_L a$ estimates, regardless of how much smaller τ_e is compared to $1/k_L a$. They emphasized that proper accounting for the electrode dynamics is needed for reliable measurements.

A comparison of methods done by Poughon et al. (2003) concluded, without explanation, that the use of the dynamic method always results in an underprediction of $k_L a$ when compared to other methods, such as the gas balance and chemical sorption methods, which are now described.

Chemical Sorption Methods. Chemical sorption methods to determine $k_L a$ are based on a chemical reaction between the absorbed gas and a chemical that is added to the liquid phase. Four of these methods will be presented here, although many others exist. The sulfite oxidation, hydrazine, and peroxide methods are applicable to systems studying oxygen transport, while the carbon dioxide absorption method, as its name implies, is for measuring dissolved carbon dioxide.

SULFITE OXIDATION METHOD. The sulfite oxidation method is based on the oxidation of sulfite to sulfate in the presence of a catalyst, where dissolved oxygen is consumed by the reaction

$$Na_2SO_3 + \frac{1}{2}O_2 \xrightarrow{\text{catalyst}} Na_2SO_4 \qquad (4.35)$$

Thus, to make this method work, the bulk fluid has to have a high concentration of sulfite and catalyst prior to aeration. Once aeration begins, any oxygen that dissolves into the fluid phase is immediately consumed by the sulfite reaction and the rate of sulfite oxidation is proportional to $k_L a$. Since the bulk fluid oxygen concentration remains at zero, $k_L a$ is estimated by

$$-\frac{dC_{\text{sulfite}}}{dt} \approx k_L a \cdot C^* \qquad (4.36)$$

The sulfite concentration in the bulk fluid is followed by taking liquid samples over a given time interval. The samples are then quenched with excess iodine and back titrated with thiosulfate to determine the residual iodine concentration and, subsequently, the sample sulfite concentration (Sobotka et al., 1982). Knowing the sulfite concentration change with time, Eq. (4.36) can be used to determine $k_L a$.

Chisti (1989) and Blanch and Clark (1997) reported that this technique had severe limitations. First, there is a need for expensive high purity chemicals. Second, the chemical reaction produces a highly ionic fluid that is non-coalescing, which may alter the system hydrodynamics. Third, sample analysis is often slow and tedious (Sobotka et al., 1982). Fourth, the sulfite oxidation rate is very sensitive to fluid properties and impurities; thus, the reaction rate depends on the type of catalyst used, its concentration, trace metals, temperature, and fluid pH. Hence, $k_L a$ determination requires that the reaction conditions be carefully controlled, the sulfite concentration kept sufficiently high, and excess catalyst must be present in the bulk fluid to ensure that oxidation occurs in the bulk fluid and not at the gas–liquid interface (Chisti, 1989). Gogate and Pandit (1999b) indicated that this method is not suitable for use in systems using pure oxygen because bubble size in the system changes dramatically by the high chemical reaction rate. Van't Riet (1979) also reported that the reaction rate constant can vary in unknown ways and that this method should be avoided.

THE HYDRAZINE METHOD. The steady-state hydrazine (N_2H_4) method makes use of the following reaction (Chisti, 1989):

$$N_2H_4 + O_2 \rightarrow N_2 + 2H_2O \tag{4.37}$$

This method uses a steady flow of hydrazine into an aerated reactor. The dissolved oxygen concentration is then followed by an oxygen electrode. The intent of this method is to introduce hydrazine into the system at a rate equal to $k_L a$ which, when accomplished, keeps the electrode signal at a constant level (i.e., the rate at which hydrazine is consumed is equal to $k_L a$).

The reaction in Eq. (4.37) does not form ionic species; therefore, the system hydrodynamics are not affected during the course of the test, unlike the sulfite oxidation method (Chisti, 1989).

PEROXIDE METHOD. The peroxide method is based on the following chemical reaction where oxygen is produced in the reactor liquid:

$$2H_2O_2 \xrightarrow{\text{catalase}} 2H_2O + O_2 \tag{4.38}$$

The oxygen is transferred to a carrier gas that is used to transport the oxygen out of the system. Under steady-state conditions, the oxygen production is equal to the oxygen transfer rate. To calculate the oxygen transfer rate, only the peroxide inlet flow rate and concentration, liquid volume, carrier gas flow rate, and dissolved oxygen concentrations are needed at steady-state conditions. This method uses catalase enzymes that are known to enhance foam formation and alter the gas bubble diameter, which is a severe limitation when considering the use of the method (Gogate and Pandit, 1999b).

CARBON DIOXIDE ABSORPTION METHOD. Another commonly employed chemical technique is the absorption of carbon dioxide into a mild alkaline or an appropriately buffered solution (Andre et al., 1981). The carbon dioxide method is similar in principle and procedure to the sulfite oxidation method. Chisti (Chisti, 1989) indicated that the limitations for this method were similar to those of the sulfite oxidation method.

4.3 SUMMARY

Many different experimental techniques are available to characterize and quantify bioreactor hydrodynamics and gas–liquid mass transfer rates. Some of these techniques, including advantages and disadvantages, were outlined in this chapter. Current experimental methods are continually being modified and refined, and new techniques will always be developed. Hence, the development of experimental methods relevant to bioreactor operation is an evolutionary process and an area rich in application.

5 Modeling Bioreactors

Lab-scale and pilot-scale experimental bioreactor studies can be expensive and challenging to complete, but are needed before industrial-scale processes are implemented. To reduce the experimental costs, models of the various systems can be developed and simulations can be completed and validated with selected experimental results. In bioreactors, the hydrodynamics play a critical role, and multiphase computational fluid dynamics (CFD) models are needed to simulate the commonly encountered gas–liquid, liquid–solid, and gas–liquid–solid bioreactor mixtures. The overall bioreactor production can also be modeled through appropriate mass and energy balances combined with biological kinetic processes. This chapter will provide a general outline of multiphase flow CFD modeling, and then provide a brief overview of basic biological process modeling.

5.1 MULTIPHASE FLOW CFD MODELING

CFD can be used to simulate the hydrodynamics found in multiphase bioreactors. Although bioreactors typically involve heat and mass transfer operations, they are generally neglected in the hydrodynamic models (Joshi, 2001; Kulkarni et al., 2007). Monahan et al. (2005) have provided a summary of the various computational approaches and physical models used in gas–liquid hydrodynamic modeling. They point out that various aspects of gas–liquid hydrodynamic modeling have been considered in the literature, but the importance of various terms and their exact model form is still under debate; these include terms that address (i) bubble–bubble interactions; (ii) two-phase turbulence modeling; (iii) gas–liquid interfacial mass, momentum, and energy transfer mechanisms; and (iv) coupling between the phases. The required grid resolution and its effect on convergence have also been addressed. Comprehensive CFD overviews are available in the literature (Azzopardi et al., 2011; Delnoij et al., 1997a; Jakobsen et al., 2005a; Joshi, 2001; Kulkarni et al., 2007), and the interested reader is referred to these and other studies for additional details.

Multiphase flow CFD simulations typically employ Eulerian–Eulerian models (Monahan et al., 2005; Pan et al., 2000; Rampure et al., 2003; Sokolichin and

An Introduction to Bioreactor Hydrodynamics and Gas-Liquid Mass Transfer, First Edition.
Enes Kadic and Theodore J. Heindel.
© 2014 John Wiley & Sons, Inc. Published 2014 by John Wiley & Sons, Inc.

Eigenberger, 1994), Eulerian–Lagrangian models (Delnoij et al., 1997a, 1997b; Delnoij et al., 1997c) or various direct numerical simulation (DNS) methods (Dijkhuizen et al., 2010; van Sint Annaland et al., 2006). The Eulerian–Eulerian model treats dispersed and continuous phases as interpenetrating continua and describes the motion for both phases in an Eulerian frame of reference. In gas–liquid bioreactors, for example, the bubbles act as the dispersed phase and the liquid is the continuous phase. In the Eulerian–Lagrangian model (Delnoij et al., 1997a, 1997b; Delnoij et al., 1997c), the continuous phase is described in an Eulerian representation while the dispersed phase (e.g., bubbles) is treated as discrete particles, and each discrete particle is tracked by solving the equations of motion for individual particles. The DNS methods, including level set, volume of fluid, lattice Boltzmann, and front-tracking methods (Dijkhuizen et al., 2010; van Sint Annaland et al., 2006), solve the instantaneous Navier–Stokes equations to obtain the dispersed and continuous phase flow field with an extremely high spatial resolution, and the interface between phases is tracked.

The main advantage of the Eulerian–Lagrangian formulation comes from the fact that each individual bubble is modeled, allowing consideration of additional effects related to bubble–bubble and bubble–liquid interactions. Mass transfer with and without chemical reaction, bubble coalescence, and redispersion, in principle, can be added directly to an Eulerian–Lagrangian hydrodynamic model. The main disadvantage of the Eulerian–Lagrangian approach is that only a limited number of particles (bubbles) can be tracked, such as when the superficial gas velocity is low (Chen et al., 2005), due to computer limitations.

The Eulerian–Eulerian method is more popular because memory storage requirements and computer power demand depend on the number of computational cells considered instead of the number of particles. Hence, the Eulerian–Eulerian approach can be applied to cases for low and high superficial gas velocities. The disadvantage of using the Eulerian–Eulerian method is that the bubble–bubble and bubble–liquid interactions cannot be considered as straightforward as the Eulerian–Lagrangian method, and models for these interactions are typically applied.

The DNS methods are the most detailed; they typically advance the gas–liquid interface through the flow field in an Eulerian mesh and do not require empirical constitutive equations. However, the DNS methods are limited to a very small number of particles/bubbles (e.g., typically 10s of bubbles in the flow field) due to computational limitations. Most industrial applications require high superficial gas velocities, and therefore the Eulerian–Eulerian method is preferred (Dudukovic, 2002; Pan et al., 2000) and will be summarized later.

Note that many bioreactor CFD simulations assume that the biological content has a negligible effect on the bioreactor hydrodynamics or, if there is an effect, only the fluid rheology is modified. Hence, many of the available bioreactor CFD studies focus on gas–liquid modeling (Roy et al., 2006). A few gas–liquid–solid CFD studies have been completed (Hamidipour et al., 2012; Jia et al., 2007; Rampure et al., 2003), and the interested reader is referred to these studies for further information.

5.1.1 Governing Equations for Gas–Liquid Flows

The two-fluid Eulerian–Eulerian model represents each phase as interpenetrating continua, and the conservation equations for mass and momentum for each phase are ensemble-averaged. Bubble coalescence and/or breakup are typically neglected in the models used in the literature. The subscript c refers to the continuous (liquid) phase and the subscript d refers to the dispersed (gas bubble) phase. The continuity equations for each phase are

$$\frac{\partial}{\partial t}(\alpha_c \rho_c) + \nabla \cdot (\alpha_c \rho_c \vec{u}_c) = R_c$$

(5.1)

$$\frac{\partial}{\partial t}(\alpha_d \rho_d) + \nabla \cdot (\alpha_d \rho_d \vec{u}_d) = R_d$$

(5.2)

where the volumetric phase fractions are denoted by α_c and α_d, respectively, and sum to 1, and ρ and $\vec{u}$ are the respective phase density and velocity. Note that α is used to specify the volumetric phase fractions in this chapter to avoid confusion with the dissipation of turbulent kinetic energy ϵ. The right-hand side of Eqs (5.1) and (5.2), R_c and R_d, are zero when mass transfer is neglected, which is common in many simulations. The momentum equations for each phase are

$$\frac{\partial}{\partial t}(\alpha_c \rho_c \vec{u}_c) + \nabla \cdot (\alpha_c \rho_c \vec{u}_c \vec{u}_c)$$

$$= -\alpha_c \nabla p + \nabla \cdot \overline{\overline{\tau}}_c + \vec{K}_{dc}(\vec{u}_d - \vec{u}_c) + \vec{F}_{vm} + \rho_c \alpha_c \vec{g}$$

(5.3)

$$\frac{\partial}{\partial t}(\alpha_d \rho_d \vec{u}_d) + \nabla \cdot (\alpha_d \rho_d \vec{u}_d \vec{u}_d)$$

$$= -\alpha_d \nabla p + \nabla \cdot \overline{\overline{\tau}}_d + \vec{K}_{cd}(\vec{u}_c - \vec{u}_d) - \vec{F}_{vm} + \rho_d \alpha_d \vec{g}$$

(5.4)

The terms on the right-hand side of Eqs (5.3) and (5.4) represent, from left to right, the pressure gradient, effective stress, interfacial momentum exchange terms (drag and virtual mass forces), and the gravitational force. Additional momentum exchange forces may be included, but there has been no consensus as to which forces are the most appropriate, which may also depend on the given system of interest (Joshi, 2001; Monahan et al., 2005). The closures for turbulence modeling and interfacial momentum exchange are discussed next.

5.1.2 Turbulence Modeling

Turbulence contributions for the continuous and dispersed phases have been based on a modified form of the standard multiphase k–ϵ equations, first presented by Kashiwa et al. (1993) and described in detail by Padial et al. (2000), to calculate turbulence at the gas–liquid interface in the form of a slip-production energy term.

The modified $k-\epsilon$ equations can be used for high superficial gas velocity flows (Law et al., 2008), and the equations for a general phase i are

$$\frac{\partial}{\partial t}(\alpha_i\rho_ik_i) + \nabla \cdot (\alpha_i\rho_ik_i\overrightarrow{u}_i) = \nabla \cdot \left(\alpha_i\frac{\mu_{t,i}}{\sigma_i}\nabla k_i\right) + \alpha_iG_i - \alpha_i\rho_i\epsilon_i$$

$$+ \sum_{j\neq i}\beta_{ij}K_{ij}|\overrightarrow{u}_i - \overrightarrow{u}_j|^2 + 2\sum_{j\neq i}E_{ij}(k_j - k_i) \quad (5.5)$$

$$\frac{\partial}{\partial t}(\alpha_i\rho_i\epsilon_i) + \nabla \cdot (\alpha_i\rho_i\epsilon_i\overrightarrow{u}_i) = \nabla \cdot \left(\alpha_i\frac{\mu_{t,i}}{\sigma_\epsilon}\nabla\epsilon_i\right) + \alpha_i\frac{\epsilon_i}{k_i}(C_{1\epsilon}G_i - C_{2\epsilon}\rho_i\epsilon_i)$$

$$+ \frac{1}{\tau_{ij}}\left\{\sum_{j\neq i}\beta_{ij}K_{ij}|\overrightarrow{u}_i - \overrightarrow{u}_j|^2\right\} \quad (5.6)$$

where

$$\mu_{t,i} = \rho_iC_{\mu,i}\frac{k_i^2}{\epsilon_i} \quad (5.7)$$

$$G_i = \mu_{t,i}(\nabla\overrightarrow{u}_i + (\nabla\overrightarrow{u}_i)^T) : \nabla\overrightarrow{u}_i \quad (5.8)$$

$$C_{\mu,i} = \frac{C_\mu}{1 + (2\sum_{j\neq i}E_{ij}k_i)/(\rho_i\epsilon_i)} \quad (5.9)$$

Note that if i is the continuous phase, then j is the dispersed phase, and vice versa.

The form of Eq. (5.9) models a return-to-isotropy effect due to fluctuating interfacial momentum coupling and reduces the turbulent viscosity from that predicted by the single-phase model. The turbulence energy exchange rate coefficient E_{ij} is given by

$$E_{ij} = \alpha_i\alpha_j\left(\frac{\rho_i\rho_j}{\rho_i + \rho_j}\right)\frac{\sqrt{k_i + k_j}}{d_B}(1 + Re_B^{0.6}) \quad (5.10)$$

where $Re_B = \rho_c|\overrightarrow{u}_d - \overrightarrow{u}_c|d_B/\mu_c$ is the bubble Reynolds number based on a characteristic (effective) bubble diameter, relative velocity between the two phases, and the liquid density and dynamic viscosity.

The first three terms on the right-hand side of Eq. (5.5) account for turbulent diffusion, mean flow shear production, and decay of turbulence kinetic energy of phase i. The fourth term on the right-hand side of Eq. (5.5) accounts for production of turbulence energy from slip between phases. The coefficient β_{ij} is given by

$$\beta_{ij} = \frac{\alpha_i}{\alpha_i + \alpha_j} \quad (5.11)$$

where

$$\alpha_i = \frac{\alpha_i^{1/3}}{\rho_i + \rho_c} \tag{5.12}$$

and ρ_c is the continuous phase density. The last term in Eq. (5.5) accounts for the exchange of turbulence energy among phases.

The first three terms on the right-hand side of Eq. (5.6) account for the diffusion of turbulence dissipation, the mean flow velocity gradient production term, and the homogeneous dissipation term. The last group of terms in Eq. (5.6) describes the effect of interfacial momentum transfer on the production of turbulence dissipation. The time constant τ_{ij} is given by the following empirical correlation:

$$\tau_{ij} = \left\{ 0.01 C_{2\epsilon} \left(\alpha_i \alpha_j \right)^{0.086} \left[\frac{\rho_i \left| \vec{u}_i - \vec{u}_j \right| d_B}{\mu_i} \right]^{0.562} \frac{\left| \vec{u}_i - \vec{u}_j \right|}{d_B} \right\}^{-1} \tag{5.13}$$

This correlation was obtained by fitting predictions of turbulence kinetic energy to data from experiments on homogeneous settling and bubbly systems (Lance and Bataille, 1991; Mizukami et al., 1992; Parthasarathy and Faeth, 1990a, 1990b). The term K_{ij} is the interfacial momentum exchange coefficient discussed next. Equations (5.7) and (5.8) are closure models for the turbulent viscosity $\mu_{t,i}$ and for the production of turbulent kinetic energy G_i of phase i. The turbulent parameters are set using standard empirical values for $k-\epsilon$ turbulence modeling where $C_{1\epsilon} = 1.44$, $C_{2\epsilon} = 1.92$, $C_\mu = 0.09$, $\sigma_k = 1.0$, and $\sigma_\epsilon = 1.3$.

5.1.3   Interfacial Momentum Exchange

The interfacial momentum exchange terms in the momentum conservation equations for each phase consist of drag and virtual mass force terms. The drag force for gas and liquid is modeled, respectively, as

$$\vec{K}_{cd}(\vec{u}_c - \vec{u}_d) = \frac{3}{4} \rho_c \alpha_d \alpha_c \frac{C_D}{d_B} |\vec{u}_c - \vec{u}_d|(\vec{u}_c - \vec{u}_d)$$

$$\vec{K}_{dc}(\vec{u}_d - \vec{u}_c) = \frac{3}{4} \rho_c \alpha_d \alpha_c \frac{C_D}{d_B} |\vec{u}_d - \vec{u}_c|(\vec{u}_d - \vec{u}_c) \tag{5.14}$$

where C_D is the drag coefficient. There are many different drag coefficient models available in the literature and these have been summarized by Joshi (2001) and Monahan et al. (2005). Two particular models commonly found in commercial CFD packages were compared by Law et al. (2008). One was the model proposed by Schiller and Naumann (1933):

$$C_D = \begin{cases} 24 \left(1 + 0.15 Re_B^{0.687} \right) / Re_B & Re_B \leq 1000 \\ 0.44 & Re_B > 1000 \end{cases} \tag{5.15}$$

The second drag coefficient model was proposed by White (1974):

$$C_D = C_{D,\infty} + \frac{24}{Re_B} + \frac{6}{1 + \sqrt{Re_B}} \qquad 0 \le Re_B \le 2 \times 10^5 \qquad (5.16)$$

where $C_{D,\infty}$ is the drag coefficient when the bubble Reynolds number goes to infinity, which is set at 0.5 (Law et al., 2008).

The virtual mass force $\overrightarrow{F}_{vm}$ is modeled as

$$\overrightarrow{F}_{vm} = 0.5\alpha_d\rho_c \left(\frac{d\overrightarrow{u}_c}{dt} - \frac{d\overrightarrow{u}_d}{dt} \right) \qquad (5.17)$$

and the coefficient of 0.5 is used for a spherical bubble.

5.1.4   Bubble Pressure Model

The bubble pressure (BP) model is reported in the literature to play an important role in bubble-phase stability and represents the transport of momentum arising from bubble–velocity fluctuations, collisions, and hydrodynamic interactions. According to Spelt and Sangani (1998), the kinetic contribution comes from fluctuations in the bubble motion; the collisional contribution is attributed to bubble–bubble collisions; and the hydrodynamic contribution arises from the relative motion of the bubbles and the spatial and velocity distribution of the bubbles. As the dispersed phase void fraction (α_d) increases from zero, the bubble-phase pressure will increase from zero, reach a maximum value, and then decrease. For low dispersed phase void fractions, the gradient in the BP with respect to the void fraction ($dP_d/d\alpha_d$) is positive and proportional to the slip velocity and gas holdup, and the collisional and hydrodynamic contributions can be neglected. In this case, Spelt and Sangani (1998) suggest the BP to be of the form

$$P_B = \rho_c C_{BP}\alpha_d(\overrightarrow{u}_d - \overrightarrow{u}_c) \cdot (\overrightarrow{u}_d - \overrightarrow{u}_c) \qquad (5.18)$$

The virtual mass coefficient C_{BP} of an isolated spherical bubble is 0.5. The BP gradient is then added to the right-hand side of the gas momentum (Eq. (5.4)) and acts as a driving force for bubbles to move from areas of higher α_d to areas of lower α_d and facilitates stabilization of the bubbly flow regime. However, Sankaranarayanan and Sundaresan (2002) indicate that as α_d increases, the collisional and hydrodynamic contributions become important.

Biesheuvel and Gorissen (1990) proposed a modified BP model of the form

$$P_B = \rho_c C_{BP}\alpha_d(\overrightarrow{u}_d - \overrightarrow{u}_c) \cdot (\overrightarrow{u}_d - \overrightarrow{u}_c) \left(\frac{\alpha_d}{\alpha_{dcp}} \right) \left(1 - \frac{\alpha_d}{\alpha_{dcp}} \right) \qquad (5.19)$$

The gas holdup at close packing α_{dcp} is typically set equal to 1.0 (Law et al., 2008; Monahan et al., 2005).

5.1.5   Bubble-Induced Turbulence

Bubbles contain potential energy when they are injected into a bioreactor. As they rise, some of the potential energy of the gas is converted into kinetic energy. The remaining energy is passed to the liquid through the gas–liquid interface, where some energy is dissipated. The energy that reaches the liquid phase is eventually dissipated in the small scales found in the wakes of the bubbles (Monahan et al., 2005). Several models have been proposed to account for this bubble-induced turbulence (BIT) (Sokolichin et al., 2004). Sato and Sekoguchi (1975) proposed a BIT model proportional to the bubble diameter and slip velocity of the rising bubbles:

$$\mu_{t,c} = \rho_c C_{BT} \alpha_d d_b |\vec{u}_d - \vec{u}_c| \tag{5.20}$$

where the value of the proportionality constant C_{BT} is 0.6 (Sato et al., 1981). The BIT model yields an effective viscosity in the liquid (continuous) phase, that is, the sum of the molecular viscosity of the continuous phase and the turbulent viscosity calculated from the BIT model, whereas the effective viscosity for the dispersed phase is assumed to equal the molecular viscosity of the dispersed phase. Equation (5.20) replaces Eq. (5.7) when the BIT model is applied (Law et al., 2008). Law et al. (2008) concluded that a BP model coupled with a BIT model is needed to produce stable solutions in gas–liquid bubble column reactors when the superficial gas velocity is low.

5.1.6   Modeling Bubble Size Distribution

Bubble size is required to calculate, for example, the drag force imparted on a bubble. Most Eulerian–Eulerian CFD codes assume a single (average) bubble size, which is justified if one is modeling systems in which the bubble number density is small (e.g., bubbly flow in bubble columns). In this case, the bubble–bubble interactions are weak and bubble size tends to be narrowly distributed. However, most industrially relevant flows have a very large bubble number density where bubble–bubble interactions are significant and result in a wide bubble size distribution that may be substantially different from the average bubble size assumption. In these cases, a bubble population balance equation (BPBE) model may be implemented to describe the bubble size distribution (Chen et al., 2005).

A general population balance equation for bubbles located at position vector $\vec{x}$ with a bubble volume V_b, at time t, can be written as (Chen et al., 2005)

$$\frac{\partial}{\partial t} f(\vec{x}, V_b, t) + \nabla \cdot [\vec{u}_b(\vec{x}, V_b, t) f(\vec{x}, V_b, t)] = S(\vec{x}, V_b, t) \tag{5.21}$$

where $f(\vec{x}, V_b, t)$ is the bubble number density function that is assumed to be continuous and specifies the probable number density of bubbles at a given time in the spatial range $d\vec{x}$ about $\vec{x}$ with a bubble volume between V_b and $V_b + dV_b$. The bubbles in $f(\vec{x}, V_b, t)$ travel at a velocity $\vec{u}_b(\vec{x}, V_b, t)$. The term on the right-hand side of Eq. (5.21) is the source term described by

$$
S(\vec{x}, V_b, t) = \frac{1}{2} \int_0^{V_b} a(V_b - V_b', V_b') f(\vec{x}, V_b - V_b', t) f(\vec{x}, V_b', t) dV_b'
$$

$$
- f(\vec{x}, V_b, t) \int_0^{\infty} a(V_b, V_b') f(\vec{x}, V_b', t) dV_b'
$$

$$
+ \int_0^{\infty} m(V_b') b(V_b') P(V_b, V_b') f(\vec{x}, V_b', t) dV_b'
$$

$$
- b(V_b') f(\vec{x}, V_b, t) + S_{ph} + S_p + S_r + \cdots \qquad (5.22)
$$

where the first term on the right-hand side is the birth rate of bubbles of volume V_b due to coalescence of bubbles of volume $V_b - V_b'$ and V_b', the second term is the death rate of bubbles of volume V_b due to coalescence with other bubbles, the third term is the birth rate of bubbles of volume V_b due to breakup of bubbles with a volume larger than V_b, and the fourth term is the death rate of bubbles of volume V_b due to breakup. The additional source/sink terms are because of the bubbles being added or subtracted from the bubble class of volume V_b due to phase change (S_{ph}), pressure change (S_p), or reaction (S_r). In addition, $a(V_b, V_b')$ is the coalescence frequency between bubbles of volume V_b and V_b', $b(V_b')$ is the breakup frequency of bubbles of volume V_b', $m(V_b')$ is the mean number of daughter bubbles produced by breakup of a parent bubble of volume V_b', and $P(V_b, V_b')$ is the probability density function of daughter bubbles produced on breakup of a parent bubble of volume V_b' (Chen et al., 2005). The various source terms in Eq. (5.22) require closure models for bubble breakup and coalescence; these models are beyond the scope of this review, but can be found in the literature (e.g., Chen et al., 2005; Jakobsen et al., 2005a; Sanyal et al., 2005).

5.2 BIOLOGICAL PROCESS MODELING

Biological process (bioprocess) models mathematically focus on describing the biological system's growth and product formation. Even simple biological processes are extremely complex from cellular component operations to overall bioreactor interactions. These complex systems, however, are often described by only a few mathematical equations, and rather simple growth kinetics, because the large cell population in the bioreactor hides individual variations in their growth and product formation; this leads to a smoothed average bioreactor behavior (Bellgardt, 2000a).

As described by Nielsen et al. (2003), biological process modeling uses a set of mathematical relationships developed through physical laws and/or empirical observations that relate the input variables of the system to the output variables. The input parameters include things such as flow rate(s), pH, temperature, agitation speed, and substrate concentration, while the output may include cell/product concentration, temperature, and flow rate. Kinetic expressions, which describe the rates of input/output concentrations, coupled to system mass balances are utilized in the overall bioprocess model. The model is specified for a given control volume, defined as a region of interest where all variables of interest are uniform. In biological process modeling, the control volume is typically taken as the entire bioreactor. For extremely large systems, the bioreactor may be divided into several control volumes, each one being homogeneous but different from the adjacent control volume(s). In contrast, when using CFD to model the bioreactor hydrodynamics, the control volume is an individual grid cell with millions of grid cells comprising the entire bioreactor.

Nielsen et al. (2003) describes four general groups of biological process models. The simplest are unstructured, nonsegregated models where the biomass is described by a single variable (e.g., cell concentration) and the cell population is assumed to be homogeneous. These models can be extended to unstructured, segregated models where individual cells within a population are modeled (e.g., cell age). Structured, nonsegregated and structured, and segregated models mathematically incorporate the cellular structure and its effect on the transport characteristics in the overall process model.

5.2.1 Simple Bioprocess Models

Simple bioprocess models are typically based on batch fermentation processes where the change in cell concentration with respect to time is proportional to the current cell concentration. This first-order rate equation is mathematically described by

$$\frac{dC_c}{dt} = kC_c \tag{5.23}$$

where C_c is the cell concentration at a given instant in time and k is a proportionality constant. This description assumes an infinite substrate concentration to support cell production.

A batch process, however, typically has a finite substrate concentration C_S set at time $t = 0$. As shown in Figure 5.1, plotting experimentally observed substrate and cell concentrations as a function of time, where concentration is on a natural logarithmic scale, the entire life cycle of the batch process can be described. Cell production can be divided into four regions (Dunn et al., 2003). First, cells are inoculated into the batch reactor at a known concentration C_{c0} and a period a cell adjustment is observed during the lag period ($0 < t < t_1$). Second, cells undergo a period of exponential growth until the substrate concentration can no longer support growth because a particular substance is limited ($t_1 < t < t_2$). Third, a stationary period of constant cell concentration may be observed while the remaining substrate material is consumed ($t_2 < t < t_3$). If sufficient time is allowed, cell death may be reached because of lack of nutrients, toxicity effects on the cells, and/or cell aging ($t_3 < t$).

Solving Eq. (5.23) during the exponential growth period reveals

$$\frac{C_c}{C_{c0}} = e^{kt} = e^{\mu t} \tag{5.24}$$

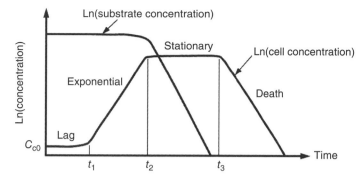

Figure 5.1 Cell life cycle in a batch fermentation process. Adapted from Dunn et al. (2003).

where the slope of linear part of Figure 5.1 ($t_1 < t < t_2$) is the growth rate per unit cell mass defined as the specific growth rate μ.

The exponential growth region can be described by a single relationship where the specific growth rate μ is a function of the substrate concentration C_S. Empirical evidence reveals that μ is a maximum when C_S is large, and μ is linear with C_S when C_S is small. The specific function that describes the entire relationship including the two limiting extremes is called the Monod equation

$$\mu = \frac{\mu_m C_S}{K_S + C_S} \tag{5.25}$$

where μ_m is the maximum specific growth rate and the saturation constant K_S is defined as the substrate concentration C_S when $\mu = 1/2\mu_m$. The Monod equation is shown in Figure 5.2 and is based on empirical observations. Although the Monod

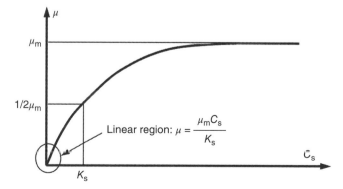

Figure 5.2 The Monod relationship describing specific growth rate as a function of substrate concentration C_S. Note that this is valid only in the exponential growth region ($t_1 < t < t_2$ in Fig. 5.1).

equation may appear to be rather simplistic, it generally describes the exponential growth rate for a wide variety of real systems (Dunn et al., 2003).

The Monod equation is not the only unstructured, nonsegregated empirical model of bioreactor cell growth. Others can be found in the summaries provided by Bellgardt (2000a), Dunn et al. (2003), or Nielsen et al. (2003).

5.3 SUMMARY

Modeling bioreactor performance is typically accomplished by focusing on either the bioreactor hydrodynamics using CFD modeling, which neglects the presence of microorganisms, or on biological process modeling assuming uniform flow condition within the bioreactor (or large sections of the bioreactor). Coupling the multiphase hydrodynamic models to the bioprocess models is needed for improved performance predictions.

6 Stirred-Tank Bioreactors

6.1 INTRODUCTION

Stirred-tank reactors (STRs) are one of the standard reactors in the chemical industry and have therefore been widely implemented for biological applications (Williams, 2002). We assume that the terms "stirred-tank reactor" and "stirred-tank bioreactor" can be used interchangeably and no distinction will be made between the two terms in this chapter; hence, the term *STR* will also be used to identify a stirred-tank bioreactor. They are used with viscous liquids, slurries, very low gas flow rates, and large liquid volumes (Charpentier, 1981; Fujasova et al., 2007; Garcia-Ochoa and Gomez, 1998). STRs are also popular because a well-mixed state is easily achieved, which aids in providing necessary substrate contact, pH and temperature control, removal of toxic by-products, uniform cell distribution, clog prevention, and particle size reduction (Branyik et al., 2005; Hoffmann et al., 2008). This chapter provides an overview of STR operation and summarizes issues related to STR gas–liquid mass transfer. Additional information can be found in the literature (Harnby et al., 1992; Linek et al., 2004; Linek et al., 1996a, 1996b; Linek et al., 1987; McFarlane and Nienow, 1995, 1996a, 1996b; McFarlane et al., 1995; Oldshue, 1983; Tatterson, 1991, 1994; Ulbrecht and Patterson, 1985).

STRs are widely applied in industry because of their low capital and operating costs (Williams, 2002). Popular applications are fermentation (Cabaret et al., 2008; Fujasova et al., 2007; Garcia-Ochoa and Gomez, 1998; Hoffmann et al., 2008; Scargiali et al., 2007; Vasconcelos et al., 2000), carbonation, oxidation (Oldshue, 1983; Scargiali et al., 2007), chlorination (Fujasova et al., 2007; Oldshue, 1983; Scargiali et al., 2007), hydrogenation (Fujasova et al., 2007; Murthy et al., 2007; Scargiali et al., 2007; Shewale and Pandit, 2006), dissolution, polymerization (Shewale and Pandit, 2006), chemical synthesis, and wastewater treatment (Cabaret et al., 2008; Ogut and Hatch, 1988; Shewale and Pandit, 2006). STRs are preferred when high gas–liquid mass transfer coefficients are needed (Bredwell et al., 1999). These reactors are usually made out of stainless steel for industrial units or out of clear materials, such as glass or certain plastics, for experimental applications (Williams, 2002).

An Introduction to Bioreactor Hydrodynamics and Gas-Liquid Mass Transfer, First Edition.
Enes Kadic and Theodore J. Heindel.
© 2014 John Wiley & Sons, Inc. Published 2014 by John Wiley & Sons, Inc.

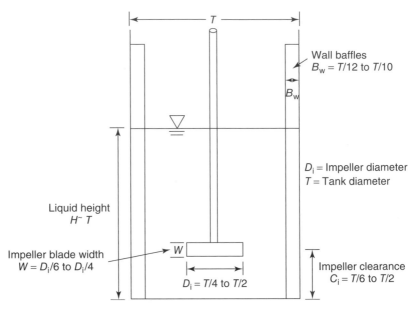

Figure 6.1   Standard single impeller stirred-tank reactor design. Adapted from Tatterson (1991).

Typical STR units (Figure 6.1) have a small height-to-diameter ratio relative to other reactor types (Charpentier, 1981). The diameter T can vary from about 0.1 m for experimental units to 10 m for industrial applications (Harnby et al., 1992). As shown in Figure 6.1, the impeller and baffle dimensions, as well as the impeller clearance, are typically a specified fraction of the tank diameter. The aspect ratio, defined as the liquid height-to-diameter ratio, is highly variable and depends on the number and arrangement of impellers and the reactor application. Single impeller systems typically have an aspect ratio of 1 (Charpentier, 1981; Tatterson, 1991), but certain exotic applications call for designs with aspect ratios up to 3 (Nielsen and Volladsen, 1993; Tatterson, 1991). Industrial multiple impeller designs are mostly limited to an aspect ratio of less than ~4 due to practical considerations (Charpentier, 1981).

Reactor shape, specifically the bottom, can vary greatly. The standard reactor design is cylindrical with a flat bottom (Ulbrecht and Patterson, 1985), but dished, conical, or curved bottoms have also been used (Harnby et al., 1992; Tatterson, 1991). The bottom shape does not seem to affect gas–liquid mass transfer or gas dispersion significantly, but the dished bottom is preferred for solid suspensions and mixing (Oldshue, 1983). Other reactor shapes, such as spherical or semispherical, are in use (Oldshue, 1983) but the standard design is preferred for gas–liquid dispersion due to operational experience and cost. Even though standard reactor designs exist in the chemical industry for liquid–liquid processes, customized STR's use for specific biological or gas–liquid applications precludes an optimized STR design for all applications (Tatterson, 1991).

6.2   STIRRED-TANK REACTOR FLOW REGIMES

Superficial gas velocity, defined as the volumetric gas flow rate divided by the STR cross-sectional area, influences gas–liquid mass transfer through two mechanisms: gas-filled cavities and gas holdup. The sweeping action of the impeller creates a low pressure void that quickly fills with sparged gas. These gas-filled cavities are the mechanism for gas dispersion and gassed power reduction (McFarlane et al., 1995). These cavities ultimately influence impeller loading, gas dispersion, and liquid circulation such that the impeller creates specific flow regimes which are of great importance for STR optimization.

6.2.1   Radial Flow Impellers

Radial flow impellers expel the fluid from the impeller region in the radial direction. The Rushton-type impeller is a standard example of radial flow impeller operation, and this impeller type is the basis for the typical flow regimes identified in radial flow impeller operation. As shown in Figure 6.2, three stable cavity groups are observed using this impeller: vortex cavities, clinging cavities, and large cavities. Vortex cavities form at constant impeller speeds and small gas flow rates. They are defined by two rolling vortices, one at the top and the other at the bottom of the impeller blade. Clinging cavities are formed with an increase in gas flow rate. They are larger than vortex cavities and cling to the blade backside, but still produce vortices at the gas tail. Large cavities, which form with another increase in the gas flow rate, are larger, smoother, and behave differently in terms of hydrodynamics (Nienow et al., 1985; Smith and Warmoeskerken, 1985).

Turbulent action forces the gas to break away from the cavity and exit the impeller zone. This breakage is the source of gas dispersion in STRs. The large cavity deserves special attention because it induces gas breaking away less violently than the other cavity types. Large cavities also have an advantage in that they hold more gas and have more surface area from which gas can break away.

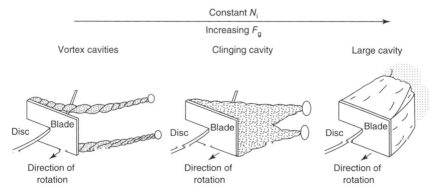

Figure 6.2   Cavity types for the Rushton-type impellers where N is the impeller speed and $Q_g(F_g)$ is the volumetric gas flow rate (Doran, 2013).

Since these cavities form at higher gas flow rates and thus superficial gas velocities, they are able to sustain a higher gas dispersion rate than the other two cavity types; however, there is a breakeven point. Cavities can become too large and hamper gas–liquid mass transfer. For example, if the cavity volume-to-surface area ratio is too large, gas dispersion will decrease (Nienow et al., 1985; Smith and Warmoeskerken, 1985).

Cavities also reduce the energy transfer between the impeller and liquid. A higher superficial gas velocity induces more gas dispersion at a lower power input. However, if too much gas is present and the cavities are too large, the energy transmission is reduced and the impeller's gas dispersion and mixing capabilities are hampered. In this case, the cavities produce an unwanted energy loss and a state described as flooding (Ogut and Hatch, 1988).

The exact nature of these events is accompanied by changes in the cavity structures for the Rushton-type turbine (Figure 6.3). Vortex and clinging cavities form symmetric structures (i.e., their size and shape are similar for every impeller blade). Although these transfer energy well, they do very little for gas dispersion making them undesirable for gas–liquid processes. If the gas flow rate is increased, a 3-3 structure is formed, which is defined by alternating large and clinging cavities. Its importance comes from its stability and gas-handling capacity. It offers the optimal gas dispersion for the lowest power input (i.e., most efficient). If the gas flow rate is increased further, the impeller is flooded. At this point, the stable 3-3 structure is replaced by a structure formed by large, unstable ragged cavities, which are inefficient for gas–liquid mass transfer and gas dispersion (Nienow et al., 1985; Smith and Warmoeskerken, 1985). The instability can also lead to varying impeller power draw that can damage the motor and gearbox system.

The observed cavity structures have inspired improvements to the Rushton-type turbine to increase the gas-handling capacity of the impeller to allow more gas and, hence, higher gas concentrations in the reactor. To accomplish this, impellers should create smaller cavities in similar structures while minimizing flooding. The concave blade disk turbine, discussed in Section 6.3, has been shown to accomplish these goals.

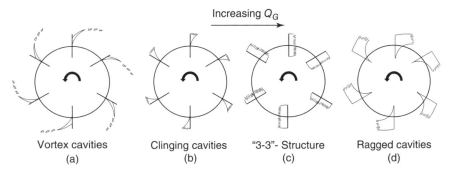

Figure 6.3 Cavity structures for the Rushton-type impeller. Adapted from Nienow et al. (1985).

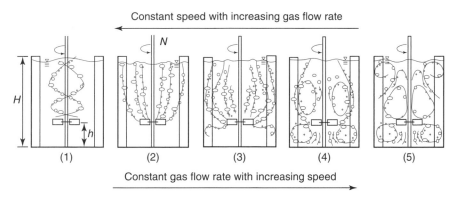

Figure 6.4 Bulk flow patterns for radial flow impellers at constant U_g. Adapted from Nienow et al. (1977).

The gas cavities on the impeller influence the gas dispersion and bulk flow regimes in an STR. For single impeller radial flow systems, five bulk flow regimes (Figure 6.4) have been defined (Nienow et al., 1977). The different regimes occur at increased impeller speeds while holding the gas flow rate constant. At low impeller speeds, the power input is very small and negligible dispersion occurs (regime (1) in Figure 6.4). Increasing the impeller speed begins to disperse the gas phase (regime (2)). The bulk flow above the impeller acts like a bubble column while the lower section is not contacted by the gas phase. Further increasing the impeller speed allows the gas to be recirculated in the upper reactor section, and some gas dispersion occurs in the lower region (regime (3)). Regime (4) is identified by gas recirculation throughout the reactor. This condition is optimal for gas–liquid mass transfer and mixing processes. At the highest impeller speeds (regime (5)), significant circulation loops and gross recirculation are observed and high turbulence at the surface promotes gas entrainment (surface aeration) (Nienow et al., 1977). The progression of these bulk flow regimes are also shown in Figure 6.5 (Nienow et al., 1985).

Nishikawa et al. (1984) determined that when negligible gas dispersion occurs, impeller type, location, or separation are irrelevant. In other words, if the total power input to the system is dominated by the sparged gas, the mixing in the STR approximates a bubble column. This effect was, however, recognized to occur at small power inputs, and the impeller power would start to dominate hydrodynamics and mass transfer at 30 W/m^3 for Rushton-type impellers (Gagnon et al., 1998). This power level is almost never observed in application since most gas–liquid dispersion with STRs occurs in the range of 500–4000 W/m^3 (Bliem and Katinger, 1988b; Bredwell and Worden, 1998; Oldshue, 1983).

The flow regime transitions can be determined from a gassed power number (or gassed-to-ungassed power ratio) versus flow number graph where the flow number is defined as

$$Fl_G = \frac{Q_G}{ND_i^3} \tag{6.1}$$

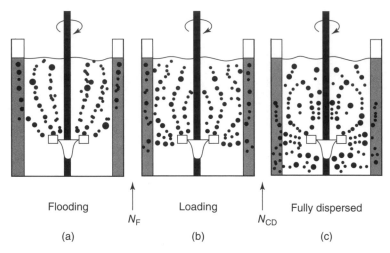

Flooding   Loading   Fully dispersed

N_F N_{CD}

(a)   (b)   (c)

Figure 6.5   Loading regimes and transitions for radial flow impellers, where N_{FL} indicates a transition from flooded to loaded regimes and N_{CD} defines the transition to completely dispersed flow regime (Jade et al., 2006).

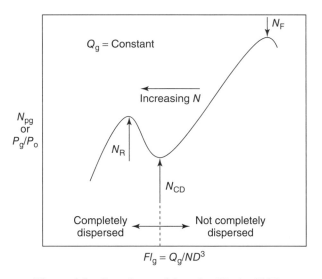

Figure 6.6   Generic transition plot (Kapic, 2005).

where Q_G, N, and D_i are the volumetric flow rate, impeller speed, and impeller diameter, respectively. Generic and experimental examples are shown in Figures 6.6 and 6.7, respectively. Flooding occurs at a local maximum of this graph represented by a flooding transition impeller speed N_F. It has also been defined as the transition point from regime (3) to regime (2) in Figure 6.4 (by decreasing N). The local minimum for the graph holds a special place. It represents the minimum power

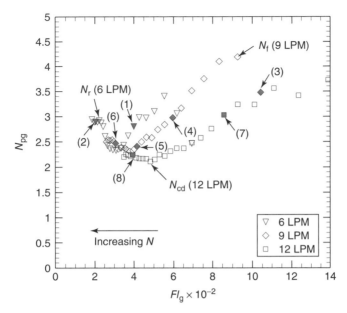

Figure 6.7   Experimental transition plot. The numbered data points represent specific conditions tested by Ford et al. (2008).

input required to achieve bulk flow regime (4), also known as complete dispersion or recirculation. The reactor should be operated such that $N > N_{CD}$; this represents the most economical operation in terms of power usage and gas utilization. Mixing time, which is defined as the time required to mix incoming fluid homogeneously into the existing liquid volume, is also optimized, but it is still higher than in an unaerated system (Hadjiev et al., 2006). If the impeller speed is increased beyond the complete dispersion impeller speed, the power use increases (Nienow et al., 1985; Smith and Warmoeskerken, 1985).

A second local maximum that follows represents the transition to the gross recirculation flow regime. This impeller speed, identified by N_R, should be avoided because it leads to inefficient gas utilization and gas entrainment. If an oxygen probe were to be positioned near the entrained gas, oxygen depletion would be a matter of time and a reduction in $k_L a$ would be recorded. In other words, the impeller should be operated between N_{CD} and N_R to achieve maximum efficiency. The transition between these flow regimes and their representative impeller speeds has been shown to be a function of impeller type, D_i/T, and scale. Given a D_i/T and an impeller type, the transition impeller speeds are scale independent, but the rate at which the transitions occur is not independent. Larger vessels have a gradual transition while smaller vessels experience transition over a limited flow number range (Nienow et al., 1985; Smith and Warmoeskerken, 1985). As shown in Figure 6.7, increasing the gas flow rate leads to smoother and delayed transitions.

6.2.2   Axial Flow Impellers

Axial flow impellers direct the fluid flow within the STR along the axis of the rotating shaft. Propeller-type impellers offer a simple example. In down-pumping axial flow impellers, the fluid is pumped downward while the gas is introduced below the impeller. The cavity propagation of axial flow impellers (Figure 6.8) is similar to the Rushton-type turbine, but gas dispersion differs in that pulsation forces gas to leave the impeller zone via the cavity tail (McFarlane et al., 1995). As shown in Figure 6.8, low impeller speeds and gas flow rates form vortex cavities at the impeller blade tip. Increasing the gas flow rate leads to the creation of a clinging cavity. The creation of larger cavities on axial flow impellers requires an increase in the gas flow rate and the impeller speed. A minimum impeller speed is needed to support larger cavities. If this minimum is not met, axial flow impellers experience vortex shedding such that the vortex detaches from the blade, leading to the next vortex creation cycle. This shedding can cause variations in torque and power draw, but usually do not cause any problems (McFarlane et al., 1995).

Incipient large cavities are formed in axial flow impellers if the gas flow rate is increased and the minimum impeller speed requirement is met. This cavity type occupies more space than the clinging cavity but disperses gas in a similar pulsating fashion. At a higher gas flow rate, large, fully developed cavities form, which occupy almost the entire blade area, but they do not extend beyond the impeller blade edge. When large cavities are present, the liquid and gas are discharged predominately in a radial direction from the impeller zone. Prior to flooding, the large cavity loses its defined boundary and blends into the flow. Once the impeller is flooded, the cavity structure is completely lost and gas simply passes through the impeller zone without any breakage or dispersion (McFarlane et al., 1995).

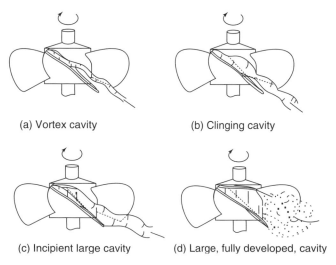

(a) Vortex cavity   (b) Clinging cavity

(c) Incipient large cavity   (d) Large, fully developed, cavity

Figure 6.8   Cavity types for axial flow impellers (McFarlane et al., 1995).

Unlike the Rushton-type turbine, cavities on axial flow impellers are rarely stable and frequently change shape, size, and identity, which can induce different gas loading regimes as well as torque and power draw variations. If gas enters the impeller zone via the impeller-induced liquid flow, the impeller is said to be loaded indirectly. In contrast, direct loading is defined by sparged gas-controlled flow into the impeller zone. In other words, the impeller-generated flow is not able to deflect the sparged gas. The transition leads to changes in the cavity size, shape, and identity, which are significant for large cavities (McFarlane et al., 1995).

Down-pumping axial flow impellers exhibit different bulk flow regimes at low and high gas flow rates due to the effects of large cavities; this phenomenon differentiates axial flow impellers from radial flow impellers; however, the major regimes (flooding, loading, and complete dispersion) have similar characteristics. McFarlane et al. (1995) used an A-315 (axial flow impeller) for visualization purposes and noted that other down-pumping axial flow impellers exhibit similar behavior. Figure 6.9 depicts A-315 behavior as described by McFarlane et al. (1995) for low gas flow rates (relative to scale and power draw). The figures progress from (a) to (d) in terms of increasing impeller speeds. In (a), the impeller is flooded at a low impeller speed. Gas rises easily through the impeller zone and behaves similar to flooded radial flow impellers. Increasing impeller speed induces minor recirculation

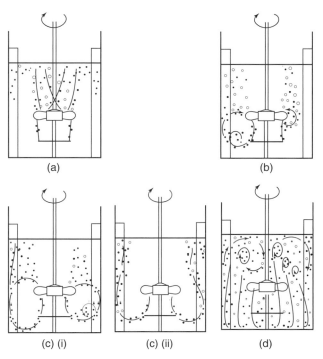

Figure 6.9 Bulk flow regimes generated by down-pumping A-315 at low gas flow rates (McFarlane et al., 1995).

loops in the impeller vicinity, and the impeller is said to be directly loaded. The transition between (b) and (c) signals the direct–indirect transition such that the impeller is partially directly and indirectly loaded. Increasing the impeller speed further leads to indirect loading (c). Cavity oscillations cause the bulk flow to vary between asymmetric (i) and occasional symmetric (ii) flow patterns. Variations in cavity type and loading are not damaging to the motor and gearbox because the flow remains mostly axial throughout the process such that the torque and power draw variations are limited. Further increase in impeller speed leads to complete dispersion as shown in (d) (McFarlane et al., 1995).

McFarlane et al. (1995) describe different flow patterns when relatively high gas flow rates are present (Figure 6.10). At higher gas flow rates and beyond a minimum impeller speed, large cavities form and have a significant impact on flow stability and bulk flow regimes. Flooding occurs in the same manner as with low gas flow rates (a). If the impeller speed is above a threshold, large cavities form causing a significant loss in pumping capacity; this is accompanied by direct loading and a significant radial and axisymmetric flow pattern as shown in (b). The impeller cavity distribution is also axisymmetric. The transition from direct to indirect loading is shown in (c). An increase in the impeller speed induces oscillations between predominantly radial (i) and axial (ii) flow. Occasional cavity shedding induces

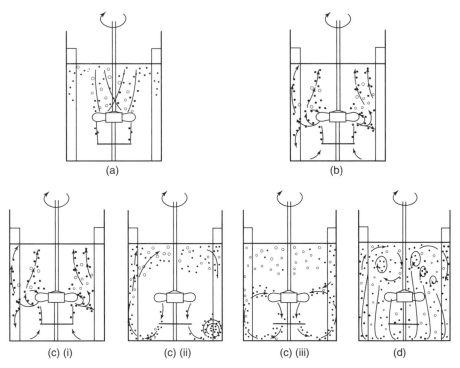

Figure 6.10   Bulk flow regimes generated by down-pumping A-315 at high gas flow rates (McFarlane et al., 1995).

impeller dominance which would force gas to the tank bottom (iii); however, this event is temporary, and flow eventually reverts back to radial or oscillating dominance. This series of events occurs over a very short impeller speed range and a subsequent increase results in proper loading and dispersion (d).

The oscillation in cavity size and flow direction (Figure 6.10c) can produce large and significant variations in power draw and torque such that the power number for the down-pumping A-315, for example, can vary between 0.84 and 1.48 at a frequency of 1–2 min. The variations in torque and power draw can have damaging effects on the motor and gear assembly (Tatterson, 1994) and can lead to vessel vibration (Sardeing et al., 2004b). In addition, large cavities tend to create asymmetric flow that can produce a significant bending moment on the impeller shaft, resulting in shaft damage. Smaller impellers at low power concentrations experience these problems much earlier because large cavities are able to form with a smaller amount of gas at these scales. Up-pumping axial flow impellers have a more stable oscillation because the sparged gas enters the impeller zone in the same direction as the fluid (McFarlane et al., 1995).

The detrimental effects of large cavity formations can be minimized by increasing the impeller speed. Reactor design improvements can be made such that large cavities require more gas or that the minimum impeller speed for large cavity formation is reduced. The goal is to reduce the impact on the normal operating range such that the event is outside normal parameters or induces fluctuations that are within the design specifications for the motor and gear assembly (McFarlane et al., 1995).

6.3   EFFECTS OF IMPELLER DESIGN AND ARRANGEMENT

The impeller provides mechanical agitation and gas dispersion. It is responsible for bubble breakup in gas-sparged STRs and for solid suspension in gas–liquid–solid STRs. Numerous impeller designs exist to meet various needs, and the economic success of a project depends on the evaluation and selection of a proper impeller in concert with reactor geometry (Ungerman and Heindel, 2007). A few standard impellers in gas–liquid dispersion are documented. Normal impeller-to-tank diameter ratio, D_i/T, is typically between 1/4 and 2/3 (Harnby et al., 1992) with the standard ratio being 1/3 (based on industrial experience) and rarely going above 1/2 (Tatterson, 1991). This geometry minimizes cost and is capable of providing a well-mixed state for the liquid phase and complete dispersion of the gas phase. An impeller with a larger D_i/T ratio proves inefficient and unnecessary. The impeller power draw is proportional to the impeller speed to the third power and the impeller diameter raised to the fifth power. It is, therefore, cheaper to operate an impeller at a faster speed than a larger diameter if more dispersion or blending is needed. Impeller clearance, defined as the distance between the impeller and tank bottom, is typically in the range $T/6–T/2$ (Tatterson, 1994) depending on the liquid viscosity, impeller type, sparger–impeller separation, and number of impellers.

Impellers usually enter the vessel from the top; however, very large or novel vessels find it useful for the impeller to enter from the bottom or side (Harnby et al.,

1992) because it can minimize the amount of structural steel needed to support the shaft and impeller (Aden et al., 2002). Impeller shafts usually have a circular cross-section and are oriented perpendicular to the reactor bottom and placed along the tank centerline. Other impeller shaft placement relative to the centerline, known as impeller eccentricity, has been practiced (Tatterson, 1991). Off-centering the impeller shaft has been shown to improve mixing, minimize the appearance of vortices (Oldshue, 1983), and produce smaller bubbles in the turbulent regime (Cabaret et al., 2008). Shaft eccentricity (changing the cross-sectional shape) minimizes vortices and can be used in place of baffles. In these applications, mixing time decreases while power input increases. Industrial applications may find this configuration undesirable due to construction and maintenance costs (Cabaret et al., 2008).

Impellers can be classified into groups based on the liquid viscosity used in the reactor (Harnby et al., 1992). Propellers, turbines, and paddles have a higher tip speed relative to other impeller types and are used for low viscosity Newtonian liquids that are encountered in most processes (Ogut and Hatch, 1988). Propellers are usually operated faster and paddles are operated slower than turbines. The standard three-bladed propeller has poor gas–liquid dispersion and contacting characteristics and will be excluded from further discussion.

6.3.1   Radial Flow Impellers

A more common impeller classification is by flow leaving the impeller zone. Impellers can be classified into radial or axial flow impellers. Some examples of radial flow impellers are the Narcissus impeller (NS), concave blade disc turbine (Figure 6.11), (Chemineer) BT-6, and the multibladed disc turbine. A six-bladed disc turbine, shown in Figure 6.12, is often referred to as a Rushton-type turbine (RT) (Ulbrecht and Patterson, 1985). The standard blade is $D_i/4$ long and $D_i/5$

Figure 6.11   Concave impeller.

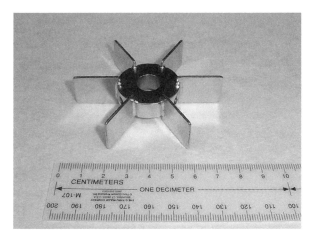

Figure 6.12   Rushton-type turbine.

wide (Oldshue, 1983). Increasing the number of blades (12 or 18) produces similar recirculation as the standard RT, but reduces the power draw drop upon gassing (Smith et al., 1977). Increasing the number of blades, however, increases the ungassed power number from ∼5 to 8–9 (Nienow, 1996), which diminishes the use of a disc turbine with higher blade numbers as a mixing device (Nienow et al., 1995). Turbine designs using retreating, angled, or hollow-faced blades are in existence, but have not been able to compete with the Rushton-type turbine for gas–liquid dispersion tasks (Harnby et al., 1992; Williams, 2002).

The Rushton-type turbine has been the most popular impeller for gas–liquid dispersion since the 1950s (Nienow, 1996). It has very good bubble breakup and gas dispersion capabilities leading to good mass transfer characteristics (Cabaret et al., 2008; Williams, 2002). It is the measuring stick to which other gas–liquid impellers are compared. The high power number of the Rushton-type turbine, which is a disadvantage for mixing purposes, is an advantage for gas–liquid dispersion. The RT is capable of creating higher maximum shear zones and produce smaller bubbles. Smaller bubbles lead to a higher interfacial area, which in turn increases the mass transfer capacity. In addition, the disc feature of the RT prevents gas bubbles from passing through the lower shear region and forces the gas flow through the high shear impeller tip region (Oldshue, 1983; Tatterson, 1991). These features give the Rushton-type turbine $k_L a$ values that are an average of 50% higher than other impellers operating under similar conditions (Sardeing et al., 2004a).

Although commonly used, Rushton-type turbines exhibit several negative traits. Some are design specific, but others are shared by all radial flow impellers. This is mainly due to the fact that newer radial impellers were designed with the goal of improving on a small number of disadvantages (Nienow, 1996). The RT experiences a power draw drop of 50–65% upon gassing (McFarlane and Nienow, 1996b). This weakness is purely operational, but forces the reactor to have a complicated gearbox design (Ulbrecht and Patterson, 1985), which adds to the installation and maintenance costs. Concave (hollow) and 12- or 18-blade disc turbines can be used to minimize this effect. They form smaller cavities on the blade backside relative to the standard RT (Smith et al., 1977). The smaller cavities lead to much smoother power curves and less variation in power draw upon gassing (Ungerman and Heindel, 2007), and hence lead to simpler gearbox designs (Ulbrecht and Patterson, 1985).

Concave blade disc turbines are of interest in gas–liquid dispersion because they are able to handle more gas than Rushton-type turbines before flooding (Smith et al., 1977; Vasconcelos et al., 2000). The mass transfer capacity for the concave blade disc turbine is very similar to the Rushton-type turbine, but Chen and Chen (1999) found that the blade curvature could be optimized for a certain power input to produce higher gas–liquid mass transfer coefficients. Unlike the Rushton-type turbine, the concave blade disc turbine requires the cup orientation to be in the direction of impeller rotation (Tatterson, 1994).

Vasconcelos et al. (2000) investigated the influence of the impeller blade shape, shown in Figure 6.13, on gas–liquid mass transfer. They concluded that the importance of blade shape was negligible for disc turbines as long as the power number, power draw drop upon gassing, and gas flow rate were similar. Other authors have come to a similar conclusion, but also stipulated that the process has to be in the turbulent regime (Nienow, 1996). Angled, concave, and lancelet blade disc turbines not only provided lower impeller power numbers but also offered smaller power draw drops upon gassing and smaller, if any, gas cavities. Hence, they are more efficient and capable of handling gas. Increasing the impeller diameter could discount the power draw disparity while allowing retrofitting to be easily accomplished. As such, the retrofitted system would provide similar gas–liquid mass transfer performance and would handle more gas, potentially allowing for an increased operational range and gas–liquid mass transfer.

Another major weakness for radial flow impellers stems from one of the strengths: the high shear rates. The power dissipation (or shear) rates are concentrated at the blade tips (Gagnon et al., 1998) and are not uniformly distributed throughout the reactor (Fujasova et al., 2007; Ungerman and Heindel, 2007). This unbalanced shear distribution can lead to stagnant zones in the outer reactor region (Bellgardt, 2000b) and higher mass transfer in the impeller stream relative to the working volume (Stenberg and Andersson, 1988a, 1988b). According to Stenberg and Andersson (1988b), 50% of the energy is dissipated in the impeller stream, 20% is dissipated in the immediate impeller vicinity, and 30% is dissipated through the rest of the reactor. This disparity leads to radial flow impellers, especially the RT, providing very poor top-to-bottom mixing (McFarlane and Nienow, 1995), particularly in more viscous fluids.

If a constant impeller speed scale-up rule is used, the impeller tip energy dissipation rate will increase due to its connection to the impeller diameter (Bliem and Katinger, 1988b). The power decay is more pronounced in larger vessels and contributes to scale-up issues (Figueiredo and Calderbank, 1979). This concentration leads to the local energy dissipation rate in the impeller vicinity being up to 270 times higher than the average. Furthermore, the local rate experiences large

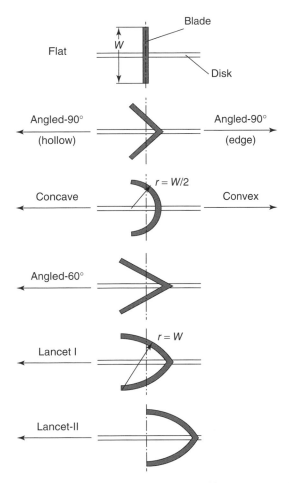

Figure 6.13   Possible blade shapes for use with disc turbines (Vasconcelos et al., 2000).

fluctuations creating problems in scale-up and reactor design comparisons. These high shear fluctuations can be harmful to some bioreactor microorganisms (Bliem and Katinger, 1988a, 1988b; Bredwell and Worden, 1998).

Finally, the fluid can experience low bulk circulation leading to low gas holdup in the bottom reactor section (Fujasova et al., 2007; Ungerman and Heindel, 2007) and gas compartmentalization (Moucha et al., 2003). Gas compartmentalization is to be avoided since it poses the danger of spent gas entrainment or gas starvation. Spent gas is inactive in production and limits the practical working volume (Fujasova et al., 2007) while gas starvation can limit the effectiveness of the microorganisms (Pollack et al., 2008).

As scale increases, a proper mixing state gains in importance that can lead to radial flow impellers, providing very poor mixing conditions in a significant portion of the reactor volume (McFarlane and Nienow, 1995). The solution has been to simply operate the impeller at a faster speed that can have detrimental effects on microorganisms, power usage, and impeller characteristics. Retrofitting a system with a larger radial flow impeller to resolve some of these problems is not usually possible because different turbine diameters produce very different torques. The higher torque of a larger diameter turbine can be damaging for the motor and drive train to the point that this practice is seen as high risk–low reward and rarely implemented (McFarlane and Nienow, 1995). The smoother power curves of the concave blade turbine are very promising for this purpose and should be seriously considered.

Chen and Chen (1999) investigated the possible replacement of the RT. The comb blade and perforated blade disc turbine were found to have higher $k_L a$ values than the standard RT at similar power inputs. They came to the conclusion that bubble breakup was not only a function of the shear-gradient magnitudes, but also the amount of time the gas phase remained in the shear field. By spatially increasing the shear field, bubbles spent longer time periods in this region and decreased the probability that larger bubbles would pass through this region without significant breakup. The results were that the comb and perforated blade disc turbine produced $k_L a$ values that were almost 12% and 30%, respectively, higher than the Rushton-type turbine at the same power input and superficial gas velocity while producing lower shear-gradient magnitudes. Currently, there is no available information of these impellers being used in processes involving microorganisms.

6.3.2 Axial Flow Impellers

Many attempts have been made to replace radial flow impellers with axial flow impellers. Examples of axial flow impellers include the Lightnin A-310 (Figure 6.14), Lightnin A-315 (Figure 6.15), pitched blade turbine (PBT) (Figure 6.16), Techmix 335 (TX), Prochem Maxflo T, SuperMIG (EKATO), marine propeller, A-3 impeller, or a multibladed paddle. The PBT and hydrofoil impellers, such as the Lightnin A-315, are the most commonly used axial flow impellers. These devices have a much lower power number than radial flow impellers that make them ideal for mixing purposes. Their blending prowess is due to the fact that the mixing time is independent of the impeller type (Nienow, 1996). Hence, the operation can be accomplished at a lower cost with axial flow impellers; however, as shown in Figure 6.17, axial flow impellers are usually inferior to radial flow impellers for mass transfer purposes. They usually produce flow in the axial direction, but can create some radial flow if $D_i/T > 0.5$ (Tatterson, 1991). This situation is usually avoided since the standard low viscosity impellers have a D_i/T ratio of about 1/3 (Harnby et al., 1992).

Axial flow impellers used in low bottom clearance tanks can also create radial flow if the direction of the flow is downward. In this case, the flow can leave the impeller zone only by flowing in the radial direction (Tatterson, 1991). This phenomenon is usually not observed since the standard bottom clearance for low viscosity impellers in gas–liquid dispersions is between one impeller diameter and one half the tank diameter (Ulbrecht and Patterson, 1985). The hydrofoil impeller

Figure 6.14   Lightnin A-310 axial flow impeller.

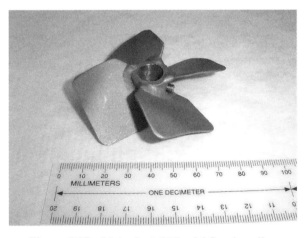

Figure 6.15   Lightnin A-315 axial flow impeller.

discharge is affected by the impeller Reynolds number.[1] If the system is operated with a viscosity such that *Re* is about 5000, the flow becomes more and more radial. This change could have a significant impact on mixing and gas dispersion and would be especially important for viscous non-Newtonian bioprocesses (McFarlane and Nienow, 1995).

The PBT discharges fluid from the impeller zone at an angle of 45–60°. The standard design has flat blades that are 45° from the horizontal and have a 1/5 blade width-to-diameter ratio. This discharge angle causes a significant radial component

---

[1]Reynolds number is defined in Eq. (3.11) in Section 3—Gas–Liquid Mass Transfer Models.

Figure 6.16   Pitched axial flow impeller.

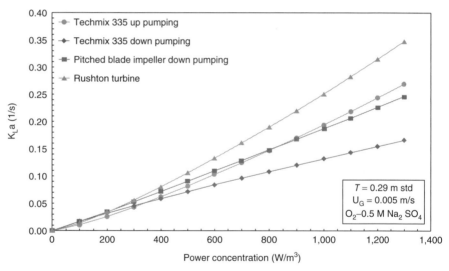

Figure 6.17   Gas–liquid mass transfer for various impeller types. Adapted from Moucha et al (2003).

regardless of the impeller size or position, which has led to the PBT to being classified as a mixed impeller in the axial family (McFarlane and Nienow, 1995). These features make the PBT an excellent mixing device. The mixing (blending) time is reduced and heat transfer is improved; however, the PBT makes a bad impeller for gas breakup. Gas bubbles are led to the blade tip where they are sliced apart, but the blade shape does not accumulate bubbles in a sufficient manner. Therefore, a large number of bubbles pass through the impeller zone without interacting with the breakup mechanism (Martín et al., 2008b). The solidity ratio, which relates the impeller blade area to the impeller swept area, is about 0.43 for the PBT (Oldshue, 1983). Low solidity ratio impellers flood before radial flow impellers and high solidity ratio impellers, so their use in gas–liquid processes in single impeller systems is atypical (Ungerman and Heindel, 2007).

In order to reduce flooding, high efficiency hydrofoil impellers, distinguished by their profiled blades (McFarlane and Nienow, 1995), were introduced. These impellers, such as the Lightnin A-315 (see, e.g., Figure 6.15), have a much higher solidity ratio of about 0.87, have a smaller power draw drop upon gassing (30–50%) (McFarlane and Nienow, 1996b), and flood later (Ungerman, 2006) than radial flow impellers. It is therefore capable of handling more gas than the radial or PBT impellers (Chen and Chen, 1999). In fact, it can handle 86% more gas than a Rushton-type turbine before flooding (Yawalkar et al., 2002b) and 40% more gas than the PBT before the onset of gas loading transition (described in Section 6.2.2) (McFarlane and Nienow, 1996b). However, the A-315 is still an axial flow impeller and produces lower shear gradients relative to the Rushton-type turbine. The advantage of using the A-315 over an RT depends upon the gas flow rates and shear sensitivity of the microorganisms. If neither of these situations is important, the RT is still the better choice. If the gas flow rates are of importance but shear sensitivity is not, the concave blade disc turbine is the more effective impeller.

The axial flow direction makes particulate suspension easier relative to radial flow impellers (Tatterson, 1991). Axial flow impellers are also better blending devices because they have a lower power number—hence power usage—and shorter blending times; however, results for gas–liquid contact have been unimpressive relative to the Rushton-type turbine, and axial flow impellers tend to have smaller gas–liquid mass transfer coefficients. In addition, large reactors using axial flow impellers have shown highly asymmetric and oscillating gas holdup distributions with periods of up to several minutes (Bakker and Oshinowo, 2004). Therefore, their use has been limited to mixing and shear normalizing in multiple impeller gas–liquid systems (Bouaifi et al., 2001) and mixing and solid suspension in single impeller systems (Oldshue, 1983).

The direction of the flow, up or down, depends on the geometry and rotation of the axial flow impeller. In general, up-pumping impellers push the gas to the surface faster (lower gas holdup) while down-pumping impellers induce recirculation and longer residence times, defined as the time a particle spends in the reactor, which leads to higher gas holdup (Moucha et al., 2003). These circulation loops are important with respect to gas holdup and contribute to multiple impeller systems offering 30% higher gas holdup values than single impeller systems (Bouaifi et al., 2001). The down-pumping impeller also offers shorter mixing times, which can lead to lower operational costs in systems where mixing is important (Gogate and Pandit, 1999a).

Some researchers, on the other hand, do not find a preference for up- or down-pumping axial flow impellers (Fujasova et al., 2007), or find that the up-pumping configuration produces more macromixing than down-pumping systems, which can lead to smaller bubble diameters (Majirova et al., 2004; Sardeing et al., 2004b). Sardeing et al. (2004b) observed that gas holdup in

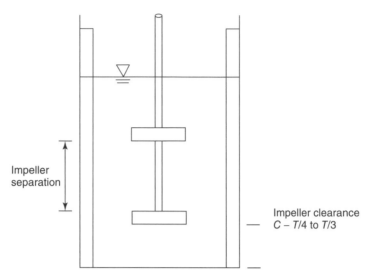

Impeller separation

Impeller clearance
— $C – T/4$ to $T/3$

Figure 6.18 Standard multiple impeller stirred-tank reactor design.

the up-pumping configuration was determined to be 10–25% higher than the down-pumping orientation. The gas holdup conclusions were drawn based on constant impeller speed data. Down-pumping axial flow impellers also have significant stability and hydrodynamic problems in certain operating ranges, which may have contributed to the up-pumping axial flow impeller's superior gas holdup performance (McFarlane and Nienow, 1995; McFarlane et al., 1995). Sardeing et al. (2004b) focused on power concentration, which they indicated to be more relevant, and the down-pumping PBT was determined to be more efficient than the up-pumping PBT, confirming similar conclusions made by others (Bouaifi and Roustan, 2001; Gogate and Pandit, 1999a; Moucha et al., 2003).

6.3.3   Multiple Impeller Systems

Multiple impeller STR designs, schematically represented in Figure 6.18, are very popular in practice (Nocentini et al., 1998) and were implemented due to short-comings of the single impeller system and industrial requirements. For example, when a single impeller system is used in an industrial-scale reactor, it may not provide proper agitation and gas dispersion in large reactors. In addition, viscous or non-Newtonian liquids do not mix well in a single impeller system. Large gas-filled cavities on the back of the impeller blades also limit the amount of gas that can be properly dispersed in a single impeller STR (Cabaret et al., 2008).

Multiple impeller systems are able to distribute energy throughout the reactor more efficiently, which leads to a more homogeneous shear rate distribution. Liquid circulation and gas dispersion are also improved, leading to longer gas-phase residence times. These factors lead to better gas utilization, higher gas–liquid mass transfer coefficients (Bouaifi and Roustan, 2001; Cabaret et al., 2008; Fujasova et al., 2007; Moucha et al., 2003; Nocentini et al., 1998), longer gas-phase residence times, better bulk flow characteristics (Shewale and Pandit, 2006), and higher gas holdup (Bouaifi et al., 2001; Bouaifi and Roustan, 2001; Shewale and Pandit, 2006) relative to single impeller systems. The industrial implementation of multiple impellers can be made to single impeller systems with minimal retrofitting and without changes to the motor and gearbox (Lines, 2000), especially if the addition is an axial flow impeller (McFarlane and Nienow, 1995).

In order to achieve a proper working condition, however, proper impeller placement has to be considered. The bottom clearance for most multiple impeller designs is between $T/4$ and $T/3$. If the impellers are too close to each other, impeller–impeller interference can lead to inefficient operation and an inability to separate operational inputs; this would add complications without enhancing efficiency. Interference also provides limited, if any, increase in $k_L a$ (Nishikawa et al., 1984). The power draw for a multiple impeller configuration will be normal upon start-up, but may decrease by about 70% of the initial value and remain at those levels throughout the process when the impellers are too close. The flow patterns of interfering impellers may also exhibit characteristics of a single impeller system (Mishra and Joshi, 1994), leading to negligible improvements and increased costs (Oldshue, 1983). For example, a second RT has been reported to increase $k_L a$ by 74% if placed correctly, but would provide no improvement if interference occurred (Nishikawa et al., 1984).

The separation between impellers depends on the impeller type, but should be a minimum of $1D_i$ (Linek et al., 1996b). The Rushton-type turbine, for example, requires a separation of $1.5D_i$, but $2D_i$ would be preferred to ensure independence (Fujasova et al., 2007). Linek et al. (1996b) found that if the impellers were acting independently, each impeller's mass transfer characteristics could be evaluated as a single impeller system. The overall mass transfer coefficient could be calculated using a weighted average, with the lower and upper sections having weights of 0.25 and 0.75, respectively (Linek et al., 1996a). However, this approach failed using non-coalescing media and was not proven at larger scales (Linek et al., 1996b).

Although common shafts are standard, dual-shaft systems have been investigated. Dual shafts allow the impellers to spin at independent speeds and directions. Although this configuration adds to setup and maintenance costs, it provides operational flexibility. Since the amount of gas that passes through each impeller zone differs, independent controls allow optimal operation for each impeller. Cabaret et al. (2008) found that the upper impeller needs to run 20% faster for homogeneous gas distribution. Dual shafts can also accommodate counter-rotation. Cabaret et al. (2008) claim that counter-rotating Rushton-type impellers in unbaffled tanks is just as efficient as standard, baffled operation. The costs of implementing dual shafts would usually be expected to outweigh any gains made using dual impellers in low viscosity liquids.

The impeller is the key component for proper STR operation, especially for multiple impeller systems. A proper selection procedure has to consider numerous options and their applicability to the particular process of interest. A mixed

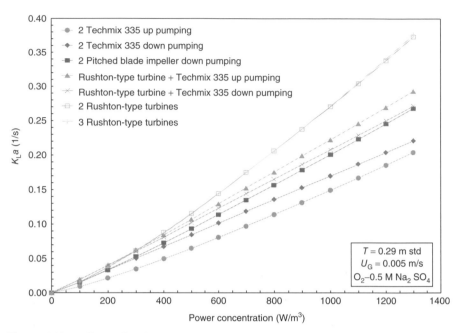

Figure 6.19   Effects of multiple impellers on gas–liquid mass transfer in an STR. Adapted from Moucha et al. (2003).

configuration using a radial and axial flow impeller is assumed to be more efficient for gas dispersion and mixing in a low viscosity Newtonian fluid than in a dual axial or radial configuration, even though the Rushton-type turbine combination provides better $k_L a$ (see, e.g., Figure 6.19). Efficiency, in this case, is defined as the capability to maximize gas–liquid mass transfer while minimizing power input (Gagnon et al., 1998). It is often advantageous to use a Rushton-type or concave blade turbine as the bottom impeller. This impeller would provide optimal bubble breakage. The upper impeller can be a down-pumping axial flow impeller to enhance gas–liquid circulation (Puthli et al., 2005).

Experiments using a radial setup are often performed to set the bar for mass transfer comparisons. Gagnon et al. (1998) contributed to this discussion by investigating the effect of adding impellers to the same reactor. A second Rushton-type turbine increased gas holdup, gas residence time, and the gas–liquid mass transfer coefficient. The addition of a third Rushton-type turbine increased these further, but at a much lower amount. They came to the conclusion that gas holdup and, subsequently, gas–liquid mass transfer does not increase linearly with the number of impellers and power drawn. Moucha et al. (2003) came to a similar conclusion when going from 1RT to 2RT but, as shown in Figure 6.19, the additional mass transfer was insignificant when the increase was made from 2RT to 3RT.

The mixed configuration efficiency and the declining increase in $k_L a$ with increasing number of turbines are determined by the impeller loading. The bottom impeller is loaded directly (by sparged gas) while the other impeller(s) are loaded indirectly (by impeller-generated flow loops). Direct loading enhances gas dispersion capabilities of the Rushton-type turbine, while indirect loading puts more emphases on liquid mixing efficacy. Impeller loading is a more important consideration in experimental-scale reactors. Larger industrial-scale reactors require more effective blending and top-to-bottom mixing than the Rushton-type turbine can provide (Bouaifi and Roustan, 2001; Fujasova et al., 2007). The Rushton-type turbine is oftentimes limited in this regard, and the conditions created in these impeller zones (cells) are more geared toward axial flow impellers (McFarlane and Nienow, 1995).

Furthermore, radial flow impellers' discharge divides the reactor volume into well-mixed systems with minimal interchange (Nienow, 1996). As a result, radial flow impellers in large-scale systems may produce compartmentalization, caverns (impeller is encased by its flow field while most of the reactor is stagnant), higher gas recirculation, and low volumetric exchange zones (Nocentini et al., 1998). For large STRs, the combination of a radial flow impeller on the bottom and a down-pumping axial flow impeller on the top enhances the reactor fluid mixing such that the reactor volume contact is maximized with minimal power input (Fujasova et al., 2007; Vasconcelos et al., 2000).

Some discrepancies and opposing suggestions in the literature can be explained by using the studies of Linek et al. (1996a) and Nocentini et al. (1998). They found unequal amounts of gas pass through each impeller section regardless of loading type, resulting in gas distribution nonuniformity that can lead to the bottom impeller being flooded far earlier than the others. Since the bottom impeller contacts the most gas and is responsible for initial bubble breakup, flooding of this impeller is severely detrimental to system operation (Nocentini et al., 1998).

Linek et al. (1996a) concluded that the bottom impeller section had gas holdup and mass transfer values that were 15% and 45%, respectively, lower than the upper section when the STR was filled with non-coalescing media (0.5 M Na_2SO_4). Similar conclusions were reached when using water, with the upper section producing higher $k_L a$ values by 15%. This also implies that impeller power consumption was not balanced, that is, one of the impellers consumed more power and created higher shear gradients than the other, producing higher k_L and a values for the upper reactor section. The problem is made worse due to the fact that gas tends to coalesce faster and easier in regions of low power (relatively speaking), creating larger bubbles that reduce the interfacial surface area and possibly lowers k_L. Coalescence, however, would be far less likely in non-coalescing media, which should explain the results of Linek et al. (1996a).

Bouaifi et al. (2001) found that the average bubble diameter was larger in the bottom section of the reactor than the upper section. They concluded that bubbles formed a distribution such that the larger bubbles were in a region outside the impeller stream and were up to four times larger than the bubbles entrained in the impeller stream. More specifically, gas in these setups would concentrate about the impeller shaft, impeller tip, and within the radial area between the impeller and reactor walls (Boden et al., 2008; McFarlane and Nienow, 1996a).

These observations were made for an axial system, but are very similar to those made by Stenberg and Andersson (1988b) for a 1RT setup, which produced a similar qualitative mass transfer behavior for these impeller types.

Bouaifi et al. (2001) also observed that a "very heterogeneous bubble distribution" would form in a dual axial flow impeller system once the bottom impeller was flooded. If the impeller was properly loaded and complete dispersion occurred, 50–60% of the bubbles had a diameter of 1–3 mm. Thus, it was more effective to operate in the loaded and completely dispersed regimes. These experiences confirm and explain the unbalanced mass transfer performance observed by Linek et al. (1996a, 1996b) and Gagnon et al. (1998) in multiple impeller systems, and by Bellgardt (2000b), Moilanen et al. (2008), and Stenberg and Andersson (1988a, 1988b) in single impeller systems.

The impeller choice in multiple impeller reactors is therefore vital. A proper selection requires a minor power increase by about 15% to produce similar $k_L a$ of a Rushton-type setup but with a much friendlier environment for microorganisms and larger scales (Fujasova et al., 2007). The required radial and axial flow impeller often depends on the operational conditions. The simplest configuration includes a Rushton-type turbine for the lower impeller and a down-pumping PBT for the upper turbine(s). Since these impellers tend to flood relatively early, it has been proposed to replace the Rushton-type turbine and down-pumping PBT to extend the operational use. For example, Pinelli et al. (2003) did not find an advantage of using two Rushton-type turbines over two BT-6 impellers (asymmetric concave blade impellers designed by Chemineer). Gas holdup and macromixing were observed to be very similar, which would imply that the concave blade disc turbine could replace a Rushton-type turbine in a single or multiple impeller system without major hydrodynamic implications while providing more gas handling capacity (Vasconcelos et al., 2000). While holding power concentration and superficial gas velocity constant, Chen and Chen (1999) observed much higher mass transfer potential and smaller bubbles by replacing the RT with a comb and perforated blade disc turbine. The A-315 could replace the down-pumping PBT if a higher gas capacity is necessary. A more homogeneous environment is also expected with this replacement at larger scales because the A-315 (e.g., Figure 6.15) provides better recirculation exchange and interaction with the other impeller(s) (Bouaifi et al., 2001).

It is common to use multiple impeller systems in operations that are expected to undergo significant changes in viscosity and rheology. These processes are operated in the laminar regime that puts more emphases on the viscous behavior of the fluid. Multiple impellers have been determined to produce better gas–liquid mass transfer in viscous fluids than the commonly used helical ribbon impeller. Most researchers, however, spend time investigating low viscosity impeller combinations for viscous non-Newtonian applications (Tecante and Choplin, 1993). These low clearance impellers can require large amounts of power, making their operation impractical, especially for very viscous non-Newtonian liquids (Gagnon et al., 1998). In these cases, the operation is simply shut down if the impellers are not capable of providing proper conditions (Ogut and Hatch, 1988).

Cabaret et al. (2008) and Gagnon et al. (1998) concluded that better mixing and higher product conversion can be achieved if a close clearance impeller, such as the helical ribbon, is used in conjunction with a radial flow impeller such as the RT in a highly viscous system. The Rushton-type turbine provides proper gas dispersion, while the close clearance impeller attempts to contact most of the reactor volume and provides proper bulk mixing, shear distribution, lower apparent viscosity, and minimal stagnant zones (Tecante and Choplin, 1993). These effects also lead to higher reactor utilization and can decrease power requirements.

6.3.4   Surface Aeration

STRs are highly turbulent mixers which can induce a high degree of surface turbulence. Although the effects of turbulence in the reactor volume are known, the interaction at the gas–liquid surface can be complicated. Unbaffled vessels can experience flow destructive vortices and solid body motion at normal operating impeller speeds. If the vessel is baffled, these vortices tend to be minor, but their influence on mass transfer can be important. Highly turbulent surfaces allow the STR to entrain head space gas, effectively adding to the sparged gas flow rate. This phenomenon is referred to as surface aeration. Therefore, direct sparging, which has been the only option discussed so far, is not required. However, this form of indirect sparging affects impeller performance and reactor hydrodynamics in the same fashion as direct sparging (Patwardhan and Joshi, 1998).

Surface aeration is used in wastewater treatment, water aeration (e.g., fishing ponds), and processes requiring the gas-phase conversion to be maximized (toxic or highly valuable gases). If the system requires the gas phase to be recycled, surface aeration allows the reactor to be capped creating a dead-end system. Hence, a recycle gas compressor is not necessary, which minimizes fixed costs (Patwardhan and Joshi, 1999). Such reactor designs limit potentially toxic exposure, increase work reliably, and have limited maintenance costs. The lack of a sparger can further extend these benefits for processes containing an excessive amount of a solid phase such as wastewater treatment (Patwardhan and Joshi, 1998).

Surface aeration is most common in multiple impeller systems and/or semibatch and batch processes (Lines, 1999). Multiple impeller designs place an impeller relatively close to the surface that can induce surface aeration at relatively low impeller speeds. Most authors do not check for this phenomenon, and it is often unclear as to which models are used for the mass balance in the gas phase. The exclusion of a dynamic gas holdup term (assuming dynamic conditions) does not affect the results if surface aeration is limited; however, if surface aeration is significant, experimental errors could be large (Figueiredo and Calderbank, 1979).

The critical impeller speed for surface aeration (N_{CSA}) can be identified using indirect sparging. A simple $k_L a$ versus N graph produces a sharp increase in $k_L a$ at N_{CSA}. Direct sparging makes this identification more difficult. Although gas may be entrained, additional gas dispersion does not occur until the impeller speed is increased by about 20% above the initial entrainment speed. Other factors controlling the surface aeration are impeller type and diameter, tank diameter, impeller clearance and submergence, baffling, and gas and liquid properties (Patwardhan and Joshi, 1998).

Surface aerators are, however, highly limited by impeller submergence and require an impeller to be very close to the surface in order to be effective. Furthermore, the operation becomes hampered with scale-up such that dead zones are common (Lines, 1999; Patwardhan and Joshi, 1998). Design and scale-up for surface aerated systems are even more challenging than a standard STR (Lines, 1999). As such, this reactor design is not capable of competing with conventional designs. However, surface aeration is an important phenomenon that occurs in STRs (Patwardhan and Joshi, 1998). Although it may not have a significant impact on gas–liquid mass transfer, it should be kept in mind especially if comparisons are to be made between competing designs. Many experimental reactors include an operating range that is much higher than traditional industrial applications (Benz, 2008). These designs may induce surface aeration and artificially increase mass transfer performance.

6.3.5 Self-Inducing Impellers

Self-inducing impellers are also used for indirect sparging purposes. The spinning action of the impeller creates a low pressure region at the impeller intake. Orifice holes are exposed to this low pressure region and connected by a hollow shaft to the atmosphere or head space (for dead-end systems) (Patwardhan and Joshi, 1999; Vesselinov et al., 2008). As such, the gas flow rate is a function of the impeller speed. The "atmospheric" pressure can be adjusted by pressurizing the shaft entrance such that additional gas is pumped into the system (Patwardhan and Joshi, 1999).

Self-inducing impellers are classified by the flow in and out of the impeller zone. Type 11 impellers are shown in Figure 6.20a and are defined by the gas being the only phase at the inlet and the outlet. The most common and simplest design is a hollow pipe with orifice holes at the ends. The hollow pipe impeller induces gas flow through Bernoulli's equation. Gas induction occurs once the pressure differential is large enough to overpower the liquid hydrostatic head. Type 12 impellers (Figure 6.20b) have a gas phase at the inlet and a gas–liquid flow at the outlet. Gas induction occurs in a similar manner to Type 11, but Type 12 impellers mix the phases in the impeller zone through some intricate designs. Type 22 impellers (Figure 6.20c) are intricate devices that have a two-phase inlet and outlet composition. The impeller induces a large vortex until the impeller is able to induce phase interface ("surface") aeration. The design calls for an axial gas–liquid inlet and radial outlet. An optional impeller hood prevents gas outlet in the axial direction, inducing a pressure differential so that the liquid is pumped into the impeller zone from the bottom reactor section (Patwardhan and Joshi, 1999).

These impellers have both advantages and disadvantages. Type 12 and 22 impellers force the gas phase to travel through a high shear volume, creating smaller average bubble diameters and higher interfacial area in the outlet flow.

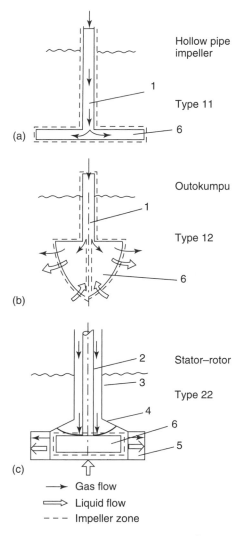

Hollow pipe
impeller

Type 11

Outokumpu

Type 12

Stator–rotor

Type 22

→ Gas flow
⇒ Liquid flow
--- Impeller zone

Figure 6.20   Self-inducing impeller types (Patwardhan and Joshi, 1999).

These designs also allow for a reduction in power usage, but this usually comes at a cost of reduced residence time that can have a negative impact on conversion. Since dead-end zones are feasible, this design provides similar advantages to surface aerators. The disadvantages of self-inducing impellers are usually their complicated designs, diminished control (with power input often being the only control variable), and gas–liquid mass transfer performance comparable to "standard" STRs. Unless the gas phase is highly toxic or the gas recycling system is too expensive, self-inducing impellers have little to offer in microorganism systems.

6.4 SUPERFICIAL GAS VELOCITY

The superficial gas velocity U_G is a description of the amount of gas present in the reactor volume and is defined as the volumetric gas flow rate per unit cross-sectional area of the reactor. This definition is easily quantifiable and has been used by many researchers as a correlation parameter for gas holdup and gas–liquid mass transfer. Most researchers cited in this chapter, with the exception of Linek et al. (2005a), have proposed a positive correlation of superficial gas velocity with gas–liquid mass transfer; however, its specific influence on mass transfer is often confusing.

If the impeller is operated below a minimum tip speed (2.25 m/s for RT), the reactor hydrodynamics are dominated by the gas flow and the reactor acts as a bubble column. At this point, gas–liquid mass transfer has an exclusive dependence on the superficial gas velocity (Charpentier, 1981). Since the intent of the STR is to provide agitation that would be superior to gas sparging alone, STRs are operated such that impeller agitation dominates the hydrodynamics (Nishikawa et al., 1981).

The superficial gas velocity is often recognized to influence gas–liquid mass transfer through gas holdup (Nocentini et al., 1993) and its influence on the interfacial surface area (Garcia-Ochoa and Gomez, 2004). It is generally assumed that the interfacial surface area can be increased by entraining more gas in the reactor, and this results in increased gas dispersion and gas–liquid mass transfer.

Bubble–bubble interaction and coalescence have to be considered when the superficial gas velocity is increased. Coalescence occurs through a three-step process. First, bubbles collide and form a liquid layer (typically $10^{-3} – 10^{-4}$ cm thick). Second, the film drains assuming that the collision force is sufficient to deform the bubble interface and that the bubbles spend enough time in contact for the film to drain. Third, assuming that a critical film layer thickness ($\sim 10^{-6}$ cm) is achieved, the film breaks and the bubbles combine (Martín et al., 2008b; Tse et al., 1998). The entire sequence of events is completed in milliseconds for coalescing liquids (Tse et al., 1998), but may take up to 15 s for liquids containing surface-active agents (Laari and Turunen, 2005). Coalescence is, therefore, influenced by the collision frequency, collision force, bubble deformation, and coalescence efficiency (Bredwell et al., 1999; Martín et al., 2008b), and is controlled by the film drainage rate (Tse et al., 1998).

Superficial gas velocity has an influence over the collision frequency. If more gas is present, there is a higher probability of collision (Martín et al., 2008b). Coalescence efficiency and drainage rate depend on the film properties which are a function of the liquid properties. The collision force, however, is the controlling factor because the bubble diameter is a function of the power input (Bouaifi et al., 2001; Nocentini et al., 1993).

Thus, increasing the superficial gas velocity may initially increase $k_L a$ because there will be more gas bubbles and a larger gas–liquid interfacial area. However, further increasing superficial gas velocity could lead to bubble coalescence, which would increase the average bubble diameter and bubble rise velocity and lower the gas residence time (Moilanen et al., 2008). All of these factors would lower gas–liquid mass transfer (Charpentier, 1981). Under these conditions, the impeller fails to disperse the gas properly, and bubbles form a heterogeneous distribution (Bouaifi et al., 2001) and rise easily (Ford et al., 2008). Even if a higher $k_L a$ is achieved by increasing U_G, a lower residence time can still lead to lower fermentation conversion levels (Bredwell et al., 1999).

The effect of increasing gas holdup on the gas–liquid interfacial area (a) is often ambiguous. This is a special situation for the STR because the power input determines the bubble diameter and hydrodynamics (Bouaifi et al., 2001; Nocentini et al., 1993), while the gas flow rate has a driving influence on bubble dynamics in other reactor designs (to be discussed in other chapters). In other words, gas holdup information does not necessarily contain any (quantitative or qualitative) information about bubble diameter and interfacial surface area for STRs such that an increase in gas holdup does not necessarily increase a (Moilanen et al., 2008). However, gas holdup is still reported as an indicator of hydrodynamic performance, gas distribution (Boden et al., 2008), and gas–liquid contacting (Garcia-Ochoa and Gomez, 2004).

Gas–liquid mass transfer correlations, however, typically fail to reflect STR hydrodynamics and predict an increase in mass transfer with an increase in the superficial gas velocity. If these correlations are used improperly, such as during scale-up or outside their representative size and operating conditions, inaccurate $k_L a$ estimates will result. In other words, the correlation form is simply incapable of representing the hydrodynamic situation and fails to decouple events occurring due to the superficial gas velocity and those occurring due to the power concentration. The current state-of-the-art $k_L a$ correlations with respect to STR conditions will be described in detail in Section 6.9, but it is important to realize that a correlation that is capable of communicating a more complete hydrodynamic picture still remains elusive.

6.5   POWER INPUT

Power dissipation has a direct impact on the gas–liquid mass transfer in STRs. As the power dissipation increases, the bubble diameter decreases (Bouaifi et al., 2001; Nocentini et al., 1993), which, in turn, increases the interfacial surface area. Bubbles break apart because the surface tension force is overcome by a higher power density. Coalescence behavior is reduced because bubbles are not allowed enough contact time for film drainage between adjacent bubbles. At the same time, however, a higher power density implies that the collision force is also increased that would enhance coalescing. Thus, an equilibrium point is reached. As the bubble diameter is reduced to $d_B < 1$ mm, the effect of increased power concentration is decreased. These small bubbles tend to have immobile interfaces that are more resistant to mass and energy transfer. These diminishing returns cause low viscosity Newtonian fluids to have an optimal bubble diameter distribution of 1–3 mm (Bouaifi et al., 2001) unless surfactant stabilized microbubbles are produced (Bredwell and Worden, 1998; Worden and Bredwell, 1998).

The power dissipation influence on the liquid-phase mass transfer coefficient (k_L) is highly debated in STRs, especially at higher power densities. The slip velocity model and eddy turbulence model have been used to explain mass transfer, but they come to different conclusions with respect to power. The slip velocity model predicts a decrease in mass transfer with increasing power dissipation while the eddy turbulence model predicts an increase. Linek et al. (2004) postulate that the main reason for the confusion stems from the miscalculation of k_L. They investigated different measurement methods and models used by others and concluded that the slip velocity models were underestimating $k_L a$ and, hence, k_L.

It was recognized that both models represent parallel mass transfer processes, but that STRs were prone to induce higher rates of eddy diffusion and turbulence (rather than molecular diffusion and surface rigidity control) over surface renewal and mass transfer. Hence, shear rates in STRs regulate surface renewal during normal operating conditions. A similar conclusion could be drawn from the border diffusion layer model (Azbel, 1981). It stipulates, for example, that the eddy diffusion coefficient is three orders of magnitude higher than the molecular diffusivity in the viscous sublayer. If power rates were increased, the viscous sublayer would decrease, which in turn would limit the effect of molecular diffusion on mass transfer, and the liquid-phase mass transfer coefficient would increase.

Other researchers have taken an engineering approach. Nocentini et al. (1993) concluded that k_L changes relatively little with power dissipation with respect to a. Thus, any changes in $k_L a$ during an operation are dominated by changes in a and k_L is of little interest (Hoffmann et al., 2007; Stenberg and Andersson, 1988b). STR flow patterns should also be considered and neglecting them can lead to a fivefold underestimation in gas–liquid mass transfer (Linek et al., 2004).

One may conclude that power dissipation is the only variable that definitely and directly influences gas–liquid mass transfer in STRs. It has direct control over bubble diameter (interfacial area) and the liquid-phase mass transfer coefficient; however, increasing the impeller speed to achieve higher gas–liquid mass transfer rates can be inefficient. STRs also operate under an economical constraint. The impeller speed is related to the power draw by an exponential factor of 3. Along with operational limitations such as the motor and gearbox assembly and microorganism shear constraint, most gas–liquid mass transfer processes operate in the power dissipation range of 3000–4000 W/m^3 (Bredwell and Worden, 1998; Oldshue, 1983). Certain specialized processes require extreme power dissipations of 40,000–100,000 W/m^3 (Gezork et al., 2000, 2001). Operation below 30 W/m^3 is not practical with an STR (Gagnon et al., 1998).

Superficial gas velocity does not necessarily control the interfacial area or liquid-phase mass transfer in STRs directly, but influences the gas dispersion efficiency and power dissipation rate. It is very difficult to disconnect the superficial gas velocity from the power concentration in STRs, even under experimental settings. Thus, most $k_L a$ correlations include the power concentration and superficial gas velocity as variables with the power concentration having a larger role than the superficial gas velocity; this will be discussed in more detail in Section 6.9.

6.6   BAFFLE DESIGN

Whirlpool effects are often troublesome for STR operation with low viscosity liquids because they diminish mixing and dispersion (Williams, 2002). Multiphase systems that are mixed with impellers can experience a central vortex that causes recirculation, low power consumption, minimal mixing (Patwardhan and Joshi, 1998), and phase separation (Tatterson, 1994). If the impeller speed is increased even further, this vortex can reach the impeller (Patwardhan and Joshi, 1998). Baffles are often used to alleviate these effects in low viscosity Newtonian liquids, but are avoided with viscous and/or non-Newtonian liquids (Cabaret et al., 2008; Harnby et al., 1992; Williams, 2002). A secondary advantage is that baffles reduce the liquid velocity forcing a larger differential with the impeller velocity. The consequence is that the ungassed power draw is higher in baffled than in unbaffled vessels, potentially increasing eddy turbulence and gas–liquid mass transfer. These events are exclusive to turbulent operation, and baffles are unnecessary in the laminar regime (Tatterson, 1994). Gassed power draw can be expected to be similar in baffled and unbaffled vessels while the impeller is loaded (Gagnon et al., 1998). A less common application for baffles in gas–liquid processes is as fins for heat transfer purposes or as gas- or liquid-feeding device.

Baffle designs are quite numerous and include half, finger, triangle, partition, and bottom baffles with vertical, horizontal, or spiral direction and a wide array of cross-sectional shapes. The basic design, however, is by far the most common because the advantages of non-standard designs are often limited and because the standard design has been widely researched with easily accessible information, scale-up data, and design specs (Ungerman, 2006). The standard design specifies four to eight equally spaced, vertical plates (Williams, 2002) with a width between $T/12$ to $T/10$ (Oldshue, 1983; Tatterson, 1991, 1994). Baffles are often offset from the reactor wall to discourage dead zones. The standard wall clearance is about $1.5\% \ T$ (Paul et al., 2004). Baffles are either welded or bolted to the vessel (Bakker and Oshinowo, 2004). If the vessel is made from or lined by a fragile material, baffles can be supported by a ring placed on top of the tank. The mounting limits these vessels to one or two baffles (Torré et al., 2007).

Prefixes are often used to describe the length or shape. For example, full baffles describe standard baffles having a length equal to the reactor height, half baffles have a length equal to half the height, etc. The shape is often distinguished by a prefix affiliated with a common object. For example, beavertail baffles have a shape that looks like a beavertail (wide in the bottom and tappers off at the top) and C-baffles have a semi-circular cross-sectional shape.

Several STR baffle arrangements have been tested to optimize gas–liquid mass transfer or process time. Surface baffles, which are half baffles located in the upper half of the reactor, limit vortex formation while increasing surface turbulence (Patwardhan and Joshi, 1998). This baffle design induces small vortices which entrain gas more effectively, increasing gas holdup and $k_L a$. The limitation of baffle usage in the lower portion of the reactor volume allows for higher turbulence and enhances sparged gas and power utilization (Gagnon et al., 1998); however, Sivashanmugam and Prabhakaran (2008) noted that such nonstandard baffles also lead to lower impeller power draw, which may decrease mass transfer in that portion of the tank. Regardless, this setup has limited application for batch and semibatch operations.

Lines (2000), however, came to a different conclusion in addressing similar goals of enhanced operation for batch and semibatch modes. He concluded that half baffles in the lower portion of the reactor volume allow for more efficient operation that would induce surface aeration without allowing large vortices to reach deep into the vessel and influence the impeller. The limitation, of course, is the liquid height. The baffles break even at a liquid height-to-reactor diameter ratio of 1.1 for single impeller systems. A two-beavertail baffle system can be used to extend the range to a ratio of 1.5 and any further increase in reactor height relative to its diameter negates this advantage.

These two examples help to point to a common occurrence: the prescription of opposing systems. Both researchers are correct, and their suggestions are useful for their particular task, but they are not universal. Each system and process has an optimal design—including baffles—that may differ from the standard. It is recommended that a proper baffle design be investigated in the initial design stages and should be compared to the standard option; however, most prefer to avoid this situation and use full baffles due to their simplicity, reduced design costs, and known operation compared to nonstandard designs (Oldshue, 1983; Tatterson, 1991, 1994).

6.7   SPARGER DESIGN

Spargers are used to input gas into the reactor and affect impeller power consumption (Ni et al., 1995), critical impeller speed for complete dispersion, and critical impeller speed for suspension (Murthy et al., 2007). Spargers are typically located underneath the impeller with the distance and size being dependent on impeller type. The most common placement is $1D_i$ below the impeller. The sparger diameter is usually smaller than the impeller diameter (Birch and Ahmed, 1997). The orifice holes are usually placed on the sparger bottom. This placement prevents the holes from getting plugged during processes that use high viscosity fluids, a solid phase, or fine particles (Patwardhan and Joshi, 1998). The conditions in the sparger orifice have to be such that the gas flow Reynolds number is in excess of 2100 to ensure that all holes are operational (Rewatkar and Joshi, 1993).

Standard sparger designs include ring, single orifice (point), pipe, porous, and membrane spargers. A ring sparger with equally spaced holes is the most commonly used design for STRs since it provides the most consistent results, uses less power, causes a smaller power drop upon gassing, and is thought to provide higher gas holdup and gas–liquid mass transfer coefficients than other sparger types. To be more exact, a small ring sparger is preferred because it produces higher $k_L a$ values than the large ring sparger or quadruple pipe sparger (Bakker, 1992).

6.7.1 Axial Flow Impellers

The selection of the sparger design has to include a discussion of the accompanying impeller type. An improper selection is often used to explain inconsistencies in published data (Ungerman and Heindel, 2007). The most common source of error is to simply use the standard ring sparger design for all impeller types. This may lead to problems with the down-pumping axial flow impellers (Birch and Ahmed, 1996, 1997; McFarlane and Nienow, 1995; McFarlane et al., 1995; Sardeing et al., 2004b). The standard ring sparger forces gas to flow in the opposite direction of the impeller flow field and induces direct loading such that variations in torque and power draw are easily realized. Larger orifice diameters exasperate this problem by increasing the rate at which cavities are allowed to grow on the impeller blades (Murthy et al., 2007). Replacement of a ring sparger with a pipe sparger can reduce gas holdup by 25% (Rewatkar et al., 1993) and does not address the flow issue since gas is sparged in a similar manner. A single orifice sparger enhances the problem since most of the gas is sent into the impeller center promoting cavity formation (McFarlane et al., 1995).

The influence on axial flow impellers can, however, be negated by using a large ring sparger such that gas exhausts in the impeller periphery. Direct loading is avoided and power draw and torque variations are reduced (McFarlane and Nienow, 1996a) because the gas is sparged into a strong downward stream such that the probability of indirect loading is increased. This regime provides the down-pumping axial flow impeller with steady operation (McFarlane et al., 1995), a lower critical impeller speed for complete dispersion and suspension, while flooding is avoided. Particle suspension is achieved with less power and gas is distributed more uniformly for this arrangement (Murthy et al., 2007). Increasing the impeller diameter has also proven to increase stability (Birch and Ahmed, 1996) and suspension efficiency (Sardeing et al., 2004a). Large axial flow impellers solve the problem due to their stable vortex formations and inherent "periodic vortex shedding" (McFarlane and Nienow, 1995).

Another solution for the loading problem, which is not always practical, is to increase the distance between the sparger and impeller, which makes it more likely that the gas is diverted away from the axial flow impeller by the flow stream, leading to more indirect loading (McFarlane et al., 1995). Increasing the distance also tends to delay the power drop upon gassing, making the separation distance an important design parameter (Garcia-Ochoa and Gomez, 1998). The advantage, however, decreases with increasing viscosity (McFarlane and Nienow, 1996a).

Sparger placement for down-pumping axial flow impellers is suggested to be $0.8D_i$ below the impeller. Larger distances increase the flow instabilities and affect the operating range (McFarlane et al., 1995), whereas smaller distances increase the frequency at which impeller loading fluctuates (McFarlane and Nienow, 1996a). The exact sparger position is often determined by other equipment and the impeller position requirement such that most designs place the sparger $1D_i$ below the impeller and optimize the proper sparger size and type to provide maximum gas holdup and minimum power drop upon gassing (Birch and Ahmed, 1996; Ungerman, 2006).

6.7.2 Radial Flow Impellers

One may think that sparger design may have an important role in the initial bubble diameter and, as such, will influence the interfacial surface area and gas–liquid mass transfer coefficient (Bouaifi et al., 2001; Garcia-Ochoa and Gomez, 1998). Designers should choose a sparger that would offer a smaller initial bubble diameter to increase the efficiency of the operation. However, it has been shown that Rushton-type impellers control the bubble diameter and dispersion such that the sparger choice is noncritical (Garcia-Ochoa and Gomez, 1998). Even if the bubble diameter originating from the sparger is very large, the bubbles are broken apart by the time they reach the impeller and do not affect cavity size, impeller loading, gas–liquid interfacial area, and gas–liquid mass transfer. However, in order to ensure that bubbles pass through the high shear impeller zone, the sparger diameter is suggested to be smaller than the impeller diameter with a standard of $0.8D_i$ (McFarlane et al., 1995; Tatterson, 1991).

This discussion sheds light into the superior performance of a multiple impeller system with a radial–axial flow impeller combination, with a radial flow impeller in the lower position and an axial flow impeller in the upper position. The radial flow impeller is not affected by the sparger type and is able to efficiently disperse small bubbles. The upper impeller is loaded indirectly by the flow field, which it generates, and is able to provide proper mixing conditions. As such, the sparger choice does not affect the performance of the other impellers. If the impellers operate independently, impellers are optimally loaded for gas dispersion and liquid mixing such that progressive reduction in $k_L a$ is minimized and the desired process time can be reduced by >30% (Lines, 2000).

6.8 MICROBIAL CULTURES

Microbial cultures are used as catalysts in bioreactors. Bacteria are the most commonly used culture, but animal, plant, or insect cells have also been implemented (Bliem and Katinger, 1988a). STRs are popular with microorganisms (Vazquez et al., 1997) because STRs enhance feedstock contact, provide pH and temperature uniformity, and maximize mixing (Hoffmann et al., 2008). Their impact on reactor hydrodynamics is mostly indirect. Occasionally, microorganisms retard turbulence if the organic holdup is above 11–15% depending on the species. The other possibility is that the microorganisms produce surface-active agents (van der Meer et al., 1992); however, their most common impact on hydrodynamics is that reaction kinetics may be limited by the environment such that the operational range (power concentration, superficial gas velocity, etc.) may be reduced. As such, it is more constructive to concentrate on the impact that hydrodynamics have on microorganisms.

The most influential sensitivity is for shear gradients that most commonly hinders productivity regardless of the mass transfer situation (Bliem and Katinger, 1988a, 1988b; Hoffmann et al., 2008). Shear gradients damage microorganisms using several mechanisms. The simplest one is cell wall (physical) damage.

This mechanism also separates animal and plant cell applications from the bacterial applications. Bacteria are usually smaller and have stronger cell walls relative to their size than animal or plant cells such that bacterial processes use a power range of $1-5$ W/kg (comparable to chemical processes) while cellular processes use $0.0005-0.1$ W/kg (Bliem and Katinger, 1988a). In other words, smaller cultures are usually able to withstand higher shear gradients because the most damaging eddies have to be on the order of the cell size. As such, animal cell growth rate has been found to reduce with eddies smaller than 130 μm (Bliem and Katinger, 1988b).

Shear gradients may also interfere with cell-to-cell interaction, cell-to-substrate adhesion, and microbial competition. In addition, certain microorganisms prefer to flocculate. Hoffmann et al. (2008)[2] concluded that bacteria, which tended to form elongated filaments, were more prone to shear-induced damage than those which formed cocci (spherical formations). Although the elongated filaments were more advantageous for food collection during calmer operation, the introduction of strong turbulence provided a competitive advantage for cocci-forming bacteria such that those dominated the population at the end of the experiment.

The bacteria's spatial juxtaposition (awareness relative to other bacteria) may also be hindered by turbulence. In the worst-case scenario, the bacteria are not able to make significant contact and are not able to achieve the necessary cell density for optimal operation (Bliem and Katinger, 1988b) or are not able to make syntropic relationships with other bacterial cultures (Hoffmann et al., 2008). The result is that start-up performance is very poor with minimal or insignificant conversion while long-term performance is not hindered in a bacterial mixture that allows competition and has at least one shear-tolerant species. Conditioning with feast and famine cycles improved recovery time and tolerance to feed and shear shocks (Hoffmann et al., 2008).

Thus, bioreactors using shear-sensitive microorganisms have to minimize cellular damage, maximize feedstock transfer to microorganisms, and maximize mixing. The latter requirements are important because the bacterial structure may change during starvation mode to make the culture even more susceptible to cell wall degradation. This situation is true for mycelia (fungi) and may be applicable to other branching bacteria. A healthy specimen, shown in Figure 6.21a, has relatively thick branches without vacuoles (empty pockets). As the bacteria starves (Figure 6.21b), it reduces the number of branches and starts to consume its internal reserves, which leads to the formation of vacuoles. As the number and size of vacuoles increases, the cell wall strengthens and its ability to resist environmental stresses decrease. As starvation is extended, the specimen will consume as much of its own mass as it can (which depends on the species) and vacuoles will dominate its structure, as can be seen in Figure 6.21c. At this point, the microorganism is easily and significantly

---

[2]The conclusions are based on a particular set of microbial species and have not been verified by other researchers. According to their published article, Hoffmann et al. (2008) experimented with the same sized vessels at a different impeller speed. Since turbulence and power concentration grow nonlinearly with impeller speed, their results and conclusions may not be universally applicable or practical at larger scale.

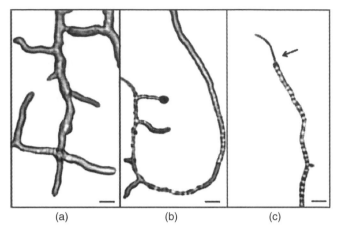

Figure 6.21 Bacteria starvation; (a) a healthy specimen, (b) a bacterium under starvation conditions, and (c) extensive starvation with the formation of many vacuoles (empty pockets) (Pollack et al., 2008).

damaged by shear gradients (Hoffmann et al., 2008). Energy and mass are diverted to the tip, as pointed out in Figure 6.23c, in order to search for a food source. This tip is of solid construction relative to the main body. Insufficient mixing can have similar effects in that the reactor volume is partially in starvation mode and not producing an optimal amount (if any) of product in those regions (Shewale and Pandit, 2006).

Microorganisms and their reaction kinetics may start out being gas–liquid mass transfer limited, but the process and changing environment may change the limiting factor to temperature or pH level. Bacteria are classified by their temperature preference into mesophilic or thermophilic families. Mesophilic bacteria operate optimally at about 30 °C with a sharp drop-off in efficiency as temperature approaches 50 °C. These cultures are used more widely because they are easier to control and produce a more consistent product, but are generally able to convert only 40% of the biological matter in 30–40 days. Thermophilic bacteria, on the other hand, prefer temperatures of about 60 °C and have proven conversion rates up to 48% in just 10 days (Demirel and Yenigün, 2002; Ros and Zupancic, 2002). Acidity is quite variable (although not for a specific bacterial culture) and can range from pH 4.3 to pH 7.9 for anaerobic bacteria (Demirel and Yenigün, 2002). Output can be maximized for acid-sensitive processes using syntropic relationships (i.e., volatile fatty acid-oxidizing bacteria and hydrogen-utilizing methanogens) (Hoffmann et al., 2008).

Furthermore, microorganism production and conversion processes often introduce unwanted by-products or create products that negatively affect bioreactor operation. For example, protein-producing microorganisms, which are often used in pharmacokinetics, produce a mixture over time that is damaging to the bacteria aside from the surface-active agent properties of the protein. Shear is tolerated by the microorganisms in this mixture, but air–liquid interfaces, which are naturally very common in gas–liquid processes, can lead to denaturation (Titchener-Hooker and Hoare, 2008). Batch and semibatch stirred-tank bioreactors are also influenced by the accumulation of products in the volume, which can significantly change liquid-phase properties. Although the production is certainly welcome, it can lead to the process being tail dominated (e.g., process time controlled by last 20%) or creating an extremely viscous liquid phase, which, in most cases, forces the operation to cease.

Many industries, in which the stirred-tank bioreactors are being implemented, require production to be very consistent and/or the design phase to be completed quickly. For example, it is common in the biopharmaceutical industry to start the design phase once approval of a drug has been secured; however, the design process requires a significant amount of time during which the patent clock is ticking. Hence, expensive delays are very common (Titchener-Hooker and Hoare, 2008).

The need for better results has led to the implementation of process and genetic engineering. Process engineering is described throughout this section. Its goal is to optimize the conditions such that production and/or conversion are increased; however, it can be difficult to predict hydrodynamic effects on microorganisms. The answer has been to carefully test microorganism on the micro (experimental) scale and implement genetic engineering techniques to create more shear-resistant strains (Zeng and Deckwer, 1996). Process engineering, however, prevails in practice as genetic engineering has not been able to produce very resistive strains (although productivity has been increased) such that stirred-tank bioreactors are limited in their power dissipation rates.

Impellers and their arrangements also need to be modified for solid suspension. Impellers used in single impeller configurations for solid suspension are preferred to have a bottom clearance up to $D_i/3$. This clearance is dependent on the process and impeller type. In order to provide proper mixing, the tanks are shorter than standard gas–liquid designs with an aspect ratio (H/T) about 0.6–0.7, and a second impeller is warranted for a ratio above 1. Solids will settle and stack around the wall if the reactor has a diameter that is too large or an impeller that is too small relative to the tank diameter (Tatterson, 1994). The bottom impeller in a multiple impeller arrangement is more important for solid suspension because the upper impeller can only suspend as much as is fed by its partner (Oldshue, 1983); however, an improper spacing can cause the upper impeller to interfere with the bottom impeller, usually degrading suspension (Tatterson, 1994).

Certain impellers create flow patterns that induce suspension at lower impeller speeds. For example, the down-pumping PBT is more efficient at solid suspension due to its axial bottom-lifting flow than radial impellers or the up-pumping PBT. Radial impellers, such as the Rushton-type turbines, are not efficient suspension impellers (Tatterson, 1994) because the particle lifting occurs through axial flow impeller suction which usually uses about half the impeller flow. Therefore, the Rushton-type turbine (RT) requires as much as three times the power for the same level of suspension as the down-pumping PBT (Oldshue, 1983). The advantage and the use of down-pumping axial flow impellers for solid suspension places more relevance on the selection and performance of the sparger design, which is explained in Section 6.7.

Operators are potentially presented with an awkward choice of achieving better gas breakage with a radial flow impeller or better suspension with an axial flow impeller. The decision really comes down to the gas flow rate. Since most biological processes require a significant amount of gas to proceed efficiently, the potential instabilities present in the down-pumping PBT and necessary gas–liquid mass transfer have led to a close clearance Rushton-type turbine (clearance of $H/4$ and $D_i = T/2$), the safest impeller choice for three-phase systems if the impeller is operated under complete dispersion conditions (Ulbrecht and Patterson, 1985). The close clearance RT has a significant axial component while still preserving some of its gas-breakage capabilities (Harnby et al., 1992).

Baffle designs for gas–liquid stirred-tank bioreactors can be used with solids that have a similar density as the liquid. If the solid phase is denser than the liquid, baffles should be much thinner if they are used at all. This is due to the fact that the decreased level of turbulence may lead to stratification, dead zones, and/or recirculation loops near the baffles. These occurrences can cause excessive buildup of solid material and poor performance. Therefore, baffles used for this purpose may have a plate thickness as small as $T/24$ (Oldshue, 1983).

6.9 CORRELATION FORMS

The mass transfer theories from Chapter 3 have been used to define operational boundaries for STRs. The hydrodynamic complexity of gas–liquid flows in STRs has curbed the practical application of theoretical models. Currently, a universal model or correlation has not been successfully developed and applied over a wide range of system configurations, scales, operating conditions, or inputs (Garcia-Ochoa and Gomez, 2004; Kawase and Moo-Young, 1988), which is a major disadvantage of STRs. The situation is even worse in gas–liquid–solid processes due to lack of relevant data and increased hydrodynamic complexity (Garcia-Ochoa and Gomez, 2004; Murthy et al., 2007). Design and scale-up are implemented using an iterative method, where previous works are used as the first step in a trial-and-error process (Benz, 2008), which continues until the reactor produces the desired conditions (Bliem and Katinger, 1988a).

In order to streamline the process, empirical, semi-empirical, or dimensionless group correlations have been proposed of which empirical correlations are the most commonly used (Bliem and Katinger, 1988a; Garcia-Ochoa and Gomez, 2004; Kawase and Moo-Young, 1988). Variables are chosen based on a gas–liquid mass transfer model believed to dominate the process as well as practical considerations. These correlations have been used in design and scale-up, but are only valid for the particular system and operating range (Benz, 2008; Garcia-Ochoa and Gomez, 2004) and have an accuracy of about $\pm 30\%$ at best (Benz, 2008) even if they are in dimensionless form (Bliem and Katinger, 1988a).

In general, STR correlations lack the ability to account for reactor variations and the specific hydrodynamic state; however, they can be used as an estimate if the corresponding systems have geometric, hydrodynamic, and flow pattern similarity. Industrial designs based on experimental setups should be referenced to systems that properly reflect the desired industrial settings and circumstances (Bliem and Katinger, 1988b; Kapic and Heindel, 2006; Kapic et al., 2006). For example, most experimental setups are operated in a Reynolds range of 5000–10,000 while large industrial units operate in the range 10,000–100,000 (Bliem and Katinger, 1988b). The disconnect becomes more obvious when we consider that a significant number of experiments include data in a power range of 5–10 kW/m^3, while industrial units are rarely operated beyond 3 kW/m^3 (Benz, 2008) or below 500 W/m^3 (Bliem and Katinger, 1988b). This is mainly due to the theoretical models' requirement of a well-mixed state for the liquid phase while industrial units require minimal costs. Hence, mixing and process time often grows much quicker with scale than anticipated and often leads to production difficulties (Nienow, 1996).

A simple approach for finding an appropriate gas–liquid mass transfer correlation is to break $k_L a$ down into its components (a and k_L) and find separate correlations for each, after which those components would be combined to generate a $k_L a$ correlation. It is very convenient to start with the interfacial area since an applicable theoretical correlation is readily available (Figueiredo and Calderbank, 1979)

$$a = \frac{6\epsilon}{d_{SM}} \tag{6.2}$$

where ϵ is the gas holdup and d_{SM} is the Sauter mean bubble diameter.

Next, an assumption is made on the mass transfer model, and the two terms are combined. If a film model is used, k_L is assumed to be inversely proportional to d_B. Therefore,

$$k_L a = \frac{\beta \epsilon}{d_B^n} \tag{6.3}$$

where d_B is the mean bubble diameter and β and n are the fitted constants. If a penetration model is used, k_L is correlated with a power concentration, usually defined in terms of P_G/V_L:

$$k_L a = \frac{C\epsilon(P_g/V_L)^A}{d_B} \tag{6.4}$$

where A and C are fitted values, P_g represents the gassed power, and V_L is the liquid volume within the STR. It should be noted that Eqs (6.3) and (6.4) fall in the semiempirical family of correlations. Gas holdup correlations, which are presented in Table 6.1, could be substituted into Eq. (6.4) to obtain the gas–liquid mass transfer approximations.

Some researchers have found a better statistical fit by using the total power, P_{tot}, defined as the gassed impeller power plus the buoyancy power of the sparged gas (Moucha et al., 2003). This has been done because the sparged gas power has been shown to impact gas–liquid mass transfer in a similar manner to impeller power (Stenberg and Andersson, 1988b). On the other hand, Gagnon et al. (1998) came to the conclusion that the impeller transfers energy to the fluid in the impeller zone and

TABLE 6.1 Gas Holdup Correlations for Stirred-Tank Reactors

$$\varepsilon_G = C(P_G/V)^a U_G^b$$

Researchers	C	a	b	U_G Range (mm/s)	P_G/V Range (W/m³)	Gas	Liquid	Impeller Configuration	H (m)	T (m)
Bouaifi et al. (2001)	22.4	0.24	0.65	3.7–18.1	0–1000	Air	Tap water	2 (PBT, A310, A315)	0.86	0.43
	24.8	0.24	0.65					2 (RT, PBT, A310, A315)		
Figueiredo and Calderbank (1979)	0.34	0.25	0.75	n/a	n/a	n/a	n/a	n/a	n/a	n/a
Linek et al. (1994)	0.23	0.31	0.73	n/a	n/a	n/a	Water	2 (RT)	n/a	n/a
	0.05	0.49	0.58				0.5 M Na₂SO₄			
Linek et al. (1996)	0.07	0.36	0.54	0–0.00848	0–4000	Nitrogen and oxygen	Water	4 (RT)	0.19, 0.38, 0.57, 0.76	0.19
	0.29	0.30	0.73							
	0.02	0.59	0.44				0.5 M Na₂SO₄			
	0.05	0.46	0.52							
Moucha et al. (2003)	0.02	0.63	0.52					1 (TXU)		
	0.05	0.49	0.57					2 (TXU)		
	0.17	0.42	0.75					3 (TXU)		
	0.05	0.42	0.53					1 (TXD)		
	0.14	0.32	0.59					2 (TXD)		
	0.26	0.36	0.74					3 (TXD)		

Reference						Gas	Liquid	Configuration		
	0.05	0.47	0.58	2.12, 4.24, 8.48	0–1300	Oxygen	0.5 M Na$_2$SO$_4$	1 (PBD)		0.29
	0.08	0.52	0.71					2 (PBD)		
	0.35	0.29	0.80					3 (PBD)	0.29, 0.58, 0.87	
	0.19	0.44	0.75					3 (RT, 2TXU)		
	0.24	0.45	0.82					3 (RT, 2TXD)		
	0.10	0.52	0.71					3 (RT, 2PBD)		
	0.03	0.51	0.50					2 (RT, TXU)		
	0.11	0.42	0.66					2 (RT, TXD)		
	0.05	0.50	0.57					2 (RT, PBD)		
	0.02	0.62	0.57					1 (RT)		
	0.05	0.49	0.58					2 (RT)		
	0.04	0.54	0.58					3 (RT)		
Pinelli et al. (2003)	0.10	0.28	0.48	7.0–14.0	200–2,600	Air	Water	2 (BT-6)	0.96, 1.44	0.48
	0.25	0.24	0.65							
Vasconcelos et al. (2000)	0.10	0.37	0.65	n/a	n/a	n/a	Water	2 (RT)	n/a	n/a
Whitton and Nienow (1993)	1.28	0.26	0.66	11.8–33.1	1.4–1.7 rps	n/a	n/a	1 (RT)	n/a	0.61, 2.67
Yawalker et al. (2002)	0.56	0.25	0.41	4.0–15.7	0.7–8.3 rps	Air	Water	1 (PBD)	0.57	0.57
	0.52	0.25	0.40							

Note the correlation by Whitton and Nienow (1993) is as cited by Yawalkar et al. (2002b).

109

that the gas phase does not influence this energy transfer directly. Most researchers, therefore, ignore the effect of the buoyancy force and the gas expansion energy in STRs. The power dependence can also be accomplished by using the impeller speed N or a combination of the impeller diameter D_i and N. Correlations based on D_i and N are more scale dependent than the power concentration. Therefore, correlations will be compared on a power concentration basis.

Equation (6.2) has two control and measurement difficulties: gas holdup and bubble diameter. Gas holdup can have dynamic features and its measurement may be difficult to implement in a reactor control scheme. The bubble diameter, especially in heterogeneous flow, is not uniform and its measurement requires visual inspection, which is troublesome in industrial or large-scale experimental units. Therefore, the representative control inputs (power concentration and superficial gas velocity) for the bubble diameter and gas holdup can be used (Moo-Young and Blanch, 1981). The two inputs represent forces acting on the bubbles such as the drag, buoyancy, inertial, and the surface tension forces. These substitutions have led to the most widely used empirical correlation form:

$$k_L a = C\left(\frac{P_g}{\forall_L}\right)^A U_G^B \qquad (6.5)$$

where A, B, and C are fitted constants and U_G is the superficial gas velocity. Correlations based on Eq. (6.5) are presented in Table 6.2.

Although Eq. (6.5) has been widely used in practice (Kawase and Moo-Young, 1988), it conveys very little information about the particular system and mass transfer mechanism (Moucha et al., 2003). For example, impeller-operating regimes, flow patterns, forces, and liquid and gas properties are not accounted for in this correlation. The particular results are global representations of the system and have little hope of representing microscale effects that are vital in gas–liquid mass transfer (Bouaifi et al., 2001). As shown in Figure 6.22, these issues have caused a wide variability in the available gas–liquid mass transfer (and gas holdup) correlations and their dependence on power concentration and superficial velocity (Yawalkar et al., 2002b). For example, the power concentration exponent ranges from 0.32 (Gagnon et al., 1998) to 1.32 (Linek et al., 1996a), and the U_G exponent ranges from 0 (Linek et al., 2005a) to 0.77 (Moucha et al., 2003). Hence, depending on the particular correlation, estimated $k_L a$ values may vary by a factor of 2 or more for a fixed power concentration.

It would be fair to conclude that the choice of fitted variables is based on statistical methods with little thought to implications on mass transfer models and forces acting on the system, thus reducing the usefulness outside the experimental range. For example, the exponents depend on the reactor size (Stenberg and Andersson, 1988b), system geometry, experimental range, and experimental method (Figueiredo and Calderbank, 1979). Therefore, one may conclude that Eq. (6.5) is leaving out major variables.

A second problem with Eq. (6.5) is that the two measured variables, power concentration and superficial gas velocity, are inherently connected through hydrodynamics. Their experimental and statistical separation is a very difficult task and is

TABLE 6.2 Standard Gas–Liquid Mass Transfer Correlations Based on Eq. (6.5)

$k_L a = C(P_G/V)^A U_G^B$

Researchers	C	A	B	U_G Range (mm/s)	P_G/V Range (W/m³)	Gas	Liquid	Impeller Configuration	H (m)	T (m)
Bcuaifi et al. (2001)	0.022	0.50	0.60	3.72–18.10	0–1,000	Air	Tap water	2 (RT, PBT, A310, A315)	0.86	0.43
Fu̱asova et al. (2007)	0.006	0.74	0.53				Water			
	0.001	1.22	0.51				0.5 M Na$_2$SO$_4$			
	0.002	0.95	0.58				Sokrat 44			
	0.006	0.69	0.53				Water			
	0.001	1.25	0.57				0.5 M Na$_2$SO$_4$	3 (RT in bottom; PBD, PBU, TXD, LTN, NS)	0.29, 0.58, 0.87	0.29
	0.001	1.00	0.50	2.12, 4.24, 8.48	0–1,200	Air	Sokrat 44			
	0.006	0.69	0.53				Water			
	0.001	1.20	0.58				0.5 M Na$_2$SO$_4$			
	0.001	0.95	0.53				Sokrat 44			
	0.005	0.72	0.50				Water			
	0.000	1.22	0.45				0.5 M Na$_2$SO$_4$			
	0.002	0.95	0.57				Sokrat 44			

(continued)

TABLE 6.2 *(Continued)*

$k_L a = C(P_G/V)^A U_G^B$

Researchers	C	A	B	U_G Range (mm/s)	P_G/V Range (W/m³)	Gas	Liquid	Impeller Configuration	H (m)	T (m)
Gagnon et al. (1998)	0.5	0.01	0.86	0–1.2	0.001–30	Air	Water	1 (RT)	0.55	0.23
	0.8	0.02	0.92					2 (RT)		
	0.2	0.02	0.72					2 (RT)		
	0.3	0.03	0.79					3 (RT)		
	0.4	0.01	0.87		0.001–100			HR		
	0.5	0.06	0.88					HRB		
	12.2	0.57	0.47		30–10,000			1 (RT)		
	2.9	0.83	0.50					2 (RT)		
	3.2	0.79	0.48					2 (RT)		
	9.2	0.60	0.50					3 (RT)		
	15.4	0.32	0.50		100–10,000			HR		
	31.1	0.38	0.65					HRB		
Gezork et al. (2001)	0.005	0.59	0.27	0–130000	0–100,000	Air	Water	1 or 2 (RT)	0.29	0.29
	0.004	0.70	0.18				0.2 M Na_2SO_4			

Hickman (1988)	0.043	0.40	0.57	2–17	50–3500	Air	Water	1 (RT)	n/a	0.6
	0.027	0.54	0.68							2
Kapic and Heindel (2006)	0.040	0.47	0.60	0.5–7.2	6.7–13.3 rps	Air	Tap water	1 (RT)	0.21	0.21
Kapic et al. (2006)	0.026	0.61	0.61	0.5–7.2	6.7–13.3 rps	CO	Water	1 (RT)	0.21	0.21
	0.001	1.23	0.55			Air	$0.5\,M\,Na_2SO_4$	2 (RT)	0.21	0.21
Linek et al. (1987)	0.005	0.59	0.40	2.12, 4.24	100–3500	Air	Water	1 (RT)	n/a	0.29
	0.001	0.95	0.40				$0.5\,M\,Na_2SO_4$			
Linek et al. (1990)	0.000	1.24	0.40	n/a	n/a	Air	Water	n/a	n/a	n/a
Linek et al. (1994)	0.000	1.21	0.40	n/a	n/a	Air	$0.3\,M\,Na_2SO_4$	3 (RT)	n/a	n/a
Linek et al. (1996)	0.009	0.63	0.54	0–8.48	0–4,000	Nitrogen and oxygen	Water	4 (RT)	0.19,	0.19
	0.006	0.68	0.50						0.38,	
	0.001	1.17	0.46						0.57,	
	0.001	1.32	0.33				$0.5\,M\,Na_2SO_4$		0.76	

(continued)

TABLE 6.2 (*Continued*)

$k_L a = C(P_G/V)^A U_G^B$

Researchers	C	A	B	U_G Range (mm/s)	P_G/V Range (W/m³)	Gas	Liquid	Impeller Configuration	H (m)	T (m)
Linek et al. (2005)	0.00003	1.18	0.00	1.8, 3.6, 5.4	10–1,500	Pure oxygen or air	0.8M Na₂SO₄	1 (RT)	0.12	0.12
	0.00003	1.16	0.00				0.8M Na₂SO₄ and Sokrat 44 (3% vol)			
	0.00004	1.08	0.00				0.8M Na₂SO₄ and CMC TS.5 (0.2wt%)			
	0.00022	0.77	0.00				0.8M Na₂SO₄ and CMC TS.5 (0.6 wt%)			
	0.00013	0.73	0.00				0.8M Na₂SO₄ and Ocenol (3 ppm by volume)			
	0.00003	1.15	0.00				0.8M Na₂SO₄ and PEG 1000 (100 ppm by mass)			

Moucha et al. (2003)									
	0.001	1.25	0.63					1 (TXU)	
	0.002	1.20	0.74					2 (TXU)	
	0.002	1.20	0.70					3 (TXU)	
	0.018	0.88	0.77					1 (TXD)	
	0.006	1.01	0.69	2.12, 4.24, 8.48	0–1,300	Oxygen	0.5 MNa$_2$SO$_4$	2 (TXD)	
	0.009	1.01	0.75					3 (TXD)	
	0.002	1.05	0.46					1 (PBD)	
	0.001	1.11	0.39					2 (PBD)	0.29, 0.58, 0.87
	0.001	1.15	0.51					3 (PBD)	0.29
	0.003	1.01	0.54					3 (RT, 2TXU)	
	0.003	1.04	0.51					3 (RT, 2TXD)	
	0.001	1.14	0.46					3 (RT, 2PBD)	
	0.001	1.06	0.36					2 (RT, TXU)	
	0.002	1.02	0.47					2 (RT, TXD)	
	0.001	1.13	0.43					2 (RT, PBD)	
	0.000	1.24	0.34					1 (RT)	
	0.001	1.23	0.56					2 (RT)	
	0.001	1.24	0.47					3 (RT)	
	0.001	1.19	0.55					All Data	

(continued)

TABLE 6.2 (*Continued*)

$k_L a = C(P_G/V)^A U_G^B$

Researchers	C	A	B	U_G Range (mm/s)	P_G/V Range (W/m³)	Gas	Liquid	Impeller Configuration	H (m)	T (m)
Ni et al. (1995)	1.645	0.50	0.64	0.5 vvm	0–10,000	Air	Yeast and feed	2 (RT)	0.195 0.37	0.12 0.24 3
Nocentini et al. (1993)	0.015	0.59	0.55	0.1–0.7 vvm	100–10,000	Air	Water and glycerol	4 (RT)	0.70, 0.93	0.23
Pinelli et al. (2003)	0.018	0.37	0.29	7.0–14.0	200–2,600	Air	Water	2 (BT-6)	0.96, 1.44	0.48
	0.005	0.59	0.40							
Puthli et al. (2003)	0.0001	0.58	0.43	1.7–6.4	300–600 rpm	Air	Water	1 (RT)	0.22	0.13
	0.0001	0.61	0.43					2 (RT, PBT)		
	0.0002	0.67	0.53					3 (RT, 2PBT)		
	0.0001	0.68	0.53				0.25% (w/v)CMC	3 (RT, 2PBT)		
	0.0001	0.66	0.54				0.375% (w/v)CMC			
	0.0022	0.36	0.56				0.50% (w/v)CMC			

Riggs and Heindel (2006)	0.051	0.51	0.65	0.5–2.88	200–600 rpm	CO	Water	1 (RT)	0.21	0.21
Smith et al. (1977)	0.010	0.48	0.40	4.0–20.0	20–5,000	Air	Water	1 (RT)	0.60 0.61 0.91 1.63	0.44 0.61 0.91 1.83
	0.020	4.75	0.40	4.0–46.0	0–700	Air	0.11 Na$_2$SO$_4$			
Van't Riet (1979)	0.026	0.40	0.50	n/a	n/a	n/a	Coalescent	1 (RT)	n/a	n/a
	0.020	0.70	0.20				Non-coalescent			
Vasconcelos et al. (2000)	0.006	0.66	0.51	n/a	n/a	n/a	Water	2 (RT)	n/a	n/a
Zhu et al. (2001)	0.031	0.40	0.50	1.0–7.5	100–1,500	Air	Water	1 (RT)	n/a	0.39

Note: The correlations by Hickman (1988), Linek et al. (1987), Van't Riet (1979), and Zhu et al. (2001) are as cited by Yawalkar et al. (2002a).

117

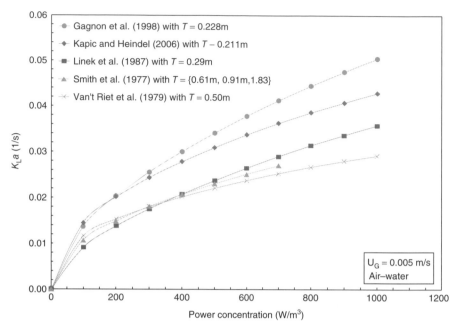

Figure 6.22 Sample mass transfer results for an STR with a single Rushton-type impeller from the correlation based on Eq. (6.5).

rarely achieved (Garcia-Ochoa and Gomez, 2004; Stenberg and Andersson, 1988b). As stated in the STR literature, questionable measurement techniques (Linek et al., 1996a) and experimental assumptions (Nocentini, 1990) make a large number of the STR correlations inadequate and conflicting (Nishikawa et al., 1984).

The need to define the system more accurately has led to several improvements on Eq. (6.5). Fujasova et al. (2007), and Moucha et al. (2003) have added a gassed power term N_{Pg} to account for the impeller type:

$$k_L a = C \left(\frac{P_{tot}}{\forall_L} \right)^A U_G^B N_{Pg}^D \qquad (6.6)$$

where A, B, C, and D are fitted constants. The (gassed) power number is defined as

$$N_{Pg} = \frac{P_g}{\rho N^3 D_i} \qquad (6.7)$$

where ρ is the liquid density and P_g is the gassed impeller power draw. Cavity effects are reflected by a decrease in the gassed power draw and hence the gassed power number. Although the power number helps to identify the particular impeller type, it does not identify the flow regime in which the impeller is operating.

Fujasova et al. (2007) and Linek et al. (1996a, 1996b) went further and attempted to account for multiple impellers. Linek et al. (1996a, 1996b) pointed to the fact that different amounts of gas pass through each impeller zone. It was suggested that one correlation should be formulated for the bottom section while the other impellers were packaged into a separate correlation. Linek et al. (1996a) presented a weighted power term that could be used in Eq. (6.5). The bottom impeller would account for 25% of the power draw while the other impellers accounted for 75%. Fujasova et al. (2007) added a gassed-to-ungassed power ratio term that would account for the different amounts of gas passing through each impeller and communicate the impeller performance upon gassing relative to its ungassed state (McFarlane and Nienow, 1996b). Hence, the total mass transfer coefficient was determined from an average of the mass transfer coefficients of each impeller section, which was calculated using

$$k_L a = C P_{tot}^A U_G^B (P_g/P_0)^D N_P^E \qquad (6.8)$$

where A, B, C, D, and E are the fitted constants and P_0 is the ungassed power.

The previous equations assume that liquid properties do not change throughout the process or hope that any viscosity changes or liquid type is reflected in the power concentration term (Ni et al., 1995). These circumstances lead Eq. (6.5) to fail when the fluid viscosity changes during the process or when the fluid exhibits non-Newtonian behavior. Several authors have suggested the inclusion of a viscosity term to account for these effects (Garcia-Ochoa and Gomez, 1998, 2004; Linek et al., 2005a; Ogut and Hatch, 1988; Tecante and Choplin, 1993):

$$k_L a = C \left(\frac{P_g}{\forall_L} \right)^A U_G^B \mu_a^C \qquad (6.9)$$

where A, B, C, and D are fitted constants and μ_a is the apparent viscosity based on the Ostwald–de Waele model. A Casson viscosity (Garcia-Ochoa and Gomez, 1998, 2004) and a liquid-to-water viscosity ratio (Nocentini et al., 1993) have also been used successfully. Unfortunately, Eq. (6.9) also shares the same disadvantages as Eq. (6.5)—it is limited to similar systems operating over similar ranges.

Flow patterns and impeller loading conditions have not been considered thus far. They are important characteristics of a system, but are identified through indirect and inefficient means. Yawalkar et al. (2002a, 2002b) and Kapic et al. (Kapic, 2005; Kapic and Heindel, 2006; Kapic et al., 2006) have attacked this prospect using a correlation based on a complete dispersion impeller speed N_{CD}. It defines the point at which complete gas dispersion is achieved at the minimal power input (e.g., see Figures 6.6 and 6.7). Yawalkar et al. (2002a) proposed

$$k_L a = C \left(\frac{N}{N_{CD}} \right)^A U_G^B \qquad (6.10)$$

where A, B, and C are fitted constants. Various correlations of this form are summarized in Table 6.3. It was noted that this $k_L a$ correlation was independent of reactor geometry, impeller type, position of the impeller, and sparger if operated at the

TABLE 6.3 Gas–Liquid Mass Transfer Correlations Based on Eq. (6.17)

$k_L a = C(N/N_{CD})^A U_G^B$

Researchers	C	A	B	U_G Range (mm/s)	N/N_{CD}	Gas	Liquid	Impeller Configuration	H (m)	T (m)
Chandrasekharan and Calderbank (1981)	2.7	1.15	0.96	3.5–18	0.85–1.56	Air	Water	1 (RT)	n/a	1.22
Calderbank (1958), van't Riet (1979)	2.8	1.14	0.97	5–36	0.67–2.65	Air	Water	1 (RT)	n/a	0.5
Hickman (1988)	4.3	1.35	1.04	2–17	066–1.61	Air	Water	1 (RT)	n/a	0.60, 2
Linek et al. (1987)	5.2	1.69	1.09	2.12–4.24	1.4–3.8	Air	Water	1 (RT)	n/a	0.29
Smith et al. (1977)	2.4	1.38	0.96	4.4–46	0.25–2.44	Air	Water	1 (RT)	0.61 0.91 1.63	0.61 0.91 1.83
Smith (1991)	6.5	1.44	1.12	n/a	n/a	Air	Water	1 (RT)	n/a	0.6, 2.7, 2.7
Smith and Warmoeskerken (1985)	12.6	1.54	1.27	n/a	n/a	Air	Water	1 (RT)	n/a	0.44
Whitton and Nienow (1993)	3.5	1.17	1.00	n/a	n/a	Air	Water	1 (RT)	n/a	0.61, 2.67
Yawalkar et al. (2002)	3.4	1.46	1.00	n/a	n/a	Air	Water	1 (RT)	n/a	
Zhu et al. (2001)	3.3	1.14	0.97	1.0–7.5	n/a	Air	Water	1 (RT)	n/a	0.69

Adapted from Yawalkar et al. (2002a)

Note: The correlations by Calderbank (1958), Linek et al. (1987), Smith and Warmoeskerken (1985), Van't Riet (1979), and Zhu et al. (2001) are as cited by Yawalkar et al. (2002a).

same N/N_{CD} ratio. For example, Kapic and Heindel (2006) correlated data from several different sources and STR sizes into a single correlation[3]:

$$\frac{k_L a}{U_G^A} = C\left(\frac{N}{N_{CD}}\right)^B \left(\frac{D_i}{T}\right)^D \tag{6.11}$$

where A, B, C, and D are fitted constants. These results are shown in Figure 6.23. This correlation does a good job of fitting the experimental data for a variety of tank sizes. However, the correlation is valid only for STRs with Rushton-type impellers operating in the completely dispersed flow regime.

In order to achieve more consistent scale-up success, full-scale and pilot reactors should have geometric and hydrodynamic similarities. The pilot reactor used for preliminary design should have operational relevance to the full-scale unit especially with respect to the impeller speed (Kapic and Heindel, 2006). Yawalkar et al. (2002a) went further to address the question of scale-up by defining Eq. (6.10) in terms of gas volume per unit liquid volume per minute (vvm):

$$k_L a = C\left(\frac{N}{N_{CD}}\right)^A (\text{vvm})(T)^B \tag{6.12}$$

where A, B, and C are fitted constants. Equation (6.12) proved more effective at predicting $k_L a$ for larger volumes. This scaling approach is, however, limited because it requires the pilot reactor to provide a vvm 10 to 20 times higher than the full-scale version (Benz, 2008).

A much simpler approach can be taken with systems having geometric and hydrodynamic similarities. It is often thought the gas–liquid mass transfer coefficient could be increased by increasing the amount of gas in the reactor volume. This idea has been in extensive use in multiple impeller systems because of the difficulty in determining k_L (Moucha et al., 2003). Total gas holdup ϵ was used to represent this concept (Moucha et al., 2003) and is defined as

$$\epsilon = \frac{\forall_g}{\forall_g + \forall_L} \tag{6.13}$$

where $\forall_g$ and $\forall_L$ are the gas and liquid volumes, respectively. A very simple form of the gas–liquid mass transfer coefficient proposed by Stenberg and Andersson (1988b) is

$$k_L a = C\epsilon \tag{6.14}$$

where C is a fitted value.

Stenberg and Andersson (1988b) found that Eq. (6.14) accounted for over 93% of their data. It was concluded that the gas–liquid mass transfer coefficient variations in a system are explained by changes in the interfacial area due to its connection to gas holdup by Eq. (6.2). They also found that a large error

---

[3] The original correlation in Kapic and Heindel (2006) inadvertently transposed the D_i/T fraction.

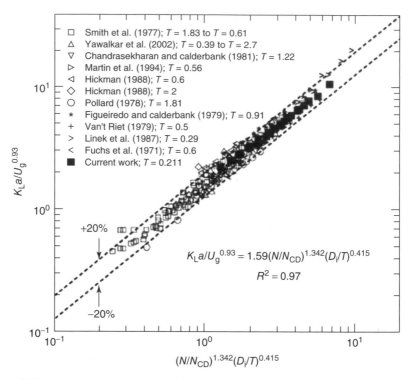

Figure 6.23 Scale-up correlations developed by Kapic and Heindel (2006). Note the original correlation by Kapic and Heindel inadvertently transposed the D_i/T fraction and has been corrected here.

in the bubble diameter would create a relatively small error in $k_L a$, which was shown using Eq. (6.3). Unfortunately, the correlation does not communicate any information regarding mass transfer process, hydrodynamics, or fluid properties and is somewhat unconventional since no other authors have presented their work in this form. The reactor design and operational uniqueness are not accounted for with Eq. (6.14). In addition, Moucha et al. (2003) found that axial flow impellers, which could provide higher gas holdup, would still underperform radial flow impellers in terms of mass transfer due to their inability to provide smaller bubble diameters. Therefore, Eq. (6.14) will fail during scale-up, but could be used as a first approximation for a similar design and size.

The scale-up problems arise from the fact that all STR gas–liquid mass transfer correlations are empirical. They are, for the most part, unable to account for hydrodynamic or liquid property changes with scale and time. Extensive attempts have been made in using nondimensional groups, especially toward solving gas–liquid processes involving non-Newtonian liquids. These correlations tend to be more complicated and require numerous static, but only few dynamic, inputs. One of the simplest correlations is presented by Ogut and Hatch (1988), which involves four dimensionless groups and requires six inputs. One of the more complicated forms, proposed by Nishikawa et al. (1981), uses 12 dimensionless groups because the model tries to explain operation during low power input leading to the STR behaving like a bubble column. A general word of warning would be that if a correlation is based on statistical fitting, it runs the risk that the fit is achieved by probability rather than causality. The result could be that the correlation predicts improbable outcomes when extended beyond the operating range (Stenberg and Andersson, 1988b).

6.10 SUMMARY

STRs are one of the standard bioreactor designs used in biological applications because gas–liquid mass transfer can be easily increased through faster impeller speed and higher gas flow rates. STRs come in many different flavors and scales. Small and experimental scale STRs can be serviced by a single impeller while most industrial applications use multiple impellers. The best choice for single impeller STRs is a radial disc impeller, but other designs may be more advantageous depending on the application. For example, retrofitting with some radial flow impeller, such as the Rushton-type turbine, may be difficult due to its high torque; however, replacement or addition of an axial flow impeller, such as the A-315, is easily achieved without further stress on the motor and gearbox. Therefore, a multiple impeller design with a non-interfering radial flow impeller (bottom) and axial flow impeller(s) is preferred.

Theoretical models explaining mass transfer are in place, but a $k_L a$ equation, which is applicable over a wide array of designs and operational range, is still not available. The interaction between the impeller and the gas and liquid phases is very complicated and a lot of ambiguity and controversy exists. Literature is often filled with contradictory suggestions and explanations. Hence, practical design procedures, such as scale-up or retrofitting, can be very complicated and results are often hard to predict correctly. Furthermore, microorganisms add another level of complication because they may be sensitive to the reactor conditions so that production declines even if gas–liquid mass transfer, the supposed limiting factor, increases. These interactions in turn have led to industrial applications often being dominated (in terms of time) by the last 10–20% of conversion (relative).

Therefore, great care has to be used in designing, operating, or retrofitting STRs for biological applications. Scale-up may be more successful if the experimental and industrial scale designs have hydrodynamic and geometric similarities because the probability of similar conditions is increased. In such a case, it is expected that production would also meet predicted values. In addition, the implemented scale-up strategy has to account for increased turbulence with scale. A common error is to use a constant impeller speed scale-up strategy, which leads to much higher levels of turbulence and power costs at the industrial scale. In fact, a global scale-up strategy does not exist, and any designs and modifications have to be thoroughly tested. The trial-and-error method cannot be avoided, but hydrodynamic and geometric similarities reduce the time and cost required to achieve satisfactory results.

7 Bubble Column Bioreactors

7.1 INTRODUCTION

Bubble columns (BCs) belong to a family of pneumatic bioreactors. These bioreactors do not have any mechanical or otherwise moving parts. Compressed air, which is used for mixing purposes, is injected into the base of a cylindrical vessel. This approach provides a cheap and simple method to contact and mix different phases (Díaz et al., 2008). The liquid phase is delivered in batch or continuous mode, which can be either countercurrent or cocurrent. The batch BC is the more common form, but the cocurrent version, shown in Figure 7.1, is also encountered. Countercurrent liquid flow is rarely used in industry as it provides minor, if any, advantages and multiple complications (Deckwer, 1992), with separation by evaporation being one of the few exceptions (Ribeiro Jr. and Lage, 2005).

The gas throughput has a significant impact on the column design. Superficial gas velocity in BCs is limited to $U_G = 0.03 - 1\,\text{m/s}$ with most applications operating at the lower end. The exact value is scale and flow regime dependent and only large industrial projects would use $U_G \approx 1\,\text{m/s}$. Such flow rates lead to very fast-rising bubbles (Deckwer, 1992; Krishna et al., 2001).

BCs tend to be tall vessels with a large aspect ratio (H/D_R). The height is an important design variable because of its influence on the process and residence times, especially for batch and semibatch operations (Roy and Joshi, 2008). Biochemical processes require an aspect ratio between 2 and 5 even for experimental work. Industrial applications require much taller vessels with an aspect ratio of at least 5 (Kantarci et al., 2005), and it is fairly common to have vessels with an aspect ratio greater than 10 (Bellgardt, 2000b). An aspect ratio greater than 5 is also preferred because it does not influence BC hydrodynamics (Ribeiro Jr., 2008). It also allows for the breakup and coalescence mechanism to stabilize and reach steady state.

The upper section of the BC is often widened to encourage gas separation. BC volume is dependent on the application. The chemical production industry uses columns with volumes on the order of $100-200\,\text{m}^3$, whereas the biotechnology industry and wastewater treatment use columns that may be up to 3000 and $20,000\,\text{m}^3$, respectively (Deckwer, 1992).

An Introduction to Bioreactor Hydrodynamics and Gas-Liquid Mass Transfer, First Edition.
Enes Kadic and Theodore J. Heindel.
© 2014 John Wiley & Sons, Inc. Published 2014 by John Wiley & Sons, Inc.

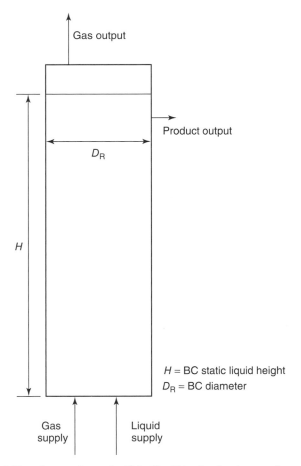

Figure 7.1   Bubble column schematic; if the liquid is also flowing continuously, the bubble column would be identified as cocurrent.

BCs require very little maintenance or floor space and have low operating costs compared to other reactor types (Ribeiro Jr., 2008). The low operating and maintenance costs are due to the lack of moving parts. Compressed gas is capable of producing a friendlier and uniform environment, which is important for processes involving shear-sensitive microorganisms or pressure-sensitive catalysts (Kantarci et al., 2005). Compressed gas is also a more effective power source for very large reactor volumes (up to $500\,m^3$) (Bellgardt, 2000b). The pneumatic power source typically produces lower energy dissipation rates compared to stirred-tank bioreactors. Furthermore, BC designs allow for online modification of microorganism concentrations (Kantarci et al., 2005) and handling of materials that may cause erosion or plugging (Ribeiro Jr., 2008).

The above advantages make BCs ideally suited for a variety of process industries including the chemical, petrochemical, biochemical, pharmaceutical, food, environmental, and metallurgical industries, and are used in operations such as oxidation, chlorination, alkylation, polymerization, and hydrogenation. BCs are also widely used for the treatment of wastewater and manufacture of synthetic fuels (through fermentation), enzymes, proteins, and antibiotics (Kantarci et al., 2005).

Ultimately, the BC is preferred by many because it is easily applied to problems and applications; however, the BC has innate complexities (Huang and Cheng, 2011). The major disadvantage is the difficulty in controlling the complex hydrodynamics found in the reactor, which have a controlling effect on the transport and mass transfer characteristics (Kantarci et al., 2005). The flow patterns, which are not well defined (Bellgardt, 2000b), create considerable backmixing and a large pressure drop through the column. These phenomena are due to the complex bubble interactions and their coalescence behavior (Martín et al., 2008a), which limit the designer's ability to control reactor performance (Dhaouadi et al., 2008; Roy and Joshi, 2008; Vial et al., 2001). The result is that BC behavior is fairly unknown, especially if reactor geometries, liquid properties, or operating ranges are varied in parallel (Godbole and Shah, 1986; Vial et al., 2001). Hence, design and scale-up are difficult (Ribeiro Jr., 2008; Vial et al., 2001) and require a tedious and iterative process (Godbole and Shah, 1986). Additional information on BCs is provided in the literature (Beenackers and Van Swaaij, 1993; Deckwer, 1992; Godbole and Shah, 1986; Han and Al-Dahhan, 2007; Kantarci et al., 2005; Lau et al., 2004; Ribeiro Jr. and Lage, 2005).

7.2   FLOW REGIMES

Gas–liquid mass transfer behavior in BCs is closely tied to gas holdup through the various flow regimes identified in Figure 7.2. The two principal and industrially useful flow regimes are the homogeneous and heterogeneous flow regimes (Mena et al., 2005). At low gas flow rates, the bubbly or homogeneous flow regime develops (Figure 7.2a). The regime is characterized by small bubbles that are a few

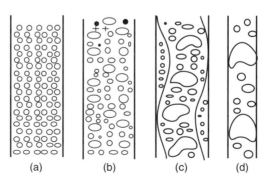

    (a)   (b)   (c)   (d)

Figure 7.2   (a) Homogeneous, (b) transition, (c) heterogeneous, and (d) slug flow regimes (Kantarci et al., 2005).

millimeters in diameter and uniformly distributed in the radial direction and, therefore, rise uniformly. The bubble diameter in the homogeneous regime tends to be controlled by the sparger design and liquid properties. Bubble–bubble interaction is limited in this regime, and bubble coalescence and breakup are negligible. So, if the sparger is capable of producing smaller bubbles, these bubbles tend to stay stable at the smaller diameter.

As the gas flow rate increases, more bubbles are created without affecting the bubble diameter or distribution significantly. Hence, the interfacial area and gas holdup increase almost linearly (Kantarci et al., 2005). Mixing is minimal, which leads to very fast average bubble rise velocity (20–30 cm/s) and short residence times, even in very tall BCs (Deckwer, 1992). Therefore, better performance can be achieved at the subsequent flow regimes, which have much better mixing features; however, a significant amount of research has been directed towards the stability of the homogeneous flow regime because certain biochemical processes require the calm environment experienced in this regime. The size of most industrial units makes it difficult for the homogeneous flow regime to be used because the amount of gas throughput and mixing is often inadequate (Kantarci et al., 2005). Additional geometric restraints exist for reactors using the homogeneous flow regime. These restrictions will be discussed in Section 7.3.

As the gas flow rate is increased, the flow evolves into an unstable structure, referred to as the transition regime (Figure 7.2b). The flow regime transition in terms of gas holdup as a function of superficial gas velocity is shown in Figure 7.3. Bubbles collide and distinct bubble classes are formed. The nature of the transition can occur through two different paths. The first is described by line "a" in Figure 7.3. The gas holdup and interfacial area are still increasing within the transition regime up to a local maximum. This effect is mainly due to the balancing act between more gas being present in the reactor and coalescence phenomenon, which lead to larger bubbles and faster bubble rise velocities, leading to a local gas

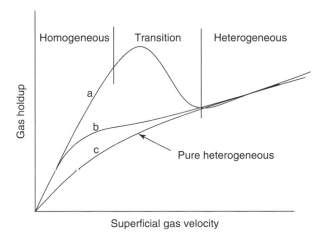

Figure 7.3 Flow regime progression (Su and Heindel, 2005b).

holdup maximum (Kantarci et al., 2005). Another reason for the increase is that the larger bubbles tend to breakup and coalesce frequently, adding to turbulence (Dhaouadi et al., 2008).

After the maximum has been reached, gas holdup and interfacial area decrease as the larger bubbles start to take control of gas holdup behavior (Kantarci et al., 2005). The second path, described by line "b" in Figure 7.3, occurs much faster and is identified by a continuous increase in gas holdup, albeit at a smaller rate, once the flow reaches the heterogeneous flow regime. Line "c" in Figure 7.3 represents a pure heterogeneous flow regime, which may occur with viscous liquids, large orifices, and/or small BC diameters (Ruzicka et al., 2003; Ruzicka et al., 2001b).

The instability that triggers the flow regime transition is mainly due to the bubble size and shape—smaller bubbles with rigid interfaces stabilize the flow because these bubbles have an inherent aversion to coalescence and breakup (León-Becerril and Liné, 2001). If the colliding bubbles are deformable, the collision is inelastic and will have an easier time forming a liquid channel to drain the liquid film, which separates the two bubbles. If, on the other hand, the colliding bubbles are nondeformable (rigid), the bubbles have a higher probability of simply bouncing off each other or separating quickly (Martín et al., 2008a). Hence, experimental results show that spherical bubbles correlate to a transition from homogeneous flow at about $U_G = 5\,cm/s$ while ellipsoidal bubbles have a transition at a lower superficial gas velocity of $U_G = 3\,cm/s$. Therefore, the moment of transition is dependent on bubble behavior in the homogeneous regime. If the homogeneous flow regime is more stable, the transition regime will be more defined, shorter, and occur at a higher superficial gas velocity (León-Becerril et al., 2002), which are beneficial side effects for industrial applications. In practice, the transition superficial gas velocity also tends to deviate with the column dimensions, sparger design, and liquid properties (Kantarci et al., 2005). In general, coalescence starts to occur at $U_G \sim 2\,cm/s$ while the actual transition to heterogeneous flow occurs at $U_G = 4\text{--}5\,cm/s$ (Deckwer, 1992).

Even though the transition regime may offer a maximum for the gas holdup and interfacial area, it is not desired for industrial processes due to its unstable and erratic nature. The instability has made the exact identification of the transition point nearly impossible. Although computational fluid dynamics and other methods are capable of predicting the other flow regimes, these methods usually have a difficult time predicting the transition point or the hydrodynamic behavior near it (Olmos et al., 2003). Hence, even if the operator wanted to work with the transition regime, it would be nearly impossible to achieve consistent results.

The evolution to the next flow regime, the churn-turbulent or heterogeneous flow regime (Figure 7.2c), is signaled by an increase in gas holdup and interfacial area with increasing gas flow rate. This growth is fairly small (less than linear) and tends to trail off. The mean bubble diameter, which tends to be on the order of a few centimeters, is controlled by coalescence and breakup mechanisms in the center section of the BC. Even though the gas–liquid mass transfer coefficient is lower in the heterogeneous regime than in the homogeneous regime, most industrial units operate in this flow regime (Ruzicka et al., 2001a) because it offers significantly better mixing and acceptable gas holdup and interfacial area values. Furthermore, selectivity and productivity requirements force a large number of industrial operations to be highly turbulent, which is an advantage of the heterogeneous regime (Jakobsen et al., 2005b).

Phase backmixing is a characteristic of the heterogeneous flow regime and represents a major disadvantage for certain types of operations. It causes very complex hydrodynamic behavior, which leads to significant problems for the design and scale-up of BCs. The backmixing and recirculation are induced by a differential static pressure between the central and wall regions (Zahradnik et al., 1997). The basic description of backmixing and circulation has changed significantly over the decades. The view was that circulation in BCs occurred mainly in upward moving cells; however, more recent studies have shed light into stationary or periodic cell behavior (Huang and Cheng, 2011).

As the flow transitions to the heterogeneous flow regime, two different bubble classes emerge, small and large, and behave differently with a unique influence on gas holdup. In the transition regime, the gas holdup for large bubbles increases much faster than the gas holdup contribution due to the small bubbles. This trend continues until the coalescence rate of smaller bubbles increases so that the growth in gas holdup and interfacial area caused by large bubbles cannot account for the decrease caused by the shrinking number of small bubbles.

As the heterogeneous flow regime is entered, the gas holdup due to the large bubble classes consistently increases, but at ever decreasing rates while the gas holdup due to small bubble classes is relatively constant (Kantarci et al., 2005). This progression can be seen in Figure 7.4; it is due to the very high bubble rise velocity of the large bubble class ($U_R \approx 160\,\text{cm/s}$ at $U_G = 20\,\text{cm/s}$) while the rise velocity for the small bubble class remains relatively unchanged ($U_R \approx 21\,\text{cm/s}$ at $U_G = 20\,\text{cm/s}$) (Schumpe and Grund, 1986).

For small column diameters ($D_R < 0.15\,\text{m}$), a fourth flow regime is feasible: slug flow (Figure 7.2d). It is characterized by a train of large bubbles, which spans the entire column diameter, dominating the flow. This flow regime is not practical and is not achieved in industrial units. Another flow regime that is not often encountered is the foaming flow regime, which is present under high superficial gas velocities, viscosity, and pressure (Kantarci et al., 2005; van der Schaaf et al., 2007).

Since most industrial processes are performed in the heterogeneous flow regime, studies have been directed towards its macroscopic flow patterns. Tzeng et al. (1993) classified these macroscopic flow structures into four regions using a 2D BC: descending flow, vortical flow, fast bubble flow, and central plume. The central region of the BC is made up of a central plume through which relatively small bubbles ascended. This central plume is surrounded by a fast bubble flow that is made of larger bubbles. At the edge of this motion, vortices form that trap bubbles and liquid to form the vortical flow region. These vortices direct bubbles near the column wall to descend (descending flow region). Chen et al. (1994) arrived at similar macroscopic structures using a 3D BC. The fast and descending

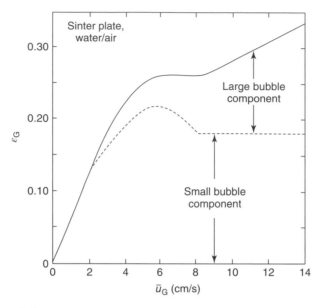

Figure 7.4   Bubble class contributions to gas holdup (Deckwer, 1992).

bubble flows, however, flowed in a spiral pattern. In addition, the homogeneous flow regime also displayed a descending flow structure near the wall region. A graphic depiction of the heterogeneous flow regime described by Chen et al. (1994) is presented in Figure 7.5.

The progression and behavior of these flow regimes is often quite complicated and depends on the superficial gas velocity, liquid properties, column dimensions, operating temperature and pressure, sparger design, and the solid phase properties (if present) (Kantarci et al., 2005). This dependence is derived from the controlling factors determining the bubble diameter.

Gas holdup and superficial gas velocity effects on gas–liquid mass transfer are analyzed based on the previously mentioned bubble interactions and flow regime progression. Global gas holdup is assumed to be directly correlated with the gas–liquid mass transfer coefficient because local variations in the interfacial area and gas holdup coincide. Having reviewed the flow regimes, it can be concluded that the transition regime offers a dilemma. Its hydrodynamics are not linear and do not follow general behavior in the homogeneous and heterogeneous flow regimes (Chaumat et al., 2005). Therefore, errors in the gas–liquid mass transfer coefficient are expected to be greater around the transition region, and better approximations are expected for correlations that take into account the regime identification.

Flow regime identification is often assumed to be based on the superficial gas velocity (all else being equal). If a correlation is based on a certain design with a specific process in mind, adjustments could be incorporated into operation while

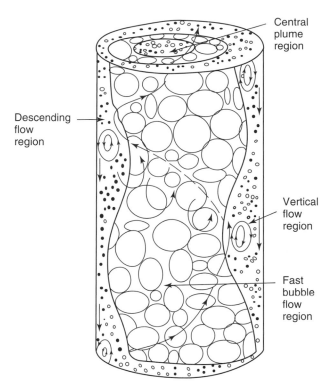

Central
plume
region

Descending
flow
region

Vertical
flow
region

Fast
bubble
flow
region

Figure 7.5 Macroscopic flow structure in the heterogeneous flow regime (Chen et al., 1994).

assuming that the results will be close to the predicted values. At the same time, most correlations neglect any changes in the liquid-phase mass transfer coefficient. If operation is expected to include rheological changes, which may have a significant effect on the liquid-phase film resistance, the correlation could fail. It would produce additional variations in the gas–liquid mass transfer correlation that could not be accounted for by variations in the interfacial area and its codependent variations in gas holdup. Therefore, a proper gas–liquid mass transfer study should attempt to separate the approximation of the liquid-phase mass transfer coefficient and the interfacial area under such circumstances.

7.3 COLUMN GEOMETRY

7.3.1 Column Diameter

The gas–liquid mass transfer coefficient and gas holdup tend to decrease with increasing BC diameter in the homogeneous and transition flow regime (Zahradnik et al., 1997) up to a critical value (Deckwer, 1992; Shah et al., 1982; Zahradnik et al., 1997), which is usually cited to be $D_R = 0.15\,m$ (Kantarci et al., 2005);

however, gas holdup has been influenced by column diameters greater than 0.15 m in the homogeneous and heterogeneous flow regimes (Ruzicka et al., 2001a; Vandu and Krishna, 2004), and the critical diameter is more accurately described to lie in the range of 0.1–0.2 m (Zahradnik et al., 1997).

The diameter dependence is created by several factors such as wall effects (Lau et al., 2004), which are negligible for water-filled columns larger than 0.10 m (Kantarci et al., 2005; Lau et al., 2004), and the flow and mixing conditions (Krishna et al., 2001; Zehner, 1989). The wall effect on gas holdup is summarized in Figure 7.6. Figure 7.6a and 7.6c represent radial gas holdup profiles in small BCs ($D_R \leq 0.060$ m), while Figure 7.6b and 7.6d represent more realistic effects in larger BCs. Hence, smaller BCs distort gas holdup behavior and may also misrepresent bubble diameter measurements (if done visually). Column diameter also has a strong influence on flow stability, defined by a critical gas holdup and gas flow rate at which the onset of the transition flow regime occurs. A larger diameter causes instability and earlier transition while a smaller diameter would induce the opposite behavior (Ruzicka et al., 2001a).

Although the slug flow regime is defined by low gas holdup and is observed in small BCs, small BCs do not necessarily have low gas holdup. The smaller BC diameter may limit the bubble size distribution, which can lead to smaller bubbles, increase stability, and sustain a higher gas holdup. Once the column diameter is

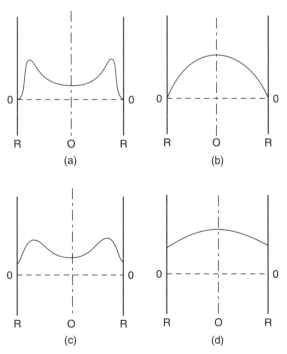

Figure 7.6   Radial gas holdup profiles for small (a and c) and large (b and d) bubble column diameters (Veera and Joshi, 1999).

larger than 0.10–0.15 m, the bubble size distribution is controlled by the coalescence and breakup mechanisms (Lau et al., 2004). This effect has been confirmed by Bouaifi et al. (2001) who investigated different gas distributors and column sizes; their results are shown in Figure 7.7.

It is often observed that the smaller bubble class is not affected by BC diameter while the larger one is (Kantarci et al., 2005; Li and Prakash, 2000). Larger BCs experience a larger degree of recirculation, which may lead to a higher degree of coalescence, higher bubble rise velocities, and lower gas holdup values (Krishna et al., 2001; Zehner, 1989). It is also feasible to observe lower gas holdup values in a smaller BC if it is operated in the slug flow regime, which is dominated by very large and fast-rising bubbles. Larger columns cannot maintain this mode of operation so that a direct comparison at a higher gas flow rate may lead to the conclusion that larger columns have larger gas holdup values (Daly et al., 1992). This effect can be seen in Figure 7.8 where the BC with the largest diameter shows the best gas holdup performance. A fairer comparison would be made on a gas flow rate per unaerated liquid volume basis.

Research work by Krishna et al. (2001) and Zehner (1989) concluded that most studies did not account for the flow regime dependence. For example, the heterogeneous flow regime, for which significant column geometry research exists, has a positive effect due to the higher degree of liquid recirculation. The homogeneous flow regime, on the other hand, has limited liquid recirculation by design. Instead, the larger BC diameter leads to faster bubble rise velocities, especially in the central region, which leads to lower gas holdup and gas–liquid mass transfer. Unfortunately, a significant number of gas holdup and gas–liquid mass transfer studies fail

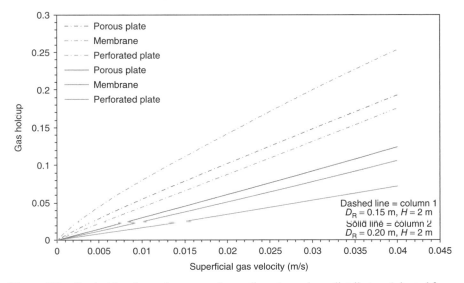

Figure 7.7   Gas holdup dependence on column diameter and gas distributor. Adapted from Bouaifi et al. (2001).

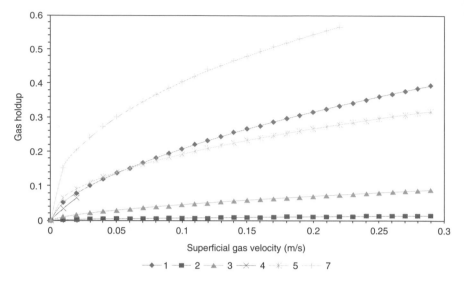

Figure 7.8   Gas holdup correlation sample using tap water at 15 °C as liquid phase: (1) Anabtawi et al. (2003) ($H = 0.60$, $D_R = 0.074$ m), (2) Deckwer (1992) ($D_R = 0.14$ m), (3) Godbole et al. (1982), (4) Hammer (1984), (5) Hikita and Kikukawa (1974) ($D_R = 0.10$ m), (6) Hughmark (1967) ($D_R = 0.0254–0.10$ m), and (7) Reilley et al. (1986) ($D_R = 0.30$ m).

to identify the flow regime in which the BC is operating, which makes meaningful comparisons often difficult.

The effect of column diameter on bubble size and flow regime has led to the introduction of flow regime maps, such as the one presented in Figure 7.9. These maps plot the regimes depending on the superficial gas velocity and BC diameter and attempt to predict the gas holdup; however, the maps are only examples and cannot be applied universally. Flow regimes and bubble behavior are also influenced by the gas distributor design and physiochemical properties of the liquid phase. Changes in these factors bring about significant variations, which are not accounted for by flow regime maps. Nonetheless, these maps may be used as operating instructions for a reactor design that has been extensively studied.

7.3.2 Unaerated Liquid Height

The BC height, defined by the static liquid level, regulates the residence time over which the bubbles are allowed to go through the coalescence and breakup process. In other words, the column height may allow enough time for equilibrium to be reached. Hence, shorter columns often experience smaller bubble size distributions and higher gas holdup values (Kantarci et al., 2005; Wilkinson, 1991; Zahradnik et al., 1997). A height greater than 1–3 m or an aspect ratio greater than 5 usually ensures that equilibrium has been established. Even though the gas holdup behavior indicates that operation with shorter BCs is more effective (Kantarci et al.,

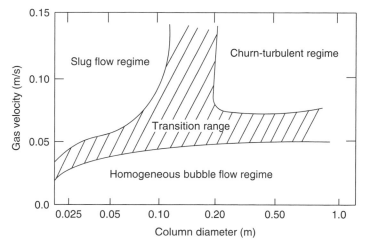

Figure 7.9   Flow regime dependence on column diameter (Kantarci et al., 2005).

2005), the practical application is limited. The residence time is simply too short and would require extensive gas recycle and a very large column footprint for meaningful BC volumes; however, the idea can be advantageous and has been applied to microreactors.

The column height may lead to bubble behavior stratification. For example, the top of the BC may experience high gas holdup due to foaming. The volume surrounding the sparger behaves differently depending on the gas distributor design. If the gas distributor is capable of producing smaller bubbles, the gas holdup is higher; however, this effect is limited only to the immediate area. The reactor bulk (middle section) usually behaves according to coalescence and breakup principles and the prevailing flow regime. Hence, the bulk region would be expected to have smaller gas holdup values (Ruzicka et al., 2001a). If the column is tall enough, such as a height greater than 1–3 m or an aspect ratio greater than 5, the sparger and foaming effects on gas holdup become negligible and the bulk region dominates reactor hydrodynamics (Veera and Joshi, 1999; Wilkinson et al., 1992; Zahradnik et al., 1997); however, foaming should still be avoided since it may damage any microorganisms and block gas disengagement.

7.3.3   Aspect Ratio

BCs are assumed to decrease gas holdup with an increase in size, but an aspect ratio above 5 does not seem to affect hydrodynamics significantly and is usually ignored. Comparisons based on aspect ratio have failed in the past due to large gas holdup data scatter. A good fit could not be obtained such that scale-up rules still need to be based on column diameter and height individually (Ruzicka et al., 2001a). Aspect ratios below 5 are rarely used in industry and are often ignored in experimental settings.

7.4 OTHER OPERATING CONDITIONS

7.4.1 Pressure

Pressure affects bubble dynamics and, therefore, has an important influence on gas holdup and gas–liquid mass transfer. It should be noted that correlations often fail to account for pressure and provide poor predictions if pressurized systems are used, as is the case in industrial applications (Dhaouadi et al., 2008). Generally, an increase in pressure is accompanied by a decrease in the bubble surface tension and an increase in bubble inertia (Kantarci et al., 2005; Luo et al., 1999) and gas solubility (Dhaouadi et al., 2008). These factors decrease the average bubble diameter, which allows for higher interfacial area and gas holdup values (Kojima et al., 1997). Thus, the higher interfacial area leads to higher gas–liquid mass transfer coefficients (Lau et al., 2004). In addition, as the bubble diameter decreases with increasing pressure, the bubble rise velocity also decreases, leading to an increased gas residence time and a more efficient gas–liquid mass transfer performance.

The effect on the liquid-phase mass transfer coefficient is most likely neutral to positive. Past experience has been that the liquid-phase mass transfer coefficient is only dependent on the phase data (Gestrich et al., 1978), but current research efforts have presented contradictory evidence (Han and Al-Dahhan, 2007), most likely due to the method used in calculating the liquid-phase mass transfer. It is probable that the effect is negligible at lower pressures due to the much more important changes in the bubble diameter and interfacial area; however, the solubility dependence on pressure could be significant, especially at higher pressure (Kojima et al., 1997).

The extent of the effect often depends on the pressure increase, liquid properties, and gas flow rates. It is often cited that pressures below 1 MPa have a negligible impact on bubble size and gas holdup; however, the references usually point to older articles that used stirred-tank reactors and have significantly different power characteristics than BCs (Stegeman et al., 1995). More recent work, on the other hand, suggests that pressure, even below 1 MPa, has a significant impact on gas holdup and the interfacial area. Furthermore, pressurization has been determined to have a more significant impact on viscous liquids, slurries, and the heterogeneous flow regime (Lau et al., 2004).

As the liquid becomes more viscous, bubbles have a tendency to coalesce more readily. The increased pressure serves as a detriment to coalescence and decreases the average bubble diameter. Hence, systems with a slurry phase benefit from pressurization (Luo et al., 1999). Bubble characteristics in the heterogeneous flow regime are defined by bubble breakage and coalescence frequencies. Once again, higher pressure tends to suppress coalescence and decrease the average bubble diameter. For example, Lau et al. (2004) varied pressure and gas flow rate in their study and found that increasing the pressure from 0.1 to 2.86 MPa led to an increase in $k_L a$ of 130% at a gas flow rate of 10 cm/s. When the same pressure increase was implemented at a gas flow rate of 20 cm/s, an increase by 187% was observed. These results can be seen in Figure 7.10. It should be noted that the gas holdup failed to predict gas–liquid mass transfer increase as pressure increased.

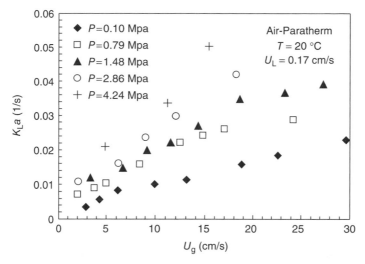

Figure 7.10 Pressure effects on gas–liquid mass transfer in a 10.16 cm bubble column with a single-nozzle gas distributor (Lau et al., 2004).

This effect can be clearly seen when a comparison is made between Figures 7.10 and 7.11, which related gas holdup to pressure.

The gas holdup data of Lau et al. (2004) are basically parallel while the gas–liquid mass transfer data have different slopes. Furthermore, the gas holdup

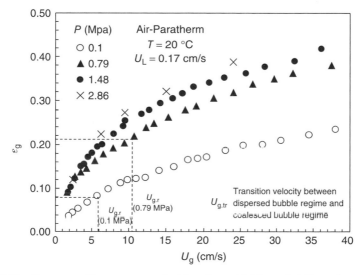

Figure 7.11 Pressure effects on gas holdup in a 10.16 cm bubble column with a single-nozzle gas distributor (Lau et al., 2004).

increase at a pressure of 1.48 MPa versus 2.86 MPa is modest at about 10% (assuming minimal experimental error) while the gas–liquid mass transfer is more impressive, especially at higher gas flow rates. These observations would lead to the conclusion that the interfacial area increased significantly from 0.1 to 1.48 MPa while the liquid-phase mass transfer coefficient had a stronger influence in the increase from 1.48 MPa to 2.86 MPa. More data points for gas–liquid mass transfer would give a larger resolution and more confidence in this conclusion.

Even more surprising was the increase in $k_L a$ with liquid velocity at high pressures observed by Lau et al. (2004). Usually, the superficial liquid velocity is ignored and is cited as having an insignificant or negative effect on gas–liquid mass transfer because a higher superficial liquid velocity is thought to decrease the gas residence time and, therefore, $k_L a$ (Chaumat et al., 2005). At a pressure of 2.86 MPa, Lau et al. (2004) observed a $k_L a$ increase by 30% when the superficial liquid velocity was increased from $U_L = 0.17$ to 0.26 cm/s.

Vibrations within the liquid phase could be introduced to improve gas holdup and gas–liquid mass transfer performance. This goal has been accomplished using mechanical vibration devices, but an easier solution would be to introduce sinusoidal pressure variations at a frequency on the order of 100 Hz and amplitude of 0.0025–0.01 mm, which lead to a reduction in the bubble diameter by 40–50% and an increase by 100–300% in gas holdup and 200% in the gas–liquid mass transfer coefficient. These variations reduced the bubble rise velocity, destabilized the bubble surface, and reduced surface tension forces, which improved the liquid-phase mass transfer coefficient (Ellenberger and Krishna, 2002).

7.4.2 Temperature

Temperature is a much more contentious issue. Generally, it is thought that higher temperatures reduce the liquid viscosity and surface tension, which would lead to a higher stability of the small bubble population and lead to higher interfacial area. In addition, the liquid-phase mass transfer (k_L) is, therefore, thought to increase with an increase in temperature due to the lower viscosity according to Calderbank's slip velocity model. At the same time, liquid-phase mass transfer coefficient is thought to decrease due to lower surface tension and turbulence. Hence, the liquid-phase mass transfer coefficient could go in either direction according to a balancing act between the two forces with increasing temperature.

Lau et al. (2004) also varied temperature with the gas flow rate and concluded that higher temperatures lead to higher gas–liquid mass transfer coefficients at constant gas flow rates. For example, an increase in temperature from 25 to 92 °C at a superficial gas velocity of 20 cm/s led to an increase in the gas–liquid mass transfer coefficient by 470% while gas holdup increased by only 25%. Although this temperature increase is huge, Lau et al. (2004) noticed a significant increase in the gas–liquid mass transfer coefficient with even a few degree Celsius change in temperature. At the same time, gas holdup did not show any significant variation.

A secondary effect could be that the rate of reaction is often directly correlated to temperature and (partial) pressure. Most rates of reaction are expected to increase with an increase in temperature or pressure. Hence, a similar assumption is made with bioreactors. This assumption can be used as a rule of thumb, but it cannot be applied generally. The microorganism and its metabolism are the ultimate judge of the rule's applicability.

The introduction of higher temperatures and pressures leads to a potential problem. The control schema for bioreactors is generally tied to the gas flow rate. The idea is that the gas flow rate would impact gas holdup, which would, in turn, control gas–liquid mass transfer. Significant temperature changes, however, decouple gas–liquid mass transfer from gas holdup and add new operational variables. Most correlations do not include temperature and pressure effects directly, but instead attempt to quantify these effects by altering or introducing effective liquid properties, which are much harder to cost-effectively measure in an industrial setting.

The biological reaction, which may occur at the interface, is often ignored during experimental measurements and correlation formulations as these are not apparent in pure liquids without the presence of bacteria; however, industrial processes may experience a significant degree of interfacial reaction. Therefore, industrial processes would be expected to perform better than experiments might suggest due to the application of higher temperatures and pressures. In effect, the use of pure liquids may provide an additional engineering factor. Alternatively, microorganisms have to play an important role as they may limit the operating range.

7.4.3 Viscosity

The liquid viscosity determines the degree to which bubbles can deform. As the viscosity increases, bubbles become more deformable and the steady-state bubble diameter increases (Martín et al., 2008a), while bubble breakup is suppressed (Zahradnik et al., 1997). Deformable bubbles allow for the bubble interface to drain much easier and allow for coalescence to occur in a shorter amount of time. First, the bubble breakup is suppressed because the higher viscosity tends to reduce turbulence. A second negative effect is that the larger bubbles have a higher rise velocity and lead to a shorter residence time for the gas phase (Zahradnik et al., 1997). Hence, higher viscosity liquids are observed to have larger bubbles and smaller gas holdups and interfacial areas (Li and Prakash, 1997; Zahradnik et al., 1997).

Viscosity also affects the liquid-phase mass transfer coefficient through the Stokes–Einstein effect. Einstein proposed that the diffusion coefficient could be expressed as

$$D_L = \frac{RT}{N_A} \frac{1}{6\pi \mu_l \, r_B} \tag{7.1}$$

where R, T, N_A, μ_L, and r_B are the gas constant, temperature, Avogadro's constant, viscosity, and bubble radius, respectively (Sharma and Yashonath, 2006). Hence, a higher viscosity could significantly reduce the diffusivity, which in turn would decrease the liquid-phase mass transfer coefficient (Chaumat et al., 2005; Lau et al., 2004; Sharma and Yashonath, 2006; Waghmare et al., 2008); however, the decrease in the interfacial area is expected to dominate the decrease in the liquid-phase mass transfer coefficient.

7.4.4   Surface Tension and Additives

Surface tension is a liquid property that tends to counter bubble deformation and encourages bubble breakup (Akita and Yoshida, 1974; Mehrnia et al., 2005; Walter and Blanch, 1986). The result is a more stable bubble interface that leads to smaller bubble diameters, a more stable flow regime (Lau et al., 2004; Schäfer et al., 2002), and higher gas holdups and interfacial areas (Kluytmans et al., 2001). It is also thought that a lower surface tension leads to a higher contact time because the liquid flow over the bubble surface is slowed (Lau et al., 2004).

Surface tension is influenced by the presence of surfactants (Kantarci et al., 2005). Surfactants attach themselves to the bubble interface and form a hydrophilic boundary at the bubble surface. The result is a much smaller bubble size and a more rigid surface. This surface, in turn, lowers the bubble surface tension and further reduces the bubble rise velocity.

Electrolytes have been shown to increase gas holdup and decrease bubble diameter (Kantarci et al., 2005) even at high concentrations where they result in a surface tension higher than that of pure water (Levin and Flores-Mena, 2001). At relatively low concentrations, electrolytes decrease surface tension and reduce the film drainage speed, leading to higher gas holdup (Kluytmans et al., 2001). Antifoam agents, on the other hand, cause a decrease in gas holdup due to higher surface tension and enhance coalescence leading to larger bubble diameters (Veera et al., 2004). Other impurities, especially of organic origin, tend to increase gas holdup and create immobile bubble interfaces; however, most research has been conducted using inorganic mixtures or pure liquids (Chaumat et al., 2005).

Alcohols create smaller bubbles and higher gas holdups because they are either amphiphilic or have a lower surface tension so that the aqueous mixture supports smaller bubbles. This behavior has been observed for a large number of alcohols in an aqueous saccharose solution. The only exception was proved to be methanol. The theory, which was successfully tested by Zahradnik et al. (1999b), is that the longer carbon chains increase the effectiveness (of alcohols as coalescence suppressants) and cause methanol, the simplest alcohol, to lose its usefulness at a much lower concentration relative to the other alcohols (Zahradnik et al., 1999a; Zahradnik et al., 1999b).

7.5   GAS DISTRIBUTOR DESIGN

Gas distributors used in BCs include (i) sintered, perforated, or porous plates; (ii) membrane or ring-type distributors; (iii) arm spargers; or (iv) single-orifice nozzles. The sintered plate (Figure 7.12a), which is usually made out of glass or metal, produces very small bubbles. Comparative experimental works are expected to find the sintered plate to perform the best using gas holdup as the criteria due to its effectiveness of creating small bubble populations; however, it is rarely used for industrial applications because it can plug very easily and requires cocurrent operation and significant maintenance. Perforated plates (Figure 7.12b), which are normally made out of rubber or metal, usually have holes 1–5 mm in diameter and are the most common gas distributor for BCs. The total aeration open area is usually maintained between 0.5% and 5% of the cross-sectional area of the reactor. Single-orifice nozzles (Figure 7.12c) are simple tubes, which are able to produce uniform gas flow far from the injection point. The flow, however, tends to be somewhat unstable. BCs may also use ring spargers (Figure 7.12d). Ring spargers are able to produce uniform and stable flow, but are usually unable to produce small bubbles (Deckwer, 1992) and lead to an earlier transition to the heterogeneous flow regime than the perforated plate (Schumpe and Grund, 1986).

The effects of gas distributor design and its extent depend on the superficial gas velocity and flow regime in which the BC operates. In the case of heterogeneous flow, the sparger has a negligible influence on the bubble size and gas–liquid mass transfer because the bubble dynamics are determined by the rate of coalescence and breakup, which are controlled by the liquid properties and the nature and frequency of bubble collisions (Chaumat et al., 2005). Hence, the sparger effect is more pronounced at lower superficial gas velocities ($U_G < 0.15\,\mathrm{m/s}$) while it is much less important at $U_G > 0.20$ m/s and nonexistent at $U_G > 0.30\,\mathrm{m/s}$ (Han and Al-Dahhan, 2007). Viscous liquids are also not affected by the gas distributor design if the column is sufficiently tall (Zahradnik et al., 1997).

Furthermore, the mechanisms that dominate gas holdup (e.g., surface tension, particle wettability, ionic force of surfactant, viscosity, and density) require consideration. If the liquid undergoes viscosity or density changes through, for example, particle addition, the initial bubble diameter does not affect gas holdup in the heterogeneous flow regime. If, on the other hand, the other mechanisms are affected

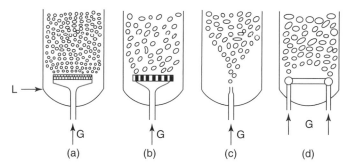

Figure 7.12   Sample bubble column aerators: (a) sintered plate, (b) perforated plate, (c) orifice nozzle, and (d) ring sparger (Deckwer, 1992).

or adjusted, such as by the addition of surfactants or electrolytes, the heterogeneous flow regime is affected by the initial bubble diameter up to a relatively high superficial gas velocity (of about 0.5 m/s), and hence, gas distributor design is important. Once a high enough superficial gas velocity is reached, the probability and frequency of bubble collision increases greatly and a precipitous decrease in gas holdup is observed (Kluytmans et al., 2001).

The homogeneous and transition regimes are highly influenced by the gas distributor design. A sparger, which is able to produce smaller initial bubble diameters, is able to produce a higher number of bubbles at the same flow rate and a more stable homogeneous flow regime (Álvarez et al., 2008). If the sparger is able to produce smaller bubbles in the homogeneous regime, the sparger would also be able to produce higher gas holdups and interfacial areas (Bouaifi et al., 2001). Hence, these spargers are also able to produce higher gas–liquid mass transfer coefficients under the homogeneous flow regime (Verma and Rai, 2003).

The distributor effect can be quite significant such that the gas–liquid mass transfer correlation can vary by up to a factor of 2 (Lau et al., 2004). The extent to which the gas distributor affects gas holdup and bubble dynamics depends on the BC geometry and superficial gas velocity. The taller the column is, the smaller the influence of the initial bubble diameter will be on the global gas holdup. A higher superficial gas velocity increases the probability and frequency of bubble collisions and decreases the effect of the initial bubble diameter and gas distributor design.

The assumption made up to this point is that all spargers are indeed capable of producing the entire range of flow regimes; however, Wilkinson (1991) noted that the discussion was irrelevant if the orifice diameter was larger than 1–2 mm because it would create bubbles that are too large and would be affected by the macroscopic flow pattern alone. In order to create significant influence, the aeration holes have to be smaller than 1 mm. Zahradnik et al. (1997), however, investigated the performance of a perforated plate and concluded that the hole diameter has to be smaller than 0.5 mm for a significant effect. A perforated plate with an orifice diameter of even 0.6 mm only created the heterogeneous flow regime using a variety of liquids.

The open area ratio, defined as the total aeration orifice area divided by the column cross-sectional area, was investigated and led to many conflicting conclusions. Different researchers approached the subject from different directions. One camp has chosen to adjust the number of holes while keeping the orifice diameter constant, whereas the other camp increases the orifice diameter while keeping the number of holes equal. Both approaches are potentially troublesome. If the open area ratio is increased by increasing the number of holes and keeping the orifice diameter constant, the holes could be spaced too close such that neighboring orifices start acting as a single sparger hole at higher gas flow rates. This sparger behavior is caused by immediate coalescence from neighboring bubbles as they are created and prior to their disengagement from the sparger orifice (Martín et al., 2008a; Su and Heindel, 2005a). If the holes are made larger while the number is held constant, the homogeneous flow regime could be skipped entirely (Zahradnik et al., 1997). Regardless, the agreement exists that if the open area ratio decreases, gas holdup increases, but the homogeneous flow regime stability may not necessarily be affected (Su and Heindel, 2005a). Logically, the open area ratio does not affect gas holdup in the fully developed heterogeneous flow regime (Kantarci et al., 2005; Su and Heindel, 2005a).

Bubble formation and orifice activity are two important factors determining stability. Synchronous bubble formation, where almost all holes are active instantaneously, tends to produce a uniform bubble and gas holdup distribution. The uniform bubble distribution leads to a more stable homogeneous flow regime, less liquid recirculation, and higher gas holdup and gas–liquid mass transfer. Asynchronous orifice operation is often accompanied by alternating or oscillating orifice activity, which leads to flow instability. The instability creates more bubble–bubble interaction and leads to lower gas holdup and gas–liquid mass transfer. Hence, the gas distributor affects the critical superficial gas velocity at which the transition regime is detected.

Perforated plates are defined by a critical flow rate above which the orifice operation is asynchronous and the liquid flow in the sparger region is relatively unstable. As the hole spacing decreases, the critical flow rate decreases as well. At the same time, perforated plates require a minimum pressure drop in order to achieve uniform orifice activity. In other words, a critical flow rate is also created at the lower end such that a lower flow rate would lead to instability as well (Kang et al., 1999; Ruzicka et al., 2003; Su and Heindel, 2005a). This effect would produce additional complications in making comparative analysis between research works using different open area ratio adjustment methods.

Vial et al. (2001) compared different gas distributor designs and made several interesting observations. The single-orifice nozzle tended to produce highly nonuniform flow while the porous plate and multiple-orifice sparger produced a fairly uniform flow pattern. This led the single-orifice sparger to always operate in the heterogeneous flow regime. Furthermore, the multiple-orifice sparger proved to be the most dependable. It produced the homogeneous flow regime until the superficial gas velocity reached about 4 cm/s. The heterogeneous flow regime would fully develop at 11–12 cm/s. The porous plate, on the other hand, was sensitive and provided different results depending on the start-up procedure.

Kluytmans et al. (2001) compared initial bubble diameters produced by different gas distributors and found that a 30-μm porous plate produced much smaller bubble diameters (0.2–0.5 mm) than the 0.5 mm perforated plate (1–2 mm). The higher gas holdup performance of the porous plate was attributed to the creation of these smaller bubbles.

Bouaifi et al. (2001) used two different columns (Column 1 with $D_R = 0.15$ m; Column 2 with $D_R = 0.20$ m; $H = 2$ m) and found that the porous plate generally produced higher gas holdup, followed by the membrane distributor and perforated plate. They agreed that smaller bubbles would lead to higher gas holdup values, and also concluded that the power consumption can vary significantly. For example, the membrane gas distributor would create very small bubbles with 80% of the bubble population being in the 3.5–4.5 mm range, which compared well with the results of the porous plate and its range of 2.5–4.5 mm. The membrane gas sparger, however, used much more power to obtain similar results. This is usually not a concern in an experimental setting, but may be of concern in (larger) industrial operations. Bouaifi et al. (2001) are the only researchers to account for the power usage of the different gas distributor designs. This is most likely due to the often ignored or relatively minor cost of compressed gas for experimental vessels.

7.6 CORRELATIONS

Correlations attempt to reflect the reactor environment as closely as possible. This goal is most often achieved through empirical data fitting (Martín et al., 2009). Analytical expressions are rarely used because they are burdensome with regard to the required data (Dhaouadi et al., 2008). A single correlation based on first principles does not exist for BCs due to the complexities discussed above. For example, Tang and Heindel (2006b) described what is needed to develop a dimensionless correlation for gas holdup in a BC based on the Buckingham-Pi Theorem, and they identified nine dimensionless parameters containing 13 variables representing three basic dimensions:

$$\epsilon = f\left\{ R_A, \frac{H}{D_R}, \frac{d_0}{D_R}, \frac{\rho_G}{\rho_L}, \frac{\mu_G}{\mu_L}, \frac{U_G}{U_L}, Re_L, Fr, We \right\} \tag{7.2}$$

where $R_A, H, D_R, d_0, \rho_G, \rho_L, \mu_G, \mu_L, Re_L, Fr,$ and We are the gas distributor open area ratio, BC height, BC diameter, gas distributor orifice diameter, gas- and liquid-phase densities, gas- and liquid-phase viscosities, gas- and liquid-phase superficial velocities, liquid-phase Reynolds number, Froude number, and Weber number, respectively.

Unfortunately for industrial settings, a majority of gas holdup and gas–liquid mass transfer correlations require inputs that are not easily collected for large tanks used in mass production settings. For example, commonly used inputs for gas holdup are the average bubble diameter and superficial gas velocity. Although superficial gas velocity is easily estimated, real-time average bubble diameter data are very hard to obtain. Large reactors have a great deal of spatial variation. In addition, industrial tanks are made out of steel and include nontransparent liquids. Visual bubble diameter observations, which are commonly used in experimental settings, would be unlikely to yield appropriate bubble size approximations. The changing environment would also provide many problems for process automation. Hence, it is rarely used in industrial-sized reactors.

Furthermore, many of the inputs are interdependent. The superficial gas velocity cannot be changed without impacting the average bubble diameter. This leads to the requirement of either a second data stream or an approximation for the bubble diameter, which could be built into the model from the very beginning. Another example would be that a significant number of correlations include the diffusivity and liquid properties as inputs. If the liquid is expected to change rheologically, which often occurs in biological processes, measurement or approximation of these inputs would be needed. Once again, this is information that is not easily obtained, even in an experimental setting.

A minor inconvenience is that many BC review, and a few original articles, do not include proper classification of variables and units, which is particularly troublesome when the correlation is not dimensionless. The inputs are not categorized as being in English or SI units. Moreover, some correlations use standard units, such as Pascal-second, while others use nonstandard units, such as milliPascal-second, without acknowledgement. Many review articles also do not define the reactor or the phases involved. The classification and definition of variables are made more difficult if the correlation is based on a theoretical derivation. These correlations often include terms that are left as constants, and further work is required to define these more accurately for practical applications. Temperature or pressure readings are usually not included as parameters, but are significant in industrial practice.

These problems can be dealt with if the correlation is selected and fit for the process in question. For example, one could easily use a gas holdup correlation such as (Guy et al., 1986)

$$\epsilon = 0.386 N_o \left(\frac{g D_R^3}{v_L^2} \right)^{0.025} \left(\frac{U_o}{\sqrt{g d_o}} \right)^{0.84} \left(\frac{d_o}{D_R} \right)^{2.075} \tag{7.3}$$

where $\epsilon, N_o, g, D_R, v_L, U_o$, and d_o are gas holdup, a sparger-dependent constant, gravitational acceleration, BC diameter, liquid viscosity (kinematic), gas velocity in the sparger orifice, and sparger orifice diameter, respectively. A separate gas–liquid mass transfer correlation may then be used, such as (Jordan and Schumpe, 2001)

$$\frac{k_L a d_B^2}{D_L} = A \left(\frac{v_L}{D_L} \right)^{0.5} \left(\frac{g \rho_L d_B^2}{\sigma} \right)^{0.34} \left(\frac{g d_B^3}{v_L^2} \right)^{0.27} \left(\frac{U_G}{\sqrt{g d_B}} \right)^{0.72}$$

$$\left[1 + 13.2 \left(\frac{U_G}{\sqrt{g d_B}} \right)^{0.37} \left(\frac{\rho_G}{\rho_L} \right)^{0.49} \right] \tag{7.4}$$

where $k_L a, d_B, D_L, A, v_L, g, \rho_L, \rho_G, \sigma$, and U_G are gas–liquid mass transfer coefficient, average bubble diameter, liquid-phase diffusion coefficient, sparger-dependent coefficient, liquid viscosity (kinematic), gravitational acceleration, liquid-phase density, gas-phase density, surface tension, and superficial gas velocity, respectively. These correlations include almost all required information to define the reactor environment except pressure and temperature.

In contrast, an industrial design has several choices. First and most complicated would be to use the experimental approach, which would be very intensive and costly. Second, a simplified correlation for gas holdup and gas–liquid mass transfer coefficient could be used:

$$\epsilon = C_2 U_G^x \tag{7.5}$$

$$k_L a = 0.467 U_G^{0.82} \tag{7.6}$$

where ϵ is the gas holdup, $k_L a$ is the gas–liquid mass transfer coefficient, U_G is the superficial gas velocity, and C_2 and x are the constants. Equations (7.5) and (7.6) were obtained by Bouaifi et al. (2001) and Shah et al. (1982), respectively. The effect of system-specific variables would have to be included in the constants. This approach would be not only highly practical and simple, but also nontransferable or limited, especially if rheological changes occur.

Lastly, a cook book approach could be used. As such, a simplified correlation would be used and time adjusted based on experience and parameter variables such as concentrations and temperature/pressure. This approach has a high upfront cost for correlation development, but would not require many data inputs and would have a low variable cost. It would represent a compromise between industrial practicality and scientific reality.

Ultimately, these problems have led to a large degree of variation in results and correlations that can be seen in Table 7.1 for gas holdup, Table 7.2 for the liquid-phase mass transfer coefficient, and Table 7.3 for the volumetric gas–liquid mass transfer correlation. The presented correlations, for example, show a wide degree and range of dependencies and variables. The result, as seen in Figure 7.13, is that some systems have anemic performance while others seem to be superstars of efficiency. Hence, the end user and designer have a great deal of work to ensure logical application of existing data and proper design constraints.

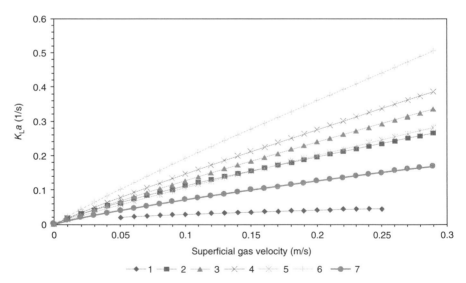

Figure 7.13 Gas–liquid mass transfer by (1) Behkish et al. (2002) (CO–hexane mixture without solids, $D_R = 0.316$ m), (2) Cho and Wakao (1988) (air–aqueous solutions, porous plate; $D_R = 0.115$ m), (3) water, (4) ethanol (96%), (5) 1-butanol, (6) toluene from Jordan et al. (2002) ($D_R = 0.115$ m), and (7) Shah et al. (1982).

TABLE 7.1 Gas Holdup Correlations for Bubble Columns

Reference	System	Conditions	Correlation
Akita and Yoshida (1973)[a]	Air–H_2O, O_2–H_2O, He–H_2O, CO_2–H_2O, air–glycol, air–aqueous glycol solution, Air–methanol	$\alpha = 0.2$ for pure liquids, 0.25 for salt solutions $D_R = 0.152$ m $H = 4.0$ m $d_0 = 0.5$ mm (SO) $U_G = 0.5$–33 cm/s P up to 0.1 MPa T up to 293K	$\dfrac{\varepsilon_G}{(1-\varepsilon_G)^4} = \alpha \left(\dfrac{D_R^2 \rho_L g}{\sigma} \right)^{1/8} \left(\dfrac{D_R^3 \rho_L^2 g}{\mu_L^2} \right)^{1/12} \dfrac{U_G}{\sqrt{g D_R}}$
Anabtawi et al. (2003)[b]	Air–(light oil, machine oil, five different engine oils)	$D_R = 74$ mm $d_0 = 10$ mm $H = 0.12$–0.60 m $U_G = 0.18$–29 cm/s $\rho_L = 906$–928 kg/m³ $\mu_L = 63$–320 mPa-s $\sigma = 24.8$–35 mN/m	$\varepsilon_G = 0.362 \dfrac{U_G^{0.60}}{H^{0.38} \mu_L^{0.24}}$
Bach and Pilhofer (1978)[c]	Air–(water, butandiol–1,3, ethylene glycol, n-octanol, tetrabromomethane, cyclohexane, cyclohexanol, ethanol, propanol, ethyl acetate)	$D_R = 0.10$ m $d_0 = 5$ mm (PP) $U_G = 1$–20 cm/s $\rho_L = 0.8$–2.98 g/cm³ $\nu_L = (0.7$–$124) \times 10^2$ cm²/s, $\sigma = 21.7$–72 mN/m	$\dfrac{\varepsilon_G}{1-\varepsilon_G} = 0.115 \left(\dfrac{U_G^3}{\nu_L g \Delta \rho / \rho_L} \right)^{0.23}$

Reference	System	Conditions	Correlation
Bouaifi et al. (2001)	Air–water	$U_G = 0.25$–4.0 cm/s Column 1 $D_R = 0.15$ m $H = 2$ m Column 2 $D_R = 0.20$ m $H = 2$ m	$\varepsilon_G = C_2 U_G^x$

$\varepsilon_G = C_2 U_G^x$

	Porous Plate		Membrane		Perforated Plate	
	C2	x	C2	x	C2	x
Column1	3.62	0.91	4.25	0.99	3.66	0.83
Column2	3.43	1.03	3.12	1.05	2.2	1.06

(continued)

147

TABLE 7.1 (*Continued*)

Reference	System	Conditions	Correlation
Deckwer (1992)	Air-CMC concentration 0.8–1.8% wt	$D_R = 0.14$ m $d_0 = 2$ mm (PP) and 0.15 or 0.2 mm (SP) $\mu_L > 50$ mPa-s	$\varepsilon_G = 0.0322 U_G^{0.674}$ for slug flow regime ($U_G > 3$ cm/s) $\varepsilon_G = 0.0908 U_G^{0.85}$ for homogeneous flow regime ($U_G < 2$ cm/s)
Gestrich and Rahse (1975)[b]	Air–organic liquids (methanol, ethanol, n-butanol, ethyl acetate, glycol, methylethylketone, $C_2H_4Cl_2$)	$H/D_R = 0.26–20$ $U_G = 1–8$ cm/s $\dfrac{\rho_L \sigma^3}{g\mu_L^4}$ $8\times10^4–5\times10^{10}$	$\varepsilon_G = 0.89 \left(\dfrac{H}{D_R}\right)^{0.035(-15.7+\log K)} \left(\dfrac{d_B}{D_R}\right)^{0.3} \left(\dfrac{U_G^2}{gd_B}\right)^{0.025(2.6+\log K)} K^{0.047} - 0.05$ $K = \dfrac{\rho_L \sigma^3}{g\mu_L^4}$ and $d_B = 3$ mm
Godbole et al. (1982)[a]	Air–viscous media		$\varepsilon_G = 0.239 U_G^{0.634} D_R^{-0.5}$
Grover et al. (1986)[a]		$D_R = 0.10$ m $d_0 = 100–120$ μm $U_G = 0.1–4.5$ cm/s $T = 303–353$K	$\varepsilon_G = \left[\left(\dfrac{1+b_1 P_V}{b_2 P_V}\right) \left(\dfrac{U_G \mu_L}{\sigma}\right)^{0.76} \left(\dfrac{\mu_L^4 g}{\rho_L \sigma^3}\right)^{-0.27} \left(\dfrac{\rho_G}{\rho_L}\right)^{0.09} \left(\dfrac{\mu_G}{\mu_L}\right)^{0.35}\right]$ $b_1 = 1.1\times10^{-4}$ and $b_2 = 5–10^{-4}$
Guy et al. (1986)[b]	Air–water Air–aqueous solutions (glycerol, carboxyl-methyl cellulose)	$D_R = 0.254$ m $d_0 = 1$ mm $U_G = 1–8$ cm/s $\dfrac{U_{b\infty} d_e}{\nu_L} > 4.02 \dfrac{g\mu_L^4}{\rho_L \sigma^3}$ $\dfrac{gd_e^2 \rho_L}{\sigma} >> 8$	$\varepsilon_G = 0.386 N_0 \left(\dfrac{gD_R^3}{\nu_L^2}\right)^{0.025} \left(\dfrac{U_0}{\sqrt{gd_0}}\right)^{0.84} \left(\dfrac{d_0}{D_R}\right)^{2.075}$

Hammer (1984)[b]	N_2–organic liquids (methanol, cyclohexane, cyclehexanol, n-octanol, $C_2H_4Cl_2$)	$d_0 = 0.15\text{--}2$ mm $U_G = 0.5\text{--}2$ cm/s $\rho_L = 780\text{--}1250$ kg/m^3 $\mu_L = 0.31\text{--}22$ mPa-s $\sigma = 21.8\text{--}31.9$ mN/m	$\varepsilon_G = 0.20 \left(\dfrac{U_G^2}{g D_R} \right)^{0.46} \left(\dfrac{D_R^3 g}{\nu_L^2} \right)^{0.08} \left(\dfrac{d_0}{D_R} \right)^{-0.17}$
Haque et al. (1985)[b]	Air–aqueous carboxy methyl cellulose solutions	$D_R = 0.10\text{--}1.0$ m $d_0 = 0.3\text{--}2.0$ mm $U_G > 3$ cm/s	$\varepsilon_G = 0.171 U_G^{0.6} \left[K(5000 U_G)^{n-1} \right]^{-0.22} D_R^{-0.15}$ (SI)
Hikita and Kikukawa (1974)[b]	Air–water Air–aqueous solutions (methanol, sucrose)	$D_R = 0.10$ and 0.19 m $d_0 = 13.0\text{--}36.2$ mm $U_G = 4.3\text{--}33.8$ cm/s $\rho_L = 911\text{--}1233$ kg/m^3 $\mu_L = 1\text{--}19.2$ mPa-s $\sigma = 38.2\text{--}75.5$ mN/m	$\varepsilon_G = 0.505 U_G^{0.47} \left(\dfrac{\sigma_w}{\sigma} \right)^{2/3} \left(\dfrac{\mu_w}{\mu_L} \right)^{0.05}$ (SI)
Hikita et al. (198_)[c]	$D_R = [0.10, 0.19]$ m $H = [1.5, 2.4]$ m $d_0 = 1.1$ cm (SO) $U_G = 4.2\text{--}39$ cm/s P up to 0.1 MPa $T = 283\text{--}303$K Non-electrolytes $\rho_L = 0.79\text{--}1.24$ g/cm^3 $\rho_G = (0.0837\text{--}1.84)\times 10^{-3}$ g/cm^3 $\mu_L = 0.66\text{--}17.8$ mPa-s $\mu_G = (0.8\text{--}1.81)\times 10^{-2}$ mPa-s $\sigma = 30\text{--}76$ mN/m Electrolytes $\rho_L = 1.01\text{--}1.17$ g/cm^3 $\mu_L = 0.9\text{--}1.87$ mPa-s $\sigma = 71.9\text{--}79.6$ mN/m	$\varepsilon_G = 0.672 f \left(\dfrac{U_G \mu_L}{\sigma} \right)^{0.578} \left(\dfrac{\mu_L^4 g}{\rho_L \sigma^3} \right)^{-0.131} \left(\dfrac{\rho_G}{\rho_L} \right)^{0.062} \left(\dfrac{\mu_G}{\mu_L} \right)^{0.107}$ where: $f = 1$ for electrolytes, $10^{0.0414 I}$ for ion strength $I < 1$ g-ion/L, 1.1 for $I > 1$ g-ion/L

(continued)

TABLE 7.1 *(Continued)*

Reference	System	Conditions	Correlation
Hughmark (1967)[a]	Air–(water, Na_2CO_3 solution, light oil, glycerol, $ZnCl_2$, and Na_2SO_3 solution)	$D_R = 0.0254–1.07$ m (PP) $U_G = 0.4–45$ cm/s $U_L = 0–12$ cm/s $\rho_L = 777–1698$ kg/m³ $\mu_L = 0.9–152$ mPa-s $\sigma = 25–76$ mN/m	$\varepsilon_G = \dfrac{1}{2 + \left(\dfrac{0.35}{U_G} \right) \left[\rho_L \dfrac{\sigma}{72} \right]^{1/3}}$
Idogawa et al. (1987)[b]	(Air, H_2, He)–water Air–organic liquids (acetone, ethanol, methanol) Air–aqueous solutions (ethanol, isoamyl alcohol)	$D_R = 0.05$ m $d_0 = 1$ mm $U_G = 0.5–5$ cm/s $\rho_L = 791–1000$ kg/m³ $\rho_G = 0.084–120.8$ kg/m³ $\mu_L = 0.35–3.0$ mPa-s $\sigma = 22.6–72.1$ mN/m	$\dfrac{\varepsilon_G}{1-\varepsilon_G} = 0.059 U_G^{0.8} \rho_G^{0.17} \left(\dfrac{\sigma}{\sigma_w} \right)^{-0.22\,\exp(-P)}$
Ityokumbul et al. (1994)[b]	Air–water	$D_R = 0.06$ m $d_0 = 134$ µm $U_G = 0–2.5$ cm/s	$\varepsilon_G = 5.9 U_G$ (SI)
Jean and Fan (1987)[b]	Air–water	$D_R = 0.0762$ (PP) $U_G = 0–12.08$ cm/s $U_L = 3.27–32.18$ cm/s	$\varepsilon_G = 0.00164 U_G^{0.988} (28.821 + 0.564 U_L - 0.221 \times 10^5 \, U_L^2)$ for $U_G \leq 5.19$ cm/s $\varepsilon_G = \dfrac{0.2933 U_G^{0.34}}{\exp(-0.248 U_L U_G^{-0.648}) + \exp(-0.243 U_L U_G^{-0.648})}$ for $U_G \geq 5.19$ cm/s

150

Reference	System	Conditions	Correlation
Jordan and Schumpe (2001)[a]	(Air, N_2, He, CO_2, H_2)–23 different organic liquids	$D_R = 0.095-0.115$ m $d_0 = 1.0-4.3$ mm $U_G = 1-21$ cm/s $\dfrac{g\rho_L d_b^2}{\sigma} = 825-1.55\times10^6$ $\dfrac{U_G}{\sqrt{gd_b}} = 0.06-1.22$ $\dfrac{\rho_G}{\rho_L} = 9.3\times10^{-5}-0.059$	$$\frac{\varepsilon_G}{1-\varepsilon_G} = B\left(\frac{g\rho_L d_b^2}{\sigma}\right)^{0.16}\left(\frac{gd_b^3}{\nu_L^2}\right)^{0.04}\left(\frac{U_G}{\sqrt{gd_b}}\right)^{0.52}\left(\frac{\rho_G}{\rho_L}\right)^{0.70}$$ $$\left[1+27.0\left(\frac{U_G}{\sqrt{gd_b}}\right)^{0.52}\left(\frac{\rho_G}{\rho_L}\right)^{0.58}\right]$$ B is sparger dependent
Jordan et al. (2002)	(Air, N_2)–(water, ethanol (96%), 1-butanol, toluene)	$D_R = 0.115$ m $H = 1.370$ m $U_G = 0.01-0.15$ m/s $\rho_L = 759-884$ kg/m³ $\mu_L = 0.58-2.94$ mPa-s $\sigma = 19.9-32.5$ mN/m D_L (10^{-9} m²/s) = 1.60–4.38 $P = 1-10$ bar $T = 293$K	$\varepsilon_G = B(U_G)^{0.86}\,(\rho_G)^{0.24}$ Values for B Water = 1.69 Ethanol (96%) = 1.62 1-Butanol = 1.34 Toluene = 1.95
Josh and Sharma (1979)[a]			$\varepsilon_G = \dfrac{U_G}{0.3+2U_G}$
Kang et al. (1999)	Air–(water, aqueous solutions of carboxy methyl cellulose)	$D_R = 0.152$ m $d_0 = 1.0$ mm $U_G = 2-20$ cm/s $\mu_L = 1-38$ mPa-s $P = 0.1-0.6$ MPa	$$\varepsilon_G = 0.910\times10^{-2.10}\left(\frac{D_R U_G \rho_G}{\mu_L}\right)^{0.254}$$
Kato and Nishiwaki (1972)[b]	Air–water	$D_R = 0.066-0.214$ m $d_0 = 1.0-3.0$ mm $U_G = 0-30$ cm/s $U_L = 0-1.5$ cm/s	$$\varepsilon_G = \frac{U_G}{31+\beta U_G^{0.8}\,(1-\exp(\gamma))}$$ where: $\beta = 4.5-3.5\exp(-0.064D_R^{1.8})$ $\gamma = -0.18U_G^{1.8}\beta^{-1}$

(continued)

TABLE 7.1 (*Continued*)

Kawase and Moo-Young (1987)[a]	$\varepsilon_G = 1.07\, Fr^{1/3}$ where: $Fr = \dfrac{U_G}{\sqrt{g D_R}}$
Kawase et al. (1992)[a]	$\dfrac{\varepsilon_G}{1-\varepsilon_G} = 0.0625 \left(\dfrac{U_G}{\nu_L g}\right)^{1/4}$
Kim et al. (1972)[b]	$D_R = 0.049$ m $d_0 = 3.175$ mm $U_G = 0\text{–}26$ cm/s $U_L = 1.4\text{–}10.2$ cm/s $1-\varepsilon_G = 1.02 \left(\dfrac{\mu_L^2 \rho_G}{D_R g \rho_L}\right)^{-0.009} \left(\dfrac{U_G \rho_L}{D_R g \rho_G}\right)^{-0.036} \left(\dfrac{U_G U_L D_R^2}{\nu_G \mu_L}\right)^{-0.015}$
Koide et al. (1979)[a]	$\varepsilon_G = \dfrac{U_G}{31+\beta(1-e)\sqrt{U_G}}$ where: $\beta = 4.5-3.5\,\exp(-0.064 D_R^{1.3})$ and $e = \dfrac{-0.18 U_G^{1.8}}{\beta}$
Koide et al. (1984)[a]	Air–water Air–aqueous solutions (electrolytes, glycol, glycerol) $D_R = 0.10\text{–}0.30$ m $d_0 = 0.5\text{–}2.5$ mm (MO) $U_G = 1\text{–}18$ cm/s $\rho_L = 997\text{–}1178$ kg/m^3 $\rho_S = 2500$ and 8700 kg/m^3 $\mu_L = 0.894\text{–}17.6$ mPa-s $\sigma = 51.5\text{–}73$ mN/m $c_S = 0\text{–}200$ kg/m^3 $d_S = 47.5\text{–}192$ μm $\dfrac{\varepsilon_G}{(1-\varepsilon_G)^4} = \dfrac{A \left(\dfrac{U_G \mu_L}{\sigma}\right)^{0.918} \left(\dfrac{g \mu_L^4}{\rho_L \sigma^3}\right)^{-0.252}}{1+4.35 \left(\dfrac{C_S}{\rho_s}\right)^{0.748} \left[\dfrac{\rho_s-\rho_L}{\rho_L}\right]^{0.88} \left(\dfrac{U_G D_R}{\nu_L}\right)^{-0.168}}$ where: $A = 0.277$ for water and non-electrolyte solutions, $A = 0.364$ for electrolyte solutions

Kojima et al. (1997)[b]	(N₂, O₂)–water (N₂, O₂)–aqueous solutions (enzyme, Na₂HPO₄, citric acid) $D_R = 0.045$ m $d_0 = 1.38–4.03$ mm $U_G = 0.5–15$ cm/s $\rho_L = 1000–1025$ kg/m³ $\mu_L = 0.890–1.075$ mPa-s $\sigma = 63.36–71.96$ mN/m $P = 0.1–1.1$ MPa	$\varepsilon_G = \varepsilon_0 = 1.18 U_G^{0.679}\left(\dfrac{\sigma}{0.076}\right)^{-0.546}$ for $P = 101$ kPa $\varepsilon_G = \varepsilon_0 = \exp\left[A\left(\dfrac{\pi D_R^4 \rho_L U_G^2}{16 d_0^3 \sigma}\right)\left(\dfrac{P}{P_{atm}}\right)^B\right]$ for $P > 101$ kPa where: $A = 1.27\times10^{-4}$ and $B = 1.0$
Kumar et al. (1976)[a]	Air–H₂O Air–kerosene Air–40% glycerol Air–2M NaOH solution $D_R = 0.05–0.10$ m $d_0 = 0.087–0.309$ cm (PP) $U_G = 0.2–14$ cm/s $\rho_L = 0.78–1.11$ g/cm³ $\mu_L = 0.88–11.5$ mPa-s $\sigma = 31.2–74.5$ mN/m	$\varepsilon_G = 0.728 U_G^* - 0.485 U_G^{*2} + 0.0975 U_G^{*3}$ where: $U_G^* = U_G\left(\dfrac{\sigma\Delta\rho g}{\rho_L}\right)^{1/4}$
Lau et al. (2004)	Air–water $D_R = 0.045–0.45$ m $H/D_R > 5$ $d_0 = 1.5$ mm $U_G = 2.8–67.8$ cm/s $U_L = 0–0.089$ cm/s P up to 22 MPa T up to 250°C $\rho_L = 790–1580$ kg/m³ $\rho_G = 0.97–33.4$ kg/m³ $\mu_L = 0.36–38.3$ mPa-s $\sigma = 23.3–72.6$ mN/m	$\dfrac{\varepsilon_G}{(1-\varepsilon_G)^4} = \dfrac{2.9\left(\dfrac{U_G^4 \rho_G}{\sigma_G}\right)^\alpha \left(\dfrac{\rho_G}{\rho_m}\right)^\beta}{\left[\cosh(Mo_m^{0.054})\right]^{4.1}}$ Where: $\alpha = 0.21 Mo_m^{0.0079}$, $\beta = 0.096 Mo_m^{-0.011}$, $Mo_m = \dfrac{(\xi\mu_L)^4 g}{\rho_m \sigma^3}$, $\ln\xi = 4.6\varepsilon_s\left\{5.7\,\varepsilon_s^{0.58}\sinh\left[-0.71\exp(-5.8\varepsilon_s)\ln Mo^{0.22}\right]+1\right\}$, and $Mo = \dfrac{g\mu_L^4}{\rho_L\sigma^3}$
Lee et al. (2000)[b]	Air–water $D_R = 0.083–0.15$ m $d_0 = 4$ mm $U_G = 0–270$ cm/s $U_L = 0–23$ cm/s	$\dfrac{\varepsilon_G}{1-\varepsilon_G} = 0.759 U_G^{0.685} U_L^{-0.116}$ (SI)

(continued)

TABLE 7.1 (*Continued*)

Lockett and Kirkpatrick (1975)[a]	$U_G(1-\varepsilon_G)+U_L\varepsilon_G = V_B\varepsilon_G(1-\varepsilon_G)^{2.39}(1+2.55\varepsilon_G^3)$	
Luo et al. (1999)[b]	$D_R = 0.1-0.61$ m $U_G = 5-69$ cm/s $U_L = 0$ m/s $\rho_L = 668-2965$ kg/m³ $\rho_s = 2200-5730$ kg/m³ $\rho_G = 0.2-90$ kg/m³ $\mu_L = 0.29-30.0$ mPa-s $\sigma = 19.0-73.0$ mN/m $\varphi_s = 0,\ 0.081, 0.181$ $T = 301K,\ 351K$ $P = 0.1-5.62$ MPa	(Air, He, CO₂, N₂)–water Air–organic liquids (methanol, glycol, glycerol, C₂H₃Br₃, n-octanol, heptanes, n-butanol, 1,3-butanediol, trichloroethylene) Air–Isopar G, Paratherm F liquids with and without suspended solids $\dfrac{\varepsilon_G}{(1-\varepsilon_G)} = 2.9\left(\dfrac{U_G^4\rho_G}{\sigma_G}\right)^{\alpha}\left(\dfrac{\rho_G}{\rho_L}\right)^{\beta}\left[\cosh(Mo_s^{0.054})\right]^{-4.1}$ Where: $\alpha = 0.21Mo_s^{0.079}$, $\beta = 0.096Mo_s^{-0.011}$, $\rho_s = \varphi_s\rho_s+\varphi_L\rho_L$, $Mo_s = \dfrac{(\xi\mu_L)^4 g}{\rho_s{}^l\sigma^3}$, $\ln\xi = 4.6\varphi_s\left\{5.7\,\varphi_s^{0.58}\sinh\left[-0.1562\exp(-5.8\varphi_s)\ln\dfrac{g\mu_L}{\rho_L\sigma^3}\right]+1\right\}$
Mashelkar (1970)[b]	Air–water Air–electrolyte aqueous solutions	$\varepsilon_G = \dfrac{U_G}{\rho_L(30+2U_G)}\left(\dfrac{\sigma_w}{\sigma}\right)^{-1/3}$
Mersmann (1977)[b]		$\varepsilon_G(1-\varepsilon_G)^n = 0.14U_G\left[\dfrac{\rho_L^2}{\sigma(\rho_L-\rho_G)g}\right]^{1/4}\left(\dfrac{\rho_L}{\rho_G}\right)^{5/72}\left[\dfrac{\rho_L}{\rho_L-\rho_G}g\right]^{1/3}\left[\dfrac{\sigma^3\rho_L^2}{\mu_L^4(\rho_L-\rho_G)g}\right]^{1/24}$
Mok et al. (1990)[b]	$D_R = 0.14$ m $d_0 = 0.3$ mm $U_G = 0.96-5.04$ cm/s $\rho_L = 997.1-998.2$ kg/m³ $\sigma = 70.5-72$ mN/m	$\varepsilon_G = 1.07\times10^{-5}\left(\dfrac{D_R U_G}{\nu_G}\right)^{1.09}\left(\dfrac{gD_R^3}{\nu_L^2}\right)^{0.096}\left(\dfrac{d_0}{D_R}\right)^{-0.19}$

Reference	Conditions	Correlation
Mauza et al. (2005)[b]	Air–water Air–aqueous solutions (glycerin, n-butanol) Square column with Width = 0.10 m d_0 = 20 and 40 μm $U_G < U_{trans}$ ρ_L = 991–1173 kg/m³ μ_L = 0.9–22.5 mPa-s σ = 48.0–72 mN/m	$$\varepsilon_G = 0.001\left[\frac{U_G}{\sqrt{gD_R}}\left(\frac{D_R^3\rho_L^2 g}{\mu_L^2}\right)^{0.1}\left(\frac{D_R^2\rho_L g}{\sigma}\right)^{2.2}\frac{D_d}{D_R}\right]^{2/3}$$
Reilley et al. (1986)[a]	D_R = 0.30 m d_0 = 1.5 mm U_G = 0.77–21.7 cm/s ρ_L = 771–1482 kg/m³ μ_L = 0.489–1.29 mPa-s σ = 28.3–72 mN/m ρ_G = 0.185–1.48 kg/m³	$\varepsilon_G = 0.009 + 296 U_G^{0.44}\rho_L^{-0.98}\sigma_L^{-0.16}\rho_G^{0.19}$ can replace ρ_L with ρ_s
Roy et al. (1953)[a]		$$\varepsilon_G = 3.88\times10^{-3}\left[Re_T\left(\frac{\sigma_w}{\sigma_L}\right)^{1/3}(1-\nu_s)^3\right]^{-0.44}$$ for $Re_T > 500$ where: $$\nu_s = \frac{W_s/\rho_s}{(W_s/\rho_s)+(W_L/\rho_L)}$$
Sada et al. (1984)[a]	(N₂, He)–(Water, molten NaNO₃, molten LiCl-KCl) N₂–Methanol N₂–Aqueous solutions (glycerol, Na₂SO₄) D_R = 0.073 m d_0 = 1.5–5.7 mm U_G = 0.56–8.3 cm/s ρ_L = 788–1888 kg/m³ μ_L = 0.45–3.658 mPa-s σ = 21.5–130 mN/m	$$\varepsilon_G = 0.32\,(1-\varepsilon_G)^4\,Bo^{0.21}\,Ga^{0.086}\,Fr\left(\frac{\rho_G}{\rho_L}\right)^{0.068}$$ Where: $Bo = \dfrac{\rho_L g D_R^2}{\sigma}$, $Ga = \dfrac{g D_R^3}{\nu_L^2}$, and $Fr = \dfrac{U_G}{\sqrt{gD_R}}$ $$\frac{\varepsilon_G}{(1-\varepsilon_G)^3} = 0.019 U_\infty^{1/16}\nu_s^{-0.215}U_\infty^{-0.16}U_G$$

(continued)

TABLE 7.1 (*Continued*)

Salvacion et al. (1995)[b]	$D_R = 0.14–0.30$ m $d_0 = 0.5–2.5$ mm $U_G = 2–15$ cm/s $\rho_L = 995–997$ kg/m³ $\rho_s = 1000–1087$ kg/m³ $\varphi_s = 0–0.20$ $\mu_L = 0.863–0.894$ mPa-s $\sigma = 55.6–72.0$ mN/m	$$\frac{\varepsilon_G}{(1-\varepsilon_G)^4} = \varepsilon_0\left[\frac{1+0.00468\,(D_R^{\,2}g\rho_L\sigma^{-1})^{0.465}f(A)}{1+2.34\varphi_s^{0.799}(D_R^{\,2}g\rho_L\sigma^{-1})^{-0.0464}}\right]$$ where: $A = 0$, $$\varepsilon_0 = 0.277\left(\frac{U_G\mu_L}{\sigma}\right)^{0.918}\left(\frac{g\mu_L}{\rho_L\sigma^3}\right)^{-0.252}$$ $$f(A) = A^{0.223}(1+0.0143A^{0.466})^{-1}$$
Schumpe and Deckwer (1987)[a]	$Bo = 1.4\times10^3–1.4\times10^5$ $Ga = 1.2\times10^7–6.5\times10^{10}$ $Fr = 3\times10^{-3}–2.2\times10^{-1}$	$\varepsilon_G = 0.2Bo^{-0.13}\,Ga^{0.11}\,Fr^{0.54}$ For highly viscous media
Smith et al. (1983)[a]		$$\varepsilon_G = \left[2.25+\frac{0.379}{U_G}\left(\frac{\rho_L}{72}\right)^{0.31}\mu_L^{0.016}\right]^{-1}$$ μ_L can be replaced with μ_s, except $\mu_s = \mu_L\,exp\left[\frac{(5/3)\nu_s}{1-\nu_s}\right]$
Sotelo et al. (1994)[b]	(Air, CO_2)–water (Air, CO_2)–aqueous solutions (sucrose, ethanol glycerin) $D_R = 0.04–0.08$ m $d_0 = 30–150$ μm $U_G = 0.64–4.9$ cm/s $\rho_L = 928–1147$ kg/m³ $\rho_G = 1.18–1.83$ kg/m³ $\mu_L = 1.0–4.17$ mPa-s $\mu_G = 0.0148–0.0191$ mPa-s $\sigma = 29.8–73.4$ mN/m	$$\varepsilon_G = 129\left(\frac{U_G\mu_L}{\sigma}\right)^{0.99}\left(\frac{\mu_L^{\,4}g}{\rho_L\sigma^3}\right)^{-0.123}\left(\frac{\rho_G}{\rho_L}\right)^{0.187}\left(\frac{\mu_G}{\mu_L}\right)^{0.343}\left(\frac{d_0}{D_R}\right)^{-0.089}$$
Syeda et al. (2002)[b]	Air–(Water, methanol, glycerol, i-propanol) $D_R = 0.09$ m $d_0 = 5$ mm	$$\varepsilon_G = 1.1855\left(\frac{\rho_L d_b}{\sigma}\right)^{0.016}\left(\frac{U_G\mu_L}{\sigma}\right)^{0.578}\left(\frac{\mu_L g}{\rho_L\sigma^3}\right)^{-0.131}\left(\frac{\rho_G}{\rho_L}\right)^{0.062}\left(\frac{\mu_G}{\mu_L}\right)^{0.107}$$

Reference	Conditions	Correlation
Ulbrecht and Patterson (1985)[b]	N_2–(Tellus oil, aqueous glucose solution) $D_R = 0.15$ and 0.23 m $d_0 = 0.5$–1.5 mm $U_G = 0$–25 cm/s $\rho_L = 867$–1380 kg/m^3 $\mu_L = 70$–550 mPa-s $\sigma = 31.0$–76.0 mN/m $P = 0.1$–1.0 MPa	$\varepsilon_G = 0.21\, \dfrac{U_G^{0.58}\rho_G^{0.3}\exp(-9\mu_L)}{D_R^{0.18}\mu_L^{0.12}}$
Zahradnik and Kastanek (1979)[b]	Air–water $D_R = 0.152$ and 0.292 m $d_0 = 0.87$–3 mm $U_G = 3.1$–27.6 cm/s	$\varepsilon_G = \dfrac{U_G}{0.3 + 2.0 U_G}$ (SI)
Zahradnik et al. (1997)	Air–water $D_R = 0.29$ m $H_0 = 1.5$ m $U_G = 0.4$–16.6 cm/s	$\varepsilon_G = 2.81 U_G^{0.9}$
Zehner (1989)		$\varepsilon_G = \dfrac{U_G/\alpha}{\sqrt{1 + 4\left(\dfrac{U_G/\alpha}{\alpha}\right)^{2/3}\dfrac{V_L(0)}{\alpha}}}$ where $\alpha = 1.4\left(\dfrac{\sigma}{\rho_L}\dfrac{\rho_L - \rho_G}{\rho_L}g\right)^{1/4}$ and $V(0) = \left[\dfrac{1}{2.5}\dfrac{\rho_L - \rho_G}{\rho_L}U_G g D_R\right]^{1/2}$
Zou et al. (1988)[a]	$D_R = 0.10$ m $d_0 = 10$ mm $\rho_L = 748.3$–997.1 kg/m^3 $\rho_G = 1.833$–2.161 kg/m^3 $\mu_L = 0.295$–0.894 mPa-s $\sigma = 18.77$–71.97 mN/m $T = 298.15$–369.65K	$\varepsilon_G = 0.17283\left(\dfrac{\mu_L^4 g}{\rho_L\sigma_L^3}\right)^{-0.15}\left(\dfrac{U_G\mu_L}{\sigma_L}\right)^{0.15}\left(\dfrac{P + P_{sat}}{P}\right)^{1.61}$

[a]As cited by Kantarci et al. (2005)

[b]As cited by Ribeiro Jr. and Lage (2005)

[c]As cited by Deckwer (1992)

157

TABLE 7.2 Liquid-Phase Mass Transfer Correlations for Bubble Column

Reference	System	Conditions	Correlation
Akita and Yoshida (1974)[a]	(Air, O_2, He, CO_2)–water Air–organic liquids (glycol, methanol) Air–aqueous solutions (glycol, methanol, glycerol, electrolyte)	$D_R = 0.07$–0.600 m $d_0 = 0.01$–5.0 mm $U_G = 4.0$–33 cm/s $\rho_L = 790$–1165 kg/m³ $\mu_L = 0.58$–21.14 mPa·s $\sigma = 22.3$–74.2 mN/m D_L (10^{-9} m²/s) $= 0.26$–4.22	$$\frac{k_L d_b}{D_L} = 0.5 \left(\frac{\nu_L}{D_L}\right)^{1/2} \left(\frac{d_b^3 g}{\nu_L^2}\right)^{1/4} \left(\frac{g d_b^2 \rho_L}{\sigma}\right)^{3/8}$$
Calderbank and Moo-Young (1961)[a]		$\rho_L = 698$–1260 kg/m³ $\mu_L = 0.371$–8.7 mPa·s $\lambda_L = 0.60$–27.92 W/(mK) D_L (10^{-4} m²/s) $= 1.9\times10^{-6}$–0.26	$$\frac{k_L d_b}{D_L} = 2.0 + 0.31 \left(\frac{d_b^2(\rho_L - \rho_G)g}{\mu_L D_L}\right)^{1/3} \quad \text{for } d_b < 2.5 \text{ mm}$$ $$k_L \left(\frac{\nu_L}{D_L}\right)^{1/2} = 0.42 \left[\frac{(\rho_L - \rho_G)\mu_L g}{\rho_L^2}\right]^{1/3} \quad \text{for } d_b > 2.5 \text{ mm}$$
Cockx et al. (1995)[a]			$$\frac{k_L}{U_G} = \frac{4.47\times10^{-2}}{\varepsilon_G} \left(\frac{\nu_L}{D_L}\right)^{-1/2}$$
Fair (1967)[a]			$$\frac{k_L d_b}{D_L} = 2.0 \left[1 + 0.276 \left(\frac{d_b U_b}{\nu_L}\right)^{1/2} \left(\frac{\nu_L}{D_L}\right)^{1/3}\right]$$
Fukuma et al. (1987)[a]		$U_G = 0.30$–1.5 cm/s $\dfrac{g d_b^3}{\nu_L} = 3.2\times10^{-5}$–$2.6\times10^8$ $\dfrac{g \rho_L d_b^2}{\sigma} = 1.4$–$120$ $\dfrac{\nu_L}{D_L} = 480$–1600	$$\frac{k_L d_b}{D_L} = 4.5\times10^{-4} \left(\frac{\nu_L}{D_L}\right)^{1/2} \left(\frac{g \rho_L d_b^2}{\sigma}\right)^{-0.2} \left(\frac{g d_b^3}{\nu_L^2}\right)^{0.8}$$

Reference	System	Conditions	Correlation
Gestrich et al. (1978)[a]	(CO$_2$, O$_2$)–water (CO$_2$, O$_2$)–aqueous solutions (glycerine, glycerol, electrolyte)		$k_L = 2.23\times10^{-4}\left(\dfrac{g\mu_L^4}{\rho_L\sigma^3}\right)^{-0.180} + 3.85\times10^{-3}\left(\dfrac{H_0}{D_R}\right)^{-0.605} U_G^{\,0.65+0.0335\frac{H_0}{D_R}}$
Gestrich et al. (1978)[a]	(CO$_2$, O$_2$)–water (CO$_2$, O$_2$)–aqueous solutions (glycerine, glycerol, electrolyte)		$k_L = 2.23\times10^{-4}\left(\dfrac{g\mu_L^4}{\rho_L\sigma^3}\right)^{-0.180} + 3.85\times10^{-3}\left(\dfrac{H_0}{D_R}\right)^{-0.605} U_G^{\,0.65+0.0335\frac{H_0}{D_R}}$
Hughmark (1967)[a]	Air–(water, Na$_2$CO$_3$ solution, light oil, glycerol, ZnCl$_2$ and Na$_2$SO$_3$ solution)	$D_R = 0.0254-1.07$ m $U_G = 0.4-45$ cm/s $U_L = 0-12$ cm/s $\rho_L = 777-1698$ kg/m^3 $\mu_L = 0.9-152$ mPa-s $\sigma = 25-76$ mN/m	$\dfrac{k_L d_b}{D_L} = 2.0 + 0.0187\left[\left(\dfrac{U_G d_b}{\varepsilon_G \nu_L}\right)^{0.484}\left(\dfrac{\nu_L}{D_L}\right)^{0.339}\left(\dfrac{d_b g^{1/3}}{D_L^{2/3}}\right)^{0.072}\right]^{1.61}$
Kawase et al. (1987)[a]			$k_L = \dfrac{2}{\sqrt{\pi}}\sqrt{D_L}\left(\dfrac{U_G g}{\nu_L}\right)^{1/4}$
Kawase and Moo–Young (1992)[a]		for small bubbles	$k_L = \dfrac{0.28 D_L^{2/3} g^{1/3}}{\nu_L^{1/3}}$
		for large bubbles	$k_L = \dfrac{0.47 D_L^{1/2} g^{1/3}}{\nu_L^{1/6}}$
Lirek et al. (2005a)	O$_2$–water O$_2$–aqueous solutions (Na$_2$SO$_3$ pure and with the addition of Sokrat 44 and carboxy–methyl–cellulose)	$D_R = 0.19$ m $U_G = 0.18-0.54$ cm/s $\rho_L = 1000-1093$ kg/m^3 $\mu_L = 1.11-3.14$ mPa-s	$k_L = 0.463(g U_G \nu_L)^{1/4}\left(\dfrac{D_L}{\nu_L}\right)^{1/2}$

(continued)

TABLE 7.2 (*Continued*)

Reference	System	Conditions	Correlation
Miller (1974)[a]	Air–water	$D_R = 0.152–0.686$ m $d_0 = 1.6–6.4$ mm $U_G = 0.762–15.2$ cm/s	$k_L = 1366 d_b^{1.376} \sqrt{\dfrac{D_L U_b}{\pi d_b}}$
Nedeltchev (2003)[a]	N_2–gasoline, toluene	$D_R = 0.316$ m $U_G = 5–25$ cm/s $\rho_L = 692–866$ kg/m³ $\mu_L = 0.46–0.59$ mPa-s	$k_L = 2.43\times10^7 d_b^{2.928} \sqrt{\dfrac{D_L U_b}{\pi d_b}}$ (SI)
Ruckenstein (1964)[a]			$\dfrac{k_L d_b}{D_L} = 1.252 \left[\dfrac{(1-\epsilon_G)^{5/3}}{2+3\left(\epsilon_G^{5/3}-\epsilon_G^{1/3}-2\epsilon_G^2\right)} \right]^{1/3} \left(\dfrac{U_b d_b}{D_L} \right)^{1/3}$
Sokolov and Aksenova (1983)[a]	(Air, CO_2)–water (Air, CO_2)–aqueous solutions (sucrose, carbonate–bicarbonate buffer)	$D_R = 0.05–0.60$ m $d_0 = 2.25–40$ mm $U_G = 0.05–4.0$ cm/s $\dfrac{\nu_L}{D_L} = 310–8320$ $\dfrac{U_L}{\mu_L \sqrt{\rho_L \sigma / g}} = 5–740$	$\dfrac{k_L \sigma}{D_L \rho_L g} = 0.05 \left(\dfrac{\nu_L}{D_L}\right)^{0.5} \dfrac{U_G}{U_L} \left(\dfrac{\rho_L \sigma}{g}\right)^{0.5}$
Vázquez et al. (2000)		$D_R = 0.113$ m $d_0 = 0.04–0.20$ mm (MO) $U_G = 0.085–0.181$ cm/s $\nu_L = 1.22\times10^{-6}$ m²/s $\sigma = 66.00–75.01$ mN/m	$k_L = A\sigma^{1.35} U_G^{0.5}$ (SI) where:

Equivalent pore diameter (10^6 m)	A
150–200	0.17587
90–150	0.18233
40–90	0.18689

[a] As cited by Kantarci et al. (2005)
[b] As cited by Ribeiro Jr. and Lage (2005)
[c] As cited by Deckwer (1992)

TABLE 7.3 Gas–Liquid Mass Transfer Correlations for Bubble Columns

Reference	System	Conditions	Correlation
Akita and Yoshida (1973)[a]	Air–H$_2$O O$_2$–H$_2$O He–H$_2$O CO$_2$–H$_2$O Air–glycol Air–aqueous glycol solution Air–methanol	$D_R = 0.15$–0.600 m $H = 4.0$ m $d_0 = 0.5$ mm $U_G = 0.4$–32.8 cm/s P up to 0.1 MPa T up to 293 K $\dfrac{U_G}{\sqrt{gD_R}} = 0.0024$–$0.135$ $gD_R^2\rho_L = 3050$–0.45×10^4 $\dfrac{v_L}{D_L}\,\sigma = 173$–$7.25\times10^3$	$$\frac{k_L a D_R^2}{D_L} = 0.6 \left(\frac{v_L}{D_L}\right)^{0.5} \left(\frac{gD_R^2\rho_L}{\sigma}\right)^{0.62} \left(\frac{gD_R^3}{v_L^2}\right)^{1.1} \varepsilon_G^{1.1}$$

Álvarez et al. (2000)	CO$_2$–water
	CO$_2$–aqueous solutions (sucrose, sodium lauryl sulfate)
		$D_R = 0.113$ m
		$d_0 = 0.04$–0.20 mm (MO)
		$U_G = 0.085$–0.150 cm/s
		$\rho_L = 997$–1030 kg/m³
		$\mu_L = 0.896$–1.135 mPa-s
		$\sigma = 65.20$–72.49 mN/m
		D_L (10^{-9} m²/s) = 1.549–1.897
			$$k_L a = A\,\frac{U_G^{2/3}\,\sigma^{3/4}\,\rho_L^{3/2}}{\mu_L^{3/4}}$$ Where:
			
			  | Equivalent pore diameter (10^6 m) | A (10^7) |
			  | 150–200 | 1.924 |
			  | 90–150 | 1.969 |
			  | 40–90 | 2.079 |

Belkish et al. (2002)	(H$_2$, CO, N$_2$, CH$_4$)–(hexane mixture, Isopar-M) with and without solids (iron oxides catalyst and glass beads)	$D_R = 0.316$ m $U_G = 5$–25 cm/s $\rho_L = 680$–783 kg/m³ $\mu_L = 0.32$–2.7 mPa-s $\sigma = 27.0$–23.0 mN/m D_L (10^{-9} m²/s) = 1.33–9. $40\varphi_s = 0$–0.36	$$k_L a = 0.18 \left(\frac{v_L}{D_L}\right)^{0.16} \left(\frac{\rho_L}{\rho_G}\right)^{2.84} (\rho_G U_G)^{0.49}\exp(-266\varphi_s)$$
Cho and Wakao (1988)[b]	N$_2$–aqueous solutions (benzene, CCl$_4$, CHCl$_3$, C$_2$H$_4$Cl$_2$, C$_2$H$_3$Cl$_2$)	$D_R = 0.11$ m $U_G = 0.71$–5.0 cm/s D_L (10^{-9} m²/s) = 0.99–2.42	$$k_L a = A D_L^{0.5} U_G^{0.81}$$ where: $A = 6500$ for single nozzle sparger or $A = 23000$ for porous plate

(continued)

TABLE 7.3 (*Continued*)

Hikita et al. (1981)[c]	$D_R = [0.10, 0.19]$ m $H = [1.5, 2.4]$ m $d_0 = 1.1$ cm (SO) $U_G = 4.2–39$ cm/s P up to 0.1 MPa $T = 283–303$ K Nonelectrolytes $\rho_L = 0.79–1.24$ g/cm³ $\rho_G = (0.0837–1.84)*10^{-3}$ g/cm³ $\mu_L = 0.66–17.8$ mPa-s $\mu_G = (0.821.81)*10^{-2}$ mPa-s $\sigma = 30–76$ mN/m Electrolytes $\rho_L = 1.01–1.17$ g/cm³ $\mu_L = 0.9–1.87$ mPa-s $\sigma = 71.9–79.6$ mN/m	$k_L a = \dfrac{14.9 g f}{U_G} \left(\dfrac{U_G \mu_L}{\sigma}\right)^{1.76} \left(\dfrac{g \mu_L^4}{\sigma^3}\right)^{-0.284} \left(\dfrac{\mu_G}{\mu_L}\right)^{0.243} \left(\dfrac{\mu_L}{\rho_L D_L}\right)^{-0.604}$ where $f = 1$ for electrolytes, $10^{0.0414I}$ for ion strength $I<1$ g-ion/L, 1.1 for $I>1$ g-ion/L
Jordan and Schumpe (2001)	$D_R = 0.095–0.115$ m $d_0 = 1.0–4.3$ mm (SO, MO) $U_G = 0.80–21$ cm/s $\dfrac{g \rho_L d_b^2}{\sigma} = 1.21–5.39$ $\dfrac{g d_b^3}{\nu_L^2} = 825–1.55\times10^6$ $\dfrac{U_G}{\sqrt{g d_b}} = 0.06–1.22$ $\dfrac{\rho_G}{\rho_L} = 9.3\times10^{-5}–0.059$ $\dfrac{\nu_L}{D_L} = 71–6.89\times10^4$	$\dfrac{k_L a d_b^2}{D_L} = A \left(\dfrac{\nu_L}{D_L}\right)^{0.5} \left(\dfrac{g \rho_L d_b^2}{\sigma}\right)^{0.34} \left(\dfrac{g d_b^3}{\nu_L^2}\right)^{0.27} \left(\dfrac{U_G}{g d_b}\right)^{0.72} \left[1+13.2 \left(\dfrac{U_G}{\sqrt{g d_b}}\right)^{0.37} \left(\dfrac{\rho_G}{\rho_L}\right)^{0.49}\right]$ where A is sparger dependent

Jordan et al. (2002)	(Air, N₂)–(water, ethanol (96%), 1-butanol, toluene)	$D_R = 0.115$ m $H = 1.370$ m $U_G = 0.01–0.15$ m/s $\rho_L = 759–884$ kg/m³ $\mu_L = 0.58–2.94$ mPa-s $\sigma = 19.9–32.5$ mN/m $D_L (10^{-9}$ m²/s) = 1.60–4.38 $P = 1–10$ bar $T = 293$K	$k_L a = C(U_G)^{0.91}(\rho_G)^{0.24}$ $k_L a = D(\epsilon_G)^{1.06}$ | Liquid	C	D
Water	0.99	0.56
Ethanol (96%)	1.14	0.67
1-Butanol	0.83	0.60
Toluene	1.49	0.72	
Kang et al. (1999)	Air–(water, aqueous solutions of carboxymethyl cellulose)	$D_R = 0.152$ m $d_0 = 1.0$ mm $U_G = 2–20$ cm/s $\mu_L = 1–38$ mPa-s $P = 0.1–0.6$ MPa	$k_L a = 0.930\times10^{-3.08}\left(\dfrac{D_R U_G \rho_G}{\mu_L}\right)^{0.254}$
Kawase and Moo-Young (1937)[a]			$\dfrac{k_L a D_R^2}{D_L} = 0.452 \left(\dfrac{\nu_L}{D_L}\right)^{1/2}\left(\dfrac{D_R U_G}{\nu_L}\right)^{3/4}\left(\dfrac{g D_R^2 \rho_L}{\sigma}\right)^{-3/5}\left(\dfrac{U_G^2}{D_R g}\right)^{7/60}$
Khudenko and Shpirt (1986)[b]	Air–water	$R_{dt} = 0.125–1$ $\dfrac{H_0}{H_b} = 0.25–1$ $U_G = 0.055–2.1$ cm/s $T = 17–24$ °C	$k_L a = 0.041 R_{dt}^{0.18}\left(\dfrac{U_G}{H_b}\right)^{0.67}$
Koide et al. (1984)[b]	Air–water Air–aqueous solutions (electrolytes, glycol, glycerol)	$D_R = 0.10–0.30$ m $d_0 = 0.5–2.5$ mm (MO) $U_G = 1–18$ cm/s $\rho_L = 997–1178$ kg/m³ $\rho_s = 2500$ and 8770 kg/m³ $\mu_L = 0.894–17.6$ mPa-s $\sigma = 51.5–73$ mN/m $c_s = 0–200$ kg/m³ $d_s = 47.5–192$ μm	$\dfrac{k_L a\sigma}{\rho_L D_{Lg}} = \dfrac{211\left(\dfrac{\nu_L}{D_L}\right)^{0.5}\left(\dfrac{g\mu_L}{\rho_L \sigma^3}\right)^{0.5}\epsilon_G^{1.18}}{1+1.47\times10^4\left(\dfrac{C_s}{\rho_s}\right)^{0.612}\left(\dfrac{U_P}{\sqrt{D_{Rg}}}\right)^{0.612}\left(\dfrac{\rho_L g D_R^2}{\sigma}\right)^{-0.477}\left(\dfrac{\rho_L U_G D_R}{\mu_L}\right)^{-0.345}}$

(continued)

TABLE 7.3 (*Continued*)

Lau et al. (2004)	(Air, N₂)-(water, Paratherm NF heat-transfer fluid) (Air, N₂, CO, He, H₂)-9 different organic liquids	$D_R = 0.045$–0.45 m $H/D_R > 5$ $d_0 = 1.5$ mm (PP) $U_G = 2.8$–67.8 cm/s $U_L = 0$–0.089 cm/s $\rho_L = 790$–1580 kg/m³ $\rho_G = 0.97$–33.4 kg/m³ $\mu_L = 0.36$–38.3 mPa-s $\sigma = 23.3$–72.6 mN/m P up to 22 MPa T up to 250 °C	$k_L a = 1.77\,\sigma^{-0.22}\exp(1.65 U_L - 65.3\mu_L)\epsilon_G^{1.2}$
Nakanoh and Yoshida (1980)[b]	Air–water Air–aqueous solutions	$D_R = 0.14$ m $d_0 = 4$ mm $U_G < 10$ cm/s $\rho_L = 995$–1230 kg/m³ $\nu_L\ (10^{-6}\ \text{m}^2/\text{s}) = 0.804$–$8.82$ $D_L\ (10^{-9}\ \text{m}^2/\text{s}) = 1.06$–$2.6$	$\dfrac{k_L a D_R^2}{D_L} = 0.09 \left(\dfrac{\nu_L}{D_L}\right)^{1/2} \left(\dfrac{D_R^3 g}{\nu_L^2}\right)^{0.39} \left(\dfrac{g D_R \rho_L}{\sigma}\right)^{0.75} \dfrac{U_G}{\sqrt{g D_R}}$
Ozturk et al. (1987)[b]	(Air, N₂, CO₂, He, H₂)-organic liquids	$D_R = 0.095$ m $d_0 = 3$ mm (SO) $U_G = 0.80$–10 cm/s $\dfrac{g\rho_L d_b^2}{\sigma} = 1.2$–$5.4$ $\dfrac{g d_b^3}{\nu_L^2} = 830$–$1.5\times10^6$ $\dfrac{U_G}{\sqrt{g d_b}} = 0.043$–$0.6$ $\dfrac{\rho_G}{\rho_L} = 9.3\times10^{-5}$–$2.0\times10^{-3}$ $\dfrac{\nu_L}{D_L} = 32$–1.5×10^5 $P = 0.1$ MPa $T = 293$ K	$\dfrac{k_L a d_B^2}{D_L} = 0.62 \left(\dfrac{\mu_L}{\rho_L D_L}\right)^{0.5} \left(\dfrac{g\rho_L d_B^2}{\sigma}\right)^{0.33} \left(\dfrac{g\rho_L^2 d_B^3}{\mu_L^2}\right)^{0.29} \left(\dfrac{U_G}{\sqrt{g d_B}}\right)^{0.68} \left(\dfrac{\rho_G}{\rho_L}\right)^{0.04}$

Reference	System	Conditions	Correlation
Salvacion et al. (1995)[b]	Air-water, Air-aqueous solutions (methanol, ethanol, n-butanol, n-hexanol, n-octanol)	$D_R = 0.14-0.30$ m $d_0 = 0.5-2.5$ mm (MO) $U_G = 2-15$ cm/s $\rho_L = 995-997$ kg/m³ $\rho_s = 1000-1087$ kg/m³ $\varphi_s = 0-0.20$ $\mu_L = 0.863-0.894$ mPa-s $\sigma = 55.6-72.0$ mN/m	$$\frac{k_L a \sigma}{\rho_L D_{Lg}} = 12.9 \left(\frac{\nu_L}{D_L}\right)^{1/2} \left(\frac{g\mu_L^2}{\rho_L \sigma^3}\right)^{-0.159} \left(\frac{g\rho_L D_R^2}{\sigma}\right)(1+0.62\varphi_s)\,f(B)$$ $$f(B) = 0.47 + 0.53\exp\left(-41.4\,\frac{B\varepsilon_G}{U_G}\sqrt{\frac{\varepsilon_G}{d_b\mu_L\rho_L U_G}}\right)$$ Where: $B = 0$ for water
Schumpe and Grund (1986)[a]	Air-water	$D_R = 0.30$ m $H = 4.4$ m $d_0 = 1$ mm $U_G < 20$ cm/s $T = 20$ °C	$$k_L a = K U_G^{0.82}\,\mu_{eff}^{-0.39}$$ where: $K = 0.063$ (water/salt solution) or 0.042 (water/0.8M Na$_2$SO$_4$ solution)
Shah et al. (1982)[c]			$k_L a = 0.467 U_G^{0.82}$
Shpirt (1981)[b]	Air-aqueous solutions of NH$_4$Cl	$D_R = 0.107$ m $H_0 = 0.30-1.0$ m $U_G = 0.094-0.75$ cm/s $T = 20$ °C	$$\frac{k_L a d_b^2}{D_L} = A\left(\frac{U_G d_b}{\nu_L}\right)^{0.45}\left(\frac{H_0}{d_b}\right)^{-0.30}$$ where: $A = 0.007$ for heterogeneous flow regime or $A = 0.0136$ for homogeneous flow regime
Sotelo et al. (1994)[b]	CO$_2$-water, CO$_2$-aqueous solutions (sucrose, ethanol, glycerin)	$D_R = 0.04-0.08$ m $d_0 = 30-150$ μm (MO) $U_G = 0.064-0.49$ cm/s $\rho_L = 998-1147$ kg/m³ $\mu_L = 1.00-4.17$ mPa-s $\mu_G = 1.48-1.53$ mPa-s $\sigma = 29.8-73.4$ mN/m $D_L (10^{-9}$ m²/s$) = 0.51-1.78$	$$\frac{k_L a U_G}{g} = 16.9\left(\frac{U_G \mu_L}{\sigma}\right)^{2.14}\left(\frac{\mu_L^4 g}{\rho_L \sigma^3}\right)^{-0.518}\left(\frac{\mu_G}{\mu_L}\right)^{0.074}\left(\frac{\nu_L}{D_L}\right)^{-0.038}\left(\frac{d_0}{D_R}\right)^{-0.98}$$
Vasquez et al. (1953)[b]	CO$_2$-water, CO$_2$-aqueous solutions (glycerin, glucose, sucrose)	$\rho_L = 997-1151$ kg/m³ $\mu_L = 0.896-6.42$ mPa-s $\sigma = 66.9-73.4$ mN/m $D_L (10^{-9}$ m²/s$) = 0.40-1.92$	$$k_L a = 0.779 D_L^{0.5}\mu_L^{-0.2} Fl_G^{1/6} \quad (SI)$$

[a] As cited by Kantarci et al. (2005)
[b] As cited by Ribeiro Jr. and Lage (2005)
[c] As cited by Deckwer (1992)

7.7 NEEDED BUBBLE COLUMN RESEARCH

Although a lot of research exists on BC gas holdup and gas–liquid mass transfer, little focuses specifically on bioreactors when microorganisms are present. Controlled experiments containing actual microorganisms as well as materials able to mimic microorganism behavior, such as fiber suspensions or other liquid additives, need to be incorporated into more comprehensive studies. For example, microorganism flocculation and its effects on hydrodynamics are not commonly studied or simulated, but obviously have serious consequences on bioreactor performance. The goal would be twofold. The first would be to quantify possible surface reactions and other phenomena when microorganisms are present in the reactor environment. The second goal would be to quantify the quality of currently used substitutes to simulate microorganisms and attempt to identify newer organic possibilities. In connection to these goals, more study is needed on the effect of higher temperature, pressure, and viscosity on microorganisms as well as bioreactor hydrodynamics.

A major study should be attempted to compare the different bioreactor performance characteristics. Comparisons between BCs and airlift reactors are available, but a wider array is lacking, but needed. Research towards this end, such as Bouaifi et al. (2001) who compared stirred-tank reactors and BCs, is sparse. Different bioreactor designs have quite unique scaling abilities and associated costs, and the economic benefits and decisions would be better understood if such studies would be more common.

7.8 SUMMARY

Bubble interactions are tightly connected to hydrodynamics so that gas holdup is usually capable of representing gas–liquid mass transfer trends fairly accurately and, more importantly, predicting reactor hydrodynamics. Thus, any BC experiment (or series of experiments) starts with a gas holdup study. The literature provides a wide array of gas holdup information; however, the more detailed experiments, especially those that investigate gas–liquid mass transfer, are much fewer in number and smaller in scope.

Most studies, for example, are based on air–water interaction in an isothermal setting, even though most industrial bioprocesses have thermal interactions and complex fluid properties. It is also common for industrial processes to be performed at relatively high temperatures and/or pressure. These settings still lack experimental coverage (Lau et al., 2004). Furthermore, gas holdup and gas–liquid mass transfer are decoupled under thermal operation, and pressure variances are rarely accounted for in currently available correlations.

Gas–liquid mass transfer correlations and their design applications have to be handled very carefully. Bubble–bubble interactions are very complicated processes, which still have not been mastered and are not easily represented with the current set of tools. Although a change in gas holdup can predict the direction of the change in gas–liquid mass transfer, it cannot predict the amount due to the hydrodynamic complexity (Chaumat et al., 2005). Scale-up, design, and application require patience and due diligence. Since most gas–liquid mass transfer correlations use gas holdup data, hydrodynamic and geometric similarities should be attempted in order to maximize the probability of successful prediction. Approximation of gas–liquid mass transfer using third-party gas holdup correlations or data has a low probability of success and should be used as a first iteration for scale-up or design. Follow-up investigations are then strongly advised.

8 Airlift Bioreactors

8.1 INTRODUCTION

The airlift reactor (ALR) is a pneumatic device that attempts to reconcile bubble column shortcomings and provide more control to the operator. The term *ALR* is also used to identify an airlift bioreactor. Two general families of airlift bioreactors exist: internal-loop airlift bioreactors and external-loop airlift bioreactors (ILALRs and ELALRs, respectively). The internal-loop variant is sectioned by a baffle (Figure 8.1a) or draught tube (Figure 8.1b), which allows for internal liquid recirculation. The external-loop airlift bioreactor (Figure 8.2) connects the up- and down-flowing regions with additional piping, thereby separating the flow paths. These two designs make up the basic idea, and extensive modification can be implemented to create a wide array of application-specific flow conditions (Ribeiro Jr. and Lage, 2005).

Airlift bioreactor construction is very simple and similar to that of a bubble column (Al-Masry, 1999; Blazej et al., 2004c). There are four basic sections: riser, gas separator, downcomer, and base. The riser is the up-flowing section of the airlift bioreactor. The gas sparger is oriented such that gas is injected into the riser causing upward fluid motion. The gas sparger location may be within the riser or the base, which is simply the region that connects the downcomer to the riser. The gas separator is located at the top of the bioreactor where gas is allowed to disengage from the liquid phase (or slurry). The downcomer is defined as the region in which down-flowing phases are present.

Airlift bioreactors tend to be larger vessels. Industrial units may have a height and diameter up to $10-40$ m and $2-10$ m, respectively. The specific dimensions are a function of process requirements. For example, industrial scale units may operate with a liquid circulation velocity up to 1 m/s and gas residence time of approximately 1 min (Giovannettone et al., 2009; van Benthum et al., 1999a; van Benthum et al., 1999b). The liquid circulation is a function of the gas flow rate and disengagement properties of the separator. Hence, the height is used to adjust the gas residence time to achieve the necessary conversion (van Benthum et al., 1999b). Pilot-scale bioreactors tend to have volumes of $0.05-0.30$ m^3. Biological applications use industrial volumes of approximately 10 m^3, but certain applications, such

An Introduction to Bioreactor Hydrodynamics and Gas-Liquid Mass Transfer, First Edition.
Enes Kadic and Theodore J. Heindel.
© 2014 John Wiley & Sons, Inc. Published 2014 by John Wiley & Sons, Inc.

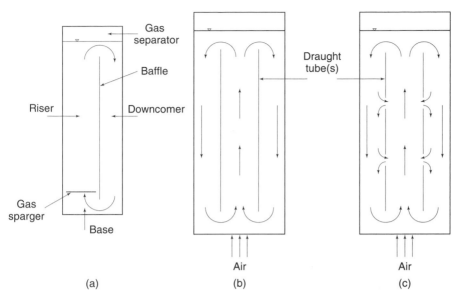

Figure 8.1 Internal-loop airlift bioreactor with (a) a baffle separating the riser and down-comer, (b) a continuous draught tube separating the riser and downcomer, and (c) a sectioned draught tube separating the riser and downcomer.

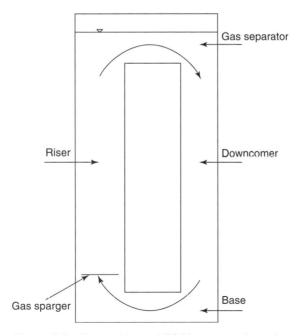

Figure 8.2 External-loop airlift bioreactor schematic.

as wastewater treatment, may use bioreactors up to 1000 m³ in size (Joshi et al., 1990).

Biological applications use smaller scale vessels because turbulence increases with scale, which leads to an increase in shear stresses as well. Airlift bioreactors have become popular in mammalian cell suspension applications for which shear stresses become important. Increasing the bioreactor size leads to an increase in the mechanical damage and lower cell densities (Martin and Vermette, 2005).

Airlift bioreactors can be viewed from two different perspectives. One is that the airlift bioreactors are variations of the bubble column. The bubble–bubble interactions, forces, construction, and bioreactor applications are very similar to those of the bubble column. On the other hand, airlift bioreactor hydrodynamics are strongly biased on the interactions between the riser and downcomer gas holdup. The gas separator, in conjunction with gas injection in the riser section, generally leads to the gas holdup in the riser section being larger than in the downcomer. This effect creates a hydrodynamic pressure difference, which leads to the liquid–gas mixture circulating in a fairly controlled manner. This mechanism is a source of many advantages unique to the airlift bioreactor.

Another advantage is that the liquid- and gas-phase flow rates may be controlled independently of each other if a control valve is placed in the downcomer (Williams, 2002). This ability introduces much more control, making the airlift bioreactor ideal for fine-tuning industrial applications. The phases are circulated by design, which allows much higher gas and liquid flow rates over that of bubble columns (Figure 8.3) without the requirement of a complicated recycle system. The difference between Figure 8.3a and b is that Figure 8.3a can be achieved through a throttling device. In other words, mixing, heat transfer, and residence time can be optimized while providing the ability to protect any microorganism or catalyst.

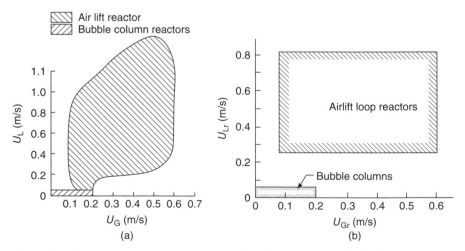

Figure 8.3 Comparison between superficial liquid and gas velocities in bubble columns and airlift bioreactors from (a) Merchuk (1986) and (b) Chisti (1989).

Dead or high shear zones are limited in airlift bioreactors because the influence of sparged gas is limited and the gas-phase distribution tends to be practically homogeneous throughout the riser volume, although a nonuniform phase distribution may be observed for approximately the first meter in the riser, which introduces variability in relatively short columns ($h_R < 4$ m) (Giovannettone et al., 2009). The only possible high shear areas are caused by turns (Merchuk and Gluz, 1999), but these can be eliminated or minimized.

The price for the advantages of ALRs is that the capital investment is usually larger relative to bubble columns. The higher costs can be offset with a liquid and/or gas input increase that is not feasible in a bubble column, which would lead to higher output; however, operation at larger flow rates in airlift bioreactors can lead to increased variable costs associated with additional gas throughput and pressurization. Foaming also becomes more likely at high gas and liquid flow rates, which leads to inefficient gas separation and recirculation of spent gas bubbles (Wei et al., 2000; Williams, 2002). Complete mixing typically requires four to nine cycles (the number of times that the liquid circulates in the ALR) (Karamanev et al., 1996), which would cause further problems if continuous operation is required. Luckily, external-loop airlift bioreactors can be designed to have much better mixing performance.

8.2   CIRCULATION REGIMES

The airlift bioreactor and bubble column have very similar bubble–bubble interactions and behavior, which leads to almost identical gas flow regimes and progression. These have been covered in detail in Section 7.2; however, more attention is placed on liquid flow behavior in airlift bioreactors since the liquid phase is a significant source of momentum and gas recirculation.

The process of gas entrainment and circulation is complicated and not easily quantified. Problems arise from an abstract relationship between the liquid and gas phases. On the one hand, the gas flow rate affects the liquid flow rate through the gas holdup and hydraulic pressure differential relationship. As the gas flow rate increases, larger bubbles rise faster and increase the circulation velocity. A higher circulation velocity, in turn, would decrease the slip velocity and make entrainment easier. On the other hand, if the liquid velocity is higher than the bubble rise velocity, bubbles would experience a drag (lift) force, which would aid entrainment.

Quantification of this process becomes complex very quickly, and a theoretical measure or technique does not exist. Researchers, instead, have to rely on empirical techniques. Nevertheless, a rough understanding of the circulation process exists and is very helpful in the empirical understanding of airlift bioreactors. Airlift bioreactor circulation can be sectioned into three general regimes, as shown in Figure 8.4. At very low gas flow rates, which correspond to $U_{Gr} < 0.012$ m/s, the induced liquid circulation velocity is not strong enough to entrain gas bubbles into the downcomer. Note that U_{Gr} is the superficial gas velocity in the riser. The gas phase is able to almost completely disengage from the liquid phase (regime 1). This

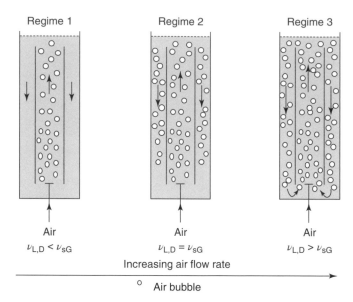

Figure 8.4 Circulation regime progression in a draught tube internal-loop airlift bioreactor (van Benthum et al., 1999b), where $v_{L,D}$ is the downcomer liquid velocity and v_{sG} is the gas slip velocity.

regime, referred to as the bubble-free regime, is usually not significantly influenced by the liquid properties simply because the amount of gas present in the system is still fairly low. In order for the liquid properties to become more important, a higher degree of bubble–bubble interaction is needed.

The liquid is capable of entraining only very small bubbles ($d_B < 1$ mm) in regime 1. The resulting downcomer gas holdup is usually small with a maximum of about 3%. Once these small bubbles are entrained in the downcomer, they are not transported far and are expected to reach depths up to 30% of the downcomer height (Albijanic et al., 2007). Any increase in the superficial gas velocity leads to a significant increase in the liquid circulation velocity (Blazej et al., 2004c; van Benthum et al., 1999b). The basic guideline is that the bubble-free regime exists as long as the downcomer superficial liquid velocity is lower than the average slip velocity (van Benthum et al., 1999a).

Once the gas is in the downcomer, the liquid has to flow even faster to cause circulation. Gas bubbles are still lighter than the liquid and have a buoyant force, which propels them to rise against the flow. The liquid-phase momentum has to provide the power to overcome the buoyant force and create a net downward force in order to cause forward motion and eventual circulation. In effect, a superficial liquid velocity exists at which gas bubbles can be suspended or are stagnant in the downcomer (regime 2 in Figure 8.4). Hence, this circulation regime is referred to as the transition regime.

When the liquid velocity in the downcomer is approximately equal to the average slip velocity, an approximate stationary bubble behavior can be observed. In addition, bubble–bubble interactions become more frequent, and liquid properties start to have more influence on bioreactor performance (Albijanic et al., 2007). Unlike regime 1, the liquid velocity does not deviate significantly and stays approximately constant in regime 2, but the transition regime is not very stable and minor variations in the superficial gas velocity can lead to regime transition. Unfortunately, a theoretical prediction of gas holdup is very challenging and experimental values vary over a wide range (van Benthum et al., 1999b). Regime 1 and the beginning of regime 2 can be described as homogeneous, while the later portion of regime 2 occurs in the transition flow regime described in Section 7.2 (Merchuk et al., 1998; van Benthum et al., 1999b).

If the downcomer liquid velocity is larger in magnitude than the bubble rise velocity, the bubble will circulate with the liquid (Albijanic et al., 2007). This minimum superficial liquid velocity usually occurs at $U_{Gr} = 3.5 - 5.0 \, \text{cm/s}$ (Jones, 2007; Wei et al., 2008) and is described by thorough gas bubble circulation (complete bubble circulation regime—regime 3 in Figure 8.4). It should be noted that regime 3 is by far the most commonly used circulation regime. The required gas flow rate for pilot and industrial scale bioreactors requires very high superficial gas velocities, which all but guarantee circulation (Blazej et al., 2004c; Chisti, 1989). Bubble-free (regime 1) and transition (regime 2) regimes are usually avoided because they have poor phase contacting, mixing, and selectivity (Wei et al., 2008). In addition, special attention and effort are required to keep the flow in bubble-free and transition regimes for an industrial-scale bioreactor.

Bioreactors operating in the bubble-free or possibly early transition regimes are usually found when cultivating mammalian cell structures or highly shear-sensitive microorganisms for which shear stress becomes the limiting operational factor; however, designers have found it easier to operate in nonstandard designs, which at least partially circulate (see, e.g., van Benthum et al. (1999a)).

The transition from regime 2 to regime 3 is not well understood. It is suspected that the transition is initiated by the gas phase entering the heterogeneous or transition flow regime, which is synonymous to bubble column flow regimes. These flow regimes would produce larger bubbles with faster bubble rise velocities, which disengage easier, decrease the downcomer gas holdup, and increase the circulation driving force. This behavior is often confirmed with circulation starting and flow entering the transition flow regime at a superficial riser gas velocity of 0.045 m/s (Joshi et al., 1990; Merchuk et al., 1998). The faster liquid circulation velocity eventually surpasses the gas slip velocity in the downcomer, and gas bubbles are entrained in the downcomer flow. The circulating gas adds to the riser gas holdup, potentially reinforcing the trend.

Complications arise when/if the gas disengagement leads to a smaller riser gas holdup, such that the driving force is not heavily influenced. The gas disengagement process has some geometric influences that cause the transition to regime 3 to occur relatively early in the transition flow regime or well into the heterogeneous flow regime. A second complication is that the recirculated gas can lead to more frequent bubble collisions and coalescence so that the riser gas holdup may decrease early in regime 3 until the flow structure stabilizes. This case requires the downcomer gas holdup to decrease at a faster rate than the riser gas holdup. Otherwise, the circulation would stay in regime 2. This behavior makes circulation regime transition highly variable. Interestingly, the transition to regime 3 occurs at a gas holdup of 10–12% regardless of the bubble flow regime, and for reasons and through mechanisms that are not well understood at this time (van Benthum et al., 1999b). The liquid circulation velocity resumes its relationship in regime 3 and increases with the riser superficial gas velocity. In general, the maximum downcomer gas holdup can be as high as ~20% (van Benthum et al., 1999b).

The riser superficial gas velocity at which transition occurs depends significantly on liquid properties. Generally, liquids containing surfactants or alcohols tend to experience circulation relatively early ($U_{Gr} \approx 0.035\,\text{m/s}$) while water experiences it later ($U_{Gr} \approx 0.045\,\text{m/s}$) (Albijanic et al., 2007; Jones, 2007; van Benthum et al., 1999b). Large bubbles, 3–5 mm in diameter, are not entrained in the downcomer in a water system until $U_{Gr} \approx 0.20\,\text{m/s}$. Large, ellipsoidal bubbles at these velocities also form significant wakes that limit the effective interfacial area. This factor is almost always ignored in theoretical gas–liquid mass transfer coefficient correlations or models (Talvy et al., 2007).

A simple measure to correlate liquid circulation velocity with the superficial gas velocity would be through a power law such as (Bello et al., 1984; Merchuk, 1986):

$$U_{Lr} = \alpha U_{Gr}^{\beta} \tag{8.1}$$

where U_{Lr} and U_{Gr} are the riser superficial liquid and gas velocities, and α and β are the fitted constants. The coefficient α depends on the bioreactor geometry and liquid properties while the exponent β is a function of the flow regime and bioreactor geometry. This base equation can be further enhanced, and terms can be added to account for specific properties or bioreactor design (Chisti, 1989). The fitted terms are usually positive, implying that the liquid circulation velocity increases with increasing superficial gas velocity; however, if slugging, choking, or throttling occurs, the liquid circulation velocity can actually decrease with an increase in superficial gas velocity (Jones, 2007).

A more detailed correlation could be arrived at by accounting for each force. For example, one would balance the hydrostatic pressure difference with bioreactor-specific pressure drops (e.g., head losses due to wall friction, elbows, and bends). This approach would be based on a theoretical foundation, but attempts have not been very successful due to the specific geometric dependence, and empirical correlations are still the most practical approach (Albijanic et al., 2007; Chisti, 1989).

Analysis could be skewed to favor riser versus downcomer data due to the riser cross-sectional area usually being larger than that of the downcomer. A better strategy would be to analyze each section separately. One approach is to assume a large degree of independence, which would work relatively well if the controlling factors in each section are independent of the other. A large effective riser diameter would not impact the riser gas holdup and would cause the riser gas holdup behavior to be a function of just the gas flow rate and liquid properties. A small downcomer diameter, on the other hand, could cause the downcomer to enter the slug flow regime, which would limit circulation. Even larger downcomer diameters can be affected by trapped gas, which can grow with superficial gas velocity. The effect could be a reduction in the effective downcomer cross-sectional area and possible flow choking (Chisti, 1989; Jones, 2007). A single correlation does not take these interactions into account.

8.3   CONFIGURATION

The specific bioreactor configuration has a significant effect on gas holdup and gas–liquid mass transfer performance, as can be seen in Figure 8.5. Other factors, such as downcomer-to-riser cross-sectional area ratio or the gas separator design, can also play significant roles. For example, external-loop airlift bioreactors use gas separators that allow more time for the gas phase to disengage than in internal-loop airlift bioreactors. Hence, the downcomer in ELALRs has a negative effect on global or total gas holdup and gas–liquid mass transfer.

Furthermore, airlift bioreactor studies often use an effective riser diameter that is less than 0.15 m, which has been shown to influence the bubble size distribution in bubble columns (Kantarci et al., 2005). Not surprisingly, research work that compares ALR designs based on the bioreactor diameter uses smaller riser and downcomer columns and finds a strong bioreactor diameter dependence, noting

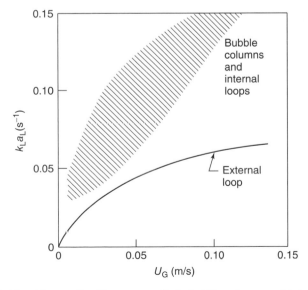

Figure 8.5 Achievable gas–liquid mass transfer in bubble columns and internal loops versus external loops (Chisti, 1989).

that bubble diameters increase with an increase in bioreactor size (Ruen-ngam et al., 2008). The bioreactor (riser) diameter effect is expected to be very similar to that of bubble columns, which has been reviewed in Section 7.3.1.

The influence of the bioreactor base is usually not discussed. The understanding is that as long as the base does not interfere with flow and increase frictional losses, one does not pay attention to it; however, some evidence exists that curved or filled bottoms may promote this limited impact (Chisti, 1989). With this in mind, the base height has a minimal impact on ALR operation as well. The only restriction is that a short base height can cause a large pressure drop, which would increase the operational cost. In such a case, gas holdup is expected to increase with an increase in bottom clearance (Al-Masry, 1999). Luckily, for designers, a restrictive bottom clearance in the experimental scale is often self-corrected during scale-up such that minimal problems are encountered in the pilot or industrial stage. None of the presented correlations at the end of this chapter attempt to reflect the bottom designs in any form.

8.3.1 Bioreactor Height

The bioreactor height can be defined in two ways. The first definition is often termed the effective bioreactor height and is defined as the distance between the base and the bottom of the gas separator. The second is the unaerated liquid height, which is defined as the distance from the base to the fluid surface prior to aeration. These definitions are shown in Figure 8.6 for the internal-loop and external-loop airlift bioreactors.

The effective bioreactor height influences the circulation path, which has many hydrodynamic implications as implied by the liquid circulation velocity defined by Blenke (1979):

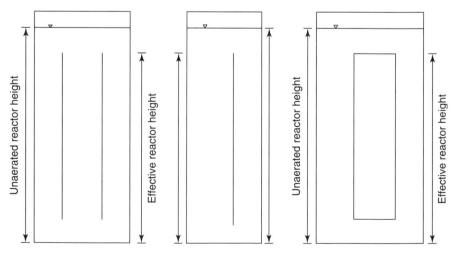

Figure 8.6 Airlift bioreactor component definitions.

$$U_{Lc} = \frac{x_c}{t_c} \tag{8.2}$$

where U_{Lc}, x_c, and t_c are the liquid circulation velocity, the circulation path length, and the circulation time, respectively. By changing the bioreactor height, the circulation path length also changes. The circulation path length influences the liquid circulation velocity, gas disengagement, and the hydraulic pressure difference, which drives the circulation flow. As the circulation path length increases, the liquid circulation velocity is expected to increase. The result would be an increase in gas disengagement and a decrease in gas holdup and gas–liquid mass transfer. At the same time, a long circulation path or time could be dangerous because it could lead to spent gas or minimal surface renewal in the downcomer (Talvy et al., 2007), possibly leading to microorganism starvation. It should be noted that the circulation time is largely influenced by the riser and downcomer residence times, while the separator and base residence times are oftentimes ignored (Joshi et al., 1990).

Bentifraouine et al. (1997a) varied the effective bioreactor height from 1 to 1.6 m and observed a significant increase in the superficial liquid velocity, which was attributed to an increase in the hydrostatic pressure differential. They also concluded that the increase in superficial gas velocity caused a decrease in the gas-phase residence time and, hence, gas holdup; however, the other possibility is that the faster liquid velocity is able to prevent the gas from disengaging and may lead to higher gas residence time. This was observed by Snape et al. (1995) who concluded that gas holdup increased and superficial liquid velocity varied minimally with an increase in bioreactor height. In other words, the effective bioreactor height's influence is dependent on the liquid flow regime and liquid velocities; however, gas–liquid mass transfer is usually seen as increasing with the effective bioreactor height (Siegel and Merchuk, 1988) unless gas holdup decreases significantly as well. It should be noted that the risk of recirculating used gas increases with higher liquid circulation velocity.

The effect of the unaerated liquid height is more complicated. If the unaerated liquid height was less than the effective bioreactor height, an increase in the unaerated liquid height would lead to an increase in the superficial liquid velocity and a decrease in gas holdup (Bentifraouine et al., 1997a, 1997b; Wei et al., 2008). If, on the other hand, the unaerated liquid height was equal to or greater than the effective bioreactor height, gas holdup, superficial liquid velocity, and, hence, ALR operation are not affected (Bentifraouine et al., 1997a, 1997b; Snape et al., 1995).

Wei et al. (2008) concluded that the unaerated liquid height had a negligible influence in the transition from the bubble-free (regime 1) to the transition regime (regime 2), but had a very important influence in the shift from the transition to complete bubble circulation regime (regime 3). Bubbles were entrained into the downcomer at approximately the same superficial gas velocity, regardless of unaerated liquid height; however, the flow resistance decreased with increased unaerated liquid height such that bubble penetration also increased, which led to an earlier transition into the complete bubble circulation regime.

The bioreactor height is often presented in the form of the riser aspect ratio (h_R/d_R). The effect of riser aspect ratio is expected to depend on the gas flow and gas circulation rate. An increase in the riser aspect ratio does not significantly change the liquid circulation velocity. Therefore, the circulation path length and circulation time increases with increasing riser aspect ratio, which allows the gas phase to achieve its equilibrium bubble diameter more efficiently.

If the superficial gas velocity and gas circulation rate are low, such that bubble–bubble interaction is low as well, an increase in the riser aspect ratio is expected to increase the average bubble diameter, decrease the gas holdup, and decrease the gas–liquid mass transfer coefficient. This effect occurs until the bioreactor is tall enough to achieve equilibrium. Bubble columns achieve this at an aspect ratio of approximately 5 and height of 1–3 m. The airlift bioreactor, especially the external-loop variant, is expected to require a higher riser aspect ratio due to the limited axial velocity variations. Evidence suggests that the superficial gas velocity is an important variable in this determination and a decrease in gas holdup may be observed up to a riser aspect ratio of 20–40. Therefore, it is fairly common to observe bubble size variations with bioreactor height in both the draught tube and external-loop airlift bioreactor (Ruen-ngam et al., 2008). If, on the other hand, the superficial gas velocity and gas circulation rate are relatively fast, a riser aspect ratio much less than 20 shows a negligible influence (Joshi et al., 1990).

8.3.2 Area Ratio

The area ratio is defined as the ratio of downcomer-to-riser cross-sectional area, but can also be represented by a downcomer-to-riser hydraulic diameter ratio. It is a simple representation of the flow restriction that exists in the bioreactor design since wall friction is of little importance unless viscosity is significantly increased (Chisti, 1989).

The area ratio influences the liquid circulation rate by adjusting the flow restriction. If the downcomer diameter or area ratio is reduced too much, the liquid flow becomes restricted and the liquid circulation velocity is expected to decrease. According to Eq. (8.2), the gas phase could follow for the same circulation path length (x_c) but the circulation time (t_c) could increase, which would lead to higher gas–liquid mass transfer (Blenke, 1979). Mere scale can have a perceived influence as well, and studies using area ratios higher than 0.25 on the pilot and industrial scales do not observe any restrictive effects. In general, frictional losses decrease with increasing scale such that the losses in a 10.5-l ILALR are four times higher than those observed in a geometrically similar 200-l ILALR (Blazej et al., 2004c). Generally, correlations model the area ratio with a negative exponent, meaning that an increase in the downcomer-to-riser area ratio leads to a decrease in gas holdup and the gas–liquid mass transfer coefficient (Garcia-Ochoa and Gomez, 2009); however, very large or very small area ratios often do not follow this rule.

An excessive reduction in the downcomer diameter (very low area ratio) would be too restrictive and reduce the liquid velocity such that the gas phase could easily disengage, which would reduce the amount of available gas (gas holdup), the driving force, and gas–liquid mass transfer. In other words, a restrained system would decrease the liquid circulation velocity, driving force, gas holdup, and gas–liquid mass transfer by decreasing the area ratio. In general, research work has not focused on area ratios below 0.10 with the exception of Jones (2007) who used an ELALR with an area ratio of 0.063. Popovic and Robinson (1984) confirmed this by increasing the area ratio from 0.11 to 0.44, which increased the superficial liquid velocity about fourfold. Airlift bioreactors are defined by the riser and very large area ratios result in bubble column-like performance; hence, area ratios beyond 4.0 are rarely used.

Valves are occasionally used to adjust the area ratio for experimentation, such as the research by Bendjaballah et al. (1999). They closed the valve from fully to 40% open without major effects on gas holdup. This would lead to the conclusion that an optimum area ratio exists, which does not impact gas holdup and also minimizes cost. The optimum ratio may not be easily determined since it depends on the scale and operating range, and the installation of a restrictive valve in the downcomer (assuming the optimum area ratio is below the valve-free area ratio) would provide the necessary flexibility and means to get there (Weiland, 1984). Additional losses and dynamics would be introduced by the valve and must also be addressed.

If circulation is a major design goal, Joshi et al. (1990) concluded that an area ratio of about 1 maximizes the liquid circulation rate in a 10 m³ volume bioreactor regardless of sparger location. If the bioreactor volume is increased to 100 m³, the optimum area ratio increases by 2. When the area ratio is adjusted, the circulation velocity monotonically decreases with increasing area ratio. In Joshi's work, a bioreactor with $V_L = 10\,\text{m}^3$, $h_R/d_R = 40$, and $P/V_L = 0.3\,\text{kW/m}^3$ experienced a liquid circulation velocity decrease from 3.0 to 0.5 m/s when the area ratio increased from 0.25 to 4.0. The higher liquid velocities at lower area ratio also introduced circulation instabilities, which disappeared at $A_d/A_r = 0.3$ at 10 m³ and $A_d/A_r = 0.5$ at 100 m³. Hence, a stable and optimum area ratio can be achieved with a riser-to-downcomer diameter ratio of about 1.41 (Joshi et al., 1990). Variations exist for the draught tube and external-loop airlift bioreactor, which depend on the maximizing characteristic and are presented in Sections 8.3.4 and 8.3.5, respectively.

8.3.3   Gas Separator

The gas separator is an important design feature that is often ignored. The simple reason is that internal-loop airlift bioreactors have only a few options, and the design is essentially the same: a vented headspace, which is similar to the tank separator shown in Figure 8.7c. The external-loop airlift bioreactor, however, is presented with additional design options (shown in Figure 8.7a and b), which provides the external-loop airlift bioreactor with some advantages for certain processes.

The tank separator (Figure 8.7c) is the simplest and most common design and is used for both the ELALR and ILALR. It is usually a simple, rectangular vented

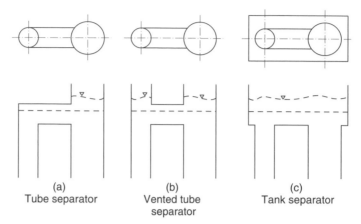

Figure 8.7   Common gas separator designs for external-loop airlift bioreactors (Jones, 2007).

box that connects the riser to the downcomer. The tank separator, however, has a negative effect in the ELALR. It increases the available gas separation area and gas separator residence time relative to a comparable ILALR design. If a comparison is made between an ELALR and ILALR on the basis of unaerated liquid height, the ELALR will have a shallower separator liquid height and higher rate of gas disengagement. Hence, the ELALR is often cited to have not only lower gas holdup and gas–liquid mass transfer performance, as shown in Figure 8.5, but also higher liquid circulation velocity (Chisti, 1989).

The unaerated liquid level is an important feature. Liquid circulation can usually be achieved by injecting more gas and ensuring a sufficiently high aerated fluid level; however, in some extreme cases, additional gas would not cause additional circulation or could actually suppress circulation to the point of no liquid circulation. This may occur when the aerated liquid level is too shallow in the gas separator such that complete gas disengagement is promoted. Therefore, increasing the liquid level to a minimal operational level would allow gas-phase circulation (Merchuk and Gluz, 1999). The operator is expected to have the ability to add enough liquid to have control, and, naturally, research has not gone toward quantifying the necessary unaerated liquid height.

More importantly for operation, the gas residence time in the separator controls disengagement such that decreasing the gas residence time in this region would lead to an increase in the downcomer gas holdup, which could decrease the hydraulic pressure differential and liquid circulation velocity (Siegel et al., 1986). The negative effect would be if the gas residence time in the separator is too long such that gas easily disengages and causes a decrease in gas holdup and gas–liquid mass transfer (Al-Masry, 1999).

One cannot easily quantify the critical fluid level since it is highly dependent on liquid properties, but a simple guideline can be used. If the gas separator residence time is longer than the minimum separation time (time required for the average bubble to rise from the top of the riser to the liquid surface), gas disengagement is encouraged (Al-Masry, 1999). In other words, the separator would encourage gas detachment if bubbles are allowed to spend enough time in the gas separator to rise to the surface given their bubble rise velocity. Hence, increasing the superficial gas velocity could increase the bubble rise velocity and decrease the minimum separation time. Luckily, for ELALR operators, this design allows liquid circulation velocity adjustments (Chisti, 1989) so that the described situation may be offset. The gas separator, however, is sometimes designed to decrease the liquid velocity such that spent gas disengages more readily. Such a system might require a second sparger in the downcomer in order to provide sufficient gas for microorganism viability (Merchuk and Gluz, 1999).

A system-specific operational rule could be introduced based on a ratio between tank separator volume and the total bioreactor volume. An optimum volume ratio is expected to minimize power input without affecting downcomer gas holdup, which would lower operational cost and optimize output. Al-Masry (1999) varied separator volume in air–water and air–glycerol systems from 0–37%. He found that the optimum ratio, based on hydrodynamics and power input, was 11% for the air–water system. Increasing the volume ratio did not lead to significantly higher riser or downcomer gas holdup. Decreasing the volume ratio produced higher gas holdup, but it also led to much lower liquid circulation velocity. The air–glycerol system proved more difficult due to coalescence, and a volume ratio of 0% failed to lead to circulation until $U_{Gr} = 0.08\,\mathrm{m/s}$. Hence, the airlift bioreactor acted as a bubble column. Increasing the volume ratio beyond 7% would cause negative effects such that gas holdup was minimal at 37%. The liquid circulation velocity in the air–glycerol system was much lower than in the air–water system, which made the air–water system hydrodynamically superior.

The other gas separator designs offer an alternative such that the gas separator residence time is limited. The (closed) tube (Figure 8.7a) and vented tube (Figure 8.7b) represent the same idea. The tube separator is a simple design that basically increases the separator liquid height (assuming comparable volume as with a tank separator), which increases the amount of time required for bubbles to reach the surface and disengage from the liquid phase; however, in reality, it works too well. Gas often fails to disengage completely and tends to build in the connector and eventually in the downcomer. This accumulation leads to flow choking and reduces the liquid circulation velocity. Another way to interpret the gas accumulation is that it reduces the effective downcomer-to-riser area ratio. Making the situation worse, the downcomer can enter the plug flow regime. Gas holdup increases, but the representative bubbles are very large and gas–liquid mass transfer suffers greatly. Ultimately, the connector and downcomer could be completely choked, at which point the airlift bioreactor acts just like a bubble column (Jones, 2007).

The vented tube connector minimizes the problem of gas accumulation, and gas is allowed to separate fairly efficiently. Unfortunately for some processes, the additional separator volume increases the gas separation efficiency relative to an internal-loop airlift bioreactor for the same reasons as for the tank separator;

however, the liquid circulation velocity keeps increasing with superficial gas velocity without a local maximum (Bentifraouine et al., 1997a; Choi, 2001), which may be advantageous for some processes. The vented tube separator is also susceptible to the risk of operating in the slug flow regime and the formation of very large bubbles, which can choke the downcomer flow (Jones, 2007).

The vented tube separator, in turn, can be improved with the introduction of a valve, which allows the overhead pressure to be varied. The valve would introduce an additional circulation control mechanism. As the valve closes, the overhead pressure would increase, and the gas would not disengage as readily. The assumption is that gas disengages too easily in the vented state, while it almost fails to disengage in the tube separator (Jones, 2007; van Benthum et al., 1999a), which a closed valve would imitate. Hence, the optimum would be somewhere in the middle, and a valve adjustment would allow this operational point to be reached. If the initial assumption does not hold true for the design, the valve addition to the vented tube separator would not improve operation.

The connector length (vented or closed) is expected to decrease the separator liquid height and minimum separation time and increase the separation efficiency (Choi and Lee, 1993). Hence, the liquid circulation velocity increases with a longer connector length, which leads to a decrease in downcomer gas holdup. The gas–liquid mass transfer coefficient also decreases with a decrease in downcomer gas holdup for similar reasons (Choi, 2001). Frictional losses would also increase, but wall losses can be ignored for water-like substances and would only be important in highly viscous liquids (Chisti, 1989).

The flow in the separator is much different than the other airlift components. The riser and downcomer have very well-defined flow without backmixing. The separator, on the other hand, has defined up-flow and down-flow regions. As with the bubble column, the up-flow region was present in the central region while the down-flow region occurs near the wall. The results are that a stagnant region is present between these flow regions, which may lead to bubble coalescence, and a lower liquid velocity in the separator than in the top of the riser (Lo and Hwang, 2003).

8.3.4 Internal-Loop Airlift Bioreactor

The internal-loop airlift bioreactor is a very simple design that presents some advantages for gas–liquid mass transfer. The two basic variants, explained earlier, include a vessel separated by a full baffle and a draught tube internal-loop airlift bioreactor. Regardless of the variant, the internal airlift bioreactor has very similar gas holdup characteristics to that of a bubble column, as long as the comparison is made within the bubble column's feasible superficial gas velocity range. This similarity is due to the fact that the downcomer gas holdup is 80–95% of the riser gas holdup, and the riser gas holdup is comparable to that of the bubble column. Hence, the global or total ILALR gas holdup is very similar to that of the bubble column (Bello et al., 1984; van Benthum et al., 1999b).

The draught tube ILALR (DT-ILALR) is more efficient than the baffled ILALR at gas–liquid mass transfer, and much more research has been directed toward this variant. This imbalance is due to the advantage that the gas residence time may be up to twice as long in the DT-ILALR than in the baffled ILALR. As a matter of fact, a majority of the correlations for ILALRs, which are presented in Tables 8.1 and 8.2, have been developed using draught tubes. The baffled ILALR is usually used to study hydrodynamic behavior. Industrial application seems to be dominated by draught tube ILALRs as well. Baffled ILALRs are a little cheaper, but the advantages of the draught tube far outweigh the costs.

Since the gas phase spends more time in contact with the liquid phase, gas–liquid mass transfer in the ILALR is usually higher than the ELALR (Shariati et al., 2007). Gas–liquid mass transfer coefficients in DT-ILALRs are even slightly higher than in bubble columns due to higher operating gas flow rates. The draught tube diameter can be optimized to minimize costs and maximize a desired variable. General guidelines have been developed for different applications using a draught tube to column diameter ratio (d_D/d_R). A ratio of 0.8–0.9 maximizes mixing and mass transfer. A range of 0.5–0.6 maximizes circulation while a range of 0.6–1.0 minimizes the mixing time (Chisti, 1989). The problem for microorganism applications and the mass transfer guideline ($d_D/d_R = 0.8 - 0.9$) is that the downcomer provides minimal contribution to global gas–liquid mass transfer or gas holdup, which runs the risk of starving the microorganisms. If microorganism gas consumption in the downcomer is important, the operator has basically two solutions. First, circulation has to become more important. A tradeoff would have to be struck and the diameter ratio (d_D/d_R) be reduced. The result would be that the microorganisms could spend less time in the oxygen-deprived environment or the circulation could be designed to entrain more gas in the downcomer. Second, the microorganisms could be suspended. If the microorganisms are sensitive to shear, they can be suspended in the downcomer. If, on the other hand, the microorganisms require lots of gas, they could be suspended in the riser. Suspension could involve packed particles, porous media, and so on.

8.3.5   External-Loop Airlift Bioreactor

The external-loop airlift bioreactor has a wide array of variants ranging from the fairly simple, shown in Figure 8.2, to quite complex multistage designs. An important advantage is that the design allows access to all major bioreactor components, which is a great benefit for troubleshooting bioreactor performance or visualization (Merchuk and Gluz, 1999); however, this advantage is often not enough, and the ELALR is typically limited for use with shear-sensitive cells, photosynthetic microorganisms like algae, or processes requiring fluid recirculation. For example, mammalian cell structure can usually tolerate shear stresses in the range of 0.05 – 500 N/m² (Chisti, 1989), but the sensitivity is highly variable with cell structure and density such that cells could be highly shear sensitive at low cell density and somewhat resistant at higher densities (Martin and Vermette, 2005). Hence, the bioreactor operating conditions need to be flexible enough to adjust from very low shear conditions and still potentially operate with a high degree of turbulence (high shear stress).

Stirred-tank bioreactors (discussed in Chapter 6) usually create shear stresses much larger than those that can be tolerated by mammalian cells, and bubble column bioreactors (discussed in Chapter 7) and ILALRs may reach the high end of the mammalian cell spectrum at best. Bubble–bubble interactions, especially bubble bursts or breakup, create high local shear stresses, which have a negative impact on mammalian cell growth. ELALRs, on the other hand, can maintain low shear rates while still providing a respectable oxygen transfer of $0.6–1.0$ mmol/(l min) (Chisti, 1989), which is sufficient even for human skin (0.0011 mmol/(l min) at 10^6 cells/ml) and liver cells (0.005 mmol/(l min) at 10^6 cells/ml) (Martin and Vermette, 2005). This is doable using minimal circulation in the downcomer and cell suspension on packed material in the lower portion of the downcomer. The cells have minimal bubble–bubble interactions and usually have enough oxygen for growth and liquid flow for waste disposal.

Nonetheless, cell density can become a major problem. For example, mammalian cells are usually $100\,\mu m$ within a blood capillary for oxygen transfer. Therefore, nature has provided a design limitation. Cells can only be $150–200\,\mu m$ away from an oxygen source, such as dissolved oxygen in a liquid, because oxygen has a maximum diffusion depth of about $240\,\mu m$ for cellular material. This may limit the cell density and, in turn, the operational gas flow or local shear rate. Some production problems of critical cells are mitigated by cellular design. Connective tissue cells are elongated and form low density cell structures, while some critical ones, such as liver or kidney cells, operate at high density, and also form many more blood capillaries (Martin and Vermette, 2005). In other words, the external-loop airlift bioreactor provides the possible production of a wide array of mammalian cell structures as well as shear-sensitive microorganism by-products.

The ELALR can be used for these processes because gas disengagement is very efficient. The bubbles have a relatively fast rise velocity and slow radial velocity. Hence, bubble–bubble interactions are diminished in the external-loop variant relative to the bubble column or stirred-tank bioreactor, which, in turn, leads to higher gas holdup sensitivity to liquid property variations in bubble columns than in ELALRs (Chisti, 1989; Joshi et al., 1990; Shariati et al., 2007). In other words, the bubble–bubble collision frequency is lower in ELALRs, which makes coalescence-adjusting liquid properties, such as viscosity, surface tension, or ionic strength, less important. So, while bubble column and internal-loop airlift bioreactor gas holdup are usually similar, the downcomer gas holdup in an external-loop airlift bioreactor is only $0–50\%$ of the riser gas holdup (Bello et al., 1984), which leads to much lower global gas holdup in ELALRs.

The bioreactor geometry effects in an ELALR can be quite complex and dynamic. As the area ratio increases, the liquid circulation velocity decreases. Hence, the gas-phase circulation time decreases and gas holdup increases. The increase in gas holdup leads to an increase in the interfacial area. Some bubble dynamics are reflected in the growth, but, due to the lower bubble–bubble interactions in ELALRs, the increase is fairly continuous, but at a relatively slow rate. For example, Joshi et al. (1990) showed that by increasing the area ratio from 0.25 to 1.0 using a 10-m^3 ELALR at $0.3\,kW/m^3$ yielded a negligible increase in the interfacial area while a further increase in area ratio to 4.0 increased the interfacial area by about 30%.

The area ratio effects on the liquid-phase mass transfer coefficient are more difficult to predict. Area ratio effects are usually studied by keeping the bioreactor volume equal, which requires the effective bioreactor height to be adjusted. As the height is increased, the interfacial solute gas concentration increases as well, which decreases the gas solubility and, in turn, the liquid-phase mass transfer coefficient. In addition, an increase in the area ratio decreases the liquid circulation rate, which increases gas holdup, but may decrease surface renewal. The greater height also raises the pressure drop and power consumption, which increases surface renewal and the liquid-phase mass transfer coefficient. The extent of these effects is dependent on the operational scale and power level, and it is hard to predict which will dominate.

For example, at a power consumption of $0.3 \, kW/m^3$ in an ELARL, Joshi et al. (1990) observed a local maximum when $V_L = 10 \, m^3$; this led to the conclusion that the decrease in gas solubility and surface renewal due to the liquid circulation velocity decline was not dominant until the area ratio was increased by about 0.5. The decline, though, was somewhat gradual. At a power consumption of $0.6 \, kW/m^3$, a local maximum was not observed, which suggested the solubility effect was stronger; however, the gas–liquid mass transfer coefficient was about 50% higher at $0.6 \, kW/m^3$ than at $0.3 \, kW/m^3$, which was attributed to a doubling of the interfacial area. Hence, Joshi et al. (1990) concluded that the solubility decrease must have played a more significant role.

An increase in viscosity has a much stronger effect on external-loop airlift bioreactors than internal-loop airlift or bubble column bioreactors. At higher viscosities, the lower gas holdup in an internal-loop airlift bioreactor is attributed to the lower bubble rise velocity, which causes the bubbles to have a longer circulation time. At lower increases in viscosity, gas entrapment into the downcomer actually becomes easier and may increase gas holdup slightly. This behavior has been observed with a 25% glycerol concentration (Hallaile, 1993). It has also been reported that a viscosity increase in the range of 1.54–19.5 mPa-s has a minimal effect on liquid circulation rate and mixing time inside a draught tube airlift bioreactor (Molina et al., 1999). The experience in the external-loop airlift bioreactor, however, is opposite. The bubble disengagement efficiency increases because the bubble residence time in the gas separator increases significantly. The result is that if the viscosity is increased to 14 mPa-s by the addition of glycerol, the bioreactor experiences a severe gas holdup decrease (McManamey and Wase, 1986; Shariati et al., 2007).

8.4   SPARGER DESIGN

Gas distributors used in airlift bioreactors are very similar to those used in bubble columns and include sintered, perforated, or porous plates, membrane or ring-type distributors, arm spargers, and single-orifice nozzles (Figure 7.12). The sintered plate (Figure 7.12a) is usually made out of glass or metal and produces very small bubbles. However, sintered plates are rarely used for industrial applications because they can easily plug and require cocurrent operation; they are preferred for laboratory use. Porous plates experience a similar fate. In addition, porous plates suffer from hysteresis and start-up dependent behavior, although some of this has been attributed to liquid impurities buildup (Merchuk et al., 1998). Perforated plates (Figure 7.12b), which are normally made out of rubber or metal, usually have holes 1–5 mm in diameter and are the most common gas distributor for bubble columns and airlift bioreactors. The total aeration open area is usually maintained between 0.5% and 5% of the cross-sectional area of the bioreactor. Single-orifice nozzles (Figure 7.12c) are simple tubes, which are able to produce uniform gas flow far from the injection point. The flow, however, tends to be somewhat unstable. Airlift bioreactors may also use ring spargers (Figure 7.12d) that are used with stirred-tank bioreactors and bubble columns. Ring spargers are able to produce uniform and stable flow, but are usually unable to produce small bubbles (Deckwer, 1992) and lead to an early transition to heterogeneous flow regime when compared to perforated plates (Schumpe and Grund, 1986). If the sparger is placed within the riser in an airlift bioreactor, the sparger is often a perforated tube or ring that attempts to mimic the perforated plate performance.

The sparger effect is more pronounced for $U_G < 0.15$ m/s while it is much less important at $U_G > 0.20$ m/s and nonexistent at $U_G > 0.30$ m/s (Han and Al-Dahhan, 2007). Hence, in the case of heterogeneous flow, the sparger has a negligible influence on the bubble size, gas holdup, or gas–liquid mass transfer because the bubble dynamics are determined by the rate of coalescence and breakup, which are controlled by the liquid properties and the nature and frequency of bubble collisions (Chaumat et al., 2005; Merchuk et al., 1998). Since airlift bioreactors are operated most commonly above 0.15–0.20 m/s, as seen in Figure 8.3, and using rheologically complex fluids, most research work ignores the sparger effect (Merchuk et al., 1998). Of the reviewed work in Tables 8.1 and 8.2, almost all (with very few exceptions) use perforated plate or tube spargers.

The commonly observed high gas flow rates result in correlations that can be fitted across data sets without accounting for sparger design (Kawase and Moo-Young, 1986b). Even a correlation for the liquid circulation velocity can be formed without regard to the sparger, and the results are still within 10%, even for very short vessels (Miyahara et al., 1986). The gas spargers usually do not provide enough kinetic energy from the gas jets to influence the liquid circulation rates (Chisti and Moo-Young, 1987). Hence, the preference and selection for spargers is based on operational restrictions. The perforated plates and tubes provide the cheapest and most reliable operation, while the others have major drawbacks (Chisti, 1989).

Merchuk et al. (1998) investigated the use of different sparger designs in a draught tube airlift bioreactor using $U_{Gr} = 0 - 0.20$ m/s. Their results concurred with the bubble column experience. If a sparger created a smaller initial bubble diameter, it would lead to higher gas holdup in the homogeneous and transition flow regime and longer mixing times. This dependence diminished with gas flow rate and practically disappeared when $U_{Gr} > 0.15$ m/s, which was determined to correspond to the transition into the heterogeneous flow regime. A cylindrical perforated-pipe sparger created additional mixing due to a more radial bubble distribution, which interfered with bubble detachment and led to a higher rate of coalescence, especially in the sparger zone, and easier transition to the heterogeneous flow regime. The sparger zone in airlift bioreactors is defined as the liquid volume up to one column diameter above the sparger location (Giovannettone et al., 2009).

Although the sparger type does not appear to significantly affect ALR performance, sparger position is important and can be optimized. If the gas sparger is left at the bottom of the bioreactor, the incoming flow tends to interfere and cause a flow imbalance, which in turn leads to accumulation of gas and possible coalescence, as shown in Figure 8.8a and b. If the sparger is placed just within the riser, the incoming fluid flow does not interfere with the incoming gas flow, and the sparged gas is not affected. Elimination of dead zones in the aeration region, as shown by the shaded regions in Figure 8.8c and d, would also be beneficial. Moving the sparger too close to the surface, however, could lead to surface aeration and a higher rate of gas separation such that gas recirculation could be in troublesome. Thus, placing the sparger too far beyond the downcomer inlet does not provide additional benefits.

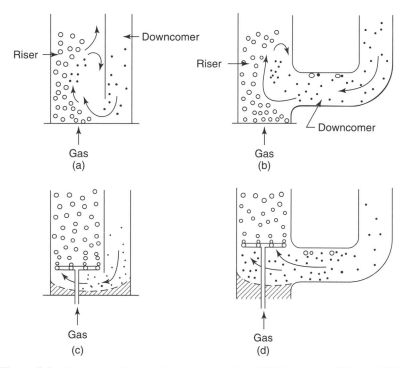

Figure 8.8   Location behavior of gas spargers in airlift bioreactors (Chisti, 1989).

Placing a sparger in the downcomer can alleviate some problems with oxygen suffocation of microorganisms and circulation, improve control, and reduce energy usage (Merchuk and Gluz, 1999). In effect, the complete bubble circulation regime could be induced earlier and with finer bubbles, which could potentially decrease the bubble diameter and increase the interfacial area; however, downcomer sparging introduces circulation instability, which can only be solved by placing the down-comer sparger at a critical height (h_{DC}).

The critical height can also be used to compare the stability between the different designs. The system is more stable if the critical height is higher. The unfortunate aspect of downcomer sparging is that it is also dependent on the riser and down-comer gas flow rates, which makes the location of the critical height highly variable. A higher riser gas flow rate increases the critical height, but the rate of increase decreases with higher downcomer gas flow. Placing the riser sparger farther into the riser may also increase the critical height (Joshi et al., 1990).

According to Joshi et al. (1990), system stability can be optimized by adjusting certain bioreactor geometries (aspect and area ratio) such that the critical height ratio (h_{DC}/h_R) is the highest possible. The area ratio was easily optimized by Joshi et al. (1990) for this purpose with a critical value of approximately 0.75. The sparger location was then dependent on the power consumption such that the critical height ratio was 0.70 at $0.6\,kW/m^3$ and 0.60 at $0.3\,kW/m^3$. In other words, the critical height determined by Joshi et al. (1990) was scale dependent. As a geometrically similar bioreactor gets larger, the critical height will increase as well.

The aspect ratio is a bit more difficult. Joshi et al. (1990) showed that the criti-cal height rose quickly with an aspect ratio up to 10. A change in the aspect ratio from 10 to 40 yielded a relatively minor change in the critical height ratio (from 0.7 to 0.75) with a liquid volume of 10 m^3. In order to achieve $h_{DC}/h_R = 0.80$, the necessary aspect ratio was about 80, which is usually not used for biological appli-cations. A similar trend was observed with bioreactor scale (using an aspect ratio of 40). The critical height ratio increased quickly up to 10 m^3 and changed slightly (from 0.7 to 0.75) with a volume increase from 10 to 100 m^3. A volume of 1000 m^3 was required to achieve $h_{DC}/h_R = 0.80$ (Joshi et al., 1990).

8.5   CORRELATIONS

Approaches used to develop gas–liquid mass transfer correlations in bubble columns have been ported over to airlift bioreactors, and, unfortunately, they have brought some issues along with them. Airlift bioreactor correlations are highly empirical, and a unifying development method does not exist. Some correlations attempt to be very specific and use multiple inputs, which are hard to quantify in an industrial setting, or use inputs that are not independent of each other. The suggestion is similar as with the bubble column: experimental units can use these more complicated correlations, but pilot or industrial scale units will have to depend on empirical and design specific correlations.

Further problems are presented by the downcomer and riser. Conditions in the two different sections can be very different, either by design or operation. Gas holdup correlations, which are summarized in Table 8.1, often reflect the different conditions, and two different approaches have been developed. The first approach is to develop correlations and models for the riser. The downcomer is then modeled based on the riser. One such approach has been taken by Bello et al. (1985a) by suggesting that

$$\varepsilon_{Gd} = 0.89\varepsilon_{Gr} \tag{8.3}$$

where ε_{Gd} and ε_{Gr} are the downcomer and riser gas holdups, respectively. The other approach is to develop correlations for the riser and downcomer independently. The correlations would use similar inputs such as the riser superficial gas velocity. The assumption is that the downcomer hydrodynamics are ultimately controlled by the riser conditions and its inputs. This approach has been taken by Li et al. (1995)

$$\varepsilon_{Gr} = 0.441 U_{Gr}^{0.841} \mu_a^{-0.135} \tag{8.4}$$

$$\varepsilon_{Gd} = 0.297 U_{Gr}^{0.935} \mu_a^{-0.107} \tag{8.5}$$

where U_{Gr} and μ_a are the riser superficial gas velocity and the apparent viscosity.

Interestingly, gas–liquid mass transfer correlations for ALRs, presented in Table 8.2, do not differentiate between the bioreactor components and are displayed in global form. The situation can get troublesome for processes employing microorganisms. Gas holdup correlations have to be used and relied upon to predict the gas–liquid mass transfer conditions in the downcomer, where the danger of starvation is the highest. Although this relationship oftentimes is intact, industrial processes commonly create surfactants, temperatures, and pressures, which may decouple the relationship.

Gas–liquid mass transfer coefficients follow the same gas holdup trends. As shown in Figure 8.9, the gas–liquid mass transfer coefficient increases monotonically with riser superficial gas velocity. The correlations by Chisti et al. (1988b) (as cited by Murchuk and Gluz (1999) and Popovic and Robinson (1984) were developed using external-loop airlift bioreactors, while the others used the draught tube internal-loop airlift bioreactor. The DT-ILALR has much better performance than the ELALR. It is unfortunate to note that gas–liquid mass transfer correlations are much fewer in number than their gas holdup counterpart.

The gas holdup correlations also have a high degree of variability, as shown in Figure 8.10. Correlations are able to reflect the consensus that the internal-loop airlift bioreactor has better gas holdup performance than the external-loop airlift bioreactor. This advantage is reflected in the power law presented in the correlations in Table 8.1. The internal-loop airlift bioreactors consistently have a higher constant and stronger dependence on the superficial gas velocity. As a matter of fact, the ILALR usually has a superficial gas velocity power close to 1 while the ELALR is usually less than 0.5. A similar observation is made for the gas–liquid mass transfer correlations in Table 8.2. The advantage of the airlift bioreactor relative to the bubble column also comes through. The bubble column correlations show a leveling effect at higher superficial gas velocities. This trend is not present in the airlift bioreactor correlations.

TABLE 8.1 Gas Holdup Correlations for Airlift Bioreactors

Reference	Bioreactor Type(s)	Bioreactor parameters	Correlations
Albijanic et al. (2007)	DT-ILALR	Air-(water, 1 wt% aqueous solutions of methanol, ethanol, n-propanol, isopropanol, and n-butanol) $U_G = 0-0.07$ m/s	$\varepsilon_G = 1.65 U_G^{0.97}\left[1+\left(\dfrac{d\sigma}{dC_A}\right)^{0.20}\right]^{1.52}$
Bello (1981)[a]	ELALR	$\varepsilon_{Gr} = 0.16\left(\dfrac{U_G^2}{U_{Lr}}\right)^{0.56}\left(1+\dfrac{A_d}{A_r}\right)$ $\varepsilon_{Gd} = 0.89\varepsilon_{Gr}$	
Bello et al. (1985a)[b]	ELALR	Air-(water, NaCl solution) $C_{NaCl} = 0.15$ kmol/m³ d_R or $d_t = 0.152$ m $h_D = 1.8$ m $U_{Gr} = 0.0137-0.086$ m/s $A_d/A_r = 0.11-0.69$	$\varepsilon_{Gr} = 0.16\left(\dfrac{U_{Gr}}{U_L}\right)^{\alpha}\left(1+\dfrac{A_d}{A_r}\right)$ $\varepsilon_{Gd} = 0.89\varepsilon_r$ for DT-ILALR $\varepsilon_{Gd} = 0.79\varepsilon_r - 0.057$ for ELALR $\alpha = 0.56$ for water $\alpha = 0.58$ for salt solution
Bello et al. (1985b)[b]	ELALR DT-ILALR BC	Air-(water, NaCl solution) $C_{NaCl} = 0.15$ kmol/m³ d_R or $d_t = 0.152$ m $h_D = 1.8$ m $U_{Gr} = 0.0137-0.086$ m/s $A_d/A_r = 0.11-0.69$ (ELALR) $A_d/A_r = (0.13, 0.35, 0.56)$ (ILALR) $A_d/A_r = 0$ (BC)	$\varepsilon_G = 3.4\times10^{-3}\left(1+\dfrac{A_d}{A_r}\right)^{-1}\left(\dfrac{P_G}{V_D}\right)^{2/3}$
Bentifraouine et al. (1997a)[c]	?	$U_G = 0.002-0.06$ m/s $U_{Lr} = 0-0.2$ m/s	$\varepsilon_{Gr} = 2U_G^{0.88}\left(1 - 0.97U_{Lr}^{0.49}\right)$

Blazej et al. (2004c)	ILALR	Air–water For 10.5 l: $d_R = 0.108$ m $d_r = 0.070$ m $h_R = 1.26$ m $h_D = 1.145$ $A_d/A_r = 1.23$ $h_c = 0.030$ m $h_R/d_R = 11$ For 32 l: $d_R = 0.157$ m $d_r = 0.106$ m $h_R = 1.815$ m $h_D = 1.710$ $A_d/A_r = 0.95$ $h_c = 0.046$ m $h_R/d_R = 12$ For 200 l: $d_R = 0.294$ m $d_r = 0.200$ m $h_R = 2.936$ m $h_D = 2.700$ $A_d/A_r = 1.01$ $h_c = 0.061$ m $h_R/d_R = 10$	For 10.5 l: $\varepsilon_{Gr} = 0.829 U_{Gr}^{0.505}$ $\varepsilon_{Gd} = 0.857\varepsilon_{Gr} - 0.0095$ in regime 2 $\varepsilon_{Gd} = 0.432\varepsilon_{Gr} - 0.0139$ in regime 3 For 32 l: $\varepsilon_{Gr} = 0.815 U_{Gr}^{0.449}$ $\varepsilon_{Gd} = 0.885\varepsilon_{Gr} + 0.0065$ in regime 2 $\varepsilon_{Gd} = 0.670\varepsilon_{Gr} - 0.0037$ in regime 3 For 200 l (only regime 3 was presented): $\varepsilon_{Gr} = 0.792 U_{Gr}^{0.427}$ $\varepsilon_{Gd} = 0.967\varepsilon_{Gr} + 0.0068$
Cai et al. (1992)[a]	ILALR		$\varepsilon_{Gr} = 2.47 U_{Gr}^{0.97}$
Chakravarty et al. (1973)[b]	DT-ILALR	Air–(water, sulphate, glycerol, and iso-butyl alcohol solutions) $d_R = 0.10$ m $d_D = (0.074, 0.59, 0.45)$ m $L_D = 0.40$ m $L_c = 0.026$ m	$\varepsilon_{Gr} = \left[(\mu_L - \mu_W)^{2.75} + 161 \left[\dfrac{74.2 - \sigma}{79.3 - \sigma} \right] \dfrac{73.3 - \sigma}{79.3 - \sigma} \right] \times 10^{-4} \times U_{Gr}^{0.88}$ $\varepsilon_{Gr} = 1.23 \times 10^{-2} \mu_L^{0.45} \left(\dfrac{A_d}{A_r} \right)^{1.08} U_{Gr}^{0.88}$

(continued)

TABLE 8.1 *(Continued)*

Chisti and Moo-Young (1986)[b]	ILALR (rectangular)	Air-(water, aqueous salt solution)+1-3 dry wt/vol. % KS-1016 cellulose fiber	Homogeneous flow regime: $\varepsilon_G = (1.488-0.496C_s)U_G^{0.892\pm0.075}$ Heterogeneous flow regime: $\varepsilon_G = (0.371-0.089C_s)U_G^{0.430\pm0.015}$
Chisti et al. (1988a)[a]	ILALR	$\varepsilon_{Gr} = 0.65U_{Gr}^{(0.603+0.078C_0)}\left(1+\dfrac{A_d}{A_r}\right)^{-0.258}$ $\varepsilon_{Gd} = 0.46\varepsilon_r - 0.0244$	
Chisti (1989)	ILALR	$U_{Gr} = 0.026-0.21$ m/s $A_d/A_r = 0.25-0.44$	$\varepsilon_{Gr} = 0.65\left(1+\dfrac{A_d}{A_r}\right)^{-0.258}U_{Gr}^{0.603}$ $\varepsilon_{Gr} = 2.4U_G^{0.97}$
Choi (2000)[c]	?	$U_G = 0.02-0.18$ m/s $h_R = 0.04-0.20$ m $A_d/A_r = 0.11-0.53$	$\varepsilon_{Gr} = 0.2447U_G^{0.5616}\left(\dfrac{A_d}{A_r}\right)^{-0.2779}h_R^{-0.0130}$
Choi (2001)[c]	ELALR	$U_G = 0.02-0.18$ m/s $L_c/L_h = 0.1-0.5$ m $A_d/A_r = 0.11-0.53$	$\varepsilon_{Gr} = 0.431U_G^{0.580}\left(\dfrac{A_d}{A_r}\right)^{-0.040}\left(\dfrac{L_c}{L_h}\right)^{-0.042}$
Ghirardini et al. (1992)[a]	ELALR	$\varepsilon_G = 0.55U_{Gr}^{0.78}\left(\dfrac{U_{Ls}}{U_L}\right)^{0.2}d_r^{0.42}$	
Hills (1976)[c]	?	$U_G = 0.4-3.2$ m/s $U_{Lr} = 0-2.5$ m/s	$\varepsilon_{Gr} = \dfrac{U_G}{0.21 + 1.35(U_G+U_{Lr})^{0.93}}$

Li et al. (1995)[a]	ILALR	$\varepsilon_{Gr} = 0.441 U_{Gr}^{0.841}\,\mu_a^{-0.135}$ $\varepsilon_{Gd} = 0.297 U_{Gr}^{0.935}\,\mu_a^{-0.107}$
Kawase et al. (1995)[a]	ILALR	$$\frac{\varepsilon_{Gr}}{1-\varepsilon_{Gr}} = \frac{U_{Gr}^{n+2/2(n+1)}}{2^{3n+1/n+1}\,n^{n+2/2(n+1)} \left(\dfrac{K}{\rho_L}\right)^{1/2(n+1)} g^{n/2(n+1)} \left(1+\dfrac{A_d}{A_r}\right)^{3(n+2)/4(n+1)}}$$
Kawase and Moo-Young (1986a)[a]	ILALR	$\varepsilon_G = 0.24 n^{-0.6} Fr^{0.84-0.14n} Ga$
Kawase and Moo-Young (1986b)[b]	DT-ILALR BC	Air-(water, pseudoplastic fluids) $U_G = 0.008-0.285$ m/s $d_R = 0.14-0.35$ m $n = 0.28-1$ $K = 0.001-1.22$ Pa-s^n $\varepsilon_G = 0.24 n^{-0.6} \left(\dfrac{U_G}{\sqrt{gd_R}}\right)^{0.84-0.14n} \left(\dfrac{gd_R^3\rho_L^2}{\mu_a^2}\right)$
Kawase and Moo-Young (1987)[a]	ELALR	$\varepsilon_{Gr} = 1.07 \left(\dfrac{U_G^2}{gd_r}\right)^{0.333}$
Kamblowski et al. (1993)[a]	ELALR	$\varepsilon_{Gr} = 0.203 \dfrac{Fr^{0.31}}{Mo^{0.012}} \left(\dfrac{U_{Gr} A_r}{U_{Lr} A_d}\right)^{0.74}$ where: $Mo = \dfrac{g(\rho_L-\rho_G)}{\sigma\rho_L^2} K^4 \left(\dfrac{8U_{Gr}}{d_r}\right)^{4(n-1)} \left(\dfrac{3n+1}{4n}\right)^{4n}$ $Fr = \dfrac{(U_{Lr}+U_{Gr})^2}{gd_r}$

(continued)

TABLE 8.1 (*Continued*)

Reference	Config.	Equation / Conditions
Koide et al. (1983a)[a]	ILALR	$\dfrac{\varepsilon_G}{(1+\varepsilon_G)^4} = 0.16\left(\dfrac{U_G\mu_L}{\sigma}\right) Mo^{-0.283}\left(\dfrac{d_t}{d_R}\right)^{-0.222}\left(\dfrac{\rho_L}{\Delta\rho}\right)^{0.283}\left[1-1.61\left(1-e^{0.00565Ma}\right)\right]^{-1}$
Koide et al. (1985)[a]	ILALR	$\dfrac{\varepsilon_G}{(1+\varepsilon_G)^4} = \dfrac{0.124\left(\dfrac{U_G\mu_L}{\sigma}\right)^{0.996}\left(\dfrac{\rho_L\sigma^3}{g\mu_L^4}\right)^{0.294}\left(\dfrac{d_t}{d_R}\right)^{0.114}}{1-0.276\left(1-e^{-0.0368Ma}\right)}$
Koide et al. (1988)[a]	ILALR	$\varepsilon_{Gr} = \dfrac{Fr}{0.415+4.27\left(\dfrac{U_{Gr}+U_{Lr}}{\sqrt{gd_t}}\right)\left(\dfrac{g\rho_L d_R^2}{\sigma}\right)^{-0.188}+1.13\,Fr^{1.22}Mo^{0.0386}\left(\dfrac{\Delta\rho}{\rho_L}\right)^{0.0386}}$
Merchuk (1986)[c]	?	$\varepsilon_{Gr} = 0.047U_{Gr}^{0.59}$ $\quad U_G = 0.002\text{–}0.5$ m/s
Merchuk et al. (1998)	DT-ILALR	Air-seawater from Almeria Bay; $U_G = 0\text{–}0.21$ m/s; $h_R = 2$ m; $d_R = 0.096$ m; $h_D = 1.5$ m; $A_r = 0.00283$ m²; $A_d = 0.00280$ m²; $\varepsilon_{Gr} = \alpha\left(\dfrac{U_{Gr}}{U_{Lr}}\right)^{\beta}$

Sparger	α	β
PP with 30×d_0 = 1 mm	0.48	1.03
PP with 30×d_0 = 0.5 mm	0.37	0.79
CS with d_0 = 120 μm	0.34	0.69
CS with d_0 = 60 μm	0.33	0.68
PS with d_0 = 120 μm	0.48	1.13
PS with d_0 = 60 μm	0.40	0.99
PS with d_0 = 30 μm	0.38	0.88

| Miyahara et al. (1986)[b] | DT-ILALR | Air-non-Newtonian CMC solutions

$\rho_L = 952–1168$ kg/m^3
$\mu_L = 1.0–14.9$ mPa-s
$\sigma = 34.1–72.0$ mN/m
$d_R = 0.148$ m
$L_d = 1.0$ m
$h_R = 1.20$ m
$d_0 = 0.0005–0.0015$ m
$A_d/A_r = 0.128–0.808$ | $$\varepsilon_{Gr} = \frac{0.4\sqrt{Fr}}{1 + 0.4\sqrt{Fr}}\left(1 + \frac{U_L}{U_{Gr}}\right)$$
For
$\varepsilon_{Gr} < 0.0133$
$$\varepsilon_{Gd} = 4.51 \times 10^6\, Mo^{0.115}\left(\frac{A_r}{A_d}\right)^{-1.32}\left(\frac{A_r}{A_d}\right)^{4.2}\varepsilon_{Gr}^{4.2}$$
For
$\varepsilon_{Gr} > 0.0133$
$$\varepsilon_{Gd} = 0.05\, Mo^{-0.22}\left(\frac{A_r}{A_d}\right)^{-1.32}\left[\left(\frac{A_r}{A_d}\right)^{0.5}\varepsilon_{Gr}\right]^{0.31 Mo^{-0.073}}$$ |
| Shariati et al. (2007) | DT-ILALR | Air-(distilled water, aqueous isomax diesel)

$U_G = 0–0.07$ m/s
$d_R = 0.14$ m
$h_D = 1.1$ m
$d_r = 0.1$ m
$h_c = 0.024$ m
$A_d/A_r = 0.906$
$h_0 = 25\times1$ mm (PP)
$\nu_L = 46.142\times10^{-6}$ m^2/s | $\varepsilon_G = 4.92\, U_{Gr}^{1.066}\,\nu_L^{-0.355}$
$\varepsilon_{Gd} = 0.788\varepsilon_{Gr} - 0.004$ |

(continued)

195

TABLE 8.1 (*Continued*)

Trilleros et al. (2005)	DT-ILALR	Air–water $h_R = 1.25$ m $d_R = 0.42$ m $d_{pg} = 0.25$ and 1 mm $\rho_{pg} = 2.6$ g/cm^3 $d_{pc} = 3$ mm $\rho_{pg} = 1.0$ g/cm^3 $d_0 = 12 \times 1$ mm (PP) $d_D = (0.044, 0.082, 0.125, 0.240)$ m $h_D = 0.630$ and 1.050 m $\varepsilon_s = 1$–7%	Large glass particles: $\dfrac{U_{Gr}}{\varepsilon_{Gr}} = \left[0.43\left(U_{Gr} + U_{Lr} + U_{Sr}\right)\right] + 0.26$ $U_{Gr} = 0.02$–0.10 m/s $(U_{Gr} + U_{Lr} + U_{Sr}) = 0.80$–14.05 m/s $\varepsilon_{Gr} = 0.013$–0.14 Small glass particles: $\dfrac{U_{Gr}}{\varepsilon_{Gr}} = \left[0.35\left(U_{Gr} + U_{Lr} + U_{Sr}\right)\right] + 0.15$ $U_{Gr} = 0.006$–0.12 m/s $(U_{Gr} + U_{Lr} + U_{Sr}) = 1.05$–14.75 m/s $\varepsilon_{Gr} = 0.005$–0.17 Polystyrene particles: $\dfrac{U_{Gr}}{\varepsilon_{Gr}} = \left[0.22\left(U_{Gr} + U_{Lr} + U_{Sr}\right)\right] + 0.15$ $U_{Gr} = 0.005$–0.12 m/s $(U_{Gr} + U_{Lr} + U_{Sr}) = 0.50$–15.20 m/s $\varepsilon_{Gr} = 0.005$–0.16
Posarac and Petrovic (1988)[a]	ELALR		$\varepsilon_{Gr} = \dfrac{0.6\,\rho_G^{0.062}\,\rho_L^{0.069}\,\mu_G^{0.107}}{\mu_L^{0.053}\,S_L^{0.185}}\,\dfrac{U_{Gr}^{0.936}}{(U_{Gr} + U_{Lr})^{0.474}}$
Popovic and Robinson (1984)[a]	ELALR BC	Air–non-Newtonian CMC solutions $\mu_L = 0.015$–0.5 Pa·s $A_d/A_r = (0, 0.11, 0.25, 0.44)$	$\varepsilon_{Gr} = 0.465 U_{Gr}^{0.65}\left(1 + \dfrac{A_d}{A_r}\right)^{-1.06}\mu_a^{-0.103}$
Vatai and Tekic (1986)[a]	ILALR		$\varepsilon_{Gr} = (0.491 - 0.498\,)U_{Gr}^{0.706}\left(\dfrac{A_d}{A_r}\right)^{-0.254}d_{rr}\mu_a^{0.0684}$

[a]As cited by Merchuk and Gluz (Merchuk and Gluz, 1999)
[b]As cited by Chisti (Chisti, 1989)
[c]As cited by Jones (Jones, 2007)

TABLE 8.2 Gas–Liquid Mass Transfer Coefficient Correlations for Airlift Bioreactors

Reference	Bioreactor Type(s)	Bioreactor Parameters	Correlations
Abijanic et al. (2007)	DT-ILALR	Air–(water, 1 wt% aqueous solutions of methanol, ethanol, n-propanol, isopropanol, and n-butanol) $U_G = 0–0.07$ m/s	$k_L a = 0.28 U_G^{0.77} \left[1 + \left(\dfrac{d\sigma}{dC_A} \right)^{0.15} \right]^{0.71}$
Bello et al. (1985a)[a]	ELALR	Air–(water, NaCl solution) d_R or $d_r = 0.152$ m $h_D = 1.8$ m $U_{Gr} = 0.0137–0.086$ m/s $A_d/A_r = 0.11–0.69$	$\dfrac{k_L a h_r}{U_L} = 2.28 \left(\dfrac{U_{Gr}}{U_L} \right)^{0.90} \left(1 + \dfrac{A_d}{A_r} \right)^{-1}$
Bello et al. (1985b)[a]	ELALR DT-ILALR BC	Air–(water, NaCl solution) d_R or $d_r = 0.152$ m $h_D = 1.8$ m $U_{Gr} = 0.0137–0.086$ m/s $A_d/A_r = 0.11–0.69$ (ELALR) $A_d/A_r = (0.13, 0.35, 0.56)$ (ILALR) $A_d/A_r = 0$ (BC)	$k_L a = 0.76 \left(1 + \dfrac{A_d}{A_r} \right)^{-2} U_{Gr}^{0.8}$ $k_L a = 5.5 \times 10^{-4} \left(1 + \dfrac{A_d}{A_r} \right)^{-1.2} \left(\dfrac{P_G}{V_D} \right)^{0.8}$

(continued)

197

TABLE 8.2 *(Continued)*

Reference	Bioreactor Type(s)	Bioreactor Parameters	Correlations
Blazej et al. (2004b)	DT-ILALR	(Air, pure oxygen, nitrogen)–(Na_2SO_3 aqueous solution) $d_R = 0.157$ m $d_D = 0.106$ m $V_L = 40$ dm^3 $d_0 = 25 \times 1$ mm (PP)	$k_L a = 0.91 \varepsilon_{Gr}^{1.39}$
Chisti et al. (1988b)[b]	ELALR	$k_L a = (0.349 - 0.102 C_s) U_{Gr}^{0.837} \left(1 + \dfrac{A_d}{A_r}\right)^{-1}$	
Freitas and Teixeira (2001)	DT-ILALR	Low density solids: $d_P = 2.131 \pm 0.102$ mm $\rho_s = 1023 \pm 1$ kg/m^3 High density solids: $d_P = 2.151 \pm 0.125$ mm $\rho_s = 1048 \pm 1$ kg/m^3 $\varepsilon_s = 0$–30% $U_{Gr} = 0.01$–0.5 m/s	Water/low density solids $k_L a = (-0.93 U_{Gr}^2 + 1.33 U_{Gr} - 0.012)(-0.0000016 \varepsilon_s^2 - 0.00099 \varepsilon_s + 0.054)$ Water/high density solids $k_L a = (-0.33 U_{Gr}^2 + 0.43 U_{Gr} - 0.0064)(-0.000080 \varepsilon_s^2 - 0.0056 \varepsilon_s + 0.17)$ Ethanol (10 g/l)/low density solids $k_L a = (-0.95 U_{Gr}^2 + 1.34 U_{Gr} - 0.021)(-0.000072 \varepsilon_s^2 - 0.00079 \varepsilon_s + 0.075)$ Ethanol (10 g/l)/high density solids $k_L a = (-0.78 U_{Gr}^2 + 1.20 U_{Gr} - 0.021)(-0.000074 \varepsilon_s^2 + 0.0035 \varepsilon_s + 0.081)$

Kawase and Moo-Young (1986a)[b]	ILALR	$Sh = 0.68n^{-0.72} Fr^{0.38n+0.52} Sc^{0.38-0.14n}$
Kawase and Moo-Young (1986b)[a]	DT-ILALR BC	Air–(water, pseudoplastic fluids) $U_G = 0.008$–0.084 m/s $d_R = 0.14$–0.305 m $n = 0.543$–1 $K = 0.00089$–2.82 Pa·s^n $\qquad \dfrac{k_L a d_R^2}{D_L} = 0.68n^{-6.72} \left(\dfrac{d_R U_G \rho_L}{\mu_a}\right)^{0.38n+0.52} \left(\dfrac{\mu_a}{D_L \rho_L}\right)^{0.38n-0.14}$
Kade et al. (1983a)[a]	DT-ILALR	Air–Newtonian fluids $\mu_L = 0.9$–13 mPa-s $\sigma = 51$–73 mN/m^3 $D_L\ (10^9) = 0.18$–2.42 m^2/s $U_G = 0.0098$–0.156 m/s $d_R = 0.10$–0.30 m $d_D = 0.06$–0.19 m $L_d = 0.70$–2.10 m $L = 0.84$–2.24 m $d_0 = 0.001$–0.015 m (SO) $d_0 = 7\times0.001$ m (PP) $\qquad \dfrac{k_L a d_R^2}{D_L} = 0.477 \left(\dfrac{\mu_L}{\rho_L D_L}\right)^{0.5} \left(\dfrac{g d_R^2 D_L}{\sigma}\right)^{0.837} \left(\dfrac{g d_R^3 D_L^2}{\mu_L^2}\right)^{0.257} \left(\dfrac{d_D}{d_R}\right)^{-0.54} \varepsilon_G^{1.36}$ $3.69\times10^2 \le \dfrac{\mu_L}{\rho_L D_L} \le 5.68\times10^4$ $1.36\times10^3 \le \dfrac{g d_R^2 D_L}{\sigma} \le 1.22\times10^4$ $2.27\times10^8 \le \dfrac{g d_R^3 D_L^2}{\mu_L^2} \le 3.32\times10^{11}$ $0.471 \le \dfrac{d_D}{d_R} \le 0.743$ $0.037 \le \varepsilon_G \le 0.21$

(continued)

TABLE 8.2 (*Continued*)

Reference	Bioreactor Type(s)	Bioreactor Parameters	Correlations
Koide et al. (1983b)[a]	DT-ILALR	Air–Newtonian fluids $\mu_L = 0.9$–13 mPa-s $\sigma = 51$–73 mN/m^3 $D_L\,(10^9) = 0.18$–2.42 m^2/s $U_G = 0.0098$–0.156 m/s $d_R = 0.10$–0.30 m $d_D = 0.06$–0.19 m $L_d = 0.70$–2.10 m $L = 0.84$–2.24 m $d_0 = 0.001$–0.004 m (SO)	$$\frac{k_L a \sigma}{D_L g \rho_L} = 2.25 \left(\frac{\mu_L}{\rho_L D_L}\right)^{0.5} \left(\frac{\rho_L \sigma^3}{g \mu_L^{\,4}}\right)^{0.136} \left(\frac{d_0}{d_R}\right)^{-0.0905} \varepsilon_G^{1.26}$$ $3.71\times10^2 \leq \dfrac{\mu_L}{\rho_L D_L} \leq 6.00\times10^4$ $1.18\times10^6 \leq \dfrac{\rho_L \sigma^3}{g \mu_L^{\,4}} \leq 5.93\times10^4$ $0.471 \leq \dfrac{d_D}{d_R} \leq 0.743$ $7.41\times10^{-3} \leq \dfrac{d_0}{d_c} \leq 2.86\times10^{-2}$ $0.0302 \leq \varepsilon_G \leq 0.305$
Koide et al. (1985)[b]	ILALR		$Sh = 2.66 Sc^{0.5} Bo^{0.715} Ga^{0.25} \left(\dfrac{d_r}{d_R}\right)^{-0.429} \varepsilon_G^{1.34}$
Li et al. (1995)[b]	ILALR		$k_L a = 0.03343\, U_{Gr}^{0.524}\, \mu_a^{-0.255}$
Merchuk et al. (1994)[b]	ILALR		$Sh = 30000 Fr^{0.97} M^{-5.4} Ga^{0.045} \left(1 + \dfrac{A_d}{A_r}\right)^{-1}$

Muthukumar and Velan (2006)	DT-ILALR	Air–(isoamyl alcohol, benzoic acid, propanol, CMC solution) $\varepsilon_s = 5\text{–}20\%$ $d_R = 0.19$ m $d_t/d_R = (0.34, 0.44, 0.49, 0.59, 0.76)$ $h_R = 2$ m $h_D = 1.22$ m $h_c = 0.05$ m $d_0 = 37\times1$ mm (PP)
		For three-phase systems that enhance $k_L a$: $$\frac{k_L a d_R^2}{D_L} = 0.7 C_s^{-0.214} \left(\frac{U_G^2}{g d_R}\right)^{0.49} \left(\frac{d_t}{d_R}\right)^{-0.46} \left(\frac{g d_R^3 \rho_L^2}{\mu_L^2}\right)^{-0.209} \left(\frac{g d_R^2 \rho_L}{\sigma}\right)^{2.515} \left(\frac{\rho_p - \rho_L}{\rho_L}\right)^{-0.144} \left(\frac{U_t \mu_L}{\sigma}\right)^{-0.1}$$ For three-phase systems that decrease $k_L a$: $$\frac{k_L a d_R^2}{D_L} = 0.64 \varepsilon_s^{-0.214} \left(\frac{U_G^2}{g d_R}\right)^{0.49} \left(\frac{d_t}{d_R}\right)^{-0.46} \left(\frac{g d_R^3 \rho_L^2}{\mu_L^2}\right)^{0.576} \left(\frac{U_t \mu_L}{\sigma}\right)^{-0.343}$$ $k_L a = 0.678 U_G^{0.907} \mu_{eff}^{-0.086}$ where: $\mu_{eff} = (5000 U_G)^{n-1}$
Nicolella et al. (1998)	DT-ILALR	Air–water–biofilm particles $U_G = 0.002\text{–}0.027$ m/s $\varepsilon_s = 5, 10, 15\%$ $d_s = (0.47, 0.91, 1.67, 1.95)$ mm $d_R = 0.060$ m $d_t = 0.043$ m $h_R = 0.700$ m $h_D = 0.360$ m $k_L a = 0.6\varepsilon_G$

(continued)

TABLE 8.2 *(Continued)*

Reference	Bioreactor Type(s)	Bioreactor Parameters	Correlations
Popovic and Robinson (1984)[a,b]	ELALR BC	Air–non-Newtonian CMC solutions $\mu_L = 0.015$–0.5 Pa-s $A_d/A_r = (0, 0.11, 0.25, 0.44)$	$k_L a = 1.911\times10^{-4} U_{Gr}^{0.525} \left(1+\dfrac{A_d}{A_r}\right)^{-0.853} \mu_a^{-0.89}$ $k_L a = 0.24 U_{Gr}^{0.837} \left(1+\dfrac{A_d}{A_r}\right)^{-1}$
Ruen-ngam et al. (2008)	DT-ILALR	Air–(tap and sea water) Salinity = (0, 15, 30, 45) ppm $U_G = 0.01$–0.07 m/s $h_R = 1.2$ m $L_D = 1$ m $h_c = 0.05$ m $d_R = 0.137$ $A_d/A_r = (0.067, 0.443, 0.661, 1.008)$ $V_L = 1.51$ $d_0 = 30\times1$ mm (PR)	When salinity = 0 ppm $Sh = 0.41+1.05 Gr^{0.48}$ When salinity = 15–45 ppm $Sh = 0.41+1.04 Gr^{0.16} Sc^{0.3}+0.13 Re^{0.46} Sc^{0.06}$
Siegel and Merchuk (1988)[b]	ELALR	$k_L a = 913 \left(\dfrac{P}{V_L d_R}\right)^{1.04} U_L^{-0.15}$	

[a] As cited by Chisti (1989).
[b] As cited by Merchuk and Gluz(1999).

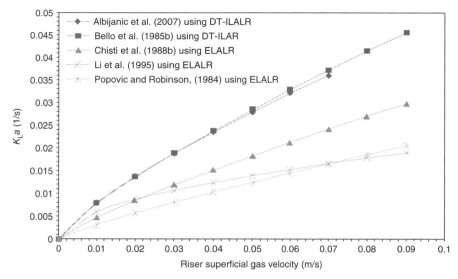

Figure 8.9   Sample gas–liquid mass transfer coefficients for ALRs.

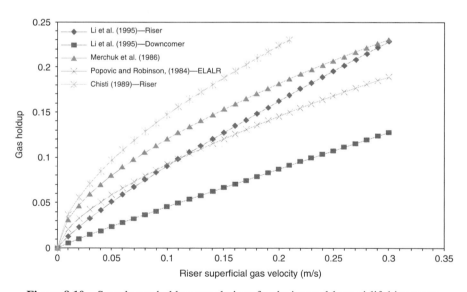

Figure 8.10   Sample gas holdup correlations for the internal-loop airlift bioreactor.

A significant problem with a large number of available correlations is that they require knowledge of the liquid phase behavior. This need is quite logical, but replication often requires a correlation for the liquid velocity, and additional error would be introduced in the generated gas holdup and gas–liquid mass transfer data.

A few measurement and assumption issues are built into the existing correlations. Gas–liquid mass transfer correlations, which are calculated at high gas flow rates, may be 15% lower than the system experiences because an incorrect assumption of ideal phase mixing is often made (Blazej et al., 2004b). A further problem is that a normal bubble size distribution is often assumed in the analysis, but the truth of this or any other assumption has not been thoroughly verified (Fadavi et al., 2008).

8.6   NEEDED RESEARCH

The research on airlift bioreactors is relatively young compared to stirred-tank and bubble column bioreactors (Figures 8.11 and 8.12). The data have been collected using EngineeringVillage2 (EV2), which allows access to several engineering-related databases and services. The search terms used in EV2 are shown in the legends of Figures 8.11 and 8.12. A moving 3-year average was used to smooth the data. This simple search reveals some interesting trends. An equal amount of research is performed in stirred-tank bioreactors and bubble columns, and it is currently preferred to airlift bioreactors. One reason for this is that many experiences from bubble columns are ported to airlift bioreactors. The result has been that the research depth for stirred-tank bioreactors and bubble column bioreactors is great, but specific studies concentrating on airlift bioreactor behavior has waned. Also, there is more mass transfer research performed in bubble

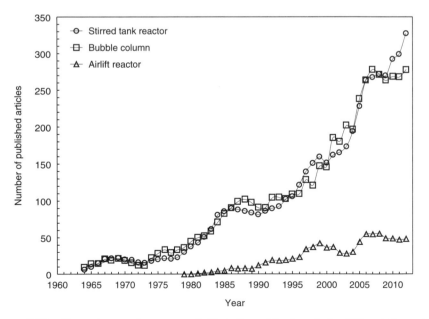

Figure 8.11   General research papers available in the public domain for different bioreactor types. Acquired 07/01/2013.

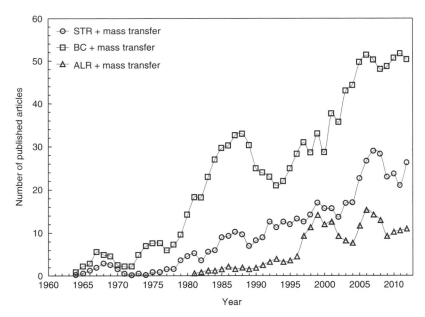

Figure 8.12   Research papers available in the public domain specific to mass transfer for different bioreactor types. Acquired 07/01/2013.

columns than in stirred-tank reactors, which is larger than ALRs. One possible explanation is that gas–liquid mass transfer operations have more applications in bubble columns than in stirred-tank reactors.

A great deal of enthusiasm surrounded airlift bioreactors stemming from the increased attention that the bioreactor had received since the mid-1990s. The progression seemed quite natural. Research in stirred-tank bioreactors had made that bioreactor's shortcomings were obvious and the advent of the bubble column was a natural transition. As research into the bubble column matured, some of its issues also came to light, and the solution, the airlift bioreactor, was seen as the next stage. It may also be concluded that the number of papers related to airlift bioreactors has been suppressed because research in bubble columns has been successfully ported to the airlift bioreactor even though major flow differences exist.

A secondary force in airlift bioreactor research development has been bioreactor design procedure. The question is not which microorganism fits the bioreactor, but which bioreactor fits the microorganism. So, the natural starting point for process research and development is to start with the microorganism and its metabolic production and then select the necessary hardware. Hence, research has become more inclusive of process engineering and biological processes (Zhong, 2010), which can spread resources thin.

One of the potential outcomes is for airlift bioreactor research to slow down for two main reasons. First, bioreactor research seems to be more mature than its biological counterpart. The rate of return on improvements in the metabolic performance of microorganisms should be higher than the rate of return on bioreactor mass transfer improvements. Second, the continued and successful application of bubble column data to airlift bioreactors should continue to incentivize bubble column researchers to practically undercut new airlift bioreactor research entrants. It is currently easier to make the case that research money serves a multipurpose when applied to bubble columns, whereas research funds on airlift bioreactors offer more specific applications.

That is not to say that airlift bioreactor research is doomed, but bioprocessing could use more specific attention. Research efforts in airlift bioreactors need to go in a direction where they would enable new processes, such as mammalian tissue growth, or incorporate the biological element of bioreactors to a larger degree. For example, current effects of variable local conditions on cellular biochemistry are not understood. As such, cells could be grown to mimic the macroscopic behavior of the desired cells, but still miss the molecular, cellular, and biochemical interactions necessary to mimic the operations of the target cells. Research where bubble column bioreactor similarities break down, such as incorporating a solid phase, are scarce and could also be successful.

Proper start-up procedures, such as initial cell density and distribution, and tracking important or toxic by-products are also troublesome (Martin and Vermette, 2005). Fiber suspensions have not been widely applied in airlift bioreactors, but their application would be of interest for suspended cell growth and density propagation. The incorporation of microorganisms in hydrodynamic studies have not been widely attempted even though microorganisms are known to increase the gas–liquid mass transfer coefficient due to a higher rate of oxygen consumption (Garcia-Ochoa and Gomez, 2009), especially at and near the surface.

More focus on consistent variable testing is needed. Oftentimes, multiple variables are adjusted while the analysis is focused on just one. The conclusions are based on one variable, and the changes in the others are ignored. Particle size comparisons are often done based on scale. One group is often referred to as large with a diameter on the millimeter scale, while the small particles are defined by a diameter less than 1 mm. In addition, these particles are often constructed of differing materials, such that a density variance is introduced, or are used in different concentrations. These facts make hydrodynamic and gas–liquid mass transfer conclusions highly variable in the literature.

This mistake may be made for scale comparison studies as well. Blazej et al. (2004c), for example, compared vessels with volumes of 10.5, 32, and 200 l. The 10.5 and 32 l versions can be classified as experimental scale, while the 200 l ALR may be considered pilot scale (Garcia-Ochoa and Gomez, 2009). Hence, the results, shown in Figure 8.13, could be very helpful for scale-up. Each vessel, however, was geometrically different with varying downcomer-to-riser area and aspect ratio and bottom clearance. Although the differences were not great and comply with the scale-up procedure of Garcia-Ochoa and Gomez (2009), it introduces an additional dilemma. Which effects are related to the scale and which to the downcomer-to-riser area ratio, or is the difference even significant? If a control

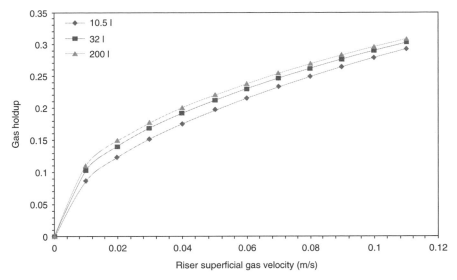

Figure 8.13   Airlift bioreactor scale effects. Adapted from Blazej et al. (2004c).

study with constant ratios is not done, these questions are much harder to answer and conclusions are based on the experimenters' interpretation and inference.

8.7   SUMMARY

Bubble column bioreactors are the main competitor to airlift bioreactors, and deployment decisions are usually based on a competitive basis that excludes the stirred-tank bioreactor. Hence, studies rarely compare the stirred-tank bioreactor, bubble column, and airlift bioreactor in a more comprehensive manner. Usual comparisons are made either between the stirred-tank bioreactor and the bubble column or between the bubble column and the airlift bioreactor. The first stage in identifying a bioreactor usually involves a comparison between the stirred-tank bioreactor and bubble column bioreactor, which clarifies the bioreactor and process requirements. If the bubble column is found to be competitive, the airlift bioreactor is introduced to the discussion.

The airlift bioreactor is often used in cases for which the bubble column lacks the operational flexibility or requirements. In other words, the stirred-tank bioreactor does not have the necessary features, and the bubble column has traits that could have negative effects on the production process. These negative traits often include extensive phase backmixing, undefined flow paths, limited gas flow rates, and potentially damaging shear rates. The airlift bioreactor can successfully address these issues and can provide further control, such as a downcomer sparger, and access to major bioreactor components.

The negative aspect of airlift bioreactors is that research is in its early stages in many respects. A basic hydrodynamic understanding is lacking. For example, the onset of liquid circulation has been reported to vary between $U_{Gr} = 0.035 - 0.050\,\text{m/s}$. It is occasionally reported to coincide with the introduction of the heterogeneous flow regime while the homogeneous flow regime is prevalent in other cases of circulation onset.

Bubble–bubble contacts are regulated by the same forces and interactions as those found in bubble columns, which would imply that the basic behavior should be very similar to bubble columns. This assumption is often invoked, and bubble column research is used in these cases to set expectations and explain outcomes; however, many assumptions have not been tested. The fact that bubble collisions do not occur as frequently in the airlift bioreactor as in the bubble column leads to the conclusion that certain aspects of the bubble column may not be applied to airlift hydrodynamics as easily or at all.

These issues, positive and negative, are reflected in the available correlations. These correlations are both highly useful and also limited. Some are useful because the inputs are easily measured and adjusted as needed; however, correlations are mostly empirical or semi-empirical, which means that they are not widely applicable but, rather, are bioreactor design dependent at best. Hence, geometric similarity is very important. Furthermore, most studies are performed in air–water systems while most industrial processes use much more complicated and time-variant liquids. In other words, the airlift bioreactor correlations have similar problems as those for stirred-tank bioreactors and bubble columns and are due to the fact that they share the "problem" source: bubble–bubble interactions. Bubble–bubble interactions are highly variable and lead to hydrodynamics which, in turn, are difficult to quantify and predict. Hence, the result has been that the airlift bioreactor correlations and models are either system dependent or not adequately constrained.

9   Fixed Bed Bioreactors

9.1   INTRODUCTION

Fixed bed reactors are three-phase systems for which the solid phase is structurally fixed. There are two basic categories of fixed bed reactors that are defined by the phase flow directions. The first class is the packed bed reactors (PBRs), schematically represented in Figure 9.1. PBRs are defined by countercurrent phase flow, which can lead to reactor stability and safety issues during exothermic reactions (Yakhnin and Menzinger, 2008). The liquid phase is sprayed from the top while the gas phase is fed from the bottom. The second class is trickle bed reactors (TBRs), shown in Figure 9.2, which have very similar designs, but each component is adjusted to conform to particular phase flow patterns. TBRs are defined by cocurrent, downward phase flow—the liquid and gas phases are fed from the top (Maiti and Nigam, 2007; Medeiros et al., 2001). The gas phase is pressurized to improve process efficiency. If the flow of the gas and liquid phases occurs upward, the reactor is referred to as flooded bed reactor (FBR). Although the differences may seem minor, the dynamics can be very different. For example, FBRs operate with the packed material being almost or completely submerged while the TBR is usually operated with minimal flooding (Al-Dahhan et al., 1997; Kolev, 2006).

FBRs ensure that the entire packing surface is wetted, which is highly important for liquid-phase limited processes and reactions; however, most biological applications are limited to gas phase, which would make the FBR less useful in its current state (as used by the chemical industry). This reactor design could be useful for biological applications if the microorganisms require significant substrate flow or are sensitive to gas–liquid interfaces; however, FBRs are typically not used in industry because they are difficult to design and manage (Maiti and Nigam, 2007).

The exact reactor choice is highly dependent on the type of reaction required by the process. Since biological applications are not very fast, the choice is going to be more dependent on the microorganism's environmental requirements. For biological applications, fixed bed reactors have traditionally been used for shear-sensitive microbes and cells (especially mammalian). In contrast, shear-resistive strains would be expected to experience better results in a bubble column or airlift reactor since these devices operate at much higher gas and liquid flow rates.

An Introduction to Bioreactor Hydrodynamics and Gas-Liquid Mass Transfer, First Edition.
Enes Kadic and Theodore J. Heindel.
© 2014 John Wiley & Sons, Inc. Published 2014 by John Wiley & Sons, Inc.

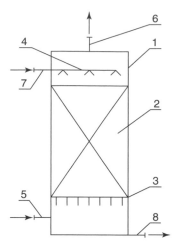

Figure 9.1   Packed bed reactor internals: (1) vessel, (2) packed material, (3) support plate, (4) liquid distributor, (5) gas input, (6) gas output, (7) liquid input, and (8) liquid output (Kolev, 2006).

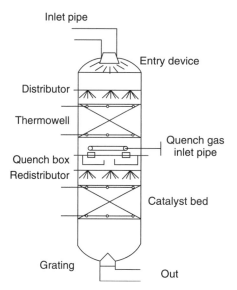

Figure 9.2   Trickle bed reactor schematic (Maiti and Nigam, 2007).

Fixed bed reactors, especially TBRs, are heavily used in industrial practice. They are used by the petroleum, petrochemical, and chemical industries for waste treatment and processing, biochemical and electrochemical processes, and hydrotreatment. Catalysts provide a mechanism to accelerate and channel very complex processes, which would normally require high pressures and/or temperatures and long processing times, into a more manageable reaction and reactor environment. The fixed bed reactor is the perfect reactor of choice to achieve these goals such that the TBR has been used to process approximately 1.6 billion metric tons of products by the petroleum industry in 1991 alone (Al-Dahhan et al., 1997). Hence, large amounts of research and development have gone, and will continue to go, into these reactors because even marginal improvements can yield significant profitability increases.

The negative aspects of fixed bed reactors are often tied to the performance, behavior of the particular catalyst, and potentially difficult design and scale-up (Nacef et al., 2007). If the catalyst is deactivated or spent fairly quickly, construction of the fixed bed reactors makes the processes impractical. The catalyst would have to be replaced too frequently, leading to high labor and packing costs. One of the main objectives is to ensure that the packing material experiences optimal wetting. Catalysts that do not wet provide inefficient reaction sites. Hence, the flow has to stay within an acceptable pattern. Flow channeling is quite problematic since it can cause inactive catalyst and performance and economic losses (Doan et al., 2008; Maiti and Nigam, 2007). On the extreme, flow channeling can lead to runaway reactions and potentially to combustion/explosion (Al-Dahhan et al., 1997).

A related issue is that identifying flow problems or catalyst utilization is often hard to accomplish in an industrial setting. There are two basic approaches. The first involves measuring pressure drop and/or temperature differentials in the reactor; however, this technique is useful only if flow channeling is significant to cause variations, and it often does not account for wetting problems. The second is to visually inspect the catalyst, but this is often done once the catalyst is spent, which can take up to 2–3 years (Maiti and Nigam, 2007). After the inspection and determination of catalyst inactivity, a plan is made to correct the flow, but the soundness of those corrections cannot be determined until the next inspection, which may be in another 2–3 years. There are other methods to investigate liquid maldistribution, but these can often be complicated or expensive (see, e.g., Llamas et al. (2008)).

Each reactor type has specific operational parameters. TBRs are usually operated adiabatically at high pressure (20–30 MPa) and temperature and employ $U_G < 30$ cm/s and $U_L < 1$ cm/s (Al-Dahhan et al., 1997; Attou et al., 1999; Nigam and Larachi, 2005) although research work is often conducted at atmospheric pressures (Attou et al., 1999). PBRs use similar phase velocity ranges, but those can be increased up to $U_G < 3.5$ m/s and $U_L \sim 0.11$ m/s for some special reactor configurations (Kolev, 2006). Biological operation, however, may call for much lower velocities. If the fixed bed reactor is used as a biofilter, $U_G < 0.001$ m/s and $U_L < 0.005$ m/s are common (Maldonado et al., 2008).

9.2   COLUMN GEOMETRY AND COMPONENTS

The PBRs and TBRs have very similar construction. Both devices are made of a cylindrical vessel, but the internal construction varies greatly between the designs, engineering firm, or even application. The basic components consist of a support mechanism for the packed material and phase distributors and extractors. The major difference for TBRs is that they tend to use packaged gas–liquid distributors rather than separate units. Other parts are basically the same although the FBR has a different collection system.

The PBRs consist of support plates, hold-down plates, liquid (re)distributors, gas or vapor distributors, gas deflection plates, and screen mesh. The identity and type of the particular parts are dependent largely on reactor scale. The gas deflection plates and screen mesh are used in a cross-flow cascade packed column, which is only useful for chemical processes. The support plates are primarily for structural integrity. They simply hold the packed material in place and allow the gas and liquid to pass through. Since the PBR is sectioned into multiple stages in industrial applications, as shown in Figure 9.1, the support plates also have to prevent phase maldistribution, which is usually accomplished with proper plate leveling. Uneven packing and loading may cause support plate bending and flow maldistribution (Kolev, 2006; Maiti and Nigam, 2007). As with aerated plates in bubble columns, the uniform liquid distribution objective is important, difficult, and often not completely achieved.

The design direction often involves using the lightest and least obstructive support plates in order to achieve packed material and flow stability. Since the PBRs are mainly used by the chemical industry, fouling and erosion are important side effects of the reactants. Hence, the chemical industry requires ready access to the packed material and support plates that are easily removed and changed. This requirement would most likely be necessary for biological applications as well. The support plate is usually fastened to a supporting ring, which is welded to the vessel, due to possible pressure surges, which could potentially lift the support plate and the supported packing.

Support plates are classified according to the type of packing (random or structured) used in the reactor. Support plates used for random packing have inclined walls to support free solid particles and ensure that these stay static. Some examples include the SP 1 (Figure 9.3), 2, and 3 multibeam support plates. The SP 1 is used for packed bed columns with a diameter larger than 1200 mm while the SP 2 and 3 (Figure 9.4) are used for diameters of 100–300 mm and 300–1200 mm, respectively.

The column diameter and geometry of the support plates also impose a minimum packing size. The support plate height is dependent on material construction. For example, carbon and stainless steel call for a height of 265 mm while thermoplastic support plates have a height of 300 mm. Owing to material properties, larger columns (with larger amounts of supported packing) require further strengthening of the support plates.

The downside of these multibeam support plates is that they are plagued by fouling and uncontrolled biological growth, which can lead to improper flow distribution and inefficient reactor operation. Hence, biological processes might find it more advantageous to use a more open support plate such as the hexa-grid (SP-HG) or cross-flow-grid (SP-CF) support plate; however, these plates are usually flat and

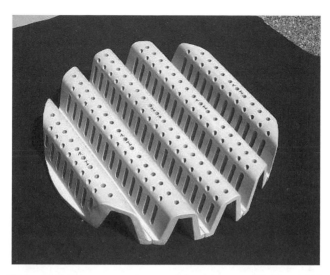

Figure 9.3   SP 1 multibeam support plate used for columns with $D_R > 1200$ mm (Raschig Jaeger Technologies, 2006).

Figure 9.4   SP 2 and 3 multibeam support plates used for $100 < D_R < 300$ mm and $300 < D_R < 1200$ mm, respectively (Raschig Jaeger Technologies, 2006).

are more applicable to structured packings. The SP-HG, shown Figure 9.5, is preferred for mass transfer processes while the SP-CF, as shown in Figure 9.6, ensures a more uniform gas distribution. Other possibilities include the flat bar plate (exclusively for structured packing), vapor distributing packing (used to reduce vapor velocity), or ceramic packing (used for corrosive processes) support plates.

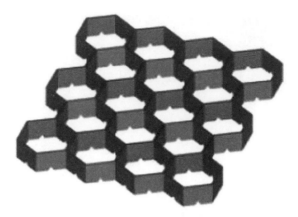

Figure 9.5 SP-HG support plate (Raschig Jaeger Technologies, 2006).

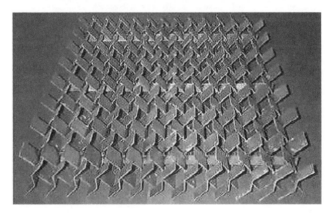

Figure 9.6 SP-CF support plate (Raschig Jaeger Technologies, 2006).

Hold-down plates are used to secure packing in case of flooding or pressure surges. In these situations, it is possible that the packing would wash away and enter the output stream. The standard hold-down plate (HP-1) is constructed from a frame with a metal screen. This hold-down plate is not meant to interfere with the liquid flow, but is also not designed to stabilize the packed material. Hence, more fragile packing, such as ceramics or carbon steel, cannot use the HP-1 plate without a breakage or grinding risk. As such, fragile packing should use the HP-2 hold-down plates, which has a grid-like structure that inhibits packing movement. Structured packing typically utilize hold-down plates (HP-P), which are bolted to the vessel and are designed to prevent movement within the bed or packing washout. An example of the hold-down plate is the grid-like Raschig grid (RG) or HP-1, shown in Figure 9.7. Plates are generally intended for larger vessels ($D_R > 3000$ mm), but can also be designed for and installed in smaller vessels ($D_R = 500 - 3000$ mm).

Figure 9.7   HP-1 hold-down plate (Raschig Jaeger Technologies, 2006).

Since packed bed columns are mainly designed for the chemical industry, a lot of attention is paid to liquid-phase distribution, because the liquid distributor design is crucial while gas-phase distribution gets less attention. If the liquid distribution operates or performs poorly, the packing is not going to be wetted effectively, which is important for soluble gases. Nonwetted packing leads to reactor underperformance (Maiti and Nigam, 2007). Although biological systems would also need significant attention, these processes could mostly do with simpler options.

There are two basic options for liquid distribution: single stream and spray distributors. Single stream distributors basically feed the liquid phase through a perforated pipe to channels, which, in turn, distribute the liquid over the cross-sectional area of the reactor, which is shown in Figure 9.8. Some examples of these are the shower-type (perforated) distributor, trough distributor, distributor with weirs, bottom-hole distributor, splash-plate distributor, channel-type distributor, distributor with gas risers, or liquid pipe distributors. Spray distributors are used for processes requiring large amounts of liquid surface, gas cooling, or homogeneous liquid distribution. Since the droplets are smaller, the spray distributors are also better at wetting particles, which produces more efficient packing operation.

Liquid distributors are often used to correct liquid flow to a more uniform pattern and/or to add more liquid to the reactor. Liquid distributors are also often used to redistribute the liquid phase as well, but a simpler option, shown in Figure 9.9, would be a perforated plate with directional facing. Liquid collectors are used to channel the liquid into the liquid outlet in order to prevent converted/used liquid to stay in the column bottom. The hardest portion of designing a collector is to not interfere with gas distribution, and the available designs are quite wide and diverse. TBRs, on the other hand, tend to use liquid collectors, which are only used for collecting and extracting liquid.

Figure 9.8 Example of liquid distributor (Raschig Jaeger Technologies, 2006).

Figure 9.9 Example of a liquid distributor (Raschig Jaeger Technologies, 2006).

Gas distributors are also an important component of packed bed columns and influence phase distribution. The design issue is to ensure a uniform gas flow distribution through the reactor or a necessary gas flow ratio relative to the liquid phase. The liquid–gas ratio is only important if the process requires a chemical reaction of the liquid phase induced by the gas phase. The packing, on the other hand, is responsible for the pressure drop profile and, hence, the downstream gas-phase distribution.

The simplest gas distributor is a perforated pipe. Since the pipe creates a significant amount of flow irregularity in the distributor region, the packed material is usually placed a distance away from the gas inlet to ensure the gas flow becomes more uniformly distributed upon entering the packed region. The separation distance is dependent on the column diameter such that a column of 1 m requires a distance of about 0.4 m. Columns with $D_R = 1 - 2$ m require a separation distance of 0.7 m while larger columns should have a distance of about 1 m (Kolev, 2006). An actual rule, which is often linear, can be developed to quantify a column diameter to separation distance ratio, but this rule would be dependent on the level of turbulence. This turbulence can be represented by a ratio of inlet to column gas velocity.

Other gas distributor options include more complicated systems such as a guided vane gas distributor, which attempts to create a more uniform gas distribution through higher turbulence, or a liquid collector–gas distributor combination; this design is more appropriate for high pressure reactors, especially if multiple packed beds are necessary. For example, the process may require the product to be extracted and more liquid/gas phase to be input into the reactor. Hence, such a device would be necessary. TBRs share this basic design with PBRs and make more extensive use of combination devices. Examples of combination distributors for the TBRs include perforated plate (Figure 9.10), chimney tray, bubble cap tray, and vapor-assist lift distributors.

A simple choice is the perforated plate while the chimney is useful if vapors are present. If the perforated plate is used and flow uniformity is desired, the number of drip points should be as large and as close together as possible. The problem with most perforated devices is that coking or accumulation of other materials may clog the holes and lead to the initial liquid flow maldistribution. Then the perforated plate requires a certain minimal liquid loading. If the minimal liquid amount is not present, some perforations will stay dry and once again lead to initial liquid maldistribution. The perforated plate cannot be designed for some minimal and universally used value because the pressure drop at nominal operation has to be minimized. Hence, the plate is usually designed for a normal state of operation while off-design conditions may cause operational problems. A quick summary of available devices for TBRs is presented in Table 9.1 (Maiti and Nigam, 2007).

The negative aspects and age of perforated plate distributors have led to improvements over the years. A historical performance summary is presented in Figure 9.11. The chimney-type distributors attempt to improve gas–liquid contacting and prevent clogging issues. The liquid is injected as jets while the air

TABLE 9.1   Distributor Comparison

Type	Spacing Density	Level Sensitivity	Liquid Rangeability	Vapor–Liquid Flexibility	Liquid–Vapor Mixing
Perforated plate	Best	Worst	Worst	Worst	Worst
Chimney	Average	Poor	Poor	Poor	Poor
Multiport chimney	Average	Average	Average	Average	Poor
Bubble cap	Worst	Average	Good	Good	Best
Gas-lift assisted	Best	Best	Best	Best	Best

Adapted from Maiti and Nigam (2007).

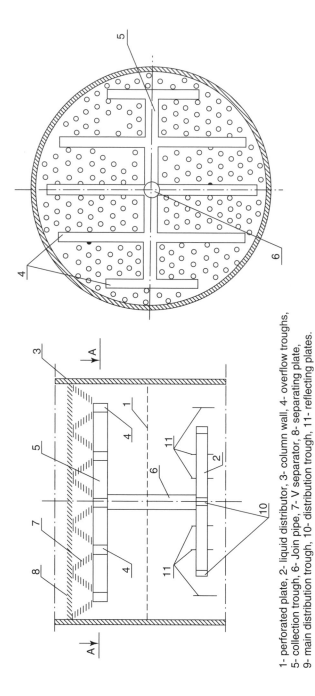

Figure 9.10 Example of a combination of liquid collector and gas distributor system (Kolev, 2006).

1- perforated plate, 2- liquid distributor, 3- column wall, 4- overflow troughs,
5- collection trough, 6- Join pipe, 7- V separator, 8- separating plate,
9- main distribution trough, 10- distribution trough, 11- reflecting plates.

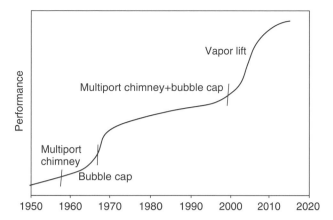

Figure 9.11   Historical distributor performance changes (Maiti and Nigam, 2007).

is used to break up the liquid into smaller droplets. Unfortunately, chimney-type distributors are highly sensitive to the liquid level and have a limited number of liquid entry points. Bubble caps are distributors that do not use the liquid hydraulic head to perform liquid distribution duties. Bubble caps, in turn, use the gas flow to distribute the liquid. Therefore, the bubble cap trays allow much wider liquid flow rates, but the design requires a larger diameter, which leads to even fewer drip points than the chimney-type distributors. Vapor-lift tubes, as the name implies, use the gas phase to push the liquid from an established level through a U-tube and distribute droplets onto the packing. The biggest advantages of the vapor-lift tubes are that they are smaller, simpler, and cheaper to construct; provide more wall coverage; and increase wetting efficiency. Hence, vapor-lift tubes are expected to become dominant for chemical processes, but a perforated plate might be good enough for biological applications. Graphical examples of combination devices for TBRs are shown in Figure 9.12 (Maiti and Nigam, 2007).

9.3   FLOW REGIME

PBRs and TBRs, respectively, share some basic flow characteristics, but major differences exist, which lead to differing reactor performance and application. At very low superficial gas velocity, the superficial liquid velocity has a significant impact on the relative gas velocity for both PBR and TBR operations. Furthermore, liquid holdup is solely a function of the superficial liquid velocity (Alix and Raynal, 2008). Under PBR operation, increasing the superficial liquid velocity causes an increase in the relative gas velocity. This effect, in turn, increases the pressure drop of wetted to dry packing. TBR operation, on the other hand, leads to the opposite effect.

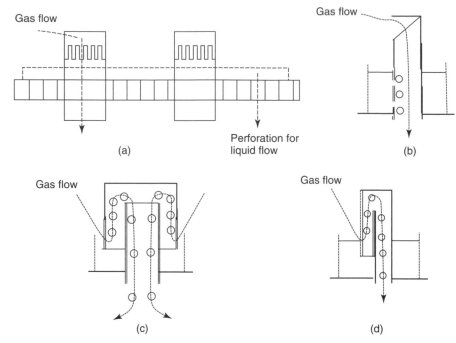

Figure 9.12 Examples of (a) perforated plate, (b) chimney, (c) bubble cap, and (d) vapor-lift tube used in TBRs (Maiti and Nigam, 2007).

By increasing the superficial gas velocity in PBRs, a critical superficial gas velocity is reached at which the friction between the gas and liquid phases leads to an increase in liquid holdup. This critical point also represents the inflection shown in Figure 9.13 and is caused by the emerging and powerful influence of the superficial gas velocity on the liquid holdup. This point is referred to as the loading point.

As the superficial gas velocity is increased, the liquid holdup increases as well until another critical point, the flooding point, is reached. At the flooding point, the slope of the pressure drop versus superficial gas velocity line becomes exponential. The drag force of the gas phase becomes the dominating influence on the pressure drop (Kolev, 2006; Stichlmair et al., 1989). The effect of the loading and flooding points and the transition on the pressure drop and slope can be seen in Figure 9.14. The identification of these boundaries allow for optimal operation, which is observed between the loading and flooding points for most industrial processes, such that mass transfer is maximized while operating costs are minimized.

The operating range between the loading and flooding points is referred to as the loaded regime. The loaded regime is important for mass transfer purposes. One way to judge mass transfer efficiency in fixed bed reactors is with the height equivalent to a theoretical plate (HETP). A theoretical plate is an abstract stage in which two phases are capable of establishing an equilibrium. Thus, having an actual height that

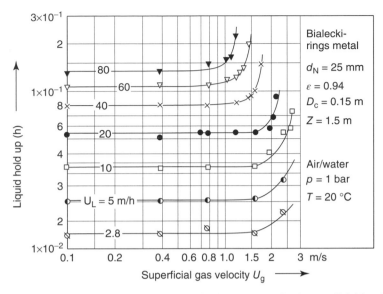

Figure 9.13 Representation of the loading point in packed bed columns (Stichlmair et al., 1989).

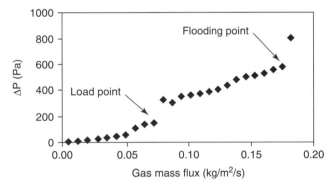

Figure 9.14 Example illustrating loading and flooding points (Breijer et al., 2008).

is lower at the same superficial gas velocity implies a more efficient mass transfer process. In the same line of reasoning, having the same equivalent height with a higher superficial gas velocity implies that more mass has been transferred, and the reaction occurs under more optimal circumstances.

Figure 9.15 represents the effect of increasing the liquid rate in direct proportion with the vapor (gas) rate. The important markers are points C, E, G, and F. Point A illustrates a minimal liquid loading or a minimal liquid and vapor rate at which the column converts the feed. If the vapor rate is lower than the critical value at point A, the HETP increases such that a practical column is not operable. Points C and F

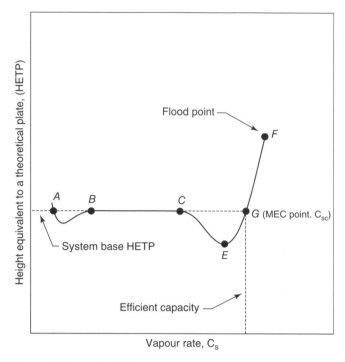

Figure 9.15 Packed bed reactor operating regimes (Kolev, 2006).

represent the loading and flooding points, respectively. The region between points B and C defines the constant separation efficiency region, where the interaction between the liquid and vapor is weak. Point E represents the minimal HETP, which would yield the operating condition for the shortest and smallest reactor; however, a more efficient point can be obtained at point G, which represents the maximal efficiency capacity (MCE). The MCE provides the most stable operations while also maximizing the mass transfer potential. Increasing the vapor rate beyond point G yields an inefficient result such that the column requires larger and larger scale in order to achieve the same result (Kolev, 2006)—not something designers and operators look for.

Another way to define the flow in PBRs is using the Reynolds number, which is defined by the packing diameter and superficial (gas) velocity. Creeping flow is observed at Reynolds numbers <1. This region is defined by a linear pressure drop with increasing interstitial velocity. A steady laminar inertial flow is observed within a Reynolds range of 10–150 where the pressure drop develops a nonlinear relationship with interstitial velocity. An unsteady laminar inertial flow develops at Reynolds numbers of 150–300; this region is defined by laminar wake oscillations in the pores. Furthermore, vortices start developing at a Reynolds number of approximately 250. At Reynolds numbers above 300, the flow becomes unsteady and resembles turbulent flow. It should be noted that experimental scale vessels experience Reynolds numbers in the creeping and steady laminar inertial flow while industrial scales operate at higher Reynolds numbers (Schuurman, 2008).

In trickle bed operation, the frictional force acts in the flow direction, decreasing the liquid holdup. For TBR operation, the flooding point does not exist in the same sense as in a PBR. Instead, the same critical ("flooding") point in the trickle flow regime is defined by a decreasing slope relative to dry packing (Kolev, 2006; Stichlmair et al., 1989). Hence, the TBR is not affected by flooding or loading conditions and can be applied to wider gas- and liquid-phase flow ranges (Breijer et al., 2008). The trickle flow reactor is only limited by the pressure drop, which causes economic constraints for TBR application. The downside is that the PBR is necessary for equilibrium reactions, which are not easily carried out in a TBR due to the cocurrent nature of the phase flow directions (Al-Dahhan et al., 1997; Kolev, 2006).

The general PBR flow structure is highly dependent on the initial liquid phase flow distribution, liquid velocity, particle shape, particle size, and packing method. The fact that the packing method is influential can be somewhat troubling with random packing because results are harder to replicate once the packing is replaced. On a smaller scale, the flow structure is affected by start-up procedure, wettability, flow modulation, and particle coordination number (number of touching neighbors per particle). The level of the nonuniformity is largely controlled by reactor internals. A properly designed and operated reactor is expected to have uniform flow distribution or at least become stable relatively quickly (Doan et al., 2008; Maiti and Nigam, 2007).

Nonuniform flow has several causes. The first cause is that the initial liquid distribution is not uniform such that the liquid does not enter the packing volume uniformly (shown in Figure 9.16a). If the liquid is not well-distributed and adjustments are not effective or easily implemented, a layer of inert packing particles on top of the reactive packing may lead to some flow improvement. The more common cause is that the packing has not been packed well, damaged, or that its shape makes the flow unstable (shown in Figure 9.16b). The shape may lead to liquid maldistribution in the case that the reactor-to-particle diameter is too small (i.e., the particles are too large relative to the reactor diameter). For example, if the ratio is <20, the liquid starts to show maldistribution near the wall region at higher superficial liquid velocities. Hence, the reactor-to-particle diameter should be at least 20 to ensure proper liquid distribution for nominal liquid velocities (Metaxas and Papayannakos, 2008). If axial dispersion is problematic, a ratio >50 minimizes the influence of axial variations (Schuurman, 2008). It should be noted that liquid flow at the wall requires a bed height-to-diameter ratio of at least 4.0 (Doan et al., 2008). It should be noted that a small reactor to particle diameter ratio, or a small bed height-to-diameter ratio, typically results in packing not properly being wetted (shown in Figure 9.16c), and the mistake is often not realized until the packing is replaced, which may not occur for at least 2–3 years (Maiti and Nigam, 2007).

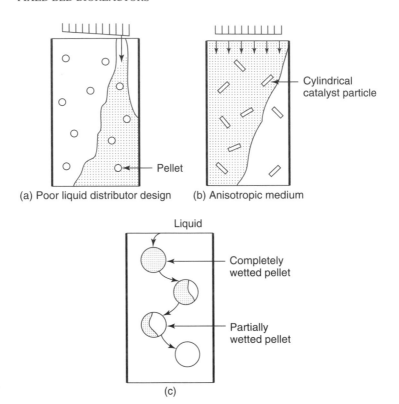

Figure 9.16 (a) Initial liquid maldistribution and (b) restrictive packing leading to (c) different levels of wetted pellets (Maiti and Nigam, 2007).

Packing manufacturers develop and test their packing extensively, but oftentimes do not have the expertise or infrastructure to test packing flow behavior. Engineering firms have the expertise, but are missing the infrastructure and incentives to test packing. Hence, the customer and end-user are expected to experience some trouble that usually requires on-site adjustment (Maiti and Nigam, 2007).

The TBR hydrodynamics are influenced by the operating conditions, reactor design, reactor internals, distributor design, and phase properties (Nacef et al., 2007). There are four basic flow regimes in a TBR. At low superficial liquid and gas velocities, the liquid trickles onto the packing, forming streams and films, while the gas phase flows through the residual voids; this is called the trickle regime. The TBR performance in the trickle regime is dependent on the pressure gradient and liquid saturation. A significant portion of the packing is unwetted in the trickle regime, which is usually minimized by increasing the superficial liquid velocity. The extent of unwetted packing may be used to identify flow nonuniformity and its causes (Liao et al., 2008) since the packing should be completely wetted during the transition from the trickle to pulsing flow regime. The liquid holdup can be

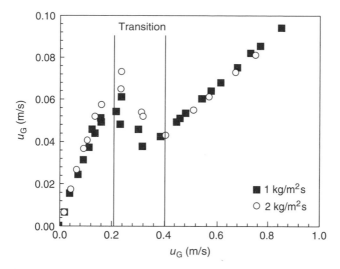

Figure 9.17   Flow regime detection in trickle bed reactors (Nacef et al., 2007).

increased by increasing the liquid flow rate, while increasing the gas flow rate leads to a decrease in the holdup (Burghardt et al., 1995).

At higher superficial liquid velocity, the pulsing regime is formed because the liquid blocks the flow paths and forces alternating liquid- and gas-rich regions in the reactor volume. This is very similar to the slug flow regime in bubble columns. An interesting behavior in the pulsing flow regime is that the liquid holdup is mostly independent of the liquid flow rate although it is still negatively affected by an increase in the gas flow rate (Burghardt et al., 1995). The identification of the pulsing regime is achieved by using the transition between the trickling and pulsing regimes. This transition is defined by a temporary decrease in drift flux (velocity) as the superficial gas velocity is increased, as shown in Figure 9.17. Correlations have been developed that attempt to model the velocity at which the transition may occur; however, these correlations are limited to only a few important parameters (phase velocities, liquid viscosity, particle diameter, and bed height) (Nacef et al., 2007) such that more complicated processes, such as those involving non-Newtonian liquids, are almost impossible to predict.

At high superficial gas and low superficial liquid velocities, a spray regime occurs that is defined by the liquid phase being turned into droplets by the continuous gas phase. At low superficial gas and higher superficial liquid velocities, the dispersed bubble (gas) regime is observed and defined by a continuous liquid phase, which entrains the gas phase as bubbles (Attou et al., 1999). The dispersed bubble is most often used in the TBR, but pulsing regime operation is also common (Burghardt et al., 1995). Biological applications would benefit the most from the dispersed bubble regime while the pulsing regime may cause damage to the microorganisms as well as potential exposure to the gas–liquid interface. A generalized regime map has not been developed because the regime transitions are dependent on packing material properties such as wettability, size, and shape (Maldonado et al., 2008).

9.4 LIQUID PROPERTIES

Liquid properties influence the behavior of the bubble interface and, consequently, have a strong effect on both the liquid-phase mass transfer coefficient and the interfacial area. Research using organic liquids, which would be very useful for bioreactor and gas–liquid mass transfer optimization, is almost nonexistent for fixed bed reactors.

It should be noted that PBRs operate differently than other gas–liquid reactors covered so far. In the other reactor designs, the gas phase is dispersed in the liquid phase such that bubbles are formed from which the gas phase is transferred to the liquid. The surface area through which the transfer occurs is the interfacial area of the bubbles. In PBRs, general operation yields droplets or liquid films immersed in the gas phase, which causes the main transfer surface to be the droplet–gas or film–gas interface. In order to increase the mass transfer interfacial area, smaller droplets or wavier films have to be produced. This process is often turbulent and may result in liquid breakup and coalescence, which may be destructive to some microorganisms (specifically those that are shear sensitive).

A unique property of fixed bed reactors, in general, is that the wetted particles also become part of the mass transfer interface. This difference is not too important for chemical systems due to the commonly used packing designs, but biological systems may have additional problems. For example, the packing serves as a supporting mechanism for the microorganisms. Hence, the microorganisms feed and breathe through the liquid layer on top of the colonies. If the liquid droplets do not touch or interact in some way with the liquid–microbial interface, mass transfer may not occur. Another design objective has to be added to biological systems, which stipulates that the packing is refreshed with liquid and that the liquid's interaction with the microbial interface is optimized. The liquid properties play a crucial role in this process. The viscosity of the liquid determines the stickiness or the wettability of the packing. The degree to which this property is important is highly variable with the specific microbial needs, and research determining this requirement in fixed bed reactors is nonexistent; however, if trickle- or flooded-bed reactors are used, gas bubbles are the mass transfer mechanisms, and the same bubble behavior and interactions are to be expected as with (three-phase) bubble columns and airlift reactors (Nacef et al., 2007).

Unfortunately for biological applications, the influence of liquid properties is often ignored in fixed bed reactor research. Approximately half the data for fixed bed reactors have been obtained using air–water systems with glass bead packing (Nacef et al., 2007). Hence, a significant problem is apparent. The available data are biased toward a situation that does not occur in industrial settings or biological applications, even if they are for experimental purposes. To make matters worse, experimental phase flow ranges oftentimes differ significantly from industrial practice (Burghardt et al., 1995) such that those results by themselves would have little meaning for the real-world application. Furthermore, these test systems experience constant properties, such as liquid viscosity or particle density. This state is often not observed in biological systems since microorganisms change the liquid properties and consume or colonize the solid phase. Hence, the effective liquid viscosity and particle density may be unsteady.

9.5   PACKING MATERIAL

Packing is the solid and fixed phase in the reactor volume. It is often used to judge reactor performance and utilization. The wettability or wetting efficiency is the representative parameter of choice and is defined as the ratio of wetted-to-external particle surface area (Metaxas and Papayannakos, 2008). The external area is not equal to the total surface area because the external area excludes the contact areas (Burghardt et al., 1995). The wetting efficiency is a function of phase flow rates, pressure, liquid properties, and packing diameter. In most cases, the wetting efficiency varies between 0.6 and 1.0. The most common approach to measuring wettability is by using tracers and visual inspection. Reaction methods can be used, but are difficult to implement from a theoretical point of view (Nigam and Larachi, 2005). The type of packing generally does not determine wettability directly, but rather influences the liquid distribution, which may lead to non-wetting if flow maldistribution occurs.

As summarized by Nigam and Larachi (2005), wettability may be modeled using saturated pores and solid surfaces. At low wettability, the liquid volume per unit-wetted surface area is large (represented by Figure 9.18a). As the wettability increases and the liquid volume per unit-wetted surface area decreases, the contact angle and the wetting efficiency also increases (Figure 9.18b); however, this is followed by a small contact angle and a decrease in efficiency (transition between Figure 9.18b and c) and then a resumption of the nominal increases (Figure 9.18c). A further increase in wettability causes a steep contact angle and a quick increase in wetting efficiency (transition between Figure 9.18c and d), which is followed again by a contact angle and efficiency decrease (Figure 9.18d). As the wetted pores reach each other, the contact angle becomes stationary and further liquid flattening is not possible such that wetting efficiency is abnormally high (transition between Figure 9.18d and e). If there are any partially solid surfaces left, the liquid film ruptures and a quick decrease in wetting efficiency is observed (Figure 9.18f). The result of higher wetting efficiency is a thinner liquid film, which would also represent a smaller resistive mass transfer force (Liao et al., 2008; Nigam and Larachi, 2005). Hence, porosity has a significant impact on wetting efficiency and hysteresis behavior.

The start-up procedure for fixed bed reactors often involves prewetting the packing in order to limit operational variations. Interestingly, reactors may demonstrate better or worse performance due to a different start-up procedure. In other words, the pressure drop or liquid holdup are not good indicators of flow uniformity, but rather show significant dependence on the start-up and prewetting procedure. The

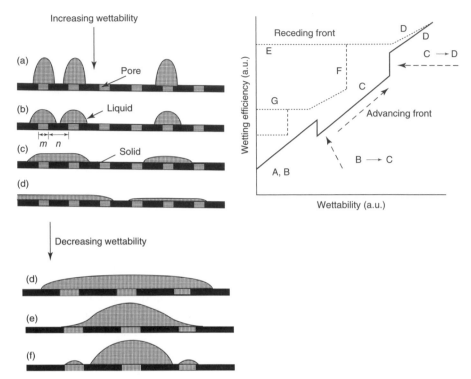

Figure 9.18   Wetting efficiency dependence on the contact angle movement (Nigam and Larachi, 2005).

possible start-up procedures include a dry, Levec, Kan liquid, Super, and Kan gas mode. The dry mode starts the process with the packing dry. The Levec mode prewets the packing for about 20 min by flooding the reactor volume and then allows the liquid to drain out after which the liquid is reintroduced and the process is allowed to start normal operation. The Kan liquid mode prewets the packing by cycling the reactor between the pulsing regime and the operating set point. The Super mode simply floods the reactor volume and then reduces the phase flow rates to the operational set points. The Kan gas mode prewets the packing by operating in the pulsing regime. The pulsing regime is achieved by increasing the gas flow to the critical pulsing point after which the phase flow rates are adjusted to the operational set point (van der Westhuizen et al., 2007). Prewetting has not been studied with biological media so that its influence is currently unknown for this application.

9.5.1 Random Packing

There are two basic types of packing materials: random and structured packings. Random packings are commonly formed from different shapes such as rings, shown in Figure 9.19, or saddles, shown in Figure 9.20. They are constructed out of ceramic, metal, plastics, or coke with the most popular metal application being

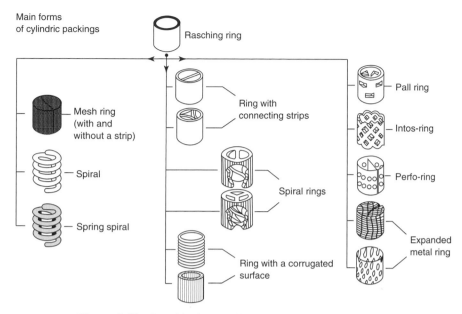

Figure 9.19   Raschig ring packing examples (Reichelt, 1974).

stainless steel. A chemical treatment may be applied to the surface in order to increase the wettability and process efficiency (Kolev, 2006; Strigle, 1994).

The simplest random packings are spherical, but are not used very often in industry. The most commonly used packing are the Raschig rings and its descendants. The historical development of random packing is summarized in Figure 9.21. Each alteration stems from the need to improve packing performance for the particular process. The most common goal is to increase the surface area available for mass transfer and reaction. The great advantage of random packing is the ease with which it can be produced and loaded into the column. Typically, the packing is simply loaded from the top onto the support plate and secured by the hold-down plates. On the other hand, the random and unstructured distribution of the packing also leads to poor phase distribution, possible flow channeling, and higher pressure drop (Kolev, 2006; Strigle, 1994). The pressure drop, on the other hand, may also be an advantage as it leads to a higher level of turbulence and, hence, higher gas–liquid mass transfer efficiency relative to structured packing (Schultes, 2003).

The first and most logical adaptation has been to add dividers, internal spirals, and corrugated surfaces to the Raschig ring. Although these rings effectively increase the area, the pressure losses and gas clogging/slugging can be significant, which may limit the operating range. The second adaptation has been to make the rings out of mesh and spirals. Finally, the rings can be perforated. Interestingly, the pall ring has been determined to fulfill most mass transfer requirements and a better random packing has not been found in the last 30 years; however, other options, such as the Hiflow ring, Ralu-Flow, IMTP, Nutter ring, and Raschig Super Ring,

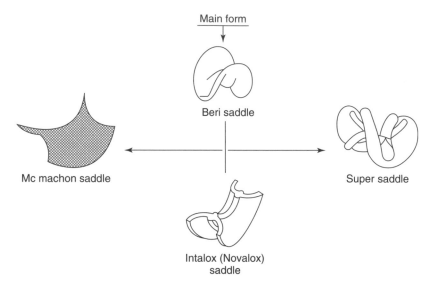

Figure 9.20 Saddle-type packing examples (Reichelt, 1974).

are available and provide satisfactory mass transfer results. Although statistics and performance measures are collected, they are highly dependent on the reactor diameter such that results presented in the literature may vary greatly and be a challenge to reproduce (Kolev, 2006).

The mass transfer requirement stipulates that the patella (connecting strip) cannot be larger than 5 mm in width. If the patella is larger than that, a droplet will form on the surface, which will effectively reduce the wetted surface available for the reaction. If the patella is less than 5 mm in width, the liquid is capable of moving over the surface without accumulating. The result is that the operating range is widened. The downside is that as the width is decreased, the pressure drop increases. This increase may limit the economics of the process.

9.5.2 Structured Packing

Structured packing is constructed to provide optimal phase channeling and uniform phase distribution. Structured packing material is very similar to the random packing and includes metals, ceramics, plastics, and other materials like wood. Structured packing is usually subdivided into smooth-walled packing, packing with turbulizers, expanded metal packing, corrugated metal sheet packing, and packing for very low superficial liquid velocities (Kolev, 2006; Strigle, 1994).

Smooth-walled packing is constructed using vertical walls, which attempts to limit the pressure losses for a given operating condition. As such, they provide the lowest pressure drop per mass transfer unit for a given volumetric mass transfer coefficient. The first generation of structured packing was fixed rings, such as Raschig rings. Although the construction was very simple, a significant problem

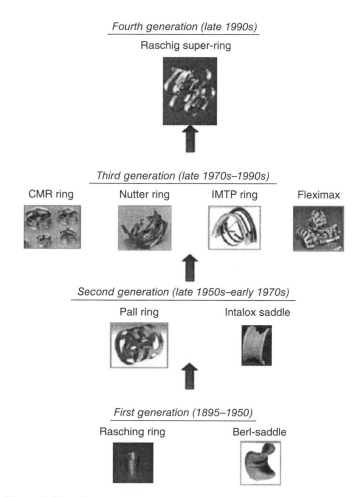

Figure 9.21   Historical random packing development (Schultes, 2003).

became apparent early on. The interior ring channel (labeled 1 in Figure 9.22) is much larger than the exterior ring channel (labeled 2 in Figure 9.22). This discrepancy leads to flow channeling and a dry packing surface. Hence, the natural evolution has been to construct structured packing using similar members and forming them into symmetrical arrangements. Some popular formations are shown in Figure 9.23. Honeycomb has quickly become very popular, largely because they are able to provide a lower pressure drop for a given volumetric mass transfer coefficient. For example, honeycomb packing (Honeycomb No. 1) can provide a pressure drop that was 8.3 times lower than the Raschig rings (Kolev, 2006).

Packing with turbulizers, referred to as Turbo-pack, is made by thermo-pressing or stamping plates such that the packing surface is ribbed, which increases turbulence near the wall. This increased turbulence leads to better heat and mass transfer

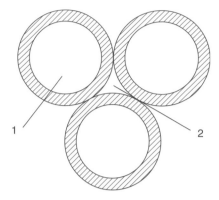

Figure 9.22   Cross section of structured (Raschig) ring packing (Kolev, 2006).

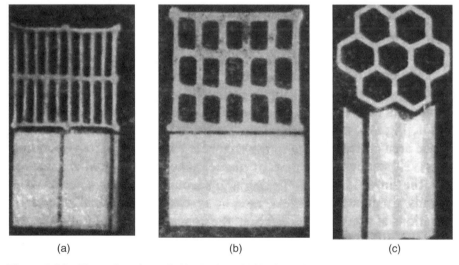

        (a)   (b)   (c)

Figure 9.23   Examples of (a) slit block, (b) grid block, and (c) honeycomb block packing structures (Kolev, 2006).

performance. Turbulizers are usually made in the horizontal direction in order to achieve axial flow uniformity. The result of using turbulizers is that the operable superficial gas and liquid velocities are increased. For example, P15-235 (type of structured packing with turbulizers) allows for superficial gas and liquid velocities up to 3.5 and 0.11 m/s (in PBRs), respectively (Kolev, 2006). The interesting behavior of these turbulizers is that at low superficial liquid velocity, the pressure drop of the wetted packing is half the pressure drop experienced by the dry packing. At high superficial liquid velocity, the pressure drop equalizes for wetted and dry packing.

Expanded metal packing is constructed by stamping or otherwise forming a channel from metal plates. The greatest advantage of this type of packing is that the channels are exclusively in the vertical direction such that the pressure drop is lower relative to the other packing types. Another derived advantage is that this also allows for much higher superficial gas velocities, potentially reaching 2.5–3.5 m/s. In order to achieve even higher gas velocities, the gas phase is input and extracted horizontally from the vessel, which allows operational superficial gas velocities up to 6 m/s. The most important disadvantage is that the vertical channel leads to the liquid preferentially wetting the leading edges. Hence, this packing requires more rigorous collection and redistribution than the other options. The exact behavior of the expanded metal packing is highly dependent on its construction, which can vary significantly. Major design considerations are the pitch, step height, and the existence and extent of perforations. The pitch, for example, can be designed so that the packing may be almost completely wet at relatively low superficial liquid velocities or able to handle an extreme (relative to other packing) amount of liquid (Kolev, 2006; Strigle, 1994).

The structured packing of corrugated sheets attempts to fix a major disadvantage of the smooth-walled packing: the possibility of free-falling liquid through the open cross-sectional area. The corrugated sheets are designed to intercept liquid (usually within one half of a wave) and enhance axial mixing (Kolev, 2006). The downside is that the pressure drop is larger relative to other structured packing options, but lower than random packing.

Certain processes, such as those relying on equilibrium absorption with low initial concentration of highly soluble gases, require very low superficial liquid velocities in the countercurrent configuration such that existent packing does not perform well. The operator has the option of using very easily wettable materials or using a highly specific packing form. This circumstance is rarely encountered in biological applications.

9.6 BIOLOGICAL CONSIDERATIONS

Biocatalysis and microorganisms have become great hopes and are seen as possible solutions to a multitude of problems. As such, research efforts are starting to turn toward suspending microorganisms on packing material to perform similar roles as reactive packing or perhaps completely new functions (Llamas et al., 2008). Biological microorganisms are typically attached to the packing surface since the microorganisms perform better when either the cell adhesion occurs or the cells are highly shear sensitive and need to be protected. Hence, fixed bed reactor performance is highly dependent on the liquid distribution uniformity. It should be noted that the liquid phase usually serves as the food source for the microorganisms. So, if the packing is not wetted, microorganisms will not colonize that section of the reactor. This, in turn, can severely limit reactor performance (Doan et al., 2008).

In addition, reactor internals may be of significant importance. Even though the perforated plate is a simple and effective device, it tends to provide minimal wall wetting and tends to channel liquid through the central region. This effect can cause interesting performance variations at different liquid flow rates and may make comparisons between different research work a challenge. If the outer regions need to have liquid exposure, a better liquid distributor should be selected. Similar channeling can also be experienced by uneven cellular growth rates within the reactor volume.

Biomass and microorganisms tend to have a positive effect on fixed bed performance. Packing is usually judged on the basis of wettability, which biomass seems to increase. Doan et al. (2008) compared the wettability of plastic spheres with the same plastic spheres in the presence of microorganisms. The result has been that the liquid holdup increased by about 20% in the presence of microorganisms.

Liquid flow distribution and gas solubility issues can be enhanced with periodic liquid flushing of the reactor volume. The flushing tends to renew interfaces and reduce the gas-phase transport resistance (Nigam and Larachi, 2005). This practice may also have additional benefits in controlling microorganism growth. If the microorganisms do very well, the colonies might grow too thick and kill the lower cell layers. In order to prevent buildup in the reactor volume, a flushing cycle could be performed to detach a specific region or amount of microbial growth.

9.7 CORRELATIONS

Correlations for fixed bed reactors, which are shown in Tables 9.2 and 9.3, are currently available for specific operating regimes and packing types. Some aspects can be generalized; however, correlations are developed for either random or structured packing even though the same principal theories are used to explain gas–liquid mass transfer and behavior (Larachi et al., 2008). Current fixed bed correlations are very design specific, and any broad correlations produce highly variable and sometimes impractical results. This has been at least partially the result of many experiments being conducted at much lower pressure than those observed during industrial application (Attou et al., 1999). Kolev (2006) collected gas–liquid mass transfer correlations for the packed bubble column for his book on packed bed columns; however, the number of correlations was small because research was limited to gas–liquid mass transfer in PBRs. The most recent correlation (other than Kolev's) is from Billet (1989). Most of the work seems to have been done in the 1950s and 1960s. Since then, correlations have been seldom formed.

Perhaps the reason for the lack of correlations is twofold. First, the more important information for chemical engineers is the pressure and temperature predictions. With these in hand, chemical engineers are able to predict the other factors of importance. Biological processes are not expected to behave like this, and more work would need to be done for this purpose.

A second reason is the complexity of the interactions, which has lead to a data mining exercise by Professor Larachi's research group at Laval University. They have been at the forefront of providing correlations capable of predicting fixed

TABLE 9.2 Gas–liquid Mass Transfer Correlations for Fixed Bed Reactors

Researcher(s)	Reactor	Packing	Correlation
Billet (1989)		RP	$k_L a = C_L \dfrac{a^{2/3}}{d_h^{0.50}} D_L^{0.50} \left(\dfrac{g \rho_L}{\mu_L} \right)^{1/6} U_L^{1/3} \dfrac{a_e}{a}$ where C_L is an experimental, packing dependent constant
Fujita and Hayakawa (1956)	PBR	Rings 5–35 mm Saddles 13–40 mm	$Sh_L = 0.025 \left(\dfrac{h_P}{\delta} \right)^{-0.19} Re_L^{0.67} Sc_L^{0.50}$
Kasatkin and Ziparis (1952)	PBR	Rings 8–20 mm	$Sh_L = 0.0021 \, Re_L^{0.75} Sc_L^{0.50}$
Hikita and Ono (1959)	PBR	Wetted single element	$Sh_L = 0.27 \left(\dfrac{d_P}{\delta} \right)^{-0.50} Re_L^{0.545} Sc_L^{0.50}$
Krevelen and Hoftijzer (1948)	PBR	Rings, coke, and others	$Sh_L = 0.00595 \, Re_L^{0.67} Sc_L^{0.33}$

(continued)

TABLE 9.2 *(Continued)*

Researcher(s)	Reactor	Packing	Correlation
Koch et al. (1949)	PBR	RP	$k_L a = 0.25 I_m{}^{0.96}$ $k_L a = 0.0085 U_L$
Kolev (1976)		RP	$Sh_L = 0.030\, Re_{Le}{}^{0.5}\, Ga_L{}^{0.28}\, Sc_L{}^{0.50}$ $Re_{Le} = \left(\dfrac{4 U_L}{a_e \nu_L} \right)$ Sh_L and Ga_L determined by d_P
Kolev and Daraktschiev (1976)		Holpack	$Sh_L = 0.00113\, Re_L{}^{0.635}\, Sc_L{}^{0.50}\, Ga_{Lh}{}^{0.366}\, Ga_L{}^{4.0}\, (s/t_P)^{0.1}$
Kolev and Semkov (1983)		RP	$Sh_L = 0.0115\, Re_L{}^{0.33}\, Ga_L{}^{0.42}\, Sc_L{}^{0.5}\, (a d_P)^{-0.37}$
Kolev and Nakov (1994)		FP with turbulizers	$Sh_L = 0.0077\, Sc_L{}^{0.5}\, Re_L{}^{0.70}\, (a h_P)^{-0.29} (a s)^{-0.19}$

Larachi et al. (2008)	Correlation Excel files available at http://www.gch.ulaval.ca/bgrandjean or http://www.gch.ulaval.ca/flarachi		
Mangers and Porter (1980)	RP		$$\frac{k_L a}{D_L} = 0.0039 \left(\frac{U_L \rho_L}{\mu_L}\right)^{\alpha} \left(\frac{\mu_L}{\rho_L D_L}\right)^{0.50} \left(\frac{\rho_L^2 g d_P^3}{\mu_L^2}\right)^{0.70} \left(\frac{\rho_L \sigma^3}{\mu_L^4 g}\right)^{0.33} \left(\frac{1}{MWR}\right)^{1.67}$$ $\alpha = 0.484 MWR^{0.108}$ $$MWR = 1.12 \left[(1-\cos\theta_\theta)^{0.60} \left(\frac{\rho_L \sigma^3}{\mu_L^4 g}\right)^{0.33}\right]$$ where θ_θ is the contact angle
Onda et al. (1961)		RP	$Sh_L = 0.01\, Re_L^{0.50}\, Sc_L^{0.50}$
Onda et al. (1958)	PBR	Rings 6–10 mm	$Sh_L = 0.00625\, Re_L^{0.50}\, Sc_L^{0.50}$
Onda et al. (1959)	PBR	Rings 6–10 mm	$Sh_L = 0.0107\, Re_L^{0.90}\, Sc_L^{0.50}$
Ramm and Chagina (1965)	PBR	FP	$k_L a = 11.6 U_L^{0.768} h_P^{-0.185}$
		Raschig rings 25 and 50 mm	$Sh_L = 0.00216\, Re_L^{0.77}\, Sc_L^{0.50}$
		Pall rings 50 mm	$Sh_L = 0.0036\, Re_L^{0.77}\, Sc_L^{0.50}$

(continued)

TABLE 9.2 *(Continued)*

Researcher(s)	Reactor	Packing	Correlation
Raschig LTD		FP	$Sh_L = 0.01019\,Re_L^{1.1-0.4\epsilon}\,Sc_L^{0.50}\,Ga_L^{-0.01}\,(ah_P)^{-0.35\epsilon}\,\epsilon^{2.9}$ $Sh_L = 0.0026\,Re_L^{0.66-0.4(1-\epsilon)}\,Sc_L^{0.5}\,Ga_L^{0.07-0.25(1-\epsilon)}\,(ah_P)^{-0.41+0.6(1-\epsilon)}\,(1-\epsilon)^{-0.2}$
Scherwood and Holloway (1940)	PBR	Rings 25–50 mm	$Sh_L = 0.00204\,Re_L^{0.78}\,Sc_L^{0.50}$
		Rings 12, 5 mm	$Sh_L = 0.00333\,Re_L^{0.65}\,Sc_L^{0.50}$
		Saddles 12.5–38 mm	$Sh_L = 0.00285\,Re_L^{0.72}\,Sc_L^{0.50}$
Shulman et al. (1955)		RP	$Sh_L = 5\left(\dfrac{F_f^{2/3}}{1-\epsilon}\right)^{0.55} Re_L^{0.45}\,Sc_L^{0.50}\,(a\delta)^{33}$ $F_f = \dfrac{483\nu_L^{2/3}}{a_s}$
Yoshida and Koyanagi (1962)	PBR	Rings 15–25 mm Saddles 12–25 mm	$Sh_L = 0.236\left(\dfrac{d_P}{\delta}\right)^{-0.50} Re_{Le}^{0.50}\,Sc_L^{0.50}$ $Re_{Le} = \left[\dfrac{4U_L}{a_e\nu_L}\right]$

Adapted from Kolev (2006).

TABLE 9.3   Liquid-Phase Mass Transfer Correlations for Fixed Bed Reactors

Researcher(s)	Reactor	Packing	Correlation
Billet (1993)		RP	$k_L = C \left(\dfrac{D_L}{4\varepsilon} \right)^{0.5} \left(\dfrac{\rho_L g}{\mu_L} \right)^{1/3} \mu_L^{1/3}$ where C is an experimental, packing dependent constant
Billet and Schultes (1993)		RP	$k_L = C_{LB} 12^{1/16} \left(\dfrac{U_L}{\varepsilon_L} \right)^{0.5} \left(\dfrac{D_L}{d_h} \right)^{0.5}$ where C_{LB} is an experimental, packing dependent constant
Kolev and Daraktschiev (1976)		Holpack	$\dfrac{k_L t_P}{D_L} = 0.00113\, Re_L^{\,0.635}\, Sc_L^{\,0.5}\, Ga_{Lh}^{\,0.366}\, (s/t_P)^{4.0}\, (a t_P)^{0.1}$ $Re_L = \dfrac{4 d_L}{a_e \nu_L} \quad$ and $\quad Ga_{Lh} = \dfrac{g t_P^3}{\nu_L^2}$

(*continued*)

TABLE 9.3 *(Continued)*

Researcher(s)	Reactor	Packing	Correlation
Norman and Sammak (1963)	Disk column	RP	$\dfrac{k_L d_d}{d_R} = 0.13 \left(\dfrac{4U_{Li}}{\mu_L}\right)^{0.61} \left(\dfrac{\mu_L}{\rho_L d_R}\right)^{0.50} \left(\dfrac{\rho_L^2 g d_d^3}{\mu_L^2}\right)^{0.17}$
Onda et al. (1959)		RP	$k_L = 0.0051 \left(\dfrac{\mu_L g}{\rho_L}\right)^{1/3} \left(\dfrac{U_L \rho_L}{a_e \mu_L}\right)^{1/3} \left(\dfrac{D_L \rho_L}{\mu_L}\right)^{1/3} (a d_P)^{0.4}$
Shi and Mersmann (288)		RP	$k_L = 0.91 \left(\dfrac{6D_L}{\pi d_P}\right)^{0.5} \dfrac{U_L^{0.19} g^{0.22} \varepsilon^{0.2} \rho_L^{0.23}}{\mu_L^{0.23} a^{0.4}} \left(\dfrac{\sigma}{\rho_L}\right)^{0.05} (1-0.93\cos\theta_\theta)^{1/3}$
Zech (1978)		RP	$k_L = C \left(\dfrac{\rho_L g d_P^2}{\sigma}\right)^{-0.15} \left(\dfrac{U_L g d_P}{3}\right)^{1/6} \left(\dfrac{6D_L}{\pi d_P}\right)^{0.5}$
			where C is an experimental, packing dependent constant

Adapted from Kolev (2006).

bed performance for a wide variety of packing and designs. Their approach has been based on developing a large database, which incorporates 861 and 4291 experiments for structured and random packings, respectively, and the use of a neural network to determine the most important factors in predicting the necessary output. The resulting average error varies between 20.9 and 29.2% among six different mass transfer parameters. They have created separate correlation Excel files for the packed, flooded, and TBRs, which can be accessed from Professor Larachi's homepage at Laval University (Larachi et al., 2008; Piché et al., 2001a; Piché et al., 2001b; Piché et al., 2001c); the link to this site is http://www.gch.ulaval.ca/bgrandjean/pbrsimul/pbrsimul.html. This is currently the best and most comprehensive resource for gas–liquid mass transfer information relative to fixed bed reactors.

9.8 NEEDED RESEARCH

Fixed bed reactors have been largely used by the chemical and related industries. Industrial biological applications are limited. Since fixed bed reactors operate at lower superficial gas and liquid velocities, it is going to be very hard for these reactor types to compete with bubble columns, airlift reactors, or membrane reactors (Gottschalk, 2008). A second competitive problem is that interfacial area in fixed bed reactors is made of liquid droplets and wetted packing. If fixed bed reactors are to be used in biological applications, more research needs to be directed toward increasing the surface area available for mass transfer. One solution that could use more investigation is the operation in the flooded regime, which would allow for more bubbles and a larger interfacial area; however, the problem with this approach is that such operation may be described as more bubble column than fixed bed reactor. So far, biological research has been focused on waste treatment, or the need to provide a support mechanisms for highly shear-sensitive media, such as mammalian cell structures. In other words, if the microorganism is tough enough for a bubble column or airlift reactor, the fixed bed reactor is not effective.

The current research effort in fixed bed reactors is directed toward the improvement of packing. The view is that the packing choice can make or break the success of the operation, and that current operations are so large that even minimal improvements would yield tremendous savings. Although the reactive function is very important given the supporting role packing has in chemical reactions, biological systems are going to be less dependent on this packing function and would be better served with packing that supports larger amounts of microorganisms.

9.9 SUMMARY

Fixed bed reactors use packed, fixed material with the purpose of achieving higher reaction rates than would be possible with two-phase interaction alone. In order to conform to different types of reactions, several styles and forms of fixed bed reactors and packing are in use. The PBR is a countercurrent model, where liquid is injected from the top of the reactor and gas from the bottom. The TBR is a downward flowing cocurrent reactor. If the flow occurs in the upward direction, it is referred to as a FBR. As far as the packing goes, the designer and operator have a wide array of options, which are mainly due to a tremendous amount of customization that has occurred over the years. This experimentation has led to dozens of different packing materials and classifications.

Generally speaking, fixed bed reactors are not able to transfer as much mass when compared to previously covered reactor types because the phase flow rates in fixed beds are usually much lower. This is due to the fact that the pressure drop across the bed would increase to a value that would make almost any operation either unprofitable or nonreactive. Hence, the general application of fixed bed reactors is to provide a support structure on which microorganisms can grow. Another application of the fixed bed reactor is to serve as a biofiltration device.

There is still a lot of work—including more basic research—left to do on fixed bed reactors for biological applications. Most of the data, conclusions, and designs have been developed for chemical processes and reactions, which usually require very high pressures and temperatures. Microorganisms often have very different requirements and interactions, which have not been explored very deeply, if at all. For example, there is not a single gas–liquid mass transfer correlation that is directly applicable to microbial design mainly because most of them have been developed in the 1940–1960s. So, if the fixed bed reactor limitations are kept in mind, the design can be adapted to serve roles currently supplied by chemical reactions or open the doors to new applications.

10 Novel Bioreactors

10.1 INTRODUCTION

Novel bioreactors are nonstandard devices, which attempt to deliver better performance, bring new features, or introduce production through a new method. Novel designs accomplish these tasks by either introducing new or significantly adjusting components in standard devices, or by introducing a completely new approach and/or concept into the production. This section will concentrate on both aspects of novel bioreactor design, but more attention will be paid to novel bioreactor design strategies. Note that new novel bioreactors are being developed all the time, and this chapter is not intended to be an exhaustive review of novel bioreactors.

10.2 NOVEL BUBBLE-INDUCED FLOW DESIGNS

Novel bubble-induced flow designs apply a plethora of mechanisms that help differentiate each specific design from other novel and standard devices. Some changes are structural and include use of different materials and internals. Others include the use of novel methods to excite the bubble interface and induce gas–liquid mass transfer. Novel methods exclude devices that are created to study specific events relating to standard devices. For example, the study by Sotiriadis et al. (2005) using a specially designed bubble column where the phases move downward to specifically study bubble behavior, bubble size, and gas–liquid mass transfer in the downcomer of airlift reactors would fall in the excluded devices.

Perhaps the simplest variation to aerated designs is to inject the phase(s) using the jetting principle. Such devices are often grouped as jet reactors. It should be noted that jet reactors do not necessarily share any other commonalities besides the phase input techniques. The reactor internals and phase flow can vary significantly. Jet reactors use liquid jets and injection devices in order to achieve high liquid velocities and turbulence. The higher turbulence generally yields to better gas–liquid mass transfer performance, which is its largest advantage. On the other hand, the shear rates, especially in the injection vicinity, tend to be much higher than the average. Microorganism growth could produce uneven reactor performance. Furthermore, the generated shear stresses are potentially too high for

An Introduction to Bioreactor Hydrodynamics and Gas-Liquid Mass Transfer, First Edition.
Enes Kadic and Theodore J. Heindel.
© 2014 John Wiley & Sons, Inc. Published 2014 by John Wiley & Sons, Inc.

microorganisms to survive. Hence, the current application of jet reactors is limited to strictly chemical processes, and research with microorganisms is limited.

Some examples of jet reactors include submerged- and plunging-jet reactors, ejector reactors, hydrocyclones, and venturi devices. Plunging-jet reactors throw a liquid jet through a gas phase, which is usually reactive, into the liquid volume. In other words, these reactors attempt to transfer the gas phase by creating gas entrainment at the liquid surface. This jet reactor requires jet velocities of up to 30 m/s (Charpentier, 1981). Submerged-jet reactors pump a liquid phase through a venturi where the liquid is combined with the gas phase. The result is a mixture that has a very bubbly appearance. The generated bubbles are very small and would be expected to have a large gas–liquid interfacial area. Varley (1995) reported a Sauter mean diameter of 0.29–1.92 mm in a 72-l submerged-jet reactor using $U_L = 3 - 12$ m/s.

The ejector reactor uses a similar injection device as the submerged-jet reactor, but the created jet is injected into an airlift-like vessel. The gas–liquid mixture is allowed to go through a riser and into a separator. Since the gas phase separates, a density difference is created and liquid recirculates into the injection zone. These reactors are capable of operating with liquid velocities of at least 20 m/s. With this kind of turbulence, the ejector reactor outperforms stirred-tank reactors at equivalent operating conditions (Charpentier, 1981).

Venturi-based reactors work similarly to submerged-jet and ejector reactors, but the big difference is that the liquid is injected into a high velocity gas-phase field, whereas the ejector reactors inject gas into a high velocity liquid-phase field. Hence, the venturi-based reactors create small liquid droplets similar to an atomizer. Venturi-based reactors are used as scrubbers or with quantities of gas phase present in the reactor volume. In contrast, ejector reactors create small bubbles that are used in liquid-phase dominated reactor volumes (Charpentier, 1981).

Jet injectors may also be combined with monolith reactors. Monoliths are usually tube reactors with channeled flow. The reaction occurs at the gas–liquid interface as well as on the channel wall, which are usually catalytic or coated with catalytic material. Monoliths can be made into vertical (similar to bubble column) or horizontal tubes, airlift devices (whereby the riser would a monolith), or even into a mechanically stirred device. Usually, however, monoliths are designed like bubble columns or airlift reactors (Broekhuis et al., 2001).

Monolith reactors could be considered a novel class on their own. The problem is that the monolith reactor relies on the catalytic properties of the wall in order to have better mass transfer and reaction performance than the other reactor types. If these properties do not exist, as they are not expected to with biological applications, the channels are too small and cause gas to slug. Hence, a monolith reactor would be limited by the amount of biological media and phase flow media, specifically the gas phase. If the microorganisms grow well, the channels could be plugged and extensive cleaning may be necessary. For most biological applications, monoliths do not provide theoretical advantages to a fixed bed reactor, bubble column, or airlift reactor. This fact is reflected in the lack of biological application of traditional monolithic reactors; however, monolith-like principles are sometimes hidden within fixed bed reactors. In these cases, the monolith-like packing is used as a support mechanism for cellular growth. For available biological studies, monolith reactors tend to compete with trickle-down (fixed bed) or membrane reactors. For example, the 2007 AIChE Annual Meeting held in Salt Lake City, UT, had a conference section dedicated to monoliths and membranes (Bauer et al., 2007). It is also a popular option to design monolith microreactors (Schönfeld et al., 2004), which are basically scaled-down versions of the large-scale reactors.

Researchers have also combined reactor types. Guo et al. (1997), for example, designed an external loop airlift reactor that incorporates a fluidized bed within the downcomer section, shown in Figure 10.1. The fluidized bed section is used to immobilize microorganisms on carrier particles in order to protect them from damage. The design is meant for the production of enzymes, biofluidization, and wastewater treatment. Although shear rates were minimized, bubbles were not entrained within the downcomer. Furthermore, the gas–liquid mass transfer coefficient was observed to increase with gas holdup. The result was that the gas–liquid mass transfer was limited due to the fact that global gas holdup for the reactor was strictly defined by the riser gas holdup without any addition by the downcomer.

Wastewater treatment plants oftentimes use vessels that combine the (slurry) bubble column with a mechanical extractor and/or a mixer. The mechanical extractor is used to scrape heavy residue at strategically located divider walls. Such a system implies that the liquid flows across the gas flow field. Siemens' proprietary Attached Growth Airlift Reactor (AGAR)-Moving Bed Bioreactor is an example of such a device.

Wastewater treatment may also require the use of UVA–UVB rays as a mechanism to kill any unwanted cellular material. Unsparged photoelectrochemical

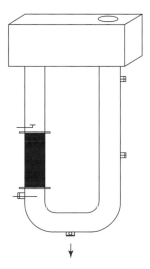

Figure 10.1 Novel external-loop airlift reactor designed by Guo et al. (1997).

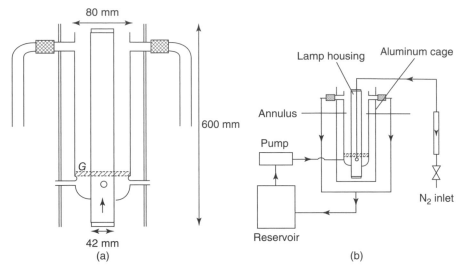

Figure 10.2 (a) Novel sparged photoelectrochemical reactor cross section and (b) schematic diagram used by Harper et al. (2001).

reactors, which resemble unsparged slurry bubble columns, are standard devices for such an operation. Harper et al. (2001), however, experimented with a modified aerated airlift reactor, shown in Figure 10.2. The design uses an UVA–UVB lamp core around which the photochemical reaction is allowed to occur. The liquid and gas can be recirculated to maximize output and conversion and optimize residence times. A similar design could be very useful for photosynthetic microorganism growth, such as algae, but no information on such a system has been found in the open literature.

Ellenberger and Krishna (2002) introduced vibrations in order to reduce the bubble rise velocity, destabilized the bubble surface, and reduced surface tension forces. These changes lead to an improved liquid-phase mass transfer coefficient and, hence, gas–liquid mass transfer coefficient. This goal has been accomplished using mechanical vibration devices in previous studies cited by the authors. In contrast, Ellenberger and Krishna (2002) introduced sinusoidal pressure variations at a frequency on the order of 100 Hz and amplitude of 0.0025–0.01 mm; this led to a reduction in the bubble diameter by 40–50% and an increase by 100–300% in gas holdup and 200% in the gas–liquid mass transfer coefficient when compared to the no pressure variation system.

Another less commonly applied adjustment with bubble columns and airlift reactors is to introduce additional turbulence by pumping or mechanically exciting the liquid phase. Such adjustments have long been viewed as being advantageous and increasing gas–liquid mass transfer, but they have also been seen as very expensive options (Lundgren and Russel, 1956). The advantage of such a system is that it allows more control of the liquid phase flow, but the cost is usually represented by the additional power required by the pump or impeller motor. The need for

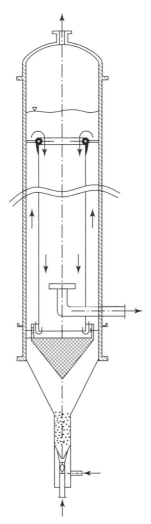

Figure 10.3 Experimental pumped circulation column proposed by Fadavi and Chisti (2005).

such a device stems from the potentially limiting suspension capabilities, liquid circulation rates, and axial nutrient gradients provided by standard airlift reactors. These issues are exaggerated with height. The pumped circulation column is shown in Figure 10.3. As expected, the high liquid flow rate ($Q_L = 2$ m^3/h) used by Fadavi and Chisti (2005) added to the turbulence and nearly doubled gas–liquid mass transfer relative to an airlift reactor. Liquid flow in an airlift reactor generally does not contribute significantly to bubble breakage, but the fast liquid flow in the pumped variant did so and led to a smaller average bubble diameter.

A mechanically induced circulation loop reactor is shown in Figure 10.4. This variant works on similar principles as the pumped circulation column.

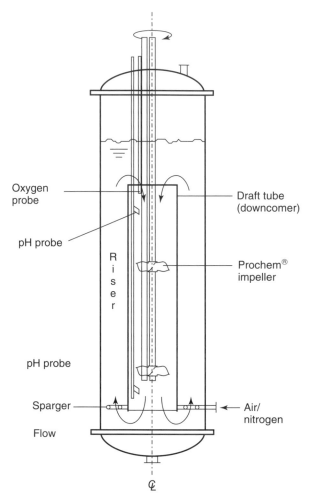

Figure 10.4 Experimental mechanically induced circulation loop reactor proposed by Chisti and Jauregui-Haza (2002).

The mechanically induced circulation loop reactor attempts to increase liquid velocity using impeller sets. Chisti and Jauregui-Haza (2002) accomplished this by imbedding a down-pumping Prochem Maxflo T hydrofoil impeller. Air is injected into the annular region of the reactor for several reasons. First, the impellers may flood with the higher gas flow rates usually applied in airlift reactors. Second, the operational set becomes easier to handle if the impellers are placed in the reactor center. Finally, the impellers are used to increase the liquid velocity rather than bubble breakage. So, the impellers used for pumping, such as the Prochem Maxflo T, do not necessarily handle gas well, and placing them in the riser would make the situation unnecessarily more complicated. In addition, the riser flow is already influenced by the gas flow rate. It should be noted that many airlift reactor issues can be solved with proper airlift reactor design and phase flow rates; however, these hybrid variants are useful when high gas flow rates are not possible or for more control. Generally, mechanically agitated airlift reactors are still seen as economically prohibitive except for a set of specialized cases.

Another area of reactor modifications has been to add or adjust column internals. For airlift reactors, some research has gone into replacing the draught tube with a net draught tube that would allow for bubble breakage near the wall region (Fu et al., 2004; Fu et al., 2003). This additional bubble breakage leads to a smaller bubble diameter in the riser, which in turn leads to higher gas–liquid mass transfer. If the process is mass transfer limited, it is expected that an increase in productivity would be observed. Although the net draught tube reactor has the potential to easily increase gas–liquid mass transfer, it also has the potential to be easily plugged by biomass growth or otherwise foul in an industrial setting.

Static mixers can be installed in bubble columns or airlift reactors to provide additional mixing efficiency. Such devices are used to break bubbles before they have a chance to coalesce. The performance of these devices range greatly, but increase in gas–liquid mass transfer by 500% has been observed (Chisti et al., 1990; Fadavi and Chisti, 2005). Such increases, however, are to be expected with highly viscous media. Viscosities that are nearer to water are going to see much smaller performance increases. Operators interested in adding static mixers to their bubble column or airlift reactor also have to consider the additional cleaning requirements that may come from such devices. In other words, static mixers may not provide enough of an improvement in low viscosity fluids to pay for their additional maintenance costs.

10.3   MINIATURIZED BIOREACTORS

Miniaturized bioreactors can be divided into two categories based on scale: microreactors and nanoreactors. These bioreactors present several fundamental advantages and open new venues. Miniaturized reactors allow for bench-scale chemical and biochemical production, which can be used by researchers. They also allow for cost-effective production when smaller quantities of a chemical are required. Other larger bioreactors are often not feasible because the production is not cost effective if the product is not very valuable or if the production is not consistent or pure enough for higher value chemicals. Miniaturized bioreactors, however, provide a great deal of control over reaction kinetics and hydrodynamics.

Another very important advantage of miniaturized bioreactors is that scale-up takes on a different form. The scale-up procedure for standard bioreactors and miniaturized bioreactors is compared in Figure 10.5. The scale-up procedure for standard bioreactors involves a complicated iteration processes. A laboratory bioreactor is designed as a proof-of-concept. Next, an experimental-scale bioreactor is designed to ensure production viability. After a few iterations, a small-scale pilot plant is constructed to test and finalize the production process, equipment placement, and economic viability. Once this step is accomplished, a large-scale

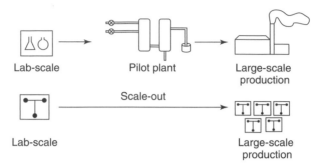

Figure 10.5 Scale-up procedure comparison (Watts and Wiles, 2007).

plant is constructed. The iteration requirement stems from the significant changes in hydrodynamics and/or reaction kinetics that are experienced as scale is increased (Watts and Wiles, 2007).

With miniaturized bioreactors, however, scale-up is a much simpler process. The procedure involves putting miniaturized bioreactors into series and/or parallel in order to produce larger output quantities. This approach keeps the reaction kinetics and hydrodynamics predictable for each component regardless of plant scale. Hence, the process is often referred to as numbering-up. In other words, the laboratory bioreactor is very similar to the industrial production, with the bioreactor quantity and controls being the most significant differences. This property keeps the start-up and development costs lower and more flexible (Ehrfeld et al., 2000; Watts and Wiles, 2007).

The numbering-up method also introduces another potential advantage. The scaled model operates as a continuous bioreactor rather than batch while providing the operator with the same control advantages of the batch operation at the same time. Therefore, the process time is expected to be shorter with miniaturized bioreactors since most standardized bioreactors use a process time that is longer than the kinetic minimum. Safety is also increased tremendously since the process can be stopped at any point in the process flow (Ehrfeld et al., 2000). These controls can be instituted automatically without the need for human supervision.

10.3.1 Microreactors

Microreactors are defined by their size rather than construction. Microreactors are miniaturized with channels between the (sub-)millimeter scale and nanometer scale. Microreactors mix the gas and liquid phases pneumatically or mechanically. The size of the complete bioreactor construction is less important. A microreactor example is shown in Figure 10.6. Microreactors are generally compounded into microreactor elements, which are placed into mixing units. These units are placed into microreactor devices, which have inputs and outputs for all the microreactor units placed within it. Microreactor devices are placed in parallel or in series in order to achieve the necessary conversion. Finally, the output is treated

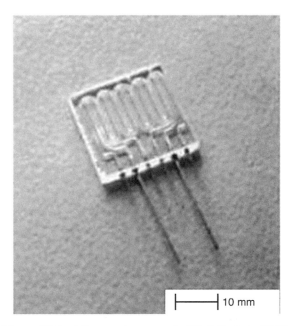

10 mm

Figure 10.6 Microreactor example (Ehrfeld et al., 2000).

(Ehrfeld et al., 2000; Watts and Wiles, 2007). A typical microreactor assembly is shown in Figure 10.7.

Microreactor construction can be accomplished using numerous tools. Costs, however, dominate the construction options. Hence, possible construction techniques are defined by channel flow scale, precision, reliability, and material selection. Naturally, as the bioreactor scale becomes smaller and precision requirements become higher, the costs tend to increase as well. An accepted approach when constructing microreactors out of metals, ceramics, or plastics is using lithography, electroplating, and molding (LIGA, Lithographie, Galvanik und Abformung). LIGA is a three-step process in which a laser, electron beam, ion beam, UV-ray lithography, or X-ray lithography are used to print microstructures. Then, electroforming is used to generate a metal layer onto the microstructure. This metal structure can be used as a mold or embossing tool for mass production (Ehrfeld et al., 2000; Hruby, 2002; Wirth, 2008). The available construction techniques are summarized in Figure 10.8. Ehrfeld et al. (2000) and Wirth (2008) are good and recent sources for the current construction techniques and their applications.

Mixing in microreactors is almost exclusively assumed to be laminar due to the small flow channel width. Laminar flow through microchannels requires the phase flow to be alternated in some fashion in order to create the mixing environment. This operation is important because mass transfer, in this case, is driven only by molecular diffusion. Hence, the creation of larger gas–liquid interfaces is the only practical course of action for gas-limited operations. Miniaturized bubble columns are able to accomplish this task well. For example, a standard reactor can create

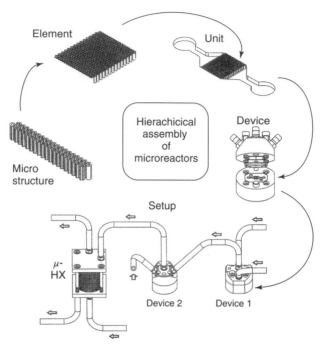

Figure 10.7 Microreactor assembly (Ehrfeld et al., 2000).

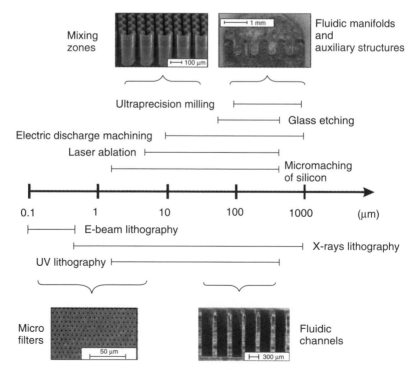

Figure 10.8 Microreactor construction techniques (Ehrfeld et al., 2000).

interfacial areas in the range of 2000 m^2/m^3, while a single-channel microbubble column and a microbubble column with channel arrays can create interfacial areas of 1700–25,300 and 5100–16,600 m^2/m^3, respectively (Ehrfeld et al., 2000).

Unfortunately, the extreme numbers are somewhat misleading. In order to properly understand their context for biological application, an understanding of microbubble column flow regimes is necessary. Bubble velocity in microreactors is defined by a gas space velocity. This velocity is similar to the superficial gas velocity, but it tends to be slower due to very significant wall effects. So, the gas space velocity is approximated using a sample bubble velocity.

Bubbles in microbubble columns are observed to be separated by a liquid film from the wall at almost all gas space velocities. At low gas space velocities, the microbubble column experiences bubbly flow. The bubbly flow is defined by microbubbles, which are spherical and as large as the channel diameter. Since bubbles of this size are highly unstable without surfactants, the gas space velocity has to be low enough to allow enough space between bubbles in order to prevent coalescing.

As the gas space velocity increases, bubbles tend to coalesce and a slug flow regime develops. The slug flow regime is defined by bubbles that are longer than the channel diameter and are still separated by some amount of liquid. Increasing the gas space velocity leads to the slugs coalescing even further, and a slug-annular flow regime is observed. The slug-annular regime experiences very long bubbles, which are separated by a very small amount of liquid. Further increasing the gas space velocity combines the elongated bubbles into an annular gas flow (annular flow regime).

The extreme interfacial values are experienced by the latter two regimes and are most likely useless for biological applications. The channel diameters are very small to begin with, and the liquid film that develops in the slug, slug-annular, and annular flow regimes is not sufficient to support microorganism growth. In order to have microorganisms survive in microbubble columns, the reactor would almost certainly have to be operated in the bubbly regime, which experiences interfacial areas of 1700 − 5100 m^2/m^3. Lower values are observed with microbubble columns with a diameter of 1100 μm, whereas the larger values are experienced with a channel diameter of approximately 300 μm (Ehrfeld et al., 2000). Even though these values are higher than those for standard equipment, the advantage for microbubble columns is not astronomical, and competition and performance are likely to be more comparable and competitive.

Although microbubble columns are popular, other microreactor mixing methods exist, such as the falling film principle. In such a system, the liquid phase would be input into the microreactor's reaction chamber from the top while the gas phase input is at the bottom. The principle is very similar to the annular flow regime in the microbubble columns. The liquid would line the reactor walls while the gas phase would move through the annular region. The difference with the falling film reactor is that the liquid is fed at a rate and way that it would guarantee flow only at the wall region and would prevent the liquid phase from mixing into the gas phase (Zhang et al., 2009).

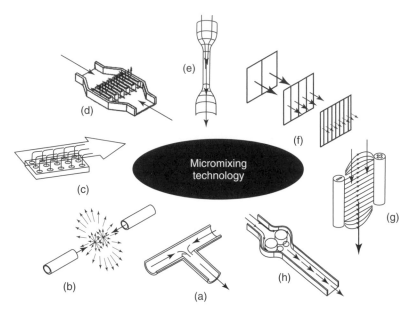

Figure 10.9   Micromixing arrangements: (a) substream contacting, (b) high energy substream contacting, (c) multiple substream injection into major stream, (d) multiple substream contacting, (e) flow restriction, (f) stream splitting, (g) forced mass transfer, and (h) periodic fluid injection (Löwe et al., 2000).

In addition to the microbubble column and falling film microreactor, several other theoretical schemes exist in order to achieve proper mixing conditions of two phases. The simplest mechanism is defined by the phases entering a tee intersection either as single streams, shown in Figure 10.9a, or as numerous substreams, shown in Figure 10.9d. Control over the mixing is provided by the phase flow rates. A more energetic method is to collide high energy (velocity) streams in order to create large interfacial areas due to atomization or spraying (Figure 10.9b). This method is most likely not useful for biological applications as it would damage the microorganisms. It could, however, be used as a phase premixer for semisuspended microorganisms. One of the phases could also be broken down into substreams, which are then injected into a larger stream of the other phase (Figure 10.9c). For example, gas could be injected into a liquid stream in order to generate bubbles with liquid intervals.

An interesting method for mass transfer problems is to channel the phase flows into a constricted area, which would increase velocity and decrease the diffusion path (Figure 10.9e). Although useful in this regard, it could create problems in biological systems by either damaging the microorganisms or not allowing enough room for the microorganisms to pass through. A more popular approach is to use splitting arrays (Figure 10.9f). The phases would be mixed into a single stream and then continuously broken apart by alternating horizontal and vertical splitters, which get finer as the mixture moves along. Although this method is popular in the chemical industry, microorganism growth would most likely cause blockage. Naturally, these methods can be combined into a single, more complex device. These mixing schemes have been applied to liquid–liquid contacting problems, but only substream tee contacting and substream injection have been tested for gas–liquid problems (Ehrfeld et al., 2000).

The natural cause for a lack of microreactor research in gas–liquid processes is that most miniaturized reactor research is concentrated on problems experienced in the chemical industry, specifically catalyzed liquid–liquid reactions. Gas–liquid, and especially biological reactions, have not been widely attempted since the field of microreactor engineering is emerging to fulfill the highest margin needs first. Biological applications are still being mastered with standardized equipment, and economic viability using microreactors seems to be some distance away. Another issue is that much of the work is still not published. For example, Ehrfeld et al. (2000) includes numerous unpublished reports and correspondences, such as the interfacial areas data. The same information for standard bioreactors, however, is easily obtained across different print media.

10.3.2   Nanoreactors

A differentiation between nanoscaled and microscaled miniaturized reactors is necessary because the naming convention is not universally applied. Nanoscaled reactors were introduced in the late 1990s and are the possible microreactor alternatives for the chemical industry (Ostafin and Landfester, 2009). An example of a nanoreactor in nature is the mitochondrion, which is the energy-producing portion in most complex cells. Nanoreactors are more accurately described as molecular reactors, and the reactor volume is defined by the number of molecules confined within the reactor rather than a standard volumetric measurement. Nanoreactor's biological applications, however, will most likely stay very limited. The nanoreactor has some critical faults and problematic characteristics for biological application. The first and largest problem is that the channel size is too small for most microorganisms of interest, such as bacteria. Nanoreactors are usually on the nanometer scale while smaller bacteria are in the micrometer scale. Hence, no additional discussion on nanoreactors will be presented.

10.4   MEMBRANE REACTOR

Some products or intermediaries in biological processes are highly shear sensitive. Standard bioreactors may not be able to protect the microorganisms sufficiently. Membrane reactors allow for semisuspension of cells by placing the cellular material within or between (semi)permeable membranes. Membranes are not novel ideas for the chemical industry for which membranes have $9 billion in component sales in 2006 (Nunes and Peinemann, 2006), but their biological application is novel. The chemical industry, however, has used membrane technology for experiments since the 1920s and for industrial applications since the 1960s, while the membrane bioreactor concept has been first defined in the literature in 1979 (Hall et al., 2001; Nunes and Peinemann, 2006). It should be noted that permeation and membrane concepts were first scientifically described in 1748 (Baker, 2004).

The initial membrane bioreactors cross-flowed the liquid phase through the membrane, which increased energy costs significantly. Currently viable membrane bioprocesses submerge the membrane in the liquid phase whereby the liquid flows parallel to the membrane matrices. This creates low pressure drops and makes the economic viability a reality. Current membrane reactors also tend to vary the volumetric membrane amount or carrier particles at 60–70% (Leiknes and Odegaard, 2007).

Membranes allow the gas phase to transfer the microorganisms without exposing the microorganisms to the bubble interface or other potentially high shear areas. This allows high cell densities which, in turn, allows for higher conversion effectiveness (Ko et al., 2008). The membrane may also allow the product to permeate away to a collection area. Since the microorganisms have little exposure to bubbles and the gas has to travel through the membrane, the membrane reactor allows for relatively small amount of gas–liquid mass transfer to occur and almost any other reactor types outperform it on that measure. Unfortunately, some microorganisms, such as animal cells, are so shear sensitive that they are able to survive only in a membrane-protected environment (Bellgardt, 2000b).

The most significant disadvantage of membrane bioreactors is that they are very costly to construct and maintain, and their long-term viability is yet to be proven (Dudukovic, 1999; Kumar et al., 2008). For example, microporous membranes have to be replaced every 3 years, which leads to operating and maintenance costs that are approximately 10 times higher than with a conventional gas treatment options (Kumar et al., 2008). Long-term viability may be problematic because the current membrane technology has permeate mass fluxes which are an order of magnitude that are too small to be competitive on a volumetric productivity basis. Furthermore, membranes provide an additional heat transfer barrier, which introduces temperature control issues that are generally not present in standard bioreactor designs (Dudukovic, 1999).

Mass transfer through a membrane occurs in several steps, which differentiate it from mass transfer in standard reactors. The first two processes are very similar. First, the gas is injected into the reactors volume. Then, the gas is transported into the liquid phase. This is usually accomplished by molecular diffusion. The mixing is not turbulent enough for bubble interface excitement. The diffused gas phase transports through the membrane, which adds a significant mass transfer resistance. Membrane transportation is a two-part process whereby the gas phase is first absorbed by the membrane and then diffuses through the membrane. This process may lead to the separation of the transfer material, which is an advantageous property of membranes for specific product removal. For example, this property would allow separate gas phases to enter the biofilm at different rates or would allow for protein separation (Gottschalk, 2008) or filtration (Ko et al., 2008). Hence, membrane reactors can be split into permselective and nonpermselective categories (Dudukovic, 1999).

The dissolved gas phase then has to diffuse through the biofilm, and the reaction is allowed to occur. The biofilm is usually attached to the membrane rather than being allowed to float (Henstra et al., 2007). Additional problems are created because any by-products, which may be toxic if they are allowed to concentrate, have to leave the reactor through the same manner. In other words, microbial operation may have significant effects on the gas phase concentrations (gas–liquid mass transfer driving force), which could become an additional restriction on total gas–liquid mass transfer (Kumar et al., 2008).

Membranes can be classified by either geometry (symmetrical or anisotropic) or construction (dense, porous, or composite). Membrane selection becomes the main design criterion for a membrane bioreactor. The membrane can be very dense and complex, which would allow for a high degree of selectivity while microporous materials would allow gas, regardless of identity, to permeate easier. Composite membranes have been developed, which attempt to allow the gas phase to permeate easily while allowing a high degree of selectivity. Biological applications make extensive use of microporous membranes, and some attempts have been made to incorporate composite membranes. The most commonly encountered problem with microporous membranes is that they easily plug, which leads to even higher maintenance costs (Kumar et al., 2008). In such a case, the operation becomes limited by the net material accumulation at or within the membrane. This accumulation can be controlled if the fouling is reversible by backwashing; however, some fouling is irreversible and requires chemical treatment, which would force the membrane bioreactor to be shut down and then restarted after the cleaning procedure (Leiknes and Odegaard, 2007).

Hence, issues have led the membrane bioreactor to be a perfect choice for a select few problems. Interestingly, even with its advantages for those shear-sensitive microorganisms, their industrial viability and application have yet to be proven. In practical terms, membrane bioreactors have to compete with fixed-bed bioreactors, hybrid systems (fixed bed incorporated into an airlift), bubble columns, and airlift reactors, all of which are able to provide much higher gas–liquid mass transfer rates. In other words, production could be optimized within an environment that is less friendly than the membrane bioreactor simply because the higher productivity due to higher amounts of gas–liquid mass transfer could offset the productivity losses due to microorganism shear damage. More information on membrane technology for chemical and biological applications can be found in Baker (2004), Nunes and Peinemann (2006), and Peinemann and Nunes (2008a, 2008b).

10.5 SUMMARY

Novel bioreactors can be represented by many different designs and variations, but the most novel and promising approach may turn out to be miniaturized reactors.

Novel mechanical or bubble-induced flow designs are not trendsetters nor do they solve many of the and gas–liquid mass transfer problems. Miniaturized reactors, however, could decrease process design and implementation significantly. The numbering-up method for these reactors reduces the time and amount of work necessary for scale-up; the process is determined for one experimental unit and then the unit is copied multiple times. The rest of the work is spent on the industrial and economic problems rather than hydrodynamic and gas–liquid mass transfer issues commonly found in scale-up issues for other bioreactors.

11 Figures of Merit

Figures of merit are quantities used to compare reactor performance across all reactor types or just the different designs of stirred-tank bioreactors. This section summarizes some reactor-specific figures of merit and problems that have prevented meaningful and significant figures of merit being developed as well as some possibilities that could be considered for further research. The figures of merit are also oriented toward stirred-tank reactors, bubble columns, and airlift reactors since these are the most common gas–liquid and gas–liquid–solid bioreactors. Fixed bed reactors use a much smaller gas and liquid flow rate so that they are not able to compete with the other reactors unless the microorganisms have to be suspended or otherwise protected.

Figures of merit for stirred-tank reactors are especially difficult because of the wide variety of equipment and arrangements used and the high degree of phase interaction complexity. Although the impeller is always used for gas breakup, mixing, and dispersion, the effects of different impellers, setups, inputs, or microorganisms make universal conclusions on hydrodynamics almost impossible. The varieties of microorganisms also make predictions on production or conversion very difficult. Furthermore, a practical model describing conditions for stirred-tank reactors across wide operational ranges and scales is currently unavailable. Since this underlying information is not available, a figure of merit, which is applicable across a wide variety of designs, control variables, inputs, and microorganisms, is also not easily within reach and rarely addressed in available research.

The task, however, is not impossible. A practical representation of the mass transfer performance is $k_L a/(P_G/\forall_L)$. It represents the mass transfer rate per unit power input. This figure of merit does not attempt to predict output or conversion, but that outcome is implied if the process is still gas–liquid mass transfer limited. If other microorganisms' performance measures are important, a time-dependent representation of output in terms of systems size (e.g., units of mass/liter-hour) can be used. Its downfall is that design or control variables are not accounted for, but it can still be used in tandem with the previous mass transfer-based figure of merit or as a long-term assessment tool.

These measures have not been extended to the different reactor designs, but they certainly could be. A few fundamental issues exist. The first arises from the fact that variable costs differ significantly between the reactor designs. For example, the

An Introduction to Bioreactor Hydrodynamics and Gas-Liquid Mass Transfer, First Edition.
Enes Kadic and Theodore J. Heindel.
© 2014 John Wiley & Sons, Inc. Published 2014 by John Wiley & Sons, Inc.

power draw in the stirred-tank reactor varies with impeller diameter raised to the fifth power and impeller speed raised to the third power. Hence, increasing the scale of a stirred-tank reactor increases the power usage at a fast pace. At the same time, larger mechanical systems, such as the motor and gear systems, may require more or more expensive maintenance. These properties ultimately lead to the operational cost growing faster than with the bubble column or airlift reactor; however, the stirred-tank reactor may be able to produce smaller bubbles if it is properly designed and setup. As scale increases, a theoretical inflection point may exist at which the advantages of the stirred-tank reactor are surpassed by the power usage and the bubble column or airlift reactor becomes a superior economic alternative. This inflection point has not been studied, and such comparisons between stirred-tank reactors, bubble columns, and/or airlift reactors are very rare. So, such a study could produce a map that could be microorganism specific, which shows the preferred reactor for a given scale and gas or liquid flow rate(s). Many variations and variables would not be included but a process-specific guide could be developed.

A second issue with figures of merit is more of a derived problem. The scale-up procedure for any given reactor design is highly variable. The stirred-tank reactor alone has numerous alternatives such as constant power density, impeller velocity, impeller tip speed, or a variation/combination of these variables. Bubble column scale-up might be even harder since the gas and liquid hydrodynamics are not easily controlled or predicted. The airlift reactor scale-up is perhaps easier due to the more controlled gas and liquid flow patterns. Regardless, the scale-up for each reactor design varies significantly, and, more importantly, the reactors do not have a standard, optimal, or commonly applied and agreed upon scale-up procedure. For example, bubble columns have to be scaled up by increasing the experimental reactor by a factor of 5–10 for each iteration. By approximately 300 l, the reactor is supposed to be scale independent and a pilot plant can be seriously considered. The airlift reactors seem to show this property at a much earlier point, 32 l (see Figure 8.14). The stirred-tank reactor, on the other hand, may never lose its scale dependence (after accounting for the power usage). In other words, comparison of industrial bubble columns and airlift reactors could be well defined, but the experimental and pilot-scaled reactors may not be. Since most research is done with experimental (small) reactors, figures of merit are often not attempted.

It is clear that a defined winner among the reactor designs does not exist. More importantly, the reactors may not be interchangeable and design competition is not clearly defined. The previously discussed issues shed light on the lack of interchangeability. The specific project or process helps to further explain the lack of figures of merit in published research. Many of the reactor applications are process or project specific. Without having an idea of the microorganism requirements or its influence on the liquid properties over time, it may be difficult to predict which reactor would be better without experience. At best, some reactors could be ruled out.

An example of the problem difficulty can be simulated by assuming economic product qualities. Let us assume that the product of interest is highly valuable, and that its quality impacts the earned price significantly. The bioreactor for such a product should be chosen on its ability to consistently produce the same product quality. In other words, the bioreactor design that can accomplish this task can get away with costing more (feed, maintenance cost, variable cost, capital cost, etc.) on an annualized output basis. Specifically, it can cost more on a marginal or variable basis since the marginal product price would be able to at least cover those expenses. Such a product also implies a certain degree of pricing power on the part of the producer. An example of such production is often encountered with patented products (proteins, antibiotics, medicines, etc.). Such a producer may be perfectly happy employing stirred-tank reactors regardless of scale.

A commoditized product, on the other hand, requires that the marginal product cost be optimize using whatever method works. The producer does not have significant pricing power. If the cost of capital is low or the economies of scale are large, the bioreactor of choice may be quite complicated and the facility may be very large such as many refineries in the United States. If the cost of capital is high, input prices are variable, and the price risks are not easily accounted for, producers may be interested in smaller facilities with a smaller initial investment and higher variable cost such as many of the early ethanol facilities. The bioreactor choice becomes dependent on preference, economic resourcefulness, and a reflection of the producer's risk appetite. The bioreactor selection for such a producer becomes much more important. It should be noted that competitive market theory calls for commodity producers to optimize the marginal cost rather than the average.

Some figures of merit, however, could be defined without accounting for the economic impact and trying to judge a bioreactor's efficiency and feasibility for a given process. The scale of interest for such figures of merit would not be industrial. By the time a process reaches even the pilot stage, the designer has to start considering the economic impact, and the economic decisions become more and more important. The efficiency and feasibility of production has to be decided early in the experimental and scale-up portion of the project. Hence, figures of merit are going to be highly dependent on the bioreactor scale-up procedures and knowledge thereof. So, more research would need to be directed into producing a unified or at least more uniform scale-up procedure set in order to predict and compare the potential output gains between the bioreactors.

Cost analysis is usually implemented based on a previous project that used similar components. A size ratio is introduced to relate the base design to the current design. A scaling factor is used to properly represent the equipment relative to its scale basis. The scale basis is determined by the type of equipment and process and commonly includes flow, volume, area, and power. An installation multiplier is introduced that is based on the material cost and type of equipment. This factor hopes to account for the necessary machining, design, and installation costs. For example, a carbon steel agitator has a multiplier of 1.3 while a stainless steel version has one of 1.2. If different equipment has a higher design and installation cost, such as heat exchangers, they carry a higher multiple (2.1 in this case) (Aden et al., 2002; Wooley et al., 1999).

In order to account for inflation in the monetary supply and fluctuations in equipment costs due to technology and commodity prices, an equipment index, such as the Chemical Engineering Purchased Equipment Index (CE Index), is used. This index, in particular, is popular because bioreactors are completely based on chemical reactor principles such that the index properly reflects fluctuations for bioreactor equipment. Proper usage should project prices to the year when purchasing decisions and commitments are to be made. A formula for the practice can be summarized by (Aden et al., 2002; Wooley et al., 1999)

$$\text{Cost} = \text{InstalledBaseCost}^* \left(\frac{\text{Base}}{\text{Current}} \right)^{\text{ScalingExp}}$$

$$^*\text{InstallationFactor}^* \left(\frac{\text{CEIndexBase}}{\text{CEIndexCurrent}} \right) \qquad (11.1)$$

Equation (11.1) assumes economies of scale, which is only true for relatively cheap materials such as mild steel. If the material is more expensive, this relationship breaks down. For medium material costs ($\sim$\$10/kg), a break-even point (diminishing returns) develops such that larger reactors prove too expensive in terms of capital, but smaller reactors are too energy intensive such that the variable costs are higher. As the material becomes more expensive, the break-even point shifts toward a smaller scale. The opposite is true for cheaper material costs. Capital and input (labor, electricity, or feed) costs also induce similar behavior. As the cost of capital increases, operators prefer smaller scales such that continued operation is used to pay for variable costs that occur during the same period (may be offset due to payment arrangements). Therefore, lower capital and input costs induce larger scales (Patwardhan and Joshi, 1999).

Economies of scale also explain industrial practices for processes that turn from low viscosity Newtonian to viscous and/or non-Newtonian liquids. If feed costs are reduced, the process is more likely to yield a positive return with scale (Ogut and Hatch, 1988) even though the amount of unused volume and discarded feed would increase as well. The rate of diminishing returns would be dependent on product price(s), material costs, and other input costs (electricity, labor) such that the operator may be inclined to halt operation (Ogut and Hatch, 1988) once the power draw becomes economically unbearable. On the other hand, if input costs are too high, the reactor design would require maximum conversion and recycling, leading to diminishing returns such that the amount of equipment may limit the scale of operation.

Equipment selection and recommendations are dependent on a proper cash flow analysis that includes costs and price shocks to inputs, capital costs (debt facilities, lines of credit), and products (Aden et al., 2002; Patwardhan and Joshi, 1999; Wooley et al., 1999). The ultimate goal hinges on investment criteria. Operation can be optimized to meet a certain level of output, maximize profit, maximize the return on equity/capital, or a combination of those that would depend on the investors and their priorities.

12 Concluding Remarks

The interesting aspect of the bioreactor design developments has been that the different designs are more complementary than competitive in nature. For example, the stirred-tank bioreactor is capable of producing small bubbles, but costs too much to operate at a large scale. The fix has been the bubble column bioreactor. Under certain conditions, the bubble column has too much backmixing. The solution is the airlift reactor. If suspension is necessary, one can use a fixed bed reactor. Ultimately, these relationships may be a significant cause that research work rarely attempts to describe more than two reactor designs. In addition, any comparisons that are done have been accomplished at the experimental scale where the stirred-tank bioreactor weaknesses are not as apparent.

For gas–liquid mass transfer purposes, the stirred-tank bioreactor ranks better than the bubble column or airlift bioreactor, and the bubble column outperforms the airlift bioreactor most of the time. The fixed bed bioreactors cannot compete unless the microorganisms or biological material is highly shear sensitive. On a cost basis, the bubble column is the cheapest followed by the airlift bioreactor. The stirred-tank bioreactors have high power costs associated with impeller scaling, whereas the fixed bed bioreactors tend to have significant pressure drops and packing and maintenance costs. However, the airlift bioreactor is the easiest to scale based on hydrodynamics. The bubble column is harder to scale effectively while the stirred-tank bioreactor has the largest variety of published scale-up procedures. These procedures also tend to have variable success as defined by the process and depend on the reference and final scale.

The ultimate success of the reactor will depend on the comfort and well-being of the microorganisms in the bioreactor. In addition to gas–liquid mass transfer, the bioreactor also has to provide efficient mixing, a friendly shear environment, and proper pressure and temperature controls so that the microorganism output can be optimized. Hence, bioreactor design can have significant variation depending on the microorganism employed by the process.

The information that is available for the different bioreactor designs provides a good understanding of each bioreactor operation, construction, advantages, and disadvantages. These aspects have been summarized qualitatively in Table 12.1 for each reactor design where the relative benefit of a particular reactor type (indicated by the number of "+" signs) or challenge (−) is provided. A summary of some of the less common or novel bioreactor options is presented in Table 12.2.

An Introduction to Bioreactor Hydrodynamics and Gas-Liquid Mass Transfer, First Edition.
Enes Kadic and Theodore J. Heindel.
© 2014 John Wiley & Sons, Inc. Published 2014 by John Wiley & Sons, Inc.

TABLE 12.1 Bioreactor Comparison with Relative Benefit (+) or Challenge (−)

	Reactor Type	Brief Description	Economics	Operations/ Maintenance	Hydro-dynamics	Design	Scalability	Comments
Gas-liquid	Bubble column (BC)	Column reactor in which the medium is aerated and mixed by gas introduced at its base.	+++	++	++	+++	---	Lack of control; high shear gradients; reactor of choice for gas conversion; no moving parts
	Airlift reactor (ALR)	Modified BC with a channel for up- and downflow. Driving force is supplied by a density difference between these sections. More control than BC. Two major variants.						Flow defined by reactor design; minimum fluid level; modifications possible leading to further improvements
	Internal loop airlift reactor (ILALR)	ILALR has an internal flow separator creating channels for up- and downflow.	++	++	+++	++	++	Limited flow control
	External loop airlift reactor (ELALR)	ELALR has distinct conduits for fluid flow which are connected at the base and gas separator sections.	+	+++	+++	++	++	Even more control and design flexibility; better hydrodynamics than ILALR
	Stirred tank reactor (STR)	Mechanically agitated and mixed vessel with flow dependent on impeller design. High power consumption.	--	-	+++	++	+	High shear gradients; ideally mixed; limited economical operating range
	Packed bed reactors	Reactor in which liquid flows over immobilized solid material also known as packing material. Solids cannot be present in input or product.						Large units possible; usually counter current configuration which is limited by flooding
	Trickle bed reactor (TBR)	Liquid is sprayed over packing with product extraction at the bottom of the reactor.	+	--	+	-	+	
	Packed-bubble column (PBC)	Packed BC.	+	--	+	-	+	Liquid flow is negligible
Gas-liquid-solid	Slurry bubble column (SBC)	G–L–S or G–S bubble column. The G–S SBC has analogous hydrodynamic behavior as the G–S bubbling FBR.	+++	++	+	+++	---	
	Fluidized bed reactor (FBR)	Reactor in which the solid phase is suspended in medium.						Many variants are available; usually operated as gas–solid system. Typically not used as bioreactors.
	Bubbling	BC variant that can also be operated in G–S mode.	+++	++	+	+++	---	
	Circulating	ALR variant.	++	++	+++	++	++	Internal or external

TABLE 12.2   Summary of Less Important and Novel Bioreactors

	Description	Comments
Others:		
Jet reactors	Reactor in which the liquid (submerged-jet reactor) or gas (ejector reactor) is introduced at a high velocity. Another variant, the venturi scrubber reactor, injects liquid into a high velocity gas stream causing the liquid to atomize.	Large interfacial areas allowing high mass and heat transfer; not applicable to most organic substances
Membrane reactor	G–L–S or G–L reactor in which liquid is diffused through a membrane and converted by the bioflim, which is attached to the membrane, to the final product. Biofouling or clogging is possible. High construction and operating cost.	Four membrane types: microporous, porous, dense, and composite.
Plate columns	Liquid is channeled by plates (such as a muffler and run counter currently with the gas phase. Able to handle large variations in flow rates and high pressures.	Two major variants: bubble-cap and sieve plates
Tube Reactors	Cocurrent BC variant with pipelines or coils serving as guidelines. Very similar to heat exchanger.	horizontal, vertical or coiled; variety of flow regimes
Spray towers/column	Usually treated as a gas–liquid reactor. Liquid is sprayed counter currently to gas flow. Used for corrosive and liquids containing substantial amount of solid materials.	Higher energy usage and capital investment
Torus reactor	Mechanically agitated loop reactor. Lower power consumption, better mixing, and good heat transfer capacity than STR.	Many variants based on different cross-sectional shapes
Wetted wall reactor	Verticalreactor with liquid phase entering at the top and flowing along its wall with gas flowing through its core.	
Novel bioreactors:		
Magnetically stabilized FBR	FBR in which a magnetic field is used to stabilize magnetic particles in the fluidized bed.	
Dually injected turbulent separation FBR	FBR variant in which large and small particles are injected separately allowing for separate residence times.	Size and cost of the reactor are minimized
Continuous centrifugal bioreactor	FBR variant in which cells are fluidized using centrifugal forces. Allows high density cell cultivation.	May not be suitable for three phase fermentation
Inverse FBR	FBR variant in which low density particles with a biofilm are used. The biofilm causes a change in its thickness over time causing the bed to growing downward.	Superior to ALR for aerobic waste water treatment with certain cultures

(*continued*)

TABLE 12.2 *(Continued)*

	Description	Comments
Blenke-cascade reactor	Baffled tower separating the reactor into different sections being mixed by upward flowing gas. Similar to plate columns.	Liquid or liquid–solid mixture can be operated co- or countercurrently
Contained FBR	FBR variant that uses a retaining mesh or grid to contain solids and allow a liquid-only product.	
Double-entry FBR	FBR variant that uses top and bottom inlets to minimize gas logging (gas phase lifts solids to reactor surface).	

13 Nomenclature

a	Gas–liquid interfacial area (per unit liquid volume) (m^2/m^3)
$a(V_b, V_b')$	Coalescence frequency between bubbles of volume V_b and V_b' (s^{-1})
a_e	Effective surface area (per unit liquid volume) (m^2/m^3)
a_s	Surface area of a single element (m^2)
A	Constant $(-)$
A_d	Downcomer cross-sectional area (m^2)
A_r	Riser cross-sectional area (m^2)
Abs	Absorption value $(-)$
$b(V_b')$	Breakup frequency of bubbles of volume V_b' (s^{-1})
b_1	Constant $(-)$
b_2	Constant $(-)$
B	Constant $(-)$
B_i	Wall baffle length (m)
B_{ij}	Turbulent modeling coefficient $(-)$
B_W	Baffle width (m)
c_s	Solid concentration (kg/m^3)
C	Experimental constant; dissolved gas concentration in the liquid phase $(-; mol/m^3)$
C_B	Experimental constant $(-)$
C_{BP}	Virtual mass coefficient $(-)$
C_{BT}	Bubble-induced turbulence constant $(-)$
C_c	Cell concentration (kg/m^3)
C_{c0}	Initial cell concentration (kg/m^3)
C_{CO}	Carbon monoxide concentration (mol/m^3)
C_D	Drag coefficient $(-)$
C_E	Gas concentration at the electrode (mol/m^3)
C_i	Impeller clearance from the bottom of an STR (m)
C_L	Experimental constant $(-)$
C_o	Initial liquid-phase gas concentration (mol/m^3)

An Introduction to Bioreactor Hydrodynamics and Gas-Liquid Mass Transfer, First Edition.
Enes Kadic and Theodore J. Heindel.
© 2014 John Wiley & Sons, Inc. Published 2014 by John Wiley & Sons, Inc.

C_p	Myoglobin concentration (mol/m³)
C_s	Substrate concentration (kg/m³)
$C_{1\epsilon}, C_{2\epsilon}, C_{\mu,i}\ k - \epsilon$	Turbulence model parameters (−)
C_2	Constant (−)
C^*	Gas concentration in the gas phase (mol/m³)
C_∞	Dissolved gas concentration at a steady state (mol/m³)
d_B	Average bubble diameter (m)
d_d	Downcomer (hydraulic) diameter (m)
d_{disc}	Disc diameter of the disc column (m)
d_D	Draught tube diameter (m)
d_h	Hydraulic diameter of the packing (m)
d_m	Membrane thickness (m)
d_p	Average particle diameter (m)
d_P	Packing diameter (m)
d_R	Reactor diameter; riser diameter (m; m)
d_{SM}	Sauter mean bubble diameter (m)
d_o	Diameter of sparger orifice (m)
D	Diffusivity; constant (m²/s; −)
D_d	Gas distributor diameter (m)
D_c	Column internal diameter (m)
D_i	Impeller diameter (m)
D_L	Liquid diffusivity; axial dispersion coefficient (m²/s; m²/s)
D_m	Membrane diffusion coefficient (m²/s)
D_R	Reactor diameter (m)
D_T	Tank diameter (m)
e	Constant (−)
E	Constant (−)
E_{ij}	Turbulent energy exchange rate coefficient (kg/m³ s)
$f(\vec{x}, V_b, t)$	Bubble number density function (m⁻³)
F	Constant; volumetric flow rate (−; m³/s)
F_f	Coefficient (−)
$\vec{F}_{vm}$	Virtual mass force (N/m³)
g	Gravitational acceleration (m/s²)
G_i	Production of turbulent kinetic energy (m²/s³)
h_c	Downcomer clearance height (m)
h_{DC}	Critical downcomer sparger distance (m)
h_P	Packing height (m)
h_R	Effective reactor height (m)
H	Static or unaerated liquid height (m)
H_b	Liquid height with bubbling (m)
H_0	Ungassed liquid height (m)
I	Ion strength (g-ion/L)
J	Mass flux (kg/s m²)
k_G	Gas-phase mass transfer coefficient (m/s)

k_L	Liquid-phase mass transfer coefficient (m/s)
$k_L a$	Volumetric gas–liquid mass transfer coefficient (s^{-1})
k	Turbulent kinetic energy per unit mass; rate constant (m^2/s^2; s^{-1})
K	Constant (−)
$\vec{K}_{dc}$	Drag force (N/m^3)
K_{ij}	Interfacial momentum exchange coefficient (kg/m^3)
K_s	Saturation constant (kg/m^3)
L_c	Tube connector length (−)
L_D	Draught tube length (−)
L_h	Reactor height; used by Choi (2001) (−)
L_m	Liquid mass superficial velocity (kmol/m^2 s)
m	Parameter group (−)
$m(V_b')$	Mean number of daughter bubbles produced by breakup of a parent bubble of volume V_b' (−)
n	Constant; exponent (− ; −)
N	Impeller speed (rev/min)
N_A	Avogadro's number (mol^{-1})
N_{CD}	Impeller speed at which a transition in flow regime occurs from loaded to completely dispersed (rev/min)
N_{CSA}	Critical impeller speed for surface aeration (rev/min)
N_F	Flooded regime impeller speed (rev/min)
N_{FL}	Impeller speed at which a transition in flow regime occurs from flooded to loaded (rev/min)
N_R	Gross recirculation regime impeller speed (rev/min)
N_o	Constant (−)
Δp	Pressure drop (Pa)
P	Pressure; power (Pa; W)
$P(V_b, V_b')$	Probability density function of daughter bubbles produced upon breakup of a parent bubble of volume V_b' (m^{-3})
P_B	Bubble pressure (Pa)
P_o	Ungassed impeller power draw (W)
P_g	Gassed impeller power draw (W)
P_{sat}	Saturation pressure (Pa)
P_{TOT}	Total power (W)
P_v	Energy dissipation rate (kW/m^3)
qX	Microbial gas consumption rate (mol/s)
Q_G	Volumetric gas flow rate (m^3/s)
r_B	Bubble radius (m)
r_c	Gas-phase reaction rate at the liquid interface (mol/m^3 s)
r_U	Radius of the gas particle (m)
R	Universal gas constant; source term (kJ/K kmol; kg/m s)

R_A	Gas distributor open area ratio (−)
s	Surface renewal rate; thickness of the sloped metal sheet (s; m)
S_p	Source/sink term due to bubbles being added/subtracted from the bubble class of volume V_b due to pressure change (m^3/s)
S_{ph}	Source/sink term due to bubbles being added/subtracted from the bubble class of volume V_b due to phase change (m^3/s)
S_r	Source/sink term due to bubbles being added/subtracted from the bubble class of volume V_b due to a reaction (m^3/s)
SS	Percent match to saturated carbon monoxide spectrum (%)
t	Time (s)
t_c	Circulation time (s)
t_P	Thickness of expanded metal sheets (m)
T	Temperature; STR tank diameter (°C; m)
T_o	Initial temperature (°C)
T_∞	Final temperature (°C)
$\vec{u}$	Velocity (m/s)
$\vec{u}_b(\vec{x}, V_b, t)$	Local bubble velocity function (m/s)
U_b	Bubble terminal velocity (m/s)
U_c	Mean circulation velocity (m/s)
U_G	Superficial gas velocity (m/s)
U_{Gr}	Riser superficial gas velocity (m/s)
U_L	Superficial liquid velocity (m/s)
U_{Lc}	Liquid circulation velocity (m/s)
U_{Ld}	Superficial liquid velocity in the downcomer (m/s)
U_{Ll}	Peripheral liquid flow rate (kg/m s)
U_{Lr}	Superficial liquid velocity in the riser (m/s)
U_{Ls}	Liquid velocity in the separator (m/s)
U_o	Gas velocity through orifice (m/s)
U_P	Superficial velocity losses to the surroundings (m/s)
U_R	Bubble rise velocity (m/s)
U_{sG}	Average slip velocity (m/s)
U_t	Terminal bubble velocity (m/s)
U_{trans}	Superficial gas velocity at which flow transitions out of the homogeneous flow regime into the transition flow regime (m/s)
U_∞	Terminal rise velocity (m/s)
vvm	Gas volume per unit liquid volume per minute $(m^3/m^3 \text{ min})$
V_b	Bubble volume (m^3)
V_L	Liquid linear velocity; liquid volume $(m/s; m^3)$
$V_{L,D}$	Downcomer liquid velocity (m/s)

$V_L(0)$	Center-line liquid velocity; used by Zehner (1989) (m/s)
$\forall_g$	Gas volume (m^3)
$\forall_L$	Liquid volume (m^3)
Vol_L	Liquid volume (m^3)
Vol_S	Volume of dissolved carbon monoxide liquid sample (m^3)
Vol_T	Total cuvette liquid volume (m^3)
W	Impeller blade width (m)
W_L	Found in Kantarci et al. (2005) but not defined ($-$)
W_s	Found in Kantarci et al. (2005) but not defined ($-$)
x	Constant ($-$)
$\vec{x}$	Position vector (m)
x_c	Circulation path (m)
y	Distance normal to surface (m)
z	Axial location (m)
Δz	Axial distance (m)

Abbreviations

ALR	Airlift reactor
A-310	Lightnin A-310 axial flow impeller (hydrofoil)
A-315	Lightnin A-315 axial flow impeller (hydrofoil)
BC	Bubble column
BIP	Bubble-induced pressure
BP	Bubble pressure
BPBE	Bubble population balance equation
BT-6	Concave blade disc turbine (radial flow impeller)
Carbopol	Carboxypolymethylene
CFD	Computational fluid dynamics
CMC	Carboxymethylcellulose
CS	Cylindrical sintered sparger
DGD	Dynamic gas disengagement
DNS	Direct numerical simulation
DR	Dilution ratio
DSF	Decision support framework
DT-ILALR	Draught tube airlift reactor
ELALR	External-loop airlift reactor
FBR	Flooded bed reactor; fluidized bed reactor
FP	Fixed packing
GTR	Gas transfer rate
HETP	Height equivalent to a theoretical plate
ILALR	Internal-loop airlift reactor
MCE	Maximum efficiency capacity
NS	Narcissus impeller (radial flow impeller)
PBD	Down-pumping pitched blade turbine
PBR	Packed bed reactor

PBT	Pitched blade disc turbine
PBU	Up-pumping pitched blade turbine
PP	Perforated plate or tube
PS	(Perforated) plane sparger (as cited by Merchuk et al. (1998))
RP	Random packing
RT	Rushton-type impeller (radial flow impeller)
RTD	Residence time distribution
STR	Stirred-tank reactor
TBR	Trickle bed reactor
TX	Techmix 335 (axial flow impeller)
TXD	Down-pumping Techmix 335
TXU	Up-pumping Techmix 335

Greek Symbols

α	Constant; volumetric phase fraction $(-; -)$
a_{dcp}	Gas holdup at close packing $(-)$
β	Constant $(-)$
γ	Shear rate (s^{-1})
δ	Diffusion layer thickness; modified film thickness, $\delta = (\mu_L^2/\rho_L^2 g)^{1/3}$ (m; m)
δ_{eff}	Liquid-phase film layer (m)
δ_o	Viscous sublayer thickness (m)
ϵ	Overall gas holdup or void fraction; turbulent energy dissipation rate $(-; \mathrm{m}^2/\mathrm{s}^3)$
ϵ_b	Baseline gas holdup used by Salvacion et al. (1995) $(-)$
ϵ_d	Downcomer gas holdup $(-)$
ϵ_G	Gas holdup $(-)$
ϵ_L	Liquid void fraction $(-)$
ϵ_m	Protein extinction coefficient (mol/m)
ϵ_r	Riser gas holdup $(-)$
ϵ_s	Solid holdup $(-)$
ϵ_0	Gas holdup at atmospheric pressure used by Kojima et al. (1997) $(-)$
$\bar{\epsilon}$	Average gas holdup $(-)$
θ_θ	Contact angle of wettability (degrees)
λ	Cuvette path length (m)
λ_L	Liquid thermal conductivity $(\mathrm{W/m\,K})$
μ	Liquid dynamic viscosity; specific cellular growth rate $(\mathrm{Pa\text{-}s}; \mathrm{s}^{-1})$
μ_a	Apparent dynamic viscosity (Pa-s)
μ_G	Gas dynamic viscosity (Pa-s)
μ_L	Liquid dynamic viscosity (Pa-s)
μ_m	Maximum specific cellular growth rate (s^{-1})

μ_t	Turbulent viscosity (Pa-s)
μ_w	Dynamic viscosity of tap water (Pa-s)
ν_L	Liquid kinematic viscosity (m^2/s^2)
ρ	Density (kg/m^3)
ρ_G	Gas density (kg/m^3)
ρ_L	Liquid density (kg/m^3)
ρ_m	Slurry density in the absence of gas (kg/m^3)
ρ_s	Density of solids (kg/m^3)
$\Delta\rho$	Phase density difference (kg/m^3)
σ	Liquid-phase surface tension (mN/m or dyn/cm)
σ_i	Turbulent parameters in $k-\epsilon$ equations (N/m)
σ_w	Surface tension of tap water (mN/m or dyn/cm)
τ	Contact time (s)
$\tau\delta$	Dead time (s)
τ_e	Electrode time constant (s)
τ_g	Gas-phase residence time (s)
$\tau_{i\phi}$	Time constant in turbulent dissipation (s)
τ_w	Wall shear stress (N/m^2)
$\overline{\tau}_w$	Average wall shear stress (N/m^2)
$\overline{\overline{\tau}}$	Effective shear tensor (N/m^3)
φ_s	Volumetric solid fraction (−)

Dimensionless Numbers

Bo	Bond number, $Bo = \rho_L g D_R^2 / \sigma$
Fl_G	Gas flow number, $Fl_G = Q_G / ND^3$
Fr	Froude number, $Fr = U_G / \sqrt{g D_R}$
Ga	Galileo number, $Ga = g D_R^3 / v_L^2$
Ga_L	Galileo number, $Ga_L = g \rho_L^2 / a^3 \mu_L^2$
Ga_{Lh}	Galileo number, $Ga_{Lh} = g t_p^3 / v_L^2$
G	Grashof number, $Gr = d_{Bs}^3 \rho_L \Delta \rho g / \mu_L^2$
Ha	Hatta number, ratio of species absorption with and without reaction
Mo	Morton number, $Mo = g \mu_L^4 / \rho_L \sigma^3$
Mo_m	Modified Morton number, $Mo_m = (\xi \mu_L)^4 g / \rho_m \sigma^3$ where $\ln \xi = 4.6\epsilon_s \{5.7\epsilon_s^{0.58} \sinh[-0.71 \exp(-5.8\epsilon_s) \ln Mo^{0.22}] + 1\}$
N_{Pg}	Gassed power number, $N_{Pg} = P_g / \rho N^3 D^5$
N_{Po}	Ungassed power number, $N_{Po} = P_o / \rho N^3 D^5$
Re	Reynolds number, $Re = \rho ND^2 / \mu$
Re_B	Bubble Reynolds number, $Re_B = \rho_c |\vec{u}_d - \vec{u}_c| d_B / \mu_c$
Re_L	Liquid-phase Reynolds number, $Re_L = \rho_L U_L D_R / \mu_L$
Re_{Le}	Reynolds number, $Re_{Le} = 4 U_L / a_e v_L$
Re_P	Reynolds number used by Mena et al. (2005); defined in terms of particle diameter

Re_T	Reynolds number used by Kantarci et al. (2005); not defined
Sc or Sc_L	Schmidt number, $Sc = \nu_\mathrm{L}/D_\mathrm{L}$
Sh	Sherwood number, usually $Sh = k_\mathrm{L} a d_\mathrm{R}/D_\mathrm{L}$
Sh_L	Sherwood number, usually $Sh_\mathrm{L} = k_\mathrm{L} d_\mathrm{P}/D_\mathrm{L}$
We	Weber number, $We = \rho_\mathrm{L} U_\mathrm{L}^{2} D_\mathrm{R}/\sigma$

Bibliography

Abashar, M.E., Narsingh, U., Rouillard, A.E., and Judd, R. (1998), "Hydrodynamic flow regimes, gas holdup, and liquid circulation in airlift reactors," *Industrial & Engineering Chemistry Research*, 37: 1251–1259.

Aden, A., Ruth, M., Ibsen, K., Jechura, J., Neeves, K., Sheehan, J., Wallace, B., Montague, L., Slayton, A., and Lukas, J. (2002), *Lignocellulosic biomass to ethanol process design and economics utilizing co-current dilute acid prehydrolysis and enzymatic hydrolysis for corn stover*, Golden, Colorado, National Renewable Energy Laboratory.

Aiba, S., Ohashi, M., and Huang, S.Y. (1968), "Rapid determination of oxygen permeability of polymer membranes," *Industrial & Engineering Chemistry, Fundamentals*, 7(3): 497–502.

Akita, K., and Yoshida, F. (1973), "Gas holdup and volumetric mass transfer coefficient in bubble columns," *Industrial & Engineering Chemistry, Process Design and Development*, 12(1): 76–80.

Akita, K., and Yoshida, F. (1974), "Bubble size, interfacial area, and liquid-phase mass transfer coefficient in bubble columns," *Industrial & Engineering Chemistry, Process Design and Development*, 13(1): 84–91.

Al-Dahhan, M.H., Larachi, F., Dudukovic, M.P., and Laurent, A. (1997), "High-pressure trickle-bed reactors: A review," *Industrial & Engineering Chemistry Research*, 36(8): 3292–3314.

Al-Masry, W.A. (1999), "Effect of liquid volume in the gas-separator on the hydrodynamics of airlift reactors," *Journal of Chemical Technology & Biotechnology*, 74(10): 931–936.

Al-Masry, W.A. (2001), "Gas holdup in circulating bubble columns with pseudoplastic liquids," *Chemical Engineering & Technology*, 24(1): 71–76.

Albijanic, B., Havran, V., Petrovic, L., Duric, M., and Tekic, M.N. (2007), "Hydrodynamics and mass transfer in a draft tube airlift reactor with dilute alcohol solutions," *AIChE Journal*, 53(11): 2897–2904.

Alix, P., and Raynal, L. (2008), "Liquid distribution and liquid hold-up in modern high capacity packings," *Chemical Engineering Research & Design*, 86(6): 585–591.

Álvarez, E., Gomez-Diaz, D., Navaza, J.M., and Sanjurjo, B. (2008), "Continuous removal of carbon dioxide by absorption employing a bubble column," *Chemical Engineering Journal*, 137(2): 251–256.

An Introduction to Bioreactor Hydrodynamics and Gas-Liquid Mass Transfer, First Edition.
Enes Kadic and Theodore J. Heindel.
© 2014 John Wiley & Sons, Inc. Published 2014 by John Wiley & Sons, Inc.

Álvarez, E., Sanjurjo, B., Cancela, A., and Navaza, J.M. (2000), "Mass transfer and influence of physical properties of solutions in a bubble column," *Chemical Engineering Research & Design*, 78(6): 889–893.

Anabtawi, M.Z.A., Abu-Eishah, S.I., Hilal, N., and Nabhan, M.B.W. (2003), "Hydrodynamic studies in both bi-dimensional and three-dimensional bubble columns with a single sparger," *Chemical Engineering and Processing*, 42(5): 403–408.

Andre, G., Moo-Young, M., and Robinson, C.W. (1981), "Improved method for the dynamic measurement of mass transfer coefficient for application to solid-substrate fermentation," *Biotechnology and Bioengineering*, 23(7): 1611–1622.

Anonymous (2005), "4500-o oxygen (dissolved)," in *Standard Methods for the Examination of Water and Wastewater* 21st Edition, M.A.H. Franson, Editor, New York, American Public Health Association, American Water Works Association, & Water Pollution Control Federation: 4.136–4.143.

Antonini, E., and Brunori, M. (1971), "Memoglobin and myoglobin in their reactions with ligands," in *Frontiers in Biology*, A. Neuberger, and E.L. Tatum, Editors, Amsterdam-London, North-Holland Reserach Monographs. **21**.

Ascanio, G., Castro, B., and Galindo, E. (2004), "Measurement of power consumption in stirred vessels—a review," *Chemical Engineering Research and Design: Transactions of the Institution of Chemical Engineers Part A*, 82(A9): 1282–1290.

Attou, A., Boyer, C., and Ferschneider, G. (1999), "Modelling of the hydrodynamics of the cocurrent gas-liquid trickle flow through a trickle-bed reactor," *Chemical Engineering Science*, 54(6): 785–802.

Azbel, D. (1981), *Two-Phase Flows in Chemical Engineering*, Cambridge University Press, New York.

Azzopardi, B.J., Mudde, R.F., Lo, S., Morvan, H., Yan, Y., and Zhao, D. (2011), *Hydrodynamics of Gas-Liquid Reactors: Normal Operation and Upset Conditions*, West Sussex, UK, John Wiley & Sons.

Bach, H.F., and Pilhofer, T. (1978), "Variation of gas hold-up in bubble columns with physical properties of liquids and operating parameters of columns," *German Chemical Engineering*, 1(5): 270–275.

Baker, R.W. (2004), *Membrane Technology and Applications*, John Wiley & Sons, Ltd., West Sussex, England.

Bakker, A. (1992), "Hydrodynamics of stirred gas-liquid dispersion," PhD Thesis, Delft University of Technology, Delft, The Netherlands.

Bakker, A., and Oshinowo, L.M. (2004), "Modelling of turbulence in stirred vessels using large eddy simulation," *Chemical Engineering Research & Design*, 82(9 SPEC ISS): 1169–1178.

Barakat, T.M.M., and Sorensen, E. (2008), "Simultaneous optimal synthesis, design and operation of batch and continuous hybrid separation processes," *Chemical Engineering Research & Design*, 86(3): 279–298.

Bauer, T., Haase, S., Al-Dahhan, M.H., and Lange, R. (2007), "Paper 559d - Monolithic reactor and particle-packed monolithic reactor for three-phase catalytic reactions", in *The 2007 Annual Meeting*, Salt Lake City, UT, AIChE.

Beck, M.S., and Williams, R.A. (1996), "Process tomography: A European innovation and its applications," *Measurement Science and Technology*, 7: 215–224.

Beenackers, A.A.C.M., and Van Swaaij, W.P.M. (1993), "Mass transfer in gas-liquid slurry reactors," *Chemical Engineering Science*, 48(18): 3109–3139.

Behkish, A., Men, Z., Inga, J.R., and Morsi, B.I. (2002), "Mass transfer characteristics in a large-scale slurry bubble column reactor with organic liquid mixtures," *Chemical Engineering Science*, 57(16): 3307–3324.

Bellgardt, K.-H. (2000a), "Bioprocess models," in *Bioreaction Engineering: Modeling and Control*, K. Schugerl, and K.-H. Bellgardt, Editors, Berlin, Springer: 44–105.

Bellgardt, K.-H. (2000b), *Bioreaction Engineering: Modeling and Control*, Springer, New York City, New York.

Bellgardt, K.-H. (2000c), "Introduction," in *Bioreaction Engineering: Modeling and Control*, K. Schugerl, and K.-H. Bellgardt, Editors, Berlin, Springer: 1–18.

Bello, R.A. (1981), "A characterization study of airlift contactors for applications to fermentation," PhD Thesis, University of Waterloo, Ontario, Canada.

Bello, R.A., Robinson, C.W., and Moo-Young, M. (1984), "Liquid circulation and mixing characteristics of airlift contactors," *Canadian Journal of Chemical Engineering*, 62(5): 573–577.

Bello, R.A., Robinson, C.W., and Moo-Young, M. (1985a), "Gas holdup and overall volumetric oxygen transfer coefficient in airlift contactors," *Biotechnology and Bioengineering*, 27(3): 369–381.

Bello, R.A., Robinson, C.W., and Moo-Young, M. (1985b), "Prediction of the volumetric mass transfer coefficient in pneumatic contactors," *Chemical Engineering Science*, 40(1): 53–58.

Bendjaballah, N., Dhaouadi, H., Poncin, S., Midoux, N., Hornut, J.-M., and Wild, G. (1999), "Hydrodynamics and flow regimes in external loop airlift reactors," *Chemical Engineering Science*, 54(21): 5211–5221.

Bentifraouine, C., Xuereb, C., and Riba, J.-P. (1997a), "Effect of gas liquid separator and liquid height on the global hydrodynamic parameters of an external loop airlift contactor," *Chemical Engineering Science*, 66: 91–95.

Bentifraouine, C., Xuereb, C., and Riba, J.-P. (1997b), "An experimental study of the hydrodynamic characteristics of external loop airlift contactors," *Journal of Chemical Technology & Biotechnology*, 69: 345–349.

Benz, G.T. (2008), "Piloting bioreactors for agitation scale-up," *Chemical Engineering Progress*, 104(2): 32–34.

Beyerlein, S.W., Cossmann, R.K., and Richter, H.J. (1985), "Prediction of bubble concentration profiles in vertical turbulent two-phase flow," *International Journal of Multiphase Flow*, 11(5): 629–641.

Biesheuvel, A., and Gorissen, W.C.M. (1990), "Void fraction disturbances in a uniform bubbly fluid," *International Journal of Multiphase Flow*, 16(2): 211–231.

Billet, R. (1989), "Packed column analysis and design," PhD Thesis, Ruhr-Universität Bochum, Bochum, Germany.

Billet, R. (1993), "Process engineering evaluation of packings and the limits of their development," *Chemie-Ingenieur-Technik*, 65(2): 157–166.

Billet, R., and Schultes, M. (1993), "Physical model for the prediction of liquid hold-up in two-phase countercurrent columns," *Chemical Engineering & Technology*, 16(6): 370–375.

Birch, D., and Ahmed, N. (1996), "Gas sparging in vessels agitated by mixed flow impellers," *Powder Technology*, 88(1): 33–38.

Birch, D., and Ahmed, N. (1997), "The influence of sparger design and location on gas dispersion in stirred vessels," *Chemical Engineering Research & Design*, 75(5): 487–496.

Bisang, J.M. (1997), "Modelling the startup of a continuous parallel plate electrochemical reactor," *Journal of Applied Electrochemistry*, 27(4): 379–384.

Blanch, H.W. and Clark, D.S. (1997), *Biochemical Engineering*, Marcel Dekker, New York.

Blazej, M., Annus, J., and Markos, J. (2004a), "Comparison of gassing-out and pressure-step dynamic methods for $k_l a$ measurement in an airlift reactor with internal loop," *Chemical Engineering Research and Design*, 82(10): 1375–1382.

Blazej, M., Jurascik, M., Annus, J., and Markos, J. (2004b), "Measurement of mass transfer coefficient in an airlift reactor with internal loop using coalescent and non-coalescent liquid media," *Journal of Chemical Technology & Biotechnology*, 79(12): 1405–1411.

Blazej, M., Kisa, M., and Markos, J. (2004c), "Scale influence on the hydrodynamics of an internal loop airlift reactor," *Chemical Engineering and Processing*, 43(12): 1519–1527.

Blenke, H. (1979), "Loop reactors," *Advances in Biochemical Engineering*, 13: 121–214.

Bliem, R., and Katinger, H. (1988a), "Scale-up engineering in animal cell technology: Part I," *Trends in Biotechnology*, 6(8): 190–195.

Bliem, R., and Katinger, H. (1988b), "Scale-up engineering in animal cell technology: Part II," *Trends in Biotechnology*, 6(9): 224–230.

Boden, S., Bieberle, M., and Hampel, U. (2008), "Quantitative measurement of gas hold-up distribution in a stirred chemical reactor using X-ray cone-beam computed tomography," *Chemical Engineering Journal*, 139(2): 351–362.

Bouaifi, M., Hebrard, G., Bastoul, D., and Roustan, M. (2001), "A comparative study of gas hold-up, bubble size, interfacial area, and mass transfer coefficients in stirred gas-liquid reactors and bubble columns," *Chemical Engineering and Processing*, 40(2): 97–111.

Bouaifi, M., and Roustan, M. (2001), "Power consumption, mixing time and homogenisation energy in dual-impeller agitated gas-liquid reactors," *Chemical Engineering and Processing*, 40(2): 87–95.

Boyer, C., Duquenne, A.M., and Wild, G. (2002), "Measuring techniques in gas-liquid and gas-liquid-solid reactors," *Chemical Engineering Science*, 57: 3185–3215.

Branyik, T., Vicente, A.A., Dostalek, P., and Teixeira, J.A. (2005), "Continuous beer fermentation using immobilized yeast cell bioreactor systems," *Biotechnology Progress*, 21(3): 653–663.

Bredwell, M.D., Srivastava, P., and Worden, R.M. (1999), "Reactor design issues for synthesis-gas fermentations," *Biotechnology Progress*, 15(5): 834–844.

Bredwell, M.D., and Worden, R.M. (1998), "Mass-transfer properties of microbubbles: I. Experimental studies," *Biotechnology Progress*, 14(1): 31–38.

Breijer, A.A.J., Nijenhuis, J., and van Ommen, J.R. (2008), "Prevention of flooding in a countercurrent trickle-bed reactor using additional void space," *Chemical Engineering Journal*, 138(1–3): 333–340.

Bridgwater, A.V. (1995), "The technical and economic feasiblity of biomass gasification for power generation," *Fuel*, 74: 631–653.

Broder, D., and Sommerfeld, M. (2003), "Combined PIV/PTV-measurements for the analysis of bubble interactions and coalescence in a turbulent flow," *The Canadian Journal of Chemical Engineering*, 81(3–4): 756–763.

Broekhuis, R.R., Machado, R.M., and Nordquist, A.F. (2001), "The ejector-driven monolith loop reactor: Experiments and modeling," *Catalysis Today*, 69(1–4): 87–93.

Brown, R.C. (2005), "Biomass refineries based on hybrid thermochemical/biological processing – An overview," in *Biorefineries, Biobased Industrial Processes and Products*, B. Kamm, P.R. Gruber, and M. Kamm, Editors, Weinheim, Germany, Wiley-VCH Verlag: 1–24.

Burghardt, A., Bartelmus, G., Jaroszynski, M., and Kolodziej, A. (1995), "Hydrodynamics and mass transfer in a three-phase fixed-bed reactor with cocurrent gas-liquid downflow," *The Chemical Engineering Journal and the Biochemical Engineering Journal*, 58(2): 83–99.

Cabaret, F., Fradette, L., and Tanguy, P.A. (2008), "Gas-liquid mass transfer in unbaffled dual-impeller mixers," *Chemical Engineering Science*, 63(6): 1636–1647.

Cai, J., Nieuwstad, T.J., and Kop, J.H. (1992), "Fluidization and sedimentation of carrier material in a pilot-scale airlift internal-loop reactor," *Water Science & Technology*, 26(9–11): 2481–2484.

Calderbank, P.H. (1958), "Physical rate processes in industrial fermentation. Part I: The interfacial area in gas-liquid contacting with mechanical agitation," *Transactions of IChemE, Part A, Chemical Engineering Research and Design*, 36: 443–463.

Calderbank, P.H., and Moo-Young, M.B. (1961), "The continuous phase heat and mass-transfer properties of dispersion," *Chemical Engineering Science*, 16(1–2): 39–54.

Carroll, J.J. (1991), "What is Henry's law?," *Chemical Engineering Progress*, 87(9): 48–52.

Ceccio, S.L., and George, D.L. (1996), "A review of electrical impedance techniques for the measurement of multiphase flows," *Journal of Fluids Engineering - Transactions of the ASME*, 118: 391–399.

Chakravarty, M., Begum, S., Singh, H.D., Baruah, J.N., and Iyengar, M.S. (1973), "Gas holdup distribution in a gas-lift column," *Biotechnology and Bioengineering Symposium*, 4: 363–378.

Chakravarty, M., Singh, H.D., Baruah, J.N., and Lyengar, M.S. (1974), "Gas holdup distribution in a gas-lift column," *Indian Chemical Engineer*, 16: 17–22.

Chand Eisenmann Metallurgical, (2008), "Spargers," Retrieved August 7, 2008, from http://www.chandeisenmann.com/products/spargers.asp.

Chandrasekharan, K., and Calderbank, P.H. (1981), "Further observations on the scale-up of aerated mixing vessels," *Chemical Engineering Science*, 36(5): 818–823.

Chang, H.N., Halard, B., and Moo-Young, M. (1989), "Measurement of $k_L a$ by a gassing-in method with oxygen-enriched air," *Biotechnology and Bioengineering*, 34(9): 1147–1157.

Chaouki, J., Larachi, F., and Dudukovic, M.P. (eds) (1996), *Non-Invasive Monitoring of Multiphase Flows*, Elsevier, New York.

Chaouki, J., Larachi, F., and Dudukovic, M.P. (1997), "Noninvasive tomography and velocimetric monitoring of multiphase flows," *Industrial & Engineering Chemistry Research*, 36: 4476–4503.

Charpentier, J.-C. (1981), "Mass-transfer rates in gas-liquid absorbers and reactors," in *Advances in Chemical Engineering*, T.B. Drew, G.E. Cokelet, J.W. Hoopes, Jr., and T. Vermeulen, Editors, New York City, New York, Academic Press. **11**: 1–133.

Chaumat, H., Billet-Duquenne, A.M., Augier, F., Mathieu, C., and Delmas, H. (2005), "Mass transfer in bubble column for industrial conditions: Effects of organic medium, gas and liquid flow rates and column design," *Chemical Engineering Science*, 60(22): 5930–5936.

Chen, P., Sanyal, J., and Dudukovic, M.P. (2005), "Numerical simulation of bubble column flows: Effect of different breakup and coalescence closures," *Chemical Engineering Science*, 60: 1085–1101.

Chen, R.C., Reese, J., and Fan, L.S. (1994), "Flow structure in a three-dimensional bubble column and three-phase fluidized bed," *AIChE Journal*, 40(7): 1093–1104.

Chen, Z.D., and Chen, J.J. (1999), "Comparison of mass transfer performance for various single and twin impellers," *Chemical Engineering Research & Design*, 77(2): 104–109.

Cheremisinoff, N.P. (1986), "Measurement techniques for multiphase flows," in *Encyclopedia of Fluid Mechanics Volume 1 - Flow Phenomena and Measurement*, N.P. Cheremisinoff, Editor, Houston, Gulf Publishing Company: 1280–1338.

Chilekar, V.P., Warnier, M.J.F., van der Schaaf, J., Kuster, B.F.M., Schouten, J.C., and van Ommen, J.R. (2005), "Bubble size estimation in slurry bubble columns from pressure fluctuations," *AIChE Journal*, 51(7): 1924–1937.

Chisti, M.Y. (1989), *Airlift Bioreactors*, Elsevier Applied Science, London.

Chisti, M.Y., Halard, B., and Moo-Young, M. (1988a), "Liquid circulation in airlift reactors," *Chemical Engineering Science*, 43(3): 451–457.

Chisti, M.Y., and Moo-Young, M. (1987), "Airlift reactors: Characteristics, applications, and design considerations," *Chemical Engineering Communications*, 60: 195–242.

Chisti, Y., Fujimoto, K., and Moo-Young, M. (1988b), *Biotechnology Progress*, 5: 72–76.

Chisti, Y., and Jauregui-Haza, U.J. (2002), "Oxygen transfer and mixing in mechanically agitated airlift bioreactors," *Biochemical Engineering Journal*, 10(2): 143–153.

Chisti, Y., Kasper, M., and Moo-Young, M. (1990), "Mass transfer in external-loop airlift bioreactors using static mixers," *Canadian Journal of Chemical Engineering*, 68(2): 45–50.

Chisti, Y., and Moo-Young, M. (1986), "Disruption of microbial cells for intracellular products," *Enzyme and Microbial Technology*, 8(4): 194–204.

Cho, J.S., and Wakao, N. (1988), "Determination of liquid-side and gas-side volumetric mass transfer coefficients in a bubble column," *Journal of Chemical Engineering of Japan*, 21(6): 576–581.

Choi, J.H., and Lee, W.K. (1993), "Circulating liquid velocity, gas holdup and volumetric oxygen transfer coefficient in external-loop airlift reactors," *Journal of Chemical Technology & Biotechnology*, 56: 51–58.

Choi, K.H. (2001), "Hydrodynamic and mass transfer characteristics of external-loop airlift reactors without an extension tube above the downcomer," *Korean Journal of Chemical Engineering*, 18(2): 240–246.

Cockx, A., Roustan, M., Line, A., and Hebrard, G. (1995), "Modelling of mass transfer coefficient k_L in bubble columns," *Chemical Engineering Research & Design*, 73(A6): 627–631.

Daly, J.G., Patel, S.A., and Bukur, D.B. (1992), "Measurement of gas holdups and Sauter mean bubble diameters in bubble column reactors by dynamic gas disengagement method," *Chemical Engineering Science*, 47(13/14): 3647–3654.

Dang, N.D.P., Karrer, D.A., and Dunn, I.J. (1977), "Oxygen transfer coefficients by dynamic model moment analysis," *Biotechnology and Bioengineering*, 19(6): 853–865.

Dechsiri, C., Ghione, A., van de Weil, F., Dehling, H.G., Paans, A.M.J., and Hoffmann, A.C. (2005), "Positron emission tomography applied to fluidization engineering," *The Canadian Journal of Chemical Engineering*, 83: 88–96.

Deckwer, W.-D. (1992), *Bubble Column Reactors*, Wiley & Sons, New York, NY.

Delnoij, E., Kuipers, J.A.M., and van Swaaij, W.P.M. (1997a), "Computational fluid dynamics applied to gas-liquid contactors," *Chemical Engineering Science*, 52(21/22): 3623–3638.

Delnoij, E., Kuipers, J.A.M., and van Swaaij, W.P.M. (1997b), "Dynamic simulation of gas-liquid two-phase flow: Effect of column aspect ratio on the flow structure," *Chemical Engineering Science*, 52(21/22): 3759–3772.

Delnoij, E., Lammers, F.A., Kuipers, J.A.M., and van Swaaij, W.P.M. (1997c), "Dynamic simulation of dispersed gas-liquid two-phase flow using a discrete bubble model," *Chemical Engineering Science*, 52(9): 1429–1458.

Demirel, B., and Yenigün, O. (2002), "Two-phase anaerobic digestion processes: A review," *Journal of Chemical Technology & Biotechnology*, 77(7): 743–755.

Deshpande, N.S., Dinkar, M., and Joshi, J.B. (1995), "Disengagement of the gas phase in bubble columns," *International Journal of Multiphase Flow*, 21(6): 1191–1201.

Devanathan, N., Moslemian, D., and Dudukovic, M.P. (1990), "Flow mapping in bubble columns using CARPT," *Chemical Engineering Science*, 45(8): 2285–2291.

Dhaouadi, H., Poncin, S., Hornut, J.M., and Midoux, N. (2008), "Gas-liquid mass transfer in bubble column reactor - Analytical solution and experimental confirmation," *Chemical Engineering and Processing Process Intensification*, 47(4): 548–556.

Díaz, M.E., Montes, F.J., and Galán, M.A. (2008), "Experimental study of the transition between unsteady flow regimes in a partially aerated two-dimensional bubble column," *Chemical Engineering and Processing Process Intensification*, 47(9–10): 1867–1876.

Dijkhuizen, W., Roghair, I., van Sint Annaland, M., and Kuipers, J.A.M. (2010), "DNS of gas bubbles behaviour using an improved 3D front tracking model—model development," *Chemical Engineering Science*, 65(4): 1427–1437.

Doan, H.D., Wu, J., and Eyvazi, M.J. (2008), "Effect of liquid distribution on the organic removal in a trickle bed filter," *Chemical Engineering Journal*, 139(3): 495–502.

Domingues, L., and Teixeira, N.L.J.A. (2000), "Contamination of a high-cell-density continuous bioreactor," *Biotechnology and Bioengineering*, 68(5): 584–587.

Donati, G., and Paludetto, R. (1999), "Batch and semibatch catalytic reactors (from theory to practice)," *Catalysis Today*, 52(2–3): 183–195.

Doran, P.M. (2013), *Bioprocess Engineering Principles*, 2nd edn, Elsevier, Oxford, UK.

Drake, J.B., and Heindel, T.J. (2011), "The repeatability and uniformity of 3D fluidized beds," *Powder Technology*, 213(1–3): 148–154.

Drake, J.B., and Heindel, T.J. (2012), "Local time-average gas holdup comparisons in cold flow fluidized beds with side-air injection," *Chemical Engineering Science*, 68(1): 157–165.

Du, B., Warsito, W., and Fan, L.S. (2006), "Imaging the choking transition in gas-solid risers using electrical capacitance tomography," *Industrial & Engineering Chemistry Research*, 45(15): 5384–5395.

Dudukovic, M.P. (1999), "Trends in catalytic reaction engineering," *Catalysis Today*, 48(1–4): 5–15.

Dudukovic, M.P. (2000), "Opaque multiphase reactors: Experimentation, modeling and troubleshooting," *Oil & Gas Science and Technology*, 55(2): 135–158.

Dudukovic, M.P. (2002), "Opaque multiphase flows: Experiments and modeling," *Experimental Thermal and Fluid Science*, 26(6–7): 747–761.

Dunn, I.J., and Einsele, A. (1975), "Oxygen transfer coefficients by the dynamic method," *Journal of Applied Chemistry and Biotechnology*, 25(9): 707–720.

Dunn, I.J., Heinzle, E., Ingham, J., and Prenosil, J.E. (2003), *Biological Reaction Engineering*, Wiley-VCH, Weinheim.

Ehrfeld, W., Hessel, V., and Lowe, H. (2000), *Microreactors: New Technology for Modern Chemistry*, Wiley-VCH, New York, NY.

Ellenberger, J., and Krishna, R. (1994), "A unified approach to the scale-up of gas-solid fluidized bed and gas-liquid bubble column reactors," *Chemical Engineering Science*, 49(24B): 5391–5411.

Ellenberger, J., and Krishna, R. (2002), "Improving mass transfer in gas-liquid dispersions by vibration excitement," *Chemical Engineering Science*, 57(22–23): 4809–4815.

Fadavi, A., and Chisti, Y. (2005), "Gas-liquid mass transfer in a novel forced circulation loop reactor," *Chemical Engineering Journal*, 112(1–3): 73–80.

Fadavi, A., Chisti, Y., and Chriascarontel, L. (2008), "Bubble size in a forced circulation loop reactor," *Journal of Chemical Technology & Biotechnology*, 83(1): 105–108.

Fair, J.R. (1967), "Designing gas-sparged reactors," *Chemical Engineering*, 74(14): 67–74.

Figueiredo, M.M.L.d., and Calderbank, P.H. (1979), "The scale-up of aerated mixing vessels for specified oxygen dissolution rates," *Chemical Engineering Science*, 34: 1333–1338.

Fogler, H.S. (2005), *Elements of Chemical Reaction Engineering*, Upper Saddle River, NJ, Prentice Hall.

Ford, J.J., Heindel, T.J., Jensen, T.C., and Drake, J.B. (2008), "X-ray computed tomography of a gas-sparged stirred-tank reactor," *Chemical Engineering Science*, 63: 2075–2085.

Franka, N.P., and Heindel, T.J. (2009), "Local time-averaged gas holdup in a fluidized bed with side air injection using X-ray computed tomography," *Powder Technology*, 193: 69–78.

Fransolet, E., Crine, M., Marchot, P., and Toye, D. (2005), "Analysis of gas holdup in bubble columns with non-Newtonian fluid using electrical resistance tomography and dynamic gas disengagement technique," *Chemical Engineering Science*, 60: 6118–6123.

Freitas, C., and Teixeira, J.A. (2001), "Oxygen mass transfer in a high solids loading three-phase internal-loop airlift reactor," *Chemical Engineering Journal*, 84(1): 57–61.

Fu, C.-C., Lu, S.-Y., Hsu, Y.-J., Chen, G.-C., Lin, Y.-R., and Wu, W.-T. (2004), "Superior mixing performance for airlift reactor with a net draft tube," *Chemical Engineering Science*, 59(14): 3021–3028.

Fu, C.-C., Wu, W.-T., and Lu, S.-Y. (2003), "Performance of airlift bioreactors with net draft tube," *Enzyme and Microbial Technology*, 33(4): 332–342.

Fuchs, R., Ryu, D.D.Y., and Humphrey, A.E. (1971), "Effect of surface aeration on scale-up proceedures for fermentation process," *Industrial & Engineering Chemistry Process Design and Development*, 10(2): 190–196.

Fujasova, M., Linek, V., and Moucha, T. (2007), "Mass transfer correlations for multiple-impeller gas-liquid contactors. Analysis of the effect of axial dispersion in gas and liquid phases on "local" $k_L a$ values measured by the dynamic pressure method in individual stages of the vessel," *Chemical Engineering Science*, 62(6): 1650–1669.

Fujita, S., and Hayakawa, T. (1956), "Liquid-film mass transfer coefficients in packed towers and rod-like irrigation towers," *Kagaku Kogaku*, 20(3): 113–117.

Fukuma, M., Muroyama, K., and Yasunishi, A. (1987), "Specific gas-liquid interfacial area and liquid-phase mass transfer coefficient in a slurry bubble column," *Journal of Chemical Engineering of Japan*, 20(3): 321–324.

Gaddis, E.S. (1999), "Mass transfer in gas-liquid contactors," *Chemical Engineering and Processing*, 38(4–6): 503–510.

Gagnon, H., Lounes, M., and Thibault, J. (1998), "Power consumption and mass transfer in agitated gas-liquid columns: A comparative study," *Canadian Journal of Chemical Engineering*, 76(3): 379–389.

Gamio, J.C., Castro, J., Rivera, L., Alamilla, J., Garcia-Nocetti, F., and Aguilar, L. (2005), "Visualization of gas-oil two-phase flows in pressurized pipes using electrical capacitance tomography," *Flow Measurement and Instrumentation*, 16(2–3): 129–134.

Garcia-Ochoa, F., and Gomez, E. (2009), "Bioreactor scale-up and oxygen transfer rate in microbial processes: An overview," *Biotechnology Advances*, 27(2): 153–176.

Garcia-Ochoa, F.F., and Gomez, E. (1998), "Mass transfer coefficient in stirred tank reactors for xanthan gum solutions," *Biochemical Engineering Journal*, 1(1): 1–10.

Garcia-Ochoa, F.F., and Gomez, E. (2004), "Theoretical prediction of gas-liquid mass transfer coefficient, specific area and hold-up in sparged stirred tanks," *Chemical Engineering Science*, 59(12): 2489–2501.

Garner, A.E., and Heindel, T.J. (2000), "The effect of fiber type on bubble size," *Journal of Pulp and Paper Science*, 26(7): 266–269.

George, D.L., Torczynski, J.R., O'Hern, T.J., Shollenberger, K.A., Tortora, P.R., and Ceccio, S.L. (2001), "Quantitative electrical-impedance tomography in an electrically conducting bubble-column vessel," in *ASME 2001 Fluids Engineering Division Summer Meeting*, New Orleans, LA, ASME Press.

George, D.L., Torczynski, J.R., Shollenberger, K.A., O'Hern, T.J., and Ceccio, S.L. (2000), "Validation of electrical-impedance tomography for measurements of material distribution in two-phase flows," *International Journal of Multiphase Flow*, 26: 549–581.

Gestrich, W., Esenwein, H., and Kraus, W. (1978), "Liquid-side mass transfer coefficient in bubble layers," *International Chemical Engineering*, 18(1): 38–47.

Gestrich, W. and Rahse, W. (1975), "Der relative gasgehalt von blasenschichten," *Chemie-Ingenieur-Technik*, 47(1): 8–13.

Gezork, K.M., Bujalski, W., Cooke, M., and Nienow, A.W. (2000), "The transition from homogeneous to heterogeneous flow in a gassed, stirred vessel," *Chemical Engineering Research & Design*, 78(3): 363–370.

Gezork, K.M., Bujalski, W., Cooke, M., and Nienow, A.W. (2001), "Mass transfer and hold-up characteristics in a gassed, stirred vessel at intensified operating conditions," *Chemical Engineering Research & Design*, 79(8): 965–972.

Gharat, S.D., and Joshi, J.B. (1992), "Transport phenomena in bubble colunm reactor II: Pressure drop," *The Chemical Engineering Journal*, 48(3): 153–166.

Ghirardini, M., Donati, G., and Rivetti, F. (1992), "Gas lift reactors: Hydrodynamics, mass transfer, and scale up," *Chemical Engineering Science*, 47(9–11): 2209–2214.

Giovannettone, J.P., Tsai, E., and Gulliver, J.S. (2009), "Gas void ratio and bubble diameter inside a deep airlift reactor," *Chemical Engineering Journal*, 149(1–3): 301–310.

Glazer, B.T., Marsh, A.G., Stierhoff, K., and Luther III, G.W. (2004), "The dynamic response of optical oxygen sensors and voltammetric electrodes to temporal changes in dissolved oxygen concentrations," *Analytica Chimica Acta*, 518(1–2): 93–100.

Godbole, S.P., Honath, M.F., and Shah, Y.T. (1982), "Holdup structure in highly viscous Newtonian and non-Newtonian liquids in bubble columns," *Chemical Engineering Communications*, 16: 119–134.

Godbole, S.P., and Shah, Y.T. (1986), "Design and operation of bubble column reactors," in *Encyclopedia of Fluid Mechanics: Gas-Liquid Flows*, N.P. Cheremisinoff, Editor, Houston, Gulf Publishing Company. 3: 1216–1239.

Gogate, P.R., and Pandit, A.B. (1999a), "Mixing of miscible liquids with density differences: Effect of volume and density of the tracer fluid," *Canadian Journal of Chemical Engineering*, 77(5): 988–996.

Gogate, P.R., and Pandit, A.B. (1999b), "Survey of measurement techniques for gas-liquid mass transfer coefficient in bioreactors," *Biochemical Engineering Journal*, 4(1): 7–15.

Gonzalez, J., Aguilar, R., Alvarez-Ramirez, J., and Barren, M.A. (1998), "Nonlinear regulation for a continuous bioreactor via a numerical uncertainty observer," *Chemical Engineering Journal*, 69(2): 105–110.

Gottschalk, U. (2008), "Bioseparation in antibody manufacturing: The good, the bad and the ugly," *Biotechnology Progress*, 24(3): 496–503.

Grover, G.S., Rode, C.V., and Chaudhari, R.V. (1986), "Effect of temperature on flow regime and gas hold-up in a bubble column," *Canadian Journal of Chemical Engineering*, 64(3): 501–504.

Guo, Y.X., Rathor, M.N., and Ti, H.C. (1997), "Hydrodynamics and mass transfer studies in a novel external-loop airlift reactor," *Chemical Engineering Journal*, 67(3): 205–214.

Guy, C., Carreau, P.J., and Paris, J. (1986), "Mixing characteristics and gas hold-up of a bubble column," *Canadian Journal of Chemical Engineering*, 64(1): 23–35.

Hadjiev, D., Sabiri, N.E., and Zanati, A. (2006), "Mixing time in bioreactors under aerated conditions," *Biochemical Engineering Journal*, 27(3): 323–330.

Hall, D.W., Scott, K., and Jachuck, R.J.J. (2001), "Determination of mass transfer coefficient of a cross-corrugated membrane reactor by the limiting-current technique," *International Journal of Heat and Mass Transfer*, 44(12): 2201–2207.

Hallaile, M. (1993), *Biotechnology*, Ben-Gurion University of the Negev, Israel.

Hamidipour, M., Chen, J., and Larachi, F. (2012), "CFD study on hydrodynamics in three-phase fluidized beds—Application of turbulence models and experimental validation," *Chemical Engineering Science*, 78(0): 167–180.

Hammer, H. (1984), *Frontiers in Chemical Reaction Engineering*, Halsted Press, New Delhi, India.

Han, L., and Al-Dahhan, M.H. (2007), "Gas-liquid mass transfer in a high pressure bubble column reactor with different sparger designs," *Chemical Engineering Science*, 62(1–2): 131–139.

Haque, M.W., Nigam, K.D.P., and Joshi, J.B. (1986), "Hydrodynamics and mixing in highly viscous pseudo-plastic non-Newtonian solutions in bubble columns," *Chemical Engineering Science*, 41(9): 2321–2331.

Harnby, N., Edwards, M.F., and Nienow, A.W. (1992), *Mixing in the Process Industry*, Butterworth-Heinemann Ltd. Boston.

Harper, J.C., Christensen, P.A., Egerton, T.A., and Scott, K. (2001), "Mass transport characterization of a novel gas sparged photoelectrochemical reactor," *Journal of Applied Electrochemistry*, 31: 267–276.

Harvel, G.D., Hori, K., Kawanishi, K., and Chang, J.S. (1999), "Cross-sectional void fraction distribution measurements in a vertical annulus two-phase flow by high speed X-ray computed tomography and real-time neutron radiography techniques," *Flow Measurement and Instrumentation*, 10(4): 259–266.

Heijnen, P., and Lukszo, Z. (2006), "Continuous improvement of batch wise operation: A decision support framework," *Production Planning & Control*, 17(4): 355–366.

Heindel, T.J. (1999), "Bubble size measurements in a quiescent fiber suspension," *Journal of Pulp and Paper Science*, 25(3): 104–110.

Heindel, T.J. (2000), "Gas flow regime changes in a bubble column filled with a fiber suspension," *The Canadian Journal of Chemical Engineering*, 78(5): 1017–1022.

Heindel, T.J. (2002), "Bubble size in a cocurrent fiber slurry," *Industrial & Engineering Chemistry Research*, 41(3): 632–641.

Heindel, T.J. (2011), "A review of X-ray flow visualization with applications to multiphase flows," *Journal of Fluids Engineering - Transactions of the ASME*, 133(7): 074001 (16 pages).

Heindel, T.J., and Garner, A.E. (1999), "The effect of fiber consistency on bubble size," *Nordic Pulp and Paper Research Journal*, 14(2): 171–178.

Henstra, A.M., Sipma, J., Rinzema, A., and Stams, A.J. (2007), "Microbiology of synthesis gas fermentation for biofuel production," *Current Opinion in Biotechnology*, 2007(18): 200–206.

Herringe, R.A., and Davis, M.R. (1978), "Flow structure and distribution effects in gas-liquid mixture flows," *International Journal of Multiphase Flow*, 4(5–6): 461–486.

Hewitt, G.F. (1978), *Measurement of Two Phase Flow Parameters*, Academic Press, New York.

Hewitt, G.F. (1982), "Measurement techniques: Overall measurements - measurement of void fraction," in *Handbook of Multiphase Systems*, G. Hetsroni, Editor, New York, Hemisphere Publishing Corp.: Chapter 10.2.1.2.

Hickman, A.D. (1988), "Gas-liquid oxygen transfer and scale-up. A novel experimental technique with results for mass transfer in aerated agitated vessels," in *Proceedings of the 6th European Conference on Mixing*, Pavia, Italy, 369–374.

Hikita, H., Asai, S., Tanigawa, K., Segawa, K., and Kitao, M. (1980), "Gas hold-up in bubble columns," *Chemical Engineering Journal*, 20(1): 59–67.

Hikita, H., Asai, S., Tanigawa, K., Segawa, K., and Kitao, M. (1981), "The volumetric liquid-phase mass transfer coefficient in bubble columns," *Chemical Engineering Journal*, 22(1): 61–69.

Hikita, H., and Kikukawa, H. (1974), "Liquid-phase mixing in bubble columns: Effect of liquid properties," *Chemical Engineering Journal and the Biochemical Engineering Journal*, 8(3): 191–197.

Hikita, H., and Ono, Y. (1959), "Mass transfer into a liquid film flowing over a packing piece," *Kagaku Kogaku*, 23(12): 808–813.

Hills, J.H. (1976), "The operation of a bubble column at high throughputs. I – Gas holdup measurements," *The Chemical Engineering Journal*, 12: 89–99.

Hoffmann, A., Mackowiak, J.F., Gorak, A., Haas, M., Loning, J.M., Runowski, T., and Hallenberger, K. (2007), "Standardization of mass transfer measurements: A basis for the description of absorption processes," *Chemical Engineering Research & Design*, 85(1): 40–49.

Hoffmann, R.A., Garcia, M.L., Veskivar, M., Karim, K., Al-Dahhan, M.H., and Angenent, L.T. (2008), "Effect of shear on performance and microbial ecology of continuously stirred anaerobic digesters treating animal manure," *Biotechnology and Bioengineering*, 100(1): 38–48.

Hol, P.D., and Heindel, T.J. (2005), "Local gas holdup variation in a fiber slurry," *Industrial & Engineering Chemistry Research*, 44: 4778–4784.

Hruby, J. (2002), "Overview of liga microfabrication," in *5th Workshop on High Energy Density and High Power RF*, Snowbird, Utah (USA), American Institute of Physics, 55–62.

Huang, Z.-B., and Cheng, Z.-M. (2011), "Determination of liquid multiscale circulation structure in a bubble column by tracing the liquid flowing trajectory," *Industrial & Engineering Chemistry Research*, 50(21): 11843–11852.

Hubers, J.L., Striegel, A.C., Heindel, T.J., Gray, J.N., and Jensen, T.C. (2005), "X-ray computed tomography in large bubble columns," *Chemical Engineering Science*, 60(22): 6124–6133.

Hughmark, G.A. (1967), "Holdup and mass transfer in bubble columns," *Industrial & Engineering Chemistry Process Design and Development*, 6(2): 218–220.

Idogawa, K., Ikeda, K., Fukuda, T., and Morooka, S. (1987), "Effect of gas and liquid properties on the behavior of bubbles in a column under high pressure," *International Chemical Engineering*, 27(1): 93–99.

Ismail, I., Gamio, J.C., Bukhari, S.F.A., and Yang, W.Q. (2005), "Tomography for multi-phase flow measurement in the oil industry," *Flow Measurement and Instrumentation*, 16(2–3): 145–155.

Ityokumbul, M.T., Kosaric, N., and Bulani, W. (1994), "Gas hold-up and liquid mixing at low and intermediate gas velocities: 1. Air-water system," *The Chemical Engineering Journal and The Biochemical Engineering Journal*, 53(3): 167–172.

Jade, A.M., Jayaraman, V.K., Kulkarni, B.D., Khopkar, A.R., Ranade, V.V., and Sharma, A. (2006), "A novel local singularity distribution based method for flow regime identification: Gas-liquid stirred vessel with Rushton turbine," *Chemical Engineering Science*, 61(2): 688–697.

Jakobsen, H.A., Lindborg, H., and Dorao, C.A. (2005a), "Modeling of bubble column reactors: Progress and limitations," *Industrial & Engineering Chemistry Research*, 44: 5107–5151.

Jakobsen, H.A., Lindborg, H., and Dorao, C.A. (2005b), "Modeling of bubble column reactors: Progress and limitations," *Industrial & Engineering Chemistry Research*, 44(14): 5107–5151.

Jean, R.-H., and Fan, L.-S. (1987), "On the particle terminal velocity in a gas-liquid medium with liquid as the continuous phase," *Canadian Journal of Chemical Engineering*, 65(6): 881–886.

Jia, X., Wen, J., Feng, W., and Yuan, Q. (2007), "Local hydrodynamics modeling of a gas-liquid-solid three-phase airlift loop reactor," *Industrial & Engineering Chemistry Research*, 46(15): 5210–5220.

Jin, H., Yang, S., He, G., Guo, Z., and Tong, Z. (2005), "An experimental study of holdups in large-scale p-xylene oxidation reactors using the γ-ray attenuation approach," *Chemical Engineering Science*, 60: 5955–5961.

Jones, S.T. (2007), "Gas-liquid mass transfer in an external airlift loop reactor for syngas fermentation," PhD Thesis, Iowa State University, Ames, Iowa.

Jones, S.T., and Heindel, T.J. (2007), "A review of dissolved oxygen concentration measurement methods for biological fermentations", in *ASABE Annual Meeting*, Minneapolis, Minnesota, ASABE: 2–15.

Jordan, U., and Schumpe, A. (2001), "The gas density effect on mass transfer in bubble columns with organic liquids," *Chemical Engineering Science*, 56(21–22): 6267–6272.

Jordan, U., Terasaka, K., Kundu, G., and Schumpe, A. (2002), "Mass transfer in high-pressure bubble columns with organic liquids," *Chemical Engineering & Technology*, 25(3): 262–265.

Joshi, J.B. (2001), "Computational flow modelling and design of bubble column reactors," *Chemical Engineering Science*, 56: 5893–5933.

Joshi, J.B., Ranade, V.V., Gharat, S.D., and Lele, S.S. (1990), "Sparged loop reactors," *The Canadian Journal of Chemical Engineering*, 68(October): 705–741.

Joshi, J.B., and Sharma, M.M. (1979), "Circulation cell model for bubble columns," *Chemical Engineering Research & Design*, 57(4): 244–251.

Kak, A.C., and Slaney, M. (1988), *Principles of Computerized Tomographic Imaging*, Society of Industrial and Applied Mathematics, New York.

Kang, Y., Cho, Y.J., Woo, K.J., and Kim, S.D. (1999), "Diagnosis of bubble distribution and mass transfer in pressurized bubble columns with viscous liquid medium," *Chemical Engineering Science*, 54(21): 4887–4893.

Kantarci, N., Borak, F., and Ulgen, K.O. (2005), "Bubble column reactors," *Process Biochemistry*, 40(7): 2263–2283.

Kantzas, A. (1994), "Computation of holdups in fluidized and trickle beds by computer-assisted tomography," *AIChE Journal*, 40(7): 1254–1261.

Kapic, A. (2005), "Mass transfer measurements for syngas fermentation," MS Thesis, Iowa State University, Ames, Iowa.

Kapic, A., and Heindel, T.J. (2006), "Correlating gas-liquid mass transfer in a stirred-tank reactor," *Chemical Engineering Research & Design*, 84(3A): 239–245.

Kapic, A., Jones, S.T., and Heindel, T.J. (2006), "Carbon monoxide mass transfer in a syngas mixture," *Industrial & Engineering Chemistry Research*, 45(26): 9150–9155.

Kara, S., Kelkar, B.G., Shah, Y.T., and Carr, N.L. (1982), "Hydrodynamics and axial mixing in a three-phase bubble column," *Industrial and Engineering Chemistry Process Design and Development*, 21: 584–594.

Karamanev, D.G., Chavarie, C., and Samson, R. (1996), "Hydrodynamics and mass transfer in an airlift reactor with a semipermeable draft tube," *Chemical Engineering Science*, 51(7): 1173–1176.

Kasatkin, A.G., and Ziparis, I.N. (1952) *Chim Prom*, 7, 203.

Kashiwa, B.A., Padial, N.T., Rauenzahn, R.M., and VanderHeyden, W.B. (1993), "Cell-centered ice method for multiphase flow simulations", U.S. Department of Energy, Los Alamos National Laboratory, Report: LA-UR-93-3922.

Kato, Y., and Nishiwaki, A. (1972), "Longitudinal dispersion coefficient of a liquid in a bubble column," *International Journal of Chemical Reactor Engineering*, 12(1): 182–187.

Kawase, Y., Halard, B., and Moo-Young, M. (1987), "Theoretical prediction of volumetric mass transfer coefficients in bubble columns for Newtonian and non-Newtonian fluids," *Chemical Engineering Science*, 42(7): 1609–1617.

Kawase, Y., and Moo-Young, M. (1986a), "Influence of non-Newtonian flow behaviour on mass transfer in bubble columns with and without draft tubes," *Chemical Engineering Communications*, 40(1–6): 67–83.

Kawase, Y., and Moo-Young, M. (1986b), "Mixing and mass transfer in concentric-tube airlift fermenters: Newtonian and non-Newtonian media," *Journal of Chemical Technology and Biotechnology*, 36(11): 527–538.

Kawase, Y., and Moo-Young, M. (1987), "Heat transfer in bubble column reactors with Newtonian and non-Newtonian fluids," *Chemical Engineering Research & Design*, 65(2): 121–126.

Kawase, Y., and Moo-Young, M. (1988), "Volumetric mass transfer coefficients in aerated stirred tank reactors with Newtonian and non-Newtonian media," *Chemical Engineering Research & Design*, 66(3): 284–288.

Kawase, Y., and Moo-Young, M. (1992), "Correlations for liquid-phase mass transfer coefficient in bubble column reactor with Newtonian and non-Newtonian fluids," *Canadian Journal of Chemical Engineering*, 70(1): 48–54.

Kawase, Y., Tsujimura, M., and Yamaguchi, T. (1995), Gas hold-up in external-loop airlift bioreactors. *Bioprocess and Biosystems Engineering*, 12(1–2): 21–27.

Kawase, Y., Umeno, S., and Kumagai, T. (1992), "The prediction of gas hold-up in bubble column reactors: Newtonian and non-Newtonian fluids," *Chemical Engineering Journal*, 50(1): 1–7.

Keitel, G., and Onken, U. (1981), "Errors in the determination of mass transfer in gas-liquid dispersions," *Chemical Engineering Science*, 36(12): 1927–1932.

Kemblowski, Z., Przywarski, J., and Diab, A. (1993), "An average gas hold-up and liquid circulation velocity in airlift reactors with external loop," *Chemical Engineering Science*, 48(23): 4023–4035.

Kertzscher, U., Seeger, A., Affeld, K., Goubergrits, L., and Wellnhofer, E. (2004), "X-ray based particle tracking velocimetry - A measurement technique for multi-phase flows and flows without optical access," *Flow Measurement and Instrumentation*, 15(4): 199–206.

Khopkar, A.R., Rammohan, A.R., Ranade, V.V., and Dudukovic, M.P. (2005), "Gas-liquid flow generated by a rushton turbine in stirred vessel: CARPT/CT measurements and CFD simulations," *Chemical Engineering Science*, 60(8–9): 2215–2229.

Khudenko, B.M., and Shpirt, E. (1986), "Hydrodynamic parameters of diffused air systems," *Water Research*, 20(7): 905–915.

Kiambi, S.L., Kiriamiti, H.K., and Kumar, A. (2011), "Characterization of two phase flows in chemical engineering reactors," *Flow Measurement and Instrumentation*, 22(4): 265–271.

Kim, D.J., and Chang, H.N. (1989), "Dynamic measurement of $k_L a$ with oxygen-enriched air during fermentation," *Journal of Chemical Technology and Biotechnology*, 45(1): 39–44.

Kim, S.D., Baker, C.G.J., and Bergougnou, M.A. (1972), "Hold-up and axial mixing characteristics of two and three phase fluidized beds," *Canadian Journal of Chemical Engineering*, 50(6): 695–691.

Kluytmans, J.H.J., Kuster, B.F.M., and Schouten, J.C. (2001), "Gas holdup in a slurry bubble column: Influence of electrolyte and carbon particles," *Industrial & Engineering Chemistry Research*, 40(23): 5326–5333.

Ko, C.-H., Chiang, P.-N., Chiu, P.-C., Liu, C.-C., Yang, C.-L., and Shiau, I.-L. (2008), "Integrated xylitol production by fermentation of hardwood wastes," *Journal of Chemical Technology & Biotechnology*, 83(4): 534–540.

Koch, H.A., Stutzman, L.F., Blum, L.F., and Hutchings, H.A. (1949), "Liquid transfer coefficients for the carbon dioxide-air-water system," *Chemical Engineering Progress*, 45(11): 677–682.

Koeneke, R., Comte, A., Juergens, H., Kohls, O., Lam, H., and Scheper, T. (1999), "Fiber optic oxygen sensors for use in biotechnology, environmental, and food industries," *Chemical Engineering and Technology*, 22(8): 666–671.

Kohls, O., and Scheper, T. (2000), "Setup of a fiber optical oxygen multisensor-system and its applications in biotechnology," *Sensors and Actuators, B: Chemical*, 70(1–3): 121–130.

Koide, K., Horibe, K., Kawabata, H., and Ito, S. (1985), "Gas holdup and volumetric liquid-phase mass transfer coefficient in solid-suspended bubble column with draught tube," *Journal of Chemical Engineering of Japan*, 18(3): 248–254.

Koide, K., Kimura, M., Nitta, H., and Kawabata, H. (1988), "Liquid circulation in bubble column with draught tube," *Journal of Chemical Engineering of Japan*, 21(4): 393–399.

Koide, K., Kurematsu, K., Iwamoto, S., Iwata, Y., and Horibe, K. (1983a), "Gas holdup and volumetric liquid-phase mass transfer coefficient in bubble column with draught tube and with gas dispersion into tube," *Journal of Chemical Engineering of Japan*, 16(5): 413–418.

Koide, K., Morooka, S., Ueyama, K., Matsuura, A., Yamashita, F., Iwamoto, S., Kato, Y., Inoue, H., Shigeta, M., Suzuki, S., and Akehata, T. (1979), "Behavior of bubbles in large scale bubble column," *Journal of Chemical Engineering of Japan*, 12(2): 98–104.

Koide, K., Sato, H., and Iwamoto, S. (1983b), "Gas holdup and volumetric liquid-phase mass transfer coefficient in bubble column with draught tube and with gas dispersion into annulus," *Journal of Chemical Engineering of Japan*, 16(5): 407–413.

Koide, K., Takazawa, A., Komura, M., and Matsunaga, H. (1984), "Gas holdup and volumetric liquid-phase mass transfer coefficient in solid-suspended bubble columns," *Journal of Chemical Engineering of Japan*, 17(5): 459–466.

Kojima, H., Sawai, J., and Suzuki, H. (1997), "Effect of pressure on volumetric mass transfer coefficient and gas holdup in bubble column," *Chemical Engineering Science*, 52(21–22): 4111–4116.

Kolev, N. (1976), "Wirkungsweise von füllkörperschüttungen," *Chemie-Ingenieur-Technik*, 48(12): 1105–1112.

Kolev, N. (2006), *Packed Bed Columns: For Absorption, Desorption, Rectification and Direct Heat Transfer*, Elsevier, Amsterdam, The Netherlands.

Kolev, N., and Daraktschiev, R. (1976), "Issledobanie massoobmena w gorizontalbnoi listowoi nasadke," *Theoretical Fundamentals of Chemical Technology*, 10(4): 611–614.

Kolev, N., and Nakov, S. (1994), "Performance characteristics of a packing with boundary layer turbulizers. III. Liquid film controlled mass transfer," *Chemical Engineering and Processing*, 33(6): 437–442.

Kolev, N., and Semkov, K. (1983), "Axial mixing in the liquid phase in packed columns," *Verfahrenstechnik*, 17(8): 474–479, 488.

Krevelen, D.V.V., and Hoftijzer, P.J. (1948), "Kinetics of simultaneous absorption and chemical reaction," *Chemical Engineering Progress*, 44(7): 529–536.

Krishna, R., De Swart, J.W.A., Ellenberger, J., Martina, G.B., and Maretto, C. (1997), "Gas holdup in slurry bubble columns: Effect of column diameter and slurry concentrations," *AIChE Journal*, 43(2): 311–316.

Krishna, R., and Ellenberger, J. (1996), "Gas holdup in bubble column reactors operating in the churn-turbulent flow regime," *AIChE Journal*, 42(9): 2627–2634.

Krishna, R., Urseanu, M.I., de Swart, J.W.A., and Ellenberger, J. (2000), "Gas hold-up in bubble columns: Operation with concentrated slurries versus high viscosity liquid," *The Canadian Journal of Chemical Engineering*, 78(6): 442–448.

Krishna, R., Urseanu, M.I., van Baten, J.M., and Ellenberger, J. (1999), "Influence of scale on the hydrodynamics of bubble columns operating in the churn-turbulent regime: Experiments vs. Eulerian simulations," *Chemical Engineering Science*, 54(21): 4903–4911.

Krishna, R., van Baten, J.M., and Urseanu, M.I. (2001), "Scale effects on the hydrodynamics of bubble columns operating in the homogeneous flow regime," *Chemical Engineering & Technology*, 24(5): 451–458.

Kulkarni, A.A., Ekambara, K., and Joshi, J.B. (2007), "On the development of flow pattern in a bubble column reactor: Experiments and CFD," *Chemical Engineering Science*, 62(4): 1049–1072.

Kumar, A., Degaleesan, T.E., Laddha, G.S., and Hoelscher, H.E. (1976), "Bubble swarm characteristics in bubble columns," *Canadian Journal of Chemical Engineering*, 54(6): 503–508.

Kumar, A., Dewulf, J., and Van Langenhove, H. (2008), "Membrane-based biological waste gas treatment," *Chemical Engineering Journal*, 136(2–3): 82–91.

Kumar, S.B., and Dudukovic, M.P. (1996), "Computer assisted gamma and X-ray tomography: Applications to multiphase flow systems," in *Non-Invasive Monitoring of Multiphase Flows*, J. Chaouki, F. Larachi, and M.P. Dudukovic, Editors, New York, Elsevier: 47–103.

Kumar, S.B., Dudukovic, M.P., and Toseland, B.A. (1997), "Measurement techniques for local and global fluid dynamic quantities in two and three phase systems," in *Non-Invasive Monitoring of Multiphase Flows*, J. Chaouki, F. Larachi, and M.P. Dudukovic, Editors, New York, Elsevier: 1–45.

Kumar, S.B., Moslemian, D., and Dudukovic, M.P. (1995), "A γ-ray tomographic scanner for imaging voidage distribution in two-phase flow systems," *Flow Measurement and Instrumentation*, 6(1): 61–73.

Kundu, S., Premer, S.A., Hoy, J.A., Trent, J.T., and Hargrove, M.S. (2003), "Direct measurements of equilibrium constants for high-affinity hemoglobins," *Biophysical Journal*, 84(6): 3931–3940.

Laari, A., and Turunen, I. (2005), "Prediction of coalescence properties of gas bubbles in a gas-liquid reactor using persistence time measurements," *Chemical Engineering Research & Design*, 83(7): 881–886.

Lance, M., and Bataille, J. (1991), "Turbulence in the liquid phase of a uniform bubbly air–water flow," *Journal of Fluid Mechanics*, 222: 95–118.

Larachi, F., Levesque, S., and Grandjean, B.P.A. (2008), "Seamless mass transfer correlations for packed beds bridging random and structured packings," *Industrial & Engineering Chemistry Research*, 47(9): 3274–3284.

Lau, R., Peng, W., Velazquez-Vargas, L.G., Yang, G.Q., and Fan, L.S. (2004), "Gas-liquid mass transfer in high-pressure bubble columns," *Industrial & Engineering Chemistry Research*, 43(5): 1302–1311.

Law, D., Battaglia, F., and Heindel, T.J. (2008), "Model validation for low and high superficial gas velocity bubble column flows," *Chemical Engineering Science*, 63(18): 4605–4616.

Lee, D.-H., Grace, J.R., and Epstein, N. (2000), "Gas holdup for high gas velocities in a gas-liquid cocurrent upward-flow system," *Canadian Journal of Chemical Engineering*, 78(5): 1006–1010.

Lee, D.J., Luo, X., and Fan, L.-S. (1999), "Gas disengagement technique in a slurry bubble column operated in the coalesced bubble regime," *Chemical Engineering Science*, 54: 2227–2236.

Lee, Y.H., and Luk, S. (1983), "Aeration," *Annual Reports on Fermentation Processes*, 6: 101–147.

Lee, Y.H., and Tsao, G.T. (1979), "Dissolved oxygen electrodes," *Advances in Biochemical Engineering*, 13: 35–86.

Leiknes, T., and Odegaard, H. (2007), "The development of a biofilm membrane bioreactor," *Desalination*, 202(1–3): 135–143.

León-Becerril, E., Cockx, A., and Liné, A. (2002), "Effect of bubble deformation on stability and mixing in bubble columns," *Chemical Engineering Science*, 57(16): 3283–3297.

León-Becerril, E., and Liné, A. (2001), "Stability analysis of a bubble column," *Chemical Engineering Science*, 56(21–22): 6135–6141.

Letzel, H.M., Schouten, J.C., Krishna, R., and van den Bleek, C.M. (1999), "Gas holdup and mass transfer in bubble column reactors operated at elevated pressure," *Chemical Engineering Science*, 54(13–14): 2237–2246.

Levin, Y., and Flores-Mena, J.E. (2001), "Surface tension of strong electrolytes," *Europhysics Letters*, 56(2): 187–192.

Li, G.Q., Yang, S.Z., Cai, Z.L., and Chen, J.Y. (1995), "Mass transfer and gas-liquid circulation in an airlift bioreactor with viscous non-Newtonian fluids," *The Chemical Engineering Journal and the Biochemical Engineering Journal*, 56(2): b101–b107.

Li, H., and Prakash, A. (1997), "Heat transfer and hydrodynamics in a three-phase slurry bubble column," *Industrial & Engineering Chemistry Research*, 36(11): 4688–4694.

Li, H., and Prakash, A. (2000), "Influence of slurry concentrations on bubble population and their rise velocities in a three-phase slurry bubble column," *Powder Technology*, 113(1–2): 158–167.

Liao, Q., Tian, X., Chen, R., and Zhu, X. (2008), "Mathematical model for gas-liquid two-phase flow and biodegradation of a low concentration volatile organic compound (VOC) in a trickling biofilter," *International Journal of Heat and Mass Transfer*, 51(7–8): 1780–1792.

Lin, T.-J., Tsuchiya, K., and Fan, L.-S. (1998), "Bubble flow characteristics in bubble columns at elevated pressure and temperature," *AIChE Journal*, 44(3): 545–560.

Linek, V. (1972), "Determination of aeration capacity of mechanically agitated vessels by fast response oxygen probe," *Biotechnology and Bioengineering*, 14(2): 285–289.

Linek, V. (1988), *Measurement of Oxygen by Membrane-Covered Probes: Guidelines for Applications in Chemical and Biochemical Engineering*, Halsted Press, New York.

Linek, V., Benes, F., and Hovorka, F. (1981), "Role of interphase nitrogen transport in the dynamic measurement of the overall volumetric mass transfer coefficient in air-sparged systems," *Biotechnology and Bioengineering*, 23(2): 301–319.

Linek, V., Benes, P., and Sinkule, J. (1990), "Critical assessment of the steady-state Na_2SO_3 feeding method for $k_L a$ measurements in fermentors," *Biotechnology and Bioengineering*, 35(8): 766–770.

Linek, V., Benes, P., and Sinkule, J. (1991a), "Hollow blade agitator for mass transfer in gas-liquid bioreactors," *Food and Bioproducts Processing*, 69(3): 145–148.

Linek, V., Benes, P., Sinkule, J., and Moucha, T. (1993), "Non-ideal pressure step method for $k_L a$ measurement," *Chemical Engineering Science*, 48(9): 1593–1599.

Linek, V., Benes, P., and Vacek, V. (1979), "Oxygen probe dynamics in flowing fluids," *Industrial & Engineering Chemistry, Fundamentals*, 18(3): 240–245.

Linek, V., Benes, P., and Vacek, V. (1984), "Experimental study of oxygen probe linearity and transient characteristics in the high oxygen concentration range," *Journal of Electroanalytical Chemistry and Interfacial Electrochemistry*, 169(1–2): 233–257.

Linek, V., Benes, P., and Vacek, V. (1989), "Dynamic pressure method for $k_L a$ measurement in large-scale bioreactors," *Biotechnology and Bioengineering*, 33(11): 1406–1412.

Linek, V., Kordac, M., Fujasova, M., and Moucha, T. (2004), "Gas-liquid mass transfer coefficient in stirred tanks interpreted through models of idealized eddy structure of turbulence in the bubble vicinity," *Chemical Engineering and Processing*, 43(12): 1511–1517.

Linek, V., Kordac, M., and Moucha, T. (2005a), "Mechanism of mass transfer from bubbles in dispersions: Part II. Mass transfer coefficients in stirred gas-liquid reactor and bubble column," *Chemical Engineering and Processing*, 44(1): 121–130.

Linek, V., Moucha, T., Dousova, M., and Sinkule, J. (1994), "Measurement of $k_L a$ by dynamic pressure method in pilot-plant fermentor," *Biotechnology and Bioengineering*, 43(6): 477–482.

Linek, V., Moucha, T., and Kordac, M. (2005b), "Mechanism of mass transfer from bubbles in dispersions: Part I. Danckwerts' plot method with sulphite solutions in the presence of viscosity and surface tension changing agents," *Chemical Engineering and Processing*, 44(3): 353–361.

Linek, V., Moucha, T., and Sinkule, J. (1996a), "Gas-liquid mass transfer in vessels stirred with multiple impellers–I. Gas-liquid mass transfer characteristics in individual stages," *Chemical Engineering Science*, 51(12): 3203–3212.

Linek, V., Moucha, T., and Sinkule, J. (1996b), "Gas-liquid mass transfer in vessels stirred with multiple impellers–II. Modelling of gas-liquid mass transfer," *Chemical Engineering Science*, 51(15): 3875–3879.

Linek, V., and Sinkule, J. (1990), "Comments on validity of dynamic measuring methods of oxygen diffusion coefficients in fermentation media with polarographic oxygen electrodes," *Biotechnology and Bioengineering*, 35(10): 1034–1041.

Linek, V., Sinkule, J., and Benes, P. (1991b), "Critical assessment of gassing-in methods for measuring $k_L a$ in fermentors," *Biotechnology and Bioengineering*, 38(4): 323–330.

Linek, V., Sinkule, J., and Benes, P. (1992), "Critical assessment of the dynamic double-response method for measuring $k_L a$. Experimental elimination of dispersion effects," *Chemical Engineering Science*, 47(15–16): 3885–3894.

Linek, V., Sinkule, J., and Vacek, V. (1985), "Dissolved oxygen probes," in *Comprehensive Biotechnology*, Oxford, England, Pergamon Press, 4: 363–394.

Linek, V., Vacek, V., and Benes, P. (1987), "A critical review and experimental verification of the correct use of the dynamic method for the determination of oxygen transfer in aerated agitated vessels to water, electrolyte solutions and viscous liquids," *Chemical Engineering Journal*, 34(1): 11–34.

Lines, P.C. (1999), "Gas-liquid mass transfer: Surface aeration in stirred vessels with dual impellers," *Institution of Chemical Engineers Symposium Series*, 146: 199–216.

Lines, P.C. (2000), "Gas-liquid mass transfer using surface-aeration in stirred vessels, with dual impellers," *Chemical Engineering Research & Design*, 78(3): 342–347.

Liu, T.-J. (1997), "Investigation of the wall shear stress in vertical bubbly flow under different bubble size conditions," *International Journal of Multiphase Flow*, 23(6): 1085–1109.

Llamas, J.-D., Pérat, C., Lesage, F., Weber, M., D'Ortona, U., and Wild, G. (2008), "Wire mesh tomography applied to trickle beds: A new way to study liquid maldistribution," *Chemical Engineering and Processing Process Intensification*, 47(9–10): 1765–1770.

Lo, C.-S., and Hwang, S.-J. (2003), "Local hydrodynamic properties of gas phase in an internal-loop airlift reactor," *Chemical Engineering Journal*, 91(1): 3–22.

Lockett, M.J., and Kirkpatrick, R.D. (1975), "Ideal bubbly flow and actual flow in bubble columns," *Transactions of the Institution of Chemical Engineers*, 1975(53): 267–273.

Lopez, J.L., Rodriguez Porcel, E.M., Oller Alberola, I., Ballesteros Martin, M.M., Sanchez Perez, J.A., Fernandez Sevilla, J.M., and Chisti, Y. (2006), "Simultaneous determination of oxygen consumption rate and volumetric aoxygen transfer coefficient in pnuematically agitated airlift bioreactors," *Industrial & Engineering Chemistry Research*, 2006: 1167–1171.

Lorenz, O., Schumpe, A., Ekambara, K., and Joshi, J.B. (2005), "Liquid phase axial mixing in bubble columns operated at high pressures," *Chemical Engineering Science*, 60(13): 3573.

Löwe, H., Ehrfeld, W., Hessel, V., Richter, T., and Schiewe, J. (2000), "Micromixing technology", in *4th International Conference on Microreaction Technology*, Atlanta, GA, AIChE: 31–47.

Lundgren, D.G., and Russel, R.T. (1956), "An air-lift laboratory fermentor," *Applied and Environmental Microbiology*, 4(1): 31–33.

Luo, H.-P., and Al-Dahhan, M.H. (2008), "Local characteristics of hydrodynamics in draft tube airlift bioreactor," *Chemical Engineering Science*, 63(11): 3057–3068.

Luo, X., Jiang, P., and Fan, L.-S. (1997), "High-pressure three-phase fluidization: Hydrodynamics and heat transfer," *AIChE Journal*, 43(10): 2432–2445.

Luo, X., Lee, D.J., Lau, R., Yang, G., and Fan, L.-S. (1999), "Maximum stable bubble size and gas holdup in high-pressure slurry bubble columns," *AIChE Journal*, 45(4): 665–680.

Magaud, F., Souhar, M., Wild, G., and Boisson, N. (2001), "Experimental study of bubble column hydrodynamics," *Chemical Engineering Science*, 56(15): 4597–4607.

Maiti, R.N., and Nigam, K.D.P. (2007), "Gas-liquid distributors for trickle-bed reactors: A review," *Industrial & Engineering Chemistry Research*, 46(19): 6164–6182.

Majirova, H., Pinelli, D., Machon, V., and Magelli, F. (2004), "Gas flow behavior in a two-phase reactor stirred with triple turbines," *Chemical Engineering & Technology*, 27(3): 304–309.

Makkawi, Y.T., and Wright, P.C. (2002), "Fluidization regimes in a conventional fluidized bed characterized by means of electrical capacitance tomography," *Chemical Engineering Science*, 57: 2411–2437.

Makkawi, Y.T., and Wright, P.C. (2004), "Electrical capacitance tomography for conventional fluidized bed measurements–Remarks on the measuring technique," *Powder Technology*, 148(2–3): 142–157.

Maldonado, J.G.G., Bastoul, D., Baig, S., Roustan, M., and Hébrard, G. (2008), "Effect of solid characteristics on hydrodynamic and mass transfer in a fixed bed reactor operating in co-current gas-liquid up flow," *Chemical Engineering and Processing Process Intensification*, 47(8): 1190–1200.

Mangers, R.J., and Ponter, A.B. (1980), "Effect of viscosity on liquid film resistance to mass transfer in a packed column," *Industrial & Engineering Chemistry, Process Design and Development*, 19(4): 530–537.

Marashdeh, Q., Fan, L.S., Du, B., and Warsito, W. (2008), "Electrical capacitance tomography – A perspective," *Industrial & Engineering Chemistry Research*, 47(10): 3708–3719.

Marchot, P., Toye, D., Pelsser, A.-M., Crine, M., L'Homme, G.L., and Olujic, Z. (2001), "Liquid distribution images on structured packing by X-ray computed tomography," *AIChE Journal*, 47(6): 1471–1476.

Martín, M., García, J.M., Montes, F.J., and Galán, M.A. (2008a), "On the effect of the orifice configuration on the coalescence of growing bubbles," *Chemical Engineering and Processing Process Intensification*, 47(9–10): 1799–1809.

Martín, M., Montes, F.J., and Galan, M.A. (2008b), "Influence of impeller type on the bubble breakup process in stirred tanks," *Industrial & Engineering Chemistry Research*, 47(16): 6251–6263.

Martín, M., Montes, F.J., and Galán, M.A. (2009), "Theoretical modelling of the effect of surface active species on the mass transfer rates in bubble column reactors," *Chemical Engineering Journal*, 155(1–2): 272–284.

Martin, Y., and Vermette, P. (2005), "Bioreactors for tissue mass culture: Design, characterization, and recent advances," *Biomaterials*, 26(35): 7481–7503.

Mashelkar, R.A. (1970), "Bubble columns," *British Chemical Engineering*, 1970(15): 1297–1304.

McFarlane, C.M., and Nienow, A.W. (1995), "Studies of high solidity ratio hydrofoil impellers for aerated bioreactors. 1. Review," *Biotechnology Progress*, 11(6): 601–607.

McFarlane, C.M., and Nienow, A.W. (1996a), "Studies of high solidity ratio hydrofoil impellers for aerated bioreactors. 3. Fluids of enhanced viscosity and exhibiting coalescence repression," *Biotechnology Progress*, 12(1): 1–8.

McFarlane, C.M., and Nienow, A.W. (1996b), "Studies of high solidity ratio hydrofoil impellers for aerated bioreactors. 4. Comparison of impeller types," *Biotechnology Progress*, 12(1): 9–15.

McFarlane, C.M., Zhao, X.-M., and Nienow, A.W. (1995), "Studies of high solidity ratio hydrofoil impellers for aerated bioreactors. 2. Air-water studies," *Biotechnology Progress*, 11(6): 608–618.

McManamey, W.J., and Wase, D.A.J. (1986), "Relationship between the volumetric mass transfer coefficient and gas holdup in airlift fermentors," *Biotechnology and Bioengineering*, 28(9): 1446–1448.

Medeiros, E.B.M., Petrissans, M., Wehrer, A., and Zoulalian, A. (2001), "Comparative study of two cocurrent downflow three phase catalytic fixed bed reactors: Application to the sulphur dioxide catalytic oxidation on active carbon particles," *Chemical Engineering and Processing*, 40(2): 153–158.

Mehrnia, M.R., Towfighi, J., Bonakdarpour, B., and Akbarnejad, M.M. (2005), "Gas hold-up and oxygen transfer in a draft-tube airlift bioreactor with petroleum-based liquids," *Biochemical Engineering Journal*, 22(2): 105–110.

Mena, P.C., Ruzicka, M.C., Rocha, F.A., Teixeira, J.A., and Drahos, J. (2005), "Effect of solids on homogeneous-heterogeneous flow regime transition in bubble columns," *Chemical Engineering Science*, 60(22): 6013–6026.

Merchuk, J.C. (1985), "Hydrodynamics and hold-up in air-lift reactors," in *Encyclopedia of Fluid Mechanics*, N.P. Cheremisinoff, Editor, Houston, Texas, Gulf Publishing Company, Chapter 3: 1485–1511.

Merchuk, J.C. (1986), "Gas hold-up and liquid velocity in a two-dimensional air lift reactor," *Chemical Engineering Science*, 41(1): 11–16.

Merchuk, J.C., Contreras, A., García, F., and Molina, E. (1998), "Studies of mixing in a concentric tube airlift bioreactor with different spargers," *Chemical Engineering Science*, 53(4): 709–719.

Merchuk, J.C., and Gluz, M. (1999), "Bioreactors: Air-lift reactors," in *Encyclopedia of Bioprocess Technology: Fermentation, Biocatalysis, and Bioseperations*, M.C. Flickinger, and S.W. Drew, Editors, New York City, New York, John Wiley & Sons, Inc. **1**: 320–353.

Merchuk, J.C., Ladwa, N., Cameron, A., Bulmer, M., and Pickett, A. (1994), "Concentric-tube airlift reactors: Effects of geometrical design on performance," *AIChE Journal*, 40(7): 1105–1117.

Merchuk, J.C., and Stein, Y. (1981), "Local hold-up and liquid velocity in air-lift reactors," *AIChE Journal*, 27(3): 377–388.

Merchuk, J.C., Yona, S., Siegel, M.H., and Zvi, A.B. (1990), "On the first-order approximation to the response of dissolved oxygen electrodes for dynamic $k_L a$ estimation," *Biotechnology and Bioengineering*, 35(11): 1161–1163.

Mersmann, A. (1977), "Auslegung und maßstabsvergrösserung von blasen- und tropfensäulen," *Chemie-Ingenieur-Technik*, 49(9): 679–691.

Metaxas, K., and Papayannakos, N. (2008), "Gas-liquid mass transfer in a bench-scale trickle bed reactor used for benzene hydrogenation," *Chemical Engineering & Technology*, 31(10): 1410–1417.

Miller, D.N. (1974), "Scale-up of agitated vessels gas-liquid mass transfer," *AIChE Journal*, 20(3): 445–453.

Mishra, V.P., and Joshi, J.B. (1994), "Flow generated by a disc turbine: Part IV: Multiple impellers," *Chemical Engineering Research & Design*, 72(A5): 657–668.

Miyahara, T., Hamaguchi, M., Sukeda, Y., and Takehashi, T. (1986), "Size of bubbles and liquid circulation in a bubble column with draught tube and sieve plate," *Canadian Journal of Chemical Engineering*, 64: 718–725.

Mizukami, M., Parthasarathy, R.N., and Faeth, G.M. (1992), "Particle-generated turbulence in homogeneous dilute dispersed flows," *International Journal of Multiphase Flow*, 18(3): 397–412.

Moilanen, P., Laakkonen, M., Visuri, O., Alopaeus, V., and Aittamaa, J. (2008), "Modelling mass transfer in an aerated $0.2\,m^3$ vessel agitated by Rushton, Phasejet and Combijet impellers," *Chemical Engineering Journal*, 142(1): 95–108.

Mok, Y.S., Kim, Y.H., and Kim, S.Y. (1990), "Bubble and gas holdup characteristics in a bubble column of CMC solution," *The Korean Journal of Chemical Engineering*, 7(1): 31–39.

Molina, E., Contreras, A., and Chisti, Y. (1999), "Gas holdup, liquid circulation and mixing behaviour of viscous Newtonian media in a split-cylinder airlift bioreactor," *Food and Bioproducts Processing*, 77(1): 27–32.

Monahan, S.M., Vitankar, V.S., and Fox, R.O. (2005), "CFD predictions for flow-regime transitions in bubble columns," *AIChE Journal*, 51(7): 1897–1923.

Moo-Young, M., and Blanch, H.W. (1981), "Design of biochemical reactors mass transfer criteria for simple and complex systems," *Advances in Biochemical Engineering*, 19: 1–69.

Moucha, T., Linek, V., and Prokopova, E. (2003), "Gas hold-up, mixing time and gas-liquid volumetric mass transfer coefficient of various multiple-impeller configurations: Rushton turbine, pitched blade and techmix impeller and their combinations," *Chemical Engineering Science*, 58(9): 1839–1846.

Mouza, A.A., Dalakoglou, G.K., and Paras, S.V. (2005), "Effect of liquid properties on the performance of bubble column reactors with fine pore spargers," *Chemical Engineering Science*, 60(5): 1465–1475.

Mudde, R.F. (2010a), "Advanced measurement techniques for gls reactors," *The Canadian Journal of Chemical Engineering*, 88(4): 638–647.

Mudde, R.F. (2010b), "Time-resolved X-ray tomography of a fluidized bed," *Powder Technology*, 199(1): 55–59.

Mudde, R.F., Bruneau, P.R.P., and van der Hagen, T.H.J.J. (2005), "Time-resolved γ-densitometry imaging within fluidized beds," *Industrial & Engineering Chemistry Research*, 44: 6181–6187.

Muller, C.R., Holland, D.J., Sederman, A.J., Mantle, M.D., Gladden, L.F., and Davidson, J.F. (2008), "Magnetic resonance imaging of fluidized beds," *Powder Technology*, 183(1): 53–62.

Murthy, B.N., Ghadge, R.S., and Joshi, J.B. (2007), "CFD simulations of gas-liquid-solid stirred reactor: Prediction of critical impeller speed for solid suspension," *Chemical Engineering Science*, 62(24): 7184–7195.

Muthukumar, K., and Velan, M. (2006), "Volumetric mass transfer coefficients in an internal loop airlift reactor with low-density particles," *Journal of Chemical Technology & Biotechnology*, 81(4): 667–673.

Nacef, S., Poncin, S., Bouguettoucha, A., and Wild, G. (2007), "Drift flux concept in two- and three-phase reactors," *Chemical Engineering Science*, 62(24): 7530–7538.

Nakanoh, M., and Yoshida, F. (1980), "Gas absorption by Newtonian and non-Newtonian liquids in a bubble column," *Industrial & Engineering Chemistry Process Design and Development*, 19(1): 190–195.

Nakanoh, M., and Yoshida, F. (1983), "Transient characteristics of oxygen probes and determination of $k_L a$," *Biotechnology and Bioengineering*, 25(6): 1653–1654.

Nauman, E.B., and Buffman, B.A. (1983), *Mixing in Continuous Flow Systems*, John Wiley & Sons, New York.

Nedeltchev, S. (2003), "Correction of the penetration theory applied for prediction of mass transfer coefficients in a high-pressure bubble column operated with gasoline and toluene," *Journal of Chemical Engineering of Japan*, 36(5): 630–633.

Ni, X., Gao, S., Cumming, R.H., and Pritchard, D.W. (1995), "A comparative study of mass transfer in yeast for a batch pulsed baffled bioreactor and a stirred tank fermenter," *Chemical Engineering Science*, 50(13): 2127–2136.

Nicolella, C., van Loosdrecht, M.C.M., and Heijnen, J.J. (1998), "Mass transfer and reaction in a biofilm airlift suspension reactor," *Chemical Engineering Science*, 53(15): 2743–2753.

Nielsen, J., Villadsen, J., and Liden, G. (2003), *Bioreaction Engineering Principles*, Kluwer Academic/Plenum Publishers, New York.

Nielsen, J., and Volladsen, J. (1993), "Description and modelling," in *Bioprocessing*, H.-J. Rehm, G. Reed, A. Puehler, and P. Stadler, Editors, New York, NY, Wiley-VCH. 3: 79–102.

Nienow, A.W. (1996), "Gas-liquid mixing studies: A comparison of Rushton turbines with some modern impellers," *Chemical Engineering Research & Design*, 74(A4): 417–423.

Nienow, A.W., Kendall, A., Moore, I.P.T., Ozcan-Taskin, G.N., and Badham, R.S. (1995), "The characteristics of aerated 12- and 18-blade Rushton turbines at transitional Reynolds numbers," *Chemical Engineering Science*, 50(4): 593–599.

Nienow, A.W., Warmoeskerken, M.M.C.G., Smith, J.M., and Konno, M. (1985), "On the flooding/loading transition and the complete dispersal condition in aerated vessels agitated by a Rushton-turbine," in *Fifth European Conference on Mixing*, Wuerzburg, West Germany, BHRA, Cranfield, England: 143–154.

Nienow, A.W., Wisdom, D.J., and Middleton, J.C. (1977), "The effect of scale and geometry on flooding, recirculation, and power in gassed stirred vessels," in *Second European Conference on Mixing*, Cambridge, England: **F1**: 1–16.

Nigam, K.D.P., and Larachi, F. (2005), "Process intensification in trickle-bed reactors," *Chemical Engineering Science*, 60(22): 5880–5894.

Nishikawa, M., Nakamura, M., and Hashimoto, K. (1981), "Gas absorption in aerated mixing vessels with non-Newtonian liquid," *Journal of Chemical Engineering of Japan*, 14(3): 227–232.

Nishikawa, M., Nishioka, S., and Kayama, T. (1984), "Gas absorption in an aerated mixing vessel with multi-stage impellers," *Journal of Chemical Engineering of Japan*, 17(5): 541–543.

Nocentini, M. (1990), "Mass transfer in gas-liquid, multiple-impeller stirred vessels: A discussion about experimental techniques for $k_L a$ measurement and models comparison," *Chemical Engineering Research & Design*, 68(3): 287–294.

Nocentini, M., Fajner, D., Pasquali, G., and Magelli, F. (1993), "Gas-liquid mass transfer and holdup in vessels stirred with multiple Rushton turbines: Water and water-glycerol solutions," *Industrial & Engineering Chemistry Research*, 32(1): 19–26.

Nocentini, M., Pinelli, D., and Magelli, F. (1998), "Analysis of the gas behavior in sparged reactors stirred with multiple Rushton turbines: Tentative model validation and scale-up," *Industrial & Engineering Chemistry Research*, 37(4): 1528–1535.

Norman, W.S., and Sammak, F.Y.Y. (1963), "Gas absorption in a packed column part I: The effect of liquid viscosity on the mass transfer coefficient," *Transactions of the Institution of Chemical Engineers*, 41(3): 109–116.

Nunes, S.P., and Peinemann, K.-V. (eds) (2006), *Membrane Technology in the Chemical Industry*, Wiley-VCH Verlag GmbH & Co., Weinheim, Germany.

Ogut, A., and Hatch, R.T. (1988), "Oxygen transfer into Newtonian and non-Newtonian fluids in mechanically agitated vessels," *Canadian Journal of Chemical Engineering*, 66(1): 79–85.

Ohki, Y., and Inoue, H. (1970), "Longitudinal mixing of the liquid phase in bubble columns," *Chemical Engineering Science*, 25(1): 1–16.

Oldshue, J.Y. (1983), *Fluid Mixing Technology*, McGraw-Hill Publications Co., New York, NY.

Olmos, E., Gentric, C., and Midoux, N. (2003), "Numerical description of flow regime transitions in bubble column reactors by a multiple gas phase model," *Chemical Engineering Science*, 58(10): 2113–2121.

Onda, K., Sada, E., and Murase, Y. (1959), "Liquid-side mass transfer coefficients in packed towers," *AIChE Journal*, 5(2): 235–239.

Onda, K., Sada, E., and Otubo, F. (1958), "Liquid-side mass-transfer coefficient for a tower packed with Raschig rings," *Kagaku Kogaku*, 22(4): 194–199.

Onda, K., Sada, E., and Saito, M. (1961), "Gas-side mass transfer coefficients in packed tower," *Kagaku Kogaku*, 25(N11): 820–828.

Ostafin, A. and Landfester, K. (eds) (2009), Nanoreactor Engineering: Engineering in Medicine & Biology, Artech House, Boston.

Ozturk, S.S., Schumpe, A., and Deckwer, W.D. (1987), "Organic liquids in a bubble column: Holdups and mass transfer coefficients," *AIChE Journal*, 33(9): 1473–1480.

Padial, N.T., VanderHeyden, W.B., Rauenzahn, R.M., and Yarbro, S.L. (2000), "Three-dimensional simulation of a three-phase draft-tube bubble column," *Chemical Engineering Science*, 55(16): 3261–3273.

Pan, Y., Dudukovic, M.P., and Chang, M. (2000), "Numerical investigation of gas-driven flow in 2-D bubble columns," *AIChE Journal*, 46(3): 434–449.

Parthasarathy, R.N., and Faeth, G.M. (1990a), "Turbulence modulation in homogeneous dilute particle-laden flows," *Journal of Fluid Mechanics*, 220: 485–514.

Parthasarathy, R.N., and Faeth, G.M. (1990b), "Turbulent dispersion of particles in self-generated homogeneous turbulence," *Journal of Fluid Mechanics*, 220: 515–537.

Patel, S.A., Daly, J.G., and Bukur, D.B. (1989), "Holdup and interfacial area measurements using dynamic gas disengagement," *AIChE Journal*, 35(6): 931–942.

Patwardhan, A.W., and Joshi, J.B. (1998), "Design of stirred vessels with gas entrained from free liquid surface," *Canadian Journal of Chemical Engineering*, 76(3): 339–364.

Patwardhan, A.W., and Joshi, J.B. (1999), "Design of gas-inducing reactors," *Industrial & Engineering Chemistry Research*, 38(1): 49–80.

Paul, E.L., Atiemo-Obeng, V.A., and Kresta, S.M. (eds) (2004), *Handbook of Industrial Mixing*, John Wiley & Sons, Inc., Hoboken, New Jersey.

Peinemann, K.-V., and Nunes, S.P. (2008a), "Membranes for energy conversion," in *Membrane Technology*, K.-V. Peinemann, and S.P. Nunes, Editors, Weinheim, Germany, Wiley-VCH Verlag GmbH & Co. KGaA.

Peinemann, K.-V., and Nunes, S.P. (2008b), "Membranes for life sciences," in *Membrane Technology*, K.-V. Peinemann, and S.P. Nunes, Editors, Weinheim, Germany, Wiley-VCH Verlag GmbH & Co. KGaA.

Piché, S., Larachi, F., and Grandjean, B.P.A. (2001a), "Flooding capacity in packed towers: Database, correlations, and analysis," *Industrial & Engineering Chemistry Research*, 40(1): 476–487.

Piché, S., Larachi, F., and Grandjean, B.P.A. (2001b), "Loading capacity in packing towers: Database, correlations and analysis," *Chemical Engineering & Technology*, 24(4): 373–380.

Piché, S.R., Larachi, F., and Grandjean, B.P.A. (2001c), "Improving the prediction of irrigated pressure drop in packed absorption towers," *The Canadian Journal of Chemical Engineering*, 79(4): 584–594.

Pinelli, D., Bakker, A., Myers, K.J., Reeder, M.F., Fasano, J., and Magelli, F. (2003), "Some features of a novel gas dispersion impeller in a dual-impeller configuration," *Chemical Engineering Research & Design*, 81(A4): 448–454.

Pollack, J.K., Li, Z.J., and Marten, M.R. (2008), "Fungal mycelia show lag time before re-growth on endogenous carbon," *Biotechnology and Bioengineering*, 100(3): 458–465.

Poorte, R.E.G., and Biesheuvel, A. (2002), "Experiments on the motion of gas bubbles in turbulence generated by an active grid," *Journal of Fluid Mechanics*, 461: 127–154.

Popovic, M., and Robinson, C.W. (1984), "Estimation of some important design parameters for non-Newtonian liquids in pneumatically-agitated fermenters," in *Proceedings of the 34th Canadian Chemical Engineering Conference*, Quebec City, Canada, 258–263.

Popovic, M., and Robinson, C.W. (1988), "External-circulation-loop airlift bioreactors: Study of the liquid circulating velocity in highly viscous non-Newtonian liquids," *Biotechnology and Bioengineering*, 32(7): 301–312.

Posarac, D., and Petrovic, D. (1988), "An experimental study of the minimum fluidization velocity in a three-phase external loop airlift-reactor," *Chemical Engineering Science*, 43(5): 1161–1165.

Poughon, L., Duchez, D., Cornet, J.F., and Dussap, C.G. (2003), "$k_L a$ determination: Comparative study for a gas mass balance method," *Bioprocess and Biosystems Engineering*, 25(6): 341–348.

Powell, R.L. (2008), "Experimental techniques for multiphase flows," *Physics of Fluids*, 20(4): 040605–040622.

Prasser, H.M. (2008), "Novel experimental measuring techniques required to provide data for CFD validation," *Nuclear Engineering and Design*, 238(3): 744–770.

Prasser, H.M., Misawa, M., and Tiseanu, I. (2005), "Comparison between wire-mesh sensor and ultra-fast X-ray tomograph for an air-water flow in a vertical pipe," *Flow Measurement and Instrumentation*, 16(2–3): 73–83.

Pugsley, T., Tanfara, H., Malcus, S., Cui, H., Chaouki, J., and Winters, C. (2003), "Verification of fluidized bed electrical capacitance tomography measurements with a fiber optic probe," *Chemical Engineering Science*, 58: 3923–3934.

Puthli, M.S., Rathod, V.K., and Pandit, A.B. (2005), "Gas-liquid mass transfer studies with triple impeller system on a laboratory scale bioreactor," *Biochemical Engineering Journal*, 23(1): 25–30.

Rados, N., Shaikh, A., and Al-Dahhan, M. (2005), "Phase distribution in a high pressure slurry bubble column via a single source computed tomography," *The Canadian Journal of Chemical Engineering*, 83: 104–112.

Ramaswamy, S., Cutright, T.J., and Qammar, H.K. (2005), "Control of a continuous bioreactor using model predictive control," *Process Biochemistry*, 40(8): 2763–2770.

Ramm, W.M., and Chagina, Z.W. (1965), "Issledowanie teplootdachi pri powerxnostnom kipenii rastworow neletuchix weshestw," *Khimicheskaia Promyshlennost*, 23: 219–222.

Rampure, M.R., Buwa, V.V., and Ranade, V.V. (2003), "Modelling of gas-liquid/gas-liquid-solid flows in bubble columns: Experiments and CFD simulations," *The Canadian Journal of Chemical Engineering*, 81: 692–706.

Raschig Jaeger Technologies (2006), "Product bulletin 1100," *Raschig Jaeger Technologies Brochures*(1100): 1–20.

Raschig LTD, "Prospectus of packings", Ludwigshafen, Germany, Raschig LTD.

Reddy, G.P., and Chidambaram, M. (1995), "Near-optimal productivity control of a continuous bioreactor," *IEEE Proceedings: Control Theory and Applications*, 142(6): 633–637.

Rees, A.C., Davidson, J.F., Dennis, J.S., Fennell, P.S., Gladden, L.F., Hayhurst, A.N., Mantle, M.D., Muller, C.R., and Sederman, A.J. (2006), "The nature of the flow just above the perforated plate distributor of a gas-fluidised bed, as imaged using magnetic resonance," *Chemical Engineering Science*, 61(18): 6002–6015.

Reichelt, W. (1974), *Stromung und stoffaustausch in fullkorperapparaten bei gegenstrom einer flussigen und einer gasformigen phase*, Weinheim/Bergstr, Verlag Chemie.

Reilley, I.G., Scott, D.S., De Bruijn, T., Jain, A., and Pixkorz, J. (1986), "A correlation for gas holdup in turbulent coalescing bubble columns," *Canadian Journal of Chemical Engineering*, 1986(64): 705–717.

Rewatkar, V.B., Deshpande, A.J., Pandit, A.B., and Joshi, J.B. (1993), "Gas hold-up behavior of mechanically agitated gas-liquid reactors using pitched blade downflow turbines," *Canadian Journal of Chemical Engineering*, 71(2): 226–237.

Rewatkar, V.B., and Joshi, J.B. (1993), "Role of sparger design on gas dispersion in mechanically agitated gas-liquid contactors," *Canadian Journal of Chemical Engineering*, 71(2): 278–291.

Ribeiro Jr., C.P. (2008), "On the estimation of the regime transition point in bubble columns," *Chemical Engineering Journal*, 140(1–3): 473–482.

Ribeiro Jr., C.P., and Lage, P.L.C. (2005), "Gas-liquid direct-contact evaporation: A review," *Chemical Engineering & Technology*, 28(10): 1081–1107.

Riggs, S.S., and Heindel, T.J. (2006), "Measuring carbon monoxide gas-liquid mass transfer in a stirred tank reactor for syngas fermentation," *Biotechnology Progress*, 22(3): 903–906.

Rodgers, T.L., Gangolf, L., Vannier, C., Parriaud, M., and Cooke, M. (2011), "Mixing times for process vessels with aspect ratios greater than one," *Chemical Engineering Science*, 66(13): 2935–2944.

Ros, M., and Zupancic, G.D. (2002), "Thermophilic aerobic digestion of waste activated sludge," *Acta Chimica Slovenica*, 49(4): 931–943.

Roy, N.K., Guha, D.K., and Rao, M.N. (1963), "Fractional gas holdup in two-phase and three-phase batch-fluidized bubble-bed and foam-systems," *Indian Chemical Engineer*, 1963: 27–31.

Roy, S., Dhotre, M.T., and Joshi, J.B. (2006), "CFD simulation of flow and axial dispersion in external loop airlift reactor," *Transactions of IChemE, Part A, Chemical Engineering Research and Design*, 84(A8): 677–690.

Roy, S., and Joshi, J.B. (2008), "CFD study of mixing characteristics of bubble column and external loop airlift reactor," *Asia-Pacific Journal of Chemical Engineering*, 3(2): 97–105.

Ruchti, G., Dunn, I.J., and Bourne, J.R. (1981), "Comparison of dynamic oxygen electrode methods for the measurement of $k_L a$," *Biotechnology and Bioengineering*, 23(2): 277–290.

Ruckenstein, E. (1964), "On mass transfer in the continuous phase from spherical bubbles or drops," *Chemical Engineering Science*, 19(2): 131–148.

Ruen-ngam, D., Wongsuchoto, P., Limpanuphap, A., Charinpanitkul, T., and Pavasant, P. (2008), "Influence of salinity on bubble size distribution and gas-liquid mass transfer in airlift contactors," *Chemical Engineering Journal*, 141(1–3): 222–232.

Ruthiya, K.C., Chilekar, V.P., Warnier, M.J.F., van der Schaaf, J., Kuster, B.F.M., Schouten, J.C., and van Ommen, J.R. (2005), "Detecting regime transitions in slurry bubble columns using pressure time series," *AIChE Journal*, 51(7): 1951–1965.

Ruzicka, M.C., Drahos, J., Fialová, M., and Thomas, N.H. (2001a), "Effect of bubble column dimensions on flow regime transition," *Chemical Engineering Science*, 56(21–22): 6117–6124.

Ruzicka, M.C., Drahos, J., Mena, P.C., and Teixeira, J.A. (2003), "Effect of viscosity on homogeneous-heterogeneous flow regime transition in bubble columns," *Chemical Engineering Journal*, 96(1–3): 15–22.

Ruzicka, M.C., Zahradník, J., Drahos, J., and Thomas, N.H. (2001b), "Homogeneous-heterogeneous regime transition in bubble columns," *Chemical Engineering Science*, 56(15): 4609–4626.

Sada, E., Katoh, S., Yoshii, H., Yamanishi, T., and Nakanishi, A. (1984), "Performance of the gas bubble column in molten salt systems," *Industrial & Engineering Chemistry, Process Design and Development*, 23(1): 151–154.

Salvacion, J.L., Murayama, M., Ohtaguchi, K., and Koide, K. (1995), "Effects of alcohols on gas holdup and volumetric liquid-phase mass transfer coefficient in gel-particle suspended bubble column," *Journal of Chemical Engineering of Japan*, 28(4): 434.

Sankaranarayanan, K., and Sundaresan, S. (2002), "Lift force in bubbly suspensions," *Chemical Engineering Science*, 57: 3521–3542.

Sanyal, J., Marchisio, D.L., Fox, R.O., and Dhanasekharan, K. (2005), "On the comparison between population balance models for CFD simulation of bubble columns," *Industrial & Engineering Chemistry Research*, 44: 5063–5072.

Sardeing, R., Aubin, J., Poux, M., and Xuereb, C. (2004a), "Gas-liquid mass transfer: Influence of sparger location," *Chemical Engineering Research & Design*, 82(9): 1161–1168.

Sardeing, R., Aubin, J., and Xuereb, C. (2004b), "Gas-liquid mass transfer: A comparison of down- and up-pumping axial flow impellers with radial impellers," *Chemical Engineering Research & Design*, 82(12): 1589–1596.

Sato, Y., Sadatomi, M., and Sekoguchi, K. (1981), "Momentum and heat transfer in two-phase bubble flow I," *International Journal of Multiphase Flow*, 7(2): 167–177.

Sato, Y., and Sekoguchi, K. (1975), "Liquid velocity distribution in two-phase bubble flow," *International Journal of Multiphase Flow*, 2(1): 79–95.

Saxena, S.C., Patel, D., Smith, D.N., and Ruether, J.A. (1988), "An assessment of experimental techniques for the measurement of bubble size in a bubble slurry reactor as applied to indirect coal liquefaction," *Chemical Engineering Communications*, 63: 87–127.

Scargiali, F., D'Orazio, A., Grisafi, F., and Brucato, A. (2007), "Modelling and simulation of gas-liquid hydrodynamics in mechanically stirred tanks," *Chemical Engineering Research & Design*, 85(5): 637–646.

Schäfer, R., Merten, C., and Eigenberger, G. (2002), "Bubble size distributions in a bubble column reactor under industrial conditions," *Experimental Thermal and Fluid Science*, 26(6–7): 595–604.

Scherwood, T.K., and Holloway, F.A.L. (1940) "Performance of packed towers—Liquid film data for several packings," *Transactions of the American Institute of Chemical Engineers*, 36(1): 36–69.

Schiller, L., and Naumann, A. (1933), "Uber die grundlegenden berechnungen bei der schwerkraftaufbereitung," *Zeitung des vereins deutscher ingenieure*, 77: 318–320.

Schmit, C.E., and Eldridge, R.B. (2004), "Investigation of X-ray imaging of vapor-liquid contactors: 1 – Studies involving stationary objects and a simple flow system," *Chemical Engineering Science*, 59: 1255–1266.

Schmit, C.E., Perkins, J., and Eldridge, R.B. (2004), "Investigation of X-ray imaging of vapor-liquid contactors: 2 - Experiments and simulations of flows in an air-water contactor," *Chemical Engineering Science*, 59: 1267–1283.

Schönfeld, H., Hunger, K., Cecilia, R., and Kunz, U. (2004), "Enhanced mass transfer using a novel polymer/carrier microreactor," *Chemical Engineering Journal*, 101(1–3): 455–463.

Schultes, M. (2003), "Raschig super-ring: A new fourth generation packing offers new advantages," *Chemical Engineering Research & Design*, 81(1): 48–57.

Schumacher, J. (2000), "A framework for batch-operation analysis within the context of disturbance management," *Computers & Chemical Engineering*, 24(2–7): 1175–1180.

Schumpe, A., and Deckwer, W.D. (1987), "Viscous media in tower bioreactors: Hydrodynamic characteristics and mass transfer properties," *Bioprocess and Biosystems Engineering*, 1987(2): 79–94.

Schumpe, A., and Grund, G. (1986), "The gas disengagement technique for studying gas holdup structure in bubble columns," *The Canadian Journal of Chemical Engineering*, 64(6): 891–896.

Schuurman, Y. (2008), "Aspects of kinetic modeling of fixed bed reactors," *Catalysis Today*, 138(1–2): 15–20.

Seeger, A., Affeld, K., Goubergrits, L., Kertzscher, U., and Wellnhofer, E. (2001a), "X-ray-based assessment of the three-dimensional velocity of the liquid phase in a bubble column," *Experiments in Fluids*, 31: 193–201.

Seeger, A., Affeld, K., Kertzscher, U., Goubergrits, L., and Wellnhofer, E. (2001b), "Assessment of flow structures in bubble columns by X-ray based particle tracking velocimetry," in *4th International Symposium on Particle Image Velocimetry*, Gottingen, Germany.

Seeger, A., Kertzscher, U., Affeld, K., and Wellnhofer, E. (2003), "Measurement of the local velocity of the solid phase and the local solid hold-up in a three-phase flow by X-ray based particle tracking velocimetry (XPTV)," *Chemical Engineering Science*, 58: 1721–1729.

Shah, Y.T., Kelkar, B.G., Godbole, S.P., and Deckwer, W.-D. (1982), "Design parameters estimation for bubble column reactors," *AIChE Journal*, 28(3): 353–379.

Shaikh, A., and Al-Dahhan, M. (2005), "Characterization of the hydrodynamic flow regime in bubble columns via computed tomography," *Flow Measurement and Instrumentation*, 16(2–3): 91–98.

Shariati, F.P., Bonakdarpour, B., and Mehrnia, M.R. (2007), "Hydrodynamics and oxygen transfer behaviour of water in diesel microemulsions in a draft tube airlift bioreactor," *Chemical Engineering and Processing*, 46(4): 334–342.

Sharma, M., and Yashonath, S. (2006), "Breakdown of the Stokes-Einstein relationship: Role of interactions in the size dependence of self-diffusivity," *The Journal of Physical Chemistry. B*, 110(34): 17207–17211.

Shewale, S.D., and Pandit, A.B. (2006), "Studies in multiple impeller agitated gas-liquid contactors," *Chemical Engineering Science*, 61(2): 489–504.

Shpirt, E. (1981), "Role of hydrodynamic factors in ammonia desorption by diffused aeration," *Water Research*, 15(6): 739–743.

Shulman, H.L., Ulrich, C.E., Proulz, A.Z., and Zimmerman, L.O. (1955), "Wetted and effective: Interfacial areas, gas- and liquid-phase mass transfer rates," *AIChE Journal*, 1(2): 253–258.

Siegel, M.H., and Merchuk, J.C. (1988), "Mass transfer in a rectangular air-lift reactor: Effects of geometry and gas recirculation," *Biotechnology and Bioengineering*, 32(9): 1128–1137.

Siegel, M.H., Merchuk, J.C., and Schugerl, K. (1986), "Air-lift reactor analysis: Interrelationships between riser, downcomer, and gas-liquid separator behavior, including gas recirculation effects," *AIChE Journal*, 32(10): 1585–1596.

Simon, J., Wiese, J., and Steinmetz, H. (2006), "A comparison of continuous flow and sequencing batch reactor plants concerning integrated operation of sewer systems and wastewater treatment plants," *Water Science & Technology*, 54(11): 241–248.

Sivashanmugam, P., and Prabhakaran, S. (2008), "Simulation of an effect of a baffle length on the power consumption in an agitated vessel," *International Journal of Food Engineering*, 4(2): Article 3.

Smith, D.N., Fuchs, W., Lynn, R.J., and Smith, D.H. (1983), *Bubble Behavior in a Slurry Bubble Column Reactor Model*, ACS, Washington, DC, USA 526.

Smith, J.M. (1991), "Simple performance correlations for agitated vessels," in *Proceedings of the 7th European Congress on Mixing*, Brugge, Belgium: 233–241.

Smith, J.M., Van't Riet, K., and Middleton, J.C. (1977), "Scale-up of agitated gas-liquid reactors for mass transfer", in *Proceedings of the Second European Conference on Mixing*, H.S. Stephens and J.A. Clarke, Editors, Crafield, UK, BHRA Fluid Engineering: 51–66.

Smith, J.M., and Warmoeskerken, M.M.C.G. (1985), "Dispersion of gases in liquids with turbines," in *Fifth European Conference on Mixing*, Wuerzburg, West Germany, BHRA, Cranfield, England, 115–126.

Snape, J.B., Zahradnik, J., Fialova, M., and Thomas, N.M. (1995), "Liquid-phase properties and sparger design effects in an external loop airlift reactor," *Chemical Engineering Science*, 50(20): 3175–3186.

Sobotka, M., Prokop, A., Dunn, I.J., and Einsele, A. (1982), "Review of methods for the measurement of oxygen transfer in microbial systems," *Annual Reports on Fermentation Processes*, 5: 127–210.

Sokol, W., and Migiro, C.L.C. (1996), "Controlling a continuous stirred-tank bioreactor degrading phenol in the stability range," *The Chemical Engineering Journal and The Biochemical Engineering Journal*, 62(1): 67–72.

Sokolichin, A., and Eigenberger, G. (1994), "Gas-liquid flow in bubble columns and loop reactors: Part I. Detailed modeling and numerical simulation," *Chemical Engineering Science*, 49(24): 5735–5746.

Sokolichin, A., Eigenberger, G., and Lapin, A. (2004), "Simulation of buoyancy driven bubbly flow: Established simplifications and open questions," *AIChE Journal*, 50(1): 24–45.

Sokolov, V.N., and Aksenova, E.G. (1982), "Mass transfer in bubble columns," *Journal of Applied Chemistry of the USSR*, 55(10): 2158–2159.

Sotelo, J.L., Benitez, F.J., Beltran-Heredia, J., and Rodriguez, C. (1994), "Gas holdup and mass transfer coefficients in bubble columns. 1. Porous glass-plate diffusers," *International Chemical Engineering*, 34(1): 82–90.

Sotiriadis, A.A., Thorpe, R.B., and Smith, J.M. (2005), "Bubble size and mass transfer characteristics of sparged downwards two-phase flow," *Chemical Engineering Science*, 60(22): 5917–5929.

Spelt, P.D.M., and Sangani, A.S. (1998), "Properties and averaged equations for flows of bubbly liquids," *Applied Scientific Research*, 58(1): 337–386.

Sriram, K., and Mann, R. (1977), "Dynamic gas disengagement: A new technique for assessing the behavior of bubble columns," *Chemical Engineering Science*, 32: 571–580.

Stegeman, D., Ket, P.J., Kolk, K.A.v.d., Bolk, J.W., Knop, P.A., and Westerterp, K.R. (1995), "Interfacial area and gas holdup in an agitated gas-liquid reactor under pressure," *Industrial & Engineering Chemistry Research*, 34(1): 59–71.

Stenberg, O., and Andersson, B. (1988a), "Gas-liquid mass transfer in agitated vessels–I. Evaluation of the gas-liquid mass transfer coefficient from transient-response measurements," *Chemical Engineering Science*, 43(3): 719–724.

Stenberg, O., and Andersson, B. (1988b), "Gas-liquid mass transfer in agitated vessels–II. Modelling of gas-liquid mass transfer," *Chemical Engineering Science*, 43(3): 725–730.

Stichlmair, J., Bravo, J.L., and Fair, J.R. (1989), "General model for prediction of pressure drop and capacity of countercurrent gas/liquid packed columns," *Gas Separation & Purification*, 3(March): 19–28.

Strigle, R.F.J. (1994), *Packed Tower Design and Applications: Random and Structured Packings*, Gulf Publishing Company, Houston, TX.

Su, X., and Heindel, T.J. (2003), "Gas holdup in a fiber suspension," *The Canadian Journal of Chemical Engineering*, 81: 412–418.

Su, X., and Heindel, T.J. (2004), "Gas holdup behavior in Nylon fiber suspensions," *Industrial & Engineering Chemistry Research*, 43(9): 2256–2263.

Su, X., and Heindel, T.J. (2005a), "Effect of perforated plate open area on gas holdup in Rayon fiber suspensions," *ASME Journal of Fluids Engineering - Transactions of the ASME*, 127(4): 816–823.

Su, X., and Heindel, T.J. (2005b), "Modeling gas holdup in gas-liquid-fiber semibatch bubble columns," *Industrial & Engineering Chemistry Research*, 44(24): 9355–9363.

Su, X., Hol, P.D., Talcott, S.M., Staudt, A.K., and Heindel, T.J. (2006), "The effect of bubble column diameter on gas holdup in fiber suspensions," *Chemical Engineering Science*, 61(10): 3098–3104.

Syeda, S.R., Afacan, A., and Chuang, K.T. (2002), "Prediction of gas hold-up in a bubble column filled with pure and, binary liquids," *Canadian Journal of Chemical Engineering*, 80(1): 44–50.

Talvy, S., Cockx, A., and Liné, A. (2007), "Modeling of oxygen mass transfer in a gas-liquid airlift reactor," *AIChE Journal*, 53(2): 316–326.

Tang, C., and Heindel, T.J. (2004), "Time-dependent gas holdup variation in an air-water bubble column," *Chemical Engineering Science*, 59(3): 623–632.

Tang, C., and Heindel, T.J. (2005a), "Effect of fiber type on gas holdup in a cocurrent air-water-fiber bubble column," *Chemical Engineering Journal*, 111(1): 21–30.

Tang, C., and Heindel, T.J. (2005b), "Gas-liquid-fiber flow in a cocurrent bubble column," *AIChE Journal*, 51(10): 2665–2674.

Tang, C., and Heindel, T.J. (2006a), "Estimating gas holdup via pressure difference measurements in a cocurrent bubble column," *International Journal of Multiphase Flow*, 32(7): 850–863.

Tang, C., and Heindel, T.J. (2006b), "Similitude analysis for gas-liquid-fiber flows in cocurrent bubble columns," in *Proceedings of FEDSM06: ASME Joint US-European Fluids Engineering Summer Meeting*, Miami, Florida.

Tang, C., and Heindel, T.J. (2007), "Effect of fiber length distribution on gas holdup in a cocurrent air-water-fiber bubble column," *Chemical Engineering Science*, 62(5): 1408–1417.

Tatterson, G.B. (1991), *Fluid Mixing and Gas Dispersion in Agitated Tanks*, McGraw-Hill, Inc., New York.

Tatterson, G.B. (1994), *Scaleup and Design of Industrial Mixing Processes*, McGraw-Hill, Inc., New York.

Tecante, A., and Choplin, L. (1993), "Gas-liquid mass transfer in non-Newtonian fluids in a tank stirred with a helical ribbon screw impeller," *Canadian Journal of Chemical Engineering*, 71(6): 859–865.

Terasaka, K., Hullmann, D., and Schumpe, A. (1998), "Mass transfer in bubble columns studied with an oxygen optode," *Chemical Engineering Science*, 53(17): 3181–3184.

Thatte, A.R., Ghadge, R.S., Patwardhan, A.W., Joshi, J.B., and Singh, G. (2004), "Local gas holdup measurement in sparged and aerated tanks by γ-ray attenuation technique," *Industrial & Engineering Chemistry Research*, 43(17): 5389–5399.

Titchener-Hooker, N.J., and Hoare, P.D.M. (2008), "Micro biochemical engineering to accelerate the design of industrial-scale downstream processes for biopharmaceutical proteins," *Biotechnology and Bioengineering*, 100(3): 473–487.

Tobajas, M., and Garcia-Calvo, E. (2000), "Comparison of experimental methods for determination of the volumetric mass transfer coefficient in fermentation processes," *Heat and Mass Transfer/Waerme- und Stoffuebertragung*, 36(3): 201–207.

Torré, J.-P., Fletcher, D.F., Lasuye, T., and Xuereb, C. (2007), "An experimental and computational study of the vortex shape in a partially baffled agitated vessel," *Chemical Engineering Science*, 62(7): 1915–1926.

Tortora, P.R., Ceccio, S.L., O'Hern, T.J., Trujillo, S.M., and Torczynski, J.R. (2006), "Quantitative measurement of solids distribution in gas-solid riser flows using electrical impedance tomography and gamma densitometry tomography," *International Journal of Multiphase Flow*, 32(8): 972–995.

Toye, D., Fransolet, E., Simon, D., Crine, M., L'Homme, G., and Marchot, P. (2005), "Possibilities and limits of application of electrical resistance tomography in hydrodynamics of bubble columns," *The Canadian Journal of Chemical Engineering*, 83: 4–10.

Toye, D., Marchot, P., Crine, M., and L'Homme, G. (1996), "Modelling of multiphase flow in packed beds by computer-assisted X-ray tomography," *Measurement Science and Technology*, 7(3): 436–443.

Tribe, L.A., Briens, C.L., and Margaritis, A. (1995), "Determination of the volumetric mass transfer coefficient ($k_L a$) using the dynamic 'gas out-gas in' method: Analysis of errors caused by dissolved oxygen probes," *Biotechnology and Bioengineering*, 46(4): 388–392.

Trilleros, J., Díaz, R., and Redondo, P. (2005), "Three-phase airlift internal loop reactor: Correlations for predicting the main fluid dynamic parameters," *Journal of Chemical Technology & Biotechnology*, 80(5): 515–522.

Tse, K., Martin, T., McFarlane, C.M., and Nienow, A.W. (1998), "Visualisation of bubble coalescence in a coalescence cell, a stirred tank and a bubble column," *Chemical Engineering Science*, 53(23): 4031–4036.

Turner, A.P.F., and White, S.F. (1999), "Process monitoring," in *Encyclopedia of Process Technology: Fermentation, Biocatalysis, and Bioseperation*, M.C. Flickinger and S.W. Drew, Editors, New York, John Wiley. 4: 2056–2070.

Tzeng, J.-W., Chen, R.C., and Fan, L.S. (1993), "Visualization of flow characteristics in a 2-D bubble column and three-phase fluidized bed," *AIChE Journal*, 39(5): 733–744.

Ueyama, K., Tsuru, T., and Furusaki, S. (1989), "Flow transition in a bubble column," *International Chemical Engineering*, 29(3): 523–529.

Ulbrecht, J.J., and Patterson, G.K. (1985), *Mixing of Liquids by Mechanical Agitation*, Gordon and Breach Science Publishers, New York.

Ungerman, A.J. (2006), *"Mass transfer enhancement for syngas fermentation,"* MS Thesis, Iowa State University, Ames, Iowa.

Ungerman, A.J., and Heindel, T.J. (2007), "Carbon monoxide mass transfer for syngas fermentation in a stirred tank reactor with dual impeller configurations," *Biotechnology Progress*, 23(3): 613–620.

Utomo, M.B., Warsito, W., Sakai, T., and Uchida, S. (2001), "Analysis of distributions of gas and TiO_2 particles in slurry bubble column using ultrasonic computed tomography," *Chemical Engineering Science*, 56: 6073–6079.

Van't Riet, K. (1979), "Review of measuring methods and results in nonviscous gas-liquid mass transfer in stirred vessels," *Industrial and Engineering Chemistry Process Design and Development*, 18(3): 357–364.

Van't Riet, K. and Tramper, J. (1991), *Basic Bioreactor Design*, Marcel Dekker, Inc., New York, New York.

van Benthum, W.A.J., van den Hoogen, J.H.A., van der Lans, R.G.J.M., van Loosdrecht, M.C.M., and Heijnen, J.J. (1999a), "The biofilm airlift suspension extension reactor. Part I: Design and two-phase hydrodynamics," *Chemical Engineering Science*, 54(12): 1909–1924.

van Benthum, W.A.J., van der Lans, R.G.J.M., van Loosdrecht, M.C.M., and Heijnen, J.J. (1999b), "Bubble recirculation regimes in an internal-loop airlift reactor," *Chemical Engineering Science*, 54(18): 3995–4006.

van Dam-Mieras, M.C.E., de Jue, W.H., de Vries, J., Currell, B.R., James, J.W., Leach, C.K., and Patmore, R.A. (1992), *Operational Modes of Bioreactors*, Elsevier Science & Technology Books, San Diego.

van der Meer, A.B., Beenackers, A.A.C.M., Burghard, R., Mulder, N.H., and Fok, J.J. (1992), "Gas/liquid mass transfer in a four-phase stirred fermentor: Effects of organic phase hold-up and surfactant concentration," *Chemical Engineering Science*, 47(9–11): 2369–2374.

van der Schaaf, J., Chilekar, V.P., van Ommen, J.R., Kuster, B.F.M., Tinge, J.T., and Schouten, J.C. (2007), "Effect of particle lyophobicity in slurry bubble columns at elevated pressures," *Chemical Engineering Science*, 62(18–20): 5533–5537.

van der Westhuizen, I., Du Toit, E., and Nicol, W. (2007), "Trickle flow multiplicity: The influence of the prewetting procedure on flow hysteresis, " *Chemical Engineering Research & Design*, 85(12): 1604–1610.

van Elk, E.P., Knaap, M.C., and Versteeg, G.F. (2007), "Application of the penetration theory for gas-liquid mass transfer without liquid bulk: Differences with systems with a bulk," *Chemical Engineering Research & Design*, 85(4): 516–524.

van Sint Annaland, M., Dijkhuizen, W., Deen, N.G., and Kuipers, J.A.M. (2006), "Numerical simulation of behavior of gas bubbles using a 3-D front-tracking method," *AIChE Journal*, 52(1): 99–110.

Vandu, C.O., and Krishna, R. (2004), "Influence of scale on the volumetric mass transfer coefficients in bubble columns," *Chemical Engineering and Processing*, 43(4): 575–579.

Vardar, F., and Lilly, M.D. (1982), "The measurement of oxygen-transfer coefficients in fermentors by frequency response techniques," *Biotechnology and Bioengineering*, 24(7): 1711–1719.

Varley, J. (1995), "Submerged gas-liquid jets: Bubble size prediction," *Chemical Engineering Science*, 50(5): 901–905.

Vasconcelos, J.M.T., Orvalho, S.C.P., Rodrigues, A.M.A.F., and Alves, S.S. (2000), "Effect of blade shape on the performance of six-bladed disk turbine impellers," *Industrial & Engineering Chemistry Research*, 39(1): 203–213.

Vasquez, G., Antorrena, G., Navaza, J.M., Santos, V., and Rodriguez, T. (1993), "Adsorption of CO_2 in aqueous solutions of various viscosities in the presence of induced turbulence," *International Chemical Engineering*, 1993(4): 649–655.

Vasquez, G., Cancela, M.A., Riverol, C., Alvarez, E., and Navaza, J.M. (2000), "Determination of interfacial areas in a bubble column by different chemical methods," *Industrial & Engineering Chemistry Research*, 39: 2541–2547.

Vatai, G., and Tekic, M.N. (1986), "Effect of pseudoplasticity on hydrodynamic characteristics of air-lift loop contactor," *Rheologica Acta*, 26: 271.

Vatanakul, M., Zheng, Y., and Couturier, M. (2004), "Application of ultrasonic technique in multiphase flows," *Industrial & Engineering Chemistry Research*, 43: 5681–5691.

Vázquez, G., Cancela, M.A., Riverol, C., Alvarez, E., and Navaza, J.M. (2000), "Application of the Danckwerts method in a bubble column: Effects of surfactants on mass transfer coefficient and interfacial area," *Chemical Engineering Journal*, 78(1): 13–19.

Vazquez, G., Cancela, M.A., Varela, R., Alvarez, E., and Navaza, J.M. (1997), "Influence of surfactants on absorption of CO_2 in a stirred tank with and without bubbling," *Chemical Engineering Journal*, 67(2): 131–137.

Veera, U.P., and Joshi, J.B. (1999), "Measurement of gas hold-up profiles by gamma ray tomography: Effect of sparger design and height of dispersion in bubble columns," *Chemical Engineering Research & Design*, 77(4): 303–317.

Veera, U.P., Kataria, K.L., and Joshi, J.B. (2004), "Effect of superficial gas velocity on gas hold-up profiles in foaming liquids in bubble column reactors," *Chemical Engineering Journal*, 99(1): 53–58.

Veera, U.P., Patwardhan, A.W., and Joshi, J.B. (2001) "Measurement of gas hold-up profiles in stirred tank reactors by gamma ray attenuation technique," *Transactions of IChemE, Part A Chemical Engineering Research and Design*, 79(6): 684–688.

Vega, J.L., Antorrena, G.M., Clausen, E.C., and Gaddy, J.L. (1989), "Study of gaseous substrate fermentations: Carbon monoxide conversion to acetate. 2. Continuous culture," *Biotechnology and Bioengineering*, 34(6): 785–793.

Verma, A.K., and Rai, S. (2003), "Studies on surface to bulk ionic mass transfer in bubble column," *Chemical Engineering Journal*, 94(1): 67–72.

Vermeer, D.J., and Krishna, R. (1981), "Hydrodynamics and mass transfer in bubble columns operating in the churn-turbulent regime," *Industrial and Engineering Chemistry Process Design and Development*, 20: 475–482.

Vesselinov, H.H., Stephan, B., Uwe, H., Holger, K., Günther, H., and Wilfried, S. (2008), "A study on the two-phase flow in a stirred tank reactor agitated by a gas-inducing turbine," *Chemical Engineering Research & Design*, 86(1): 75–81.

Vial, C., Bendjaballah-Lalaoui, N., Poncin, S., Wild, G., and Midoux, N. (2003), "Comparison, combination, and validation of measuring techniques for flow and turbulence analysis in bubble columns and airlift reactors," *The Canadian Journal of Chemical Engineering*, 81(3–4): 749–755.

Vial, C., Camarasa, E., Poncin, S., Wild, G., Midoux, N., and Bouillard, J. (2000), "Study of hydrodynamic behaviour in bubble columns and external loop airlift reactors through analysis of pressure fluctuations," *Chemical Engineering Science*, 55(15): 2957–2973.

Vial, C., Laine, R., Poncin, S., Midoux, N., and Wild, G. (2001), "Influence of gas distribution and regime transitions on liquid velocity and turbulence in a 3-D bubble column," *Chemical Engineering Science*, 56(3): 1085–1093.

Waghmare, Y.G., Rice, R.G., and Knopf, F.C. (2008), "Mass transfer in a viscous bubble column with forced oscillations," *Industrial & Engineering Chemistry Research*, 47(15): 5386–5394.

Wallis, G.B. (1969), *One-Dimensional Two-Phase Flow*, McGraw-Hill, New York.

Walter, J.F., and Blanch, H.W. (1986), "Bubble break-up in gas-liquid bioreactors: Break-up in turbulent flows," *Chemical Engineering Journal*, 32(1): B7–B17.

Warsito, W., and Fan, L.-S. (2003a), "3D-ECT velocimetry for flow structure quantification of gas-liquid-solid fluidized beds," *The Canadian Journal of Chemical Engineering*, 81(3–4): 875–884.

Warsito, W., and Fan, L.-S. (2003b), "ECT imaging of three-phase fluidized bed based on three-phase capacitance model," *Chemical Engineering Science*, 58: 823–832.

Warsito, W., and Fan, L.-S. (2005), "Dynamics of spiral bubble plume motion in the entrance region of bubble columns and three phase fluidized beds using 3D ECT," *Chemical Engineering Science*, 60: 6073–6084.

Watts, P., and Wiles, C. (2007), "Micro reactors: A new tool for the synthetic chemist," *Organic and Biomolecular Chemistry*, 5(5): 727–732.

Wei, C., Xie, B., and Xiao, H. (2000), "Hydrodynamics in an internal loop airlift reactor with a convergence-divergence draft tube," *Chemical Engineering & Technology*, 23(1): 38–45.

Wei, Y., Tiefeng, W., Malin, L., and Zhanwen, W. (2008), "Bubble circulation regimes in a multi-stage internal-loop airlift reactor," *Chemical Engineering Journal*, 142(3): 301–308.

Weiland, P. (1984), "Influence of draft tube diameter on operation behavior of airlift loop reactors," *German Chemical Engineering*, 7(6): 374–385.

Weiland, P., and Onken, U. (1981), "Fluid dynamics and mass transfer in an airlift fermenter with external loop," *German Chemical Engineering*, 4(1): 42–50.

Westerterp, K.R., and Molga, E.J. (2006), "Safety and runaway protection in batch and semi-batch reactors: A review," *Chemical Engineering Research & Design*, 84(A7): 543–552.

White, F.M. (1974), *Viscous Fluid Flow*, McGraw-Hill, New York.

Whitton, M.J., and Nienow, A.W. (1993), "Scale-up correlations fr gas hold-up and mass transfer coefficients in stirred tank reactors" in *Proceedings of the 3rd International Conference on Bioreactor and Bioprocess Fluid Dynamics*, 135–149.

Wilkin, R.T., McNeil, M.S., Adair, C.J., and Wilson, J.T. (2001), "Field measurement of dissolved oxygen: A comparison of methods," *Ground Water Monitoring and Remediation*, 21(4): 124–132.

Wilkinson, P.M. (1991), "Physical aspect and scale-up of high pressure bubble column," PhD Thesis, University of Groningen, Groningen, The Netherlands.

Wilkinson, P.M., Spek, A.P., and van Dierendonck, L.L. (1992), "Design parameters estimation for scale-up of high-pressure bubble columns," *AIChE Journal*, 38(4): 544–554.

Williams, J.A. (2002), "Keys to bioreactor selection," *Chemical Engineering Progress*, 98(3): 34–41.

Williams, R.A., and Beck, M.S., (eds) (1995), *Process Tomography: Principles, Techniques and Applications*, Butterworth-Heinemann Ltd, Oxford.

Wirth, T. (ed.) (2008), Microreactors in Organic Synthesis and Catalysis, *Weinheim*, Wiley-VCH Verlag GmbH & KGaA, Germany.

Wooley, R., Ruth, M., Sheehan, J., Ibsen, K., Majdeski, H., and Galvez, A. (1999), "Lignocellulosic biomass to ethanol process design and economics utilizing co-current dilute acid prehydrolysis and enzymatic hydrolysis current and futuristic scenarios", National Renewable Energy Laboratory, Golden, Colorado.

Worden, R.M., and Bredwell, M.D. (1998), "Mass transfer properties of microbubbles: II. Analysis using a dynamic model," *Biotechnology Progress*, 14(1): 39–46.

Worden, R.M., Bredwell, M.D., and Grethlein, A.J. (1997), "Engineering issues in synthesis-gas fermentations," in *ACS Symposium Series: Fuels and Chemicals from Biomass*, B.C. Saha, and J. Woodward, Editors, Washington, DC, American Chemical Society. 666: 320–335.

Yakhnin, V.Z., and Menzinger, M. (2008), "Estimating spectral properties of the thermal instability in packed-bed reactors," *Chemical Engineering Science*, 63(6): 1480–1489.

Yawalkar, A.A., Heesing, A.B.M., Versteeg, G.F., and Pangarkar, V.G. (2002a), "Gas-liquid mass transfer coefficient in stirred tank reactors," *The Canadian Journal of Chemical Engineering*, 80: 840–848.

Yawalkar, A.A., Pangarkar, V.G., and Beenackers, A.A.C.M. (2002b), "Gas hold-up in stirred tank reactors," *The Canadian Journal of Chemical Engineering*, 80: 158–166.

Yin, F., Afacan, A., Nandakumar, K., and Chuang, K.T. (2002), "Liquid holdup distribution in packed columns: Gamma ray tomography and CFD simulation," *Chemical Engineering and Processing*, 41: 473–483.

Yoshida, F., and Koyanagi, T. (1962), "Mass transfer and effective interfacial areas in packed columns," *AIChE Journal*, 8(3): 309–316.

Zahradnik, J., Fialová, M., and Linek, V. (1999a), "The effect of surface-active additives on bubble coalescence in aqueous media," *Chemical Engineering Science*, 54(21): 4757–4766.

Zahradnik, J., Fialova, M., Ruzicka, M., Drahos, J., Kastanek, F., and Thomas, N.H. (1997), "Duality of the gas-liquid flow regimes in bubble column reactors," *Chemical Engineering Science*, 52(21/22): 3811–3826.

Zahradnik, J., and Kastanek, F. (1979), "Gas holdup in uniformly aerated bubble column reactors," *Chemical Engineering Communications*, 3(4–5): 413–429.

Zahradnik, J., Kuncová, G., and Fialová, M. (1999b), "The effect of surface active additives on bubble coalescence and gas holdup in viscous aerated batches," *Chemical Engineering Science*, 54(13–14): 2401–2408.

Zech, J.B. (1978), "Liquid flow and mass transfer in an irrigated packed column," PhD Thesis, TU Munchen, Munchen, Germany.

Zehner, P. (1989), "Mehrphasenstromungen in gas-flussigkeits-reaktoren," *Dechema-Monograph*, 1989(114): 215–233.

Zeng, A.-P., and Deckwer, W.-D. (1996), "Bioreaction techniques under microaerobic conditions: From molecular level to pilot plant reactors," *Chemical Engineering Science*, 51(10): 2305–2314.

Zhang, H., Chen, G., Yue, J., and Yuan, Q. (2009), "Hydrodynamics and mass transfer of gas-liquid flow in a falling film microreactor," *AIChE Journal*, 55(5): 1110–1120.

Zheng, Y., and Zhang, Q. (2004), "Simultaneous measurements of gas and solid holdups in multiphase systems using ultrasonic technique," *Chemical Engineering Science*, 59: 3505–3514.

Zhong, J.-J. (2010), "Recent advances in bioreactor engineering," *Korean Journal of Chemical Engineering*, 27(4): 1035–1041.

Zhu, H., Shanks, B.H., and Heindel, T.J. (2008), "Enhancing CO-water mass transfer by functionalized MCM41 nanoparticles," *Industrial & Engineering Chemistry Research*, 47(20): 7881–7887.

Zhu, H., Shanks, B.H., and Heindel, T.J. (2009), "Effect of electrolytes on CO-water mass transfer," *Industrial & Engineering Chemistry Research*, 48(6): 3206–3210.

Zhu, Y., Bandopadhayay, P.C., and Wu, J. (2001), "Measurement of gas-liquid mass transfer in an agitated vessel-A comparison between different impellers," *Journal of Chemical Engineering of Japan*, 34(5): 579–584.

Zou, R., Jiang, X., Li, B., Zu, Y., and Zhang, L. (1988), "Studies on gas holdup in a bubble column operated at elevated temperatures," *Industrial & Engineering Chemistry Research*, 27(10): 1910–1916.

Index

An Introduction to Bioreactor Hydrodynamics and Gas-Liquid Mass Transfer, First Edition.
Enes Kadic and Theodore J. Heindel.
© 2014 John Wiley & Sons, Inc. Published 2014 by John Wiley & Sons, Inc.